Constitutive Modeling of Frictional Materials

The Advantage, Simplicity and Elegance of Using Stress Invariants

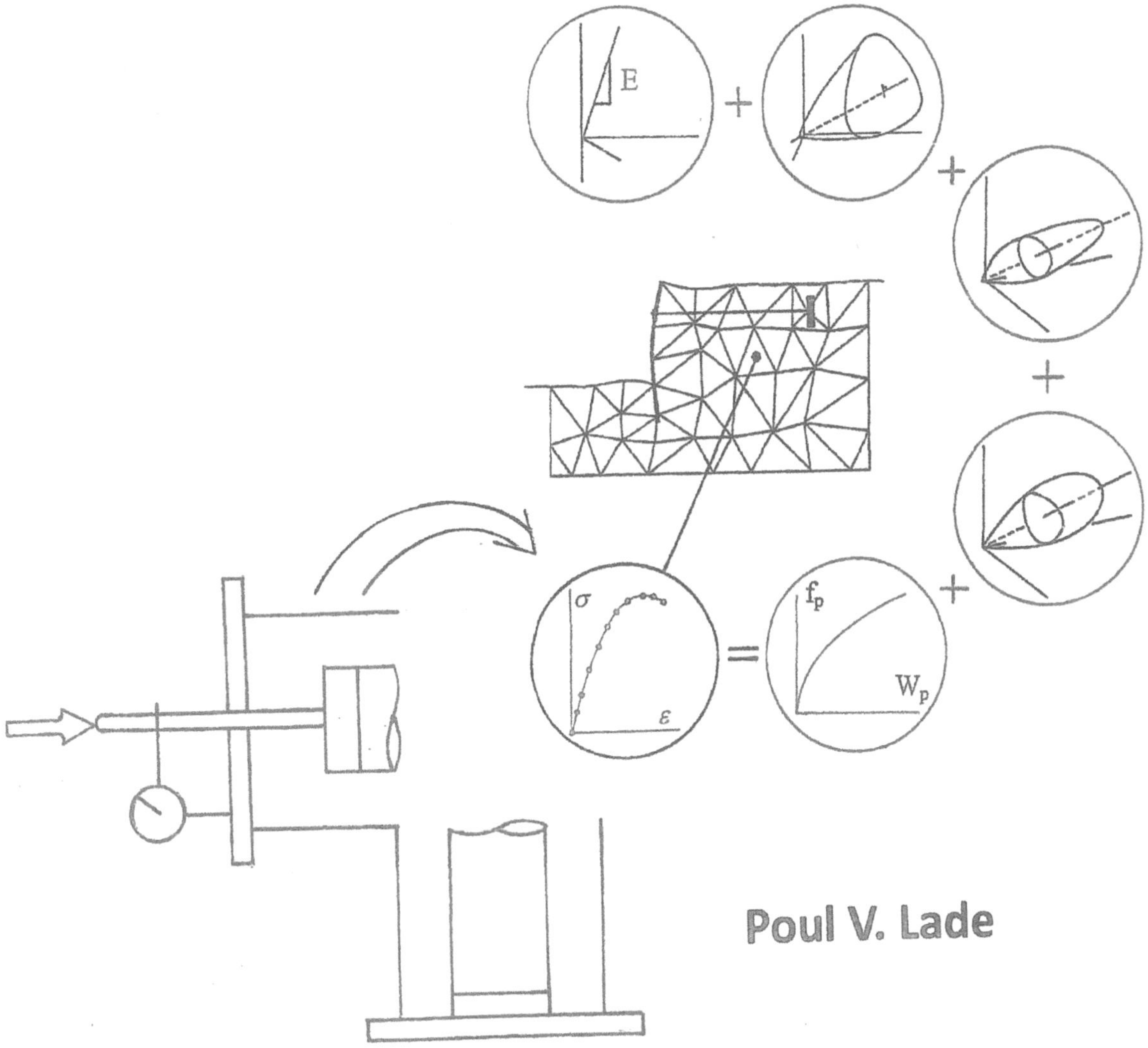

Poul V. Lade

Constitutive Modeling of Frictional Materials

The Advantage, Simplicity and Elegance of Using Stress Invariants

Poul V. Lade

Department of Civil, Environmental and Infrastructure Engineering

George Mason University

Fairfax, Virginia, USA

ISBN: 978-1-970528-46-6

2026

Contents

Constitutive Modeling of Frictional Materials

Foreword

This book presents the development of a three-dimensional, work-hardening plasticity theory for frictional materials. It is based on peer-reviewed articles presented in journals and conference proceedings tied together in a coherent presentation with comments, lecture notes and additional references given in prefaces and/or postscripts to each chapter. Since each of these articles had to stand alone, some repetition and overlap cannot be avoided.

The author's research philosophy has always been to perform the experiments first to find out how the material (most often sand or clay) behaves, and then model the observed behavior. Therefore, experimental equipment was designed and constructed to study the three-dimensional stress-strain and strength behavior of soils during monotonic loading and during large, three-dimensional stress reversals as well as principal stress rotation. The equipment consisted of (1) a cubical or true triaxial apparatus in which three different principal stresses can be applied in fixed directions, and (2) a torsion shear apparatus in which two of the three principal stresses can be rotated.

Based on the observed three-dimensional behavior, constitutive models were developed for frictional materials such as soil, rock and concrete employing elasticity and work-hardening plasticity theory. Using stress invariants for development of constitutive models has several advantages: the stress invariants are expressible both in principal stresses and in general stresses (normal and shear stresses). All material parameters can be derived from triaxial compression test, in which only principal stresses are applied. The performance of these tests has been explained in detail in a book by the author on "Triaxial Testing of Soils", published by Wiley-Blackwell in 2016. The model can be used for predicting the behavior of the frictional material for any stress path. Simple expressions can be developed for all the model components: failure surface, yield surface, and plastic potential surface, which capture the relatively complex behavior of frictional materials in a very simple and elegant fashion.

Plasticity theory for modeling the behavior of materials was originally developed for solid metals, and certain rules, postulates and conventions were consequently established. However, all materials, except solid metals, are frictional in nature, indicated by increasing shear strength with mean normal stress. This is due to collapse of pores, which causes friction between the internal surfaces, and this phenomenon increases with mean normal stress. Therefore, the original developments for solid metals have been reevaluated, mostly on the basis of experimental evidence for sands, but these reevaluations are true for all frictional materials.

In developing the constitutive models presented here, the author has adhered to Einstein's Principle of Simplicity: "A model should be as simple as possible, but not simpler." In other words, maintaining simplicity of the model without sacrificing essential accuracy is desirable. Any simplifying assumptions in the constitutive model have been evaluated relative to their practical importance and implications.

Chapter I provides an introduction in which the reasons for and the applications of an advanced constitutive model for the behavior of frictional materials are presented. An advanced constitutive model is too complex to be used for closed-form solutions to any geotechnical problem. The 31 models reviewed all require the use of finite element or finite difference computer programs to achieve the desired prediction of the behavior of the geotechnical structure in question. This introduction sets the stage for the presentation of the Single Hardening Model for Frictional Materials, the central topic in this book.

Chapter II sets out the general framework for elasticity and work-hardening plasticity theories as well as the mechanics principles which must be obeyed for legitimate modeling of all materials. These principles were originally stated by Prager (1949) in connection with development of plasticity theory for solid metals, which preceded the evolution for other materials.

To highlight the difference in behavior between solid metals and frictional materials, **Chapter III** presents a comparison of solid metals and frictional materials, which all includes voids that collapse with increasing mean normal stress, thus providing additional strength due to increased friction between internal surfaces. Frictional materials also exhibit volume changes, either dilative or contractive, during shearing. The volume changes play important roles in the stress-strain and strength behavior exhibited by frictional materials.

Chapter IV shows the development of the expressions for the 3 stress invariants and the 3 deviator stress invariants. The 3 stress invariants are used for all stress-dependent expressions for the elastic modulus, the failure surface, the yield surface and the non-associated plastic potential surface. The shapes of the stress invariants are shown in the principal stress space.

The development of an expression for the variation of the elastic modulus in terms of the first and the second stress invariants is presented in **Chapter V**. This modulus is based on the condition of zero work development or dissipation in a closed stress loop. It requires Poisson's ratio to be constant, and this condition is shown to be true for several series of triaxial compression tests on sands over large ranges of confining pressures.

Chapter VI presents the development of a three-dimensional failure criterion in terms of the first and the third stress invariants. This criterion is shown to capture the strength of all materials, frictional as well as non-frictional solid metals. Comparisons of experimental results for sands, clays, rocks, concrete, cast iron, powder metals and solid metals show that the strengths of these materials are well represented by this failure criterion. The required parameters can be obtained from simple experiments such as triaxial compression tests or simple tensions tests.

The early controversy of using non-associated or associated plastic flow for frictional materials is discussed in **Chapter VII**. Experiments were performed on sand to demonstrate that non-associated flow was observed, whether the sand volume change behavior was dilative or contractive during shearing. No combination of yield surfaces with associated plastic flow can replace the non-associated plastic flow observed in these experiments.

Chapter VIII presents the effects of non-associated plastic flow in dense and loose sands. The two types of instability observed in these sand deposits, both in terms of their physical behavior and in terms of calculations in computer programs are discussed. The type of instability observed and predicted for dense sand that tends to dilate is expressed in terms of shear banding, which occurs past peak failure in triaxial compression. For loose sand, the instability occurs in the

hardening regime inside the failure surface. In comparison, solid metals are stable all the way up to peak failure, at which condition shear banding develops.

Following these preliminary observations, **Chapter IX** presents the development of the Single Hardening Model for Frictional Materials. Based on experimental observations, the non-associated plastic potential is expressed in terms of stress invariants. The yield surface, also expressed in terms of stress invariants, is uniquely associated with the plastic work, such that a yield surface is formed by the location of all points with the same constant plastic work as measured from the stress origin. All material parameters can, in principle, be derived from any type of experiment, but they are most easily derived from triaxial compression tests. The model has been employed to predict the behavior of different types of materials as demonstrated in this chapter.

Chapter X shows the effects of large three-dimensional stress reversals and the development of a rotational model to capture this behavior. Upon large stress reversals, elastic and also plastic strains develop and these may be captured by yield and plastic potential surfaces that rotate around the stress origin in the principal stress space. This portion of the model predicts the effects of small and large stress reversals with good accuracy.

Principal stress rotation occurs in the field during any loading conditions of footings, retaining walls, etc., and this condition is discussed in detail in **Chapter XI**. Experimental evidence from torsion shear tests as well as theoretical proof of coincidence of plastic strain increments in physical space with total stress directions are predicted well by the constitutive model.

Chapter XII explains the incrementalization procedure necessary to implement the developed constitutive model in finite element and finite difference computer programs. Following the incrementalization, the complete implementation of the Single Hardening Model is presented, and the model is now ready to be programmed in a subroutine to be used with e.g. ABAQUS (finite element) or FLAC (finite difference) computer programs.

Time effects such as creep and stress relaxation are discussed in **Chapter XIII**. It is shown that these effects are due to grain crushing in sands, and they may be included in the Single Hardening Model by augmenting the work-hardening relationship by an appropriate multiplier as demonstrated for a particular granular material. While the time effects in clays are caused by a different mechanism, the overall behavior may be captured by similar model components as those for sands.

While all developments have been made with the assumption that the frictional materials exhibit isotropic behavior, real soils are most often cross-anisotropic with the vertical direction as the material axis. **Chapter XIV** shows how to model the cross-anisotropic failure conditions, and it indicates the method for handling cross-anisotropic stress-strain and volume change behavior.

Chapter XV presents a summary and conclusions for the development and use of the Single Hardening Model for Frictional Materials.

hardening regime inside the failure surface. In comparison, solid metals are stable all the way up to peak failure, at which condition shear banding develops.

Following these preliminary observations, **Chapter IX** presents the development of the Single Hardening Model for Frictional Materials. Based on experimental observations, the non-associated plastic potential is expressed in terms of stress invariants. The yield surface, also expressed in terms of stress invariants, is uniquely associated with the plastic work, such that a yield surface is formed by the location of all points with the same constant plastic work as measured from the stress origin. All material parameters can, in principle, be derived from any type of experiment, but they are most easily derived from triaxial compression tests. The model has been employed to predict the behavior of different types of materials as demonstrated in this chapter.

Chapter X shows the effects of large three-dimensional stress reversals and the development of a rotational model to capture this behavior. Upon large stress reversals, elastic and also plastic strains develop and these may be captured by yield and plastic potential surfaces that rotate around the stress origin in the principal stress space. This portion of the model predicts the effects of small and large stress reversals with good accuracy.

Principal stress rotation occurs in the field during any loading conditions of footings, retaining walls, etc., and this condition is discussed in detail in **Chapter XI**. Experimental evidence from torsion shear tests as well as theoretical proof of coincidence of plastic strain increments in physical space with total stress directions are predicted well by the constitutive model.

Chapter XII explains the incrementalization procedure necessary to implement the developed constitutive model in finite element and finite difference computer programs. Following the incrementalization, the complete implementation of the Single Hardening Model is presented, and the model is now ready to be programmed in a subroutine to be used with e.g. ABAQUS (finite element) or FLAC (finite difference) computer programs.

Time effects such as creep and stress relaxation are discussed in **Chapter XIII**. It is shown that these effects are due to grain crushing in sands, and they may be included in the Single Hardening Model by augmenting the work-hardening relationship by an appropriate multiplier as demonstrated for a particular granular material. While the time effects in clays are caused by a different mechanism, the overall behavior may be captured by similar model components as those for sands.

While all developments have been made with the assumption that the frictional materials exhibit isotropic behavior, real soils are most often cross-anisotropic with the vertical direction as the material axis. **Chapter XIV** shows how to model the cross-anisotropic failure conditions, and it indicates the method for handling cross-anisotropic stress-strain and volume change behavior.

Chapter XV presents a summary and conclusions for the development and use of the Single Hardening Model for Frictional Materials.

hardening regime inside the failure surface. In comparison, solid metals are stable all the way up to peak failure, at which condition shear banding develops.

Following these preliminary observations, **Chapter IX** presents the development of the Single Hardening Model for Frictional Materials. Based on experimental observations, the non-associated plastic potential is expressed in terms of stress invariants. The yield surface, also expressed in terms of stress invariants, is uniquely associated with the plastic work, such that a yield surface is formed by the location of all points with the same constant plastic work as measured from the stress origin. All material parameters can, in principle, be derived from any type of experiment, but they are most easily derived from triaxial compression tests. The model has been employed to predict the behavior of different types of materials as demonstrated in this chapter.

Chapter X shows the effects of large three-dimensional stress reversals and the development of a rotational model to capture this behavior. Upon large stress reversals, elastic and also plastic strains develop and these may be captured by yield and plastic potential surfaces that rotate around the stress origin in the principal stress space. This portion of the model predicts the effects of small and large stress reversals with good accuracy.

Principal stress rotation occurs in the field during any loading conditions of footings, retaining walls, etc., and this condition is discussed in detail in **Chapter XI**. Experimental evidence from torsion shear tests as well as theoretical proof of coincidence of plastic strain increments in physical space with total stress directions are predicted well by the constitutive model.

Chapter XII explains the incrementalization procedure necessary to implement the developed constitutive model in finite element and finite difference computer programs. Following the incrementalization, the complete implementation of the Single Hardening Model is presented, and the model is now ready to be programmed in a subroutine to be used with e.g. ABAQUS (finite element) or FLAC (finite difference) computer programs.

Time effects such as creep and stress relaxation are discussed in **Chapter XIII**. It is shown that these effects are due to grain crushing in sands, and they may be included in the Single Hardening Model by augmenting the work-hardening relationship by an appropriate multiplier as demonstrated for a particular granular material. While the time effects in clays are caused by a different mechanism, the overall behavior may be captured by similar model components as those for sands.

While all developments have been made with the assumption that the frictional materials exhibit isotropic behavior, real soils are most often cross-anisotropic with the vertical direction as the material axis. **Chapter XIV** shows how to model the cross-anisotropic failure conditions, and it indicates the method for handling cross-anisotropic stress-strain and volume change behavior.

Chapter XV presents a summary and conclusions for the development and use of the Single Hardening Model for Frictional Materials.

Chapter I

Introduction – Overview of Constitutive Models for Soils

Preface

The reasons for and the applications of a constitutive model for the behavior of frictional materials are presented here. The requirements for a work-hardening plasticity model are discussed and outlined in terms of experiments to study the soil behavior as well as the mathematical tools based on sound theoretical frameworks such as elasticity and plasticity theories. The use of so-called non-associated plastic flow for frictional materials is required for realistic models, for which the material parameters can be determined from simple experiments such as triaxial compression tests. The number of material parameters and their physical significance are discussed along with the capabilities of the corresponding models. Finally, a detailed review of and evaluation of 31 existing models are presented.

OVERVIEW OF CONSTITUTIVE MODELS FOR SOILS

Poul V. Lade[1], Member, Geo-Institute

ABSTRACT: Numerous constitutive models have been developed over the past forty years for modeling the stress-strain behavior of soils. These models are to be used with finite element and/or finite difference calculations of soil structures and soil/structure interaction problems under axisymmetric, plane strain, and/or general three-dimensional conditions. Simple as well as advanced models have been formulated on the basis of principles of mechanics, some more rigorous than others, some based on experimental evidence, and others based on theoretical principles. The capabilities and the shortcomings of these models are not always easy to ascertain, and the requirements for determination of parameters are not uniform. It is consequently difficult to determine which model to select for a particular task.

Presented here is an overview of the principles, main characteristic features, and components of existing, widely available constitutive models. Their requirements in terms of experimental data for calibration are presented, and their capabilities are reviewed.

INTRODUCTION

With the development of numerical methods such as finite element and finite difference methods it has become feasible to analyze and predict the behavior of complex soil structures and soil/structure interaction problems. Such analyses depend considerably on the representation of the relations between stresses and strains for the various materials involved in the geotechnical structure. In numerical computations the relations between stresses and strains in a given material are represented by a so-called constitutive model, consisting of mathematical expressions that model the behavior of the soil in a single element, as indicated in Fig. 1. Because soils are most often the weakest materials involved in common geotechnical problems, they

[1] Professor, Department of Civil Engineering, The Catholic University of America, Washington, D.C. 20064, U.S.A., Lade@cua.edu

Lade, P.V. "Overview of Constitutive Models for Soils," *ASCE Geotechnical Special Publication No. 128, Soil Constitutive Models: Evaluation, Selection, and Calibration*, Edited by J.A. Yamamuro and V.N. Kaliakin, January 2005, pp. 1-34.

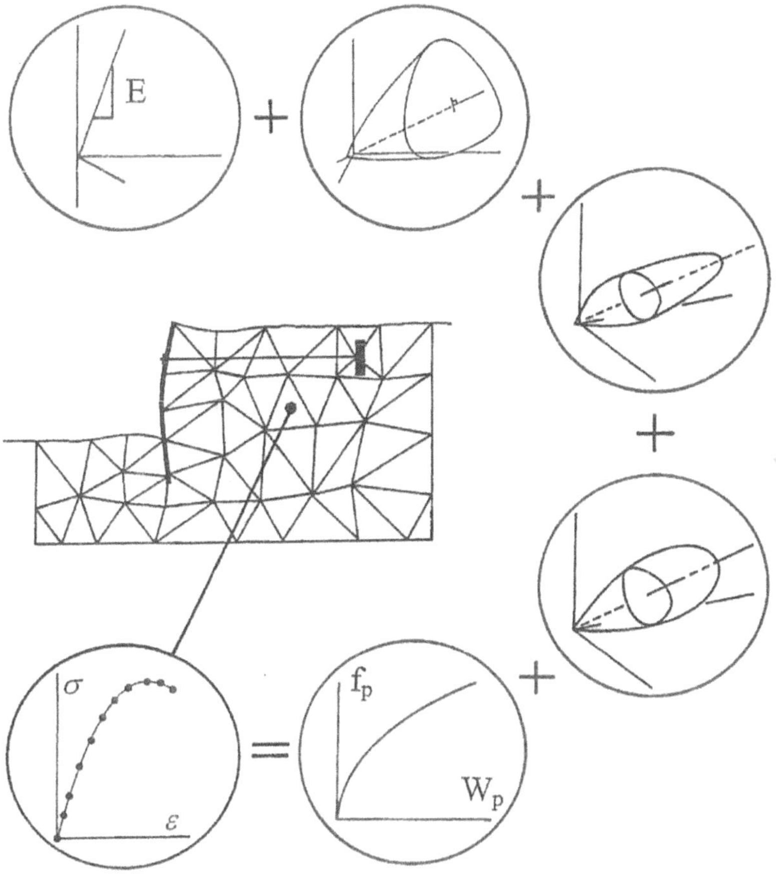

FIG. 1. Soil structure divided into a finite number of elements each of which is represented by a constitutive model based on elasticity and plasticity theories.

determine the deformations and the possibility of failure of the structure, and it is therefore important to characterize these materials accurately over the entire range of stresses and strains to which they will become exposed. Other construction materials such as concrete and steel may remain stiff in comparison with the soils, and it may be sufficient to characterize these materials as elastic or as elastic-perfectly plastic. Thus, the purpose of a constitutive model is to simulate the soil behavior with sufficient accuracy under all loading conditions in the numerical computations.

Significant developments of constitutive models have occurred over the past four decades. Naturally, the initial models were relatively simple, and a progression in complexity and capabilities of the models have lead to much improved abilities to capture the behavior of soil structures under complex loading conditions. Simple as well as advanced models have been formulated on the basis of principles of mechanics, some more rigorous than others, some based on experimental evidence, and others based on theoretical principles. Characterization of soil behavior can become quite involved, because the stress-strain relations are nonlinear in nature, the soils are fundamentally frictional materials, and volume changes occur during drained shearing.

It is clear that constitutive models have already gone through substantial improvements, and more will come. Thus, development of new and improved constitutive models is a continuing endeavor over a foreseeable future. However, the principles, main characteristic features, and components in the constitutive models for soils that are currently widely available to practitioners and researchers in packaged numerical software and in the technical literature are briefly presented and reviewed here. Important aspects of these models are the requirement for experimental data for calibration and the ease with which material parameters are determined.

This presentation is not intended as a highly in-depth discussion of each model. Virtually no particular models are mentioned in the text other than when used as examples in general discussions of e.g. the number of parameters in the models. Simple comparisons of the major components and capabilities of models are all presented in tables with no further comments.

CONSIDERATIONS IN MODEL DEVELOPMENT

Soils are complex materials consisting of a solid skeleton of grains in contact with each other and voids filled with gas (air) and/or water or other fluid. The soil skeleton transmits normal and shear forces at the grain contacts, and this skeleton of grains behaves in a very complex manner that depends on a large number of factors, void ratio and confining pressure being among the most important. However, the overall behavior of the soil skeleton may be captured within principles of continuum mechanics (solid mechanics). Interspersed in the voids is water (incompressible fluid) and gas (compressible fluid), each of which obeys its own physical laws. The mixture of grains, water, and air produces a material that, in comparison with other engineering materials, is one of the most difficult to characterize.

The Effective Stress Principle

A key principle in the behavior of soils was realized by Terzaghi (1923), namely that soils deform in response to changes in effective stresses, and these are calculated from:

$$\sigma' = \sigma - \eta \cdot u \tag{1}$$

in which σ' is the effective stress, σ is the total stress, u is the pore pressure, and η is a coefficient whose value depends on the compressibilities of the grain skeleton and the grains themselves. For all common geotechnical problems the value of η can be taken as unity. Only for extremely high stresses, encountered under very exceptional circumstances, does the value of η become smaller than unity (Lade and de Boer 1997).

Since soils deform in response to changes in effective stresses, theoretically sound constitutive models are expressed in terms of effective stresses. Nevertheless, models have also been expressed in terms of total stresses, but they suffer from the same shortcomings as any total stress analysis of a soil structure: They are only valid for the

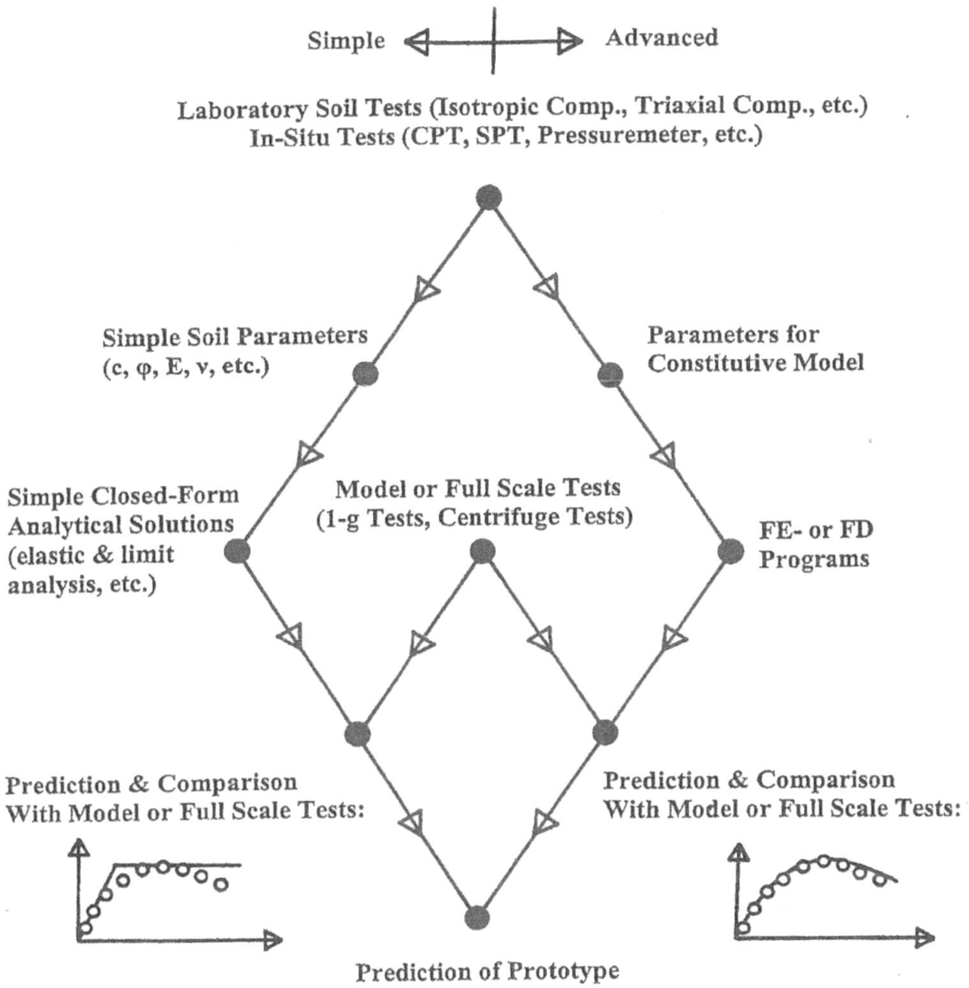

FIG. 2. Simple and advanced procedures for predicting the behavior of prototype structures.

particular conditions for which they have been calibrated. Variations in pore pressure or drainage conditions, different from those developing in the experiment from which parameter values are determined, cannot be modeled (see any textbook in geotechnical engineering, e.g. Lambe and Whitman 1969), so their usefulness is limited.

Simple versus Advanced Analysis - What is Gained?

Fig. 2 shows a simplified diagram of the analysis procedures generally used in geotechnical engineering. In the simple procedures, represented on the left hand side, simple and easily recognizable soil parameters (e.g. c, φ, E, ν) are derived from laboratory and/or in-situ tests and utilized in closed-form solutions for the particular boundary value problem under consideration. Figs. 3 and 4 illustrate some simple

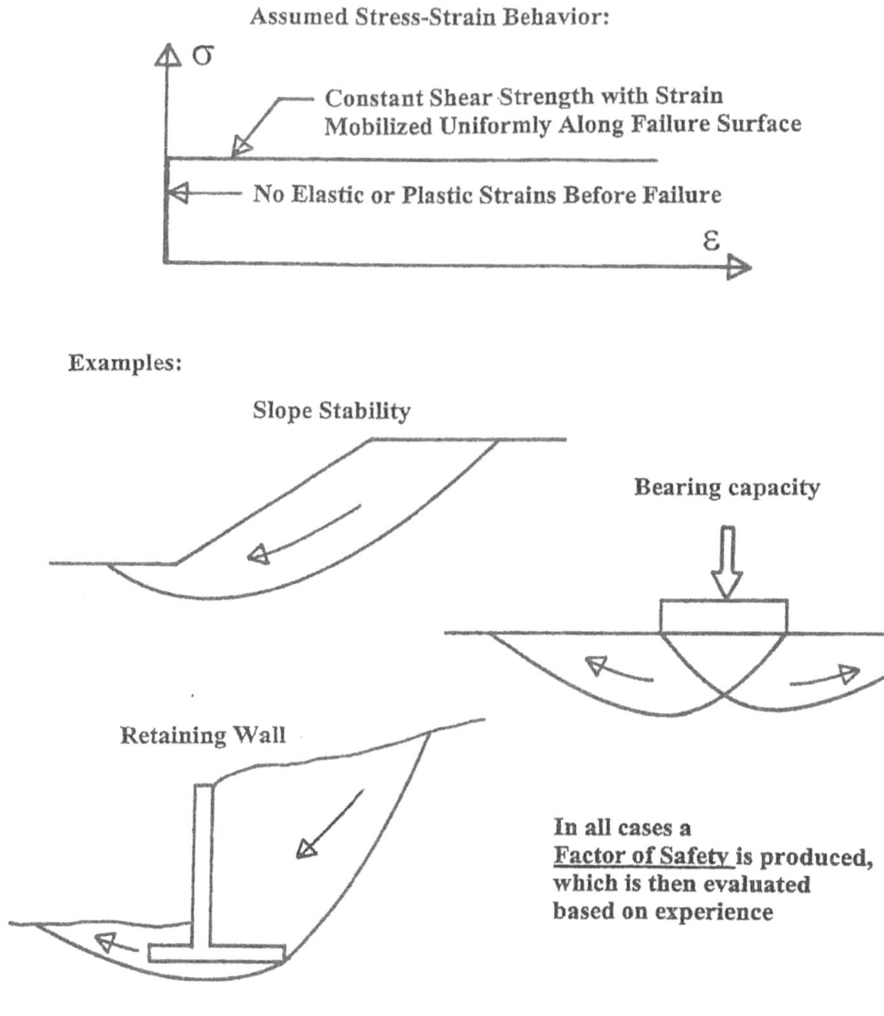

FIG. 3. Examples of failure or stability of soil structures for which closed-form solutions are available.

cases for which closed-form solutions are readily available. Usually considerations of failure or stability are made first using formulas from which the factor of safety may be calculated from solutions for slope stability, bearing capacity, and/or stability of retaining walls, as shown in Fig. 3. In these analyses strains are not considered, which corresponds to an assumed stress-strain relation as shown in the upper part of Fig. 3.

Once the stability is ensured, the deformations or settlements may be evaluated using closed-form solutions based on elasticity theory, such as that indicated in Fig. 4 for the settlement of a footing. No matter how high the bearing pressures used in the elastic formulation, failure considerations are not part of the calculations from such formulas.

The results of these classical procedures may be verified by prediction and comparison with model or full scale tests of elements of the prototype structure (e.g.,

Assumed Elastic Behavior:
Immediate deformations and/or settlements are calculated as linear elastic using closed-form solutions for elastic boundary value problems with an "elastic" modulus estimated from a stress-strain curve

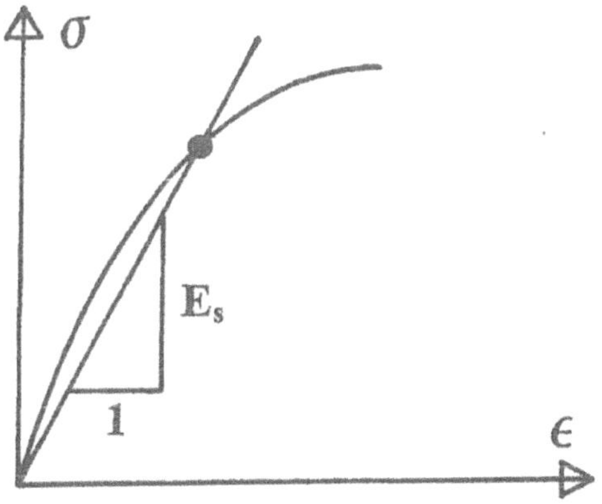

Example:

Width of Foundation

Shape Factor for Foundation (based on small scale tests)

$$S = q \cdot B \cdot \left(\frac{1-\nu^2}{E_s} \right) \cdot I_w$$

Settlement **Bearing Pressure** **"Elastic" Properties**

FIG. 4. Example of deformation consideration represented by settlement calculation of a footing for which a closed-form solution is available.

one pile). Finally, prediction of the behavior of the prototype may be performed. The simple procedures predict simplified responses such as linear elastic settlements and failure, but prediction of the entire load-deformation relation for a prototype structure is often inaccurate if not impossible to perform, especially in the load range where failure is a distinct possibility.

The real behavior of a footing is indicated in Fig. 5. As loading progresses, the stresses in soil elements located along the potential failure surface in the ground, shown in the upper part of Fig. 5, advance along the stress-strain curve as indicated in the lower part of Fig. 5. The initial state of stress in the ground may be characterized by a K_0-condition, and as loading progresses, the stresses increase up to peak failure and beyond into the softening region of the stress-strain curve. Thus, a highly nonlinear relation between stresses and strains are encountered in the real behavior of the footing.

In the advanced procedure, indicated on the right hand side of Fig. 2, a constitutive model is used to capture the entire stress-strain relation obtained from laboratory and/or in-situ tests. Incorporating the constitutive model in numerical methods, the behavior of model or full-scale tests may be predicted and serve to verify the

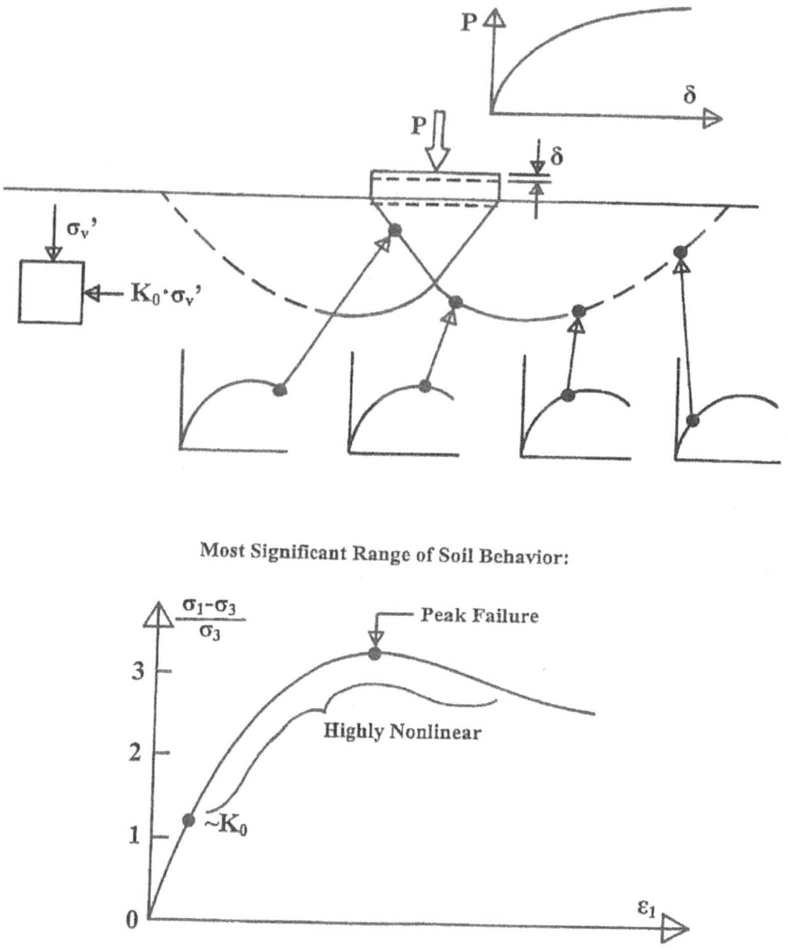

FIG. 5. Real behavior of a footing involving highly nonlinear soil stress-strain behavior.

capability of the constitutive model and the numerical method. Finally, the behavior of the prototype may be calculated with better overall accuracy.

Using finite element (or finite difference) methods, a geotechnical structure of any axisymmetric, plane, or three-dimensional shape may be divided into a finite number of elements (nodes), as indicated in Fig. 1. A constitutive model for each element then simulates the soil behavior, and the numerical method ensures equilibrium and compatibility between the soil elements. Thus, much more complex earth structures and soil/structure interaction problems can be analyzed, and their behavior in the form of deformations and stress conditions everywhere in the structure may be obtained for the entire loading history. Such results could not be obtained with closed-form solutions, which are typically only developed for rather simple boundary value problems.

Requirements of Constitutive Models

One of the critical elements in the advanced procedure is the constitutive model. It is paramount to employ realistic constitutive models that can reproduce the important aspects of the soil stress-strain behavior under various loading conditions. To develop such models requires advanced experiments to study the soil behavior under various loading conditions, and it requires employment of mathematical tools based on sound theoretical frameworks such as, e.g., elasticity and plasticity theories.

Some of the advanced experiments available today are the torsion shear, directional shear, and true triaxial tests. These are suitable for studying soil behavior under three-dimensional stress conditions with and without stress reversals and with and without rotation of principal stresses. While the model should be able to simulate the soil behavior observed in these tests, they should not be required for calibration of the constitutive models. The constitutive model should be such that the required soil parameters can be obtained from relatively simple tests.

Sequential Development of Models

Most practical models, applicable in numerical methods, are formulated in terms of some form of theories of elasticity and plasticity. They have been developed based on observed behavior in laboratory experiments, and it is the responsibility of the research community to demonstrate that they do, in fact, capture the behavior of soils as they are purported to do. Thus, as realization about the complex behavior of soils is expanding, the models are continuously being improved and new modeling principles have been developed.

For example, the following sequence of observations of soil behavior have lead to improved models: Effects of virgin (primary) loading and unloading-reloading, confining pressure, stress path (stress history), intermediate principal stress, instability (leading to shear banding in dense sand and overconsolidated soils and to liquefaction in loose silty sands), anisotropy, and stress rotation (note: this list of phenomena is not exhaustive). While all of these (and many more) effects have been known to exist for some time, some of them are still to be researched and understood in detail before they may be modeled correctly. Thus, the simple models can only handle the first mentioned phenomena, while the more advanced models may handle additional phenomena. It is difficult to say that any one of these phenomena has definitively been researched, understood, and included in any given model. The latter two topics (effects of anisotropy and effects of stress rotation) are in the forefront of current research, and practically applicable models that correctly capture these effects are still to be developed.

Shortcomings of Elasticity Theory

The behavior of soils may be predicted with reasonable accuracy by models based on elasticity for stress states not approaching failure. However, phenomena such as irreversibility of a portion of the strains, stress-path dependency, coupling effects such as volume changes due to shear stresses, referred to as shear-dilatancy, rotation of principal stress axes, and most behavior patterns near and beyond failure cannot be handled by elasticity theory. Two simple examples will illustrate the shortcomings of Hooke's law for elastic behavior.

Fig. 6 shows that zeroes occupy the sub-matrices (1, 2) and (2, 1) of the elastic material matrix that relates plastic strain increments to stress increments. The zeroes in sub-matrix (1, 2) imply that there are no relations between normal strains and shear stresses. However, Example (1) in Fig. 6 shows a dense sand specimen acted upon by a normal stress and a shear stress, as for example in a simple shear test. If the shear stress is increased, then increments in shear strains will result, but observations of real soil indicates that the soil will also dilate such that the specimen height increases.

? (1)

$$\begin{Bmatrix} \Delta\varepsilon_x \\ \Delta\varepsilon_y \\ \Delta\varepsilon_z \\ \Delta\gamma_{yz} \\ \Delta\gamma_{zx} \\ \Delta\gamma_{xy} \end{Bmatrix} = \left[\begin{array}{ccc|ccc} \frac{1}{E} & -\frac{\nu}{E} & -\frac{\nu}{E} & 0 & 0 & 0 \\ -\frac{\nu}{E} & \frac{1}{E} & -\frac{\nu}{E} & 0 & 0 & 0 \\ -\frac{\nu}{E} & -\frac{\nu}{E} & \frac{1}{E} & 0 & 0 & 0 \\ \hline 0 & 0 & 0 & \frac{1}{G} & 0 & 0 \\ 0 & 0 & 0 & 0 & \frac{1}{G} & 0 \\ 0 & 0 & 0 & 0 & 0 & \frac{1}{G} \end{array}\right] \begin{Bmatrix} \Delta\sigma_x \\ \Delta\sigma_y \\ \Delta\sigma_z \\ \Delta\tau_{yz} \\ \Delta\tau_{zx} \\ \Delta\tau_{xy} \end{Bmatrix}$$

? (2)

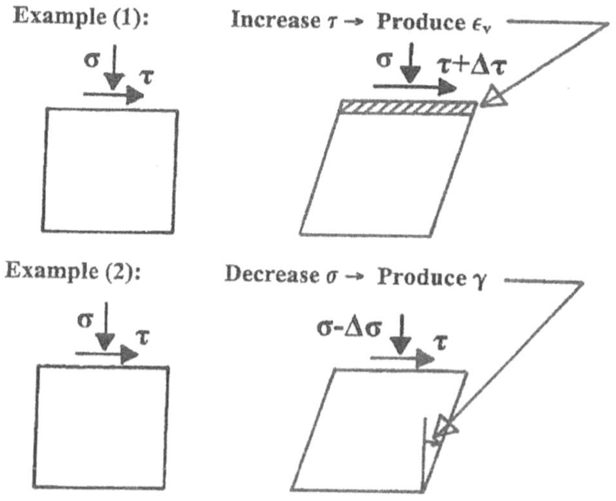

FIG. 6. Examples to indicate that Hooke's law for elastic behavior does not correctly handle coupling effects: (1) normal (and volumetric) strain increments due to changes in shear stress, and (2) shear strain increments due to changes in normal stress.

Thus, normal strain increments result from increments in shear stress. Since zeroes are present in sub-matrix (1, 2) of the elastic material matrix, no normal strain increments can be predicted from increments in shear stress. Therefore, Hooke's law is not able to model this behavior of soils. The zeroes in sub-matrix (2, 1) of the elastic material matrix may be checked by the experiment in Example (2) in Fig. 6. Here the initial condition is the same as in Example (1), but now the normal stress on the specimen is reduced. Observations of real soil behavior shows that the specimen will produce shear strain increments for such a reduction in normal stress, and sufficient reduction will result in failure. However, the zeroes in sub-matrix (2, 1) preclude any prediction of shear strain increments due to increments in normal stresses. These two examples clearly demonstrate some serious shortcomings of elasticity theory for modeling behavior of soils for stress conditions approaching failure. In fact, all frictional materials exhibit behavior of the types exemplified in Fig. 6.

Why Using Hardening Plasticity Theory?

Models based on hardening plasticity have the potential to predict the behavior reviewed above. In addition, elasticity and plasticity also differ in predictions of material behavior during gradual stress rotation, as indicated in Fig. 7. The theoretical framework for incremental elasticity implies that the increment of elastic strain coincides in direction with the increment of stress, as shown in Fig. 7(a). This differs clearly from the predictions of isotropic hardening plasticity theory, which has built into its framework the condition that the increment of plastic strain coincides in direction with the total stress, as shown in Fig. 7(b). Observations of soil behavior show that far away from failure the behavior favors the elastic predictions, while the behavior observed closer to failure resembles the behavior simulated by plasticity theory.

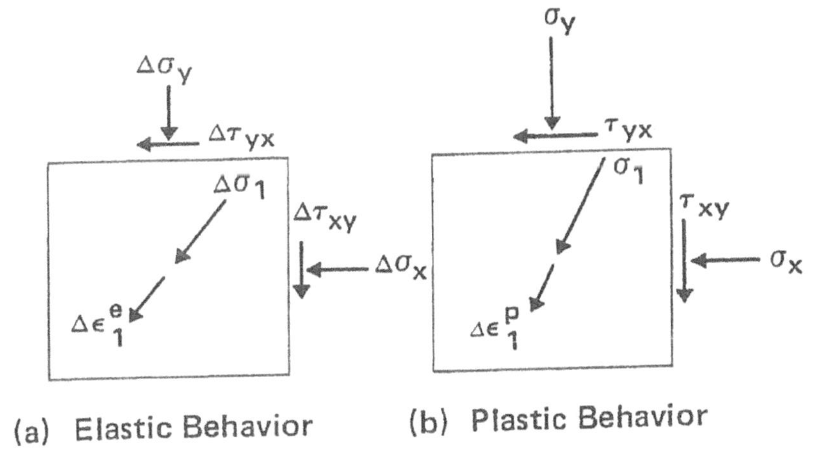

FIG. 7. (a) Elastic, and (b) plastic behavior of soil element during rotation of stress axes.

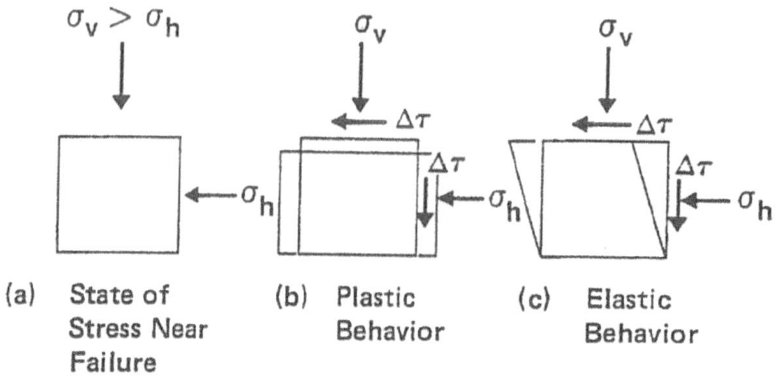

FIG. 8. (a) Stresses on a soil element near failure, its (b) plastic behavior (Saint Venant's principle), and its (c) elastic behavior.

Another way of indicating the difference between elasticity and plasticity theories is shown in Fig. 8. A soil specimen is acted upon by vertical and horizontal normal stresses such that $\sigma_v > \sigma_h$, but failure has not been reached, as shown in Fig. 8(a). An increment in shear stress is added to the soil element, as shown in Figs. 8(b) and (c). Plastic behavior, shown in Fig. 8(b), entails vertical compression of the specimen, because the direction of the much larger total stress determines the direction of the plastic strain increment, independent of whether the stress increment is a normal or a shear stress increment. In comparison, elastic behavior results in strain increments in proportion to the applied stress increment and is independent of the already existing stress state.

These two types of behavior are both observed in soils. The elastic type behavior is observed at low stress levels near the isotropic stress state and immediately upon unloading and/or reloading where plastic behavior is absent or minimal, while plastic behavior is observed at higher stress levels closer to failure. Elastic strain increments are always present when the state of stress changes, while plastic strain increments may be present or completely absent.

Thus, realistic constitutive models require both elastic and hardening/softening plastic behavior in their framework. All realistic constitutive models for soils involve these two components of deformation.

Associated versus Nonassociated Plastic Flow

Early considerations in development of plasticity theory required the relative magnitudes of the plastic strain increments to be such that the direction of the plastic strain increment vector, when superimposed on stress space, would be normal to the plastic yield surface. That implied that this direction (or the relative magnitudes of the plastic strain increments) could be determined from the derivative of the mathematical expression for the yield surface, which was therefore also referred to as the plastic potential surface. Thus, the plastic flow rule (i.e. the expressions for the plastic strain increments) could be derived from the expression for the yield surface, i.e. plastic flow was associated with the plastic yield surface. A constitutive model based on this concept was using an "associated flow rule."

Employing "associated flow" resulted in a great advantage, because knowing either the yield surface (from experimental determination of stress conditions at plastic yielding) or knowing the plastic potential surface (from experimental determination of the relative magnitudes of plastic strain increments and therefore the direction of the plastic strain increment vector at different stress points) would automatically result in knowing both surfaces. This simplified the mathematical framework for the model of the plastic behavior. Besides, violation of the "normality rule" was considered to have serious implications in terms of stability of the material, and constitutive models based on so-called nonassociated plastic flow rules, for which the plastic potential surface would be different from the yield surface, could result in unstable material behavior.

The concept of associated plastic flow works well for solid metals, but frictional materials do not seem to fit well with this concept. Much too high rates of volume dilation are predicted if associated flow is used in connection with models for frictional materials. The difference between observed soil behavior and behavior predicted on the basis of associated flow is most pronounced for frictional materials with high effective friction angles (e.g. dense sand), while materials with less pronounced frictional characteristics, such as clays with lower effective friction angles, may be modeled with some approximation by associated flow rules.

Later theoretical considerations have shown that the normality rule does not have to be fulfilled, and observations of sand behavior clearly indicate that experimentally determined plastic strain increment vectors are not perpendicular to the yield surface. Further, the instabilities that might occur in materials that exhibit nonassociated flow actually have been observed in the form of shear bands occurring in the hardening regime (the ascending branch of the stress-strain relation) under plane strain conditions and in the form of instability (decrease in load bearing capacity) leading to liquefaction of loose silty sand.

While it is generally understood at this time that frictional materials obey nonassociated flow rules, some constitutive models developed for clay are still based on associated plastic flow. The characteristics of frictional materials are less pronounced for these materials.

Prediction of Pore Water Pressures

Pore water pressures are rigorously determined from a constitutive model on the basis that a fully saturated soil element cannot change volume during shearing. Thus, the sum of all volumetric strain components must add up to zero during any change in stress. This requires that for each increment in loading, plastic dilation is balanced by an equal and opposite amount of elastic contraction or vice versa: plastic contraction is balanced by an equal and opposite amount of dilation. This results in a unique effective stress path from which the pore pressure can be obtained as the difference between the total stresses and the effective stresses for each increment of loading.

In case that only one component of volumetric strain is present, e.g. the elastic component inside the yield surface, then the effective stress path will remain in one octahedral plane corresponding to a constant mean normal stress until the yield surface is reached again.

If a constitutive model contains only one component of volumetric strain, then realistic pore pressures cannot be predicted from the model. On occasion, a separate pore pressure generation model has then been employed to predict pore pressures under undrained conditions. However, such models are usually not compatible with the performance of the constitutive model.

WHICH PROBLEMS CAN BE SOLVED?

Many different types of geotechnical problems may be analyzed using advanced constitutive models and numerical methods. Duncan (1994) reviewed the literature ten years ago to study the variety of problems that had been investigated using the advanced approach, and to look into the constitutive models that had been employed in the numerical analyses. He organized the review into sections and pointed out that the following types of geotechnical applications had been encountered: Soil structure interaction, soil reinforcement and anchorage, dams, embankments and settlement due to fluid extraction, tunnels, and natural and unbraced cut slopes. Thus, a large variety of problems benefit from insight provided by advanced analyses of stresses and deformations in the soil structures. In fact, it has become feasible to analyze and predict the behavior of any type of complex soil structure and soil/structure interaction problem.

The greatest advantages and insights are obtained for problems in which the soil behavior plays an important role. These include problems in which the interaction between stresses and soil volume changes plays a dominant role. Fig. 9 gives some examples of soil/structure interaction in which the interaction between confinement and volume changes, i.e. the real soil behavior is essential in prediction of the behavior of the structure.

Another type of problem demonstrating the importance of understanding and capturing the soil behavior correctly is shown in Fig. 10. Here the rockfill material in the upstream shell of a rockfill dam may collapse during first wetting due to filling of the water reservoir. This produces displacements of the crest that resemble those observed due to imminent upstream slope failure. Is the dam failing or are these displacements to be expected? An advanced analysis with appropriate modeling of the soil behavior can answer the question.

Fig. 11 illustrates some problems in which the self-weight of the soil plays an important role in the stability analyses of the soil structure. For example, in slope stability analyses, the weight of the soil is the direct cause of potential instability. A classical slope stability analysis produces an accurate estimate of failure, because the normal stresses on the failure plane are proportional to the weight of the soil and independent of the soil behavior. Thus, the normal stress, and therefore the shear strength available along the failure plane, does not change due to e.g. a tendency for the soil to change volume. The normal stress remains proportional with the soil weight, even if the soil tends to dilate. Thus, the calculated factor of safety remains unaffected in any advanced analyses that may include more accurate soil modeling. In such problems, there is limited advantage in an advanced analysis for the purpose of

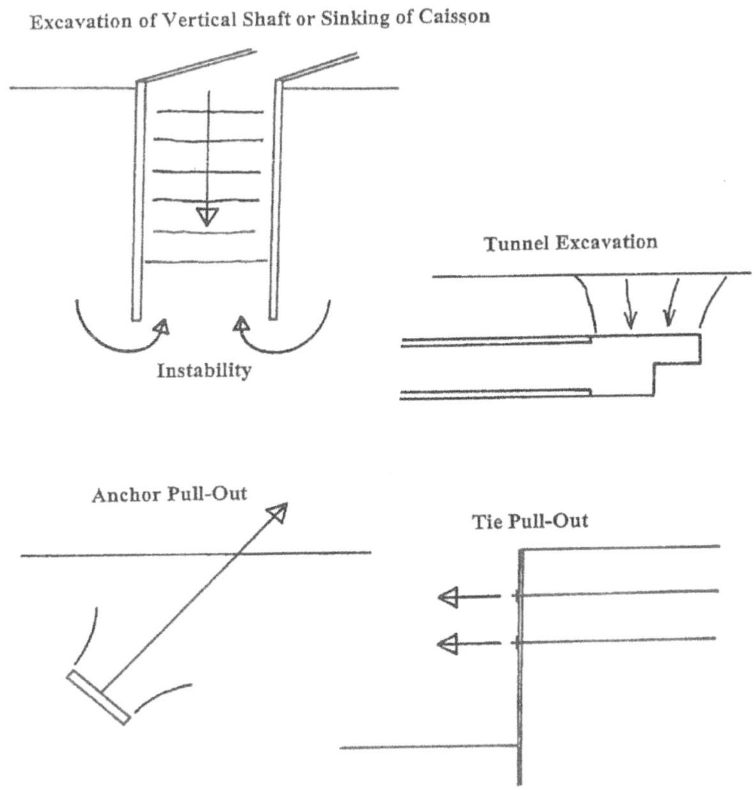

FIG. 9. Examples of soil/structure interaction in which the soil behavior plays an important role in the deformation and stability of the structure.

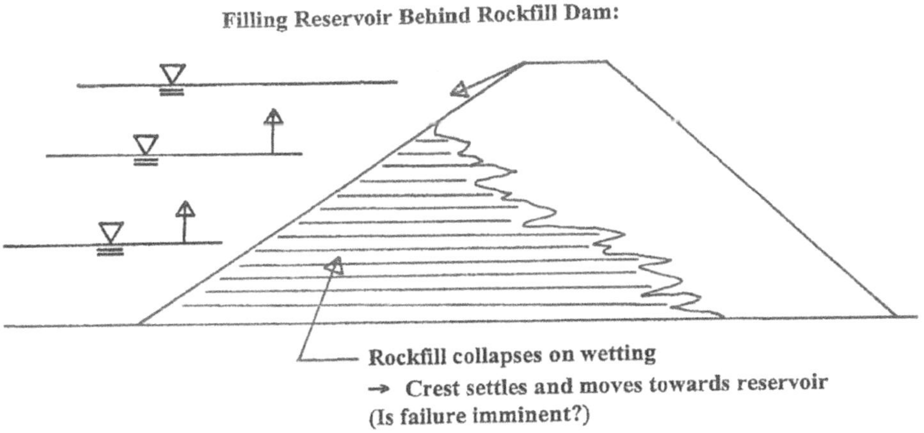

FIG. 10. Filling of water reservoir behind a rockfill dam with resulting crest displacement due to first wetting and collapse of rockfill material in upstream shell.

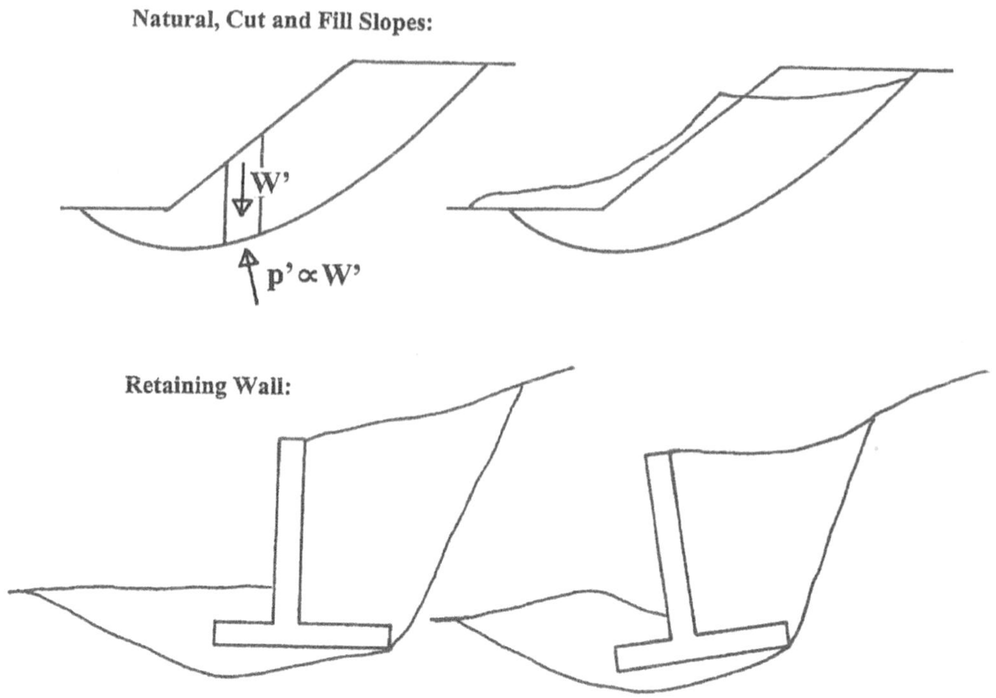

Stability considerations produce accurate estimate of failure, because normal stresses on failure plane are proportional to weight of soil and independent of soil behavior, ...
but <u>deformations</u> are not elastic
→ simple procedures cannot be used to calculate deformations

FIG. 11. Examples of slope and retaining wall problems in which soil self-weight plays a dominant role in their stability.

determining the stability. However, deformations cannot be obtained from a classical slope stability analysis, while an advances numerical analysis will be able to produce such deformations.

IMPLEMENTATION IN NUMERICAL COMPUTER PROGRAMS

Many of the commercially available finite element and finite difference programs, e.g. ABAQUS, PLAXIS, and FLAC, allow implementation of most simple as well as advanced constitutive models. Many of these programs have become increasingly user-friendly, and some have been specialized for geotechnical engineering problem solving. Unfortunately, the perception that advanced models may be too complicated for implementation and for easy parameter determination reduces the usefulness of such numerical programs. Therefore, the present exposure of constitutive models for soils should be useful for more frequent applications of advanced analysis procedures in the future.

PARAMETER DETERMINATION

Testing Requirements

To predict deformations as well as strengths of soils, the constitutive model should include expressions for the soil behavior observed during compression and during shearing. The modes of deformation are different under these two loading conditions: One involves continuous compression with no failure, while the other leads to failure and beyond. Consequently, the model should contain components that simulate the corresponding two modes of deformation, and experiments that contain information regarding these two modes of deformation should be available for determination of the material parameter values (calibration) that characterize the particular soil in question.

In principle, the parameters for a given model should be derivable from any set of experiments that contains information regarding compression and shearing behavior. This is because the mathematical expressions in the ideal constitutive model may be specialized to correspond to the particular tests available for parameter determination. But most often, the simplest and most convenient experimental results that contain the required information are those from conventional triaxial compression tests and isotropic compression tests.

Typically the results of three triaxial compression tests and one isotropic compression test contain the minimum required information, and such test results are often available for parameter determination. It may be possible to use the results from uniaxial compression tests (K_0-consolidation) in place of isotropic compression tests. However, it becomes much more complicated if a model requires results from shear tests that cannot be performed in a triaxial apparatus, because more advanced equipment is seldom available for soil testing. On the other hand, results from simpler shearing tests such as those performed in a direct shear box may not be appropriate for parameter determination for any but the most simple models.

The three triaxial compression tests and the isotropic compression test should be performed with the following important points in mind:

1) The specimens are tested with the assumption that they represent the average soil behavior at the location of the soil samples in the ground (there is no interest in the lowest ("safe") strength for the advanced analysis).

2) The tests are performed by simulating the conditions in the ground as closely as possible. This includes

   a) range of stresses (due to variable, nonlinear behavior)

   b) drainage conditions in the field (but typically drained tests are performed on sands and undrained tests are performed on clays)

   c) density and soil fabric (undisturbed soil specimens are required in most cases)

3) The test performance should include primary loading as well as unloading-reloading cycles (to evaluate the elastic parameters), and all physical quantities such as loads, pressures, deformations, volume changes or pore pressures must be measured for a complete characterization of the stress-strain behavior of the soil. Every portion of the experiments is required for parameter determination.

Because the results of advanced tests such as true triaxial tests and torsion shear tests are not available on a routine basis, the constitutive model must have built-in effects of the intermediate principal stress, anisotropy, and stress rotation for which necessary parameters, if any, must come from conventional triaxial tests.

Once the parameters for a model have been determined, they do not change value during the computations, and they should apply to all types of geotechnical problems in which the soil is present. With regard to the latter point, there are two types of constitutive models in existence: For the first and most common type, samples are taken out of the ground and parameters are determined for the particular void ratio present in the soil sample. The constitutive model is then calibrated to this particular initial void ratio, and volume changes (or changes in void ratio) as well as strength can be predicted without updating the parameter values during the computations. For example, the friction angle for a sand at a given void ratio may be one of the parameter values in a model of the first type. While the friction angle for a sand is known to vary considerably with the void ratio, the value determined for the given initial void ratio (e.g. the void ratio of the soil sample recovered from the ground) is assumed to apply at failure for the sand being loaded from that initial void ratio, even though the void ratio changes during shearing. This type of model is suitable for the particular problem at hand for which the parameters have been determined and associated with the particular initial void ratio. Most geotechnical analysis procedures have been worked out for this approach. Clearly, if a different initial void ratio is present in the soil sample, then another set of parameter values is obtained, including that for the friction angle.

For the second type of constitutive model, the parameters are not associated with any particular initial void ratio. These parameters may be used to calculate changes due to variation in void ratio during loading, and this type of model does not change parameter values with initial void ratio. While such a model may seem advantageous, only relatively simple models have been devised with these attributes. Examples of this type of model are the Critical State models such as the Modified Cam Clay Model and the Nor-Sand model.

What is the Physical Significance of the Parameters?

Advanced models may contain parameters that are not immediately recognizable or do not have immediate physical meaning. These parameters may be constants in mathematical expressions that describe, e.g., the curved three-dimensional failure envelope expressed in terms of stress invariants. The friction angle, which decreases with increasing normal stress, a phenomenon that is well known in geotechnical engineering, is not directly recognizable in such equations. Nevertheless, the parameters express the same phenomenon, namely that the shear strength increases almost (but not quite) linearly with the normal stress. Thus, equations used in advanced constitutive models to express phenomena such as the curvature of failure envelopes, dependence on the third stress invariant, and volume dilation may contain parameters whose physical meaning are not immediately recognizable in a similar manner as the familiar Young's modulus, Poisson's ratio, friction angle, and cohesion. Such parameters are in all cases simply curve-fitting parameters whose usefulness is similar to the well-known parameters.

The Single Hardening Model (Kim and Lade 1988, Lade and Kim 1988a and b) is an elasto-plastic constitutive model that includes the various necessary components typical of these types of models, as indicated in Table 1. The function of each component and their physical significance are described in this table (where possible). The parameters occurring in the Single Hardening Model are employed to characterize these components, as shown in Table 2, and while the components have functions and physical significance in the overall model, it may not be possible to assign physical significance to each individual parameter. Rather, these parameters are curve-fitting parameters.

Number of Model Parameters

The number of parameters in a model is often mentioned as a consideration and a concern in relation to employment of this model. But should it really be a concern, or are there other considerations that are more important for employment of advanced constitutive models? To answer this question, three analyses methods are discussed in view of the number of parameters involved, their role in the models, and the capabilities of the methods.

Parameters are listed in Table 2 for the simple analysis approach (SA) described above in which a stability analysis is followed by a deformation or settlement analysis. The parameters used in the Modified Cam Clay (MCC) model are also listed, since this model is well known to geotechnical engineers. Finally, Table 2 contains the parameters employed in the Single Hardening (SH) model.

For the SA approach, the elastic properties are represented by one value of Poisson's ratio, ν, and one value of Young's modulus, E. Similarly, the MCC model involves two constant, elastic parameters, κ and G, such that the elastic bulk modulus, $K = p'/\kappa$, i.e. $K\ (= E/[3(1-2\nu)])$ is proportional with the mean normal stress p', and the shear modulus $G\ (= E/[2(1+\nu)])$ is constant. In the SH model, Poisson's ratio is constant, and Young's modulus increases linearly with the modulus number M, and it increases according to a power function that contains the stress state to the power λ.

Failure is described by the well-known parameters c and ϕ in the SA approach, while the MCC model involves the parameter M that indicates the straight-line critical state (= failure for normally consolidated clay). The MCC model does not involve effective cohesion (or tensile strength) unless another parameter is added. The SH model includes a curved failure surface for which η_1 indicates the opening angle (similar to the friction angle), m is the curvature parameter, and the parameter "a" characterizes the effective cohesion (or the tensile strength). For sand $a = 0$, i.e. one less parameter than indicated in Table 2.

In the MCC model Γ indicates the void ratio for which the critical state line passes through $p' = 1$, and λ relates to the slope of the compressibility line and the critical state line (which is parallel to the compressibility line) on an e–ln(p') diagram. These two parameters are used to calculate the magnitude of the plastic strain increments. In the SH model the magnitude of the plastic strain increments are controlled by the

TABLE 1. Components and Their Physical Significance in Elasto-Plastic Models.

	Component	**Function**
Elastic Behavior	Hooke's Law	Produces elastic strains whenever the stresses change
Plastic Behavior	Failure Criterion	Imposes limits on stress states that can be reached
	Plastic Potential Function	Produces relative magnitudes of plastic strain increments (similar function as Poisson's ratio for elastic strains)
	Yield Criterion	Determines when plastic strain increments occur: Only when yield surface is pushed out/in (hardening/softening)
	Hardening/Softening Relation	Determines magnitudes of plastic strain increments (similar function as Young's modulus for elastic strains)

TABLE 2. Components and Number of Parameters in Elasto-Plastic Models.

Component	Parameters for Simple **Elastic-Perfectly Plastic Approach**	Parameters for **Modified Cam Clay Model**	Parameters for **Single Hardening Model**
Hooke's Law	ν, **E**	κ, **G**	ν, **M**, λ
Failure Criterion	**c**, ϕ	**M**	η_1, **m**, **a**
Plastic Potential Function		-	μ, ψ_2
Yield Criterion		-	**h**, α
Hardening/ Softening Relation		Γ, λ	**C**, **p**

parameters C and p, which are both curve-fitting parameters used in a power function relating plastic work to the state of stress. Thus, C and p serve the same purpose in the SH model as Γ and λ in the MCC model.

Four more parameters are present in the SH model. Two of these, h and α, are used to describe the shape of the yield surface in three dimensions. In the MCC model, this is done by the value of M, which therefore controls failure as well as the shape of the yield surface. While this is very practical and reduces the number of parameters, it does not provide the same flexibility in the MCC model as found in the SH model.

Finally, two parameters, μ and ψ_2, are used in the SH model to describe the plastic potential, which is different from the yield surface, thus providing for non-associated plastic flow, as found from experiments on real soils. Associated plastic flow is assumed in the MCC model, i.e. the yield surface acts as the plastic potential, and no additional parameters are therefore required in this model.

If these three analysis procedures and models are evaluated on an equal basis, then for soils without effective cohesion, the SA approach involves 3 parameters (for c = 0), the MCC model requires 5 parameters, and the SH model requires 11 parameter values (for a = 0). However, the SA approach clearly has its shortcomings in analysis capabilities as described in a previous section, while the MCC model primarily works for "well-behaved, insensitive" clays. Analyses performed with the SA and the MCC procedures do not provide the flexibilities presented by the SH model, which works equally well for sand and clay, involves non-associated plastic flow, models 3D failure correctly, and can handle contractive as well as dilative volume changes for soils in the normally consolidated range. The SH model has also been shown to work well for materials with effective cohesion such as cemented soils, concrete, and rock (for which $a \neq 0$).

Thus, models with larger number of parameters are typically more versatile and provide more flexibility than models with fewer parameters. In addition, determination of parameters for a given model does really not represent the greatest limitation on the use of the model, because the parameter values can typically be found by computer programs or by some computer program-assisted method.

The real limitation in the employment of advanced analysis methods lies in the expense for performing the required experiments. Although the isotropic compression and the triaxial compression tests are relatively simple to perform, they most often require undisturbed samples as well as a laboratory with the necessary equipment and know-how to perform these tests. Determination of parameters represents a small effort compared to the effort and expense involved in obtaining the experimental results.

OVERVIEW AND EVALUATION OF CONSTITUTIVE MODELS

To compare many of the well-known models available today, their attributes are listed and compared in tabular form, and an evaluation of the models is attempted. Several models occur in many different versions in the literature. To the extent possible, a representative version has been used for comparison with the other models. No discussion is presented in connection with the comparison of the models.

Tables 3-6 contain the models divided into different groups with similar frameworks and each model is given a name, usually the name under which they are generally known. However, many models could fit under different frameworks, because they include approaches and/or components from different frameworks. Key references to the literature where each model has been presented first or in which it has been explained particularly well are also given. The types of soil for which the model has been developed and shown to work are listed along with the key attributes such as the shapes of the failure, yield, and plastic potential surfaces as well as the hardening parameter(s) employed.

The capabilities of each model (as well as they could be discerned from the publications) are given along with the experiments required for calibration of the model parameters. In particular, it has been noted whether difficult, non-standard experiments are required for calibration. Here it has been assumed that conventional triaxial compression tests, isotropic compression tests, and conventional K_0-compression tests (conventional oedometer tests) can be performed on a routine basis. However, experiments such as K_0-compression tests with measurements of lateral stresses and triaxial extension tests as well as other experiments usually performed for purposes of research are considered here to be relatively complex and beyond conventional testing expertise.

Finally, the tables contain the number of parameters required in each model (although this should not be of primary concern, as discussed above) along with an overall evaluation of the model.

The evaluations listed in the tables are given with the intent of providing some guidance towards realistic, practically useful, and applicable constitutive models for soils. In this evaluation it has been kept in mind that "a model should be as simple as possible, but not simpler." Thus, included in the tables are some of the early, simple models along with the most advanced models available today. Several models are "experimental" in nature, and some of them are still under development. Thus, the evaluation, which aims to indicate practically useful models, reflects these conditions.

Only three categories are used in the evaluation. These categories and the conditions for a model to fit in each of these categories are summarized as follows:

> Category 1: Model includes (1) theoretically sound framework that is (2) sufficiently transparent and accessible to anticipate and evaluate the model performance; (3) model includes effects of confining pressure, (4) model can handle 3D conditions; (5) it is straight forward to find the parameters; (6) these may be determined from conventional experiments (difficulty not beyond conventional triaxial compression tests, isotropic compression tests and K_0-compression tests (conventional oedometer tests)); and (7) the model exhibits an overall high quality of fit with the observed behavior. Note that each model is only applicable as indicated in the other columns, but within these limitations it performs well.
>
> Category 2: Model lacks one or more of the conditions mentioned above.
>
> Category 3: Model is deficient in several of the conditions mentioned above.

TABLE 3(a). Constitutive Models. Their Attributes and Their Capabilities.

Type of Model	Model	References	Types of Soil			Failure Surface	Curved Failure Surface in Triaxial Plane?	Yield Surface	Plastic Potential: Associated or Nonassociated?	Hardening Parameter
			Sand	Clay	Cemented Soil					
Elastic	Hooke's Law	Textbook	Yes	Yes	Yes	NA	NA	NA	NA	NA
Elastic	Hyperbolic	Duncan and Chang (1970)	Yes	Yes	Yes	Mohr-Coulomb	Yes, see Duncan et al. (1980)	NA	NA	NA
Simple Elastic Plastic	Drucker-Prager	Drucker and Prager (1952)	Yes	Yes	Yes	Extended von Mises	No	= Failure surface	Associated	None
Simple Elastic Plastic	Drucker's Cap	Drucker et al. (1957)	Yes	Yes	Yes	Extended von Mises	No	Spherical Cap	Associated	Plastic Work
Simple Elastic Plastic	Stress-Dilatancy	Rowe (1962)	Yes	No	No	Mohr-Coulomb +Dilation Effect	Yes	Mohr-Coulomb	Nonassociated	NA
Simple Elastic Plastic	Mohr-Coulomb	Smith & Griffiths (1982) Brinkgreve and Vermeer (1997)	Yes	Yes	Yes	Mohr-Coulomb	No	= Failure Surface	Nonassociated	None
Simple Elastic Plastic	DiMaggio-Sandler	Dimaggio and Sandler (1971) Sandler et al. (1976)	Yes	Yes	Yes	Curved Extended von Mises	Yes	Elliptical Cap	Associated	Plastic Volumetric Strain
Simple Elastic Plastic	PLAXIS Soft Soil	Brinkgreve and Vermeer (1997)	No	Yes	Yes	Mohr-Coulomb	No	Elliptical Cap	Associated	Plastic Shear & Vol. Strain
Simple Elastic Plastic	Lade and Duncan	Lade and Duncan (1975)	Yes	Yes	No	Smooth Triangular, Conical	No	Smooth Triangular, Conical	Nonassociated	Plastic Work

TABLE 3(b). Constitutive Models, Their Attributes and Their Capabilities.

Model	Does Model Predict/Include					Experiments for Parameter Determination	Difficult, Non-Standard Experiments Required	Number of Parameters	Model Category
	Drained Softening	Large Stress Reversals	Cyclic Loading	Realistic Pore Pressures	Realistic 3D Behavior				
Hooke's Law	No	No	No	No	Yes, away from Failure	Any Comp. Test	No	2	3
Hyperbolic	No	No	No	No	Yes, away from Failure	3 Triax. Comp.	No	7	3
Drucker-Prager	No	No	No	No	No	Triax. Comp.	No	4	3
Drucker's Cap	No	No	No	No	No	Triax. Comp.	No	NA	3
Stress-Dilatancy	Yes	No	No	No	No, works only for Discrete Conditions	Triax. Comp. Pl. Strain Comp., or Triax. Ext	Yes	NA	3
Mohr-Coulomb	No	No	No	No	No	3 Triax. Comp.	No	5	2
DiMaggio-Sandler	No	No	No	No	No	3 Triax. Comp., 1 Iso. Comp.	No	10	2
PLAXIS Soft Soil	No	No	No	No	Yes	3 Triax. Comp.	No	7	2
Lade and Duncan	No	No	No	No	Yes	3 Triax. Comp.	No	9	2

TABLE 4(a). Constitutive Models, Their Attributes and Their Capabilities.

Type of Model	Model	References	Types of Soil			Failure Surface	Curved Failure Surface in Triax. Plane?	Yield Surface	Plastic Potential: Associated or Nonassociated?	Hardening Parameter
			Sand	Clay	Cemented Soil					
Critical State	Modified Cam Clay	Roscoe and Burland (1968)	No	Yes	No	Extended von Mises	No	Elliptical Cap	Associated	Plastic Volumetric Strain
	Elasto-Viscoplastic	Adachi and Oka (1982)	No	Yes	No	Smooth Triangular. Conical	No	Bullet-shaped (Orig. Cam Clay)	Associated	Plastic Volumetric Strain
	Structured Cam Clay	Liu and Carter (2002)	No	Yes	No	Extended von Mises	No	Elliptical Cap	Associated	Plastic Volumetric Strain
	Anisotropic Cam Clay	Sekiguchi and Ohta (1977)	No	Yes	No	Extended von Mises	No	Bullet-shaped (Orig. Cam Clay)	Associated	Plastic Volumetric Strain
	Anisotropic Plasticity	Dafalias et al. (2003)	No	Yes	No	Rotated, Extended von Mises	No	Distorted Ellipse	Nonassociated	Plastic Vol. & Shear Strain
	Nor-Sand	Jefferies (1993)	Yes	Yes	No	Smooth Triangular, Conical	No	Bullet-shaped (Orig. Cam Clay)	Associated	Plastic Work
	Unified SMP	Matsuoka et al. (1999)	Yes	Yes	No	Smooth Triangular, Conical	No	Modified Elliptical Cap	Associated	Product of Stress Funct. & Plastic Vol. Strain
	t_{ij}-Concept	Nakai and Matsuoka (1986) Nakai (1989)	Yes	Yes	No	Smooth Triangular, Conical	No	Modified Bullet-shape (Mod. Orig. Cam Clay)	Associated in t_{ij}-Space (Nonassoc.)	Plast. Vol. Strain(clay "Plastic Work" in t_{ij}-Space (sand)

TABLE 4(b). Constitutive Models, Their Attributes and Their Capabilities.

Model	Does Model Predict/Include					Experiments for Parameter Determination	Difficult, Non-Standard Experiments Required	Number of Parameters	Model Category
	Drained Softening	Large Stress Reversals	Cyclic Loading	Realistic Pore Pressures	Realistic 3D Behavior				
Modified Cam Clay	Yes, for OC clay	No	No	Yes	No	3 Triax. Comp. & 1 Iso. Comp.	No	5	2
Elasto-Viscoplastic	Yes, for OC clay	No	No	Yes	Yes	3 Triax. Comp. & 1 Iso. Comp.	No	10	1
Structured Cam Clay	Yes, for OC clay	No	No	Yes	No	3 Triax. Comp. & 1 Iso./K_0 Comp.	No	10	2
Anisotropic Cam Clay	Yes, for OC clay	No	No	Yes	Yes	3 Triax. Comp. & 1 Iso. Comp.	No	6	2
Anisotropic Plasticity	Yes, for OC clay	No	No	Yes	No	3 CU Triax. Comp. & Ext. & 1 Iso./K_0-Comp. with Meas. Lateral Stress	Yes	8	2
Nor-Sand	Yes, for OC soil	No	Yes, Undrained	Yes	Yes	2 CD Triax. Comp. & 2 CU Triax. Comp. with Iso. Comp.	No	7	1
Unified SMP	Yes, for OC soil	No	No	Yes	Yes	1 Triax. Comp. & 1 Iso. Comp.	No	5	2
t_{ij}-Concept	Yes, for OC soil	No	No	Yes	Yes	1 Triax. Comp. & 1 Iso. Comp.	No	5 (clay) 6 (sand)	2

TABLE 5(a). Constitutive Models, Their Attributes and Their Capabilities.

Type of Model	Model	References	Types of Soil			Failure Surface	Curved Failure Surface in Triax. Plane?	Yield Surface	Plastic Potential: Associated or Nonassociated?	Hardening Parameter
			Sand	Clay	Cemented Soil					
Incrementally Nonlinear	Darve	Darve (1982)	Yes	Yes	Yes	Mohr-Coulomb	No	NA	NA	NA
	Hypoplastic	Kolymbas (1991) Gudehus (1996) Bauer (1996)	Yes	No	No	Smooth Triangular, Conical	No	NA	NA	NA
Multiple Surfaces	Nested Surfaces	Iwan(1967) Mroz(1967) Prevost (1978)	Yes	Yes	Yes	Rotated, Extended von Mises	No	Rotating, Extended von Mises	Associated	Plastic Vol. &/or Shear Strain
	Fuzzy Set Plasticity	Klisinski et al (1987) Klisinski (1988)	Yes	Yes	Yes	Smooth Triangular, Conical	No	Smooth Triangular, Conical & Oct. Plane	Nonassociated and Associated	Plastic Work
Double Hardening	Lade	Lade (1977)	Yes	Yes	No	Smooth Triangular, Conical	Yes	Smooth Triangular, Conical & Spher. Cap	Nonassociated and Associated	Plastic Work
	PLAXIS Hardening	Brinkgreve and Vermeer (1997)	Yes	Yes	Yes	Mohr-Coulomb	No	Curved Mohr-Coulomb & Ellipt. Cap	Nonassociated and Associated	Plastic Vol. & Shear Strain
	MONOT	Molenkamp (1981) Griffiths et al. (1982) Hicks (2003)	Yes	Yes	Yes	Smooth Triangular, Conical	Yes	Smooth Triangular, Conical & Spher. Cap	Nonassociated and Associated	Plastic Vol. & Shear Strain

TABLE 5(b). Constitutive Models, Their Attributes and Their Capabilities.

Model	Does Model Predict/Include					Experiments for Parameter Determination	Difficult, Non-Standard Experiments Required?	Number of Parameters	Model Category
	Drained Softening	Large Stress Reversals	Cyclic Loading	Realistic Pore Pressures	Realistic 3D Behavior				
Darve	No	Yes	Yes	No	No	Triax. Comp. & Ext., Cyclic Tests	Yes, Depends on Conditions to be Modeled	16	2
Hypoplastic	Yes	Yes	Yes	Yes	Yes	Triax. Comp. & K_0-Comp.	Yes, Depends on Conditions to be Modeled	8	2
Nested Surfaces	No	Yes	Yes	Yes	No	Iso. & Triax. Comp. & Ext., Cyclic Tests	Yes, Depends on Conditions to be Modeled	Many Discrete Values (Prevost 1979)	2
Fuzzy Set Plasticity	Yes	Yes	Yes	Yes	Yes	Triax. Comp. & Iso. Comp.	Yes, Depends on Conditions to be Modeled	13-20	2
Lade	Yes	No	No	Yes	Yes	3 Triax. Comp. & 1 Iso-Comp.	No	14	1
PLAXIS Hardening	No	No	No	Yes	Yes	3 Triax. Comp. & 1 Iso-Comp	No	11	2
MONOT	No	Yes	Yes	Yes	Yes	3 Triax. Comp. & 1 Iso-Comp	No	21	1

TABLE 6(a). Constitutive Models, Their Attributes and Their Capabilities.

Type of Model	Model	Reference(s)	Types of Soil			Failure Surface	Curved Failure Surface in Triaxial Plane?	Yield Surface	Plastic Potential: Associated or Nonassociated?	Hardening Parameter
			Sand	Clay	Cemented Soil					
Bounding Surface	Dafalias	Dafalias et al. (1982) Kaliakin and Dafalias (1989)	No	Yes	No	Smooth Triangular, Conical	No	Elliptical Cap	Associated	Plastic Volumetric Strain
	MIT-E3	Whittle (1993) Ganendra and Potts (1995)	No	Yes	No	Rotated, Extended von Mises	No	Rotated, Elliptical Cap	Nonassociated	Plastic Volumetric Strain
	MIT-S1	Pestana and Whittle (1999) Pestana et al. (2002a) Pestana et al (2002b)	Yes	Yes	No	Smooth Triangular, Conical	No	Distorted Lemniscate	Nonassociated	Plastic Vol. & Shear Strain
Single Yield Surface	Single Hardening	Kim and Lade (1988) Lade and Kim (1988a) Lade and Kim (1988b)	Yes	Yes	Yes	Smooth Triangular, Conical	Yes	Tear-Drop Shaped	Nonassociated	Plastic Work
	Sinfonietta Classica	Nova (1988) Nova et al. (2003)	Yes	Yes	Yes	Smooth Triangular, Conical	No	Tear-Drop Shaped	Nonassociated	Plastic Vol. & Shear Strain
	Disturbed State Concept/ Hierarchial	Desai et al. (1986) Desai (2001)	Yes	Yes	Yes	Smooth Triangular, Conical	No	Tear-Drop Shaped	Associated or Nonassociated	Plastic Vol. & Shear Strain
	Generalized Plasticity	Pastor et al. (1990) Ling and Liu (2003)	Yes	No	No	Smooth Triangular, Conical	Yes	Elliptical Cap	Nonassociated	Plastic Vol. & Shear Strain

TABLE 6(b). Constitutive Models, Their Attributes and Their Capabilities.

Model	Does Model Predict/Include					Experiments for Parameter Determination	Difficult, Non-Standard Experiments Required	Number of Parameters	Model Category
	Drained Softening	Large Stress Reversals	Cyclic Loading	Realistic Pore Pressures	Realistic 3D Behavior				
Dafalias	Yes, for OC clay	Yes	Yes	Yes	Yes	3 CU Triax. Comp. 3 CU Triax. Ext. 1 Iso/K_0-Comp.	Yes	12	2
MIT-E3	Yes, for OC clay	Yes	Yes	Yes	Yes	2 CK_0U Triax. Comp. 1 CK_0U Triax. Ext. 1 CD Triax. Comp. K_0-Comp. K_0-Comp. with Meas. Lateral stresses Resonant Column	Yes	15	2
MIT-S1	Yes	Yes	Yes	Yes	Yes	CU Triax. Comp. (Iso/K_0) CD Triax. Comp. . (Iso/K_0) Iso./K_0-Comp. K_0-Comp. with Meas. Lateral stresses Resonant Column/Bender Element	Yes	13	2
Single Hardening	Yes	Yes, see: Lade and Inel (1997) Inel and Lade (1997)	Yes	Yes	Yes	3 Triax. Comp. & 1 Iso. Comp.	No	12	1
Sinfonietta Classica	No	No	No	Yes	Yes	3 Triax. Comp. & 1 Iso. Comp.	No	17	1
Disturbed State Concept/ Hierarchial	Yes	No	No	Yes	Yes, requires 3D Tests	3 Triax. Comp. & 1 Iso. Comp.	Yes	15	1
Generalized Plasticity	Yes	Yes	Yes	Yes	Yes	2 CD or CU Triax. Comp. & 1 Cyclic Test	No	15	1

The early models were developed at the time when only the simplest, classical 3D failure criteria were available, experimental results of 3D tests had not been performed, concepts in hardening plasticity were still being formulated, and conditions for plastic yielding were not yet fully discovered. These models appear in category 3. The models in category 1 fulfill all the points listed above, while those in category 2 are clearly missing one or more of the important qualities given above.

It should be noted that some of the most advanced and capable models have been placed in category 2, because they require difficult, non-standard experiments for calibration. However, if performance of such experiments is not considered to be a impediment to employment of such models, then the models clearly belong in category 1.

The writer is responsible for placement of each model within one of the three categories, and he apologizes for errors to the extent that misunderstandings have resulted in incorrect evaluation of the models.

SUMMARY

Presented here is a short review of many of the considerations involved in decisions to use advanced methods in analyses and design of geotechnical structures. This includes the requirements of the models, and why the models have been developed in terms of elasticity and hardening plasticity theories. An attempt has been made to answer some of the frequently asked questions regarding the use of constitutive models in advanced numerical analyses. Tables are presented with comparison of the main characteristic features and components of existing, widely available constitutive models. Key references are given to publications in which the models are first presented or particularly well explained. The requirements of the models in terms of experimental data for calibration are presented and compared in tables, and the capabilities of the models are reviewed. To provide some guidance in selecting realistic, practically useful, and applicable constitutive models for soils to employ in advanced numerical analyses, each model has been placed in one of three categories, the conditions for which are clearly stated.

Acknowledgment

The writer has benefited from countless invaluable discussions with Dr. Jerry A. Yamamuro over many years regarding experimental techniques, soil behavior, and constitutive modeling. Much appreciation is expressed for the assistance extended in connection with producing the present article.

REFERENCES

Adachi, T., and Oka, F. (1982) "Constitutive equation for normally consolidated clays based on elasto-viscoplasticity," *Soils and Foundations*, 22: 57-70.

Bauer, E. (1996) "Calibration of a comprehensive hypoplastic model for granular materials," *Soils and Foundations*, 36(1), 13-26.

Brinkgreve, R.B.J., and Vermeer, P.A. (1997) *"PLAXIS finite element code for soil and rock analysis – Version 7,"* Balkema, Rotterdam.

Dafalias, Y.F., Herrmann, L.R., and DeNatale, J.S. (1982) "The bounding surface plasticity model for isotropic cohesive soils and its application at the Grenoble Workshop," *Proc. Int. Workshop on Constitutive Relations for Soils*, Grenoble, 273-287.

Dafalias, Y.F., Papadimitriou, A.G., and Manzari, M.T. (2003) "Simple anisotropic plasticity model for soft clays," *Proc. Int. Workshop on Geotechnics of Soft Clays: Theory and Practice*, P.A. Vermeer et al. (eds.), 189-195.

Darve, F. (1982) "An incrementally non-linear constitutive law: Assumptions and predictions," *Proc. Int. Workshop on Constitutive Relations for Soils*, Grenoble, 385-403.

Desai, C.S. (2001) *Mechanics of Materials and Interfaces: The Disturbed State Concept,* CRC Press, Boca Raton, FL, USA.

Desai, C.S., Somasundaram, S., and Frantziskonis, G. (1986) "A hierarchial approach for constitutive modeling of geologic materials," *Int. J. Numer. Anal. Meth. Geomech.*, 10(3), 225-257.

DiMaggio, F.L., and Sandler, I.S. (1971) "Material Model for Granular Soils," *J. Engr. Mech. Div.*, ASCE, 97(EM3), 935-950.

Drucker, D.C., and Prager, W. (1952) "Soil Mechanics and Plastic Analysis or Limit Design," *Quart. Appl. Math.*, 10(2), 157-165.

Drucker, D.C., Gibson, R.E., and Henkel, D.J. (1957) "Soil Mechanics and Work-Hardening Theories of Plasticity," *J. Soil Mech. Found. Engr. Div.*, ASCE, 122, 338-346.

Duncan, J.M. (1994) "The Role of Advanced Constitutive Relations in Practical Applications," *Proc. 13th Int. Conf. Soil Mech. Found. Engr.*, New Delhi, India, 5, 31-48.

Duncan, J.M., and Chang, C.Y. (1970) "Non-linear analysis of stress and strain in soils," *J. Soil Mech. Founds Div.*, ASCE, 96(SM5) 1629-1653.

Duncan, J.M., Byrne, P., Wong, K.S., Mabry, P. (1980) "Strength, Stress-Strain and Bulk Modulus Parameters for Finete Element Analyses of Stresses and Movements in Soil Masses," *Report No. UCB/GT/80-01*, University of California, Berkeley, 77 pages.

Ganendra, D, and Potts, D.M. (1995) Discussion of 'Evaluation of constitutive model for overconsolidated clays,' by A.J. Whittle, *Geotechnique*, 45(1), 169-173.

Griffiths, D.V., Smith, I.M., and Molenkamp, F. (1982) "Computer implementation of a double-hardening model for sand," *Proc. IUTAM Conf. on Deformation and Failure of Granular Materials,* Delft, 213-221.

Gudehus, G. (1996) "A comprehensive constitutive equation for granular materials," *Soils and Foundations*, 36(1), 1-12.

Hicks, M.A. (1995) "MONICA – a computer algorithm for solving boundary value problems using the double hardening constitutive law MONOT: I. Algorithm development" *Int. J. Num. Anal. Meth. Geomech.* 19: 1-27.

Hicks, M.A. (2003) "Experience in calibrating the double-hardening constitutive model Monot," *Int. J. Numer. Analyt. Meth. Geomech.*, 27(13), 1123-1151.

Inel, S., and Lade, P.V. (1997) "Rotational Kinematic Hardening Model for Sand, Part II. Characteristic Work Hardening Law and Predictions," *Computers and Geotechnics*, Elsevier, 21(3), 217-234.

Iwan, W.D. (1967) "On a Class of Models for the Yielding Behavior of Continuous and Composite Systems," *J. Appl. Mech.*, Trans. ASME, 34(E3) 612-617.

Jefferies, M.G. (1993) "Nor-Sand: a simple critical state model for sand," *Geotechnique*, 43(1), 91-103.

Kaliakin, V.N. and Dafalias, Y.F. (1989) "Simplifications to the bounding surface model for cohesive soils," *Int. J. Numer. Analyt. Meth. Geomech.*, 13, 91-100.

Kim, M.K., and Lade, P.V. (1988) "Single Hardening Constitutive Model for Frictional Materials, I. Plastic Potential Function", *Computers and Geotechnics*, 5(4), 307-324.

Klisinski, M. (1988) "Plasticity theory based on fuzzy sets," *J. Engr. Mech.*, ASCE, 114(4), 563-582.

Klisinski, M., Alawi, M.M., Sture, S., Ko, H.-Y., and Muir Wood, D. (1987) "Elasto-plastic model for sand based on fuzzy sets," *Proc. Int. Workshop on Constitutive Equations for Granular Non-Cohesive Soils*, A. Saada and G. Bianchini (eds.), Case Western Reserve University, Cleveland, 325-347.

Kolymbas, D. (1991) "Computer-aided design of constitutive models, *Int. J. Numer. Analyt. Meth. Geomech.* 15, 593-604.

Lade, P.V., (1977) "Elasto-Plastic Stress-Strain Theory for Cohesionless Soil with Curved Yield Surfaces," *Int. J. Solids Structures*,13, 1019-1035.

Lade, P.V., and de Boer, R. (1997) "The concept of effective stress for soil, concrete and rock," *Geotechnique*, 47(1), 61-78.

Lade, P.V., and Duncan, J.M. (1975) "Elastoplastic Stress-Strain Theory for Cohesionless Soil," *J. Geotech. Engr. Div.*, ASCE, 101(GT10), 1037-1053.

Lade, P.V., and Inel, S. (1997) "Rotational Kinematic Hardening Model for Sand, Part I. Concept of Rotating Yield and Plastic Potential Surfaces," *Computers and Geotechnics*, Elsevier, 21(3), 183-216.

Lade, P.V., and Kim, M.K. (1988a) "Single Hardening Constitutive Model for Frictional Materials, II. Yield Criterion and Plastic Work Contours", *Computers and Geotechnics*, 6(1), 13-29.

Lade, P.V., and Kim, M.K. (1988b) "Single Hardening Constitutive Model for Frictional Materials, III. Comparisons with Experimental Data", *Computers and Geotechnics*, 6(1), 31-47.

Lambe, T.W., and Whitman, R.V. (1969) *"Soil Mechanics,"* John Wiley & Sons, New York.

Ling, H.I., and Liu, H. (2003) "Pressure-level denpendency and densification behavior of sand through a generalized plasticity model," *J. Engr. Mech.*, ASCE, 129(8), 851-860.

Liu, M.D., and Carter, J.P. (2002) "Structured Cam Clay Model," *Can. Geotech. J.*, 39(6), 1313-1332.

Matsuoka, H. and Nakai, T. (1974) "Stress-deformation and strength characteristics of soil under three different principal stresses," *Proc. JSCE*, 232, 59-70.

Matsuoka, H., Yao, Y.P., and Sun, D.A. (1999) "The Cam-clay models revised by the SMP criterion, " *Soils and Foundations*, 39(1), 81-95.

Molenkamp, F. (1981) Elasto-plastic double hardening model MONOT," *LGM Report CO-218595*, Delft Soil Mechanics Laboratory.

Mroz, Z. (1967) "On the Description of Anisotropic Work Hardening," *J. Mech. Phys. Solids*, 15, 163-175.

Nakai, T. (1989) "An isotropic hardening elastoplastic model for sand considering the stress path dependency in three-dimensional stresses," *Soils and Foundations*, 29(1), 119-137.

Nakai, T., and Matsuoka, H. (1986) "A generalized elastoplastic constitutive model for clay in three-dimensional stresses," *Soils and Foundations*, 26(3), 81-98.

Nova, R. (1988) "Sinfonietta classica: an exercise on classical soil modeling," *Proc. Symp. Constitutive Equations for Granular Non-Cohesive Soils*, Cleveland, A. Saada & G. Bianchini, (eds.), Balkema, Rotterdam, 501-519.

Nova, R., Castellanza, R., and Tamagnini, C. (2003) "A constitutive model for bonded geomaterials subject to mechanical and/or chemical degradation," *Int. J. Num. Anal. Meth. Geomech.* 27(9), 705-732.

Pastor, M., Zienkiewicz, O.C., and Chan, A.H.C. (1990) "Generalized plasticity and the modeling of soil behavior," *Int. J. Numer. Analyt. Meth. Geomech.*, 14(3), 151-190.

Pestana, J.M., and Whittle, A.J. (1999) "Formulation of a unified constitutive model for clays and sands," *Int. J. Numer. Analyt. Meth. Geomech.*, 23(12), 1215-1243.

Pestana, J.M., Whittle, A.J., and Salvati, L.A. (2002) "Evaluation of a constitutive model for clays and sands: Part I – sand behaviour," *Int. J. Numer. Analyt. Meth. Geomech.*, 26(11), 1097-1121.

Pestana, J.M., Whittle, A.J., and Gens, A. (2002) "Evaluation of a constitutive model for clays and sands: Part II – clay behaviour," *Int. J. Numer. Analyt. Meth. Geomech.*, 26(11), 1123-1146.

Prevost, J.H. (1978) "Plasticity Theory for Soil Stress-Strain Behavior," *J. Engr. Mech. Div.*, ASCE, 104(EM5), 1177-1194.

Prevost, J.H. (1979) "Mathematical modeling of soil stress-strain-strength behavior," *Proc. Third Int.Conf. Mumerical Methods in Geomechanics*, 1, 347-361.

Roscoe, K.H. and Burland, J.B. (1968) "On the generalized behaviour of 'wet' clay," *Engineering Plasticity*, 48: 535-609.

Rowe, P.W. (1962) "The stress-dilatancy relation for static equilibrium of an assembly of particles in contact," *Proceedings of the Royal Society*, 269, 500-527.

Sandler, I.S., DiMaggio, F.L., and Baladi, G.Y. (1976) "Generalized Cap Model for Geologic Materials," *J. Geotech. Eng. Div.*, ASCE, 102(GT7), 683-699.

Sekiguchi H., and Ohta, H. (1977) "Induced anisotropy and time dependency in clays," Constitutive aequations of Soils, *Proc.Specialty Session 9, Ninth Int.Conf. Soil Mech. Found. Engr.*, Tokyo, 1: 229-238.

Smith, I.M., and Griffiths, D.V. (1982) *Programming the finite element method – Second edition*, John Wiley & Sons.

Terzaghi, K. (1923) "Die Berechnung der Durchlässigkeitsziffer des Tones aus dem Verlauf der Hydrodynamischen Spannungserscheinungen," *Sitzungber. Akad. Wiss. Wien*, 132, 125-138.

Whittle, A.J. (1993) "Evaluation of a constitutive model for overconsolidated clays," *Geotechnique*, 43(2), 289-313.

Wood, D.M. (1990) "*Soil Behaviour and Critical State Soil Mechanics*," Cambridge University Press.

Chapter II
General Framework for Work-Hardening Plasticity Theory

Preface

The general frameworks for elasticity and work-hardening plasticity theories, as well as the mechanics principles that governs the stress-strain and strength modeling, are set out here. Prager (1949) presented these principles for solid metals in a short paper that has often been quoted.

There are two types of strains, recoverable elastic strains and irrecoverable plastic strains. The total strains consist of elastic and plastic strains, and these are additive, as follows:

$$d\varepsilon_{ij} = d\varepsilon_{ij}^{e} + d\varepsilon_{ij}^{p} \tag{2.1}$$

Time-dependent strains such as creep may also be included. Creep strains are also irrecoverable and therefore plastic in nature. The time-dependent strains are not included here, but their nature and inclusion are discussed in Chapter XIII.

The elastic strains are modeled by Hooke's law for which an elastic modulus and a Poisson's ratio are required, as discussed in Chapter V. It is assumed here that the materials are isotropic in nature. The effects of cross-anisotropy are discussed in Chapter XIV.

The plastic strains are obtained from work-hardening plasticity theory, which is explained here. The general framework consists of three components:

(1) A failure criterion, presented in Chapter VI.

(2) A flow rule, which for solid metals takes the following form:
Experimental observations indicate that solid metals do not exhibit any volume changes when deformed. This results in plastic strain increments perpendicular to the hydrostatic axis in the principal stress space, as shown on the following triaxial plane:

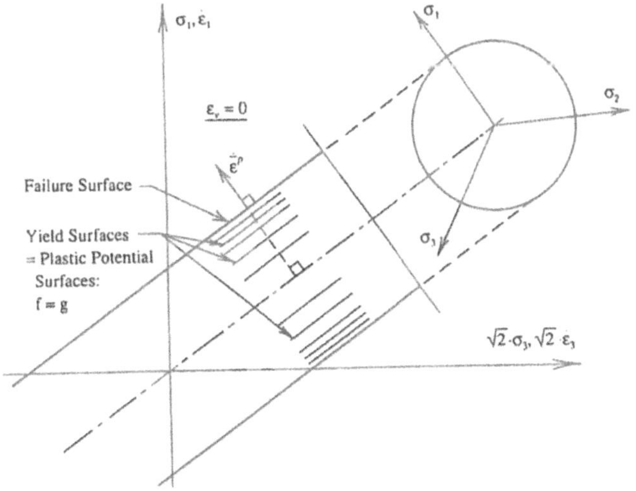

Fig. 2.1. Yield, failure, and plastic potential surfaces for solid metals, which exhibit zero volume changes during loading.

Here the yield and failure surfaces are modeled by the Von Mises criterion, which has a circular cross-section with the hydrostatic axis as the central axis. Because the superimposed plastic strain increments are perpendicular to the hydrostatic axis and therefore also perpendicular to the yield surface, the plastic strain increments may be expressed as follows:

$$d\varepsilon_{ij}^{p} = d\lambda \cdot \frac{\partial f}{\partial \sigma_{ij}} \tag{2.2}$$

which expresses that the plastic strain increments are proportional to the derivatives of the yield criterion, $f = f(\sigma_{ij})$, which are perpendicular to the yield surface. Thus, the plastic strain increments are associated with the yield surface, and this is therefore referred to as "associated plastic flow." The yield surface is therefore acting as the so-called plastic potential surface. The term $d\lambda$ is a proportionality constant that depends on the plastic work increment and is fitted to provide the magnitude of the plastic strain increments.

Frictional materials, however, do not show "associated plastic flow," because the plastic strain increments determined from experiments are not perpendicular to the yield surfaces observed for these materials. Therefore, a separate plastic potential expression is substituted for the expression for the yield surface. Thus, a non-associated plastic potential, $g = g(\sigma_{ij})$, is adopted for frictional materials. This is discussed in detail in Chapter VII, and the important consequences for frictional materials are presented in Chapter VIII. The expression for the plastic strain increments therefore becomes:

$$d\varepsilon_{ij}^{p} = d\lambda \cdot \frac{\partial g}{\partial \sigma_{ij}} \tag{2.3}$$

3) A yield criterion and a work-hardening law, which has the following form:

$$f(\sigma_{ij}, \varepsilon_{ij}{}^{p}) = 0 \tag{2.4}$$

For isotropic work-hardening, which determines the magnitude of the proportionality constant λ, this takes the following form:

$$f(\sigma_{ij}, W_p) = f'(\sigma_{ij}) - f''(W_p) = 0 \tag{2.5}$$

where the plastic work is calculated as follows:

$$W_p = \int \sigma_{ij} \cdot d\varepsilon_{ij}{}^{p} \tag{2.6}$$

Prager's Conditions:

Plasticity theory is versatile in the sense that the failure criterion, yield criterion and the plastic potential can be considered independently. However, there are some conditions that must be met by these functions to avoid violating the basic principles of continuum mechanics. Prager (1949) suggested four conditions which must be satisfied by the yield and plastic potential functions: continuity, uniqueness, consistency and irreversibility. Fig. 2.2 illustrates the stress, stress increment and strain increment vectors for the explanation of Prager's conditions. Here, the increment in stress, $d\sigma$, may be resolved in a component, $d\sigma_n$, normal to the yield surface, f_p, and a tangential component, $d\sigma_t$.

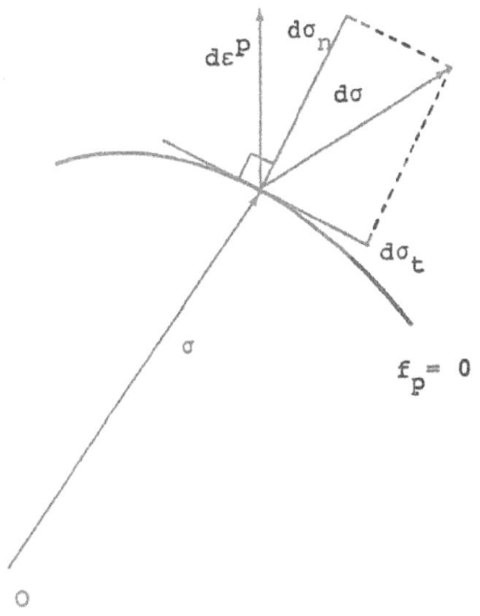

Fig. 2.2. Explanation of Prager's conditions in terms of stress, stress increments and plastic strain increment vectors on stress space.

(1) As a limiting case of transition between loading and unloading, neutral loading should not produce any plastic strains. Thus, the tangential component $d\sigma_t$ does not alter the state of plastic strains. Only the normal component results in plastic strain increments. This is called the condition of continuity.

(2) The uniqueness condition requires a unique distribution of the incremental stress and incremental strain fields for a given disturbance of the boundary values. Although Prager (1949) deduced that an associated flow rule must be used in order to satisfy this condition, Mroz (1963) showed that uniqueness can be achieved by selecting proper hardening parameters.

(3) The condition of consistency states that loading from a plastic state must again lead to a plastic state, i.e. the state after the loading must still satisfy the following equation:

$$f(\sigma_{ij}, W_p) = f'(\sigma_{ij}) - f''(W_p) = 0 \tag{2.7}$$

(4) The condition of irreversibility requires that the rate of entropy production be positive. Thus, for an isothermal process,

$$\sigma \cdot d\varepsilon^p > 0 \tag{2.8}$$

In other words, the stress vector must make an acute angle with the normal to the plastic potential surface. Prager (1949) concluded that the plastic potential surface must be convex with respect to the stress origin. Mroz (1963) showed that the convexity requirement also provides a sufficient condition for non-associated flow rules. In general, the condition of irreversibility is satisfied if the direction of increasing plastic potential, g_p, at a stress point is outward from the stress origin.

References:

Mroz, Z. (1963) "Non-associated flow laws in plasticity," Journal de Mecanique, 2(1): 21-42.

Prager, W. (1949) "Recent developments in the mathematical theory of plasticity," Journal of Applied Physics, 20(3): 235-241.

Chapter III
Comparison of Behavior of Solid Metals and Frictional Materials

Preface

Once it was realized that solid metals is a special case of materials and quite different from frictional materials, an article was written to highlight these differences. Many of the topics that are cursory mentioned in this article are further discussed in the following chapters.

INTERNATIONAL JOURNAL FOR NUMERICAL AND ANALYTICAL METHODS IN GEOMECHANICS, VOL. 12, 351–370 (1988)

THEME/FEATURE PAPER

EFFECTS OF VOIDS AND VOLUME CHANGES ON THE BEHAVIOUR OF FRICTIONAL MATERIALS

POUL V. LADE

Department of Civil Engineering, School of Engineering and Applied Science, University of California, Los Angeles, California 90024, U.S.A.

SUMMARY

Development of constitutive models for engineering materials is often based on a framework of elasticity and plasticity theories as developed for metals. Presented here is a review of some of the basic differences in the experimentally observed behaviour of metals and frictional materials. Two of the distinguishing characteristics between metals and frictional materials are the absence or presence of voids and the tendency to change or maintain constant volume during compression and shear. These differences must be recognized in order to develop reasonable constitutive models for frictional materials.

INTRODUCTION

Plasticity theory has to a large extent been developed on the basis of observed behaviour of metals, and the principles embodied in classical plasticity theory apply particularly well for such materials. Various types of materials respond in quite different ways to application of stresses. This has not always been recognized, and many of the principles of plasticity theory have often been used uncritically in the development of constitutive models for materials which exhibit patterns of behaviour different from those observed for metals. This situation may have developed due to a lack of knowledge about the behaviour of such materials under general three-dimensional stress and strain conditions. In particular, frictional materials such as soil, rock, concrete and ceramics show behaviour patterns quite different from those of metallic materials.

The fundamental information regarding the behaviour of materials must be determined from appropriate experiments. Sufficient experimental data are now available to allow recognition of the basic differences in behaviour of metals and frictional materials. The major differences as well as the similarities are listed in Table I and reviewed below. Each of the behaviour characteristics is discussed and illustrated with examples from experimental studies. Two of the distinguishing characteristics between metals and frictional materials appear to be the absence or presence of voids and the tendency to change or maintain constant volume during compression and shear.

The behaviour of metals being well established, the emphasis will be placed on illustration of the behaviour of frictional materials. The time-independent behaviour of both types of materials may be characterized by elasticity theory and work-hardening and -softening plasticity theory. However, the differences in observed behaviour result in fundamentally different constitutive models for metals and frictional materials.

0363–9061/88/040351–20$10.00

Received 27 July 1987
Revised 6 November, 1987

Table I. Similarities and differences in behaviour of metals and frictional materials

	Metals	Frictional materials
1	Material is solid throughout	Material contains pores or voids
2	Mass density can be assumed to be the same	Mass density can vary (at zero stress state)
3	Metals deform and fail in response to changes in total stresses	Frictional materials deform and fail in response to changes in effective stresses which are equal to total stresses minus pore pressures ($\sigma' = \sigma - u$)
4	Mechanical properties are constant and can often be listed in tables	Mechanical properties vary and are often characterized in relation to density or void ratio
5	Metals show initial elastic behaviour with nonlinear plastic deformations occurring only after the yield point is exceeded at high shear stresses	Frictional materials such as non-cemented soils exhibit nonlinear plastic behaviour even at very small stress changes near the isotropic stress state
6	Plastic yielding does not occur in response to isotropic stress increases (in the usual range of working stresses)	Plastic yielding (i.e. irrecoverable deformation) does occur in response to isotropic stress increases
7	Work input for deformation of metals is independent of mean normal stress	Work input for deformation of frictional materials increases with mean normal stress
8	Stiffness and strength are constant corresponding to constant density	Stiffness and strength increase with density
9	Stiffness and strength are independent of mean normal stress	Stiffness and strength increase with mean normal stress
10	Stress-path dependent behaviour is independent of mean normal stress	Stress-path dependent behaviour relates to the magnitude of the mean normal stress
11	Volume remains constant during shearing	Significant volume changes are associated with shearing under drained conditions
12	Saint-Venant's principle for plastic deformations characterizes the behaviour of metals during rotation of principal stresses	Saint-Venant's principle for plastic deformations characterizes the behaviour of frictional materials during rotation of principal stresses
13	Compressive and tensile strengths are approximately equal in magnitude	Compressive strength is greater than tensile strength
14	Shear strength is constant corresponding to no volume change	Shear strength is related to allowed volume change
15	Three-dimensional failure surface is almost circular in cross-section with central axis along hydrostatic axis	Cross-section of three-dimensional failure surface is triangular with smoothly rounded corners
16	Metals exhibit associated plastic flow	Frictional materials exhibit non-associated plastic flow
17	Metals exhibit strain- or work-hardening as well as -softening	Frictional materials exhibit strain- or work-hardening as well as -softening
18	Stability is governed by Drucker's stability postulate	Stability is related to the volume change behaviour: dilative materials are stable inside the failure surface although Drucker's stability criterion is violated; compressive materials are unstable inside the failure surface for certain stress-paths which may result in liquefaction
19	Development of shear bands (Luder's bands) occurs before peak failure and results in strengthening of the material	Development of shear planes in triaxial compression tests occurs after peak failure in dilative materials and results in decreasing shear strength

Most of the experimental results used for illustrations have been produced by the author and his co-workers. Various observed patterns of behaviour are presented and illustrated by these results, but the paper is descriptive in nature and does not represent an exhaustive review of the literature.

MATERIAL CONSTITUTION

Metals consist of atoms bonded together in crystals. Most conventional types of metals are polycrystalline, i.e. they are composed of small crystals fitting together in a tight, solid structure without macroscopic voids or pores. The internal atomic bonds in the crystals and the bonds between the crystals are much stronger than the bonds created by cementation or cohesion between mineral grains in rocks, concrete and ceramics.

Due to lack of voids or pores in most metals of technological importance, the mass density of each type of metal is constant, and the mechanical properties of metals are usually constant. Therefore, each metal with a given composition possesses properties which can be determined and tabulated in engineering handbooks once and for all.

There are some exceptions to this general pattern of behaviour for metals. Materials such as cast iron and bronze contain pores, and their mass densities as well as their engineering properties vary with porosity.

In comparison with solid metals, the composition of frictional materials includes voids or pores, and their mass or bulk densities vary with porosity. For example, non-cemented dry soils can be compacted to various degrees, and the amount of voids, or the void ratio, present in the compacted soil is directly dependent on the degree of compaction. Similarly, concrete of a given composition can be prepared at different densities before it hardens. The properties of frictional materials are variable and they are consequently not suitable for listing in engineering handbooks. However, general patterns of behaviour can be established for frictional materials, and these patterns are usually presented in relation to densities or void ratios.

THE EFFECTIVE STRESS PRINCIPLE

Porous materials in which the pores are interconnected allow flow of gas and/or fluid through their structure. The fluid in the pores can sustain hydrostatic pressures of considerable magnitudes and importance. Frictional materials deform and fail in response to changes in effective normal stresses which are equal to the total applied stresses minus the effect of the hydrostatic pressures generated in the gas and/or fluid in the pores. This principle can be expressed as

$$\sigma' = \sigma - a \cdot u \tag{1}$$

in which σ' = effective stress, σ = total stress, u = pore pressure, and a = factor that determines the relative influence of the hydrostatic pore pressure. The value of 'a' depends on the area of contact between particles per unit gross area of material, the angle of intrinsic shearing resistance, and the effective friction angle. For soils, in which the contact areas between particles are very small compared with the size of the particles, the value of 'a' is unity. For materials such as rock and concrete, in which the contact areas may be considerable in relation to the particle sizes, the value of 'a' is smaller than unity. A detailed analysis and discussion of the effective stress principle is given by Skempton.[1]

The effective stress principle applies to frictional materials for which it plays a unique and important role. It should be emphasized again that the behaviour in terms of deformation and failure of such materials is controlled entirely by effective stresses, not total stresses. A concept similar to the effective stress principle is not required for metals which are solid or without interconnected pores. Metals deform and fail in response to change in total applied stresses.

STRENGTH

The strengths of solid metals under general three-dimensional stress conditions can usually be characterized with acceptable accuracy by the von Mises failure criterion.[2] According to this criterion, failure occurs when the octahedral shear stress reaches a limiting value. In principal stress space the von Mises criterion describes a cylindrical surface whose axis coincides with the hydrostatic axis. The tensile and compressive strengths of solid metals are equal, and the magnitude of the mean normal stress has negligible influence on the strength in the normal range of engineering stress applications. This behaviour is correctly described by the von Mises criterion.

The three-dimensional strengths of frictional materials follow a different pattern. The diagram in Figure 1 shows experimental results from tests on sand[3] and concrete.[4] The failure surface for sand exhibits no strength in the unconfined state, whereas concrete possesses compressive as well as tensile strength when unconfined. Materials of geologic origin with effective cohesion have strengths in tension which are many times smaller than their strengths in compression. For example, the tensile strength of concrete is only about one-tenth of the unconfined compressive strength. Frictional materials derive their shear strengths from the applied normal stress through the friction between material constituents. Thus, the strengths of frictional materials increase with the mean normal stress. However, the rate of strength increase usually decreases with increasing confining pressure. This results in curved failure surfaces as shown in Figure 1(a) and 1(c).

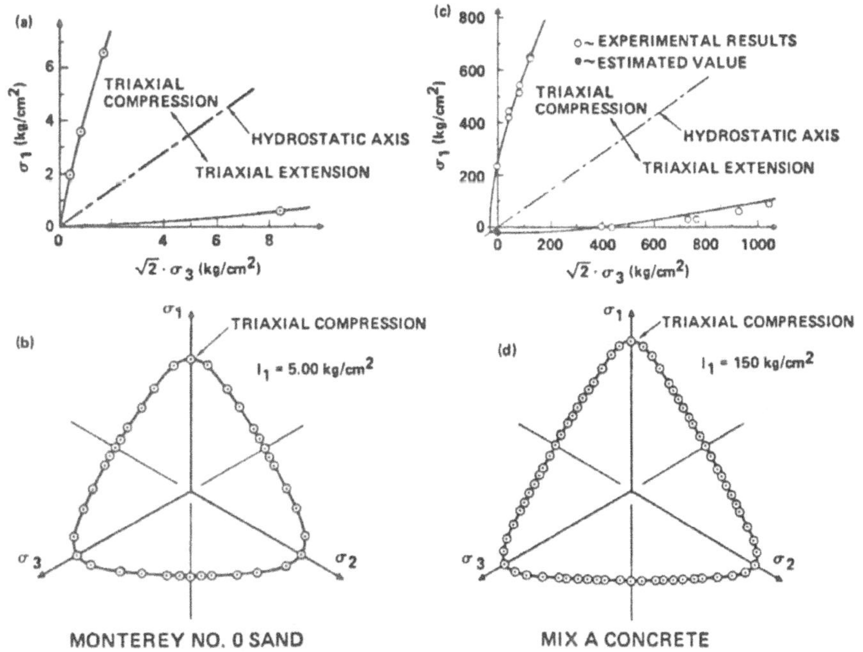

Figure 1. Failure surfaces for dense sand and concrete shown in (a), (c) triaxial planes and in (b), (d) octahedral planes

The cross-sectional shapes of the failure surfaces for sand and concrete are shown in the octahedral plane in Figure 1(b) and 1(d). Both surfaces are distinctly triangular with smoothly rounded corners. The surface for sand is more rounded than that for concrete, but description of these surfaces requires in both cases the involvement of the third stress invariant of the stress tensor. This invariant (or similar) embodies the smooth triangular shape which is so characteristic of frictional materials.

The variation of strengths of frictional materials can be directly related to density or void ratio. For example, experimental results for sands indicate that the friction angle in triaxial compression (ϕ) and the void ratio (e) may be related with reasonable accuracy through the following empirical expression[5]

$$e \cdot \tan \phi = \text{const.} \tag{2}$$

in which the value of the constant is determined experimentally for a given sand. This expression indicates that the friction angle increases with decreasing void ratio or with increasing density. Similar increases in strength with increasing density are also observed for other frictional materials.

Whether a material is of frictional nature can be discovered by comparing the strengths in compression and in tension. If the compressive strength is greater than the tensile strength, then the material exhibits behaviour characteristics in common with conventional frictional materials. For example, the tensile strength of cast iron is about one-half of its compressive strength. Further investigation reveals that cast iron possesses many other characteristics of frictional materials of the types discussed below.

It is interesting to note that a frictional material forced to maintain constant volume exhibits many behaviour characteristics in common with solid metals. For example, unconsolidated–undrained triaxial compression tests (UU-tests with $\sigma_2 = \sigma_3$) on saturated soil with the same initial density and with no allowed volume change during shear produce exactly the same stress–strain relations and strength independently of the magnitudes of the externally applied mean normal stress. This type of behaviour is therefore similar to that obtained for metals.

However, the three-dimensional behaviour for soils is different from that obtained for metals. Figure 2 shows the strengths obtained from consolidated–undrained cubical triaxial tests on normally consolidated, remoulded Grundite clay.[6] The density of the specimens is constant for each consolidation pressure, and no volume changes are allowed during testing of the specimens. Figure 2(a) shows the results interpreted in terms of total stresses, i.e. the effects of the pore water pressures inside the specimens are disregarded. Whereas solid metals exhibit a circular cross-section in the octahedral plane, the experimental results for undrained clay with constant volume form a cross-section which is closer to the hexagonal Tresca criterion. However, the Tresca criterion, which has been fitted to the strength in triaxial compression ($\sigma_2 = \sigma_3$), overestimates the undrained strength by increasing amounts as the intermediate principal stress approaches the major principal stress. In triaxial extension ($\sigma_2 = \sigma_1$), the undrained strength is about 7 per cent smaller than the strength obtained in triaxial compression for Grundite clay.

It is interesting to note that when the same test data are interpreted in terms of effective stresses, the resulting cross-sections in octahedral planes, shown in Figure 2(b), are triangular with smoothly rounded corners, i.e. similar to those for sand and concrete shown in Figure 1(b) and 1(d). It should be noted that the three cross-sections shown in Figure 2(b) correspond to three different mean normal stresses as represented by the first stress invariant I_1 in this figure. Thus, the three cross-sections represent three views of the same three-dimensional failure surface. Only one effective stress failure surface exists for a given normally consolidated clay, and it is shaped as a cone with the apex at the origin, as shown for sand in Figure 1(a).

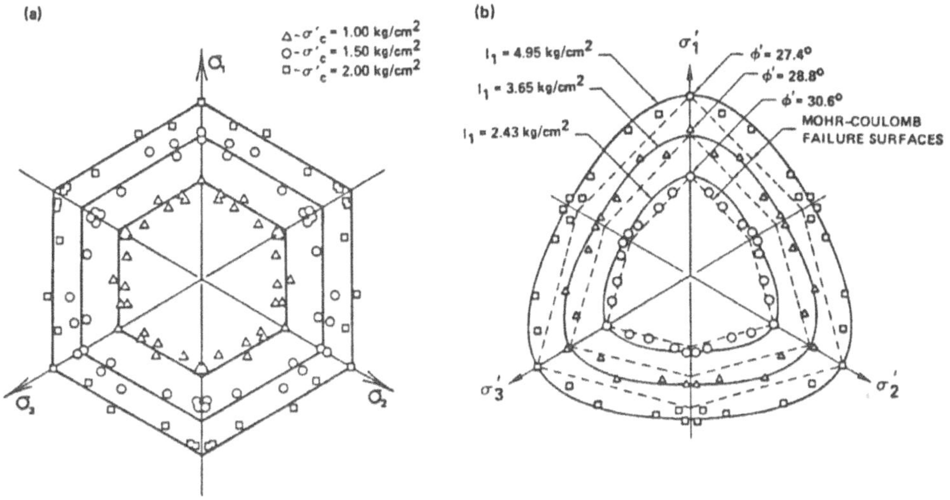

Figure 2. Cross-sectional shapes of failure surfaces from cubical triaxial CU-tests on normally consolidated, remoulded Grundite clay interpreted in terms of (a) total stresses, and (b) effective stresses

It is clear from the examples of experimental behaviour presented above for sand, clay and concrete that two of the distinguishing characteristics for metals and frictional materials are the presence of voids and the tendency to change or maintain constant volume during shear.

ELASTIC AND PLASTIC DEFORMATIONS

Different types of materials respond in different ways to applied stresses. Application of moderate tensile stresses to a specimen of a typical metal results in adjacent atoms being pulled only slightly apart. Removal of the stresses causes the metal to return to its initial size and shape. The deformation, which is recoverable and elastic in nature, is usually linear, i.e. the deformation is proportional to the applied stresses. If higher stresses are applied to the metal specimen, the atoms move further apart and they do not return to their original position upon unloading. Thus, permanent deformation of the metal has occurred. Nonlinear, permanent or plastic deformations are produced after the yield point has been exceeded. Application of even higher stresses causes additional plastic deformations and ultimately results in fracture of the metal.

Thus, metals show initial elastic behaviour with nonlinear plastic deformations occurring only after the yield point is exceeded at high shear stresses. Isotropic stress increases do not result in plastic deformation of solid metals, because they lack voids to allow compression. Only shear stresses cause plastic deformations, and yield surfaces for metals are therefore parallel to the hydrostatic axis, and they have cross-sections in the octahedral plane which form concentric circles. During the process of plastic yielding, the elastic modulus remains constant and independent of the magnitudes of normal and shear stresses.

The stress–strain behaviour of frictional materials is considerably influenced by the presence of voids and the occurrence of volume changes, both during isotropic compression and during shear. Figure 3(a) shows a diagram of typical observed variations of stress difference, $(\sigma_1-\sigma_3)$, and volumetric strain, ε_v, with axial strain, ε_1, obtained from a drained triaxial compression test performed with constant confining pressure on sand.[7] Similar results would be obtained from tests on other frictional materials. The stress–strain relationship is nonlinear, and a gradual

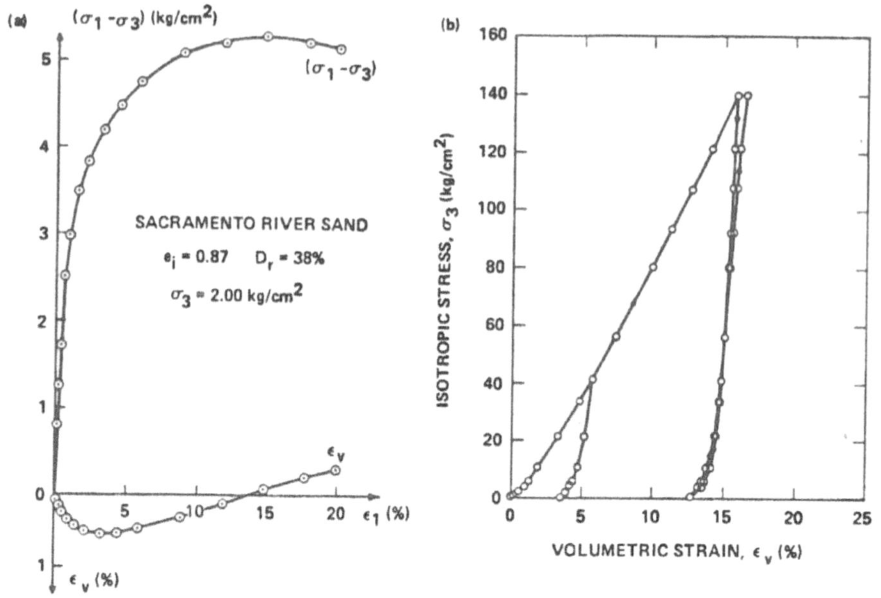

Figure 3. Stress–Strain and volume change relations from (a) drained triaxial compression test, and (b) isotropic compression test on Loose Sacramento River Sand

decrease in strength follows peak failure. The volumetric strain is initially compressive and this behaviour may be followed by expansion (as shown in Figure 3a) or by continued compression.

Both recoverable and irrecoverable deformations occur from the beginning of loading. The first yield point is therefore located on the hydrostatic axis. The deformations may be characterized as elastic and plastic, respectively. The plastic strains are initially smaller than the elastic strains, but at higher values of stress difference the plastic strains dominate the elastic strains.

Isotropic compression of sand results in both elastic and plastic deformations. The plastic deformation is allowed by the presence of voids in the material. Figure 3(b) shows that a yield point is established upon application and removal of the isotropic stress. The plastic yield surface for the sand must therefore cross the hydrostatic axis.

From these simple observations it is clear that the plastic yield surface for a material such as sand is affected by shear stresses as well as normal stresses. In fact, a combination of shear and normal stresses governs the plastic yield behaviour of frictional materials.

Experimental results show that the effect of the intermediate principal stress is to steepen the initial slope of the stress–strain relationship. This may be predicted by the elastic portion of a constitutive model. The elastic modulus varies with normal as well as shear stresses.[8] The initial density of the soil also plays an important role for the magnitude of the elastic modulus.

The stress–strain behaviour pattern reviewed above and many additional aspects of the behaviour of frictional materials may be characterized by elasticity theory and plasticity theory. These theories provide convenient frameworks for modelling the recoverable and the irrecoverable deformations in frictional materials.

YIELD SURFACES

As mentioned above, yield surfaces for metals are parallel to the hydrostatic axis, and they have cross-sections in the octahedral plane which form concentric circles. Thus, only shear stresses produce plastic yielding in metals.

The plastic behaviour of frictional materials is influenced by both normal and shear stresses. A typical pattern of yield surfaces for an isotropic granular material is shown on the triaxial plane in Figure 4. In three dimensions these yield surfaces are shaped as asymmetric tear drops with smoothly rounded triangular cross-sections similar to those of the failure surfaces shown in Figure 1(b) and 1(d). For an isotropic material the yield surfaces intersect the hydrostatic axis in a perpendicular manner, they bend backwards towards the origin, and cross the failure surface at sharp angles as indicated in Figure 4. Yield surfaces shaped approximately as those shown in Figure 4 may be represented by one continuous function (e.g. 9).

The stress-path for a conventional drained triaxial compression test performed at constant confining pressure, such as that shown in Figure 3(a), is indicated in Figure 4. As the specimen is loaded from point B on the hydrostatic axis to C on the failure surface, the inclination of the yield surface changes. At low deviator stresses near point B, the yield surface is inclined towards the outwards direction of the hydrostatic axis. As loading proceeds, the inclination of the yield surface changes gradually and it becomes inclined towards the origin as failure is approached at point C.

The yield surfaces shown in Figure 4 also indicate that isotropic compression results in plastic yielding, as presented in Figure 3(b).

WORK INPUT

The work required to deform a solid metal from the hydrostatic axis to failure is related to shear stresses only. Because plastic deformations and ultimate failure are produced due to changes in octahedral shear stresses only, the magnitude of the mean normal stress does not play any role in the amount of work input necessary to deform and fail the metal.

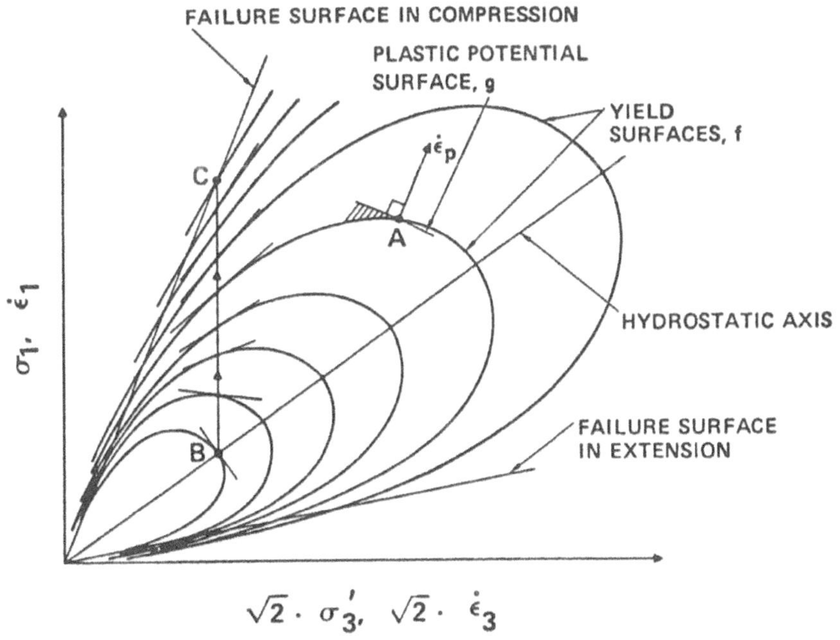

Figure 4. Pattern of yield surfaces for isotropic granular materials. Stress-path for conventional triaxial compression test (BC)

In comparison, the work input to deform a frictional material from the hydrostatic axis to failure is dependent on the initial and final states of stress. For primary (or virgin) loading, the work input varies with the mean normal stress and the applied shear stress. The higher the shear stress at failure (which in turn depends on the normal stress), the higher the work input required to bring a frictional material to failure.

STRESS-PATH DEPENDENCY

Whereas the recoverable, elastic strains are not stress-path dependent, it is the occurrence of the irrecoverable, plastic strains that accounts for the effects of stress-path. The plastic strains occurring in metals are dependent on the applied octahedral shear stress only. A given increment in octahedral shear stress will produce the same plastic strains whether applied at low or high mean normal stresses. However, the plastic behaviour of metals is dependent on the stress-path for general conditions involving primary loading, unloading and reloading.

The plastic behaviour of frictional materials is dependent on both shear stresses and normal stresses. It is therefore possible to follow different stress-paths between two stress points which produce different amounts of strain.

The dependency of the strains on the stress-path is particularly pronounced at high stress levels where the plastic strains are relatively large. Figure 5 shows two stress-paths, both with varying stress difference and varying confining pressure. One stress-path involves primary loading (stress level continuously increasing above previous maximum stress level) and unloading, whereas the other involves proportional loading along the hydrostatic axis and primary loading. Comparison of the strains occurring for the two types of stress-paths shows that the magnitudes of the strains can be considerably different.[10] The stress-path with the higher average stress level will produce larger strains than the stress-path with the lower average stress level. Thus, the strains for stress-path A-1-B are larger than the strains for stress-path A-2-B. Note that this observation applies only when one or both of the stress-paths involve unloading or reloading, or both.

The two stress-paths shown in Figure 6 both lead from point A to point B and both involve only primary loading. Experimental results have indicated[10] that any two stress-paths from point

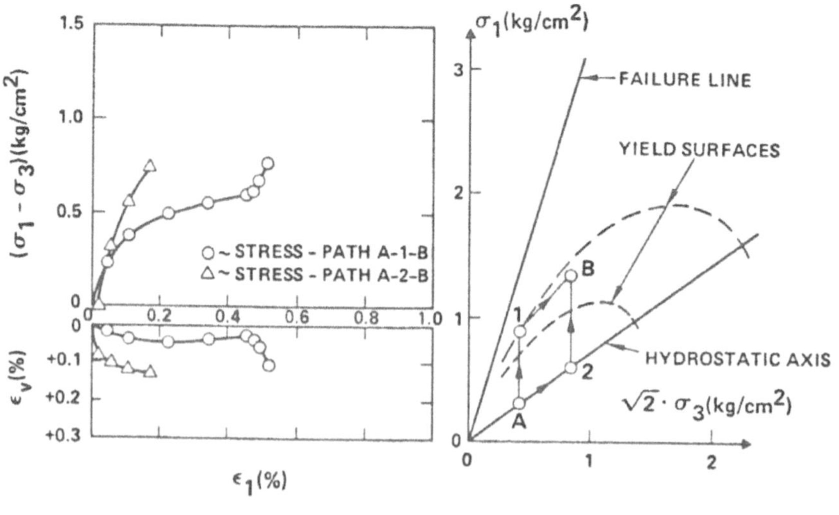

Figure 5. Comparison of stress–strain behaviour for two stress-paths, one involving primary loading and unloading (A-1-B), the other involving proportional loading and primary loading (A-2-B) for Loose Monetary No. O Sand.

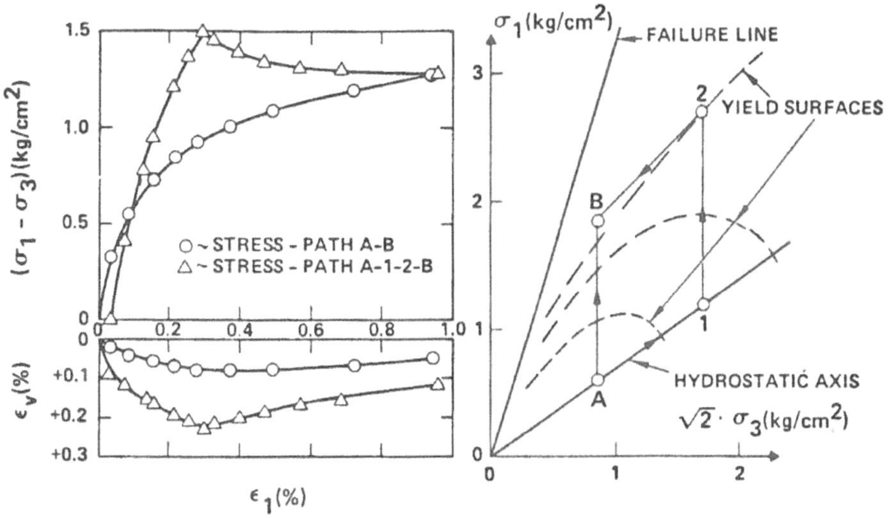

Figure 6. Comparison of stress–strain behaviour for two stress paths, one involving primary loading (A-B), the other involving proportional loading and primary loading (A-1-2-B) for Loose Monetary No. O Sand

A to point B will result in the same or very nearly the same strains, provided that both stress-paths involve only primary loading.

Thus, frictional materials, by virtue of their dependence on both normal and shear stresses, in general exhibit stress-path dependent behaviour. Both the occurrence and the magnitude of the irrecoverable strains can be modelled quite well by work-hardening plasticity theory.

ROTATION OF PRINCIPAL STRESSES

Elastic material behaviour implies that the directions of strain increment coincide with those of stress increment during rotation of principal stresses. This is illustrated in Figure 7(a), and this is the behaviour observed at low stress levels or during unloading and reloading of frictional materials.

Previous experimental evidence of the plastic behaviour of metals lead Saint-Venant[11] to propose that the principal axes of incremental plastic strain coincide with the principal stress axes during stress rotation. This is illustrated in Figure 7(b).

Experiments on soils have indicated that during rotation of the principal stress directions, the stress increment and strain increment axes coincide at low stress levels with transition to coincidence of strain increment and stress axes at high stress levels. This is illustrated by the experimental results from torsion shear tests on Santa Monica Beach Sand shown in Figure 8. Thus, the behaviour pattern observed for frictional materials at high stress levels corresponds with Saint-Venant's observed plastic behaviour of metals.

Saint-Venant's principle has another important, but more subtle implication. Consider a soil element on the verge of failing under the influence of a small shear stress, $\Delta\tau$, as shown in Figure 9(a). The theory of plasticity, as developed in part on the basis of Saint-Venant's principle, predicts that the incremental strains at failure are vertical and horizontal, i.e. they are normal strains coinciding with the directions of the principal stresses, as shown in Figure 9(b). In comparison, (nonlinear) elastic behaviour would predict the incremental shear deformation shown in Figure 9(c). Experimental evidence clearly supports the behaviour indicated in Figure 9(b).

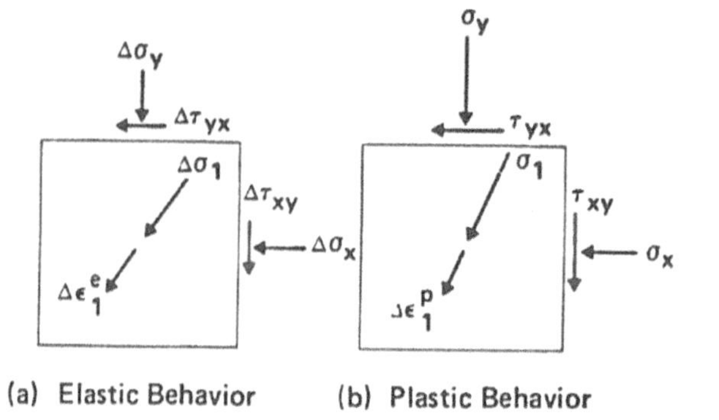

Figure 7. (a) Elastic and (b) plastic behaviour of material element during rotation of principal stress axes

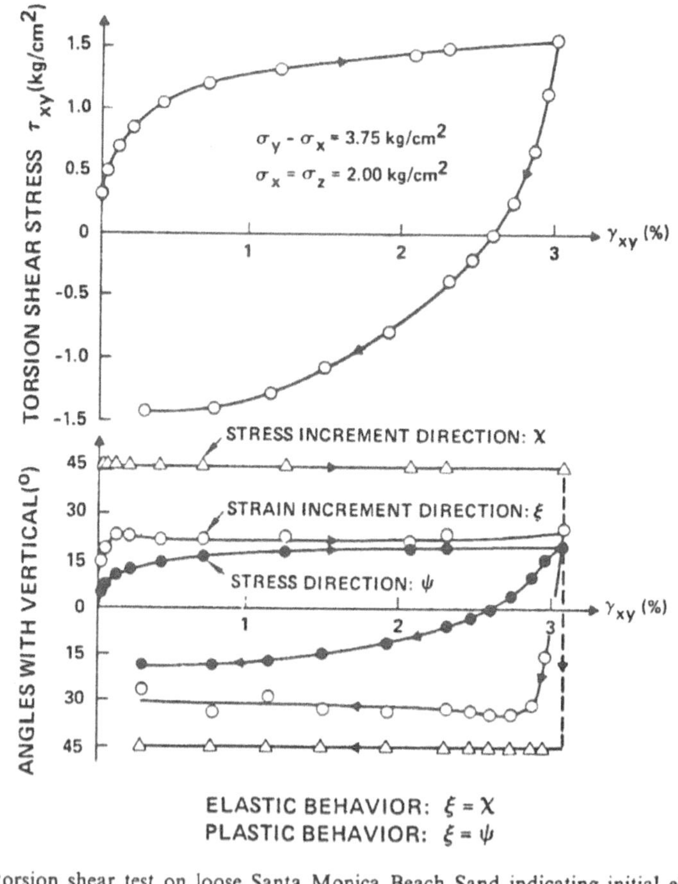

Figure 8. Results of torsion shear test on loose Santa Monica Beach Sand indicating initial elastic behaviour with gradual transition to plastic behaviour during rotation of principal stress axes

The behaviour of both metals and frictional materials during rotation of principal stress directions can be modelled quite well by a combination of elasticity and plasticity theories. Thus, an elasto-plastic work-hardening constitutive model has the ability to incorporate the characteristic behaviour observed and described above for metals and for frictional materials.

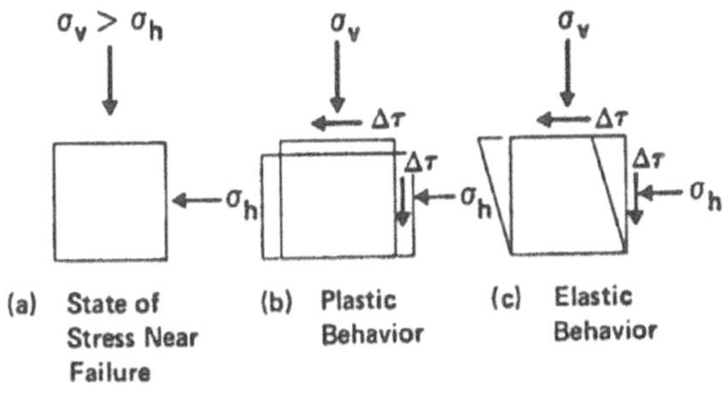

Figure 9. Effects of (b) plastic behaviour (Saint-Venant's principle) and (c) elastic behaviour of (a) material element near failure

VOLUME CHANGES

Solid metals do not exhibit macroscopic plastic volume changes during isotropic compression and shear. Their densities therefore remain almost constant (they have elastic volume changes) during any possible loading condition.

When frictional materials are loaded in compression or shear, their volumes and therefore their densities change. For example, for a given confining pressure loose sand will compress and dense sand will dilate during shear, as indicated in Figure 3(a). These volume changes, whether compression or dilation, can attain substantial magnitudes, and they account in a rather sensitive manner for many of the behaviour patterns associated with frictional materials.

To illustrate the importance of the shear–dilatancy relations, consider the behaviour of a dense sand in triaxial compression shown in Figure 10. The granular material exhibits a high rate of dilation under drained conditions and the stress–strain relation is smoothly curved. In comparison, volume changes are completely prevented in the undrained test on the granular material. Because the tendency for dilation is prevented, a tension or a decrease in pore pressure occurs. This results in an equal increase in effective confining pressure and a corresponding increase in strength. If cavitation of the pore fluid is avoided by using a high back-pressure (as is present, for example, at the ocean bottom), the effective confining pressure can increase substantially, and the undrained strength can be several times higher than the drained strength. The stress–strain relationship for the undrained test is completely different from that of the drained test.

Because the volume changes are so important for the type of behaviour exhibited by frictional materials, it is paramount that they be correctly accounted for in the constitutive model, both in terms of their rate of development and their magnitude. Plasticity theory can be employed to model the volume changes of frictional materials accurately and without violation of basic physical laws.

ASSOCIATED AND NON-ASSOCIATED FLOW

Yield surfaces for metals are shaped as circular cylinders with their axes on the hydrostatic axis of the principal stress space. Since metals do not exhibit volume changes during shear, the plastic strain increment vectors are perpendicular to the hydrostatic axis and the yield surfaces. These yield surfaces can therefore also serve as plastic potential surfaces, and the corresponding flow rule for plastic strains is consequently associated.

Experimental evidence from tests on several types of frictional materials have clearly indicated that the use of associated flow rules results in prediction of too large volumetric expansion. To

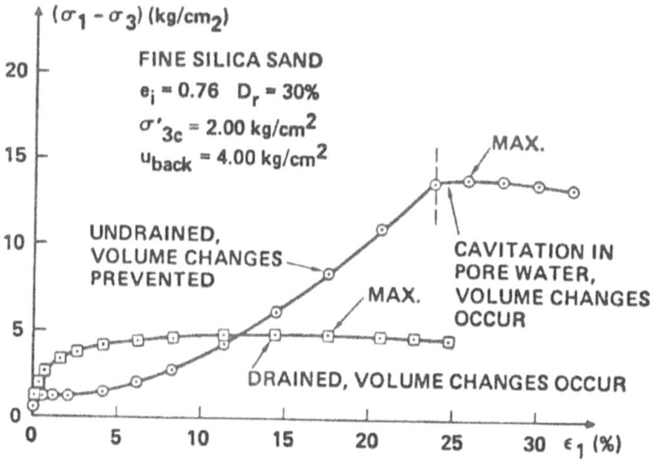

Figure 10. Comparison of stress–strain behaviour obtained from drained and undrained triaxial compression tests on fine silica sand

characterize the volume changes correctly, it is necessary to employ a non-associated flow rule. The plastic potential surfaces therefore do not coincide with the yield surfaces. Plastic potential surfaces have similar shapes as yield surfaces, and for non-associated flow the two families of surfaces cross each other. Experimental evidence for frictional materials indicates that the plastic potential surfaces have flatter rounded ends than the yield surfaces, and their apices are not at the origin, but rather located at points on the negative pressure axis. A typical plastic potential surface is shown at point A in Figure 4.

To demonstrate that non-associated plastic flow is in fact obtained for frictional materials, the results of two tests on dense Monterey No. O Sand are shown in Figure 11. One test is a conventional triaxial compression test in which the confining pressure was held constant. The

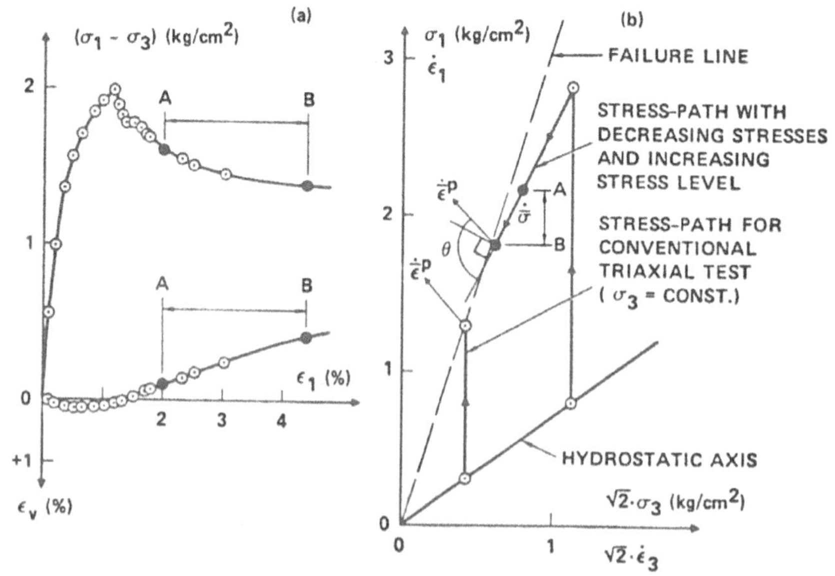

Figure 11. (a) Stress–strain behaviour, and (b) stress-path with decreasing stresses indicating non-associated flow for triaxial test on Loose Monterey No. O Sand

other test was performed with a stress-path near failure involving primary loading with decreasing stress difference $(\sigma_1 - \sigma_3)$ and decreasing confining pressure σ_3. If stresses and strain increments are plotted on the same diagram, as in Figure 11(b), the direction of the plastic strain increment vector is uniquely determined from the state of stress, and it is independent of the stress-path leading to this state of stress.[10, 12] The stress-path with decreasing stresses, shown in Figure 11(b), is so steep as to form an obtuse angle ($\theta = 110$ degrees) with the direction of the strain increment vector. Nevertheless, the two plastic strain increment vectors are parallel.

The fact that plastic yielding is occurring along the stress-path with decreasing stresses is seen from the stress–strain curve in Figure 11(a). Section AB on the stress-path is labelled similarly on the stress–strain curve which indicates large plastic strains between A and B. Since the yield surface is being pushed out, it must be steeper than the stress-path direction AB. If it were less steep, so as for example to be perpendicular to the plastic strain increment vector thereby indicating associated flow, then point B would be inside point A corresponding to unloading from A to B. This would contradict the large plastic strains shown in Figure 11(a). Consequently, the yield surface must be steeper than stress-path AB, and non-associated plastic flow is therefore clearly indicated.

STABILITY AND INSTABILITY

The stability postulate formulated by Drucker[13–15] is suitable for metals which exhibit associated flow. According to this postulate, stability requires that the second increment of work is positive or zero:

$$\Delta\sigma \cdot \Delta\varepsilon \geqslant 0 \tag{3}$$

in which $\Delta\sigma$ = increment of stress and $\Delta\varepsilon$ = resulting increment in strain. For metals, positive values of the second increment of work are always associated with the stable, ascending part of the stress–strain relationship, whereas negative values are associated with the unstable, descending part of the stress–strain curve obtained after peak failure.

Frictional materials which exhibit non-associated flow may encounter stress-paths for which equation (3) is not fulfilled, even before the failure surface has been reached. Figure 12 shows a schematic illustration of the region in which equation (3) is not fulfilled for a material with non-associated flow. The region is shaped as a wedge between the current yield surface f and the plastic potential surface g corresponding to the current stress point. All stresses, including the stress difference $(\sigma_1 - \sigma_3)$, are decreasing within the wedge between f and g, whereas the stress ratio (σ'_1/σ'_3) is increasing in this region. The stress-path with decreasing stresses shown in Figure 11(b) is located within the shaded wedge in Figure 12, and according to Drucker's stability postulate, should lead to unstable behaviour. The experiments shown in Figure 11 were performed under strain control. These experiments are therefore unable to indicate whether the soil behaviour is unstable or stable.

However, experiments with similar stress-paths have recently been performed under stress control[16] to study the stability of frictional materials with non-associated flow. These stress controlled experiments also clearly showed the plastic dilation of the sand to be non-associated. When the sand was exposed to stress-paths in the region of potential instability ($\Delta\sigma \cdot \Delta\varepsilon < 0$), none was observed. The tall cylindrical specimens with lubricated ends deformed as perfect right cylinders with no developing shear bands, no bulging or shearing in a non-uniform manner, or any other signs of instability. Little but negligible creep was present in the sand specimens. The experiments were repeated a number of times with repeatable results.

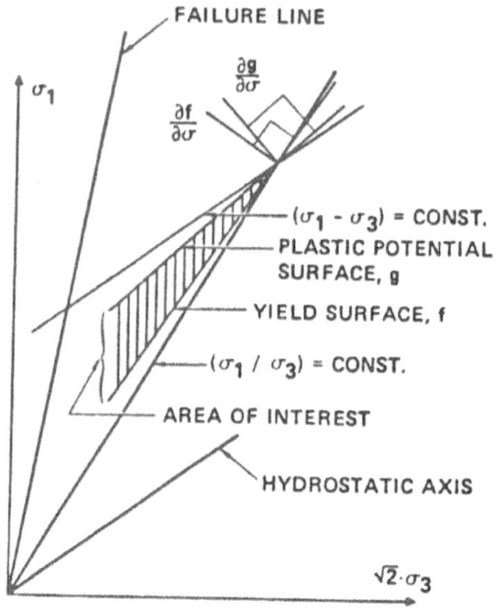

Figure 12. Wedge-shaped region of stress-paths with decreasing stresses in which granular material with non-associated flow may be unstable during hardening inside the failure surface

On the other hand, materials which compress during shear can become unstable *inside* the failure surface and this may lead to liquefaction. This instability may occur during cyclic loading (earthquakes, wave loading) and it may occur during monotonic, static loading (resulting in flow slides). The instability may occur whether the pores are filled with water or with air. In a series of stress controlled triaxial tests on granular material which tended to compress during shear,[17] the specimens were exposed to stress-paths within the shaded wedge shown in Figure 12. Figure 13 shows the actual stress-path followed in these tests. In each test the saturated specimen was first loaded under drained conditions to a preselected stress level S. The drainage valve was then closed and instability developed in each specimen due to increasing pore pressures, i.e. the specimens could not sustain the applied load. The stress-paths shown in Figure 13 were within the shaded wedge illustrated in Figure 12, the specimens exhibited non-associated flow, and instability was obtained in all cases in the hardening regime *inside* the failure surface.

It should be emphasized that instability is not synonymous with failure. Thus, a failure surface does not go through each of the points at which instability was produced. In fact, these points are well within the failure surface as obtained from drained tests. Obviously, each one of the specimens could have been loaded to higher stress differences as long as the stress-paths were outside the wedge between the yield and the plastic potential surfaces. Therefore, a failure condition expressed in terms of the maximum stress differences obtained from, for example, undrained tests initiated at the hydrostatic axis would clearly underestimate the true effective strength of the soil.

Constitutive models which employ associated plastic flow rules must assume the failure surface to go through the point at which instability is produced. This is because stability is guaranteed according to Drucker's stability condition until the failure surface is reached, at which instability ensues. But then stable loading to higher stress differences, which is clearly possible, could not be predicted by such models. Further, since instability as well as stability can be obtained at any point in the region of the tests shown in Figure 13, it is not possible to define any one consistent failure surface to be used in a model which incorporates associated flow.

The question of stability is directly related to the volume change behaviour of the frictional material. Generally, materials that dilate exhibit stable behaviour until the failure surface has

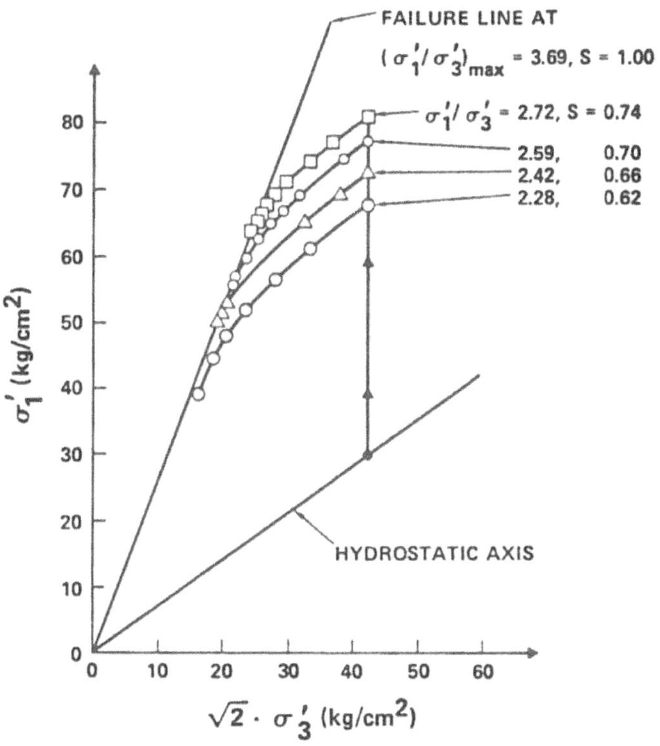

Figure 13. Effective stress-paths for stress-controlled triaxial compression tests on Loose Sacramento River Sand

been reached, whereas materials which compress may become unstable *inside* the failure surface. More detailed stability conditions for granular materials exhibiting non-associated flow were formulated on the basis of the experimental results obtained in the two studies.[16, 17] These conditions take the form of additions to Drucker's stability postulate. They are expressed in terms of the orientations of the plastic potential surface and the yield surface.

Conceptual model

The stability/instability phenomenon discussed above may be illustrated by a simple, two-dimensional, conceptual model shown in Figure 14. A block with rough surfaces is pinned against a rough vertical wall by the horizontal normal force N. The constant, vertical force F, caused by the gravitational field, acts to pull the block down. A simple analysis shows that stability is maintained as long as

$$\frac{\Delta N}{F} \geqslant \frac{-\tan v}{\tan\phi\cdot\tan(\phi+v)} \tag{4}$$

in which ΔN is a normal force increment in excess of $F/\tan\phi$ which is the value of N required to maintain equilibrium of a block along a flat wall without asperities; ϕ is the basic wall friction angle ($\phi = \arctan\mu$); and v is the angle of dilation indicated on Figure 14. The value of v is positive for dilation and negative for compression. Equation (4) shows that ΔN can always be made large enough to maintain stability as long as $\phi > |v|$. This condition is fulfilled for all real soils with the possible exception of extremely loose deposits which collapse due to any small disturbance.

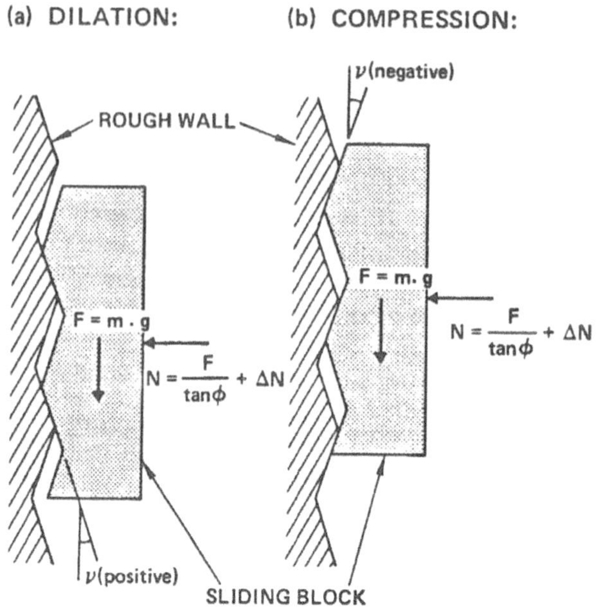

Figure 14. Simple, two-dimensional, conceptual model illustrating the stability/instability phenomenon observed in triaxial tests on granular materials

Equation (4) indicates that stability of the block in Figure 14(a) is always maintained for the case of dilation as ΔN is reduced towards zero. In fact, stability will be maintained for negative values of ΔN whose lower limit may be determined for specific values of ϕ, v and F.

For the case of compression shown in Figure 14(b), the value of $v<0$. For this case, instability will develop as ΔN is reduced towards zero. The actual value of ΔN at which instability occurs may be calculated from equation (4) for specific values of ϕ, v and F. Note, however, that this value of ΔN is still positive i.e. instability is produced due to compression for which v is negative. Note also that ΔN can be reduced to zero for the case of a flat wall without asperities for which $v=0$.

Whereas this conceptual model may be too simplistic to capture all aspects of the stability/instability phenomenon in real frictional materials, it does serve to demonstrate the influence and consequences of volume changes in such materials.

STRAIN SOFTENING

The phenomenon of strain softening following peak failure can occur in frictional materials such as sand and clay. Such materials can exhibit considerable decreases in strength after the peak stress has been exceeded, i.e. strain-softening. In a recent review paper[18] it was argued that softening is not a true material property, but represents the overall behaviour of a fragmented, non-uniformly straining, structured material subject to geometric and boundary constraints. Using the recently published results of some very carefully performed triaxial compression tests on sand, [19] it was suggested that any strain softening observed in these experiments is so small that for practical purposes it can be reasonably neglected. However, the experiments do indicate softening, and theoretical models for development of shear planes in soils (see, for example, References 19 and 20) require negative hardening moduli, i.e. softening to occur as a necessary condition for localization of plastic strains and development of shear planes.

Figure 15 shows typical results of a triaxial compression test on dense sand. During shear the dense specimen dilates, thus producing increasing void ratio and decreasing density. The dilation continues at a steady rate to large strains considerably beyond peak failure. The volume change curve as well as the specimen, which deformed as a right cylinder, does not in any way indicate that peak failure has been encountered. There is therefore no detectable difference in the specimen in the hardening regime and in the softening regime. The strain softening is associated with a decrease in density. Loose sands which compress during shear do not exhibit strain softening.

Thus, both experiments and theoretical considerations point to the existence of uniformly occurring strain softening in frictional materials. It is associated with smoothly developing decreasing material densities. This phenomenon can be modelled quite well by work-hardening and -softening plasticity theory.

DEVELOPMENT OF SHEAR PLANES

Above the yield point in metals, existing dislocations move and produce irrecoverable, plastic strains (see, for example, References 21 and 22). Dislocations move principally along primary slip planes, and these tend to concentrate in groups to form slip bands. Following some motion, the dislocations can be impeded by interaction with other dislocations, by particles of different type, by twin boundaries, or by gain boundaries. The result is that the slip planes become stronger, and deformation of the metal continues at a different location. Thus, slip bands may be distributed through a region with uniform stress. The overall result of the slip band formation is that the metal becomes stronger.

Development of slip bands or shear bands in frictional materials may be studied experimentally.[20] During the performance of the triaxial compression test shown in Figure 15, the specimen

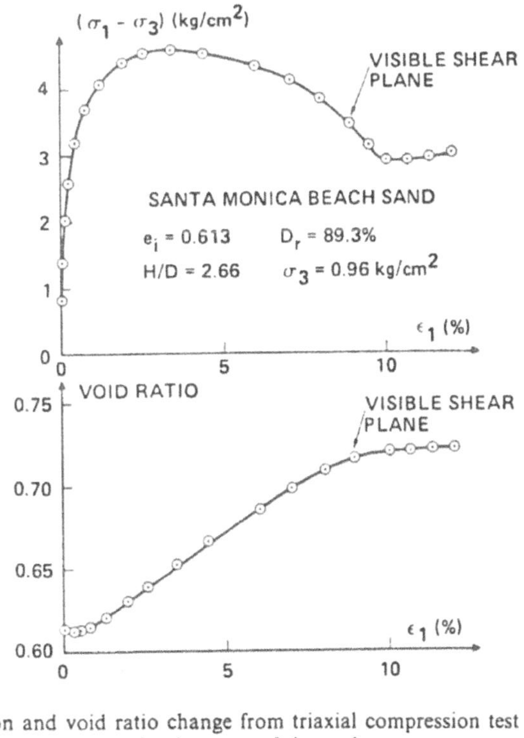

Figure 15. Stress–strain relation and void ratio change from triaxial compression test exhibiting strain-softening and development of shear plane

was carefully observed for early detection of developing shear bands. The location on the stress–strain curve at which the first observation of a shear band was made is indicated on the diagram. Considerable straining beyond the peak failure point may be required before shear bands occur in compression tests. The stress–strain curve clearly shows a drop in strength and the rate of dilation diminishes substantially immediately before the shear band becomes visible. Thus, the reduced rate of dilation appears to be associated with the occurrence of the shear band and the achievement of critical state conditions within the shear band in the compression test. Once a shear band has developed fully, the stresses and the volume changes tend to level off. It is likely that the initiation of shear banding, i.e. localization of deformation, results in elastic unloading of the specimen outside the developing shear band which is now becoming rapidly weaker than the remaining major parts of the specimen.

Development of shear planes in triaxial compression ($\sigma_2 = \sigma_3$) occurs after peak failure and results in further decrease in strength. This behaviour is the opposite of that occurring in metals and it appears to be associated with dilation of the sand. Shear bands are not observed in loose specimens which compress during shear. Loose specimens deform uniformly throughout the entire stress–strain relationship.

CONCLUSIONS

The development of constitutive models for metals and frictional materials are most often based on a framework provided by elasticity theory and work-hardening plasticity theory. Plasticity theory was developed mainly on the basis of the observed behaviour of metals. The behaviour of frictional materials is considerably different from that of metals. It is important that the differences as well as the similarities are recognized, and that principles which characterize the behaviour of frictional materials are used for development of constitutive models for such materials.

The fundamental behaviour of frictional materials can only be discovered from experimental studies. Their most important behaviour characteristics have been briefly reviewed, and the effects of voids and volume changes on the behaviour have been emphasized. It appears that two of the distinguishing characteristics between metals and frictional materials are the absence or presence of voids and the tendency to change or maintain constant volume during compression and shear. However, the behaviour of both types of materials can be characterized by elasticity theory and work-hardening plasticity theory.

REFERENCES

1. A. W. Skempton, 'Effective stress in soils, concrete and rocks', *Conf. on Pore Pressure and Suction in Soils*, London, Butterworths, 1961, pp. 4–16.
2. G. I. Taylor and H. Quinney, 'The plastic distortion of metals', *Phil. Trans. Roy. Soc.*, **A230**, 323–362 (1931).
3. P. V. Lade and J. M. Duncan, 'Cubical triaxial tests on cohesionless soil', *J. Soil Mech. Found. Div.*, A.S.C.E., **99**, 793–812 (1973).
4. L. L. Mills and R. M. Zimmerman, 'Compressive strength of plain concrete under multiaxial loading conditions', *Am. Conc. Inst. J.*, **67**, 802–807 (1970).
5. A. Caguot and J. Kerisel, *Traite de Mecanique des Sols*, 4th edn., Gauthier-Villars, Paris, 1966.
6. P. V. Lade and H. M. Musante, 'Three-Dimensional behavior of remolded clay', *J. Geotech. Engr. Div.*, A.S.C.E., **104**, 193–209 (1978).
7. K. L. Lee and H. B. Seed, 'Drained strength characteristics of sands', *J. Soil Mech. Found. Div.*, A.S.C.E., **93**, 117–141 (1967).
8. P. V. Lade and R. B. Nelson, 'Modelling the elastic behavior of granular materials', *Int. J. numer. Anal. methods geomech.*, **11**, 521–542 (1987).
9. C. S. Desai, S. Somasundaram and G. Frantziskonis, 'A hierarchical approach for constitutive modelling of geologic materials', *Int. j. numer. anal. methods. geomech.*, **10**, 225–257 (1986).

10. P. V. Lade and J. M. Duncan, 'Stress-path dependent behavior of cohesionless soil', *J. Geotech. Engr. Div.*, A.S.C.E., **102**, 51–68 (1976).
11. B. de Saint-Venant, 'Memoire sur l'etablissement des equations differentielles des mouvement interieurs operes dans les corps solides ductiles au dela des limites ou l'elasticite pourrait les ramener a leur premier etat', *Comptes Rendus Acad. Sci.*, Paris, **70**, 1323–1325 (1870).
12. H. B. Poorooshasb, I. Holubec and A. N. Sherbourne, 'Yielding and flow of sand in triaxial compression: Part I', *Can. Geot. J.*, **III**, (4) 179–190 (1966).
13. D. C. Drucker, 'A more fundamental approach to stress–strain relations', *Proc. 1st U.S. Nat. Congress Appl. Mech.*, pp. 487–491 (1951).
14. D. C. Drucker, 'On uniqueness in the theory of plasticity', *Q. Appl. Math.*, **14**, 35–42 (1956).
15. D. C. Drucker, 'A definition of stable inelastic material', *J. Appl. Mech.*, **26**, 101–106 (1959).
16. P. V. Lade, R. B. Nelson and Y. M. Ito 'Nonassociated flow and stability of granular materials', *J. Engr. Mech.*, A.S.C.E., **113**, 1302–1318 (1987).
17. P. V. Lade, R. B. Nelson and Y. M. Ito, 'Instability of granular materials with nonassociated flow', *J. Engr. Mech.*, A.S.C.E., (Accepted, December 1987).
18. H. E. Read and G. A. Hegemier, 'Strain softening of rock, soil and concrete—a review article', *Mech. Mater.*, **3**, 271–294 (1984).
19. A. Hettler and I. Vardoulakis, 'Behavior of dry sand tested in a large triaxial apparatus', *Geotech.*, **34**(2) 183–198 (1984).
20. J. F. Peters, P. V. Lade and A. Bro, 'Shear band formation in triaxial and plane strain tests', in *Advanced Triaxial Testing of Soil and Rock, ASTM 977* (Eds. R. T. Donaghe, R. C. Chaney and M. L. Silver), Am. Soc. Test. Mat., 1988.
21. J. J. Gilman, *Micromechanics of flow in solids*, McGraw-Hill, New York, 1969.
22. A. G. Guy, *Introduction to Material Science*, McGraw-Hill, New York, 1972.

Chapter IV
Stress Invariants

Preface

Augustin-Louis Chauchy (French mathematician, 1789-1857) defined stress as force per unit area, measured in Newtons per square meter, also called Pascals (N/m^2= Pa) in the International System of measurements, or in pounds per square inch (psi) in the Imperial System. Sometimes the stresses can become rather high and therefore kilo-Pascals (kPa) or Mega-Pascals (MPa) may be employed. Normal stresses (σ), i.e. stresses perpendicular to the area, may be in compression or in tension, while shear stresses (τ) are parallel to the area. To describe the complete three-dimensional stress state at a point, a stress tensor (referred to as the Cauchy stress tensor) consisting of a 3x3 symmetric matrix is employed.

Stress invariants were developed later, and they became prominent in the 1950s and 1960s in connection with analyses of elasticity problems. They have since been found to be very useful in computational mechanics, as will be shown below.

The following chapter sections are from: Lade, P.V. (2016) "Triaxial Testing of Soils", Wiley-Blackwell, Chichester, United Kingdom. 402 pages.

2.6 Three-dimensional stress states

2.6.1 General 3D stress states

The 3D state of stress at a point is completely defined if six independent quantities are known, for example three mutually perpendicular planes (defined by their normals) and the three stresses (each of which can be decomposed into one normal and two shear components) on these planes. Such a general state of stress is shown in Fig. 2.69. As for 2D conditions, the normal stresses are positive when compressive in soil mechanics. The stress components can be arranged in a symmetric matrix, called a stress tensor, as shown in Fig. 2.69.

2.6.2 Stress invariants

The principal stresses, σ_1, σ_2, and σ_3, are the normal stresses on planes on which no shear stresses act. The principal stresses can be computed from the above six independent quantities. To determine the principal stress magnitudes and their

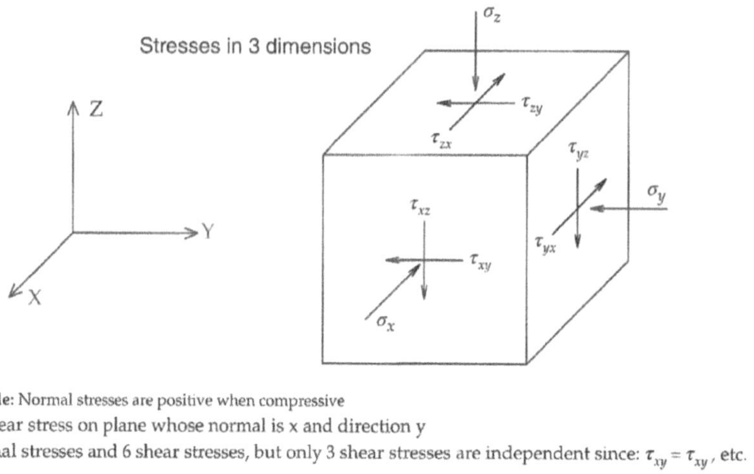

$$\begin{bmatrix} \sigma_x & \tau_{yx} & \tau_{zx} \\ \tau_{xy} & \sigma_y & \tau_{zy} \\ \tau_{xz} & \tau_{yz} & \sigma_z \end{bmatrix}$$

Figure 2.69 Three-dimensional stress state for material element.

directions, the far corner of the element shown in Fig. 2.69 is cut off by a plane ABC characterized by its normal, as indicated in Fig. 2.70. The normal to plane ABC is characterized by a unit vector $(l, m, n) = (\cos\alpha, \cos\beta, \cos\gamma)$ with the property:

$$l^2 + m^2 + n^2 = \cos^2\alpha + \cos^2\beta + \cos^2\gamma = 1 \quad (2.98)$$

The stresses on the three mutually perpendicular planes are indicated in Fig. 2.70. The stresses acting on plane ABC in the directions of the coordinate axes, p_x, p_y, and p_z, are determined by force equilibrium in the X-, Y-, and Z-directions.

To express equilibrium of forces in the coordinate directions, the areas of triangles OBC, OAC, OAB and ABC are required. They are calculated by setting the area of ABC = ω, as shown in Fig. 2.71, and the areas of the other triangles are expressed by projecting triangle ABC along each of the coordinate axes. Therefore:

$$\text{Area of OBC} = \omega \cdot l \quad (2.99)$$

$$\text{OAC} = \omega \cdot m \quad (2.100)$$

$$\text{OAB} = \omega \cdot n \quad (2.101)$$

Equilibrium of forces in the X-direction therefore gives:

$$p_x \cdot \omega = \sigma_x \cdot \omega \cdot l + \tau_{yx} \cdot \omega \cdot m + \tau_{zx} \cdot \omega \cdot n \quad (2.102)$$

and canceling ω produces:

$$p_x = \sigma_x \cdot l + \tau_{yx} \cdot m + \tau_{zx} \cdot n \quad (2.103)$$

Similar equations are obtained for the Y- and Z-directions:

$$p_y = \tau_{xy} \cdot l + \sigma_y \cdot m + \tau_{zy} \cdot n \quad (2.104)$$

$$p_z = \tau_{xz} \cdot l + \tau_{yz} \cdot m + \sigma_z \cdot n \quad (2.105)$$

These three expressions can be written in matrix form as follows:

$$\begin{Bmatrix} p_x \\ p_y \\ p_z \end{Bmatrix} = \begin{bmatrix} \sigma_x & \tau_{yx} & \tau_{zx} \\ \tau_{xy} & \sigma_y & \tau_{zy} \\ \tau_{xz} & \tau_{yz} & \sigma_z \end{bmatrix} \begin{Bmatrix} l \\ m \\ n \end{Bmatrix} \quad (2.106a\text{–}c)$$

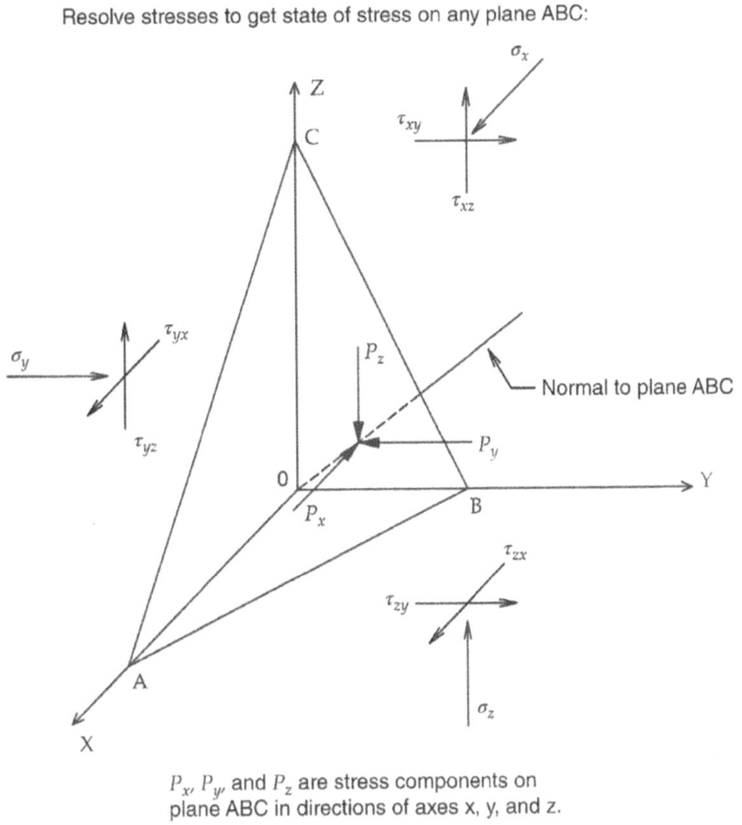

Figure 2.70 Stress state acting on plane ABC indicated by normal oriented relative to the coordinate axes.

The matrix of stresses is referred to as the stress tensor, σ_{ij}. Although the individual components of the stress tensor depend on the coordinate system, the quantity as a whole does not change. Similar to the 2D system represented by the Mohr's circle (in Section 2.1.7 and represented in Figs 2.10 and 2.11), the Mohr's circle represents a tensor in 2D such that the shear and normal components are different on each plane, yet the circle itself and the principal stresses are unchanged. A general tensor has numerical values associated with it called invariants, which for the 2D case corresponds to the values of p and q on the MIT plot. For the 3D case the principal stresses denoted by (σ_1, σ_2, σ_3) are the three solutions to a cubic equation, as shown below.

By rotating the normal to the plane as well as the plane ABC around point O in Fig. 2.71, three locations will be encountered at which no shear stresses act on plane ABC. Such a plane is called a principal plane and the normal stress acting on this plane is a principal stress.

To determine the principal stresses, assume there is one principal plane on which the shear stresses are zero and only a normal (principal) stress = σ_i exists, as shown in Fig. 2.72. Then

$$\begin{Bmatrix} p_x \\ p_y \\ p_z \end{Bmatrix} = \begin{Bmatrix} \sigma_i \cdot \cos\alpha \\ \sigma_i \cdot \cos\beta \\ \sigma_i \cdot \cos\gamma \end{Bmatrix} = \begin{Bmatrix} \sigma_i \cdot l \\ \sigma_i \cdot m \\ \sigma_i \cdot n \end{Bmatrix} \quad (2.107a\text{–}c)$$

Then, since the values of (p_x, p_y, p_z) in Eq. (2.106a–c) and in Eq. (2.107a–c) have to be equal to maintain equilibrium, subtracting the two set of equations from each other will result in zero:

$$\begin{bmatrix} (\sigma_x - \sigma_i) & \tau_{yx} & \tau_{zx} \\ \tau_{xy} & (\sigma_y - \sigma_i) & \tau_{zy} \\ \tau_{xz} & \tau_{yz} & (\sigma_z - \sigma_i) \end{bmatrix} \begin{Bmatrix} l \\ m \\ n \end{Bmatrix} = \begin{Bmatrix} 0 \\ 0 \\ 0 \end{Bmatrix} \quad (2.108a\text{–}c)$$

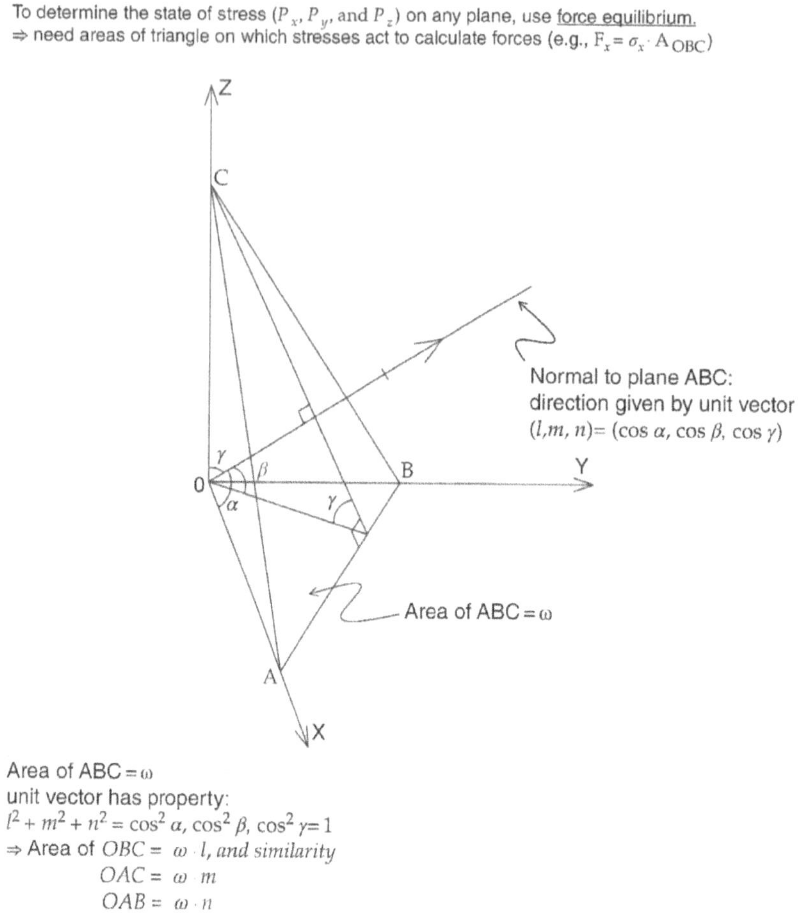

Figure 2.71 Areas of coordinate triangles, OBC, OAC and OAB, on which stresses act.

These are three linear, simultaneous equations for determination of σ_i. To obtain a nontrivial solution, the determinant of the matrix must be zero:

$$\begin{vmatrix} (\sigma_x - \sigma_i) & \tau_{yx} & \tau_{zx} \\ \tau_{xy} & (\sigma_y - \sigma_i) & \tau_{zy} \\ \tau_{xz} & \tau_{yz} & (\sigma_z - \sigma_i) \end{vmatrix} = 0 \qquad (2.109)$$

Calculating the value of the determinant produces the *characteristic equation*:

$$\begin{aligned} &\sigma_i^3 - (\sigma_x + \sigma_y + \sigma_z)\cdot\sigma_i^2 + (\sigma_x\cdot\sigma_y + \sigma_y\cdot\sigma_z + \sigma_z\cdot\sigma_x \\ &-\tau_{xy}\cdot\tau_{yx} - \tau_{yz}\cdot\tau_{zy} - \tau_{zx}\cdot\tau_{xz})\cdot\sigma_i - (\sigma_x\cdot\sigma_y\cdot\sigma_z \\ &+\tau_{xy}\cdot\tau_{yz}\cdot\tau_{zx} + \tau_{yx}\cdot\tau_{zy}\cdot\tau_{xz} - \sigma_x\cdot\tau_{yz}\tau_{zy} \\ &-\sigma_y\cdot\tau_{zx}\cdot\tau_{xz} - \sigma_z\cdot\tau_{xy}\cdot\tau_{yx}) = 0 \end{aligned} \qquad (2.110)$$

or

$$\sigma_i^3 - I_1\cdot\sigma_i^2 + I_2\cdot\sigma_i - I_3 = 0 \qquad (2.111)$$

The magnitudes of the principal stresses, σ_i ($i = 1, 2, 3$), can be found as roots of Eq. (2.111). Since for a given stress state the principal stresses are independent of the choice of coordinate system in which the normal and shear stresses are expressed, the coefficients in the cubical equation, I_1, I_2, and I_3, must also be independent of the coordinate system. These coefficients are therefore invariants with respect to change of coordinate system and they have the same values for all systems. These invariants can consequently also be expressed in terms of principal stresses:

$$I_1 = \sigma_x + \sigma_y + \sigma_z = \sigma_1 + \sigma_2 + \sigma_3 \qquad (2.112)$$

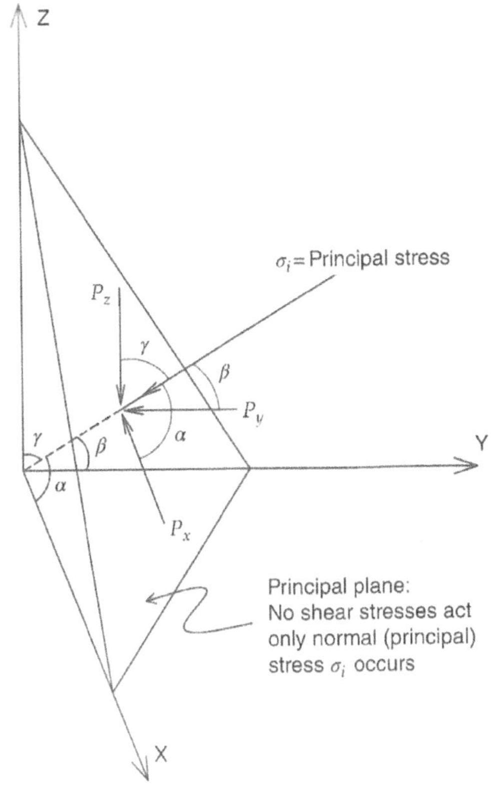

Figure 2.72 Determination of principal stresses acting on material element.

$$I_2 = \sigma_x \cdot \sigma_y + \sigma_y \cdot \sigma_z + \sigma_z \cdot \sigma_x - \tau_{xy} \cdot \tau_{yx} - \tau_{yz} \cdot \tau_{zy} - \tau_{zx} \cdot \tau_{xz} = \sigma_1 \cdot \sigma_2 + \sigma_2 \cdot \sigma_3 + \sigma_3 \cdot \sigma_1 \qquad (2.113)$$

$$I_3 = \sigma_x \cdot \sigma_y \cdot \sigma_z + \tau_{xy} \cdot \tau_{yz} \cdot \tau_{zx} + \tau_{yx} \cdot \tau_{zy} \cdot \tau_{xz} - \sigma_x \cdot \tau_{yz} \cdot \tau_{zy} - \sigma_y \cdot \tau_{zx} \cdot \tau_{xz} - \sigma_z \cdot \tau_{xy} \cdot \tau_{yx} = \sigma_1 \cdot \sigma_2 \cdot \sigma_3 \qquad (2.114)$$

The three quantities I_1, I_2, and I_3 are called the first, the second, and the third invariants of the stress tensor. Note that the sign used in the cubical equation for the coefficient to σ_i and the consequent sign of I_2 is the user's choice, but it must be used consistently thereafter. Note also that the shear stresses $\tau_{xy} = \tau_{yx}$, and so on in terms of numerical magnitude, but they remain as separate entities in the above expressions for the stress invariants, because they are different from each other with regard to differentiation when used in expressions containing these invariants.

2.6.3 Stress deviator invariants

Decomposition of stress tensor

The symmetric stress tensor, σ_{ij}, can be decomposed into two symmetric tensors, the *hydrostatic stress* (or spherical stress) tensor and the *deviatoric stress* tensor:

$$\sigma_{ij} = \sigma_m \cdot \delta_{ij} + s_{ij} \qquad (2.115)$$

where δ_{ij} is Kronecker's symbol ($\delta_{ij} = 1$ for $i = j$ and $\delta_{ij} = 0$ for $i \neq j$). The stress tensor can also be written in matrix form:

$$\begin{bmatrix} \sigma_x & \tau_{yx} & \tau_{zx} \\ \tau_{xy} & \sigma_y & \tau_{zy} \\ \tau_{xz} & \tau_{yz} & \sigma_z \end{bmatrix} = \begin{bmatrix} \sigma_m & 0 & 0 \\ 0 & \sigma_m & 0 \\ 0 & 0 & \sigma_m \end{bmatrix} + \begin{bmatrix} (\sigma_x - \sigma_m) & \tau_{yx} & \tau_{xz} \\ \tau_{xy} & (\sigma_y - \sigma_m) & \tau_{zy} \\ \tau_{xz} & \tau_{yz} & (\sigma_z - \sigma_m) \end{bmatrix} \qquad (2.116a–c)$$

The hydrostatic stress or the mean normal stress σ_m is defined as:

$$\sigma_m = \frac{1}{3} \cdot (\sigma_x + \sigma_y + \sigma_z) = \frac{1}{3} \cdot I_1 \qquad (2.117)$$

And the deviatoric stress s_{ij} is therefore:

$$s_{ij} = \sigma_{ij} - \sigma_m \cdot \delta_{ij} \qquad (2.118)$$

in which the individual components are given in the deviatoric matrix above.

As for the stress tensor, invariant quantities can be determined for the deviatoric stress tensor. The *characteristic equation* is formed as follows:

$$s_i^3 - J_1 \cdot s_i^2 - J_2 \cdot s_i - J_3 = 0 \qquad (2.119)$$

in which the invariants of the deviatoric stress tensor are expressed as follows:

$$J_1 = s_x + s_y + s_z = (\sigma_x - \sigma_m) + (\sigma_y - \sigma_m) + (\sigma_z - \sigma_m) \qquad (2.120)$$

And with σ_m from Eq. (2.117):

$$J_1 = 0 \tag{2.121}$$

$$\begin{aligned} J_2 &= \frac{1}{2}\cdot s_{ij}\cdot s_{ij} \\ &= \frac{1}{6}\cdot[(\sigma_x-\sigma_y)^2+(\sigma_y-\sigma_z)^2+(\sigma_z-\sigma_x)^2] \\ &\quad +\tau_{xy}\cdot\tau_{yx}+\tau_{yz}\cdot\tau_{zy}+\tau_{zx}\cdot\tau_{xz} \\ &= \frac{1}{6}\cdot[(\sigma_1-\sigma_2)^2+(\sigma_2-\sigma_3)^2+(\sigma_3-\sigma_1)^2] \end{aligned} \tag{2.122}$$

$$\begin{aligned} J_3 &= \frac{1}{3}\cdot s_{ij}\cdot s_{jk}\cdot s_{ki} \\ &= (\sigma_x-\sigma_m)(\sigma_y-\sigma_m)(\sigma_z-\sigma_m)+\tau_{xy}\cdot\tau_{yz}\cdot\tau_{zx} \\ &\quad +\tau_{yx}\cdot\tau_{zy}\cdot\tau_{xz}-(\sigma_x-\sigma_m)\cdot\tau_{yz}\cdot\tau_{zy} \\ &\quad -(\sigma_y-\sigma_m)\cdot\tau_{xz}\cdot\tau_{zx} \\ &\quad -(\sigma_z-\sigma_m)\cdot\tau_{xy}\cdot\tau_{yx} \end{aligned} \tag{2.123}$$

$$\begin{aligned} J_3 &= (\sigma_1-\sigma_m)(\sigma_2-\sigma_m)(\sigma_3-\sigma_m) \\ &= \frac{1}{27}\cdot(2\sigma_1-\sigma_2-\sigma_3)(2\sigma_2-\sigma_3-\sigma_1)(2\sigma_3-\sigma_1-\sigma_2) \end{aligned} \tag{2.124}$$

The solution to the characteristic equation yields the principal stress deviators s_1, s_2, and s_3. Like the stress invariants, the coefficients J_1, J_2, and J_3 are independent of the coordinate system, and they are called the first, second and third stress deviator invariants, respectively.

The stress deviator invariants may be related to the stress invariants as follows:

$$J_1 = 0 \tag{2.125}$$

$$J_2 = \frac{1}{3}\cdot I_1^2 - I_2 \tag{2.126}$$

$$J_3 = \frac{2}{27}\cdot I_1^3 - \frac{1}{3}\cdot I_1\cdot I_2 + I_3 \tag{2.127}$$

The principal deviatoric stresses, s_1, s_2, and s_3, coincide in directions with the principal stresses, and the solutions to the characteristic equations are really equivalent, except the principal deviatoric stresses are smaller than the principal stresses by the amount of σ_m.

2.6.4 Magnitudes and directions of principal stresses

Magnitudes of principal stresses

The cubical equation for the principal stresses in Eq. (2.111) [and that for the deviatoric stresses in Eq. (2.119)] has three real roots for a proper 3D stress state. These roots may be determined by graphing the characteristic equation, as indicated in Fig. 2.73, or they may be determined by solving Eq. (2.111):

For a cubical equation in the form:

$$x^3 + A\cdot x^2 + B\cdot x + C = 0 \tag{2.128}$$

the solution is provided by Korn and Korn (1961):

$$p = -\frac{A^2}{3} + B \tag{2.129}$$

$$q = 2\cdot\left(\frac{A}{3}\right)^3 - \frac{A\cdot B}{3} + C \tag{2.130}$$

$$x_1 = 2\cdot\sqrt{\frac{-p}{3}}\cdot\cos\left(\frac{\alpha}{3}\right) - \frac{A}{3} \tag{2.131}$$

$$x_{2,3} = -2\cdot\sqrt{\frac{-p}{3}}\cdot\cos\left(\frac{\alpha}{3}\pm 60°\right) - \frac{A}{3} \tag{2.132}$$

in which

$$\cos\alpha = -\frac{q}{2\cdot\sqrt{-\left(\frac{p}{3}\right)^3}} \tag{2.133}$$

For Eq. (2.111) the principal stresses are obtained as follows:

$$\sigma_1 = 2\cdot\sqrt{\frac{I_1^2}{9}+\frac{I_2}{3}}\cdot\cos\left(\frac{\alpha}{3}\right) + \frac{I_1}{3} \tag{2.134}$$

$$\sigma_2 = -2\cdot\sqrt{\frac{I_1^2}{9}+\frac{I_2}{3}}\cdot\cos\left(\frac{\alpha}{3}+60°\right) + \frac{I_1}{3} \tag{2.135}$$

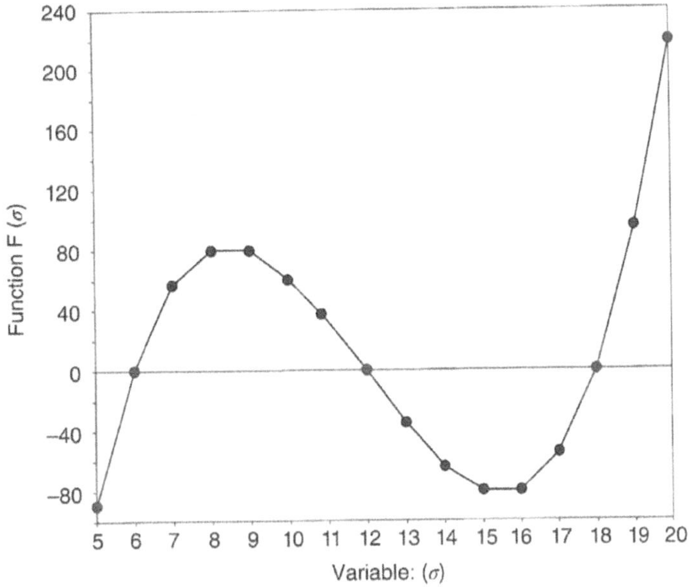

Figure 2.73 The roots of the characteristic equation (= principal stresses) may be determined by graphing the expression for the characteristic equation and finding the values of the normal stress at which this equation is zero (after Nordal 1994).

$$\sigma_3 = -2\cdot\sqrt{\frac{I_1^2}{9}+\frac{I_2}{3}}\cdot\cos\left(\frac{\alpha}{3}-60^\circ\right)+\frac{I_1}{3} \qquad (2.136)$$

where

$$\cos\alpha = \frac{2\cdot\frac{I_1^3}{27}+\frac{I_1\cdot I_2}{3}+I_3}{2\cdot\sqrt{\left(\frac{I_1^2}{9}+\frac{I_2}{3}\right)^3}} = \frac{3\sqrt{3}\cdot J_3}{2\cdot J_2^{3/2}} \qquad (2.137)$$

For a proper stress state there will always be three real roots to the cubical equation. Note that the roots come out in an ordered fashion such that $\sigma_1 = x_1$, $\sigma_2 = x_2$, and $\sigma_3 = x_3$. Note also that the deviatoric stresses, s_1, s_2, and s_3, equal the first terms in Eqs (2.134), (2.135), and (2.136).

Directions of principal stresses

The directions of the normals to the principal planes, that is the directions of the principal stresses, $(l, m, n) = (\cos\alpha, \cos\beta, \cos\gamma)$, can be found from Eq. (2.108a–c) and the fact that (l, m, n) is a unit vector as expressed in Eq. (2.98):

$$l_i = \cos\alpha = \frac{A_i}{\sqrt{A_i^2+B_i^2+C_i^2}} \qquad (2.138)$$

$$m_i = \cos\beta = \frac{B_i}{\sqrt{A_i^2+B_i^2+C_i^2}} \qquad (2.139)$$

$$n_i = \cos\gamma = \frac{C_i}{\sqrt{A_i^2+B_i^2+C_i^2}} \qquad (2.140)$$

where

$$A_i = (\sigma_y-\sigma_i)(\sigma_z-\sigma_i)-\tau_{zy}\cdot\tau_{yz} \qquad (2.141)$$

$$B_i = \tau_{zy}\cdot\tau_{xz}-\tau_{xy}\cdot(\sigma_z-\sigma_i) \qquad (2.142)$$

$$C_i = \tau_{xy}\cdot\tau_{yz}-\tau_{xz}\cdot(\sigma_y-\sigma_i) \qquad (2.143)$$

The principal deviatoric stresses, s_1, s_2, and s_3, coincide in directions with the principal stresses.

2.7 Principal stress space

To represent a general 3D state of stress, the principal stress space is most often employed. This space consists of a Cartesian coordinate system whose axes represent the three principal stresses σ_1, σ_2, and σ_3. These stresses are positive and compressive in the octant shown in Fig. 2.74(a). The stress condition in a soil element may be represented in principal stress space by a point whose coordinates are given by (σ_1, σ_2, σ_3), as illustrated in Fig. 2.74(a).

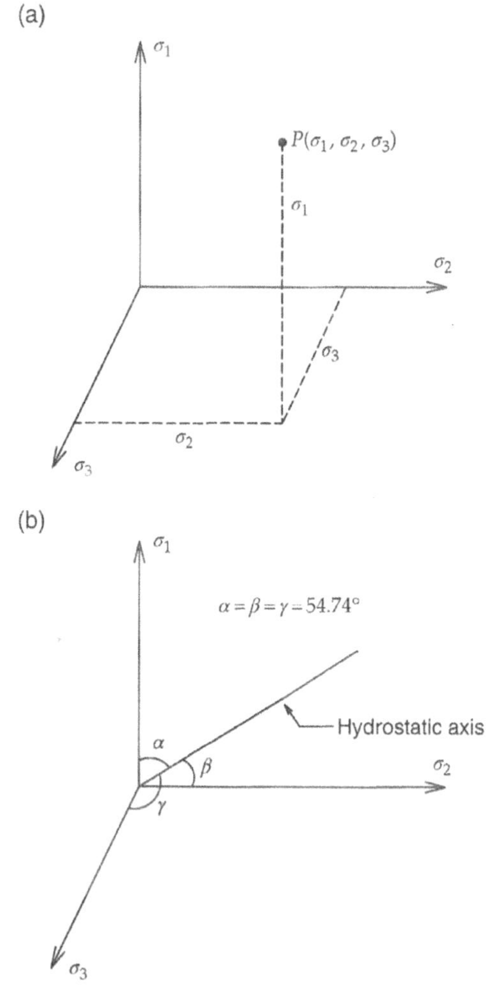

Figure 2.74 The principal stress space with (a) stress point and (b) hydrostatic axis.

The hydrostatic axis or space diagonal is the line in the coordinate system that forms equal angles with the axes, as shown in Fig. 2.74(b). Points on this line represent hydrostatic states of stress corresponding to equal values of the principal stresses ($\sigma_1 = \sigma_2 = \sigma_3$). The hydrostatic axis is characterized by the direction cosines $\bar{n} = (\cos\alpha, \cos\beta, \cos\gamma) = (\frac{1}{\sqrt{3}}, \frac{1}{\sqrt{3}}, \frac{1}{\sqrt{3}})$ and the angles between the hydrostatic axis and the three coordinate axes are therefore $\cos^{-1}(\frac{1}{\sqrt{3}}) = 54.74°$.

It is difficult to work with a 3D stress space on a routine basis. Two planes in the principal stress space are often used for plotting test results. A triaxial plane is a plane that contains the hydrostatic axes and one of the principal stress axes. There are three triaxial planes in the principal stress space. Figure 2.75(b) shows the triaxial plane that contains the σ_1-axis. An octahedral plane is a plane whose normal is the hydrostatic axis. There are eight (octa) planes that together form an octahedron, as shown in Fig. 2.75(a). There are an infinite number of octahedral planes within the octant in which all three principal stresses are positive. One octahedral plane is shown in Fig. 2.75(b).

2.7.1 Octahedral stresses

The normal and shear stresses on any octahedral plane are denoted the octahedral normal stress and the octahedral shear stress and they are expressed in terms of the principal stresses or from a general state of stress or from invariants, as follows:

$$\sigma_{oct} = \frac{1}{3} \cdot (\sigma_x + \sigma_y + \sigma_z) = \frac{1}{3} \cdot (\sigma_1 + \sigma_2 + \sigma_3) = \frac{1}{3} \cdot I_1 \quad (2.144)$$

$$\tau_{oct} = \frac{1}{3} \cdot \sqrt{(\sigma_x - \sigma_y)^2 + (\sigma_y - \sigma_z)^2 + (\sigma_z - \sigma_x)^2 + 6(\tau_{xy} \cdot \tau_{yx} + \tau_{yz} \cdot \tau_{zy} + \tau_{zx} \cdot \tau_{xz})} \quad (2.145a)$$

$$= \frac{1}{3} \cdot \sqrt{(\sigma_1 - \sigma_2)^2 + (\sigma_2 - \sigma_3)^2 + (\sigma_3 - \sigma_1)^2} \tag{2.145b}$$

$$= \frac{\sqrt{2}}{3} \cdot \sqrt{I_1^2 - 3 \cdot I_2} = \sqrt{\frac{2}{3} \cdot J_2} \tag{2.145c}$$

The octahedral stress components corresponding to the stress point $P(\sigma_1, \sigma_2, \sigma_3)$ can be found as illustrated in Fig. 2.76. The total stress vector $\overline{OP}$ can be decomposed into the components $\overline{OQ}$ on the hydrostatic axis and $\overline{QP}$ in the octahedral plane through P. The length $|\overline{OQ}|$ can be obtained as the projection of $\overline{OP}$ on the hydrostatic axis. The scalar product of $\overline{OP}$ and the unit vector $\bar{n}$ on the hydrostatic axis gives:

$$|\overline{OQ}| = \frac{1}{\sqrt{3}} \cdot \sigma_1 + \frac{1}{\sqrt{3}} \cdot \sigma_2 + \frac{1}{\sqrt{3}} \cdot \sigma_3 = \frac{\sqrt{3}}{3} \cdot (\sigma_1 + \sigma_2 + \sigma_3) = \sqrt{3} \cdot \sigma_{oct} \tag{2.146}$$

The deviatoric component $|\overline{QP}|$ of the total stress vector can be obtained as:

$$|\overline{QP}| = |\overline{OP} - \overline{OQ}| \tag{2.147}$$

where

$$\overline{OP} = (\sigma_1, \sigma_2, \sigma_3) \tag{2.148}$$

and

$$\overline{OQ} = |\overline{OQ}| \cdot \bar{n} = \frac{1}{3} \cdot (\sigma_1 + \sigma_2 + \sigma_3)(1,1,1) \tag{2.149}$$

Performing the calculation indicated in Eq. (2.147) results in:

$$|\overline{QP}| = \frac{\sqrt{3}}{3} \cdot \sqrt{(\sigma_1 - \sigma_2)^2 + (\sigma_2 - \sigma_3)^2 + (\sigma_3 - \sigma_1)^2} = \sqrt{3} \cdot \tau_{oct} \tag{2.150}$$

Figure 2.76 shows that the octahedral stress components, σ_{oct} and τ_{oct}, corresponding to the stress point $P(\sigma_1, \sigma_2, \sigma_3)$ are represented in the principal stress space as $\frac{1}{\sqrt{3}} \cdot |\overline{QP}|$ and $\frac{1}{\sqrt{3}} \cdot |\overline{QP}|$,

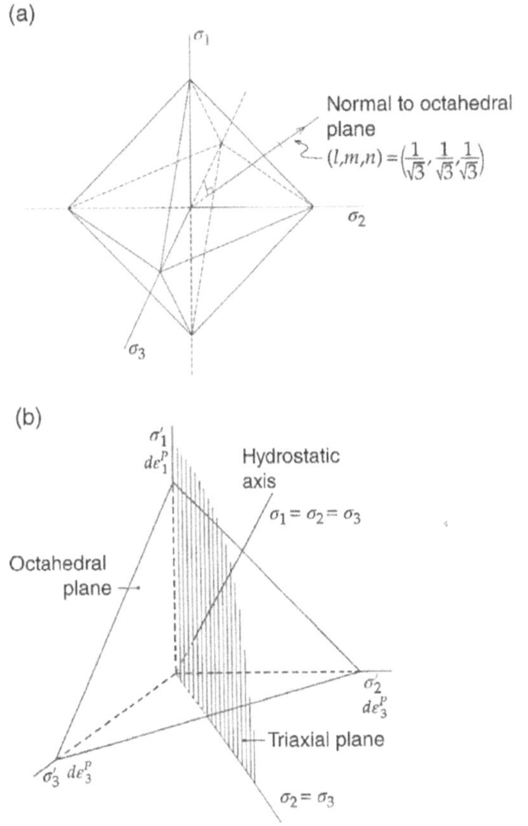

Figure 2.75 Principal stress space with (a) eight octahedral planes forming an octahedron and with (b) triaxial plane and octahedral plane in octant with positive normal stresses.

respectively. The distance from the origin to the octahedral plane that contains τ_{oct} is therefore $\frac{1}{\sqrt{3}} \cdot |\overline{OQ}|$.

Note that all points located on a circle in the octahedral plane have the same value of σ_{oct} and τ_{oct}, as shown in Fig. 2.77.

2.7.2 Triaxial plane

The triaxial plane that contains the σ_1-axis is shown in Fig. 2.78. In this plane the hydrostatic axis forms an angle of $\theta = \arctan(\frac{1}{\sqrt{2}}) = 35.26°$ with the horizontal axis. Any state of stress that can be produced in a triaxial test in which the

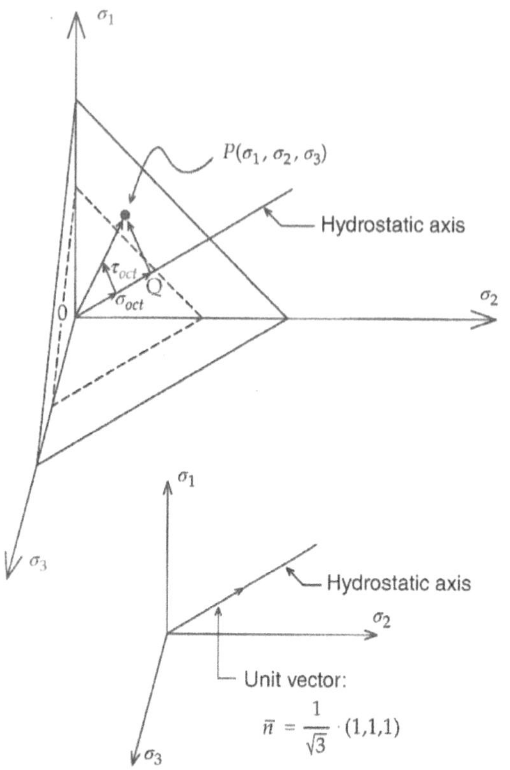

Figure 2.76 Octahedral stress vectors and octahedral planes used in location of stress point P.

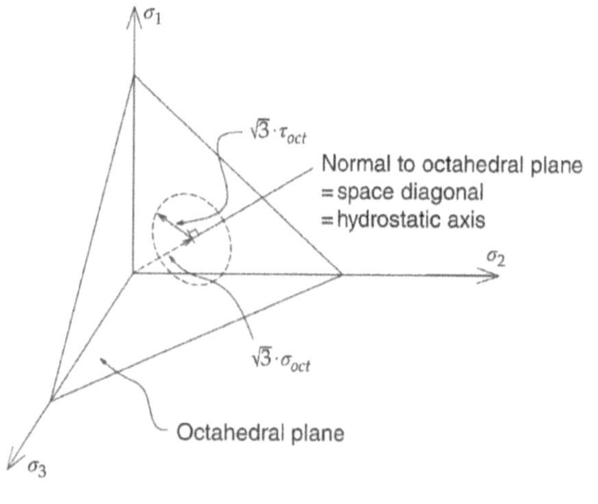

Figure 2.77 Points on a circle in an octahedral plane have the same values of σ_{oct} and τ_{oct}.

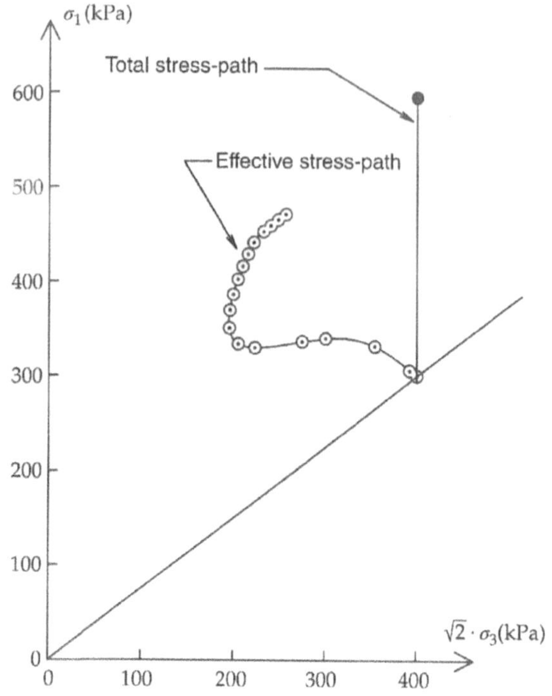

Figure 2.78 Triaxial plane with total and effective stress paths for triaxial compression test on normally consolidated, remolded Edgar Plastic Kaolinite clay.

state of stress is axisymmetric can be shown in a triaxial plane. States of stress in triaxial compression plot above the hydrostatic axis and states of stress in triaxial extension plot below the hydrostatic axis. The total and effective stress paths for an ICU test is shown in Fig. 2.78. These are the stress paths previously shown on the *p*–*q* diagrams in Fig. 2.52.

The stress paths shown in the triaxial plane and the Cambridge *p*–*q* diagram are very similar. In fact, these two diagrams are the same within a linear transformation. Figure 2.79 shows the two diagrams superimposed with the coordinates for a stress point indicated with reference to the triaxial plane. A similar direct comparison cannot be made between the triaxial plane and the modified Mohr diagram, because the abscissa of the latter cannot be changed by a linear transformation to match the distance along the hydrostatic axis.

While the Cambridge *p*–*q* diagram has the advantage of direct transformation into the

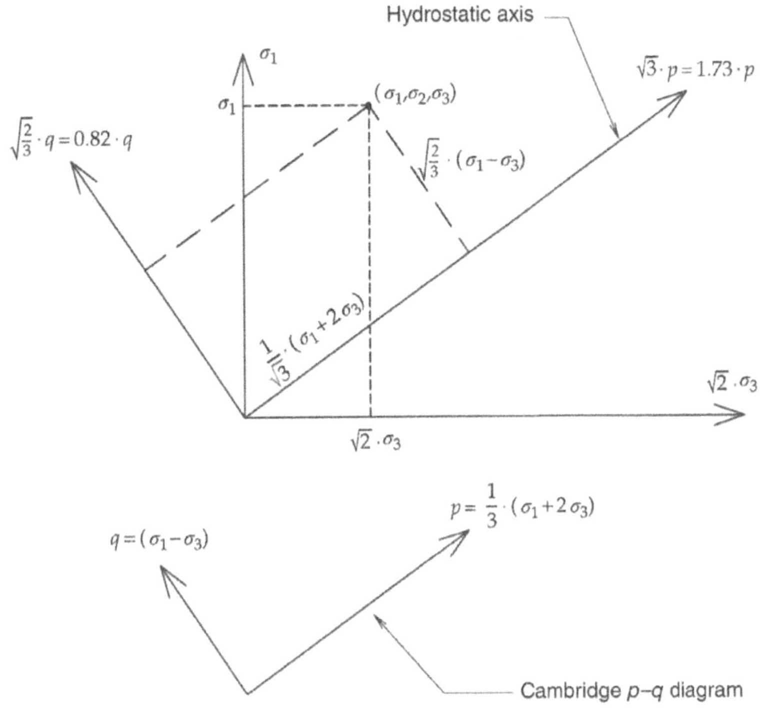

Figure 2.79 Comparison between triaxial plane and Cambridge *p*–*q* diagram.

triaxial plane and therefore has a direct relation with the principal stress space, the modified Mohr diagram (= MIT *p*–*q* diagram) is useful for stress analyses in two dimensions, as indicated in Section 2.1.7. In addition, stress paths shown in the Cambridge *p*–*q* diagram or the triaxial plane are larger, because the slope of the failure surface is steeper in this diagram than in the modified Mohr diagram.

2.7.3 Octahedral plane

The other 2D diagram in the principal stress space that is useful for presentation of test results is the octahedral plane. The results of tests with three unequal stresses are often shown on this diagram.

In the octahedral plane the center point represents the hydrostatic axis, as shown in Fig. 2.80. The three axes are 120° apart and they represent the projections of the principal stress axes on

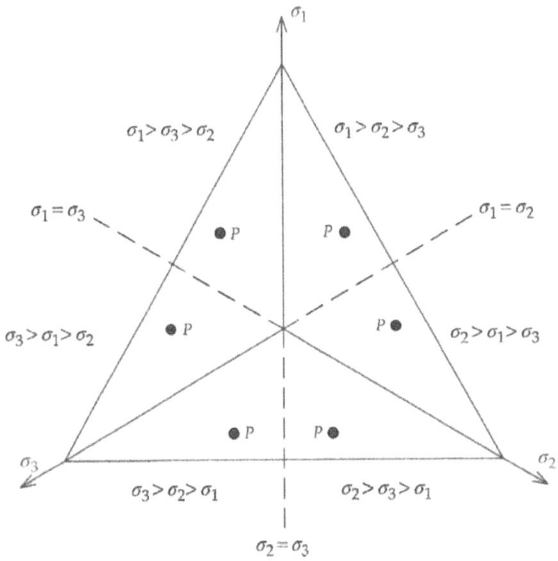

Figure 2.80 Octahedral plane with state of stress represented in each of six sectors.

the octahedral plane. These axes are lines of symmetry such that $\sigma_1 = \sigma_2$, $\sigma_2 = \sigma_3$, and $\sigma_3 = \sigma_1$ represent the traces in the octahedral plane of the three triaxial planes.

If σ_1, σ_2, and σ_3 are taken as the major, intermediate and minor principal stresses, only one-sixth of the octahedral plane is necessary for representing any state of stress. However, any one of three axes could be the major principal stress axis. Thus, interchanging the subscripts (1, 2, 3) one state of stress is represented in each of the six parts of the octahedral plane. The state of stress represented by the point P is shown in each of the six sectors of the octahedral plane in Fig. 2.80.

Figure 2.81(a) shows the Mohr–Coulomb failure criterion in the principal stress space for a material without effective cohesion. The failure surfaces form a cone with the apex at the origin. The cross-sections in the octahedral plane have shapes of irregular hexagons with acute and obtuse angles at the points corresponding to the states of stress in triaxial compression and extension, respectively. The cross-sections change shape as the magnitude of the friction angle changes, as indicated in Fig. 2.81(b). The shape approaches an equilateral triangle for friction angles approaching 90°, and it resembles a regular hexagon at very small friction angles.

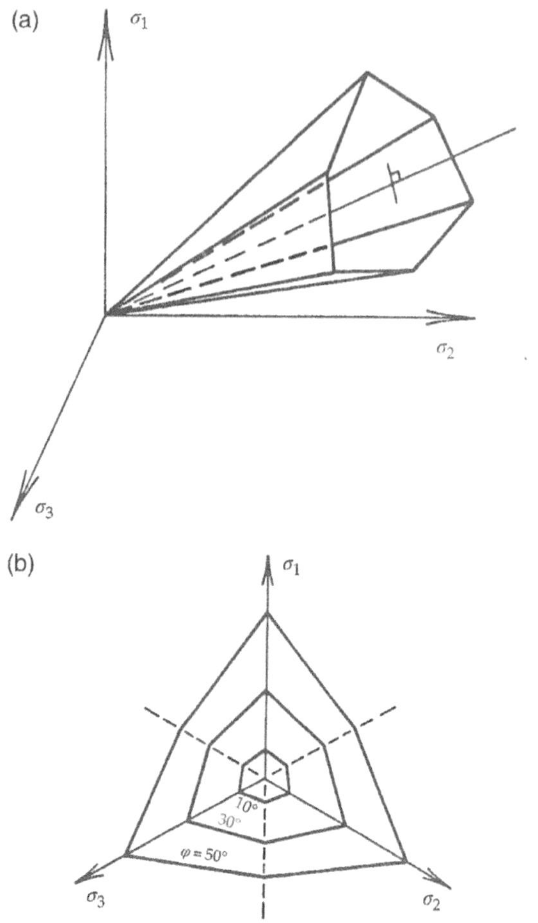

Figure 2.81 (a) Three-dimensional representation of Mohr–Coulomb failure criterion in principal stress space and (b) variation of cross-sectional shape with friction angle in an octahedral plane.

2.7.4 Characterization of 3D stress conditions

b-Value

The magnitude of the intermediate principal stress, σ_2, relative to the major and minor principal stresses, σ_1 and σ_3, may be indicated by the value of b, originally employed by Habib (1953):

$$b = \frac{\sigma_2 - \sigma_3}{\sigma_1 - \sigma_3} \tag{2.151}$$

The value of $b = 0$ when $\sigma_2 = \sigma_3$, as in a triaxial compression test, while $b = 1$ for a triaxial extension test in which $\sigma_2 = \sigma_1$. The value of b increases linearly from 0 to 1 with increasing σ_2-value from σ_3 to σ_1. This parameter may be used for example to show the variation of the friction angle with the intermediate principal stress, as exemplified in the φ–b diagram in Fig. 2.82.

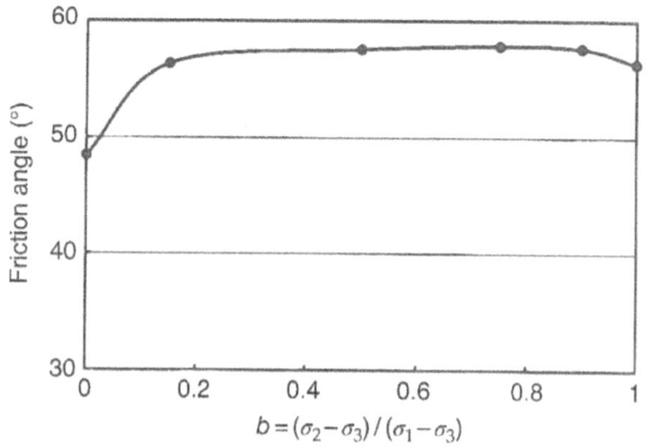

Figure 2.82 Example of φ–b diagram for dense Monterey No. 0 sand.

Lode angle

The Lode angle, θ, relates to the octahedral plane and is defined from Eq. (2.137) (Lode 1926; Nayak and Zienkiewicz 1972) as $\theta = \alpha/3$, and it is given by:

$$\cos(3\theta) = \frac{3\sqrt{3}}{2} \cdot \frac{J_3}{J_2^{3/2}} \tag{2.152}$$

For triaxial compression, $\theta = 0°$ and for triaxial extension, $\theta = 60°$. The value of θ is related to the value of b is indicated in Fig. 2.83, and as given by the following expression:

$$\cos(\alpha) = \cos(3\theta) = \frac{1}{2} \cdot \frac{(2b^3 - 3b^2 - 3b + 2)}{(b^2 - b + 1)^{3/2}} \tag{2.153}$$

An example of the use of the Lode angle is indicated by the variation of the friction angle

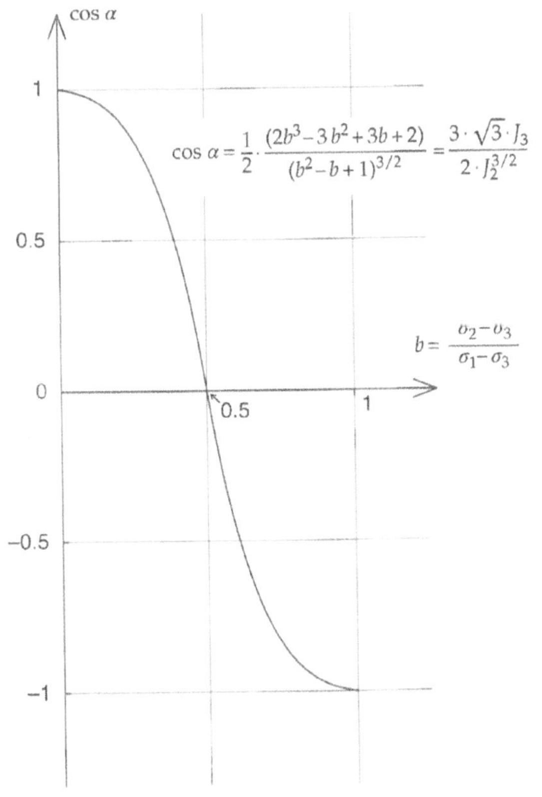

Figure 2.83 Relationship between $\cos\alpha$ and $b = (\sigma_2 - \sigma_3)/(\sigma_1 - \sigma_3)$.

in the octahedral plane as shown in Fig. 2.84. This diagram looks very similar to the φ–b diagram, but the Lode angle is rarely employed for this purpose.

b-Values and Lode angles for cross-anisotropic soils

To present results of tests on cross-anisotropic materials, it is important to clearly indicate the directions of stress and strain relative to the principal axes of the material. For this purpose a Cartesian coordinate system is employed as indicated in Fig. 2.85. The X-axis coincides with the axis of rotational symmetry for the cross-anisotropic specimens. The stresses are labeled according to this coordinate system.

The Lode angle, θ, is measured clockwise from the σ_x-axis to the stress point P (σ_x, σ_y, σ_z) as indicated on the octahedral plane in Fig. 2.85(b, c) and is calculated as follows:

$$\tan\theta = \sqrt{3} \cdot \frac{\sigma_y - \sigma_z}{(\sigma_x - \sigma_y) + (\sigma_x - \sigma_z)} \tag{2.154}$$

Values of θ are indicated on the stress axes in Fig. 2.85(c). For the ranges of θ between 0° and 60°, 60° and 120°, and 120° and 180°, the b-values are calculated as follows:

$$b = \frac{\sigma_y - \sigma_z}{\sigma_x - \sigma_z} \quad \text{for } 0° \le \theta \le 60° \tag{2.155a}$$

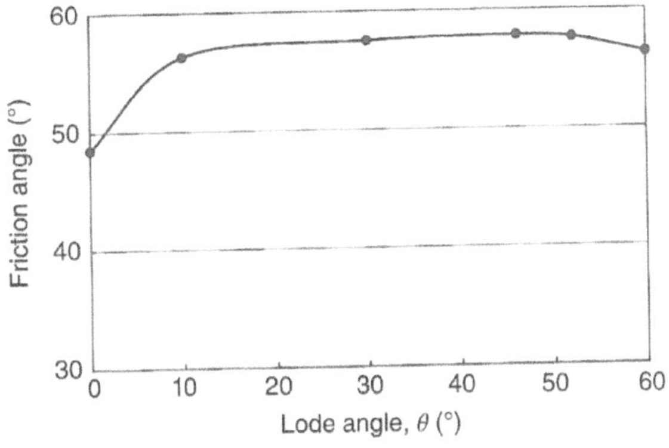

Figure 2.84 Example of ϕ–θ diagram in which θ is the Lode angle in the octahedral plane. Same data for dense Monterey No. 0 sand as shown in Fig. 2.83.

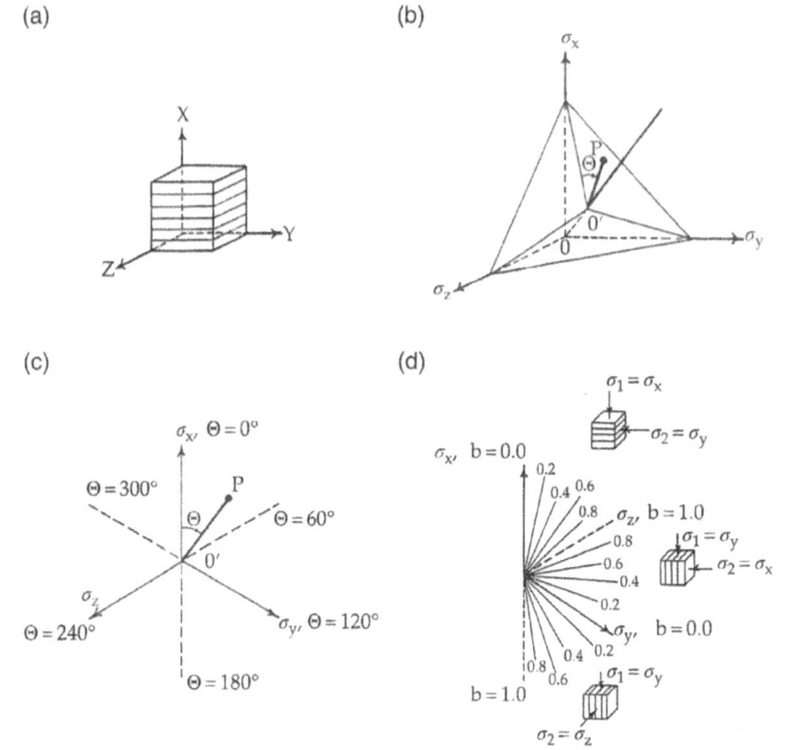

Figure 2.85 Orientation of a cross-anisotropic specimen relative to (a) a Cartesian coordinate system, (b) principal stress space, and (c and d) an octahedral plane. Reproduced from Kirkgard and Lade 1993 by permission of Canadian Science Publishing.

$$b = \frac{\sigma_x - \sigma_z}{\sigma_y - \sigma_z} \quad \text{for } 60° \le \theta \le 120° \qquad (2.155b)$$

$$b = \frac{\sigma_z - \sigma_x}{\sigma_y - \sigma_x} \quad \text{for } 120° \le \theta \le 180° \qquad (2.155c)$$

It is clear that the value of θ is sufficient to indicate the relative magnitudes of the principal stresses, and it also provides information regarding which of the normal stresses are the major, intermediate and minor principal stresses. The value of b varies from 0 to 1 in each of the six sectors of the octahedral plane, and calculation of b requires previous ordering of the principal stresses. However, the parameter b has frequently been used in studies of 3D behavior of soils, and it is convenient to use this parameter together with the parameter θ in discussions of test results.

Figure 2.86 shows experimental results from true triaxial tests on intact, cross-anisotropic San Francisco Bay Mud projected on an octahedral plane.

2.7.5 Shapes of stress invariants in principal stress space

If the stress invariants I_1, I_2, and I_3 are set equal to constant values, then each of these invariants takes very instructive shapes in the part of the principal stress space in which the principal stresses are positive. Thus, I_1 = constant is shown in Fig. 2.87. Figure 2.87(a) shows that I_1 = constant in the octahedral plane is a plane that cuts the principal stress axes at the same values of the principal stresses, while Fig. 2.87(b) indicates the edges in the triaxial plane, and Fig. 2.87(c) shows the plane in the principal stress space.

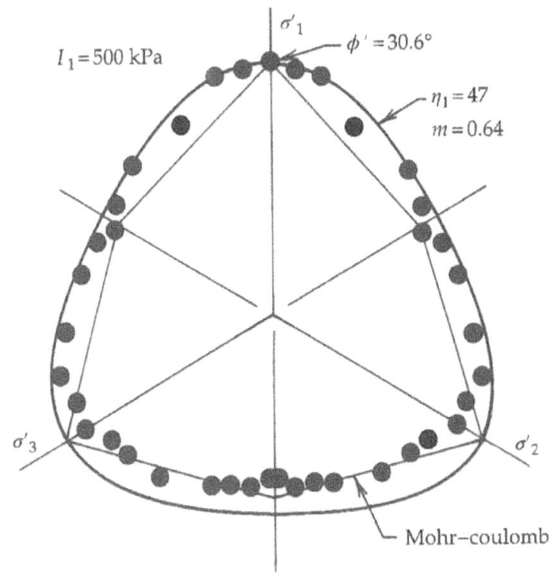

Figure 2.86 Octahedral plane with results from true triaxial tests on intact, cross-anisotropic San Francisco Bay Mud. Reproduced from Kirkgard and Lade 1993 by permission of Canadian Science Publishing.

Figure 2.88 shows the shape of I_2 = constant. Figure 2.88(a) indicates that the traces in the octahedral plane are shaped as concentric circles, whose values are shown on the diagram. Figure 2.88(b) shows the traces in the triaxial plane, which are symmetric around the hydrostatic axis and asymptotic to the major principal stress axis. Figure 2.88(c) shows the combination of these shapes in the principal stress space.

Figure 2.89 indicates the shape of I_3 = constant. The cross-sectional shapes shown in Fig. 2.89(a) are triangular with smoothly curved traces in the octahedral plane. Depending on the value of I_3, the shape is more or less rounded and near the hydrostatic axis the shape becomes circular. For lower values of I_3 the shape becomes more triangular and reaches out in the corners of the octahedral plane. Figure 2.89(b) indicates the traces of I_3 in the triaxial plane. They are asymptotic to the major principal stress axis and to the plane created by the other two principal stresses. Figure 2.89(c) shows the combination of these views in the principal stress space.

2.7.6 Procedures for projecting stress points onto a common octahedral plane

Straight failure envelope

It is often useful to study the shape of failure surfaces in the principal stress space. The cross-sectional shapes of failure surfaces are best shown on an octahedral plane. However, tests conducted with three unequal principal stresses most often do not fail at the same value of the octahedral normal stress, and the results therefore cannot be plotted directly on the same octahedral plane. For soils without cohesion and with straight failure surfaces in meridian planes (i.e., planes that contain the hydrostatic axis), the principal stresses at failure can be modified according to the following expression such that all stress points fall on one octahedral plane:

$$(\sigma_1{}^*,\sigma_2{}^*,\sigma_3{}^*) = (\sigma_1,\sigma_2,\sigma_3)\cdot\frac{\sigma_{oct}}{\frac{1}{3}(\sigma_1+\sigma_2+\sigma_3)} \qquad (2.156)$$

where σ_1, σ_2, and σ_3 are the principal stresses measured at failure, and $\sigma_{oct}\left[=\frac{1}{3}(\sigma_1{}^*+\sigma_2{}^*+\sigma_3{}^*)\right]$ is the octahedral normal stress corresponding to the octahedral plane on which the test results are to be plotted. The value of σ_{oct} is of no importance to the cross-sectional shape of the failure surface for soils without cohesion and with straight failure surfaces. Note that only the magnitudes of the principal stresses are modified, whereas the ratios between the principal stresses remain constant. Thus, the friction angle is not changing due to the modification of stresses given in Eq. (2.156).

For soils with cohesion, it is necessary to modify the normal stresses before projecting them along the straight lines going through the stress origin. This is done by translating the principal stress space along the hydrostatic axis to account for the cohesion and tensile strength that can be sustained by such materials. Thus, a constant stress, $a\cdot p_a$, is added to the normal stresses:

$$\bar{\sigma}_1 = \sigma_1 + a\cdot p_a \qquad (2.157a)$$

$$\bar{\sigma}_2 = \sigma_2 + a\cdot p_a \qquad (2.157b)$$

$$\bar{\sigma}_3 = \sigma_3 + a\cdot p_a \qquad (2.157c)$$

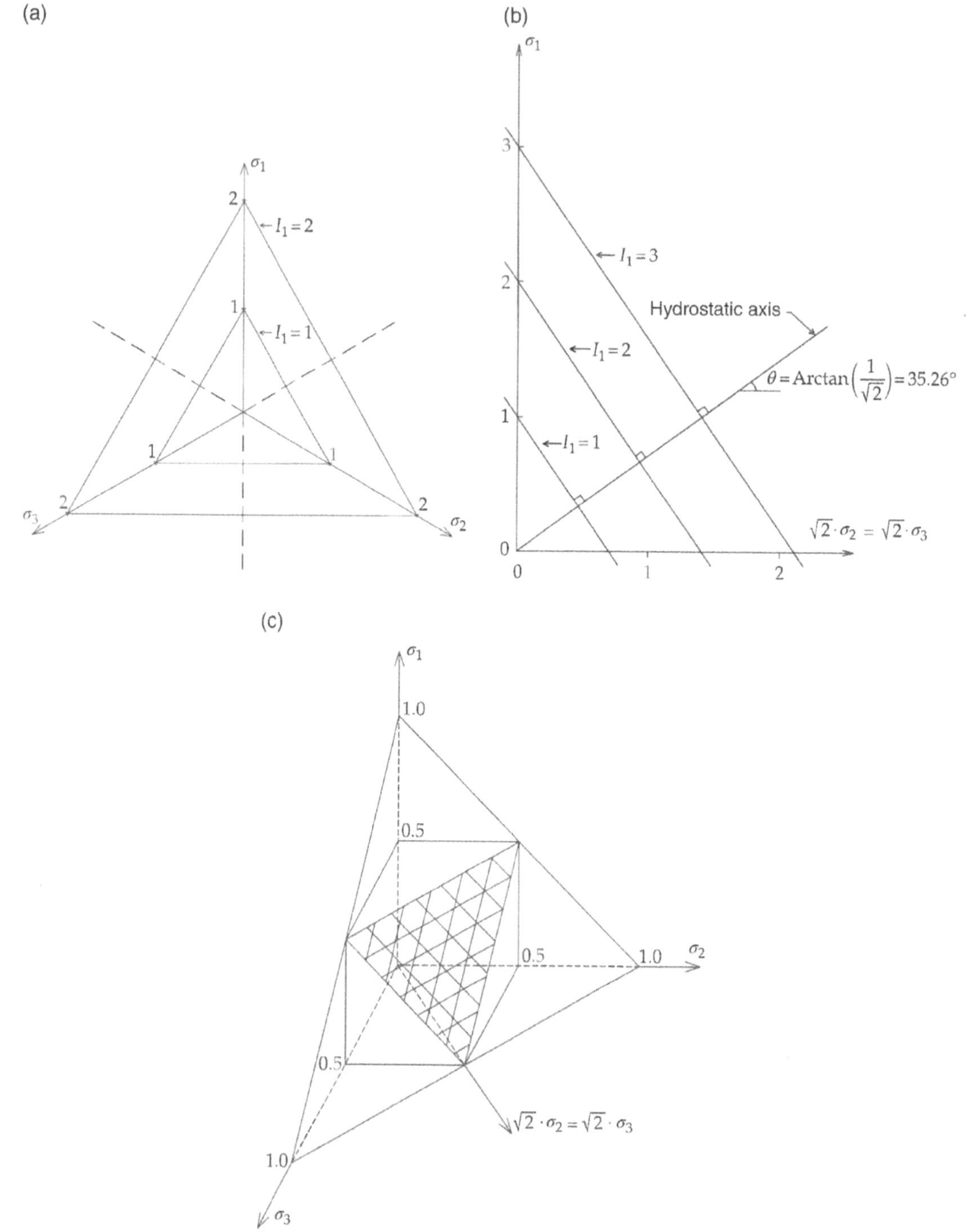

Figure 2.87 I_1 = constant plotted on (a) an octahedral plane, (b) a triaxial plane, and in (c) principal stress space.

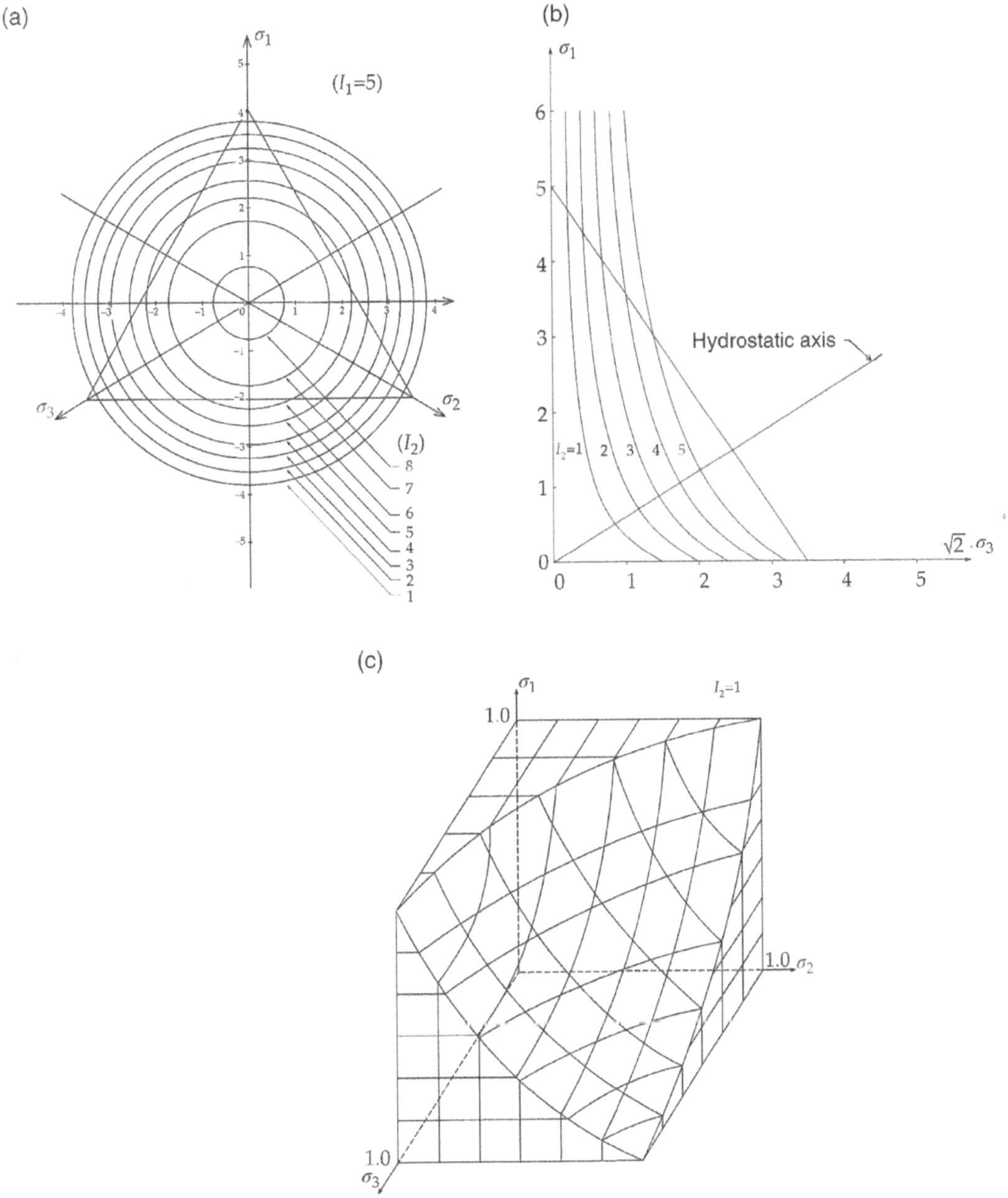

Figure 2.88 I_2 = constant plotted on (a) an octahedral plane, (b) a triaxial plane, and in (c) principal stress space.

This translation along the hydrostatic axis causes the failure surface to go through the translated stress origin, it accounts for the cohesion and tensile strength of the materials as explained by Lade (1982b), and it allows a straight line projection onto the common octahedral plane of the type indicated in Eq. (2.156).

The stress state (σ_1^*, σ_2^*, σ_3^*) for various tests, which has been modified to have a common octahedral normal stress, may now be plotted on the common octahedral plane as shown below.

For soils with curved failure surfaces in meridian planes, it is necessary to project the stress

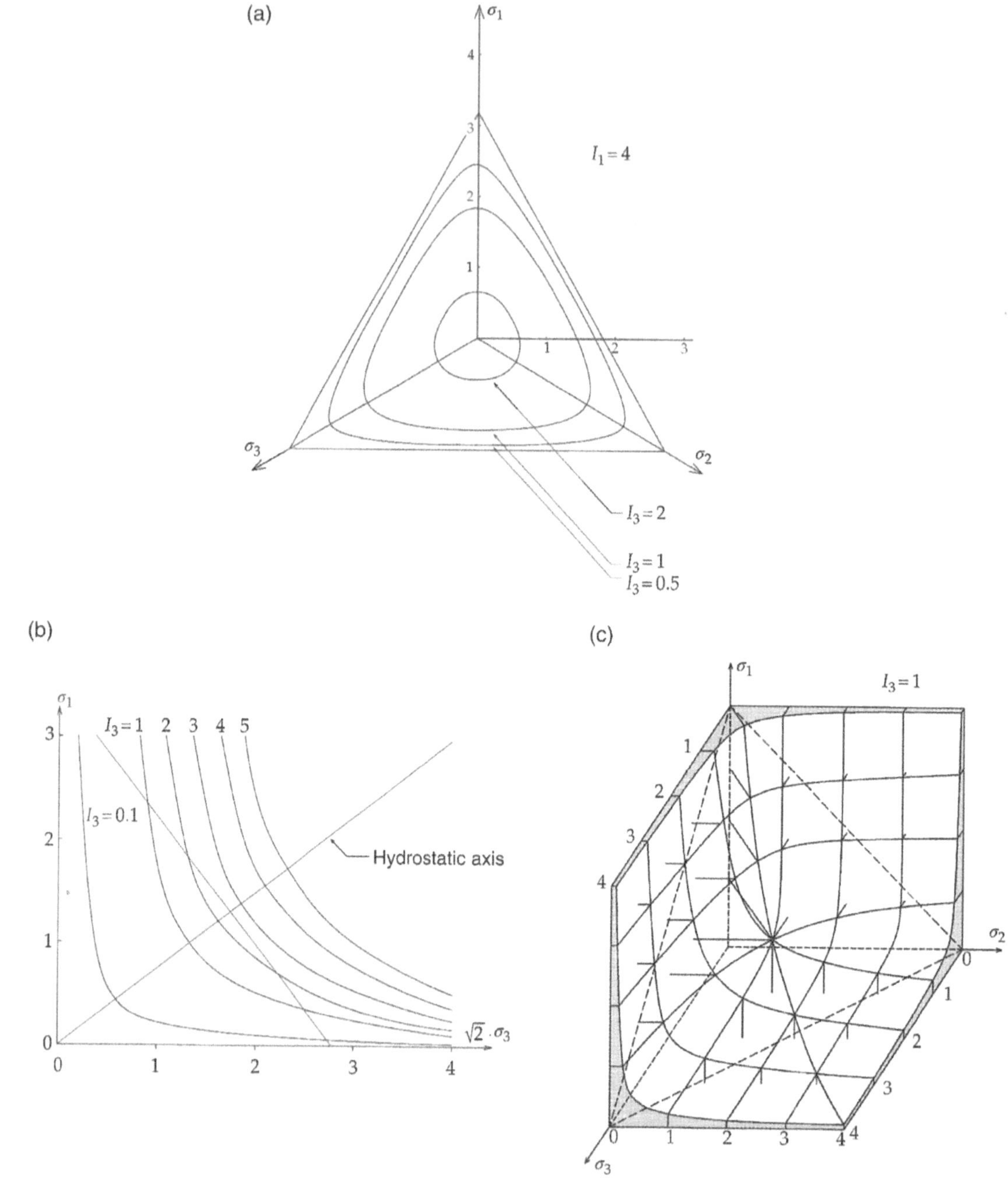

Figure 2.89 I_3 = constant plotted on (a) an octahedral plane, (b) a triaxial plane, and in (c) principal stress space.

points onto a common octahedral plane using a more complex procedure, as indicated below.

Curved failure envelope

The projection of a stress point along a curved failure surface onto a common octahedral plane is more complex, because the expression for the failure surface has to be known a priori, and it may involve solving a cubical equation as in the case presented here. The expression for the curved failure surface employed here is that proposed by Lade (1977):

$$\left(\frac{I_1^3}{I_3}-27\right)\left(\frac{I_1}{p_a}\right)^m=\eta_1 \qquad (2.158)$$

where I_1 and I_3 are the first and the third stress invariants, reviewed in Section 2.6.2, and p_a is atmospheric pressure expressed in the same units as the stresses. The parameters η_1 and m can be determined by plotting $\left(I_1^3/I_3-27\right)$ versus p_a/I_1 at failure in a log–log diagram, as shown in Fig. 2.90, and locate the best fitting straight line. The intercept of this line with $p_a/I_1 = 1$ is the value of η_1, and m is the slope of the line. For simplicity, it is assumed in the following development that the stresses have already been normalized by p_a and p_a will not be shown.

Figure 2.91 shows the known stress state (σ_1', σ_2', σ_3') at point 1, and the adjusted stress (σ_1^*, σ_2^*, σ_3^*) at point 2 is to be determined by projecting point 1 along a line with slope m to reach the desired common octahedral plane characterized by the first stress invariant I_{12}. The failure state at point 1 is defined by:

$$\left(\frac{I_1^3}{I_3}-27\right)_1\cdot\left(I_{11}\right)^m=\bar{\eta}_1 \qquad (2.159)$$

The failure state at point 2 can also be described by:

$$\left(\frac{I_1^3}{I_3}-27\right)_2\cdot\left(I_{12}\right)^m=\bar{\eta}_1 \qquad (2.160)$$

which may be rearranged to:

$$\left(\frac{I_{12}^3}{I_{32}}\right)=\frac{\bar{\eta}_1}{I_{12}^m}+27 \qquad (2.161)$$

and

$$I_{32}=\frac{1}{\dfrac{\bar{\eta}_1}{I_{12}^m}+27}\cdot I_{12}^3 \qquad (2.162)$$

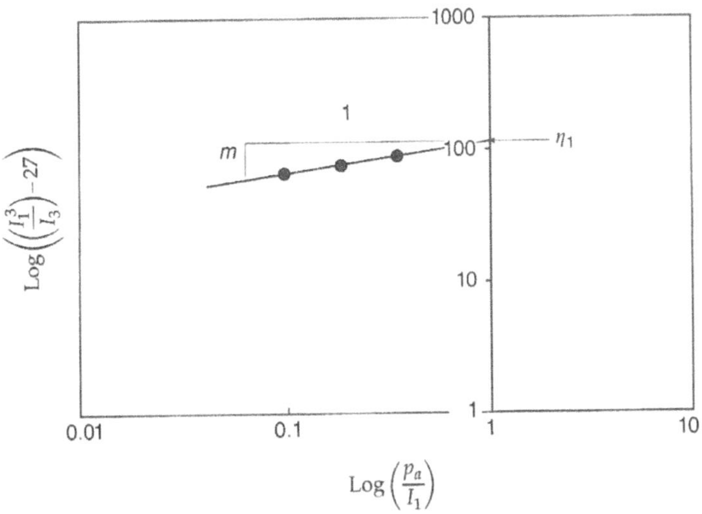

Figure 2.90 Determination of the best fit straight line on a log–log diagram for characterization of the 3D failure criterion.

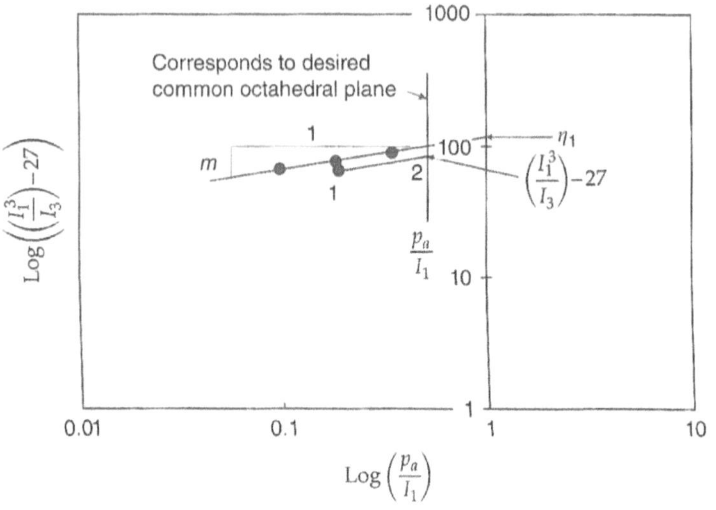

Figure 2.91 Procedure used for projection of stress point along a curved failure surface to a common octahedral plane.

The parameter b is defined as:

$$b = \frac{\sigma_2{}^* - \sigma_3{}^*}{\sigma_1{}^* - \sigma_3{}^*} \tag{2.163}$$

from which the intermediate principal stress can be expressed as:

$$\sigma_2{}^* = \sigma_3{}^* + b \cdot (\sigma_1{}^* - \sigma_3{}^*) \tag{2.164}$$

The desired first stress invariant is:

$$I_{12} = \sigma_1{}^* + \sigma_2{}^* + \sigma_3{}^* \tag{2.165}$$

With substitution of $\sigma_2{}^*$ from Eq. (2.164) into Eq. (2.165) the following is obtained:

$$\begin{aligned} I_{12} &= \sigma_1{}^* + \sigma_3{}^* + b \cdot (\sigma_1{}^* - \sigma_3{}^*) + \sigma_3{}^* \\ &= (1+b) \cdot \sigma_1{}^* + (2-b) \cdot \sigma_3{}^* \end{aligned} \tag{2.166}$$

from which the desired major principal stress can be expressed as:

$$\sigma_1{}^* = \frac{1}{(1+b)} \cdot \left[I_{12} - (2-b) \cdot \sigma_3{}^* \right] \tag{2.167}$$

The desired third stress invariant is given by:

$$I_{32} = \sigma_1{}^* \cdot \sigma_2{}^* \cdot \sigma_3{}^* \tag{2.168}$$

The expression for $\sigma_2{}^*$ in Eq. (2.164) is substituted into Eq. (2.168):

$$\begin{aligned} I_{32} &= \sigma_1{}^* \cdot \left[\sigma_3{}^* + b \cdot (\sigma_1{}^* - \sigma_3{}^*) \right] \cdot \sigma_3{}^* \\ &= \sigma_1{}^* \cdot (\sigma_3{}^*)^2 + b \cdot (\sigma_1{}^*)^2 \cdot \sigma_3{}^* - b \cdot \sigma_1{}^* \cdot (\sigma_3{}^*)^2 \\ &= (1-b) \cdot \sigma_1{}^* \cdot (\sigma_3{}^*)^2 + b \cdot (\sigma_1{}^*)^2 \cdot \sigma_3{}^* \end{aligned} \tag{2.169}$$

The expression for $\sigma_1{}^*$ in Eq. (2.167) is substituted into Eq. (2.169):

$$\begin{aligned} I_{32} = (1-b) \cdot \frac{1}{1+b} \cdot \left[I_{12} - (2-b) \cdot \sigma_3{}^* \right] \cdot (\sigma_3{}^*)^2 \\ + b \cdot \frac{1}{(1+b)^2} \cdot \left[I_{12} - (2-b) \cdot \sigma_3{}^* \right]^2 \cdot \sigma_3{}^* \end{aligned} \tag{2.170}$$

Rearrangement of Eq. (2.170) produces:

$$\begin{aligned} I_{32} = \frac{b \cdot (2-b)^2 - (1+b)(1-b)(2-b)}{(1+b)^2} \cdot (\sigma_3{}^*)^3 \\ + \frac{(1+b)(1-b) - 2b(2-b)}{(1+b)^2} \cdot I_{12} \cdot (\sigma_3{}^*)^2 \\ + \frac{b \cdot I_{12}^2}{(1+b)^2} \cdot (\sigma_3{}^*) \end{aligned} \tag{2.171}$$

Further rearrangement produces the following cubical equation:

$$[b\cdot(2-b)^2-(1+b)(1-b)(2-b)](\sigma_3{}^*)^3 + [(1+b)(1-b)-2b(2-b)]\cdot I_{12}\cdot(\sigma_3{}^*)^2 + b\cdot I_{12}^2\cdot(\sigma_3{}^*)-(1+b)^2\cdot I_{32}=0 \quad (2.172)$$

or

$$(\sigma_3{}^*)^3+\frac{(b^2-4b+1)\cdot I_{12}}{(2-b)(2b-1)}\cdot(\sigma_3{}^*)^2+\frac{b\cdot I_{12}^2}{(2-b)(2b-1)}\cdot(\sigma_3{}^*) + \frac{-(1+b)^2\cdot I_{32}}{(2-b)(2b-1)}=0 \quad (2.173)$$

This is a cubical equation in (σ_3*). It is in the form of:

$$x^3+Ax^2+Bx+C=0 \quad (2.174)$$

The solution to the cubical equation is given by Korn and Korn (1961), as reviewed in Section 2.6.4.

For $b \neq 0.5$, the solution to the cubical equation in Eq. (2.173) becomes:

$$b<0.5:\sigma_3{}^*=x_2=-2\cdot\sqrt{-\frac{p}{3}}\cdot\cos\left(\frac{\alpha}{3}+60^\circ\right)-\frac{A}{3} \quad (2.175)$$

$$b>0.5:\sigma_3{}^*=x_3=-2\cdot\sqrt{-\frac{p}{3}}\cdot\cos\left(\frac{\alpha}{3}-60^\circ\right)-\frac{A}{3} \quad (2.176)$$

in which A is the coefficient to (σ_3*)^2 in Eq. (2.173), and p and α are determined from Eqs (2.129) and (2.133), respectively.

For $b = 0.5$, Eq. (2.172) becomes a quadratic equation:

$$(b^2-4b+1)\cdot I_{12}\cdot(\sigma_3{}^*)^2+b\cdot I_{12}^2\cdot(\sigma_3{}^*)-(1+b)^2\cdot I_{32}=0 \quad (2.177)$$

The solution to Eq. (2.177) is:

$$\sigma_3{}^*=\frac{1}{3}\cdot I_{12}-\frac{1}{3}\cdot\sqrt{\frac{I_{32}}{I_{12}}}\cdot\sqrt{\frac{I_{12}^3}{I_{32}}-27} \quad (2.178)$$

Consequently, depending on the value of b, which will be retained from point 1 to point 2, the minor principal stress σ_3* can be calculated from Eq. (2.175), Eq. (2.176), or Eq. (2.178). Then the major principal stress can be calculated from Eq. (2.167), and subsequently, the intermediate principal stress can be obtained, Eq. (2.164).

Because the principal stresses depend on stress invariants, these principal stresses are themselves invariants, and it may be easier to work with them directly.

The stress state (σ_1*, σ_2*, σ_3*), which has been projected to the common octahedral plane, may be plotted on the octahedral plane as shown below.

2.7.7 Procedure for plotting stress points on an octahedral plane

The point P corresponding to the modified principal stresses σ_1*, σ_2*, and σ_3* can be placed on the octahedral plane according to the following procedure. The principal stress space with the octahedral plane that contains the stress point P is shown in Fig. 2.92(a). The procedure for finding the distances between projection of P on the principal stress axes in the octahedral plane and the hydrostatic axis will be demonstrated. The calculations all pertain to the triaxial planes, and the plane containing the σ_1-axis is shown in Fig. 2.92(b). The point in which the hydrostatic axis crosses the octahedral plane is designated O′. The distance between the origin and the projection of O′ on the σ_1-axis is equal to $\sigma_{oct}=\frac{1}{3}\cdot(\sigma_1{}^*+\sigma_2{}^*+\sigma_3{}^*)$, as may be seen from Fig. 2.92(b). The projection of P on the triaxial plane is designated P′. The distance O′P′ = a then becomes [see Fig. 2.92(b)]:

$$a=(\sigma_1{}^*-\sigma_{oct})\cdot\frac{1}{\cos\theta} \quad (2.179)$$

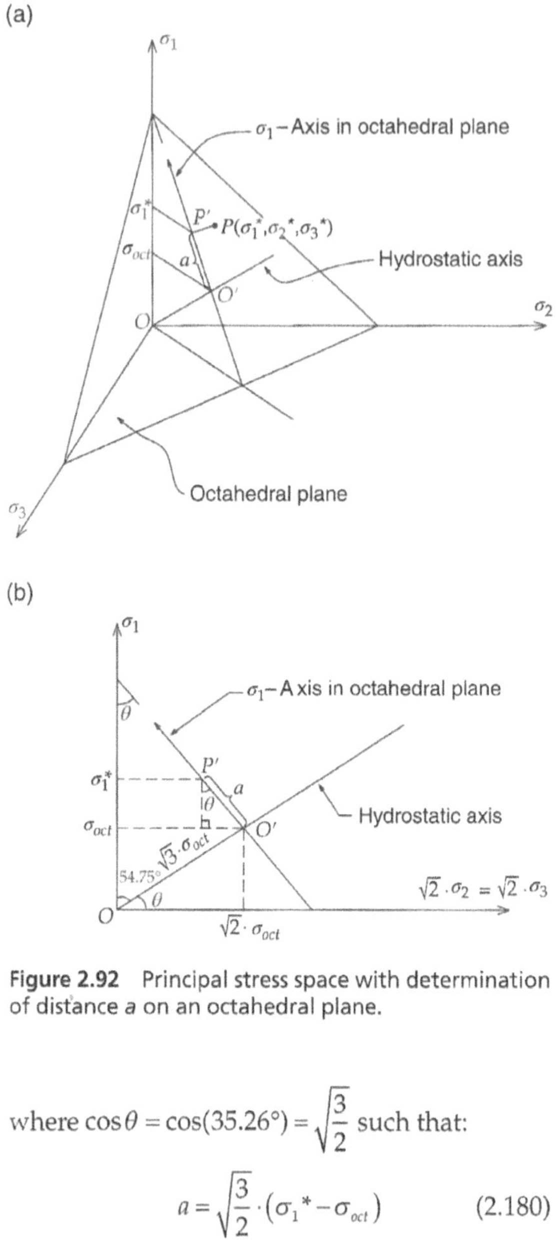

Figure 2.92 Principal stress space with determination of distance *a* on an octahedral plane.

where $\cos\theta = \cos(35.26°) = \sqrt{\frac{3}{2}}$ such that:

$$a = \sqrt{\frac{3}{2}} \cdot (\sigma_1{}^* - \sigma_{oct}) \tag{2.180}$$

Similar expressions can be obtained for the corresponding values of b and c:

$$b = \sqrt{\frac{3}{2}} \cdot (\sigma_2{}^* - \sigma_{oct}) \tag{2.181}$$

$$c = \sqrt{\frac{3}{2}} \cdot (\sigma_3{}^* - \sigma_{oct}) \tag{2.182}$$

With known values of the distances a, b, and c, the point P can be plotted on the octahedral plane, as shown in Fig. 2.93(a). The lengths a, b, and c are marked out on the axes in the octahedral plane and lines perpendicular to the axes are drawn to intersection to give the position of stress point P. It is seen that it is necessary to use only two of the three values a, b, and c.

For the purpose of simplifying the positioning of point P on the octahedral plane, the following trigonometric considerations are made so the coordinates of P are determined in the X–Y diagram shown in Fig. 2.93(b). The y-coordinate is equal to a:

$$y = \sqrt{\frac{3}{2}} \cdot (\sigma_1{}^* - \sigma_{oct}) \tag{2.183}$$

The x-coordinate is determined as shown in Fig. 2.93(b):

$$x = -c \cdot \cos 30° - (a + c \cdot \sin 30°) \cdot \tan 30°$$

$$= -c \cdot \left(\frac{\sqrt{3}}{2} + \frac{1}{2} \cdot \frac{\sqrt{3}}{3} \right) - a \cdot \frac{\sqrt{3}}{3}$$

$$x = \frac{\sqrt{3}}{3} \cdot (-a - 2c) \tag{2.184}$$

Substituting the values of a and c from Eqs (2.180) and (2.181) into Eq. (2.182) and reducing, the x-coordinate becomes:

$$x = \frac{\sqrt{2}}{2} \cdot (\sigma_2{}^* - \sigma_3{}^*) \tag{2.185}$$

Using Eqs (2.183) and (2.185), the point P(σ_1, σ_2, σ_3) can readily be located on the octahedral plane, as shown in Fig. 2.93(b).

2.7.8 Representation of test results with principal stress rotation

For experiments with rotation of principal stresses, shear stresses are applied to the surface of the specimen as is done in torsional, directional shear and direct shear tests. The rotation

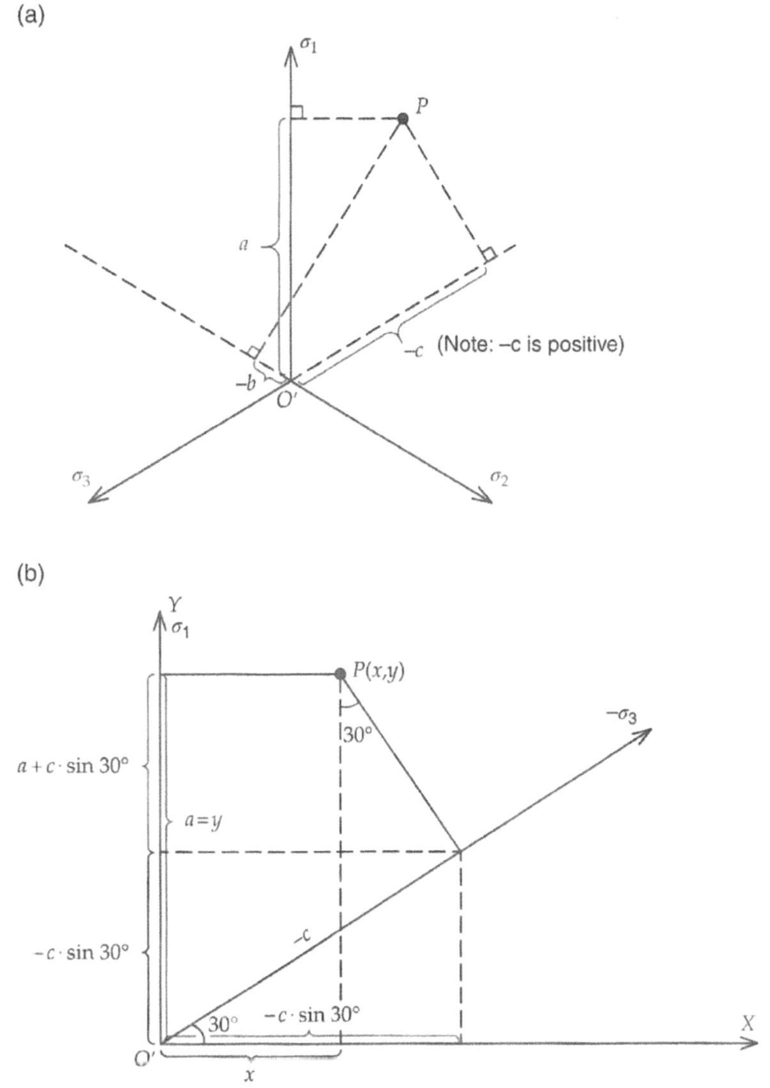

Figure 2.93 Location of stress point P on an octahedral plane from values of (a) *a*, *b*, and *c* or (b) *x* and *y*.

of principal stresses that occur in such tests cannot be indicated in the principal stress space and another diagram may be employed to show the results of such tests. In this diagram the applied shear stress $\tau_{z\theta}$ is plotted versus the stress difference ($\sigma_z - \sigma_\theta$) expressed in polar coordinates. Such diagrams are shown in Chapter 11.

Chapter V
Model for the Elastic Behavior of Frictional Materials

Preface

While the elastic strains are small relative to the plastic strains, it is important to include the elastic strains at all time, because they are used to keep track of the state of stress. Thus, zero elastic strains are not an option in the finite element or finite difference programs, even when the elastic strains are small relative to the plastic strains. On the other hand, accounting for anisotropy in the elastic strains is very complicated, both in terms of determination of the material parameters and in terms of the much more complex elastic material matrix. Since the elastic strains are typically small, the additional complication of accounting for anisotropy is practically unimportant for most geotechnical projects.

INTERNATIONAL JOURNAL FOR NUMERICAL AND ANALYTICAL METHODS IN GEOMECHANICS. VOL. 11, 521–542 (1987)

MODELLING THE ELASTIC BEHAVIOUR OF GRANULAR MATERIALS

POUL V. LADE AND RICHARD B. NELSON

Department of Civil Engineering, School of Engineering and Applied Science, University of California, Los Angeles, California 90024, U.S.A.

SUMMARY

A review of the literature indicates that the elastic behaviour of granular materials is isotropic and that Poisson's ratio is constant, whereas Young's Modulus, the bulk modulus and the shear modulus vary with the mean normal stress and the deviatoric stress. A nonlinear, isotropic model for the elastic behaviour is developed on the basis of theoretical considerations involving the principle of conservation of energy. Energy is therefore neither generated not dissipated in closed-loop stress paths or in closed-loop strain paths. The framework for the model consists of Hooke's law, in which Poisson's ratio is constant and Young's modulus is expressed as a power function involving the first invariant of the stress tensor and the second invariant of the deviatoric stress tensor. The characteristics of the model are described, and the accuracy is evaluated by comparison with experimental results from triaxial tests and three-dimensional cubical triaxial tests with a variety of stress paths. Parameter determination from unloading–reloading cycles in conventional triaxial compression tests is demonstrated, typical parameter values are given for granular materials and extension of the model to soils with effective cohesion is described.

INTRODUCTION

The time-independent stress–strain relations for soils may be characterized by elastic and plastic behaviour. Both strain components are nonlinear in nature, and the elastic strains are relatively small in the presence of plastic strains. However, the complete soil response consists of elastic strains along the initial portion of any general three-dimensional stress path involving unloading or reloading. The strains along stress paths such as proportional unloading are for practical purposes entirely elastic. Further, constitutive models must include elastic strain components in order to invert the stress–strain relations. Therefore, even though the elastic strain components may be small, they cannot be ignored.

Experimental results indicate that the elastic properties are functions of the state of the soil element, i.e. they may be expressed in terms of the soil density or void ratio and the state of stress acting on the soil. For a given void ratio, the elastic properties have been shown to vary with the mean normal stress as well as the deviatoric stress.

Presented here is an isotropic model for the elastic behaviour of soils. The model is based on a theoretical development which guarantees lack of energy generation or dissipation for any closed-loop stress path or any closed-loop strain path. Experimental data from conventional triaxial and three-dimensional cubical triaxial tests on sand are used to demonstrate the capabilities and the determination of soil parameters for the model.

0363–9061/87/050521–22$11.00

Received 15 September 1986

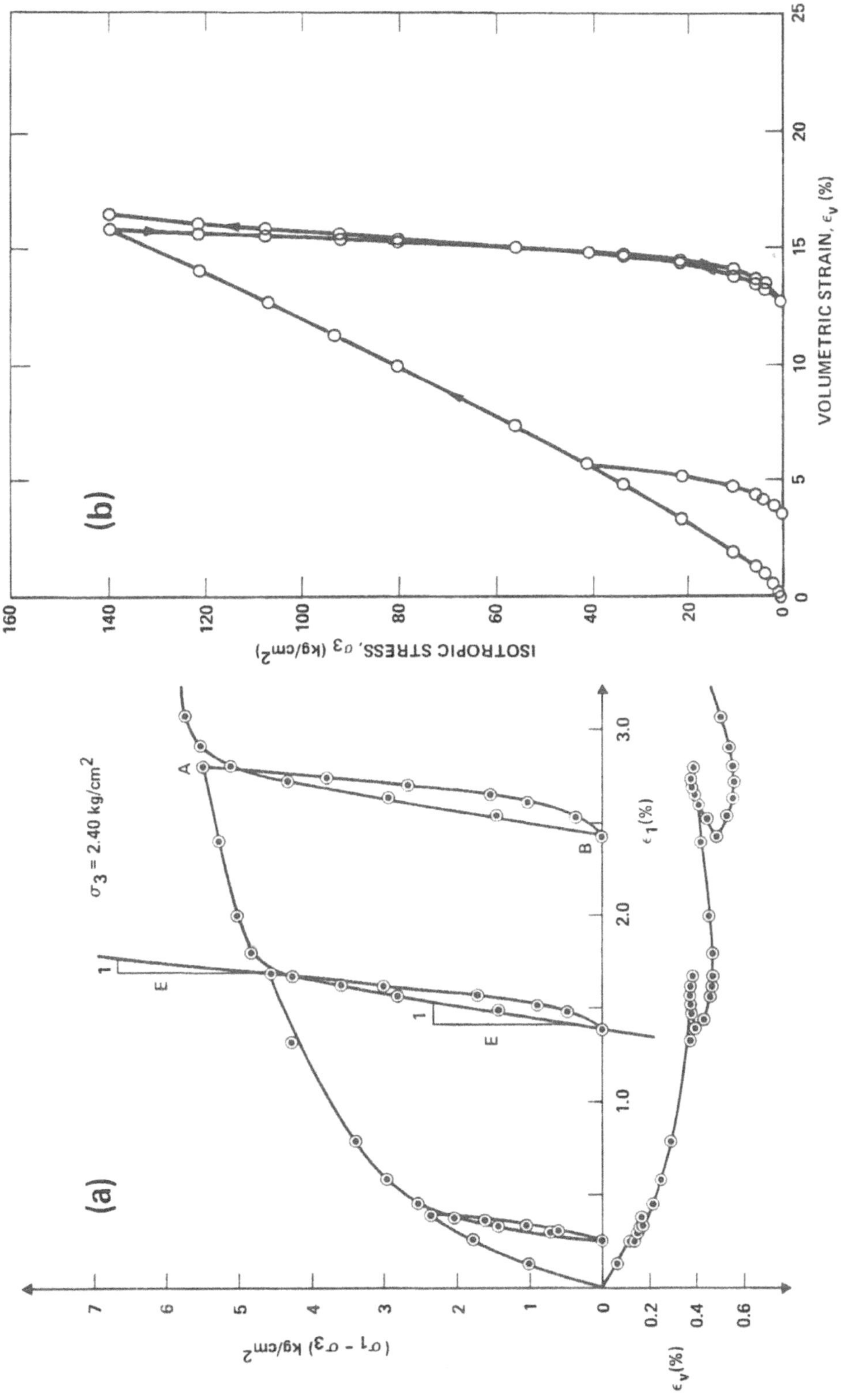

Figure 1. (a) Stress–strain and volume change behaviour of loose Santa Monica Beach Sand in triaxial compression test with several unloading–reloading cycles. (b) Isotropic compression of loose Sacramento River Sand with primary loading, unloading and reloading branches

OBSERVED BEHAVIOUR AND PREVIOUS MODELS

Basic observed behaviour

The framework for modelling the elastic behaviour of soil consists of Hooke's law, which for an isotropic material requires two independent parameters. These may be Young's modulus E and Poisson's ratio ν or the bulk modulus B and the shear modulus G. The parameters B and G are related to E and ν as follows:

$$B = \frac{E}{3(1-2\nu)} \tag{1}$$

$$G = \frac{E}{2(1+\nu)} \tag{2}$$

Experimental results indicate that E, B and G vary with the state of stress, whereas Poisson's ratio most often is assumed to be constant. Figure 1(a) shows the results of a conventional drained triaxial compression test on loose Santa Monica Beach Sand with three unloading–reloading cycles. The strains occurring immediately after each stress reversal are considered to be elastic. The elastic moduli indicated on the second cycle in Figure 1(a) are seen to vary with the deviatoric stress, since the effective confining pressure is maintained constant. Although the mean normal stress also varies in this test, an expression relating the elastic modulus to the mean normal stress is unable to capture the variation of Young's modulus correctly.[1]

Figure 1(b) shows the results of an isotropic compression test on loose Sacramento River Sand. The unloading–reloading branches, which for practical purposes are identical, indicate nonlinear elastic behaviour of the sand over a very large range of isotropic confining pressures.

Thus, both the mean normal stress and the deviatoric stress may influence the variations of the elastic moduli E, B and G. The most common models used to capture the variations of these three elastic moduli are briefly reviewed below.

Common models

Among the most widely studied modulus values is the initial tangent modulus, or initial slope of the stress–strain curve for triaxial compression. Both theoretical considerations and practical experiments[2–11] show that the initial tangent modulus may be expressed as a power function of the initially isotropic, effective confining pressure σ'_3:

$$E_i = K_i p_a \left(\frac{\sigma'_3}{p_a}\right)^n \tag{3}$$

in which p_a is atmospheric pressure expressed in the same units as E_i and σ'_3, K_i is a modulus number, and n is an exponent determining the rate of variation of E_i with σ'_3. Both K_i and n are dimensionless numbers. Janbu[8] presented results of a large number of tests. He found values of n from 0·35 to 0·55 for sands and silty sands with porosities of 35–50 per cent. Modulus numbers for the same sands had values from 50 to 500.

Because the initial slope of the stress–strain curve is often influenced by non-recoverable, plastic deformations, Duncan and Chang[12] proposed to use the slope of an unloading–reloading cycle from a triaxial compression test as the elastic modulus. This slope is determined between points A and B on the third cycle in Figure 1(a). Thus, K_i is replaced by K_{ur} in equation (3). Analyses of a large number of experiments[13] showed that K_{ur} was from 1 to 3 times higher than K_i, whereas n had the same value for initial loading and unloading–reloading.

The simple formulation in equation (3) has been widely used, because it appears to capture the elastic behaviour observed in triaxial and isotropic compression, and the material parameters can be determined from the results of conventional triaxial tests. However, Zytynski *et al.*[14] pointed out that this model for the variation of Young's modulus results in violation of the principle of conservation of energy. Thus, depending on the direction of a closed stress loop, the model will generate or dissipate energy in violation of the basic premise of elastic behaviour.

The deformation characteristics of a soil may also be expressed through the bulk modulus B, which is related to the hydrostatic stress component, and the shear modulus G, which is related to the deviatoric stress component. The advantage of using B and G is that these parameters may be evaluated independently and may more readily be related to the state of stress. Modelling of B and G may then take forms similar to that in equation (3).

Hardin and Black[15] proposed an empirical expression for the shear modulus G_{max} determined from wave propagation velocities and from small amplitude cyclic simple shear tests:

$$G_{max} = 1230\,OCR^{k}\frac{(2{\cdot}973 - e)^2}{(1 + e)}\sigma_o'^{0{\cdot}5} \tag{4}$$

in which σ_o' is the effective mean normal stress, and the exponent k applied to the over-consolidation ratio, may be related to the plasticity index of the soil. The values of G_{max} and σ_0' are expressed in psi. This expression also includes the effect of the void ratio e on the shear modulus. Many expressions with the same general form have been presented in the literature:

$$G_{max} = AF(e)\sigma_o'^{n} \tag{5}$$

in which A is a constant that may be related to OCR, $F(e)$ is a function describing the influence of void ratio, and the exponent n is often taken as 0·5. Hardin[16] has presented some of the specific forms of equation (5) together with typical values for the material parameters.

In addition to the formulations given above, other models for the elastic behaviour of soils have been presented in the literature (see, for example, References 17 and 18). These have gained less acceptance, in part due to the considerable number of material constants required and in part due to the requirement of results from more complicated or advanced tests. Further discussion of models for the elastic behaviour of soils have been given in References 19, 20 and 21.

Isotropy of elastic behaviour

Rowe[17] presented results of experiments on Mersey River Sand (4 void ratios), Silver Sand (4 void ratios), Fine Quartz Sand (2 void ratios), Feldspar (2 void ratios) and Glass Ballotini (3 void ratios) which all exhibited isotropic behaviour when unloaded, even when the strains during loading indicated anisotropic behaviour. Thus, sands with initial anisotropic fabric may exhibit anisotropic behaviour, but the elastic behaviour observed during unloading is for practical purposes isotropic. A similar conclusion was reached by Krizek,[22] who presented results of unconfined compression tests on sedimented specimens of kaolin clay with different degrees of inherent anisotropy. Results for sensitive clay studied by Wong and Mitchell[23] also showed nearly isotropic elastic behaviour, while the plastic stress–strain relations were anisotropic.

Poisson's ratio

The theoretical limits of Poisson's ratio are $-1 \leqslant \nu \leqslant 0{\cdot}5$. These limits are obtained on the basis that the elastic work $\frac{1}{2}\{\sigma\}^{T}\{\varepsilon\} \geqslant 0$ for any change in stress. Equation (1) for the bulk modulus shows that this requirement results in $\nu \leqslant 0{\cdot}5$, and equation (2) for the shear modulus provides the lower limit of $-1 \leqslant \nu$. In practice, Poisson's ratio for the elastic behaviour of soils is limited to the range from 0 to 0·5.

Experimental evidence indicates that Poisson's ratio is isotropic and practically constant for a soil at a given void ratio. The experiments on several granular materials presented by Rowe[17] showed that Poisson's ratio could be considered to be isotropic during elastic unloading.

Hardin[16] noted that accurate measurements of the elastic Poisson's ratio from wave propagation experiments was difficult, due to the insensitivity of the soil behaviour to the value of v. However, analysis of his resonant column tests on Ottawa Sand produced constant values of v in the range from 0·11 to 0·23. Based on a review of experimental data, Hardin[16] concluded that Poisson's ratio for soils lies somewhere between 0 and 0·2 and that any value within this range is accurate enough for most purposes.

Yokota and Konno[24] performed cyclic triaxial tests with a frequency of 0·25 Hz on reconstituted and undisturbed specimens of cohesive soils and alluvial sandy soils. Their investigation showed that Poisson's ratio was essentially constant at 0·5 for the cohesive soil, reflecting the undrained condition in these tests. The sandy soils produced values from individual tests in the range from 0·2 to 0·4. Their test results showed a tendency for slightly increasing Poisson's ratios with increasing magnitudes of shear strains, whereas the influence of confining pressure was found to be negligible.

El Hosri[25] conducted low amplitude cyclic triaxial tests on dry Hostun Sand with careful measurements of the axial and radial strains made directly on the specimens. He varied the average grain diameter, the uniformity coefficient, the void ratio, the effective confining pressure and the strain amplitude. Poisson's ratio was found to be essentially constant in each test and independent of confining pressure. Values of v were obtained in the range from 0·18 to 0·23 for different material compositions and void ratios. El Hosri[25] also observed that Poisson's ratio was constant within cyclic strain amplitudes up to $\pm 0{\cdot}5$ per cent.

Whereas Poisson's ratio for a soil at a given void ratio appears to be constant, experimental evidence suggests that v increases with increasing void ratio. Reanalysis of K_o-unloading branches from numerous tests performed on Ham River Sand[26] shows a consistent variation of v from 0·20 at $e = 0{\cdot}57$ to 0·31 at $e = 0{\cdot}75$. Other experiments, including those presented below for Santa Monica Beach Sand, indicate similar results.

THEORETICAL DEVELOPMENT

The experimental evidence reviewed above indicates that the elastic behaviour of soils can be captured by an isotropic elastic model in which Poisson's ratio is constant and the elastic moduli E, B and G are functions of the mean normal stress as well as the deviatoric stress. Expressing the mean normal stress in terms of the first invariant of the stress tensor I_1 and the deviatoric stress in terms of the second invariant of the deviatoric stress tensor J'_2:

$$B = B(I_1, J'_2) \tag{6}$$

$$G = K(I_1, J'_2) \tag{7}$$

in which

$$I_1 = \sigma_1 + \sigma_2 + \sigma_3 = \sigma_x + \sigma_y + \sigma_z \tag{8}$$

and

$$J'_2 = \tfrac{1}{2}(s_x^2 + s_y^2 + s_z^2 + 2s_{yz}^2 + 2s_{zx}^2 + 2s_{xy}^2) \tag{9}$$

where the s symbols denote deviatoric stress:

$$s_x = \sigma_x - (1/3)I_1 = 1/3[(\sigma_x - \sigma_y) + (\sigma_x - \sigma_z)] \tag{10a}$$

$$s_y = \sigma_y - (1/3)I_1 = 1/3[(\sigma_y - \sigma_z) + (\sigma_y - \sigma_x)] \quad (10b)$$
$$s_z = \sigma_z - (1/3)I_1 = 1/3[(\sigma_z - \sigma_x) + (\sigma_z - \sigma_y)] \quad (10c)$$

and

$$s_{yz} = \tau_{yz};\ s_{zx} = \tau_{zx};\ s_{xy} = \tau_{xy} \quad (10d,e,f)$$

Note that

$$J'_1 = s_x + s_y + s_z = 0 \quad (11)$$

Substitution of equations (10) in equation (9) yields

$$J'_2 = \tfrac{1}{6}[(\sigma_x - \sigma_y)^2 + (\sigma_y - \sigma_z)^2 + (\sigma_z - \sigma_x)^2] + \tau_{xy}^2 + \tau_{yz}^2 + \tau_{zx}^2 \quad (12)$$

The elastic work per unit volume along stress path ACB in Figure 2 is calculated as

$$W_{ACB} = \int_{ACB} dW = \int_{ACB} \{\sigma\}^T\{d\varepsilon\} \quad (13)$$

or, as shown in Appendix I, this may be written as

$$W_{ACB} = \int_{ACB} \left(\frac{I_1 dI_1}{9B} + \frac{dJ'_2}{2G}\right) \quad (14)$$

and similarly for stress path ADB:

$$W_{ADB} = \int_{ADB} \left(\frac{I_1 dI_1}{9B} + \frac{dJ'_2}{2G}\right) \quad (15)$$

According to the principle of conservation of energy, the work must be independent of the path and $W_{ACB} = W_{ADB}$. Thus, for the cycle ACBDA:

$$W_{ACB} + W_{BDA} = \int_{ACBDA} \left(\frac{I_1 dI_1}{9B} + \frac{dJ'_2}{2G}\right) \quad (16)$$

The integral in equation (16) will be zero if it can be written in the general form

$$\oint (P\,dx + Q\,dy) = \iint_A \left(\frac{\partial Q}{\partial x} - \frac{\partial P}{\partial y}\right) dA \quad (17)$$

If $P = \partial W/\partial x$ and $Q = \partial W/\partial y$, i.e. if P and Q can be derived from a potential function, then the line integral vanishes and no net work is done on the closed stress path. Thus, for the work to

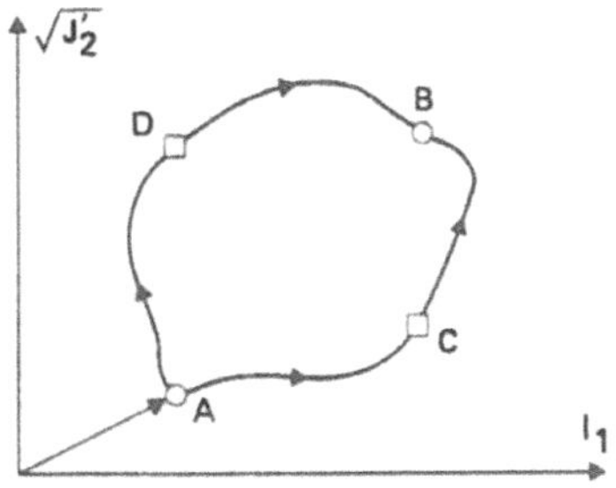

Figure 2. Stress paths from A to B shown in $I_1 - \sqrt{(J'_2)}$ diagram

be path independent, it must be possible to write the quantity $\mathrm{d}W$ in the form

$$\mathrm{d}W = \frac{I_1 \mathrm{d}I_1}{9B} + \frac{\mathrm{d}J'_2}{2G} = \frac{\partial W}{\partial I_1}\mathrm{d}I_1 + \frac{\partial W}{\partial \sqrt{(J'_2)}}\mathrm{d}\sqrt{(J'_2)} \tag{18}$$

in which I_1 and $\sqrt{(J'_2)}$ are the independent variables. Since $\mathrm{d}J'_2 = 2\sqrt{(J'_2)}\,\mathrm{d}\sqrt{(J'_2)}$, equation (18) becomes

$$\mathrm{d}W = \frac{I_1}{9B}\mathrm{d}I_1 + \frac{\sqrt{(J'_2)}}{G}\mathrm{d}\sqrt{(J'_2)} = \frac{\partial W}{\partial I_1}\mathrm{d}I_1 + \frac{\partial W}{\partial \sqrt{(J'_2)}}\mathrm{d}\sqrt{(J'_2)} \tag{19}$$

which for any $W[I_1, \sqrt{(J'_2)}]$ requires

$$\frac{\partial W}{\partial I_1} = \frac{I_1}{9B} \tag{20}$$

and

$$\frac{\partial W}{\partial \sqrt{(J'_2)}} = \frac{\sqrt{(J'_2)}}{G} \tag{21}$$

Finally, since

$$\frac{\partial^2 W}{\partial I_1 \partial \sqrt{(J'_2)}} = \frac{\partial^2 W}{\partial \sqrt{(J'_2)} \partial I_1} \tag{22}$$

the following constraints on the expressions for B and G are obtained:

$$\frac{I_1}{9B^2}\frac{\partial B}{\partial \sqrt{(J'_2)}} = \frac{\sqrt{(J'_2)}}{G^2}\frac{\partial G}{\partial I_1} \tag{23}$$

Substitution of equations (1) and (2) with $\nu = \text{constant}$ into equation (23) produces

$$\frac{1}{\sqrt{(J'_2)}}\frac{\partial E}{\partial \sqrt{(J'_2)}} = R\frac{1}{I_1}\frac{\partial E}{\partial I_1} \tag{24}$$

in which

$$R = 6\frac{1+\nu}{1-2\nu} = \frac{9B}{G} \tag{25}$$

Equation (24) has a general solution $E = E(X)$, where

$$X = I_1^2 + R[\sqrt{(J'_2)}]^2 \tag{26}$$

It is helpful to note that, in this case,

$$\mathrm{d}W = \frac{1}{9B}[I_1 \mathrm{d}I_1 + R\sqrt{(J'_2)}\mathrm{d}\sqrt{(J'_2)}] = \frac{1}{18B(X)}\mathrm{d}X \tag{27}$$

Thus,

$$W = \int_0^X \frac{1}{18B(\bar{X})}\mathrm{d}\bar{X} \tag{28}$$

and

$$\frac{\partial W}{\partial I_1} = \frac{\partial W}{\partial X}\frac{\partial X}{\partial I_1} = \frac{1}{18B(X)}2I_1 = \frac{I_1}{9B} \tag{29}$$

$$\frac{\partial W}{\partial \sqrt{(J'_2)}} = \frac{1}{18B} 2R\sqrt{(J'_2)} = \frac{\sqrt{(J'_2)}}{G} \tag{30}$$

i.e. the same expressions as in equations (20) and (21).

In terms of non-dimensional material constants and stress functions, the following power law satisfies equation (24):

$$E = Mp_a\left[\left(\frac{I_1}{p_a}\right)^2 + R\frac{J'_2}{p_a^2}\right]^\lambda \tag{31}$$

in which p_a is the atmospheric pressure expressed in the same units as E, I_1 and $\sqrt{(J'_2)}$, and the modulus number M and the exponent λ are constant, dimensionless numbers. This functional representation can also be shown to fulfil the energy conservation argument for closed *strain* paths. This is demonstrated in Appendix II.

According to equation (31) Young's modulus is constant along rotationally symmetric ellipsoidal surfaces whose long axis coincides with the hydrostatic axis and whose centre is located at the origin of the principal stress space. The magnitude of Poisson's ratio determines the shape of the ellipsoidal surface. For $\nu = 0$, the value of $R = 6$ and the surface becomes spherical, whereas for $\nu = 0{\cdot}5$, the value of $R = \infty$ and the surface degenerates into a line coinciding with the hydrostatic axis.

Figure 3 shows cross-sections of the ellipsoidal surfaces in triaxial and octahedral planes for different values of Poisson's ratio. The cross-sections in the octahedral plane are always circular, whereas the cross-sections in the triaxial plane are shaped as ellipses whose aspect ratio depends on the value of Poisson's ratio. The locations of typical failure surfaces for cohesionless soils with friction angles from 30 to 50 degrees are shown for triaxial compression and extension. Only the portions of the ellipsoids located inside the conical failure surfaces can be reached by stress states in cohesionless soils, and only these portions are of interest for such soils.

Whereas Figure 3 shows contours along which Young's moduli are constant, the actual magnitude of E is determined by the material parameters M and λ. The results of an experimental study involving triaxial compression and three-dimensional cubical triaxial tests on sand with

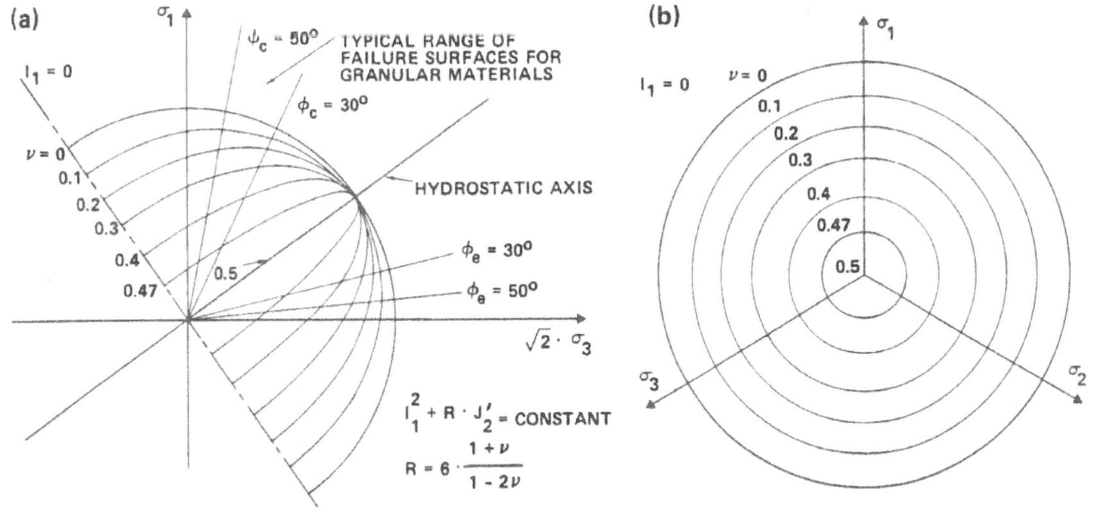

Figure 3. Contours of constant Young's modulus shown in (a) triaxial plane and (b) octahedral plane

various stress paths are used to demonstrate the determination of M and λ and to study the capability of equation (31) to capture the variation of Young's modulus.

EXPERIMENTAL STUDY

Testing program

Sand tested. All tests were performed on Santa Monica Beach Sand, which is composed of subangular to subrounded grains consisting mainly of quartz and feldspar. The characteristics of this sand are summarized as follows: mean diameter, 0·27 mm; coefficient of uniformity, 1·58; specific gravity of grains, 2·66; maximum void ratio, 0·87; and minimum void ratio, 0·58.

The tests presented here were performed on dense and loose specimens. The dense specimens had a void ratio of 0·61, corresponding to a relative density of 90 per cent, and the loose specimens had a void ratio of 0·81, corresponding to a relative density of 20 per cent. Only triaxial tests with different stress paths were performed on the dense sand. Triaxial as well as cubical triaxial tests with different stress paths were performed on the loose specimens in order to obtain relatively large strains, thus facilitating the interpretation of the results.

Specimens and measurements. The drained triaxial tests were performed on cylindrical specimens 2·8 in. (71 mm) in diameter and 7·5 in. (190 mm) high. Lubricated ends were used to minimize effects of end restraint. The initially air-dry specimens were saturated using the CO_2 method described by Lade and Duncan.[27] The vertical load, the confining pressure, the vertical deformation and the volume changes were measured during the tests. Constant or varying confining pressures in the range from 0·30 kg/cm^2 (29·4 kN/m^2) to 6·00 kg/cm^2 (588·6 kN/m^2) were employed in the triaxial tests.

The drained three-dimensional triaxial tests on loose sand were performed on cubical specimens with side lengths of 3·0 in. (76 mm). The cubical triaxial apparatus and the testing procedures used in the apparatus were described in detail by Lade.[28] A constant confining pressure of 1·20 kg/cm^2 (117·7 kN/m^2) was used in these tests, while the vertical and the horizontal deviator stresses were varied in the range from zero to 4·23 kg/cm^2 (415·0 kN/m^2) to achieve various three-dimensional stress paths.

The deformations measured in the two types of tests were corrected for sand penetration into the lubricated end plates, and the volumetric deformations were corrected for effects of membrane penetration occurring in tests with varying confining pressure.

Stress paths. Tests with simple stress paths as well as with complex stress paths were performed. Figure 4 shows the points of initiation and the directions of the stress paths involving unloading or reloading, thus resulting in primarily elastic strains. Figure 4(a) shows the stress path directions employed in the triaxial tests, and Figure 4(b) shows the stress path directions in the plane corresponding to $\sigma_3 = 1{\cdot}20$ kg/cm^2 (117·7 kN/m^2) for the cubical triaxial tests. In order to obtain an appreciation of the magnitude and composition (elastic vs. plastic) of the measured strains associated with a given stress path, each straight unloading or reloading path inside the primary yield surfaces was designed to involve relatively large stress changes.

The triaxial tests with simple stress paths were conducted with constant confining pressure. These tests included conventional triaxial compression tests with several unloading–reloading cycles at increasingly high values of stress difference, tests with compression–extension–compression cycles and tests with extension–compression–extension–compression cycles. Tests with more complex stress paths involving varying confining pressure and large unloading and reloading

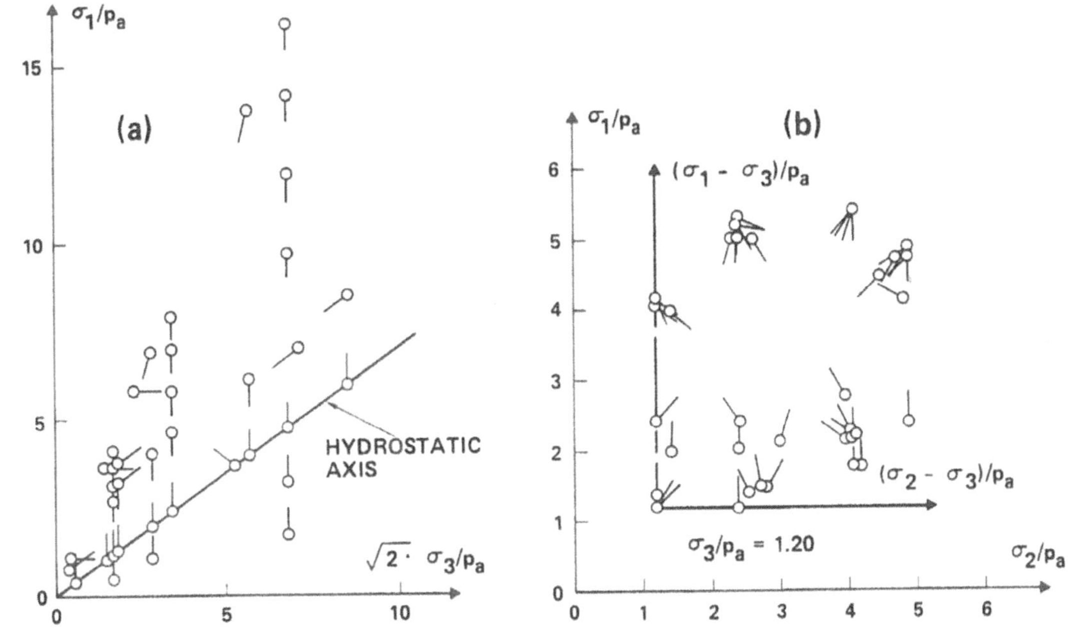

Figure 4. Points of initiation and stress path directions for tests on Santa Monica Beach Sand shown in (a) triaxial plane and (b) plane with $\sigma_3 = \text{constant}$

stress changes were also performed to study the pattern of elastic soil behaviour in relation to the current stress path and the previous stress history, if any.

The cubical triaxial tests were all performed with constant confining pressure σ_3, but the two stress differences $(\sigma_1 - \sigma_3)$ and $(\sigma_2 - \sigma_3)$ were varied to create three-dimensional stress paths from which the effects of σ_1 and σ_2 could be evaluated. These stress paths also involved large unloading and reloading stress changes to obtain reasonable amounts of strain for analysis.

Test results

The results of the tests described above form the basis for experimental verification of the theoretically developed expression for Young's modulus in equation (31). Calculation of the elastic moduli was performed using stress paths lengths of such magnitudes that at least one of the principal elastic strains was greater than $\pm 0{\cdot}015$ per cent. This was done to obtain a reasonable degree of accuracy and reliability of the moduli. Nevertheless, some scatter in the modulus values is to be expected due to the small strains encountered during unloading and reloading and due to the sensitivity of these small strains to accurate measurements and corrections. Because two of the principal stresses were changed along most of the stress paths employed in this study, the calculated elastic moduli would depend on the value of Poisson's ratio. The moduli were therefore determined in connection with regression analyses in which the constant Poisson's ratio could be varied to obtain the best overall fit. The moduli were associated with the states of stress at which unloading or reloading was initiated.

DETERMINATION OF MATERIAL PARAMETERS

Because Young's modulus is expressed in terms of stress invariants, any type of test in which all stresses and strains are measured may be used for determination of the three material parameters

M, λ and ν required to describe the elastic behaviour. In order to determine the material parameters for a given set of experimental data, a value of Poisson's ratio is estimated and the values of E are calculated from Hooke's law. In this study, a single elastic modulus was calculated for each stress path using the equation corresponding to the largest measured principal strain. The parameters M and λ in equation (31) can then be determined by plotting (E/p_a) vs. $[(I_1/p_a)^2 + R(J_2'/p_a^2)]$ in a log-log diagram, as shown in Figure 5 for dense Santa Monica Beach Sand. The intercept of the best-fitting straight line with $[(I_1/p_a)^2 + R(J_2'/p_a^2)] = 1$ is the value of M, and λ is the slope of the line.

In order to obtain the overall best-fitting parameters, regression analyses may be performed to determine the highest possible value of the coefficient of determination r^2. Figure 6 shows the effect of varying Poisson's ratio in the range from 0 to 0·5 on the values of r^2, M and λ for dense Santa Monica Beach Sand. It is interesting to note that these values are relatively insensitive to the value of Poisson's ratio in the range from $\nu = 0{\cdot}1$, $\nu = 0{\cdot}2$. The diagram shown in Figure 5 corresponds to the best-fit value of Poisson's ratio of $\nu = 0{\cdot}14$.

Poisson's ratio was also determined directly from stress paths in triaxial tests with constant confining pressures according to

$$\nu = -\frac{\dot{\varepsilon}_3}{\dot{\varepsilon}_1} = \frac{1}{2}\left(1 - \frac{\dot{\varepsilon}_v}{\dot{\varepsilon}_1}\right) \tag{32}$$

in which $(\dot{\varepsilon}_v/\dot{\varepsilon}_1)$ is the slope of the volume change curve immediately after stress reversal. Figure 7 shows values of Poisson's ratio determined from equation (32) for dense and loose Santa Monica Beach Sand plotted against I_1. The values of ν shown in Figure 7 were all determined from reloading branches initiated at the hydrostatic axis, where the best estimate of $(\dot{\varepsilon}_v/\dot{\varepsilon}_1)$ is obtained. Although there is some scatter in the test data, there are no discernible effects of the mean normal stress on Poisson's ratio. The average values, ν_{avg}, and the standard deviations, $\pm$ s, are indicated in Figure 7, and these values are also indicated in Figure 6 for dense Santa Monica Beach Sand. Figure 6 shows that the average value of $\nu_{avg} = 0{\cdot}15$, determined directly from the experimental

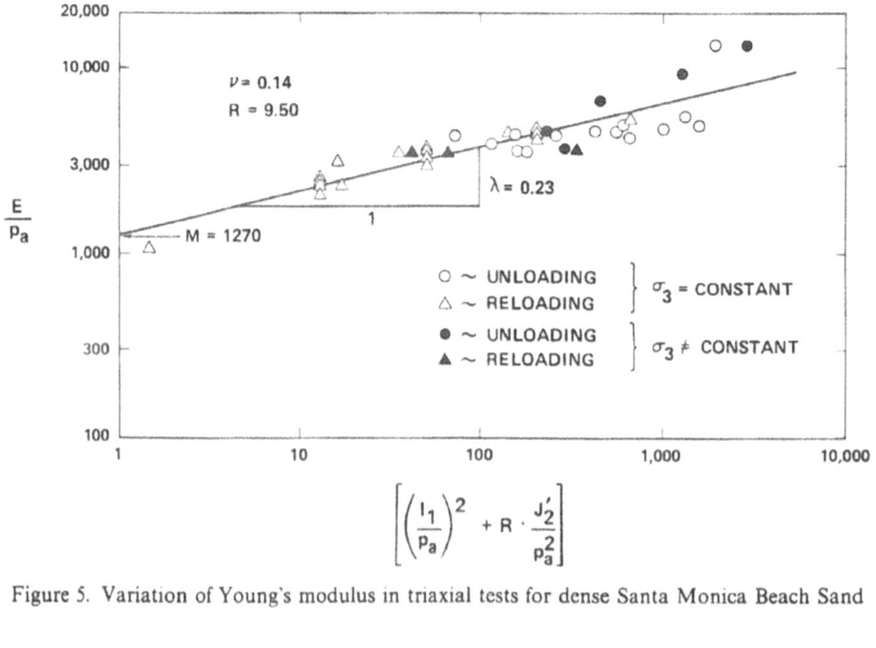

Figure 5. Variation of Young's modulus in triaxial tests for dense Santa Monica Beach Sand

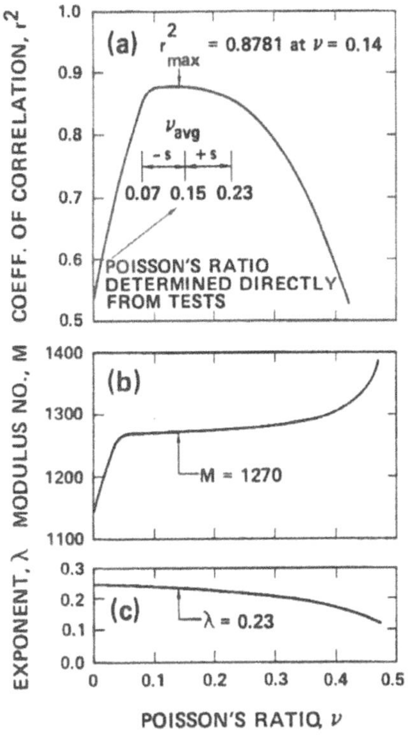

Figure 6. Variation of coefficient of correlation, modulus number and exponent with Poisson's ratio in triaxial tests on dense Santa Monica Beach Sand

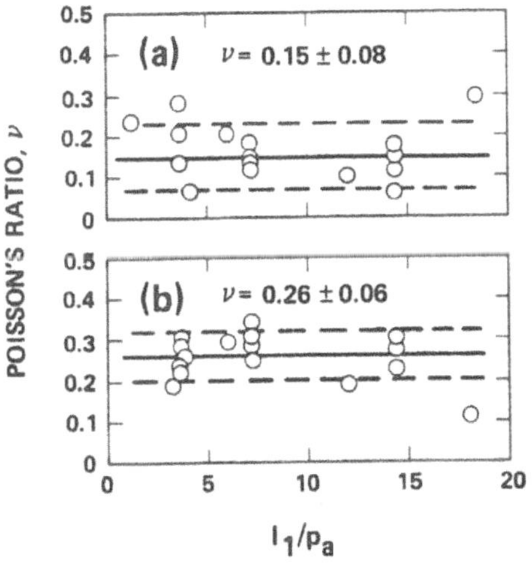

Figure 7. Poisson's ratios determined directly from results of triaxial tests on (a) dense and (b) loose Santa Monica Beach Sand

results, is located in the middle of the range in which r^2, m and λ are relatively insensitive to Poisson's ratio, and that this range is approximately indicated by the values of $\nu_{avg} \pm s$.

The results for loose Santa Monica Beach Sand are shown in Figures 8 and 9. These data include results from triaxial tests and cubical triaxial tests with various stress paths. Figure 8 shows that the results from the cubical triaxial tests fall in the range of results from the triaxial

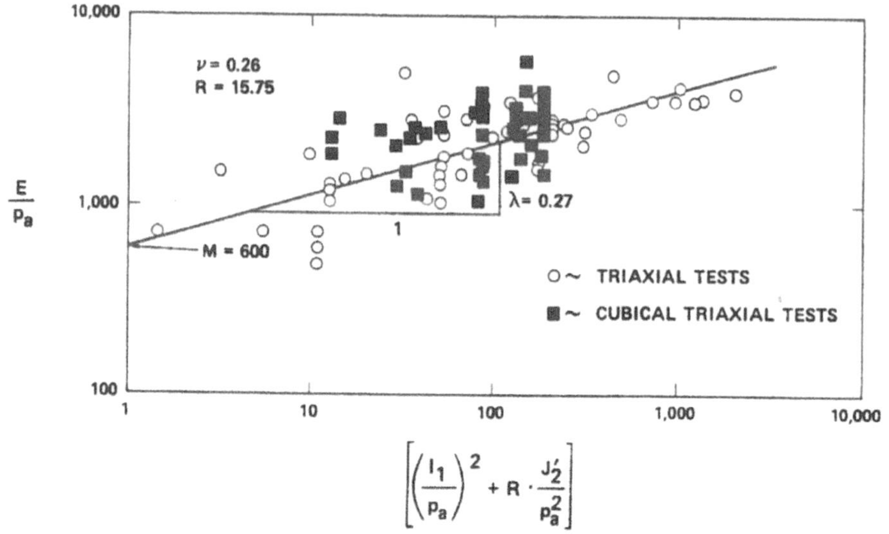

Figure 8. Variation of Young's Modulus in triaxial and cubical triaxial tests for loose Santa Monica Beach Sand

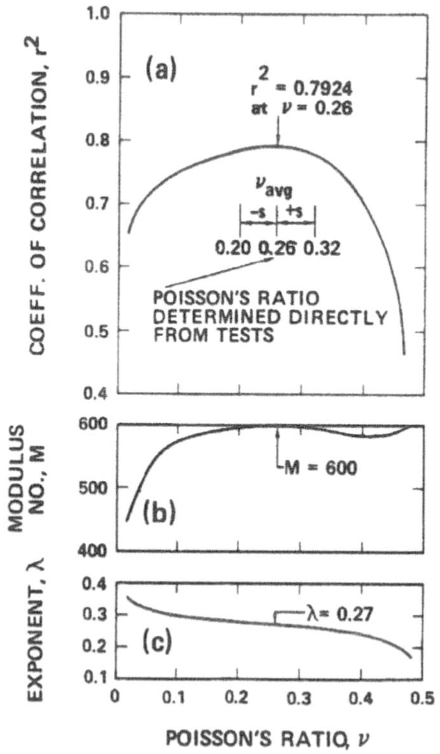

Figure 9. Variation of coefficient of correlation, modulus number and exponent with Poisson's ratio in triaxial tests on loose Santa Monica Beach Sand

tests. Thus, the values of Young's moduli determined from triaxial tests are applicable to general three-dimensional conditions. Figure 9 shows the results of regression analyses obtained by varying Poisson's ratio in the range from 0 to 0·5. The curves correspond to data from the triaxial tests. The average value and the range of Poisson's ratio determined directly from reloading branches in triaxial tests are also shown on Figure 9. Again, both ν_{avg} and $\nu_{avg} \pm s$ approximately coincide with the best-fit value and the range in which r^2, M and λ are relatively insensitive to Poisson's ratio.

PARAMETER DETERMINATION FROM SIMPLE TESTS

Based on the results and analyses presented above, it appears that the material parameters M, λ and ν required to describe the elastic behaviour may be determined from results of conventional drained triaxial compression tests which include unloading–reloading cycles. These parameters may be determined with reasonable accuracy from such simple tests, because almost equally accurate descriptions of the elastic behaviour can be achieved over a range of Poisson's ratios.

The best estimate of Poisson's ratio is obtained from the slopes of the volume change curves associated with the reloading branches initiated at or near the hydrostatic axis. Equation (32) is then used to calculate Poisson's ratio, and it is preferable to obtain an average value from several tests. Using ν_{avg}, the value of R in equation (31) can be determined. Individual modulus values may be obtained from the slopes of the unloading and the reloading stress–strain relations obtained at constant confining pressures. These modulus values are associated with the stress states at which stress reversals are initiated. The parameters M and λ can then be determined as indicated in Figure 5 without regression analysis.

MATERIAL PARAMETER VALUES

Using the procedure based on simple tests as explained above, several sets of test data were analysed. Descriptions of soils and material parameter values are given in Table I. The average Poisson's ratios were obtained from individual data sets corresponding to reloading, as explained above, as well as from unloading at constant confining pressure and from K_o-unloading. Considerable scatter in the values of Poisson's ratio were obtained as indicated by the standard deviations. However, the average values of ν may very well be within the ranges in which M and λ are relatively insensitive to Poisson's ratio. Thus, it is possible to obtain reasonable descriptions of the elastic behaviour for the soils.

The values of the modulus number M listed in Table I vary from 440 to 1270, whereas those of the exponent λ vary in the relatively narrow range from 0·22 to 0·33. Although the data base is small, these values of M and λ may be representative of granular materials in general.

It is interesting to compare the material parameters M and λ with those to be used in equation (3) and with those implied by equation (4). According to these expressions, Young's modulus varies with the confining pressure σ'_3 and with the mean normal stress σ'_o, respectively. Because Young's modulus in equation (31) varies with both mean normal stress and deviator stress, a direct comparison of material parameter values for the three expressions can only be made at the hydrostatic axis where the stress dependency can be eliminated.

Equating Young's moduli from equations (3) and (31) produces the following expression for the modulus number K and the exponent n:

$$K = 3^{2\lambda} M \tag{33}$$

and

$$n = 2\lambda \tag{34}$$

Values of K and n calculated from these expressions are also listed in Table I. It may be seen that K varies in the range from 700 to 2300. These values compare very well with values of K_{ur} obtained from unloading–reloading branches in conventional static triaxial compression tests

Table I. Parameter values for different granular materials

Soil	Description				Poisson's ratio		Parameters		Equation (33)	Equation (34)	Equation (35)
	D_{50}(mm)	C_u	e	D_r(%)	ν_{avg}	s	M	λ	K	n	M_{hb}
Santa Monica Beach Sand	0·27	1·58	0·61	90	0·15	0·08	1270	0·23	2100	0·46	1550
			0·81	20	0·26	0·06	600	0·27	1090	0·54	1160
Monterey No. 0 Sand	0·43	1·53	0·57	98	0·17	0·14	1120	0·33	2310	0·66	1340
			0·78	27	0·17	0·05	800	0·26	1420	0·52	1150
Fine Silica Sand	0·18	2·0	0·76	30	0·27	0·08	440	0·22	710	0·44	1400
Mohawk Model Soil	3·5	6·2	0·64	73	0·25	0·13	630	0·27	1140	0·54	1470
Niagara-Type A Material	4·8	1·52	0·71	95	0·23	0·12	670	0·26	1190	0·52	1340
Niagara-Type B Material	1·2	8·6	0·52	95	0·24	0·11	570	0·29	1080	0·58	1670
Sacramento River Sand	0·21	1·5	0·87	38	0·2 (estimated)		500	0·28	930	0·56	990

and listed in Reference 13. The exponent n varies from 0·44 to 0·66. These values are also directly comparable with values obtained for granular materials and listed in Reference 13.

For comparison of parameter values for equations (4) and (31), Young's modulus is obtained from equations (2) and (4) and equated with the expression in equation (31). Equation (4) was derived for mean normal stresses and shear moduli expressed in psi. This stress unit dependency is eliminated by factoring the atmospheric pressure p_a into the equation. The exponent 0·5 in equation (4) is replaced by the general parameter n, as in equation (5). For hydrostatic stresses and normally consolidated sands ($OCR = 1$), this produces

$$M_{hb} = 2(1+\nu)1230\frac{(2{\cdot}973-e)^2}{(1+e)}\frac{1}{3^{2\lambda}}\frac{1}{\sqrt{(14{\cdot}7)}} \tag{35}$$

and the exponent n is expressed by equation (34). The values of M_{hb} for Hardin and Black's expression are calculated from Poisson's ratio ν, void ratio e and exponent λ and listed in Table I. The magnitudes of M_{hb} are one to three times higher than those of M. This is most likely, because equation (4) was derived from wave propagation velocities and from small amplitude cyclic simple shear tests. These small strain tests produce higher modulus values than static tests with larger changes in stress and strain. The average value of the exponents n listed in Table I is 0·54, with a standard deviation of $\pm 0{\cdot}06$. This compares favourably with the value of 0·5 used in Equation (4).

Whereas the material parameters evaluated at the hydrostatic axis are comparable in magnitude, the stress dependency of Young's modulus is different for the three expressions. Equation (3) indicates variation of the modulus with effective minor principal stress σ'_3, whereas experiments at constant σ'_3 show an additional dependency on stress difference. Equation (4) captures the modulus variation with effective mean normal stress σ'_o, but does not include a dependency on deviator stress. In comparison, equation (31) provides for modulus variation with effective mean normal stress as well as with deviator stress according to the pattern indicated above. Both theoretical considerations and experimental evidence support the expression for Young's modulus presented here.

MATERIALS WITH EFFECTIVE COHESION

Although only cohesionless materials were used for verification and material parameter determination, it is likely that the nonlinear elastic behaviour of materials with effective cohesion may also be modelled with good accuracy by the model presented above. In order to include the effect of cohesion, a translation of the principal stress space along the hydrostatic axis is performed.[29,30] Thus, a constant stress, ap_a, is added to the normal stresses before substitution in equations (8) and (12):

$$\bar{\sigma}_x = \sigma_x + ap_a \tag{36a}$$

$$\bar{\sigma}_y = \sigma_y + ap_a \tag{36b}$$

$$\bar{\sigma}_z = \sigma_z + ap_a \tag{36c}$$

in which 'a' is a dimensionless parameter and p_a is atmospheric pressure in the same units as σ_x, σ_y and σ_z. The value of ap_a reflects the tensile strength in materials with effective cohesion, and the parameter 'a' may be determined as indicated for concrete and other frictional materials with effective cohesion.[29,30] The translation of the co-ordinate system in connection with the elastic modulus variation is required to obtain increasing values of I_1 from the origin of the (translated) stress space. This method of accounting for the modulus variation was used successfully in a study of frozen soil in which the effective cohesion plays an important role in the stress–strain behaviour.[31]

EVALUATION OF MODEL FOR ELASTIC BEHAVIOUR

The model developed above for the elastic behaviour of soil captures the nonlinear stress–strain relations commonly observed during unloading and reloading in triaxial compression and isotropic compression. Young's modulus varies with both mean normal stress and deviatoric stress and is therefore able to capture the elastic behaviour immediately upon unloading and upon reloading. Figure 10(a) shows a magnified diagram of the second unloading–reloading cycle for loose Santa Monica Beach Sand previously shown in Figure 1(a). The solid lines indicate the nonlinear elastic behaviour obtained from equation (31), with material parameter values as indicated in Figure 8. The slopes of the measured stress–strain relations immediately after stress reversals match quite well with those obtained from the model. The hysteresis loop is not modelled by equation (31). This loop is caused by the plastic strains occurring during unloading and reloading. The best-fit value of Poisson's ratio given in Figure 8 also matches well with the volume change curves shown in Figure 10(a). The predicted slope of the volume change curve is better defined for reloading than for unloading. This is because larger amounts of plastic strains occur during unloading which then obscure the elastic behaviour.

The isotropic compression test for loose Sacramento River Sand, previously shown in Figure 1(b), is shown in Figure 10(b), with the solid lines indicating the nonlinear behaviour obtained from the proposed model. Insufficient data was available for this sand to obtain overall best-fit values of the required material parameters. These were therefore obtained directly from the unloading–reloading branches shown in Figure 10(b). This figure serves to show the ability of the model to capture the nonlinear behaviour during isotropic unloading and reloading, for which only negligible plastic strains are usually observed.

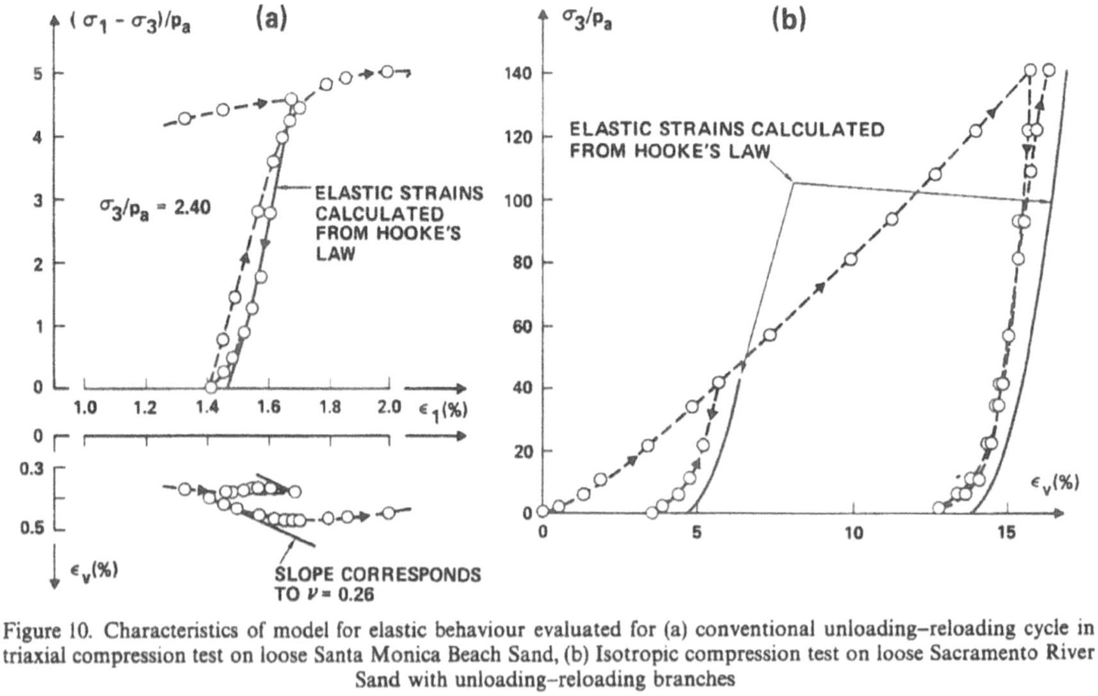

Figure 10. Characteristics of model for elastic behaviour evaluated for (a) conventional unloading–reloading cycle in triaxial compression test on loose Santa Monica Beach Sand, (b) Isotropic compression test on loose Sacramento River Sand with unloading–reloading branches

The overall fit between model and experimental results was demonstrated in Figures 5 and 8 for dense and loose Santa Monica Beach Sand. Although some scatter is present due to the small strains associated with elastic behaviour, the model appears to capture the elastic behaviour of sands with acceptable accuracy. The relative effects of the mean normal stress and the deviatoric stress are weighted by the R-value which is related to the constant value of Poisson's ratio. The model was developed on a theoretical basis, which ensures that the principle of conservation of energy is fulfilled for the nonlinear elastic behaviour of soils.

CONCLUSIONS

An isotropic model for the nonlinear elastic behaviour of granular materials has been developed. The model is based on Hooke's law in which Poisson's ratio is constant, and Young's modulus is expressed as a power function involving the first invariant of the stress tensor and the second invariant of the deviatoric stress tensor. The expression for Young's modulus is derived from theoretical considerations based on the principle of conservation of energy. Thus, energy is neither generated not dissipated along a closed-loop stress path. Experimental results from triaxial tests and cubical triaxial tests on sand with various two- and three-dimensional stress paths were used to demonstrate the ability of the model to capture the true nonlinear elastic behaviour of sands. Material parameters may be determined from unloading–reloading cycles in conventional triaxial compression tests. These parameters were shown to be applicable to general three-dimensional stress states. Although only results from tests on granular materials were employed in the present study, extension of the model to include the nonlinear elastic behaviour of materials with effective cohesion is possible, as described.

ACKNOWLEDGEMENT

The authors wish to acknowledge the valuable comments of the reviewers which led to significant improvements of the manuscript.

APPENDIX I

Equation (13) may be reduced to equation (14) as shown in the following. To distinguish between stress and strain invariants, the subscripts σ and ε are added. From equations (8) and (12):

$$dI'_{1\sigma} = d\sigma_x + d\sigma_y + d\sigma_z \tag{37}$$

and

$$\begin{aligned} dJ'_{2\sigma} = \tfrac{1}{3}[(\sigma_x - \sigma_y)(d\sigma_x - d\sigma_y) + (\sigma_y - \sigma_z)(d\sigma_y - d\sigma_z) + (\sigma_z - \sigma_x)(d\sigma_z - d\sigma_x)] \\ + 2(\tau_{yz}d\tau_{yz} + \tau_{zx}d\tau_{zx} + \tau_{xy}d\tau_{xy}) \end{aligned} \tag{38}$$

or

$$dJ'_{2\sigma} = s_x d\sigma_x + s_y d\sigma_y + s_z d\sigma_z + 2(\tau_{yz}d\tau_{yz} + \tau_{zx}d\tau_{zx} + \tau_{xy}d\tau_{xy}) \tag{39}$$

In the same manner as for stresses (equations (8) and (9)), the relevant strain invariants and strain deviatoric invariants are defined by the relations

$$I_{1\varepsilon} = \varepsilon_x + \varepsilon_y + \varepsilon_z \tag{40}$$

and

$$J'_{2\varepsilon} = \tfrac{1}{2}(e_x^2 + e_y^2 + e_z^2 + 2e_{yz}^2 + 2e_{zx}^2 + 2e_{xy}^2) \tag{41}$$

where the e symbols denote deviatoric strains:

$$e_x = \varepsilon_x - \tfrac{1}{3}I_{1\varepsilon} = \tfrac{1}{3}[(\varepsilon_x - \varepsilon_y) + (\varepsilon_x - \varepsilon_z)] \tag{42a}$$

$$e_y = \varepsilon_y - \tfrac{1}{3}I_{1\varepsilon} = \tfrac{1}{3}[(\varepsilon_y - \varepsilon_z) + (\varepsilon_y - \varepsilon_x)] \tag{42b}$$

$$e_z = \varepsilon_z - \tfrac{1}{3}I_{1\varepsilon} = \tfrac{1}{3}[(\varepsilon_z - \varepsilon_x) + (\varepsilon_z - \varepsilon_y)] \tag{42c}$$

and

$$e_{yz} = \varepsilon_{yz}; \quad e_{zx} = \varepsilon_{zx}; \quad e_{xy} = \varepsilon_{xy} \tag{42d,e,f}$$

Note that

$$J'_{1\varepsilon} = e_x + e_y + e_z = 0 \tag{43}$$

Substitution of equations (42) in equation (41) yields

$$J'_{2\varepsilon} = \tfrac{1}{6}[(\varepsilon_x - \varepsilon_y)^2 + (\varepsilon_y - \varepsilon_z)^2 + (\varepsilon_z - \varepsilon_x)^2] + \varepsilon_{xy}^2 + \varepsilon_{yz}^2 + \varepsilon_{zx}^2 \tag{44}$$

For an isotropic elastic material the stress–strain relations may be written in the stiffness form:

$$\sigma_x = (B + \tfrac{4}{3}G)\varepsilon_x + (B - \tfrac{2}{3}G)\varepsilon_y + (B - \tfrac{2}{3}G)\varepsilon_z \tag{45a}$$

$$\sigma_y = (B - \tfrac{2}{3}G)\varepsilon_x + (B + \tfrac{4}{3}G)\varepsilon_y + (B - \tfrac{2}{3}G)\varepsilon_z \tag{45b}$$

$$\sigma_z = (B - \tfrac{2}{3}G)\varepsilon_x + (B - \tfrac{2}{3}G)\varepsilon_y + (B + \tfrac{4}{3}G)\varepsilon_z \tag{45c}$$

$$\tau_{yz} = 2G\varepsilon_{yz} \tag{45d}$$

$$\tau_{zx} = 2G\varepsilon_{zx} \tag{45e}$$

$$\tau_{xy} = 2G\varepsilon_{xy} \tag{45f}$$

In view of the strain invariants and the deviatoric strain expressions, the stress–strain relations may also be written as

$$\sigma_x = BI_{1\varepsilon} + 2Ge_x \tag{46a}$$

$$\sigma_y = BI_{1\varepsilon} + 2Ge_y \tag{46b}$$

$$\sigma_z = BI_{1\varepsilon} + 2Ge_z \tag{46c}$$

$$\tau_{yz} = 2Ge_{yz} \tag{46d}$$

$$\tau_{zx} = 2Ge_{zx} \tag{46e}$$

$$\tau_{xy} = 2Ge_{xy} \tag{46f}$$

Summing equations (46a, b, c) produces

$$I_{1\sigma} = 3BI_{1\varepsilon} \tag{47}$$

Also, the following expression can be formed:

$$s_x = \tfrac{1}{3}[(\sigma_x - \sigma_y) + (\sigma_x - \sigma_z)] = \frac{2G}{3}[(e_x - e_y) + (e_x - e_z)] \tag{48}$$

which reduces to

$$s_x = 2Ge_x \tag{49}$$

This relationship holds for all deviatoric stress and strain components:

$$s_{ij} = 2Ge_{ij} \quad \text{with} \quad i, j = x, y, z \tag{50}$$

This means that

$$J'_{2\sigma} = 4G^2 J'_{2\varepsilon} \tag{51}$$

The increment of work dW shown in equation (13) is

$$dW = \sigma_x d\varepsilon_x + \sigma_y d\varepsilon_y + \sigma_z d\varepsilon_z + 2\tau_{yz} d\varepsilon_{yz} + 2\tau_{zx} d\varepsilon_{zx} + 2\tau_{xy} d\varepsilon_{xy} \tag{52}$$

Given the relations above, equation (52) may be written as

$$dW = BI_{1\varepsilon}(d\varepsilon_x + d\varepsilon_y + d\varepsilon_z) + 2G(e_x d\varepsilon_x + e_y d\varepsilon_y + e_z d\varepsilon_z + 2e_{yz} de_{yz} + 2e_{zx} de_{zx} + 2e_{xy} de_{xy}) \tag{53}$$

or

$$dW = BI_{1\varepsilon} dI_{1\varepsilon} + 2G dJ'_{2\varepsilon} \tag{54}$$

In view of equations (47) and (51) the work increment may also be written in the form

$$dW = \frac{1}{9B} I_{1\sigma} dI_{1\sigma} + \frac{1}{2G} dJ'_{2\sigma} \tag{55}$$

This is the expression given in equation (14).

APPENDIX II

Using the proposed expression for Young's modulus and constant Poisson's ratio, it may be shown that energy is conserved along a closed *strain* path. The stress invariant form in equation (26) may be written in terms of strain invariants using equation (47) and (51):

$$X = (3B)^2 I_{1\varepsilon}^2 + R4G^2[\sqrt{(J'_{2\varepsilon})}]^2 \tag{56}$$

and using the expression for R in equation (25):

$$X = (3B)^2 \left(I_{1\varepsilon}^2 + \frac{36}{R} [\sqrt{(J'_{2\varepsilon})}]^2 \right) \tag{57}$$

Assuming that the bulk modulus B is a function of X only, as presented in the text, then let

$$F(X) = \frac{X}{[3B(X)]^2} = Y \tag{58}$$

where

$$Y = I_{1\varepsilon}^2 + \frac{36}{R} [\sqrt{(J'_{2\varepsilon})}]^2 \tag{59}$$

is an invariant strain form generally similar to the invariant stress X.

Equation (58) may be inverted, at least in principle, to give

$$X = H(Y) \tag{60}$$

The work increment in equation (54) may then be written as

$$dW = BI_{1\varepsilon} dI_{1\varepsilon} + \frac{36}{R} B\sqrt{(J'_{2\varepsilon})} d\sqrt{(J'_{2\varepsilon})} \tag{61}$$

or

$$dW = B(X) \frac{dY}{2} \tag{62}$$

Thus,

$$dW = \tfrac{1}{2}B[H(Y)]\,dY \tag{63}$$

and

$$W = \int_0^Y \tfrac{1}{2}B[H(\bar{Y})]\,d\bar{Y} \tag{64}$$

Following the same argument used for the stress path, but writing the work increment dW in the form shown in equation (61), a potential function $W[I_{1\varepsilon}, \sqrt{(J'_{2\varepsilon})}]$ will exist, and the material will be conservative when exercised around closed strain paths provided that

$$\frac{\partial W}{\partial I_{1\varepsilon}} = BI_{1\varepsilon} \tag{65}$$

and

$$\frac{\partial W}{\partial \sqrt{(J'_{2\varepsilon})}} = \frac{36}{R}B\sqrt{(J'_{2\varepsilon})} \tag{66}$$

Equations (65) and (66) may be verified by differentiating equation (64).

REFERENCES

1. P. V. Lade, 'Large stress reversals in triaxial tests on sand', Proc. 4th Int. Conf. Num. Meth. Geomech., Edmonton, Canada, 1982, Vol. 1, pp. 171–182.
2. G. Wilson and J. L. E. Sutton, 'A contribution to the study of the elastic properties of sand', Proc. 2nd Int. Conf. Soil Mech. Found. Eng., Rotterdam, 1948, Vol. 1, pp. 197–202.
3. R. D. Mindlin, 'Compliance of elastic bodies in contact', *J. Appl. Mech.*, Trans. A.S.M.E., **71**, A-259–268 (1949).
4. R. D. Mindlin and H. Deresiewicz, 'Elastic spheres in contact under varying oblique forces', *J. Appl. Mech.*, Trans. A.S.M.E., **75**, A-327–344 (1953).
5. J. Duffy and R. D. Mindlin, 'Stress–strain relations and vibrations of a granular medium', *J. Appl. Mech.*, A.S.M.E., **24**(4), 585–593 (1957).
6. B. Jacobson, 'Some fundamental properties of sand', Proc. 4th Int. Conf. Soil Mech. Found. Eng., London, 1957, Vol. 1, pp. 167–171.
7. O. U. Janaradanan and A. N. Harkauli, 'Rebound characteristics of cohesionless soils', Proc. 2nd Asian Regional Conf. Soil Mech. Found. Eng., Japan, 1963, pp. 1–4.
8. N. Janbu, 'Soil compressibility as determined by odometer and triaxial tests', Proc. European Conf. Soil Mech. Found Eng., Wiesbaden, 1963, Vol. 1, pp. 19–25.
9. H. M. Makhlouf and J. J. Stewart, 'Factors influencing the modulus of elasticity of dry sand', Proc. 6th Int. Conf. Soil Mech. Found. Eng., Montreal, 1965, Vol. 1, pp. 298–302.
10. H.-Y. Ko and R. F. Scott, 'Deformation of sand in hydrostatic compression', *J. Soil Mech. Found. Div.*, A.S.C.E., **93**(SM3), 137–156 (1967).
11. H.-Y. Ko and R. F. Scott, 'Deformation of sand in shear', *J. Soil Mech. Found. Div.*, A.S.C.E., **93**(SM5), 283–310 (1967).
12. J. M. Duncan and C.-Y. Chang, 'Nonlinear analysis of stress and strain in soils', *J. Soil Mech. Found. Div.*, A.S.C.E., **96**(SM5), 1629–1653 (1970).
13. K. S. Wong and J. M. Duncan, 'Hyperbolic stress–strain parameters for nonlinear finite element analyses of stresses and movements in soil masses', *Geot. Eng. Report No. TE 74-3*, Dept. of Civ. Eng., Univ. of Calif., Berkeley (1974).
14. M. Zytynski, M. F. Randolph, R. Nova and C. P. Wroth, 'On modeling the unloading–reloading behavior of soils', *Int. j. numer. anal. methods geomech.*, **2**, 87–94 (1978).
15. B. O. Hardin and W. L. Black, Closure to 'Vibration modulus of normally consolidated clay', *J. Soil Mech. Found. Div.*, A.S.C.E., **95**(SM6), 1531–1537 (1969).
16. B. O. Hardin, 'The nature of stress–strain behavior for soils', Proc. Spec. Conf. Earthquake Eng. Soil Dyn., A.S.C.E., Pasadena, 1978, Vol. 1, pp. 3–90.
17. P. W. Rowe, 'Theoretical meaning and observed values of deformation parameters for soil', Proc. Roscoe Memorial Symp. Stress–strain Behavior of Soils, pp. 143–194 (1971).
18. M. D. Coon and R. J. Evans, 'Recoverable deformation of cohesionless soils', *J. Soil Mech. Found. Div.*, A.S.C.E., **97**(SM2), 375–391 (1971).
19. C. S. Desai and H. J. Siriwardane, *Constitutive Laws for Engineering Materials*, Prentice-Hall, Englewood Cliffs, New Jersey, 1984.

20. B. Loret, 'On the choice of elastic parameters for sand', *Int. J. Num. Anal. Methods Geomech.*, **9**, 285–292 (1985).
21. I. Nelson and M. L. Baron, 'Application of variable moduli models to soil behavior', *Int. j. solids struct.*, **7**, 399–417 (1971).
22. R. J. Krizek, 'Fabric effects on strength and deformation of kaolin clay', Proc. 9th Int. Conf. Soil Mech. Found. Eng., Tokyo, 1977, Vol. 1, pp. 169–176.
23. P. K. K. Wong and R. J. Mitchell, 'Yielding and plastic flow of sensitive cemented clay', *Geotech.*, **25**,(4), 763–782 (1975).
24. K. Yokota and M. Konno, 'Dynamic Poisson's ratio of soil', Proc. 7th World Conf. Earthquake Eng., Istanbul, 1980, Vol. 3, pp. 475–478.
25. M. S. El Hosri, 'Contribution a l'etude des proprietes mecanique des materiaux', *These d'Etat*, Univ. Paris VI, France (1984).
26. O. Daramola, 'On estimating K_o for overconsolidated granular soils', *Geotech.*, **30**(3), 310–313 (1980).
27. P. V. Lade and J. M. Duncan, 'Cubical triaxial tests on cohesionless soil', *J. Soil Mech. Found. Div.*, A.S.C.E., **99**(SM10), 793–812 (1973).
28. P. V. Lade, 'Cubical triaxial apparatus for soil testing', *Geot. Testing J.*, GTJODJ, **1**(2), 93–101 (1978).
29. P. V. Lade, 'Three-parameter failure criterion for concrete', *J. Eng. Mech. Div.*, A.S.C.E., **108**(EM5), 850–863 (1982).
30. P. V. Lade, 'Failure criterion for frictional materials', in *Mechanics of Engineering Materials* (C. S. Desai and R. H. Gallagher, Eds.), Wiley, New York, 1984, pp. 385–402.
31. P. V. Lade, H. L. Jessberger and N. Diekmann, 'Stress–Strain and volumetric behavior of frozen soil', Proc. 2nd Int. Symp. Ground Freezing, Trondheim, 1980, pp. 48–64.

Numerical Methods in Geomechanics (Innsbruck 1988), Swoboda (ed.)
© 1988 Balkema, Rotterdam. ISBN 90 6191 809 X

Model and parameters for the elastic behaviour of soils

Poul V.Lade
Department of Civil Engineering, University of California, Los Angeles, USA

ABSTRACT: A nonlinear, isotropic model for the elastic behavior of soils has been developed on the basis of theoretical considerations involving the principle of conservation of energy. Energy is therefore neither generated nor dissipated in a closed-loop stress path or strain path. The framework for the model consists of Hooke's law in which Poisson's ratio is constant and Young's modulus is expressed as a power function involving the first invariant of the stress tensor and the second invariant of the deviatoric stress tensor. The characteristics of the model are described, and the accuracy is evaluated by comparison with experimental results from triaxial tests. Parameter determination from unloading-reloading cycles in conventional triaxial compression tests is demonstrated, and parameter values are given for a variety of soils.

1 INTRODUCTION

The models employed for the elastic behavior most often assume the soil to behave as an isotropic material that obeys Hooke's law with a constant Poisson's ratio. Experimental data tend to confirm these assumptions, but the types of dependence on stress state assumed for Young's modulus, the shear modulus and the bulk modulus have shortcomings.

Presented here is an isotropic model for the nonlinear elastic behavior of soils. The model is based on a theoretical development which guaranties lack of energy generation or dissipation for any closed-loop stress path or strain path. The characteristics of the model are described, and determination of parameters from simple tests is demonstrated. Test data for fifteen different soils have been analyzed, and the parameter values are listed and evaluated.

2 COMMON ELASTIC MODELS AND THEIR SHORTCOMINGS

The two most well-known types of expressions for the elastic moduli are those given below. The first expression gives Young's modulus E as a function of the minor principal stress $\sigma_3 = \sigma_{min}$ [1,2]:

$$E = K \cdot p_a \cdot \left[\frac{\sigma_3}{p_a}\right]^n \qquad (1)$$

in which p_a is atmospheric pressure in the same units as E and σ_3. The two material parameters K and n are constant for a soil at a given void ratio. Conditions for which the minor principal stress is constant, thus producing the same value of Young's modulus, are illustrated in Figure 1. Young's modulus is constant along three perpendicular planes crossing the hydrostatic axis at the common value of minor principal stress. The expression in equation (1) has been used with constant and with variable Poisson's ratios.

For an isotropic elastic material, Young's modulus and Poisson's ratio may be replaced by the shear modulus G and the bulk modulus B in Hooke's law. These moduli have often been expressed in terms of the mean normal stress σ_0 [3,4]:

$$G \text{ and/or } B = A \cdot F(e,OCR) \cdot \sigma_0^n \qquad (2)$$

The material parameters incorporated in this class of expressions have occasionally included variations due

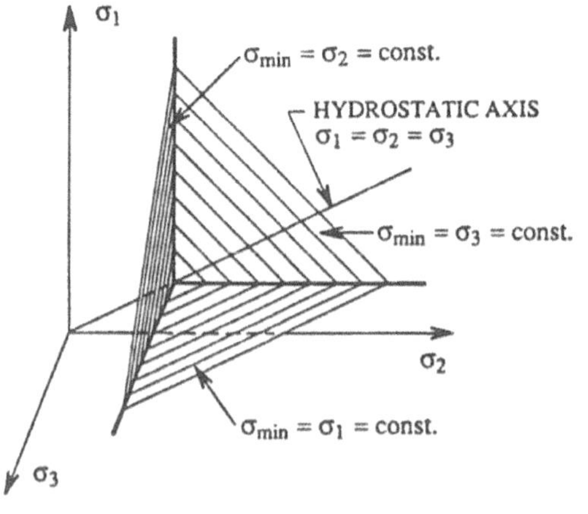

Figure 1. Three Perpendicular Planes in Which Elastic Modulus is Constant According to Equation (1).

to factors such as void ratio e and overconsolidation ratio OCR [4] as indicated in equation (2). For a soil with given e and OCR the values of A,F and n are constant. The mean normal stress and consequently the moduli given by equation (2) are constant in an octahedral plane as illustrated in Figure 2.

Theoretical reasons exist for the variation of elastic moduli with stress state [e.g. 5]. The simple formulations indicated above are empirical in nature, i.e., they have been devised on the bases of results from available, simple tests such as isotropic compression and triaxial compression. The expressions are widely used, because they appear to capture the elastic behavior observed in such tests, and because the material parameters can be determined from the results of conventional triaxial tests.

However, these models for the variation of elastic moduli result in violation of the principle of conservation of energy [6]. Thus, depending on the direction of a closed stress loop, these models will generate or dissipate energy in violation of the basic premise of elastic behavior. According to elasticity theory, a material exhibits true elastic behavior if energy is conserved at the end of a closed stress loop or strain loop.

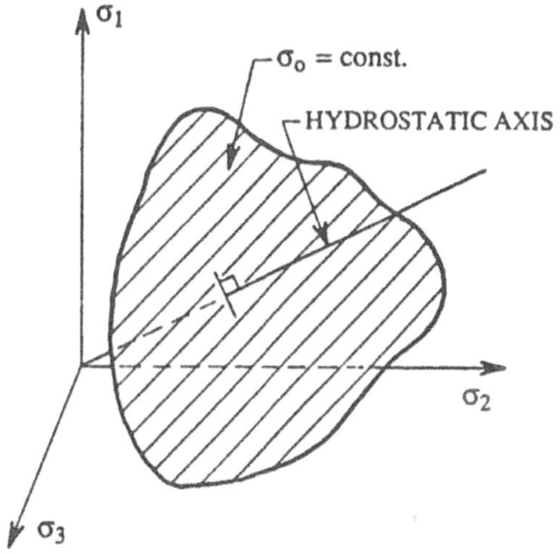

Figure 2. Octahedral Plane in Which Elastic Modulus is Constant According to Equation (2).

3 TRUE ELASTIC MODEL FOR SOILS

An isotropic model for the nonlinear elastic behavior of soils was developed by Lade and Nelson [7]. This model is based on Hooke's law in which Poisson's ratio ν = constant. The expression for Young's modulus was derived from theoretical considerations based on the principle of conservation of energy. According to this derivation, Young's modulus E can be expressed in terms of a power law involving nondimensional material constants and stress functions as follows:

$$E = M \cdot p_a \cdot \left[\left(\frac{I_1}{p_a} \right)^2 + R \cdot \frac{J_2'}{p_a^2} \right]^\lambda \qquad (3)$$

in which

$$R = 6 \cdot \frac{1 + \nu}{1 - 2\nu} \qquad (4)$$

I_1 is the first invariant of the stress tensor, and J_2' is the second invariant of the deviatoric stress tensor, given as follows:

$$I_1 = \sigma_x + \sigma_y + \sigma_z \qquad (5)$$

$$J_2' = \frac{1}{6} [(\sigma_x - \sigma_y)^2 + (\sigma_y - \sigma_z)^2 + (\sigma_z - \sigma_x)^2] \qquad (6)$$

$$+ \tau_{xy}^2 + \tau_{yz}^2 + \tau_{zx}^2$$

The parameter p_a is atmospheric pressure expressed in the same units as E, I_1, and $\sqrt{J_2'}$, and the modulus number M and the exponent λ are constant, dimensionless numbers.

According to equation (3) Young's modulus is constant along rationally symmetric ellipsoidal surfaces whose long axis coincides with the hydrostatic axis and whose center is located at the origin of the principal stress space. The magnitude of Poisson's ratio determines the shape of the ellipsoidal surface. For $\nu = 0$ the value of $R = 6$ and the surface becomes spherical, whereas for $\nu = 0.5$ the value of $R = \infty$ and the surface degenerates into a line coinciding with the hydrostatic axis.

Figure 3 shows cross-sections of the ellipsoidal surfaces in triaxial and octahedral planes for different values of Poisson's ratio. The cross-sections in the octahedral plane are always circular, whereas the cross-sections in the triaxial plane are shaped as ellipses whose aspect ratio depends on the value of Poisson's ratio. The locations of typical failure surfaces for soils with friction angles from 30° to 50° are shown for triaxial compression and extension. Only the portions of the ellipsoids located inside the conical failure surface can be reached by stress states in soils, and only these portions are of interest.

Figure 4 shows a three-dimensional ellipsoidal surface along which Young's modulus is constant according to equation (3). Comparison of this surface with those in Figures 1 and 2 indicates that the conditions for which Young's modulus are constant are quite different for the three expressions. As shown in Figure 1, the model expressed by equation (1) produces surfaces of constant moduli which are convex with regard to the origin, whereas the truly elastic model described by equation (3) results in surfaces that are concave with regard to the origin. The model described by equation (2) produces planes of constant moduli that are neither convex nor concave with regard to the origin. None of the two previous models are contained as special cases within the truly elastic model described here.

4 DETERMINATION OF PARAMETER VALUES

The material parameters M, λ, and ν required to describe the elastic behavior of soils may be deter-

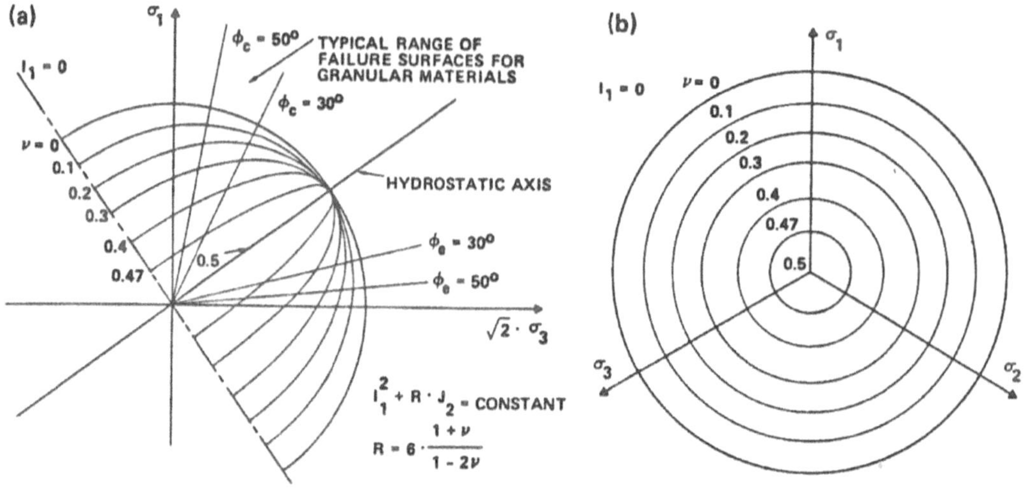

Figure 3. Contours of Constant Young's Modulus Shown in (a) Triaxial Plane and in (b) Octahedral Plane.

mined from conventional drained triaxial compression tests which include unloading-reloading cycles.

4.1 Possion's ratio

The best estimate of Poisson's ratio is obtained from the slope of the volume change curve associated with reloading initiated at or near the hydrostatic axis according to .

$$\nu = -\frac{\dot{\varepsilon}_3}{\dot{\varepsilon}_1} = \frac{1}{2} \cdot \left[1 - \frac{\dot{\varepsilon}_v}{\dot{\varepsilon}_1}\right] \tag{7}$$

in which $(\dot{\varepsilon}_v/\dot{\varepsilon}_1)$ is the slope of the volume change curve immediately after stress reversal, as indicated in the lower part of Figure 5.

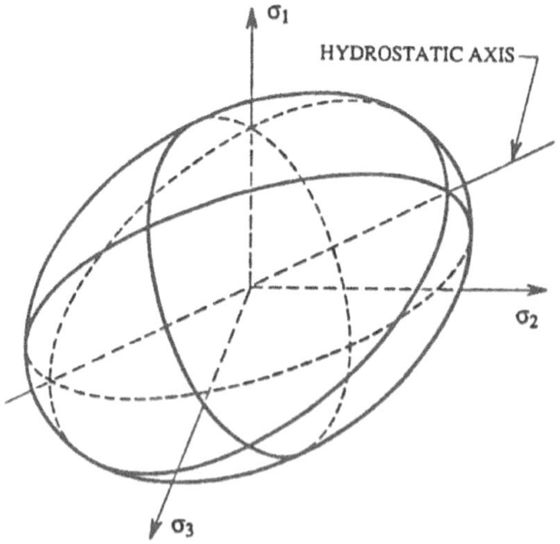

Figure 4. Ellipsoidal Surface Along Which Young's Modulus is Constant According to Equation (3).

It is preferable to obtain an average value of Poisson's ratio from several tests. Figure 6 shows values determined from equation (7) for medium dense Sacramento River Sand plotted against I_1/p_a. Although there is some scatter in the data, there is no discernible

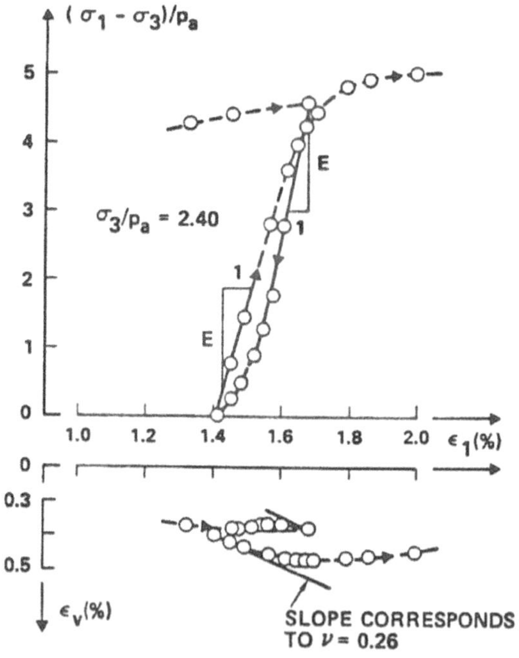

Figure 5. Young's Moduli and Slopes of Volume Change Curve for Determination of Poisson's Ratio from Unloading-Reloading Cycle in Conventional Triaxial Compression Test on Loose Santa Monica Beach Sand.

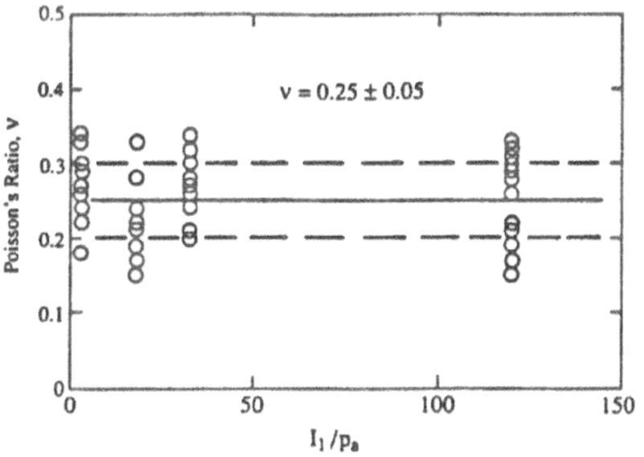

Figure 6. Possion's Ratios Determined from Results of Triaxial Tests on Medium Dense Sacramento River Sand ($e_i = 0.71$, $D_r = 78\%$)

effect of mean normal stress on Poisson's ratio. Similar results have also been obtained for several other soils [e.g. 7,8]. The average value and the standard deviation for Poisson's ratio are indicated on Figure 6.

4.2 Young's modulus

Using the average value of Poisson's ratio, the value of R can be determined from equation (4). Individual modulus values may be obtained from the slopes of the unloading and the reloading stress-strain relations as shown in the upper part of Figure 5. These modulus values are associated with the stress states at which stress reversals are initiated. The parameters M and λ in equation (3) are then determined by plotting (E/p_a) versus $[(I_1/p_a)^2 + R \cdot (J_2'/p_a^2)]$ in a log-log diagram as shown in Figure 7 for medium dense Sacramento River Sand. The intercept of the best fitting straight line with $[(I_1/p_a)^2 + R \cdot (J_2'/p_a^2)] = 1$ is the value of M, and λ is the slope of the line.

It is interesting to note that the reloading moduli obtained at the hydrostatic axis where $J_2' = 0$ essentially define the straight line in Figure 7. The positions of the unloading moduli are suitably adjusted through J_2' and R so as to fall on the same line. For this particular case, the points corresponding to isotropic compression are located slightly below the other points. However, this is usually not the case. Thus, once v and R have been determined, any type of test may, in principle, be used to obtain the parameter M and λ.

5 ELASTIC PARAMETER VALUES

Seventeen sets of data from fifteen different soils have been analyzed to study the range of parameter values obtained. Descriptions of soils and material parameter values are given in Table 1. The average Poisson's ratios were obtained from reloading, as explained above, as well as from unloading at constant confining pressure and from K_0-unloading. Considerable scatter in the values of Poisson's ratio were obtained as indicated by the standard deviations. However, it was

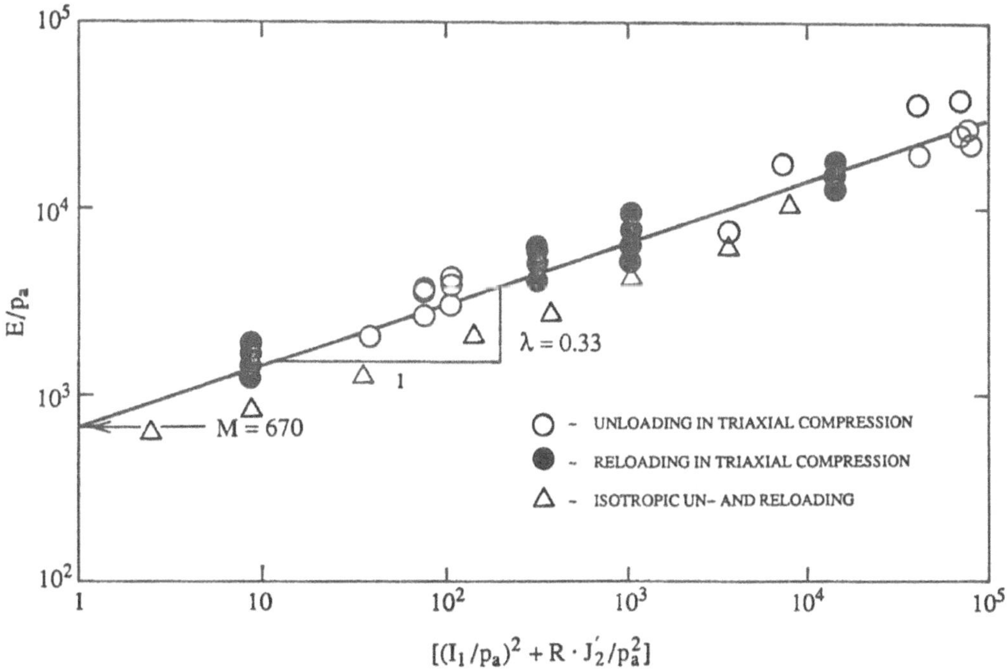

Figure 7. Determination of Parameters for Young's Modulus for Medium Dense Sacramento River Sand ($e_i = 0.71$, $D_r = 78\%$)

Table 1. Description and Elastic Parameter Values for 15 Different Soils

Soil (Reference*)	Description				Poisson's Ratio		Young's Modulus		
	D_{50}(mm)	C_u	e	D_r (%)	v_{avg}	s	M	λ	Coeff. of Correlation r
Santa Monica Beach Sand (Lade and Nelson [7])	0.27	1.58	0.61	90	0.15	0.08	1270	0.23	0.878
			0.81	20	0.26	0.06	600	0.27	0.792
Montery No. 0 Sand Lade and Duncan [9])	0.43	1.53	0.57	98	0.17	0.14	1120	0.33	0.849
			0.78	27	0.17	0.05	800	0.26	0.659
Fine Silica Sand	0.18	2.0	0.76	30	0.27	0.08	440	0.22	0.673
Mohawk Model Soil	3.5	6.2	0.64	73	0.25	0.13	630	0.27	0.782
Niagara-Type A Material	4.8	1.52	0.71	95	0.23	0.12	670	0.26	0.872
Niagara-Type B Material	1.2	8.6	0.52	95	0.24	0.11	570	0.29	0.788
Dense Silica Sand (Duncan and Chang [2])	0.25	1.8	0.50	100	0.16	-	750	0.21	0.960
Ottawa Sand (Yong and Ko, [10])	0.42	1.3	0.51	87	0.13	0.04	800	0.18	0.429
Dense Sand (Goldscheider [11])	0.42	1.8	0.56	100	0.14	-	280	0.38	0.970
Hostun Sand (Saada [12])	0.34	1.7	0.62	98	0.22	0.04	400	0.32	0.914
Reid Bedford Sand (Saada [12])	0.24	1.7	0.68	51	0.11	-	550	0.33	0.859
Intact Clayey Sand	0.20	200	0.51	65	0.25	0.03	400	0.36	0.963
Remolded Clayey Sand	0.20	200	0.51	65	0.25	0.03	390	0.36	0.987
Compacted Silt	0.013	U.S. Class. ML	0.42	RC = 95% Mod. Proctor	0.27	0.06	340	0.20	0.762
Sacramento River Sand (Lee [13])	0.21	1.5	0.71	78	0.25	0.05	670	0.33	0.975

* Those without reference are from the author's files.

shown in [7] that the average values of ν may well be within ranges in which M and λ are relatively insensitive to Poisson's ratio. Table 1 shows that Poisson's ratio varies from 0.11 to 0.27 with an average value for all soils (mostly granular) of 0.21.

The modulus number M varies from 280 to 1270, whereas the exponent λ varies from 0.18 to 0.38. The coefficient of correlation r is also listed in Table 1 to indicate the quality of fit with the model for each data set. In view of the number of data sets that have been evaluated, the magnitudes and ranges of M, λ, and ν listed in Table 1 are considered to be representative for granular materials.

6 CONCLUSION

An isotropic model for the nonlinear elastic behavior of soils has been presented. The model is based on Hooke's law in which Poisson's ratio is constant, and Yong's modulus is expressed in terms of stress invariants. The expression for Young's modulus has been derived from theoretical considerations based on the principle of conservation of energy. Material parameters may be determined from unloading-reloading cycles in conventional triaxial compression tests. Parameters were determined and studied for fifteen different soils.

REFERENCES

[1] N. Janbu, Soil Compressibility as Determined by Odometer and Triaxial Tests, Proc. Europ. Conf. Soil Mech. Found. Eng., Wiesbaden, 1, 19-25 (1963)

[2] J.M. Duncan and C.-Y. Chang, Nonlinear Analysis of Stress and Strain in Soils, J. Soil Mech. Found. Div., 96, 1629-1653 (1970)

[3] B.O. Hardin and W.L. Black, Closure to Vibration Modulus of Normally Consolidated Clay, J. Soil Mech. Found. Div., 95, 1531-1537 (1969)

[4] B.O. Hardin, The Nature of Stress-Strain Behavior of Soils, Proc. Spec. Conf. Earthquake Eng. Soil Dyn., Pasadena, 1, 3-90 (1978)

[5] R.D. Mindlin, Compliance of Elastic Bodies in Contact, J. Appl. Mech., 71, A-259-268 (1949)

[6] M. Zytynski, M.F. Randolph, R. Nova and C.P. Wroth, On Modeling the Unloading-Reloading Behavior of Soils, Int. J. Anal. Methods Geomech., 2, 87-94 (1978)

[7] P.V. Lade and R.B. Nelson, Modeling the Elastic Behavior of Granular Materials, Int. J. Anal. Methods Geomech., 11 (1987)

[8] P.W. Rowe, Theoretical Meaning and Observed Values of Deformation Parameters for Soil, Proc. Roscoe Memorial Symp. Stress-Strain Behavior of Soils, 143-194 (1971)

[9] P.V. Lade and J.M. Duncan, Cubical Triaxial Tests on Cohesionless Soils, J. Soil Mech. Found. Div., 99, 793-812 (1973)

[10] R.N. Yong and H.-Y. Ko, Pre-Workshop Package, Proc. Workshop Limit Equilibrium, Plasticity and Generalized Stress-Strain in Geotechnical Engineering, Montreal (1980)

[11] M. Goldscheider, True Triaxial Tests on Dense Sand, Constitutive Relations for Soils, Grenoble (1982), Balkema, Rotterdam, 11-54 (1984)

[12] A.S. Saada, Information Package, Proc. Int. Workshop on Constitutive Equations for Granular Non-Cohesive Soils, Cleveland (1987)

[13] K.L. Lee, Triaxial Compressive Strength of Saturated Sand Under Seismic Loading Conditions, Ph.D.-thesis, Univ. of California, Berkeley (1965)

Chapter VI
Failure Criterion for Isotropic Materials

Preface

Central and most important to development of work-hardening plasticity theory for all materials is a suitable three-dimensional failure criterion. For a given material, this criterion distinguishes between states of stress that can be reached and states of stress that cannot be reached. Analyses of three-dimensional experimental results show that the shape of the failure surfaces for frictional materials such as soils, rocks and concrete, as well as metals with pores, such as cast iron and powder metals, are influenced by the minor principal stress, the intermediate principal stress, the void ratio or density and by cross-anisotropy. The latter issue of cross-anisotropy is dealt with in Chapter XIV. Other issues such as shear banding and instability inside the failure surface are explained in Chapter VIII. Over-consolidation are also important to the failure conditions for these materials. In the limit, where friction is negligible, the proposed criterion simplifies to the Von Mises criterion which describes the failure of solid metals such as martensite with good accuracy. The required parameters can be obtained from simple experiments such as triaxial compression tests or simple tension tests. For the occasional case that the result of only a single test is available, the parameters may be estimated as shown in the following article.

The failure criterion presented here is expressed in terms of stress invariants that may be utilized for all types of isotropic materials. This three-dimensional criterion was first developed for soils with straight failure surfaces (Lade and Duncan, 1975) and subsequently for soils with curved failure surfaces (Lade,1977) and later expanded to include concrete (Lade, 1982) and rock (Kim and Lade, 1984; Lade, 1993). The criterion for soils is expressed in terms of the first and the third invariants of the stress tensor as follows:

$$(I_1^3/I_3 - 27)\cdot(I_1/p_a)^m = \eta_1 \qquad (6.1)$$

in which the stress invariants are given in terms of principal stresses as:

$$I_1 = \sigma_1 + \sigma_2 + \sigma_3 \qquad (6.2)$$

$$I_3 = \sigma_1 \cdot \sigma_2 \cdot \sigma_3 \qquad (6.3)$$

and p_a is atmospheric pressure expressed in the same units as the stresses. The value of I_1^3/I_3 is 27 at the hydrostatic axis where $\sigma_1 = \sigma_2 = \sigma_3$. The parameters η_1 and m in Eq. (1) can be determined by plotting $(I_1^3/I_3 - 27)$ versus (p_a/I_1) at failure on a log-log diagram and locating the best fitting straight line, as shown in Fig. 6.1. The intercept of this line with $(p_a/I_1) = 1$ is the value of η_1, and m is the slope of the line.

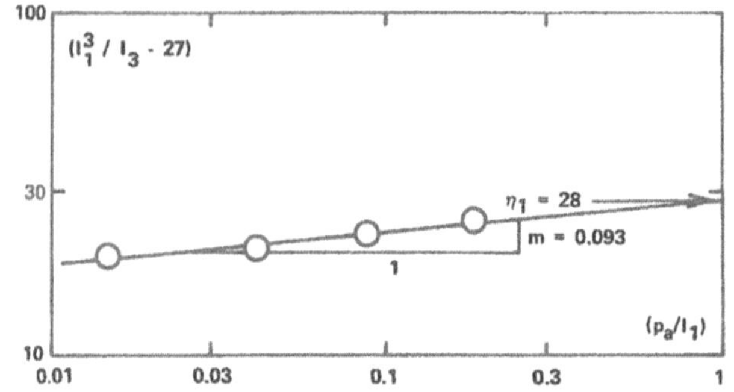

Fig. 6.1. Determination of the values of η_l and m involved in failure criterion for loose Sacramento River Sand (after Lade, 1977).

The failure surface is shaped like an asymmetric bullet with the pointed apex at the origin of the stress axes, as shown in the principal stress space in Fig. 6.2. The apex angle increases with the value of η_l. The failure surface is always concave towards the hydrostatic axis, and its curvature increases with the value of m. For constant value of m and increasing η_l-values, the cross sections in the octahedral plane changes from circular to triangular with smoothly rounded edges in a fashion that agrees with the experimental evidence.

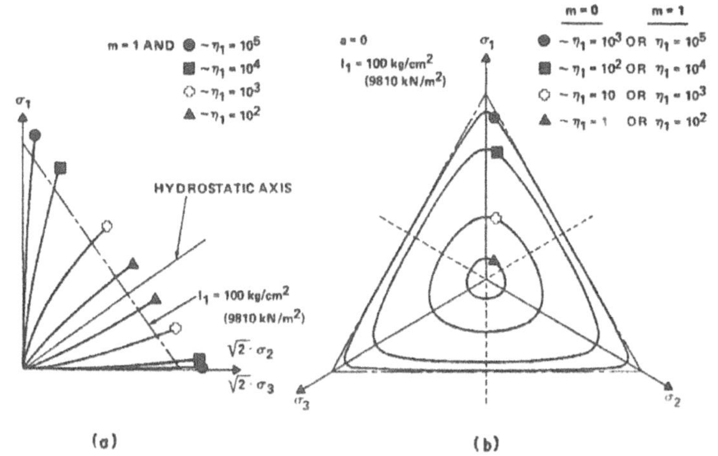

Fig. 6.2. Characteristics of failure surfaces shown in principal stress space. Traces of failure surfaces shown in (a) triaxial plane and in (b) octahedral plane (after Lade, 1982).

Frictional materials include concrete and rock, which have many characteristics in common with soils, and it may be possible to model their strengths by a criterion similar to that in Eq. (6.1). A translation of the failure surface along the hydrostatic axis in the principal stress space is performed to include the cohesion and the tensile strengths that can be sustained by concrete and rock, as illustrated in Fig. 6.3. Thus, a constant stress, $a \cdot p_a$, is added to the normal stresses before substitution in Eq. (6.1):

$$\bar{\sigma}_x = \sigma_x + a \cdot p_a \tag{6.4}$$

$$\bar{\sigma}_y = \sigma_y + a \cdot p_a \tag{6.5}$$

$$\bar{\sigma}_z = \sigma_z + a \cdot p_a \tag{6.6}$$

where a is a dimensionless parameter and p_a is atmospheric pressure in the same units as σ_x, σ_y and σ_z. The value of $a \cdot p_a$ reflects the effects of the tensile strength of the material.

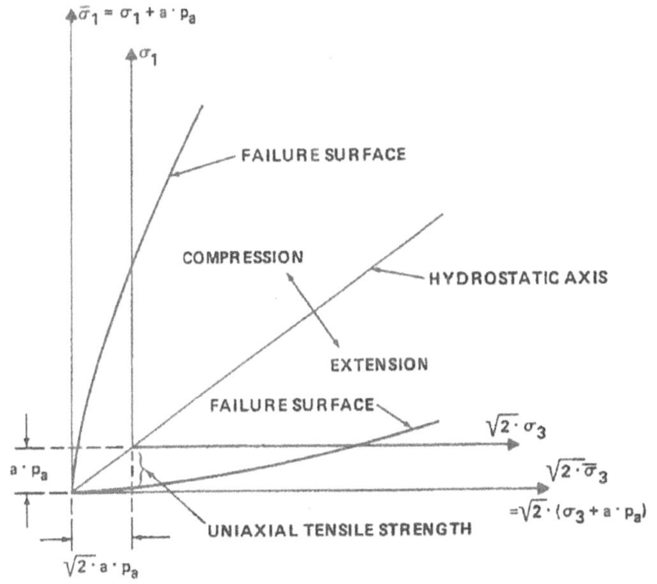

Fig. 6.3. Translation of principal stress space along hydrostatic axis to include effect of tensile strength in failure criterion (after Lade, 1982).

The three material parameters interact in calculation of, for example, the uniaxial compressive strength of a material. Thus, an infinite number of combinations of 'a', η_1, and m could result in the same value of the uniaxial compressive strength.

Combining the shapes in the triaxial, biaxial and octahedral planes produce a three-dimensional failure surface in the principal stress space, as shown in the two diagrams in Fig. 6.4. The curved shape in the triaxial plane is emphasized in Fig. 6.4(a), whereas the cross-sectional shape is exhibited more clearly in the view in Fig. 6.4(b). The compression-compression regions of the three biaxial planes are shown in Fig. 6.4(a), and the uniaxial compression and tension points are also indicated in this figure.

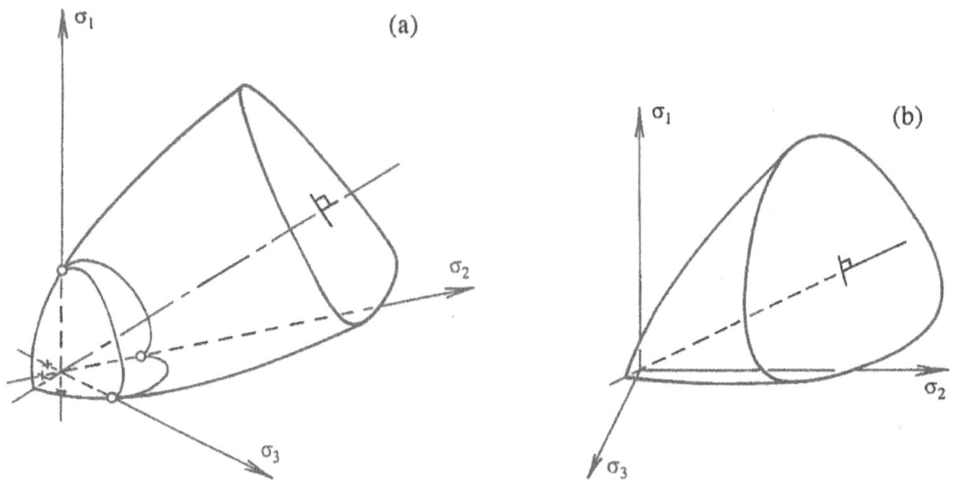

Fig. 6.4. Three-dimensional failure surface produced from experimental evidence and viewed (a) from the side to show the curved envelope in triaxial plane and in biaxial planes, and (b) from the end to show the smoothly rounded triangular shape in the octahedral plane.

For geological materials the failure surface is shaped as a pointed bullet with smoothly rounded triangular cross-sections, and it has three independent characteristics: (1) the opening angle near the origin of the principal stress space, which relates to the friction angle, (2) the curvature in planes containing the hydrostatic axis, i.e. curved meridians, and (3) the tensile strength. Each of these characteristics may be different for different types of materials. Capturing these three features therefore requires at least three independent parameters, such as included in the criterion given in Eq. (6.1).

In the limit, as for solid metals, the frictional effects disappear and cohesion dominates. This occurs for tempered martensite (4330 steel) as shown in Fig. 6.5. The results indicate equal strengths in tension and compression. Note that here again all parameters are determined from relatively simple tension tests and these are used to predict the overall three-dimensional failure conditions. They are similar to those indicated by the Von Mises failure criterion. Thus, the cross-section is a circle, as also obtained from the criterion in Eq. (6.1).

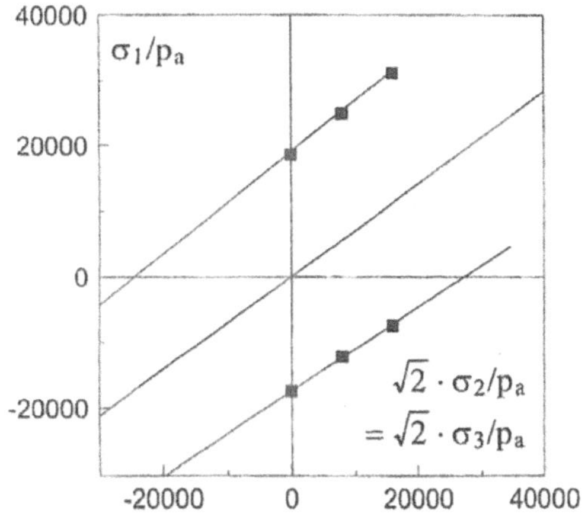

Fig. 6.5. Comparison of failure criterion with experimental results on triaxial plane for tempered martensite (4330 steel) presented by Spitzig et al. (1975).

References

Kim, M.K. and Lade, P.V. (1984) Modeling of Rock Strength in Three Dimensions. *Int. J. Rock Mech. Min. Sci. & Geomech. Abstr.*, 21(1): pp. 21-33.

Lade, P.V. and Duncan, J.M. (1975) Elastoplastic Stress-Strain Theory for Cohesionless Soil. *J. Geotech. Engr. Div.*, ASCE, 101(GT10): pp. 1037-1053.

Lade, P.V. (1977) Elasto-Plastic Stress-Strain Theory for Cohesionless Soil with Curved Yield Surfaces. *Int. J. Solids Struct.*, 13: pp. 1019-1035.

Lade, P.V. (1982) Three-Parameter Failure Criterion for Concrete, *J. Engr. Mech. Div.*, ASCE, 108(EM5): pp. 850-863.

Lade, P.V. (1993) Rock Strength Criteria: The Theories and the Evidence. In COMPREHENSIVE ROCK ENGINEERING, Principles, Practice & Projects, Editor-in-Chief: J.A. Hudson, Chapter 11 in Vol. 1: *Fundamentals* edited by E.T. Brown, Pergamon Press, Oxford, England: pp. 255-284.

Lade, P.V. (1997) "Modeling the Strengths of Engineering Materials in Three Dimensions," *Mechanics of Cohesive-Frictional Materials*, Wiley, 2: pp. 339-356.

Lade, P.V. (2006) "Assessment of Test Data for Selection of 3-D Failure Criterion for Sand" *Int. J. Num. Analyt. Meth. in Geomech.*, 30(4): pp. 307-333.

Spitzig, W.A., Sober, R.J. and Richmond, O. (1975) "Pressure dependence of yielding and associated volume expansion in tempered martensite." *Acta Metallurgica*, 23: pp. 885-893.

Mechanics of Engineering Materials
Edited by C. S. Desai and R. H. Gallagher

Chapter 20

Failure Criterion for Frictional Materials

P. V. Lade

20.1 INTRODUCTION

Failure of engineering structures, whether simple or complex, usually involves three-dimensional stress conditions in the materials. Detailed analyses of such structures therefore require knowledge of the stress–strain behaviour and the states of stress which constitute failure. However, the use of inadequate material models, especially for frictional materials, is often one of the limiting factors in the analyses procedures. In addition, the behaviour and the stress distributions are likely to be different in intact and in fractured materials. To determine the stress conditions which govern fracture of intact materials, a general three-dimensional failure criterion is required. Such a criterion is also one of the important parts of a constitutive law for frictional materials.

Frictional materials are characterized by increasing shear strengths with increasing normal stresses. Many experimental investigations have been performed, especially in recent years, to study the failure conditions for frictional materials such as sands, clays, cemented soils, concrete, mortar, rock, etc. These laboratory investigations have been characterized by utilization of increasingly improved equipment, in which the boundary conditions are well known, thus leading to reliable results of good quality. Along with these studies, formulations of failure criteria have been proposed. Some of these are aimed at practical design applications, whereas other more complex expressions have been developed for use in advanced computer codes. Most of the failure criteria proposed for three-dimensional stress states involve relatively complex expressions for which more than three material parameters are required.

These criteria have been developed to capture the experimentally determined shape of the failure surfaces as observed in the principal stress space. Several studies have shown that the failure surfaces in the principal stress space are shaped as pointed bullets with cross-sections in octahedral planes which are triangular, monotonically curved surfaces with smoothly rounded corners.

In addition to the characteristic cross-sectional shape in the octahedral plane, the three-dimensional failure surfaces for frictional materials have three independent characteristics: (1) the opening angle of the failure surface, often described by the friction angle, (2) the curvature of the failure surface in planes containing the hydrostatic axis, i.e. curved meridians, and (3) the tensile strength (which is zero for materials without effective cohesion). At least three independent parameters are necessary for description of the three separate and distinct characteristics of the failure surface for frictional materials.

A general, three-dimensional failure criterion expressed in terms of stress invariants has been developed for frictional materials. It involves three independent material parameters, each relating to one of the three characteristics mentioned above. These parameters may be determined from any type of strength tests, including the simplest possible such as uniaxial compression and triaxial compression or biaxial tests.

The procedure for determination of the three material parameters is demonstrated and comparisons between failure criterion and experimental results are presented for different frictional materials. The values of the three parameters are summarized for the various materials, and their magnitudes are compared and discussed in light of the materials to which they belong.

20.2 FRICTIONAL MATERIALS WITHOUT EFFECTIVE COHESION

Many frictional materials such as gravels, sands, silts, clays, rockfill, mine tailings, coal, feed grain, etc. do not have effective cohesion. According to Mitchell [1] 'tests over large ranges of effective stress show that the actual effective stress failure envelope is curved,' and 'that cohesion is either zero or very small, even for heavily over-consolidated clays. Thus, a significant true cohesion, if defined as strength present at zero effective stress, does not exist in the absence of chemical bonding (cementation)'. Several studies of materials without effective cohesion under two- and three-dimensional stress conditions show that these materials have many characteristics in common.

20.2.1 Evaluation of classical failure criteria

Several existing constitutive models employ the Extended Tresca criterion or the Extended von Mises (or Drucker–Prager) criterion to describe the failure condition for frictional materials. The shapes of the corresponding failure surfaces in the principal stress space are conical and they have cross-sections of a regular hexagon (Tresca) or a circle (von Mises) with their centre lines coinciding with the hydrostatic axis. According to both these criteria the material has the same strength in compression and extension for a given magnitude of mean normal stress.

In an evaluation of possible failure criteria for sands, Bishop [2] pointed out that the two failure criteria described above are in principle unable to represent

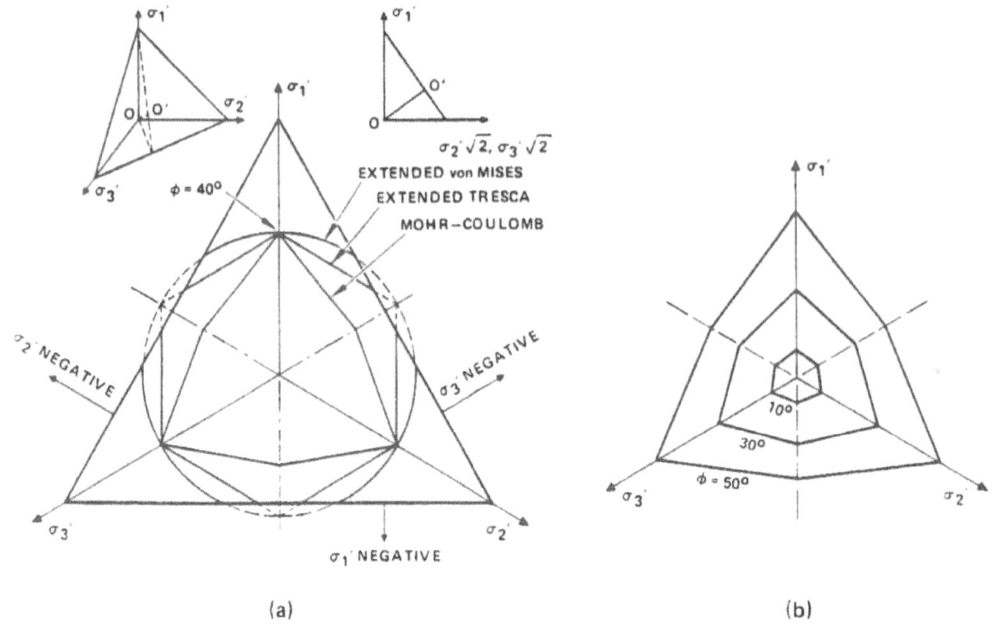

Figure 20.1 (a) Representation of classical failure criteria in principal stress space, showing boundaries of positive stress space (after Bishop [2]). (b) Cross-sections of the Mohr–Coulomb failure criterion shown for three different friction angles

the behaviour of cohesionless materials. Figure 20.1(a) indicates that the failure surfaces may extend outside the part of the principal stress space where all stresses are compressive and positive. The friction angle in triaxial compression for which both these failure surfaces are tangential to the coordinate planes is 36.9°. For higher friction angles the states of stress near triaxial extension are located in the parts of the stress space where one of the principal stresses is negative, and this is clearly unreasonable for materials without effective cohesion. Even for friction angles in triaxial compression smaller than 36.9°, the two criteria fail to model correctly the experimentally observed variation of the three-dimensional strengths of frictional materials.

The Mohr–Coulomb failure surface is also conical, and its cross-section is an irregular hexagon as indicated in Figure 20.1(a). This failure surface exhibits some of the characteristics necessary for correct modelling of failure of frictional materials. Thus, for materials without effective cohesion all principal stresses remain positive, even for very high friction angles. The shape of the cross-section of the Mohr–Coulomb failure surface resembles a regular hexagon for very small friction angles and it approaches an equilateral triangle for friction angles approaching 90°. The variation in shape for more conventional friction angles, as shown in Figure 20.1(b), is supported by experimental evidence.

The intermediate principal stress does not appear in the Mohr–Coulomb failure criterion, the failure surfaces are pointed in octahedral planes as indicated in Figure 20.1(b), and their traces in planes containing the hydrostatic axis are straight lines. However, experimental evidence clearly shows that the intermediate principal stress has an important influence on the strength of frictional materials. Furthermore, traces of experimental failure surfaces in octahedral planes are smooth throughout their lengths, and they intersect the projections of the principal stress axes at right angles. Experimental results also indicate that the failure surfaces are curved in planes containing the hydrostatic axis. The Mohr–Coulomb criterion does not model these significant aspects of failure in frictional materials.

20.2.2 Three-dimensional failure criterion

The three-dimensional failure criterion for frictional materials without effective cohesion presented here was previously developed for soils with curved failure envelopes [3]. This criterion is expressed in terms of the first and third stress invariants of the stress tensor as follows:

$$(I_1^3/I_3 - 27)\cdot(I_1/p_a)^m = \eta_1 \tag{20.1}$$

where

$$I_1 = \sigma_1 + \sigma_2 + \sigma_3 = \sigma_x + \sigma_y + \sigma_z \tag{20.2}$$

$$\begin{aligned} I_3 &= \sigma_1 \cdot \sigma_2 \cdot \sigma_3 \\ &= \sigma_x \cdot \sigma_y \cdot \sigma_z + \tau_{xy} \cdot \tau_{yz} \cdot \tau_{zx} + \tau_{yx} \cdot \tau_{zy} \cdot \tau_{xz} \\ &\quad - (\sigma_x \cdot \tau_{yz} \cdot \tau_{zy} + \sigma_y \cdot \tau_{zx} \cdot \tau_{xz} + \sigma_z \cdot \tau_{xy} \cdot \tau_{yx}) \end{aligned} \tag{20.3}$$

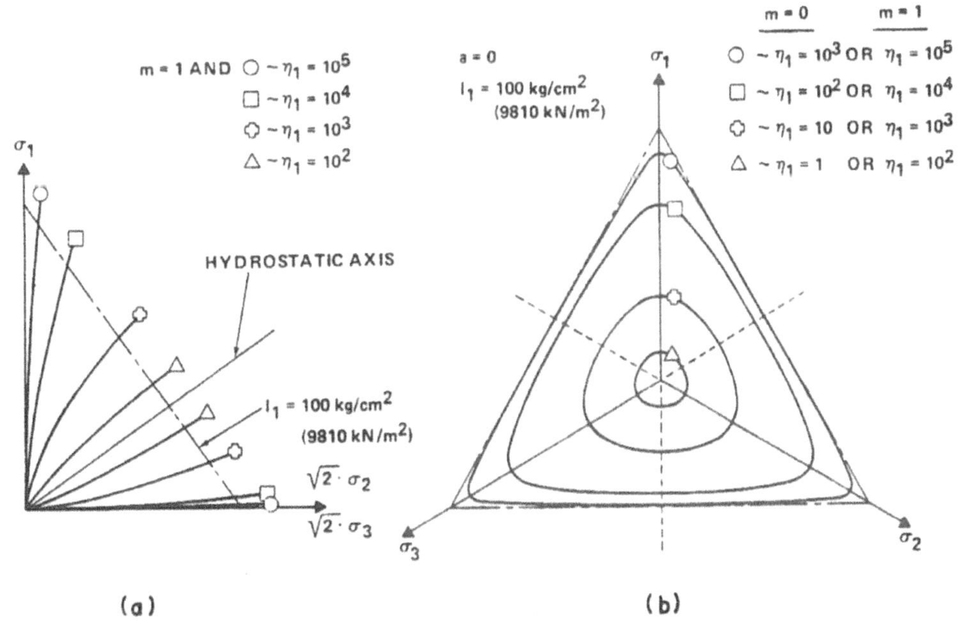

Figure 20.2 Characteristics of failure surfaces shown in principal stress space. Traces of failure surfaces shown in (a) triaxial plane, and in (b) octahedral plane

and p_a is atmospheric pressure expressed in the same units as the stresses. The value of I_1^3/I_3 is 27 at the hydrostatic axis where $\sigma_1 = \sigma_2 = \sigma_3$. The parameters η_1 and m in equation (20.1) can be determined by plotting $(I_1^3/I_3 - 27)$ versus (p_a/I_1) at failure in a log–log diagram and locate the best fitting straight line. The intercept of this line with $(p_a/I_1) = 1$ is the value of η_1 and m is the slope of the line.

In principal stress space the failure surface defined by equation (20.1) is shaped like an asymmetric bullet with the pointed apex at the origin of the stress axes as shown in Figure 20.2(a). The apex angle increases with the value of η_1. The failure surface is concave towards the hydrostatic axis, and its curvature increases with the value of m. For $m = 0$ the failure surface is straight. Figure 20.2(b) shows typical cross-sections in the octahedral plane (I_1 = const.) for $m = 0$ and $\eta_1 = 1$, 10, 10^2, and 10^3. As the value of η_1 increases, the cross-sectional shape changes from circular to triangular with smoothly rounded edges in a fashion that conforms to experimental evidence. The shape of these cross-sections does not change with the value of I_1 when $m = 0$. For $m > 0$ the cross-sectional shape of the failure surface changes from triangular to become more circular with increasing value of I_1. Similar changes in cross-sectional shape are observed from experimental studies on frictional materials. The cross-sections in Figure 20.2(b) also correspond to $m = 1$ and $\eta_1 = 10^2$, 10^3, 10^4, and 10^5.

20.2.3 Comparison of failure criterion and test data

The failure criterion described above has been shown to model the experimentally determined three-dimensional strengths of sand and clay with good

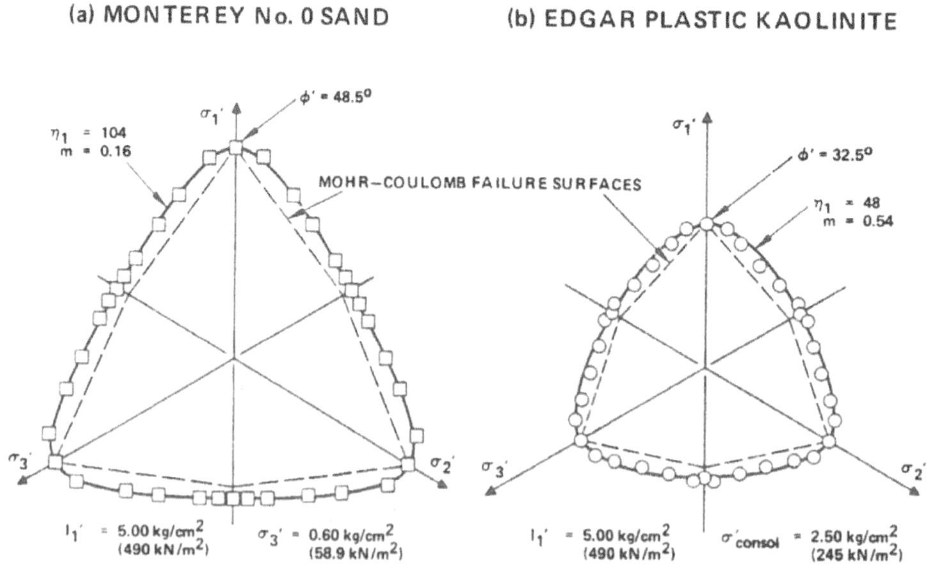

Figure 20.3 Comparison of failure criterion in octahedral planes with results of cubical triaxial tests on (a) dense Monterey No. 0 sand, and (b) normally consolidated, remoulded Edgar plastic kaolinite

accuracy in the range of stresses where the failure envelopes are concave towards the hydrostatic axis [3 to 5].

Figure 20.3 shows examples of comparisons between failure criterion and test data in terms of effective stresses for dense Monterey No. 0 sand and normally consolidated, remoulded Edgar plastic kaolinite. The values of η_1 and m suitable for description of failure in the two soils are given in Figure 20.3. The data points were projected on the common octahedral planes along curved meridians in order to provide a correct comparison between failure criterion and experimental data. It may be seen that the failure criterion models the experimentally obtained three-dimensional failure surfaces with good accuracy for both sand and clay.

20.3 FRICTIONAL MATERIALS WITH EFFECTIVE COHESION

Examples of frictional materials with effective cohesion include cemented soils (artificial and natural), frozen soils, ice, concrete, mortar, rocks, ceramics, graphite, plaster of Paris, and hydrostone. These materials exhibit tensile strengths and their shear strengths increase with increasing normal stresses.

20.3.1 General three-dimensional failure criterion for frictional materials

Because frictional materials with effective cohesion have many characteristics in common with those without effective cohesion, it has been proposed [6] that

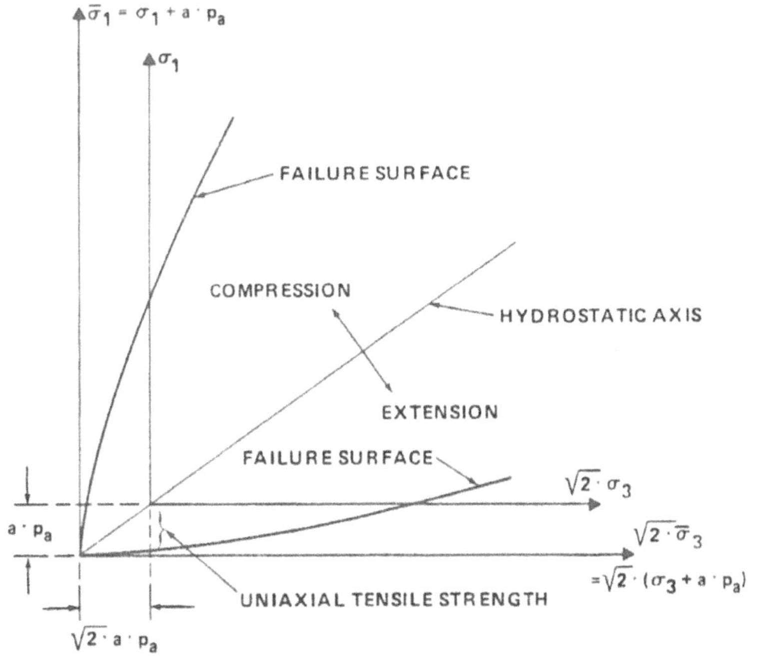

Figure 20.4 Translation of principal stress space along hydrostatic axis to include effect of tensile strength in failure criterion

their strengths can be expressed by a criterion similar to that in equation (20.1). In order to include the cohesion and the tensile strength in the failure criterion, a translation of the principal stress space along the hydrostatic axis is performed as illustrated in Figure 20.4. Thus, a constant stress $a \cdot p_a$ is added to the normal stresses before substitution in equation (20.1):

$$\bar{\sigma}_x = \sigma_x + a \cdot p_a \quad (20.4a)$$

$$\bar{\sigma}_y = \sigma_y + a \cdot p_a \quad (20.4b)$$

$$\bar{\sigma}_z = \sigma_z + a \cdot p_a \quad (20.4c)$$

where 'a' is a dimensionless parameter and p_a is atmospheric pressure in the same units as σ_x, σ_y, and σ_z. The value of $a \cdot p_a$ reflects the effect of the tensile strength of the material. Although the three material parameters describe separate characteristics of the failure surface, they do interact in calculation of, for example, the uniaxial compressive strength of the material. Thus, an infinite number of combinations of 'a', η_1, and m could result in the same value of the uniaxial compressive strength.

20.3.2 Determination of material parameters

In order to determine the values of the three material parameters for a given set of experimental data, the value of 'a' is estimated and $a \cdot p_a$ is added to the normal stresses before substitution in equation (20.1). The procedure for finding η_1 and m as described above is then followed. To facilitate the estimate of 'a', advantage may be taken of the fact that $a \cdot p_a$ must be slightly greater than the uniaxial tensile strength as indicated on Figure 20.4. If tensile tests are not part of a regular testing programme, a sufficiently accurate value of the uniaxial tensile strength may be obtained from the following approximation formula.

20.3.2.1 *Uniaxial tensile strength*

The uniaxial tensile strength, σ_t, has often been expressed as a fraction of the uniaxial compressive strength, σ_c. For example, Mitchell [7] indicates that σ_t (actually given as the flexural strength which may be higher than the true value of

Table 20.1 Values of parameters T and t for various types of frictional materials

Material	T	t	Reference
Cemented soils	−0.37	0.88	Mitchell [7]
Concrete and mortar	−0.61	0.67	Wastiels [9]
Igneous rocks	−0.53	0.70	This chapter
Metamorphic rocks	−0.00082	1.6	This chapter
Sedimentary rocks	−0.22	0.75	This chapter
Ceramics	−1.0	0.73	This chapter

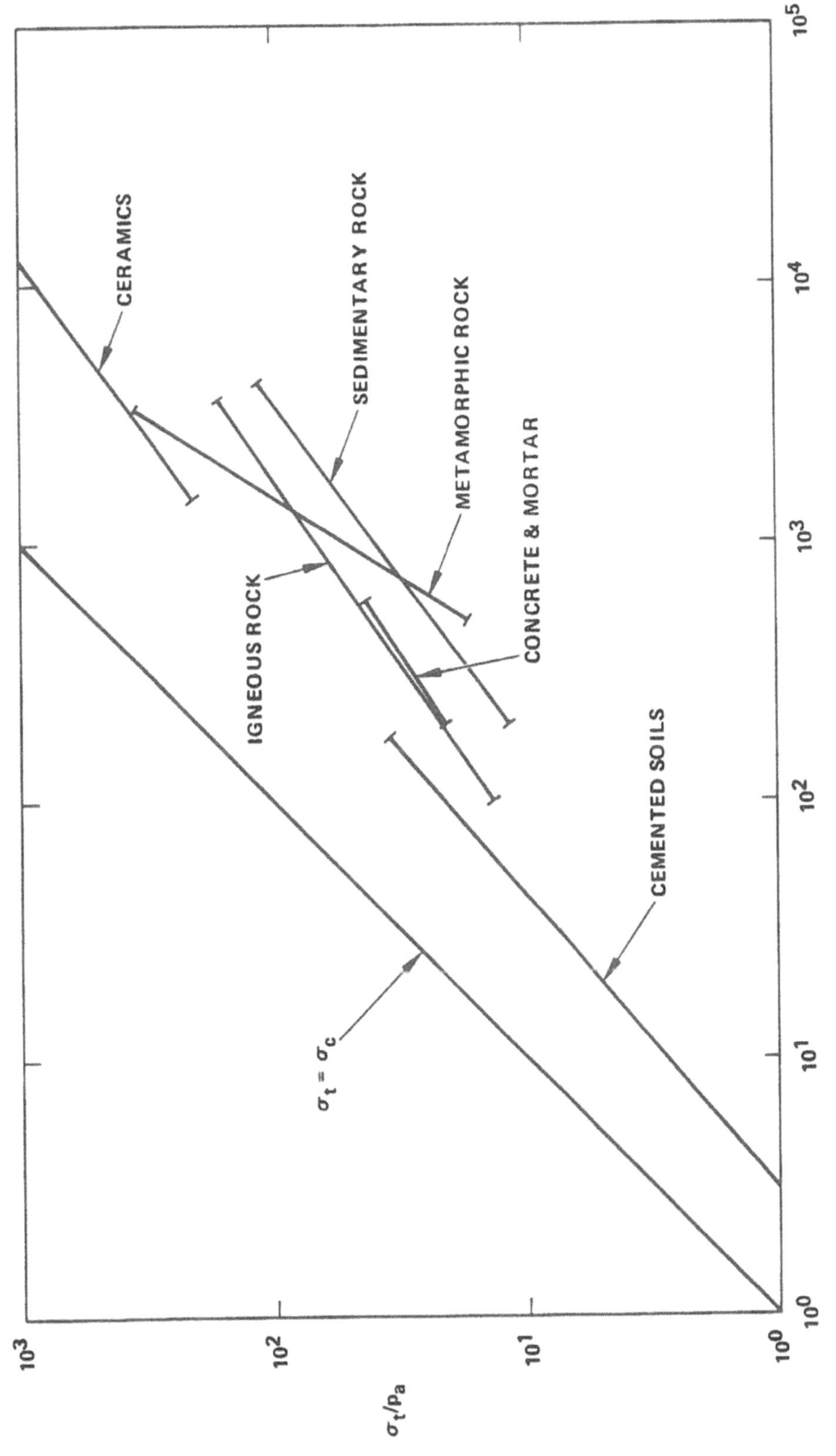

Figure 20.5 Relations between uniaxial tensile strengths and uniaxial compressive strengths for various types of frictional materials

σ_t) for cemented soils is about $\frac{1}{5}$ to $\frac{1}{3}$ of σ_c, whereas data compiled by Hannant [8] show that σ_t for concrete varies between 5 per cent and 13 per cent of σ_c. However, the values of σ_t and σ_c may be related through a power function of the following type [7, 9]:

$$\sigma_t = T \cdot p_a \cdot \left(\frac{\sigma_c}{p_a}\right)^t \tag{20.5}$$

where T and t are dimensionless numbers, and p_a is atmospheric pressure in the same units as those of σ_t and σ_c.

Values of T and t have been determined for several frictional materials and listed in Table 20.1. These values represent the best fit between experimental data and the simple expression in equation (20.5). Scatter in the test data was present for all materials, and the values of T and t given in Table 20.1 may not suitable for individual cases. The data base for ceramics was taken from only a few sources, whereas more extensive data bases were used for the other materials in Table 20.1.

Figure 20.5 shows a comparison of the relations between uniaxial tensile and uniaxial compressive strengths for the materials listed on Table 20.1. The straight lines shown on the log–log diagram span over the ranges of uniaxial compressive strengths indicated by available data. Both relatively weak and very strong frictional materials are represented in Figure 20.5. Note that the lines tend to cluster in an oblong area which slopes away from the line representing equal uniaxial tensile and uniaxial compressive strengths. Thus, the weak materials have relatively higher uniaxial tensile strengths than the strong materials.

20.3.2.2 *Regression analyses*

Because the failure criterion is expressed in terms of stress invariants, any type of test in which all stresses are measured may be used for determination of the three material parameters. However, it is advantageous to require only the simplest possible types of tests such as, for example, uniaxial compression and triaxial compression or biaxial tests for this determination, and then check whether these simple tests are sufficient for adequate characterization of the failure condition for the particular material under investigation. This may be done using various sets of data available in the literature which include both simple and more complex three-dimensional tests.

In order to obtain the overall best fitting parameters, regression analyses may be performed to determine the highest possible value of the coefficient of determination r^2. Figure 20.6 shows an example of the effect of varying the parameter 'a' on the values of r^2, η_1, and m for the tests on Mix A concrete performed by Mills and Zimmerman [10]. Only the results of the uniaxial compression and the triaxial compression tests in addition to the estimated value of the uniaxial tensile strength (from equation (20.5)) were used to

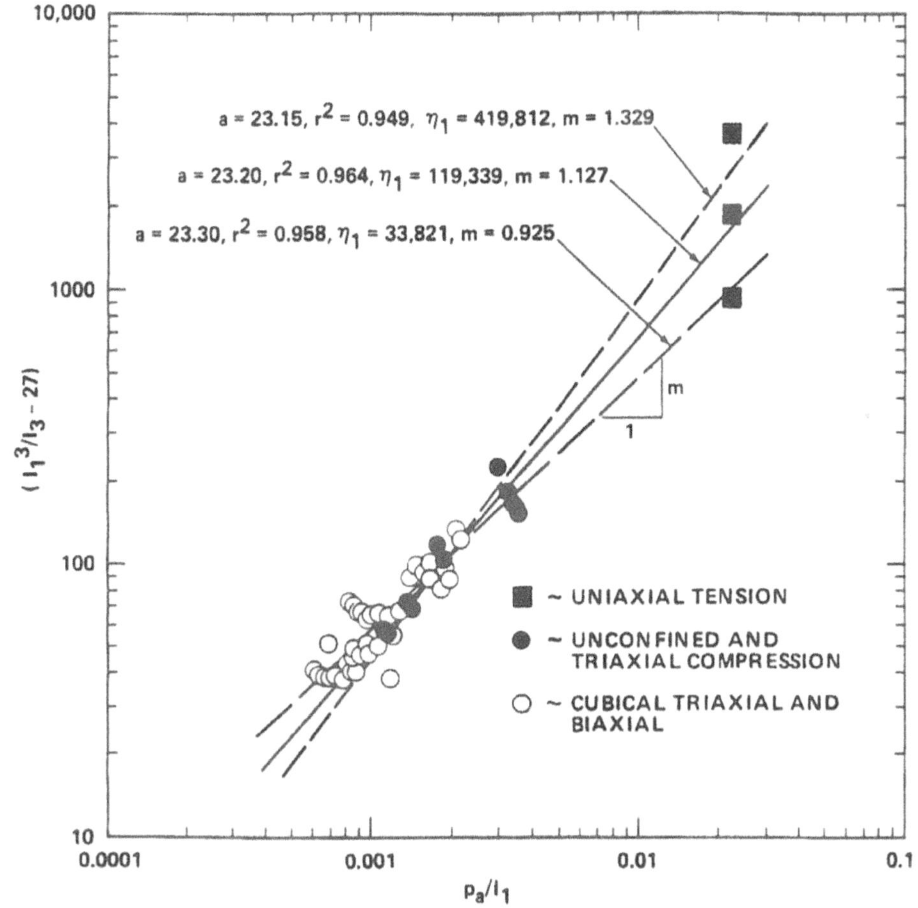

Figure 20.6 Determination of material parameters involved in failure criterion for Mix A concrete tested by Mills and Zimmerman [10]

determine the three material parameters. The uniaxial tensile strength was estimated to be $-23.1\,\text{kg/cm}^2$ ($-2266\,\text{kN/m}^2$) for Mix A concrete and the best fit value of 'a' = 23.2 resulted in $\eta_1 = 119{,}339$ and $m = 1.127$.

Except the three points corresponding to the uniaxial tensile strength on Figure 20.6, the points corresponding to the other tests do not translate enough on the diagram to show their separate locations. The points corresponding to the uniaxial tensile strength tend to influence the location of the best fit straight line. However, each of the three lines would describe the failure surface in the region of compressive stresses with reasonable accuracy. Thus, it is an advantage to incorporate the uniaxial tensile strength, even though it may be an estimate, in determination of the material parameters in order to stabilize the failure criterion in the region close to the origin and to describe the tensile strength for the material with reasonable accuracy.

The results of the cubical triaxial tests on Mix A concrete are shown on Figure 20.6 for comparison. It may be seen that some scatter of the data around the solid line does exist, but the material parameters selected on the basis of the simple tests appear to represent the data quite well.

20.3.3 Comparison of failure criterion and test data

In order to evaluate the capabilities of the three-dimensional failure criterion, comparisons have been made between experimental data and failure surfaces calculated from equations (20.1) and (20.4). Data were plotted on the biaxial plane, and data points were projected on the octahedral plane for all available data sets. Those data sets containing results of triaxial compression and extension tests were also shown on the triaxial plane. Examples of these comparisons are given below for concrete and rock.

Comparison of test data (points) and failure surface (solid line) is shown on the normalized biaxial plane in Figure 20.7(a) for the tests performed on concrete with $\sigma_c = 590\,\text{kg/cm}^2$ ($57\,880\,\text{kN/m}^2$) by Kupfer *et al.* [11]. All data, except those corresponding to tension–tension, were used to determine the material parameters. The failure critertion is seen to represent the test data with reasonable accuracy. In order to study the failure surface relative to the data in the tension–tension area, the data are shown on the enlarged diagram in Figure 20.7(b). It may be seen that the failure surface is smoothly rounded at the corner and that it corresponds very well with the data in this region.

A major investigation was performed by Mills and Zimmerman [10]. The results of their tests on Mix A concrete, which contained tests in triaxial, biaxial,

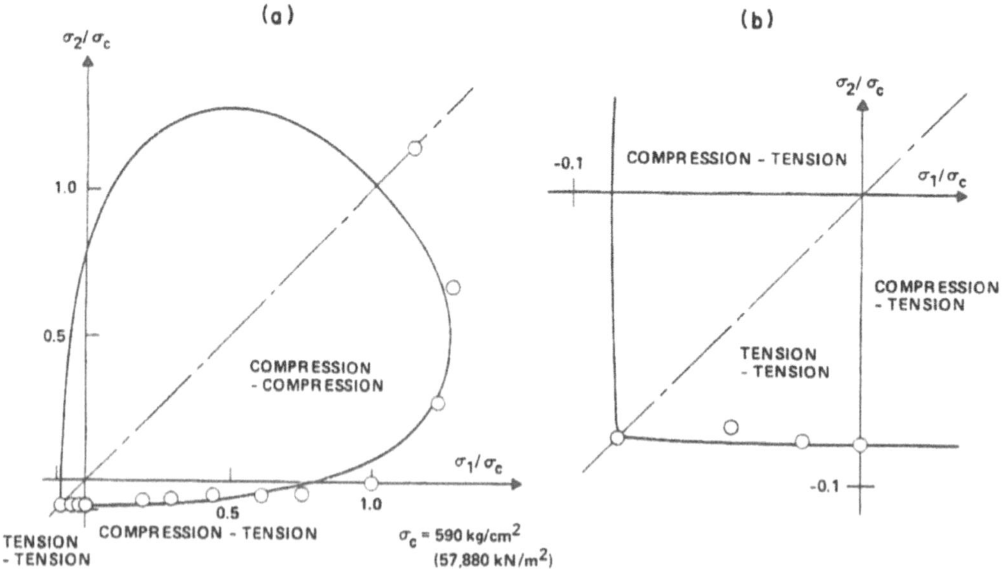

Figure 20.7 Comparison of failure criterion with results of biaxial tests performed by Kupfer, Hilsdorf, and Rusch [11] in (a) biaxial plane, and (b) enlarged tension–tension region

and octahedral planes, provided a good, coherent set of data. This set of data was used for illustration of material parameter determination in Figure 20.6. Comparisons between failure criterion and test data for Mix A concrete are shown in Figure 20.8. The material parameters for this concrete were determined on the basis of data from uniaxial compression and triaxial compression tests, and an estimated value of the uniaxial tensile strength. Thus, the good agreement shown in Figure 20.8(a) between the results of the triaxial compression tests and the failure critertion could be expected. However, the strengths obtained in triaxial extension are also well represented by the failure criterion. The data obtained in the compression–compression region of the biaxial plane are shown in Figure 20.8 (b). Although there is some scatter in the test results, the failure criterion is seen to represent the data quite well. Note that the pointed corner in the tension–tension area is actually smoothly rounded as shown in Figure 20.7(b).

Figure 20.8(a) also shows that the failure surface in extension cuts across the $\sigma_1 = 0$ plane at a very shallow angle. Therefore, any small deviation between test data and failure surface at this intersection in the triaxial plane will appear as a large deviation in the biaxial plane. Comparison of the data points for Mix A concrete indicated by arrows in Figures 20.8(a) and 20.8(b) shows that these appear to deviate somewhat from the failure surface in the biaxial plane (Figure 20.8(b)), whereas the same points in the triaxial plane are very close to the proposed failure surface. Any little amount of restraint in the testing apparatus would result in too large strength in biaxial extension, and this would show up very clearly in the biaxial plane. However, an evaluation in the triaxial plane would likely show that the test data are not that far from the actual failure surface. The natural scatter in test data could easily account for deviations of the magnitude indicated in Figure 20.8(b).

The data from cubical triaxial tests on Mix A concrete obtained by Mills and Zimmerman [10] are projected on the octahedral plane corresponding to $I_1 = 150\,\text{kg/cm}^2$ ($14\,715\,\text{kN/m}^2$) in Figure 20.8(c). Values of the minor principal stress, σ_3, of 0, $29.5\,\text{kg/cm}^2$ ($2894\,\text{kN/m}^2$), $59.11\,\text{kg/cm}^2$ ($5798\,\text{kN/m}^2$), and $88.6\,\text{kg/cm}^2$ ($8692\,\text{kN/m}^2$) were used in these tests. The points in Figure 20.8(c) corresponding to these values of σ_3 are shown separately on the octahedral plane for comparison with the failure surface. The projected data points were transferred to the common octahedral plane along the curved meridians using a technique involving the diagram in Figure 20.6. Note again that only data from uniaxial compression, triaxial compression, and uniaxial tension were used for determination of material parameters. The data from these tests are at the top of the diagrams in Figure 20.8(c). The experimental points on the octahedral plane describe a failure surface which is triangular with monotonically curved surface and smoothly rounded edges, as does the failure criterion.

A study of the three-dimensional strength of sandstone performed by Akai and Mori [12] involved uniaxial compression, triaxial compression, biaxial

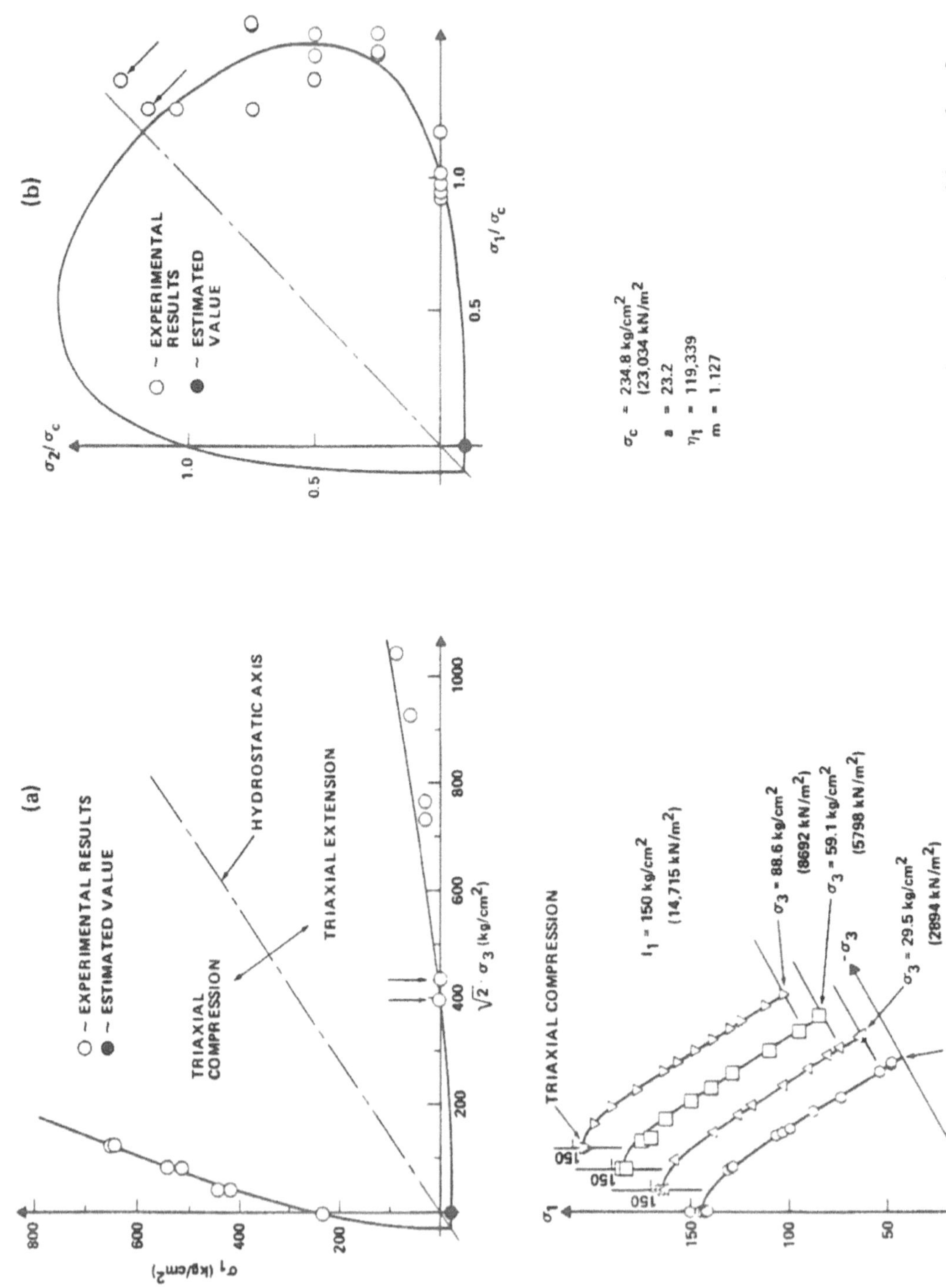

Figure 20.8 Comparison of failure criterion with results of tests on Mix A concrete performed by Mills and Zimmerman [10] in (a) triaxial plane, (b) biaxial plane, and (c) octahedral plane

compression, and cubical triaxial tests. Comparisons between failure criterion and test data for the sandstone are shown in Figure 20.9. The material parameters listed on Figure 20.9 for this rock were determined from the results of the uniaxial compression, triaxial compression, and biaxial compression tests. The value of the uniaxial tensile strength was not measured, but an estimate based on equation (20.5) with values of T and t for sedimentary rock from Table 20.1 gave $\sigma_t = -44.6$ kg/cm^2. This value was not used for material parameter determination in this case. The value of $a \cdot p_a = 37.4$ kg/cm^2 resulted in a better overall fit when the estimated value of σ_t was not employed for determination

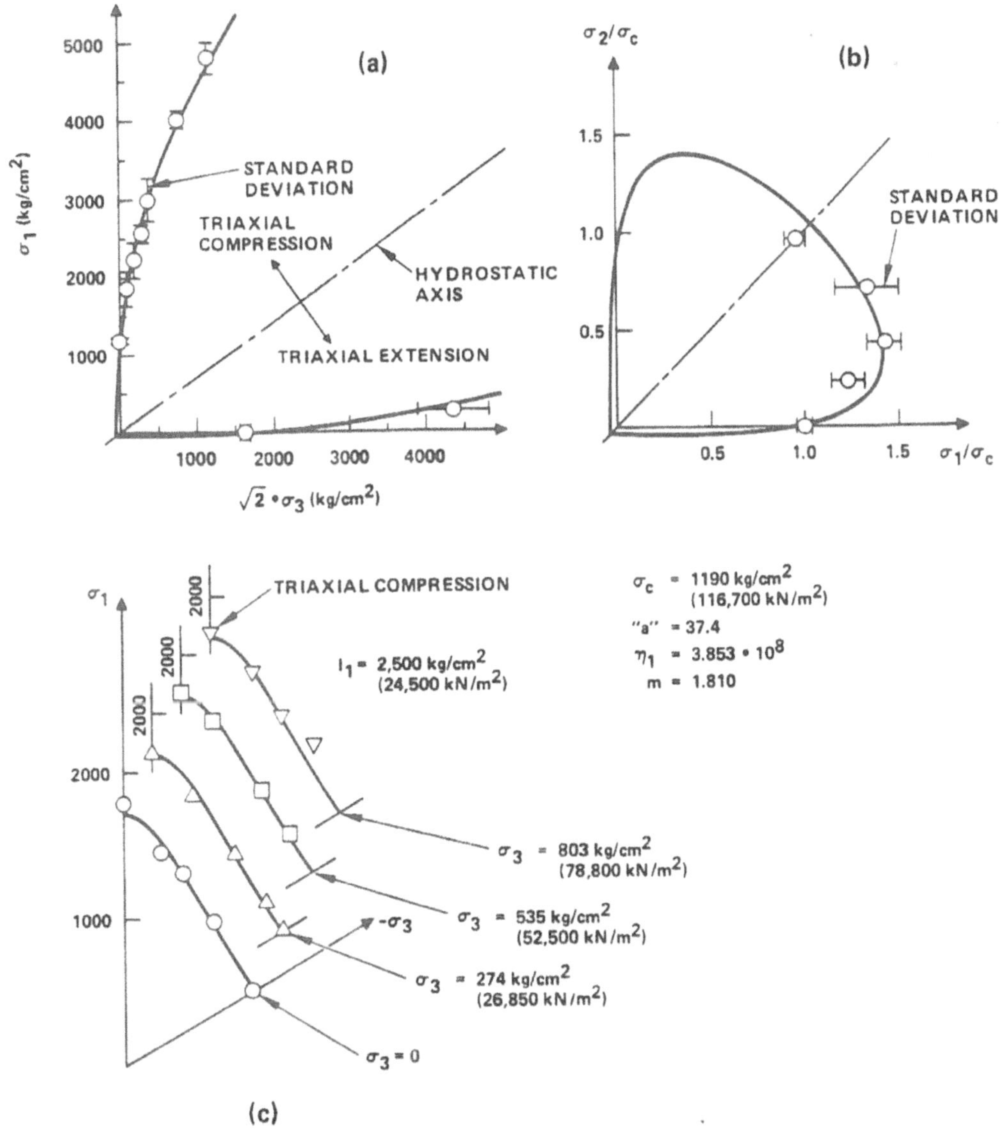

Figure 20.9 Comparison of failure criterion with results of tests on sandstone performed by Akai and Mori [12] in (a) triaxial plane, (b) biaxial plane, and (c) octahedral plane

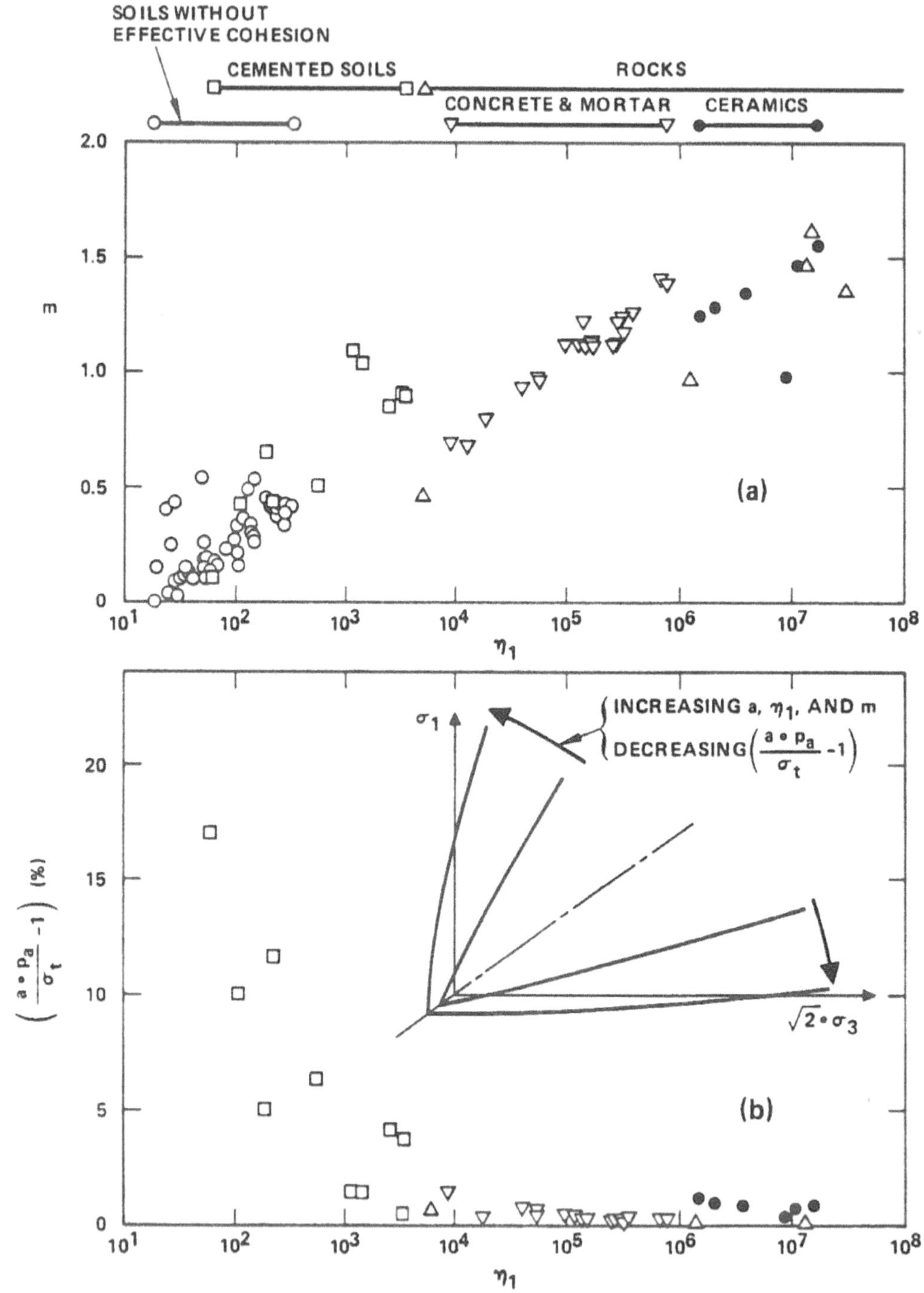

Figure 20.10 (a) Variation and ranges of the parameters η_1 and m for various frictional materials. (b) Magnitudes of the parameter 'a' relative to the uniaxial tensile strength σ_t plotted against η_1 for frictional materials with effective cohesion

of parameters. If σ_t were included, $a \cdot p_a$ would become 44.7 kg/cm^2, but these two values of $a \cdot p_a$ are easily within the scatter of the data employed to obtain T and t.

The comparison between failure criterion and test data for sandstone shown in Figure 20.9 indicates similar results as discussed in connection with Figure 20.8 for Mix A concrete. The strength of the sandstone is about five times higher than that of Mix A concrete as measured by the uniaxial compressive strengths. The values of the parameters a, m, and especially η_1 are therefore higher for the sandstone than for the concrete. The shape of the failure surface in the biaxial plane is somewhat different for the two materials, and this is correctly modelled by the failure criterion given by equations (20.1) and (20.4). The standard deviations of the major principal stresses, σ_1, as listed by Akai and Mori [12], are shown on Figures 20.9(a) and (b). The overall fit between failure surface and experimental data in Figure 20.9 is considered to be quite good and generally within the scatter of the test results.

Because emphasis was placed on obtaining the best overall fit between data and failure criterion, equal weights were placed on all test results used for parameter determination. It may therefore occur that the proposed failure criterion does not fit perfectly with the results of any one particular test. For example, the failure surface in Figure 20.7(a) does not go through the point corresponding to the uniaxial compressive strength. If it is desirable to obtain good correlation between the failure criterion and the uniaxial compressive strength, it may be necessary to apply heavier weights to this strength than to the results from other tests used for parameter determination.

20.4 PARAMETER VALUES FOR VARIOUS FRICTIONAL MATERIALS

Because the characteristic features of the three-dimensional failure surfaces have been observed for many different types of frictional materials, such as those described above, it is reasonable to believe that they exist for other frictional materials which have not yet been tested under general three-dimensional stress conditions. Therefore, the values of the three material parameters have also been determined for other frictional materials for which results of simple tests are available in the literature.

Figure 20.10(a) shows a diagram in which values of m are plotted against values of η_1 for several types of frictional materials with and without effective cohesion. The ranges of η_1 which have been obtained for these different materials are indicated by the bars in the upper part of Figure 20.10(a). It is likely that these ranges will be expanded when additional data sets for each type of material are analysed. Thus, the transition from cemented soils to rock is most likely smooth with no clear distinction between the two materials. Only a few sources of data for ceramics have been consulted for this study, and the range of the parameters for ceramics is probably larger than indicated in Figure 20.10. However, it is clear that the increased frictional strength indicated by η_1 is accompanied by an increased curvature of the failure surface as expressed by the value of m.

The parameter 'a' used for frictional materials with effective cohesion must be higher than the uniaxial tensile strength, σ_t. Values of σ_t are related to the uniaxial compressive strength, σ_c, in Figure 20.5. Thus, if uniaxial tension tests are not part of the testing programme, an approximate value of σ_t may be obtained from Figure 20.5 or from equation (20.5). Because many engineering materials are exposed to both compressive and tensile stresses in the same structure, it was proposed to include a reasonable value of the uniaxial tensile strength in the parameter determination in order to obtain an overall representative failure criterion for the material under investigation. The values of the parameter 'a' obtained in this study have been related to σ_t through the term $(a \cdot p_a - \sigma_t)/\sigma_t$ (in per cent) which expresses the percentage by which 'a' is greater than σ_t. This percentage is plotted versus η_1 in Figure 20.10(b). The high values (up to 17 per cent) of this percentage were obtained for cemented soils for which the parameter η_1 is relatively small. Values below 2 per cent were obtained for all other frictional materials with effective cohesion. In fact, a gradual decrease in percentage is obtained with increasing value of η_1.

The insert in Figure, 20.10(b) indicates the general trend for the three parameters 'a', η_1, and m. All three parameters have been found to increase with increasing strength of the frictional material.

Some materials, such as compacted, partly saturated soils, may exhibit an apparent cohesion. These materials have not been included in the present study. However, the failure conditions for such materials may be modelled by the failure criterion in equations (20.1) and (20.4) using total stressed for parameter determination and subsequent application in analysis procedures.

20.5 CONCLUSION

Failure surfaces for frictional materials with and without effective cohesion have several characteristics in common, and these are all captured by a general three-dimensional failure criterion formulated in terms of the first and the third stress invariants of the stress tensor. This failure criterion involves only three independent material parameters. Although these parameters interact with one another, each parameter corresponds to one of three failure characteristics of the frictional materials. The material parameters may be determined from simple tests such as uniaxial compression and triaxial compression or biaxial tests. For the purpose of including reasonable values of tensile strengths in the failure criterion (for frictional materials with effective cohesion), it may be necessary to include the uniaxial tensile strength in the parameter determination. A simple expression for evaluation of the uniaxial tensile strength on the basis of the uniaxial compressive strength is given. Representative sets of data for soils, concrete, and rock, for which results of three-dimensional tests were available, have been analysed, and comparisons between the failure criterion and the experimental data are made in triaxial, biaxial, and octahedral planes. Typical values of the three material parameters for various types of frictional materials have been determined, and their magnitudes are compared and discussed in light of the materials to which they belong. The ability of the general three-dimensional failure criterion to capture the characteristics of failure in frictional materials appears to be quite good with accuracies generally within the natural scatter of the test data.

ACKNOWLEDGEMENTS

Messrs E. Geiger and J. Tsai performed the tests on Edger Plastic Kaolinite, and Mr M. Kim compiled the data relating to failure of rock. Their assistance is gratefully acknowledged. Financial support for the work relating to soils has been provided under several grants from the National Science Foundation.

REFERENCES

1. J. K. Mitchell, *Fundamentals of Soils Behavior*, John Wiley & Sons, Inc., New York, NY, 1976.
2. A. W. Bishop, 'The strength of soils as engineering materials', 6th Rankine Lecture, *Geotechnique*, **16**, 2, 91–130 (1966).
3. P. V. Lade, 'Elasto-plastic stress–strain theory for cohesionless soil with curved yield surfaces', *International Journal of Solids and Structures*, **13**, 1019–1035, Pergamon Press, Inc., New York, NY, 1977.
4. P. V. Lade, 'Prediction of undrained behavior of sand', *Journal of the Geotechnical Engineering Division, ASCE*, **104**, No. GT6, Proc. Paper 13834, 721–735 (1978).
5. P. V. Lade and H. M. Musante, 'Three-dimensional behavior of remolded clay', *Journal of the Geotechnical Engineering Division, ASCE*, **104**, No. GT2, Proc. Paper 13551, 193–209 (1978).
6. P. V. Lade, 'Three-parameter failure criterion for concrete', *Journal of the Engineering Mechanics Division, ASCE*, **108**, No. EM5, Proc. Paper 17383, October 1982.
7. J. K. Mitchell, 'The properties of cement-stabilized soils', *Proc., Workshop on Materials and Methods for Low Cost Road*, pp. 365–404, Rail and Reclamation Works, Leura, Australia, September 1976.
8. D. J. Hannant, 'Nomograms for the failure of plain concrete subjected to short-term multiaxial stresses', *The Structural Engineer*, **52**, 5, 151–165 (1974).
9. J. Wastiels, 'Behaviour of concrete under multiaxial stresses—a review', *Cement and Concrete Research*, Vol. 9, pp. 35–44, Pergamon Press, 1979.
10. L. L. Mills and R. M. Zimmerman, 'Compressive strength of plain concrete under multiaxial loading conditions', *American Concrete Institute Journal*, **67**, 802–807 (1970).
11. H. Kupfer, H. K. Hilsdorf, and H. Rusch, 'Behavior of concrete under biaxial stresses', *American Concrete Institute Journal*, **66**, 656–666 (1969).
12. K. Akai and H. Mori, 'Ein Versuch über Bruchmechanismus von Sandstein under mehrachsigem Spannungszustand', *Proc. of the Second Congress of the International Society of Rock Mechanics*, Vol. II, Paper No. 3–30, Belgrade, Yugoslavia, 1970.

Estimating Parameters from a Single Test for the Three-Dimensional Failure Criterion for Frictional Materials

Poul V. Lade, M.ASCE[1]

Abstract: The three-dimensional failure criterion for soils was expanded to include concrete and used for rocks. The failure criterion is expressed in terms of the first and the third invariants of the stress tensor and it requires two parameters for characterization of the strength of soils and three parameters for concrete and rocks. For soils, one parameter is used to describe the opening angle of the failure surface, i.e., similar to the friction angle, and the second parameter is employed to describe the curvature of the failure surface in meridian planes that contain the hydrostatic axis. The third parameter, used for concrete and rocks, is slightly higher than the uniaxial tensile strength and is used to capture the cohesion and tensile strength. Typically, the results of three triaxial compression tests and the tensile strength from a Brazilian test or from an estimate are required for the determination of these three parameters. Careful inspection of parameters, determined from numerous high-quality tests on soils, concrete, and rocks presented in the literature, revealed that two of the parameters relate to each other. Therefore, it is possible to obtain the two parameters from the results of a single triaxial compression test. The relation between the two parameters is demonstrated, and it is shown how the results from a single test are used to determine the parameters to describe the three-dimensional failure surface for soils, concrete, and rocks. **DOI: 10.1061/(ASCE)GT.1943-5606.0001137.** © *2014 American Society of Civil Engineers.*

Author keywords: Concrete; Rock; Sand; Shear strength; Three-dimensional; Triaxial test.

Introduction

The three-dimensional (3D) failure criterion proposed for soils by Lade (1977) was enhanced to include concrete (Lade 1982) and also employed for rocks (Kim and Lade 1984). The failure criterion is shaped as a pointed bullet with smoothly rounded triangular cross sections, as shown in Fig. 1. It has three independent characteristics: (1) the opening angle near the origin of the principal stress space, which is often described by the friction angle; (2) the curvature in planes containing the hydrostatic axis, i.e., curved meridians; and (3) the tensile strength. Each of these characteristics may be different for different materials. A failure criterion that can capture these features, therefore, requires at least two material parameters for soils and three parameters for concrete and rocks.

The material parameters have been determined via a large number of high-quality tests on soils, concrete, and rocks presented in the literature. For concrete and rocks, one of these parameters is slightly greater than the tensile strength, which may be estimated on the basis of the uniaxial compressive strength. Close inspection reveals that the other two parameters are related to each other for each of the materials (i.e., sand, clay, concrete, and different types of rocks). Therefore, it is possible to obtain these latter two parameters from the results of a single triaxial compression test. The relation between the two parameters is demonstrated for soils, concrete, and rocks, and it is shown how the results from a single test are used to determine the parameters to describe the 3D failure surface for any frictional material.

[1]Professor, Dept. of Civil Engineering, Catholic Univ. of America, Washington, DC 20064. E-mail: Lade@cua.edu

Note. This manuscript was submitted on May 17, 2013; approved on March 27, 2014; published online on April 21, 2014. Discussion period open until September 21, 2014; separate discussions must be submitted for individual papers. This paper is part of the *Journal of Geotechnical and Geoenvironmental Engineering*, © ASCE, ISSN 1090-0241/04014038(5)/ $25.00.

3D Failure Criterion

The 3D failure criterion was developed first for soils with curved failure envelopes (Lade 1977) and later expanded to include concrete (Lade 1982) and rock (Kim and Lade 1984). The criterion for soils is expressed in terms of the first and the third stress invariants, I_1 and I_3, of the stress tensor as follows:

$$\left(\frac{I_1^3}{I_3} - 27\right)\left(\frac{I_1}{p_a}\right)^m = \eta_1 \quad (1)$$

where p_a = atmospheric pressure expressed in the same units as the stresses. The value of I_1^3/I_3 is 27 at the hydrostatic axis where $\sigma_1 = \sigma_2 = \sigma_3$.

The parameter η_1 is used to describe the opening angle of the failure surface, i.e., similar to the friction angle, and the other parameter, m, is employed to describe the curvature of the failure surface in the meridian planes. For η_1 = constant and $m = 0$, the failure surface is straight and the shape of the cross sections does not change with the value of I_1. For $m > 0$, the cross-sectional shape of the failure surface changes from triangular to become more circular with an increasing value of I_1.

To include the cohesion and the tension that can be sustained by concrete and rock, a translation of the principal stress space along the hydrostatic axis was performed, as illustrated in Fig. 2. Thus, a constant stress $a \cdot p_a$ was added to the normal stresses before substitution into Eq. (1)

$$\overline{\sigma}_x = \sigma_x + a \cdot p_a \quad (2a)$$

$$\overline{\sigma}_y = \sigma_y + a \cdot p_a \quad (2b)$$

$$\overline{\sigma}_z = \sigma_z + a \cdot p_a \quad (2c)$$

where a is a dimensionless parameter and the value of $a \cdot p_a$ reflects the effect of the tensile strength of the rock.

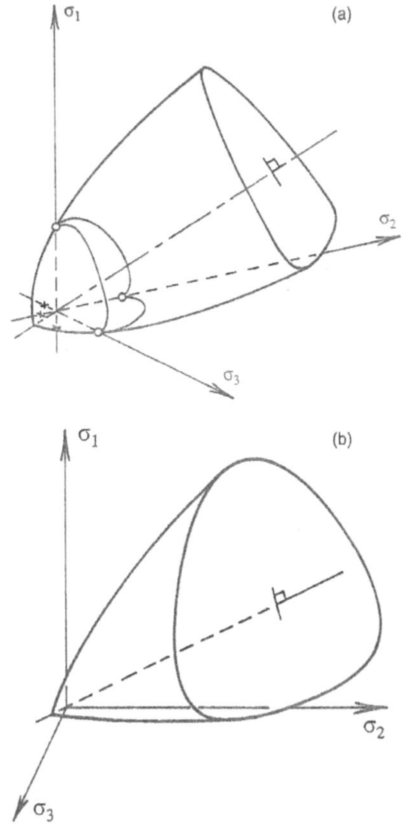

Fig. 1. 3D failure surface produced from experimental evidence and viewed from (a) the side, to show curved envelope in triaxial plane and biaxial planes; (b) the end, to show smoothly rounded triangular shape in octahedral plane

This criterion has been shown to capture the isotropic 3D failure conditions with good accuracy for soils (Lade 2006), for concrete (Lade 1982, 1984), and for rocks (Kim and Lade 1984), and it has been discussed in detail and compared with other failure criteria for frictional materials (Lade 1984, 1993, 2006).

Tensile Strength of Concrete and Rocks

It is an advantage to incorporate the uniaxial tensile strength in determination of material parameters for rock. This helps stabilize the failure criterion in the region close to the origin, and it improves the description of the failure conditions in the range of tensile stresses, where fracture occurs. If tensile tests are not part of a regular testing program, an estimated value of the uniaxial tensile strength may be obtained from an approximation formula, as given subsequently.

Kim and Lade (1984) expressed the uniaxial tensile strength using a power function of the following type:

$$\sigma_t = T \cdot p_a \cdot \left(\frac{\sigma_c}{p_a}\right)^t \tag{3}$$

where T and t = dimensionless numbers. A careful review of available test data indicates that a single relation between σ_t and σ_c for all rock types may be too crude to be useful. Kim and Lade (1984) and Lade (1993) separated the rock types according to the three great rock groups (i.e., igneous, metamorphic, and sedimentary rocks). Much experimental data were evaluated and different values of T and t were determined, as listed in Table 1.

Parameter Determination

To determine the values of the three parameters, a, η_1, and m, for a given set of experimental data, the value of a was estimated first and $a \cdot p_a$ was added to the normal stresses, as indicated in Eq. (2), before substitution into Eq. (1). Typically, the results of three triaxial compression tests are required for determination of the two parameters, η_1 and m. The expression for the failure criterion in Eq. (1) was rearranged and logs taken on both sides of the equation

$$\log\left(\frac{I_1^3}{I_3} - 27\right) = \log \eta_1 + m \cdot \log\left(\frac{p_a}{I_1}\right) \tag{4}$$

By plotting $(I_1^3/I_3 - 27)$ versus (p_a/I_1) on log-log scales, as shown in Fig. 3 for sand number D, the value of η_1 is determined as the intercept between the best-fitting straight line and the vertical line corresponding to $(p_a/I_1) = 1$. The geometric slope of the straight line is the exponent m.

To facilitate the estimate of a for concrete and rocks, advantage may be taken of the fact that $a \cdot p_a$ must be slightly greater than the uniaxial tensile strength of the rock, as indicated in Fig. 2. If tensile tests are not part of the regular testing program, an estimated value of σ_t may be obtained from Eq. (3). Fig. 4 shows an example of the effect of varying the parameter a on the values of η_1 and m as well as the coefficient of determination, r^2, for tests on Mizuho trachyte performed by Mogi (1971). The parameters corresponding to the highest value of r^2 are those preferred to model the failure conditions.

Parameter Values for Frictional Materials

The parameter values have been determined from triaxial compression tests for a large number of frictional materials and presented by Lade (1984, 1993). Fig. 5(a) shows a plot of m versus $\log(\eta_1)$ for sands. The data contain 39 sets of parameters, which form a straight line that may be expressed as

$$m = A \cdot \log\left(\frac{\eta_1}{B}\right) \tag{5}$$

For sands, the best-fitting straight line is described by values of $A = 0.33$ and $B = 16.8$. Thus, m and η_1 are related by Eq. (5).

Fig. 5(b) shows a diagram in which m is plotted against $\log(\eta_1)$ for igneous rocks. Although there is some scatter, the data consisting of 18 sets of parameters form a straight line. For igneous rocks, the best-fitting straight line is described by values of $A = 0.249$ and $B = 55.2$. Thus, the two parameters are related by Eq. (5).

Table 2 presents the values of A and B determined for various frictional materials, including sands and clays, concrete, each of the three great rock groups, and coal, because its behavior is different from the rocks in the three great rock groups. The values of a, η_1, and m determined for the rocks were presented by Lade (1993). The values for coal are quite different and the scatter is much lower for coal than for the other rock types. Combining the three great rocks groups and omitting the coal produces average values for A and B, which also are listed in Table 2.

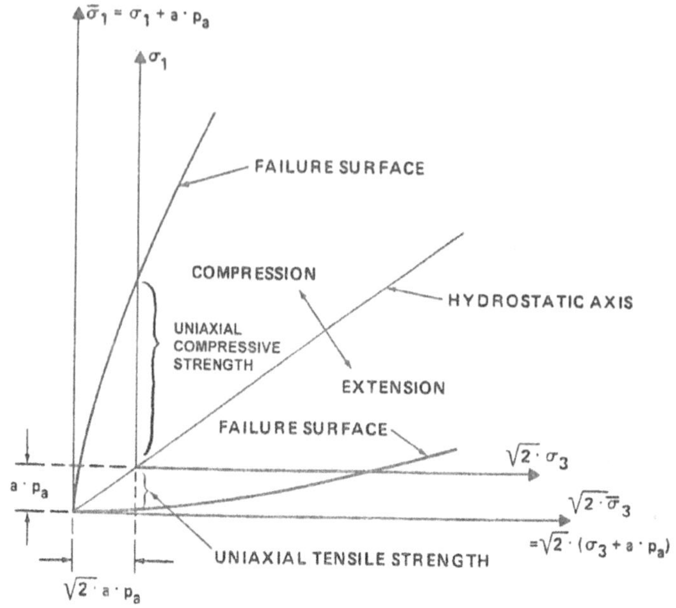

Fig. 2. Translation of principal stress space along hydrostatic axis to include effect of tensile strength in failure criterion

Table 1. Values of T and t for the Relation between Uniaxial Tensile and Compressive Strengths for Cemented Soils, Concrete, and Rocks

Type of frictional material	Reference	T	t	n_{max}	r^2	σ_{cmax}/p_a	σ_{cmin}/p_a
Cemented soils	Mitchell (1976)	−0.37	0.88				
Concrete and mortar	Wastiels (1979)	−0.61	0.67				
Igneous rock	—	−0.435	0.740	130	0.5710	5,330	42
Metamorphic rock	—	−0.0518	1.017	72	0.6985	3,360	66
Sedimentary rock	—	−0.316	0.770	174	0.5964	4,100	10
Organic rocks (coal)	—	−0.0316	1.096	14	0.6554	450	66
All rocks	—	−0.219	0.825	376	0.6513	5,330	10

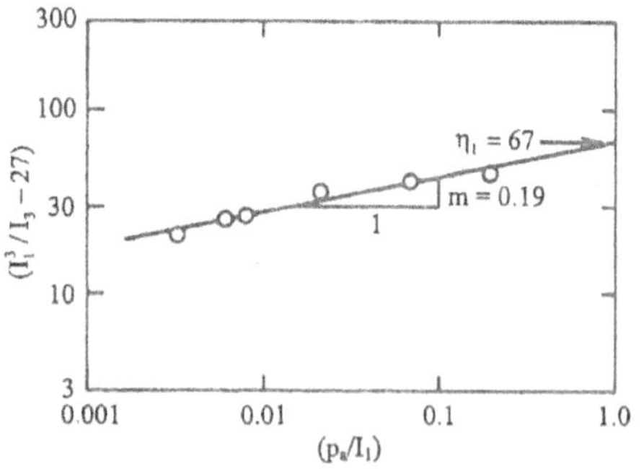

Fig. 3. Determination of material parameters involved in failure criterion for sand number D

Parameters from a Single Test

Because η_1 and m may be related to each other by the simple expression in Eq. (5), it is possible to obtain both parameter values from a single test, such as a triaxial compression test. It is clear that this test must be of high quality, because it is the basis for determining the 3D failure stresses of the frictional material in question.

Because $m > 0$ for a concave failure surface and $m = 0$ for a straight failure surface, the lowest value of B is $B = \eta_1$, according to Eq. (5). This corresponds to the lowest value of η_1 possible for a given material.

Substituting the expression for m in Eq. (5) into Eq. (4) and setting $(I_1^3/I_3 - 27) = B$ and solving for (p_a/I_1) produces

$$\log\left(\frac{p_a}{I_1}\right) = \frac{\log B - \log \eta_1}{A \cdot (\log \eta_1 - \log B)} = -\frac{1}{A} \tag{6}$$

from which

$$\left(\frac{p_a}{I_1}\right) = 10^{-1/A} \tag{7}$$

Fig. 6(a) shows that, on a log-log diagram of $(I_1^3/I_3 - 27)$ versus (p_a/I_1), this value of (p_a/I_1) indicates the point of interception between a horizontal line corresponding to $(I_1^3/I_3 - 27) = \eta_1 = B = 16.8$ for sands and all other lines describing failure surfaces in terms of η_1 and m for sands. Thus, a pattern of straight lines describing failure surfaces for all sands is established, as shown in the

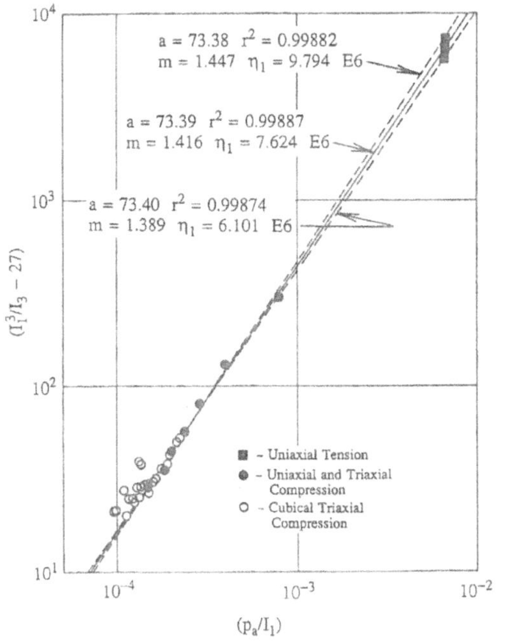

Fig. 4. Determination of material parameters involved in failure for Mizuho trachyte (data from Mogi 1971)

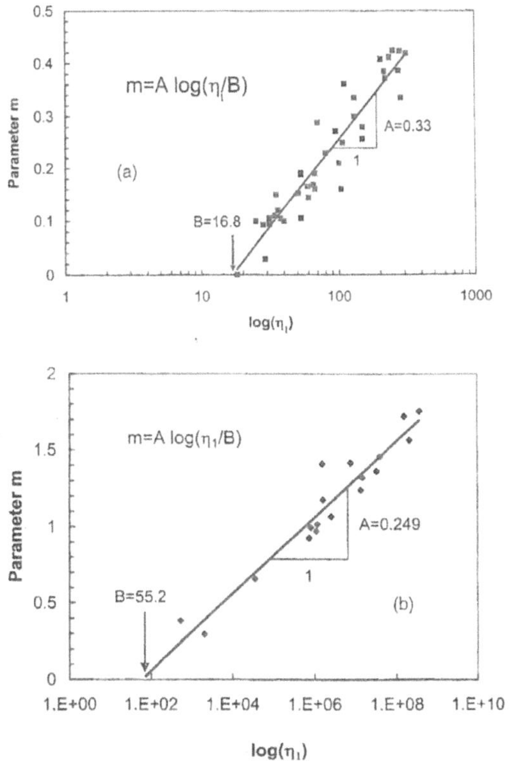

Fig. 5. Relation between m and $\log(\eta_1)$ for (a) sands; (b) igneous rocks

Table 2. Values of A and B for the Relation between η_1 and m for Various Frictional Materials

Type of frictional material	A	B	n_{max}	r^2
Sands	0.33	16.8	39	0.8914
Clays	0.49	3.7	12	0.4554
Concretes	0.354	108.4	22	0.9364
Igneous rocks	0.249	55.1	18	0.9265
Metamorphic rocks	0.257	22.7	12	0.8330
Sedimentary rocks	0.210	2.15	34	0.9011
Organic rocks (coal)	0.358	130.9	22	0.9765
All rocks (no coal)	0.229	9.62	64	0.8752

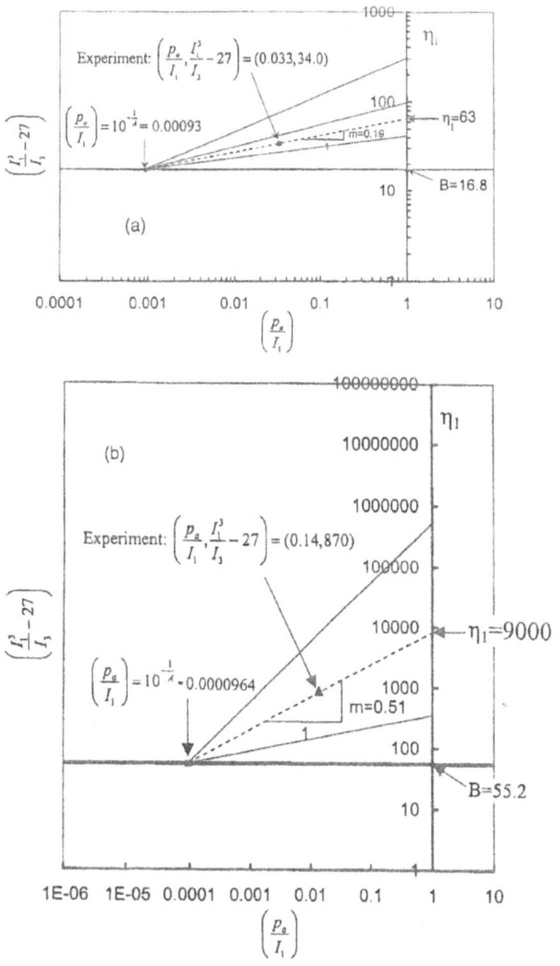

Fig. 6. Master diagram for determination of η_1 and m based on one experiment for (a) sands; (b) igneous rocks

master diagram in Fig. 6(a). This pattern is possible because the values of η_1 and m are uniquely related for sands, as indicated by Eq. (5) and shown for sands in Fig. 5(a). Similar master diagrams may be constructed for the other frictional materials for which the appropriate values of A and B are listed in Table 2. As an example, Fig. 6(b) shows the master diagram for igneous rock types.

An example of how to use the master diagram is indicated for sand in Fig. 6(a). Based on the values of the effective principal stresses at failure from a triaxial compression test on a given sand, the value of $(I_1^3/I_3 - 27) = 34.0$ is plotted versus $(p_a/I_1) = 0.033$ on the master diagram for sands. A straight line is then drawn through the corresponding experimental point and the common interception point, as indicated by the dashed line. This line crosses the η_1-axis at the value of $\eta_1 = 63$, and the geometric slope indicates the value of $m = 0.19$ for the given sand. A similar demonstration of parameter determination from a single triaxial compression test is indicated in Fig. 6(b) for igneous rocks.

Conclusions

The 3D failure criterion proposed for soils by Lade (1977), and employed for concrete (Lade 1982) and rocks (Kim and Lade 1984), models the tensile strength, the curvature in meridian planes, and the effect of the intermediate principal stress in octahedral planes with good accuracy for these frictional materials. This criterion involves three parameters: (1) a, which describes the cohesion and the tensile strength; (2) η_1, which describes the opening angle at the stress origin (similar to the friction angle); and (3) m, which describes the curvature of the failure surface. Based on numerous sets of experiments on frictional materials, it was observed that unique relations exist between η_1 and m for each material type. These relations may be described by straight lines on a semilog diagram, and they allow the determination of both parameters from a single triaxial compression test. Master diagrams can be established for each material type; determination of the values of η_1 and m from results of a single triaxial compression test was demonstrated.

References

Kim, M. K., and Lade, P. V. (1984). "Modelling rock strength in three dimensions." *Int. J. Rock Mech. Min. Sci. Geomech. Abstr.*, 21(1), 21–33.

Lade, P. V. (1977). "Elasto-plastic stress-strain theory for cohesionless soil with curved yield surfaces." *Int. J. Solids Struct.*, 13(11), 1019–1035.

Lade, P. V. (1982). "Three-parameter failure criterion for concrete." *J. Engrg. Mech. Div.*, 108(5), 850–863.

Lade, P. V. (1984). "Failure criterion for frictional materials." Chapter 20, *Mechanics of engineering materials*, C. S. Desai and R. H. Gallagher, eds., Wiley, New York, 385–402.

Lade, P. V. (1993). "Rock strength criteria: The theories and the evidence." Chapter 11, *Comprehensive rock engineering*, J. A. Hudson, ed., Vol. 1, Pergamon Press, Oxford, U.K., 255–284.

Lade, P. V. (2006). "Assessment of test data for selection of 3-D failure criterion for sand." *Int. J. Numer. Anal. Methods Geomech.*, 30(4), 307–333.

Mitchell, J. K. (1976). "The properties of cement-stabilized soils." *Proc., Workshop on Materials and Methods for Low Cost Road*, Rail and Reclamation Works, Leura, Australia, 365–404.

Mogi, K. (1971). "Fracture and flow of rocks under high triaxial compression." *J. Geophys. Res.*, 76(5), 1255–1269.

Wastiels, J. (1979). "Behaviour of concrete under multiaxial stresses—A review." *Cem. Concr. Res.*, 9(1), 35–44.

Chapter VII
Associated versus Non-Associated Plastic Flow in Soils

Preface

On account of no plastic volume changes of solid metals, Daniel C. Drucker (1951, 1956, 1959) derived the normality condition, i.e. the plastic flow direction is perpendicular to the yield surface in the principal stress space, and this behavior is referred to as associated plastic flow. This also guarantees stability of the material in physical space as well as in calculations all the way up to peak failure where the solid metals become unstable. This condition is referred to as Drucker's stability postulate. This works well for solid metals for which it was developed, but it does not correctly characterize the behavior of frictional materials, as shown in the following articles.

In the early days of development of constitutive models for frictional materials (1960s and 1970s), it was most often assumed, a priori, that the material exhibited associated flow. For example, this was clearly stated in connection with the development of the Cam Clay model (Schoefield and Wroth, 1968) and the modified Cam Clay model (Roscoe and Burland, 1968). In addition, associated plastic flow results in symmetric material matrices, and therefore allows for easier calculation procedures. While there were no experimental results available at the time, it was also assumed that the third stress invariant did not play a role in these Cambridge University models.

References

Drucker, D.C. (1951) "A More Fundamental Approach to Stress-Strain Relations," *Proceedings*, 1st U.S. National Congress on Applied Mechanics, pp. 487-491.

Drucker, D.C. (1956) "On Uniqueness in the Theory of Plasticity," *Quarterly Applied Mathematics*, 14: pp. 35-42.

Drucker, D.C. (1959) "A Definition of Stable Inelastic Materials," *Journal Applied Mathematics*, 26: pp. 101-106.

Roscoe, K.H. and Burland, J.B. (1968) "On the generalized behaviour of 'wet' clay," *Engineering Plasticity*, 48: pp. 535-609.

Schofield, A. and Wroth, P. (1968) *Critical State Soil Mechanics*, McGraw-Hill.

Nonassociated Flow and Stability of Granular Materials

By Poul V. Lade,[1] Richard B. Nelson,[2] Members, ASCE, and Y. Marvin Ito[3]

Abstract: Materials exhibiting nonassociated flow should, according to Drucker's stability postulate, become unstable when exposed to certain stress paths inside the failure surface. Results of a series of triaxial tests designed to expose the type of behavior displayed by granular materials are presented and discussed. The sand dilates during shear, it exhibits nonassociated flow, and when exposed to stress paths in the region of potential instability, none is observed. The reliability of the test results, the possible influence of viscous effects on the stability, and the effect of release of elastic energy are studied and discussed in detail. Thus, Drucker's stability postulate is not applicable to granular materials. The type of volume-change behavior displayed by the material is of great importance in the question of stability of granular materials.

Introduction

The occurrence of nonassociated plastic flow in frictional materials such as sand, clay, concrete, and rock has raised questions regarding the uniqueness and stability of such materials. The stability postulate formulated by Drucker (3–5) for time-independent materials provides a sufficient condition for stability and ensures uniqueness in dynamic as well as static problems. Drucker's postulate is satisfied provided that associated flow is employed in construction of constitutive models involving convex, plastic yield surfaces. Whereas this postulate provides a sufficient condition for stability, it has been suggested that it is not a necessary condition (13,15).

Considerations of stability have mainly been explored on a theoretical level. Experiments to study whether or not frictional materials are, in fact, unstable under certain loading conditions have not been performed. A series of tests is presented on granular materials designed to expose the type of behavior that is characteristic of frictional materials. The problem of instability and the factors that control the occurrence of unstable behavior are discussed in light of experimental results. It is shown that Drucker's stability postulate is not applicable to frictional materials in general and that the volume-change behavior (dilatancy) plays an important role in the stability of such materials.

Drucker's Stability Postulate

The condition for stability formulated by Drucker (3–5) involves two components expressed in terms of plastic work increments:

[1]Prof., Dept. of Civ. Engrg., Univ. of California, Los Angeles, CA 90024.
[2]Prof., Dept. of Civ. Engrg., Univ. of California, Los Angeles, CA 90024.
[3]Adjunct Prof., Dept. of Civ. Engrg., School of Engrg. and Applied Sci., Univ. of California, Los Angeles, CA 90024; Mgr., Engrg. Mech. Group, California Res. and Tech., Inc., Chatsworth, CA 91311.

Note.—Discussion open until February 1, 1988. To extend the closing date one month, a written request must be filed with the ASCE Manager of Journals. The manuscript for this paper was submitted for review and possible publication on August 4, 1986. This paper is part of the *Journal of Engineering Mechanics*, Vol. 113, No. 9, September, 1987. ©ASCE, ISSN 0733-9399/87/0009-1302/$01.00. Paper No. 21769.

$$dW_p - dW_p^* = (\sigma_{ij} - \sigma_{ij}^*)\,\dot{\epsilon}_{ij}^p \geq 0 \quad \text{(1a)}$$

$$d^2W = \dot{\sigma}_{ij}\dot{\epsilon}_{ij}^p \geq 0 \quad \text{(1b)}$$

where σ_{ij} = a stress state located on the yield surface; σ_{ij}^* = a stress state located inside the yield surface; and $\dot{\sigma}_{ij}$ = a stress increment initiated at σ_{ij} on the yield surface, causing the plastic strain increment, $\dot{\epsilon}_{ij}^p$. Eq. 1*a*, which is often referred to as stability in the large, requires the yield surface to be convex. Eq. 1*b* is referred to as stability in the small, and it is fulfilled, in general, if the plastic strain increment vector is perpendicular to the yield surface. This implies that the plastic potential surface coincides with the yield surface, so that the plastic strain increments produced upon yielding are perpendicular to the yield surface. Thus, Drucker's stability postulate requires constitutive laws to incorporate a so-called associated flow rule in order to guarantee that stable behavior is obtained.

It is Eq. 1*b* that will be used to define stable or unstable behavior. If this inequality is fulfilled for any $\dot{\sigma}_{ij}$, then the material will be unconditionally stable. Otherwise, according to Drucker's postulate the material is conditionally stable, with the second increment d^2W positive for some $\dot{\sigma}_{ij}$ but negative for others, or completely unstable with d^2W negative for all $\dot{\sigma}_{ij}$. Complete instability, with d^2W negative, is obtained for all materials after the failure surface has been reached, whether the material exhibits associated or nonassociated flow.

The instability of interest here occurs before the failure surface has been reached. All nonassociated flow models exhibit a region where d^2W is negative. However, it is by no means clear that the materials which are being modeled are physically unstable in this region. It is the purpose of this experimental investigation to determine the physical behavior of a real soil in this region of negative d^2W.

Occurrence of Nonassociated Flow

In relation to a previous study of stress-path–dependent behavior of cohesionless soil (10), the uniqueness of stress-strain behavior for materials exhibiting nonassociated flow was discussed. Several triaxial tests were performed on Monterey number 0 sand with stress paths involving primary loading with decreasing stress difference $(\sigma_1 - \sigma_3)$ and decreasing confining pressure, σ_3. The results of these tests were compared with results of conventional triaxial compression tests in which the confining pressure was held constant. If stresses and strain increments are plotted on the same diagram, as in Fig. 1(*b*), the direction of the plastic strain-increment vector is uniquely determined from the state of stress and is independent of the stress path leading to this state of stress. All strains are calculated as engineering strains, i.e., they refer to the original dimensions of the specimens. Since the prefailure strains of interest in this study are less than 6%, any effects of second-order terms or large strains are negligible. The failure surface indicated in Fig. 1(*b*) corresponds to peak failure conditions from drained tests in which maximum stress difference and maximum effective stress ratio occur simultaneously. The prefailure stress path with decreasing stresses shown in Fig. 1(*b*) is so steep as to form an obtuse angle (θ = 110°) with the di-

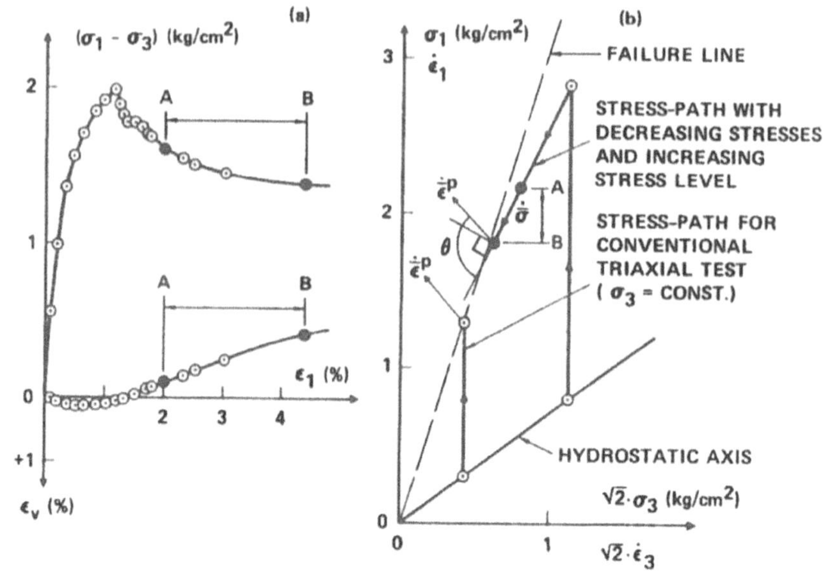

FIG. 1.—(*a*) Stress-Strain Behavior; (*b*) Stress Path with Decreasing Stresses Indicating Nonassociated Flow for Triaxial Test on Loose Monterey Number 0 Sand (e = 0.78; D_r = 27%) (1.0 kg/cm² = 98 kN/m²)

rection of the strain-increment vector. Nevertheless, the two plastic strain-increment vectors are parallel. This has also been demonstrated by Poorooshasb, et al. (16).

The fact that plastic yielding is occurring along the stress path with decreasing stresses is seen from the stress-strain curve in Fig. 1(*a*). Section AB on the stress path is labeled similarly on the stress-strain curve, which indicates large plastic strains between A and B. Since the yield surface is being pushed out, it must be steeper than the stress-path direction AB. If it were less steep, so as, e.g., to be perpendicular to the plastic strain-increment vector thereby indicating associated flow, then point B would be inside point A, corresponding to unloading from A to B. This would contradict the large plastic strains shown in Fig. 1(*a*). Consequently, the yield surface must be steeper than stress path AB and nonassociated flow is therefore clearly obtained.

Since the stress-increment vector and the strain-increment vector form an obtuse angle for the stress path considered, the scalar product of these two vectors (Eq. 1*b*) is negative. According to Drucker's stability postulate, the sand should exhibit unstable behavior along this stress path. However, the tests in Ref. 10 were all performed under strain control. This method of testing does not allow instabilities (i.e., gross collapse) of the material to be exposed. To force exposure of such instabilities, it is necessary to perform the tests under load or stress control.

It is important to recognize that the sand stress-strain curve in Fig. 1(*a*) corresponds to work hardening rather than softening. Although the stresses along stress path AB are decreasing, failure has not been reached, and softening of the material is therefore not occurring. The decreasing stresses along stress path AB are deliberately followed and not imposed

1304

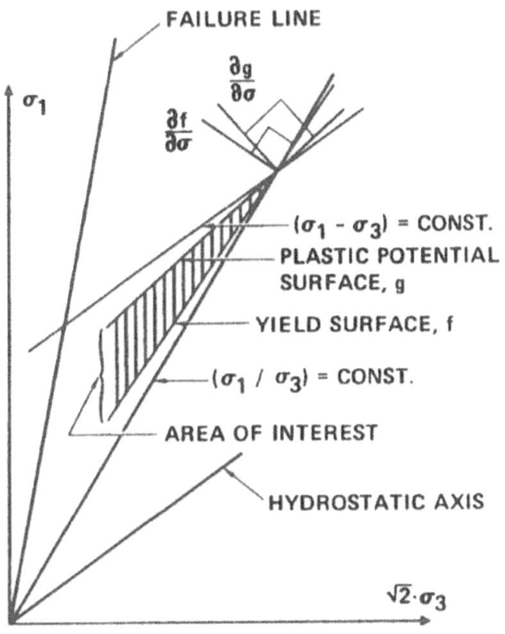

FIG. 2.—Wedge-Shaped Region of Stress Paths with Decreasing Stresses in which Granular Material with Nonassociated Flow May Be Unstable during Hardening Inside Failure Surface

by the material, whereas decreasing stresses are dictated by a softening material, because its strength has been exceeded and it is behaving in an unstable fashion.

Fig. 2 shows a schematic illustration of the region in which Eq. 1*b* is not fulfilled for a material with nonassociated flow. The region is shaped as a wedge between the current yield surface f and the plastic potential surface g corresponding to the current stress point. This wedge-shaped region is located within a larger region bounded by lines corresponding to (σ_1/σ_3) = const and $(\sigma_1 - \sigma_3)$ = const, as indicated in Fig. 2. All stresses, including the stress difference $(\sigma_1 - \sigma_3)$, are decreasing within the wedge between f and g, whereas the stress ratio (σ_1/σ_3) is increasing in this region. By performing triaxial tests with stress paths located in this region, it should be possible to obtain experimental evidence regarding the stability of materials with nonassociated flow.

Experimental Setup

In order to study experimentally the potential instabilities in soil behavior that might occur for the stress paths in the prefailure region described earlier, it is necessary to perform triaxial tests under stress control rather than strain control. This allows the triaxial specimen to exhibit any unstable behavior that may occur for the stress paths of interest.

Fig. 3 shows the experimental setup used to apply both horizontal and vertical stresses to a triaxial specimen under stress control. The minor principal stress, σ_3, was applied as a confining pressure in the triaxial

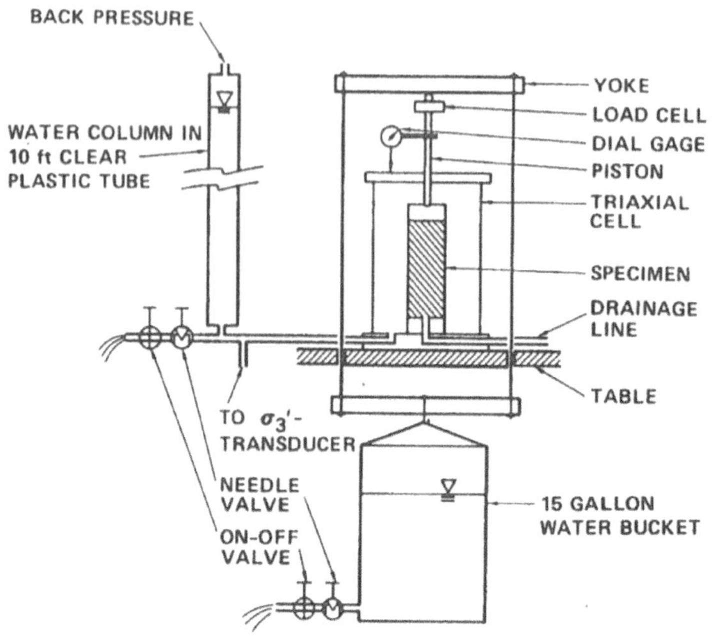

FIG. 3.—Experimental Setup for Stress Control of Triaxial Test

cell. This pressure was supplied by a column of water contained in a 10-ft tall clear platic tube. The water could be metered out through the needle valve at the bottom of the column, thus decreasing the minor principal stress on the specimen at a desired rate.

The vertical stress difference ($\sigma_1 - \sigma_3$) was applied to the specimen by the piston protruding from the top plate of the triaxial cell. The vertical load was supplied by a bucket of water hanging in a yoke below the table supporting the triaxial cell. By filling the 15 gal bucket with water at any desired rate, it was possible to increase the stress difference ($\sigma_1 - \sigma_3$) and to fail the specimen under load control with increasing vertical stress. However, all tests performed for the present study were conducted under load control with decreasing vertical stresses. This was accomplished by metering water out through a needle valve located near the bottom of the bucket. Thus, after having first loaded the specimen by filling water into the bucket, the water was let out through the needle valve at a desired rate. The actual vertical load applied to the specimen was measured with a load cell attached to the piston as seen in Fig. 3. The vertical stress rates employed during loading of the specimens averaged 0.025 kg/cm^2/min (2.5 kN/m^2/min) for increasing stresses and 0.001 kg/cm^2/min (0.1 kN/m^2/min) for decreasing stresses.

This simple experimental setup, in which the specimen was loaded under stress control, would allow any unstable behavior of the granular material to reveal itself by an inability to support the bucket of water. Thus, unstable soil behavior was expected to result in a dramatic drop of the water bucket to the laboratory floor, leaving no doubt about the nature of the soil behavior along certain stress paths of the types discussed earlier.

Sand Tested

To avoid experimental problems due to excessive effects of membrane penetration caused by changing confining pressures (11,14), the tests were performed on a fine sand. Membrane penetration into the pores of granular soils is small to negligible when the average diameter of the soil grains is smaller than 0.1–0.2 mm (6).

Fine silica sand composed of angular particles consisting mainly of quartz were used in the study. The characteristics of this sand are summarized as follows: (1) Mean diameter, 0.18 mm; (2) coefficient of uniformity, 2.0; (3) specific gravity of grains, 2.66; (4) maximum void ratio, 0.85; and (5) minimum void ratio, 0.55.

Tests were performed on dense specimens with void ratio of 0.58, corresponding to a relative density of 90%.

Specimen Preparation and Testing Procedures

All triaxial tests in the experimental program were performed on cylindrical specimens with diameter = 2.80 in. (7.1 cm). Some tests were performed on specimens with height = 2.80 in. (7.1 cm), corresponding to a height-to-diameter ratio of H/D = 1.0. Other tests were conducted on specimens with height = 7.4 in. (18.8 cm) corresponding to H/D = 2.65. The tall specimens were employed in most of the tests to allow any instabilities, e.g., development of shear planes or bands, to occur freely and uninterrupted by the end plates. In addition, lubricated end plates were used in all tests to avoid development of significant shear stresses at the cap and base.

The specimens were prepared by dry pluviation of sand and saturated using the CO_2-method (9). In addition, a nominal back pressure of 30 psi (207 kN/m^2) was applied to the specimens to ensure a high degree of saturation. B-values of 0.97 or better were measured in most tests.

The vertical load, the confining pressure, the vertical deformation, and the volume changes were measured during each test. Corrections were applied to the measured vertical load for uplift forces on the piston, piston friction, the load taken by the rubber membrane, and for the buoyed weight of the cap and the buoyed weight of the upper half of the specimen. The measured confining pressure was corrected for the additional hoop tension produced in the rubber membrane due to the volume change of the sand specimen. Corrections to linear as well as volumetric strains were found to be negligible.

Conventional Drained Triaxial Compression Tests

Four conventional, consolidated-drained, triaxial compression tests were performed to establish the stress-strain behavior and the strength of the fine silica sand. Three of the four triaxial tests were performed on specimens with H/D = 1.0 and with effective confining pressures of 0.075, 0.150, and 0.300 kg/cm^2 (7.4, 14.7, and 29.4 kN/m^2), corresponding to water column heights of 30, 59, and 118 in. (75, 150, and 300 cm). The fourth test was performed on a specimen with H/D = 2.65 and with an effective confining pressure of 0.250 kg/cm^2 (24.5 kN/m^2), correspond-

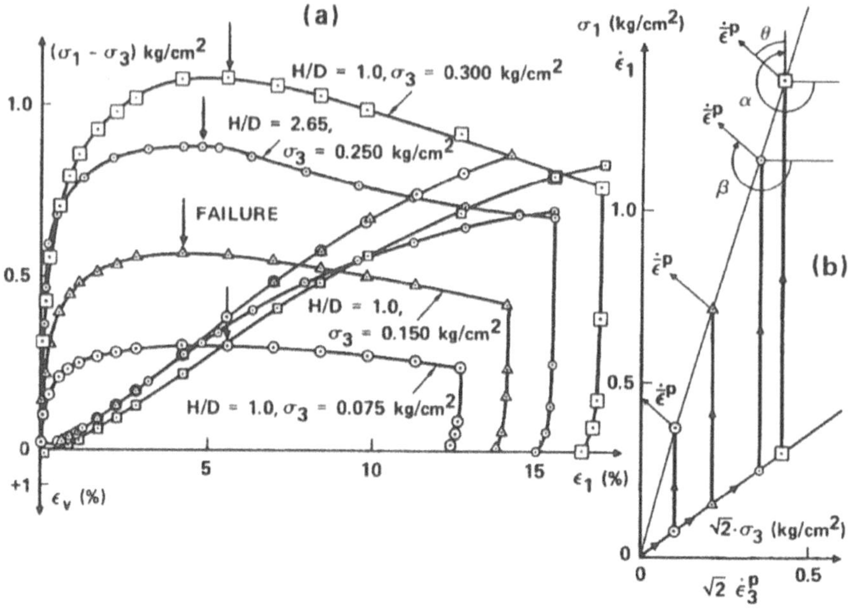

FIG. 4.—(*a*) Stress-Strain and Volume-Change Behavior; (*b*) Stress Path and Plastic Strain-Increment Vector Directions Observed in Conventional Triaxial Compression Tests on Fine Silica Sand (1.0 kg/cm² = 98 kN/m²)

ing to 100 in. (250 cm) of water. The four tests were all performed under strain control. Following failure, each specimen was unloaded to enable determination of the elastic behavior of the sand.

Stress differences and volumetric strains are plotted versus vertical strains for the four tests in Fig. 4(*a*). It may be seen that the differences in stress-strain and volumetric behavior between the short and the tall specimens are small. The initial modulus is a little higher for the tall specimen than for the short specimens, but the strain-to-peak failure is in the range of 4–6% for all specimens. The decreases in strength following peak failure also appear to form a consistent pattern. The friction angles from the four tests are listed in Table 1.

To compare these friction angles in a meaningful way, they have been adjusted to correspond to a void ratio of $e = 0.580$ and a relative density of 90.0%. The relationship $e \cdot \tan \phi = \text{const}$ is suitable for adjustment of friction angles when small differences in void ratio are present between

TABLE 1.—Results of Conventional Triaxial Compression Tests on Fine Silica Sand (1.0 kg/cm² = 98 kN/m²)

Test number (1)	*H/D* (2)	*e* (3)	σ_3 (kg/cm²) (4)	ϕ' (measured) (degrees) (5)	ϕ' (adjusted e = 0.580) (degrees) (6)	α ($\dot{\sigma}$) (degrees) (7)	β ($\dot{\epsilon}$) (degrees) (8)	θ (degrees) (9)
CD-30	1.0	0.579	0.075	41.7	41.7	270.0	218.8	51.2
CD-59	1.0	0.579	0.150	40.8	40.8	270.0	218.7	51.3
CD-100	2.65	0.580	0.250	39.6	39.6	270.0	219.9	50.1
CD-118	1.0	0.585	0.300	39.9	40.1	270.0	220.1	49.9

the specimens (1). The friction angles for the short specimens show a pattern of decreasing values with increasing confining pressure as is usually found for soils (8). The tall specimen produced a friction angle that is slightly lower than those from the short specimens. Those differences have also been observed before from similar tests on tall and short specimens (7).

Fig. 4(*a*) also shows that the volume changes observed in the four tests form a consistent pattern with only slightly decreasing rates of dilation with increasing confining pressures. The corresponding strain-increment vector directions are superimposed on the triaxial plane shown in Fig. 4(*b*). These vector directions are directly related to the rates of dilation. The vectors are shown initiating from the failure points at the ends of the stress paths for the four triaxial compression tests.

From the diagram in Fig. 4(*b*), it appears that the strain-increment vectors are parallel. To obtain a more sensitive measure of their relative directions, the angles β measured clockwise from the σ_3-axis to the positive directions of the vectors, were calculated from the original test data according to the following expression:

$$\tan(\beta - 180^\circ) = \frac{\sqrt{2}}{1 - \dfrac{\dot{\epsilon}_v}{\dot{\epsilon}_1}} \quad \text{(2)}$$

The values of β (Table 1) increase slightly with increasing confining pressure, thus reflecting the pattern of dilation rates shown in Fig. 4(*a*). The β-values vary consistently from 218.7°–220.1°, i.e., over a very narrow range of 1.40° with an average value of 219.4°.

The directions of the stress-increment vectors resulting in the aforementioned strain-increment vectors are vertical on Fig. 4(*b*). The corresponding α-values are therefore 270.0°, as listed in Table 1.

The angles θ between the stress-increment and strain-increment directions are all smaller than 90°. The scalar product given in Eq. 1*b* is therefore positive, corresponding to stable behavior in theory as well as experimentally.

Stress-Path Tests with Decreasing Stresses

A total of seven tests were performed with stress paths in the sector defined by the yield surface and the plastic potential surface. Six of these were conducted on tall specimens with $H/D = 2.65$, and one was performed on a short specimen with $H/D = 1.0$.

All tests were initiated at a confining pressure of 0.300 kg/cm^2 (29.4 kN/m^2). The specimens were first loaded vertically by filling the 15 gal bucket with water at a moderately slow, constant rate. As listed in Table 2, in five of the seven tests the specimens were loaded to a vertical stress difference of 0.98 kg/cm^2 (96 kN/m^2), whereas stress differences of 0.86 kg/cm^2 and 0.72 kg/cm^2 (84 and 71 kN/m^2) were used in the other two tests. The proximity of these stress states to the failure line may be expressed in terms of the stress level S defined as the ratio of the current to the maximum stress difference at a given confining pressure. Thus, the failure stress state corresponds to a stress level of $S = 1.00$. The three

TABLE 2.—Results of Stress Path Tests with Decreasing Stresses on Fine Silica Sand (1.0 kg/cm² = 98 kN/m²)

Test number (1)	H/D (2)	e (3)	$(\sigma_1 - \sigma_3)_{max}$ (kg/cm²) (4)	S at $(\sigma_1 - \sigma_3)_{max}$ (5)	σ_{3F} (kg/cm²) (6)	ϕ' (measured) (degrees) (7)	ϕ' (adjusted e = 0.580) (degrees) (8)	α ($\dot{\sigma}$) (degrees) (9)	β ($\dot{\epsilon}$) (degrees) (10)	θ (degrees) (11)
SP2	2.65	0.591	0.97	0.97	0.281	38.4	38.9	116.2	220.9	104.7
SP3	1.0	0.568	0.99	0.93	0.254	40.0	39.4	112.2	218.2	106.0
SP4	2.65	0.582	0.98	0.95	0.275	39.1	39.2	117.5	219.2	101.7
SP5	2.65	0.582	0.98	0.92	0.253	40.0	40.1	112.8	219.0	106.2
SP6	2.65	0.580	0.98	0.94	0.254	39.7	39.7	116.8	219.4	102.6
SP7	2.65	0.580	0.86	0.86	0.209	39.6	39.6	109.5	219.0	109.5
SP8	2.65	0.580	0.72	0.75	0.173	40.0	40.0	115.2	218.6	103.4

stress differences reached in the seven tests correspond to stress levels of 0.94, 0.86, and 0.75, respectively.

Following the initial loading, the specimens were exposed to stress paths in which all stresses, as well as the vertical stress difference, were reduced in such ratios that the specimens were observed to compress vertically. Too severe a reduction of the vertical stress difference would result in unloading and therefore vertical extension of the specimens. Thus, the stress paths imposed on the specimens were of the type that would potentially lead to unstable behavior of the sand.

Fig. 5 shows the stress-strain and volume change curves, as well as the stress path for test number SP7. These results clearly illustrate the essence of the study presented here. Fig. 5(*a*) indicates that the vertical stress difference was first increased to 0.86 kg/cm² (84 kN/m²), while the confining pressure was held constant at 0.300 kg/cm² (29.4 kN/m²). From the corresponding stress level of 0.86, the specimen was further loaded under load control along a stress path, which potentially could lead to unstable behavior of the sand. Fig. 5(*b*) shows that this second branch of the stress path is curved, although the flow rates from the two reservoirs were held constant. The curvature is caused by the accelerated increase in cross-sectional area of the specimen as failure is approached and subsequently crossed.

It is clear from the stress-strain curve that failure cannot be distinguished from observation of the vertical stress difference alone. However, the effective stress ratio will reach a maximum along the stress path indicated in Fig. 5(*b*). This maximum occurs at the point on the stress path that, when connected with the origin, forms the greatest slope of any point on the stress path. This criterion of failure is often used for granular materials and for soils without effective cohesion. Thus, for such materials and for frictional materials in general (soils, rock, concrete), it is important to consider the magnitude of all stresses when conditions of failure are to be described.

The portion of the stress path of interest in the present study is that from point B to point C. All stresses, as well as the vertical stress difference, are decreasing along this second branch. The strain increments produced along this branch correspond to dilation of the sand. Superimposed on the triaxial plane in Fig. 5(*b*) are the strain-increment vectors at points B and C. These strain-increment vectors, which are almost par-

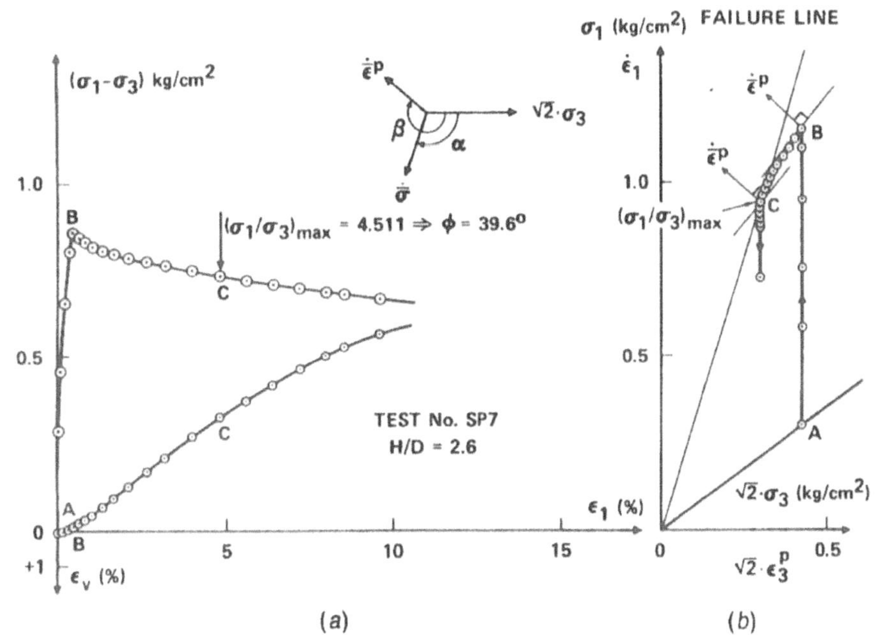

FIG. 5.—(*a*) Stress-Strain and Volume-Change Curves; (*b*) Stress Path and Plastic Strain-Increment Vectors for Test Number SP7 on Fine Silica Sand (1.0 kg/cm² = 98 kN/m²)

allel, are by definition perpendicular to the respective plastic potential surfaces indicated at these two points.

It is clear that each increment of stress along the entire length of branch BC forms obtuse angles, θ, with the corresponding strain-increment directions. Thus, the scalar product of the two vectors expressed in Eq. 1*b* is negative, and the behavior of the granular material should be unstable according to the inequality. However, each of the seven tests exhibited stable behavior before the maximum stress ratio had been reached. In fact, the specimens deformed uniformly, i.e., there was no visible evidence of bulging or shear-band formation until well beyond failure as measured by the maximum stress ratio.

The essential results of the seven tests are summarized in Table 2. Detailed results are given in Ref. 12. As for the conventional tests, the friction angles have been adjusted to correspond to a void ratio of e = 0.580. The seven specimens failed within a narrow range of confining pressures, and the adjusted friction angles are therefore directly comparable. The average value of the friction angles is 39.6° with a standard deviation of 0.46°. Thus, the strengths obtained from the seven stress-path tests are exactly the same as those obtained from conventional triaxial compression tests (ϕ' = 39.6° for test number CD-100 with H/D = 2.65). This shows that the sand specimens remained perfectly stable, and they reached the same failure surface as obtained from conventional tests. In none of the tests did shear planes or bands develop until well beyond failure, and there was consequently no difference in terms of stability and strength between the short specimens, in which shear-plane propagation may be inhibited, and tall specimens, in which shear planes are free to develop and propagate when they occur (see Ref. 7).

The directions of the strain-increment vectors are indicated by the angles β measured clockwise from the σ_3-axis to the positive directions of these vectors. The β-values are calculated from Eq. 2 and listed in Table 2. Since all specimens failed within a narrow range of confining pressures, there should not and there does not appear to be any systematic variation of the β-values in the seven tests. The average β-value is 219.2° and the standard deviation is 0.85°. Thus, the strain-increment directions obtained from the stress-path tests are the same as those obtained from the conventional tests in which the average β-value is 219.4°. The fact that the directions of the strain-increment vectors are independent of the stress path has been observed before (10,16). These important characteristics of the material behavior make it possible to express and determine the relative magnitudes of the plastic strain increments from a potential function.

The directions of the stress-increment vectors are calculated from the stresses leading up to failure. The directions are expressed in terms of the angles α, measured clockwise from the σ_3-axis to the positive directions of the stress-increment vectors. The values of α are determined from the following expression:

$$\tan(\alpha - 90°) = \frac{\sqrt{2}\,\dot{\sigma}_3}{\dot{\sigma}_1} \quad \ldots\ldots\ldots\ldots \quad (3)$$

and listed in Table 2. It may be seen that a range of α-values from 109.5°–117.5° are covered in the experiments.

The angles θ between the positive directions of the strain-increment and stress-increment directions are calculated as the difference between β and α and listed in Table 2. The values of θ range from 101.7°–109.5°, i.e., from 11.7°–19.5° greater than 90°, which corresponds to perpendicularity of the two vectors. Thus, the plastic flow of the granular material is clearly of the nonassociated type. More importantly in the present context, the seven stress paths all involve conditions for which the scalar products of the strain-increment and stress-increment vectors are negative. Therefore the material behavior should be unstable. However, each specimen behaved in a stable manner, and they all reached strengths of the same magnitudes as those obtained from conventional triaxial compression tests.

Analysis

The occurrence of nonassociated flow and the consequent problem of stability of materials that exhibit nonassociated flow behavior have been the subject of much discussion. For example, the reliability of the experimental data that provided the evidence for nonassociated flow has at times been questioned. In a similar fashion, assuming and accepting the possibility that instabilities may occur, the potential problem has been circumvented by inclusion of rate effects, e.g., viscous flow or creep, to disguise the unstable behavior and to temporarily maintain stability of the material. Release of stored elastic energy to counter dissipation of plastic work along stress paths with decreasing stresses has also been proposed. Each of these points will be considered in detail below.

Reliability of Test Data.—Because experimental data are at times subject to justified criticism and because such data may be unreliable, the occurrence of nonassociated flow occasionally has been considered to be the result of imprecise performance and measurements in tests that otherwise would exhibit associated flow.

Applied loads and pressures can be measured with good accuracy in experiments such as triaxial compression tests. The most obvious inaccuracies in these tests arise from nonuniform strains exhibited by, e.g., bulging of the specimen. This mode of deformation is due to end restraint caused by development of shear stresses between the end plates and the specimen. To avoid this problem, lubrication consisting of one or more thin rubber sheets coated with silicone grease may be used on the cap and base. By adjusting the number of rubber sheets and the amount of silicone grease, the specimen can be made to deform as a right cylinder, thereby producing uniform strains.

Lubricated ends were used in the tests performed for this study. The shape of the specimen was carefully monitored during each test to detect any signs of bulging or other nonuniform deformation patterns. The specimens in all tests deformed as perfect right cylinders until well beyond failure (maximum stress ratio). The experimental results presented here are therefore considered to be reliable and to correctly reflect the true behavior of the soil.

In spite of the care with which the present experimental program was executed, it was of interest to investigate the degree to which nonuniform strain distributions could disguise the type of behavior (associated or nonassociated flow) exhibited by the soil. For this purpose, the specimen has been assumed to deform nonuniformly and to develop from a right cylinder into the shape of a barrel. It is therefore obvious that the lateral strain, ϵ_3, is nonuniformly distributed along the vertical axis of the specimen. In the analysis, it is assumed that the contour of the barrel is shaped as a parabola with no expansion at the ends. Two simple options are available for the distribution of the vertical strain, ϵ_1: (1) Parabolic distribution with zero strains at the ends and maximum strain in the middle; and (2) uniform strain distribution. Both these assumed strain distributions are extreme on either side of the real strain distribution in a specimen with full friction on the end plates.

The results of the analyses are given in Table 3 in terms of the ratio $\dot{\epsilon}_1/(\sqrt{2}\,\dot{\epsilon}_3)$, which is equivalent to the negative value of the right-hand side of Eq. 2. The values of $\dot{\epsilon}_1$ and $\dot{\epsilon}_v$ in Table 3 are those usually calculated by assuming the specimen to deform as a right cylinder with uniform strains. The strain-increment ratio is calculated for the ends and the middle of the specimen for various combinations of uniform and parabolic strain distributions. The expression for the strain-increment ratio for uniform strain distributions (Col. 2) is that employed in Eq. 2.

For parabolic distributions of both ϵ_1 and ϵ_3 with no strains at the ends, the ratio of the strain increments in the middle (Col. 3) becomes the same as that obtained for uniform strain distributions. Thus, the type of behavior exhibited by the soil is easily recognizable, even from grossly noncylindrical specimens with most of the deformation occurring in the plastic work along stress paths with decreasing stresses has also been proposed. Each of these points will be considered in detail below.

Reliability of Test Data.—Because experimental data are at times subject to justified criticism and because such data may be unreliable, the occurrence of nonassociated flow occasionally has been considered to be the result of imprecise performance and measurements in tests that otherwise would exhibit associated flow.

Applied loads and pressures can be measured with good accuracy in experiments such as triaxial compression tests. The most obvious inaccuracies in these tests arise from nonuniform strains exhibited by, e.g., bulging of the specimen. This mode of deformation is due to end restraint caused by development of shear stresses between the end plates and the specimen. To avoid this problem, lubrication consisting of one or more thin rubber sheets coated with silicone grease may be used on the cap and base. By adjusting the number of rubber sheets and the amount of silicone grease, the specimen can be made to deform as a right cylinder, thereby producing uniform strains.

Lubricated ends were used in the tests performed for this study. The shape of the specimen was carefully monitored during each test to detect any signs of bulging or other nonuniform deformation patterns. The specimens in all tests deformed as perfect right cylinders until well beyond failure (maximum stress ratio). The experimental results presented here are therefore considered to be reliable and to correctly reflect the true behavior of the soil.

In spite of the care with which the present experimental program was executed, it was of interest to investigate the degree to which nonuniform strain distributions could disguise the type of behavior (associated or nonassociated flow) exhibited by the soil. For this purpose, the specimen has been assumed to deform nonuniformly and to develop from a right cylinder into the shape of a barrel. It is therefore obvious that the lateral strain, ϵ_3, is nonuniformly distributed along the vertical axis of the specimen. In the analysis, it is assumed that the contour of the barrel is shaped as a parabola with no expansion at the ends. Two simple options are available for the distribution of the vertical strain, ϵ_1 : (1) Parabolic distribution with zero strains at the ends and maximum strain in the middle; and (2) uniform strain distribution. Both these assumed strain distributions are extreme on either side of the real strain distribution in a specimen with full friction on the end plates.

The results of the analyses are given in Table 3 in terms of the ratio $\dot{\epsilon}_1/(\sqrt{2}\,\dot{\epsilon}_3)$, which is equivalent to the negative value of the right-hand side of Eq. 2. The values of $\dot{\epsilon}_1$ and $\dot{\epsilon}_v$ in Table 3 are those usually calculated by assuming the specimen to deform as a right cylinder with uniform strains. The strain-increment ratio is calculated for the ends and the middle of the specimen for various combinations of uniform and parabolic strain distributions. The expression for the strain-increment ratio for uniform strain distributions (Col. 2) is that employed in Eq. 2.

For parabolic distributions of both ϵ_1 and ϵ_3 with no strains at the ends, the ratio of the strain increments in the middle (Col. 3) becomes the same as that obtained for uniform strain distributions. Thus, the type of behavior exhibited by the soil is easily recognizable, even from grossly noncylindrical specimens with most of the deformation occurring in the

TABLE 3.—Values of $\dot{\epsilon}_1/(\sqrt{2}\dot{\epsilon}_3)$ for Different Assumed Strain Distributions in Triaxial Specimen

Strain conditions (1)	Uniform ϵ_1, uniform ϵ_3 (2)	Parabolic ϵ_1, parabolic ϵ_3, no strains at ends (3)	Uniform ϵ_1, parabolic ϵ_3, $\epsilon_3 = 0$ at ends (4)
At ends	$-2/(1 - \dot{\epsilon}_v/\dot{\epsilon}_1)$	$0/0 = ?\ (-\infty)$	$-\infty$
In middle	$-2/(1 - \dot{\epsilon}_v/\dot{\epsilon}_1)$	$-2/(1 - \dot{\epsilon}_v/\dot{\epsilon}_1)$	$-2/3 \cdot \sqrt{2}/(1 - \dot{\epsilon}_v/\dot{\epsilon}_1)$

middle. Although fixity in terms of ϵ_3, i.e., $\epsilon_3 = 0$, can occur at the ends, at least mathematically, the value of $\epsilon_1 = 0$ would correspond to an infinite modulus. Allowing for a modulus of finite value, the ratio of $\dot{\epsilon}_1/(\sqrt{2}\,\dot{\epsilon}_3) = -\infty$ is reasonable for full fixity at the ends of the specimen. Alternatively, the specimen might be unloading at the ends (due to reducing stresses along the stress path under consideration), in which case the soil behavior is elastic and not of interest in the present context.

If, on the other hand, the vertical strain distribution is assumed to be uniform, the results listed in Col. 4 of Table 3 are obtained. In this case, the ratio of the strain increments in the middle is multiplied by a factor of 2/3, i.e., the strain-increment vector is apparently moving towards perpendicularity to the stress path. Additional considerations are necessary to investigate whether the results of the triaxial tests now indicate associated flow, and this in turn might be the reason for the stable behavior of the granular material.

Since the ϵ_3-distribution is assumed to be parabolic, the cross-sectional area in the middle of the specimen is larger than that obtained from a perfect cylindrical specimen with the same measured vertical deformation and volume change. The cross-sectional area at the middle of the barrel-shaped specimen may be expressed as

$$A_{\text{barrel}} = A_{\text{cylinder}} \cdot \left[1 + \frac{1}{2} \cdot (\epsilon_1 - \epsilon_v)\right] \quad \text{(4)}$$

Since ϵ_1 is compressive and positive and ϵ_v is negative for dilation, the cross-sectional area of a barrel-shaped specimen is greater and the vertical stress difference is smaller in the middle than the corresponding values for a cylindrical specimen. The effective confining pressure acting at the middle of the specimen would be unaffected by the end restraint.

For the sake of investigating whether these modifications are sufficient to disguise the real soil behavior, it is assumed that the seven specimens tested in this study became barrel-shaped along the stress paths with decreasing stresses. After applying the modifications to stresses and strains described earlier, the directions of stress-increment and strain-increment vectors expressed in terms of the angles α and β are calculated and listed in Table 4 for the seven stress-path tests. The differences in the directions of stress-increment and strain-increment vectors are expressed by the angle θ. It may be seen by comparison with the results in Table 2 that the values of θ decreased in all cases towards 90°, i.e., the two vector directions approached perpendicularity. However, in none of the cases

TABLE 4.—Directions of Stress Increment and Strain Increment for Assumed Parabolic Lateral Strain Distribution and Assumed Uniform Vertical Strain Distribution in Triaxial Specimens

Test number (1)	H/D (2)	α ($\dot{\sigma}$) (degrees) (3)	β ($\dot{\epsilon}$) (degrees) (4)	θ (degrees) (5)
SP2	2.65	110.8	210.0	99.2
SP3	1.0	105.4	207.7	102.3
SP4	2.65	109.2	208.6	99.4
SP5	2.65	105.8	208.4	102.6
SP6	2.65	112.1	208.7	96.6
SP7	2.65	103.9	208.3	104.4
SP8	2.65	109.8	208.0	98.2

was perpendicularity obtained. In fact, the angles θ remained in all cases greater than 90°, and the soil therefore exhibited nonassociated flow behavior. Thus, even an unlikely combination of parabolic ϵ_3-distribution and uniform ϵ_1-distribution is insufficient to disguise the real soil behavior. The true ϵ_1-distribution in a specimen with full-end restraint is most likely in between the assumed uniform and parabolic distributions, and the true θ-values for these boundary conditions are therefore higher than those listed in Table 4.

Rate/Time Effects.—The occurrence of viscous flow or creep may maintain the stability, at least temporarily, of otherwise unstable materials. This effect often manifests itself through increasing strength with increasing loading rate. Thus, the material appears to remain stable at high magnitudes of shear stresses when these are applied at a fast rate. Examples of this behavior can be found for soils, ice, rock, concrete, etc. (see Ref. 2).

In a similar fashion, it is possible that the granular material tested for this study remained stable because the load was changed at a reasonable finite rate. To check if effects of viscious behavior were artificially maintaining stability of the sand specimens, the testing procedures for two of the seven specimens included establishment of creep curves at several stress levels. One representative set of creep curves is shown in Fig. 6 for test number SP5. The first four creep curves were established during loading of the specimen at the constant confining pressure of 0.300 kg/cm^2 (29.4 kN/m^2). This was done by simply stopping the vertical loading of the specimen and observing the deformation with time. Very little but increasing amounts of creep occur with increasing stress level. The following two creep curves were determined along the stress path involving decreasing stresses. These curves were measured after having stopped the changes in both vertical and lateral stresses. Because the stress levels were closer to the failure stress ratio (i.e., 0.96 and 0.99), greater amounts of creep strains occurred. However, in both cases the creep rates decreased to very small values after 60 min. The specimen was perfectly stable, even after the majority of viscous effects had gone away. Thus, it is clear that viscous effects cannot account for the stability of the sand.

The last creep curve was measured after the maximum (failure) stress ratio had been obtained. The stress level for this curve was initially 0.98,

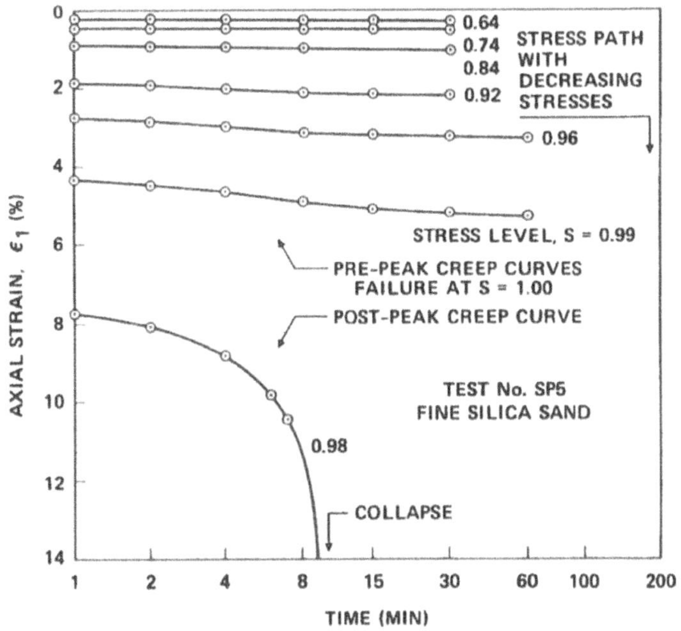

FIG. 6.—Creep Curves Obtained at Several Stress Levels in Test Number SP5 on Fine Silica Sand

i.e., the stresses had only decreased a small amount after failure due to increasing cross-sectional specimen area. Nevertheless, the specimen was clearly unstable after failure, and it moved to complete collapse with no further change in vertical load or confining pressure. The reason that instability was not observed immediately upon failure, i.e., immediately after $S = 1.00$ had been reached, was that viscous effects maintained stability over a short period of time, allowing further increase in cross-sectional area to occur before it was realized that failure had been passed and that collapse was in progress.

It can be seen that the sand specimen maintains stability before failure independent of any viscous effects exhibited by the material. However, immediately following failure, according to the maximum stress-ratio criterion, the instability of the sand is clearly evident. The deformation to complete collapse of the specimen begins slowly and then accelerates to a constant terminal velocity, after which the specimen deforms at a constant rate. Thus, the specimen did not exhibit catastrophic failure, as, e.g., in a liquefaction failure of loose sand.

Release of Stored Elastic Energy.—In order to account for the stability of granular materials during stress paths with decreasing stresses, it may be possible that elastic energy is released to counter the dissipation of plastic energy. Stability should then be maintained as long as

$$\dot{\sigma}_{ij} \cdot \dot{\epsilon}_{ij} = \dot{\sigma}_{ij} \cdot (\dot{\epsilon}^e_{ij} + \dot{\epsilon}^p_{ij}) = \dot{\sigma}_{ij} \cdot \dot{\epsilon}^e_{ij} + \dot{\sigma}_{ij} \cdot \dot{\epsilon}^p_{ij} \geq 0 \quad \quad (5)$$

in which case the sign of the elastic energy increment is the opposite of the sign of the plastic energy increment along stress paths with decreasing stresses. Thus, stability should be maintained along potentially unstable stress paths as long as

$$\dot{\sigma}_{ij}\,\dot{\epsilon}^{e}_{ij} \geq |\,\dot{\sigma}_{ij}\,\dot{\epsilon}^{p}_{ij}| \quad \ldots\ldots (6)$$

in which both sides of the unequal sign have positive values. By forming the ratio of elastic to plastic energy increments and expanding the terms, the following inequality is produced:

$$\frac{\frac{1}{E}\cdot[\dot{\sigma}_1^2 - 4\dot{\sigma}_1\dot{\sigma}_3 + 2(1-\nu)\dot{\sigma}_3^2]}{|\dot{\sigma}_1\dot{\epsilon}_1^p + 2\dot{\sigma}_3\dot{\epsilon}_3^p|} \geq 1 \quad \ldots\ldots (7)$$

The test most likely to fulfill this condition for stability is that whose value of θ in Table 2 is nearest to 90°. Test number SP4 involves the lowest value of θ = 101.7°. The left-hand side of Eq. 7 is therefore calculated for this test, using an elastic modulus of E = 400 kg/cm^2 (39,240 kN/m^2) (determined from the unloading branches in Fig. 4) and a Poisson's ratio of 0.25 (determined from the initial portion of the volume change curve during reloading). The value of the ratio of energy increments is determined to be 0.031, i.e., much less than unity. The other six tests would have even smaller values of the ratio of energy increments occurring on the left-hand side of Eq. 7. Therefore, the release of stored elastic energy accounts for a maximum of 3.1% of the plastic energy increment in the seven stress-path tests with decreasing stresses, and this mechanism cannot account for the stability of the granular material.

The tests described earlier all exhibited stable behavior until the failure surface was reached. These tests were performed on dense sand at low confining pressures, at which the material exhibits the type of behavior that is very characteristic of frictional materials. Notably, the granular material dilated under the conditions under which it was tested. On the other hand, granular materials that tend to compress may become unstable and liquefy during shear. The type of volume-change behavior exhibited by the material is of overriding significance in the question of stability. This has been discussed in further detail by the writers (12).

Summary

The stability postulate formulated by Drucker provides a sufficient condition for stability and requires that materials obey an associated plastic flow rule. Frictional materials exhibit nonassociated plastic flow, and the stability of such materials has often been questioned. To provide experimental evidence regarding the stability of granular materials, a series of triaxial tests was performed on specimens of dense sand exposed to stress paths with decreasing stresses and increasing stress levels. These tests were performed under stress control so that any instability in the material could freely develop. The results of the tests showed that the granular material dilated during shear, it followed a nonassociated flow rule, and stability was obtained in all cases until the failure surface was reached. Thus, stable behavior was observed in a region in which Drucker's stability postulate was violated. The reliability of the test results, the possible influence of viscous effects on the stability of the material, and the effect of release of elastic energy were investigated experimentally and discussed in detail. None of these effects could account for the stability of the material.

The occurrence of stable behavior found in this study is related to the dilation exhibited by the granular material. The type of volume-change behavior exhibited by the material is of great significance in the question of stability of granular materials.

Appendix.—References

1. Caquot, A., and Kerisel, J., *Traité de Mecanique des Sols*, 4th ed., Gauthier-Villars, Paris, France, 1966.
2. Casagrande, A., and Wilson, S. D., "Effects of Rate of Loading on the Strength of Clays and Shales at Constant Water Content," *Geotechnique*, Vol. 2, 1951, pp. 251–264.
3. Drucker, D. C., "A More Fundamental Approach to Stress-Strain Relations," *Proceedings, 1st U.S. National Congress on Applied Mechanics*, 1951, pp. 487–491.
4. Drucker, D. C., "On Uniqueness in the Theory of Plasticity," *Quarterly Applied Mathematics*, 1956, Vol. 14, pp. 35–42.
5. Drucker, D. C., "A Definition of Stable Inelastic Material," *Journal Applied Mathematics*, Vol. 26, 1959, pp. 101–106.
6. Frydman, S., Zeitlen, J. G., and Alpan, I., "The Membrane Effect in Triaxial Testing of Granular Soils," *Journal of Testing and Evaluation*, Vol. 1, No. 1, Jan., 1973, pp. 37–41.
7. Lade, P. V., "Localization Effects in Triaxial Tests on Sand," *Proceedings, Symposium on Deformation and Failure of Granular Materials*, International Union of Theoretical and Applied Mechanics, P. A. Vermeer and H. J. Luger, Eds., Delft, Holland, Aug., 1982, pp. 461–471.
8. Lade, P. V., "Failure Criterion for Frictional Materials," Chapter 20, *Mechanics of Engineering Materials*, C. S. Desai and R. H. Gallagher, Eds., John Wiley and Sons, Ltd., New York, N.Y., 1984, pp. 385–402.
9. Lade, P. V., and Duncan, J. M., "Cubical Triaxial Tests on Cohesionless Soil," *Journal of the Soil Mechanics and Foundations Division*, ASCE, Vol. 99, No. 10, Oct., 1973, pp. 793–812.
10. Lade, P. V., and Duncan, J. M., "Stress-Path Dependent Behavior of Cohesionless Soil," *Journal of the Geotechnical Engineering Division*, ASCE, Vol. 102, No. 1, Jan., 1976, pp. 51–68.
11. Lade, P. V., and Hernandez, S. B., "Membrane Penetration Effects in Undrained Tests," *Journal of the Geotechnical Engineering Division*, ASCE, Vol. 103, No. 2, Feb., 1977, pp. 109–125.
12. Lade, P. V., Nelson, R. B., and Ito, Y. M., "Experimental Study of Nonassociated Flow and Stability of Granular Materials," California Research and Technology, Inc., Chatsworth, Calif., Sep., 1985.
13. Mandel, J., "Conditions de Stabilite et Postulat de Drucker," *Proceedings, International Union of Theoretical and Applied Mechanics Symposium on Rheology and Soil Mechanics*, Grenoble, France, 1964, pp. 58–67.
14. Martin, G. R., Finn, W. D. L., and Seed, H. B., "Effects of System Compliance on Liquefaction Tests," *Journal of the Geotechnical Engineering Division*, ASCE, Vol. 104, No. 4, Apr., 1978, pp. 463–479.
15. Mroz, Z., "Nonassociated Flow Laws in Plasticity," *Journal de Mecanique*, Vol. 2, No. 1, 1963, pp. 21–42.
16. Poorooshasb, H. B., Holubec, I., and Sherbourne, A. N., "Yielding and Flow of Sand in Triaxial Compression: Part I," *Canadian Geotechnical Journal*, Vol. III, No. 4, Nov., 1966, pp. 179–190.

Instability of Granular Materials with Nonassociated Flow

By Poul V. Lade,[1] Member, ASCE, Richard B. Nelson,[2] Member, ASCE, and Y. Marvin Ito[3]

Abstract: Materials exhibiting nonassociated flow should, according to Drucker's stability postulate, become unstable when exposed to certain stress-paths inside the failure surface. Results of a series of triaxial tests designed to expose the type of behavior displayed by granular materials are presented and discussed. The sand tends to compress during shear, it exhibits nonassociated flow, and instability is observed along stress-paths in the region of potential instability. In a previous study of similar type in which dilation developed during shear and the sand exhibited nonassociated flow, when the specimens were exposed to stress-paths in the region of potential instability, none was observed. The type of volume change behavior displayed by the material is of great importance in the question of stability of granular materials. Stability conditions for materials with nonassociated flow are presented.

Introduction

Frictional materials such as sand, clay, concrete, and rock exhibit nonassociated plastic flow, and consequently questions have been raised regarding uniqueness and stability of such materials. Drucker's stability postulate for time-independent materials (Drucker 1951, 1956, 1959) provides a sufficient condition for stability and ensures uniqueness in dynamic as well as static problems. Drucker's postulate is satisfied provided that associated flow is employed in construction of constitutive models involving convex, plastic yield surfaces. Although Drucker's postulate provides a sufficient condition for stability, theoretical considerations have suggested that it is not a necessary condition (Mandel 1964, Mroz 1963).

Experiments to support this conclusion have been presented by the authors (Lade et al. 1987). In a series of tests designed to expose the type of behavior that is characteristic of granular materials, the sand exhibited nonassociated flow, and stability was obtained at the same time Drucker's postulate was being violated. The occurrence of stable behavior inside the failure surface (i.e., before failure had been reached) was found to be associated with the dilation or volumetric expansion exhibited by granular materials.

In the present paper a sand very similar to that investigated in Lade et al. (1987) is again investigated, but in a loading region where shear induced volumetric compression occurs, rather than dilation as in Lade et al. (1987). Again, stress-paths are explored along which Drucker's stability postulate is

[1]Prof., Dept. of Civ. Engrg., School of Engrg. and Appl. Sci., Univ. of California, Los Angeles, CA 90024.

[2]Prof., Dept. of Civ. Engrg., School of Engrg. and Appl. Sci., Univ. of California, Los Angeles, CA 90024.

[3]Adj. Prof., Dept. of Civ. Engrg., School of Engrg. and Appl. Sci., Univ. of California, Los Angeles, CA 90024; Engrg. Mech. Grp., Calif. Res. and Tech., Inc., Chatsworth, CA 91311.

Note. Discussion open until May 1, 1989. To extend the closing date one month, a written request must be filed with the ASCE Manager of Journals. The manuscript for this paper was submitted for review and possible publication on June 26, 1987. This paper is part of the *Journal of Engineering Mechanics*, Vol. 114, No. 12, December, 1988. ©ASCE, ISSN 0733-9399/88/0012-2173/$1.00 + $.15 per page. Paper No. 23046.

violated. The experimental apparatus is a load controlled device so that any instabilities will develop freely. It will be shown that, for the load paths considered herein, instabilities develop.

The problem of instability and the factors which control the occurrence of unstable behavior of granular materials are discussed in view of experimental results in the present paper. Based on the evidence presented here and in the previous publication (Lade et al. 1987), conditions for stability are formulated for granular materials. These conditions take the form of additions to Drucker's stability postulate.

Drucker's Stability Postulate

The stability condition formulated by Drucker (1951, 1956, 1959) involves two components expressed in terms of plastic work increments:

$$dW_p - dW_p^* = (\sigma_{ij} - \sigma_{ij}^*)\dot{\epsilon}_{ij}^p \geq 0 \quad (1a)$$

$$d^2W_p = \dot{\sigma}_{ij}\dot{\epsilon}_{ij}^p \geq 0 \quad (1b)$$

where σ_{ij} = a stress state located on the yield surface; σ_{ij}^* = a state located inside the yield surface; and $\dot{\sigma}_{ij}$ = a stress increment initiated at σ_{ij} on the yield surface causing the plastic strain increment $\dot{\epsilon}_{ij}^p$. Inequality 1(*a*), which is often referred to as stability in the large, requires the yield surface to be convex. Inequality 1(*b*) is referred to as stability in the small, and it is fulfilled in general if the plastic strain increment vector is perpendicular to the yield surface. This implies that the plastic potential surface coincides with the yield surface such that the plastic strain increments produced upon yielding are perpendicular to the yield surface. Thus, Drucker's stability postulate requires constitutive laws which incorporate an associated flow rule in order to guarantee that stable behavior is obtained.

It is the second Inequality 1(*b*) which defines stable or unstable behavior. If this inequality is fulfilled for any $\dot{\sigma}_{ij}$, then the material will be unconditionally stable. Otherwise, according to Drucker's postulate the material will be unconditionally stable, with the second increment d^2W_p positive for some $\dot{\sigma}_{ij}$ but negative for others, or it may be unstable with d^2W_p negative for all $\dot{\sigma}_{ij}$.

The instability of interest here occurs before the failure surface has been reached. Complete instability with d^2W_p negative, is obtained for all materials after the failure surface has been reached, whether the material exhibits associated or nonassociated flow. Inside the failure surface, but in a state of inelastic behavior (yield), all nonassociated flow mathematical models exhibit a region where d^2W_p is negative and, hence, where the material is unstable. However, it is by no means clear that the real materials which are being modeled are physically unstable in this region. It is the purpose of the experimental investigation to determine the physical behavior and the factors that influence the behavior of a real sand in this region of negative d^2W_p.

Nonassociated Flow and Possible Consequences

A typical pattern of yield surfaces for an isotropic granular material is shown on the triaxial plane in Fig. 1. In three dimensions these yield surfaces are shaped as asymmetric tear drops. For an isotropic material the yield

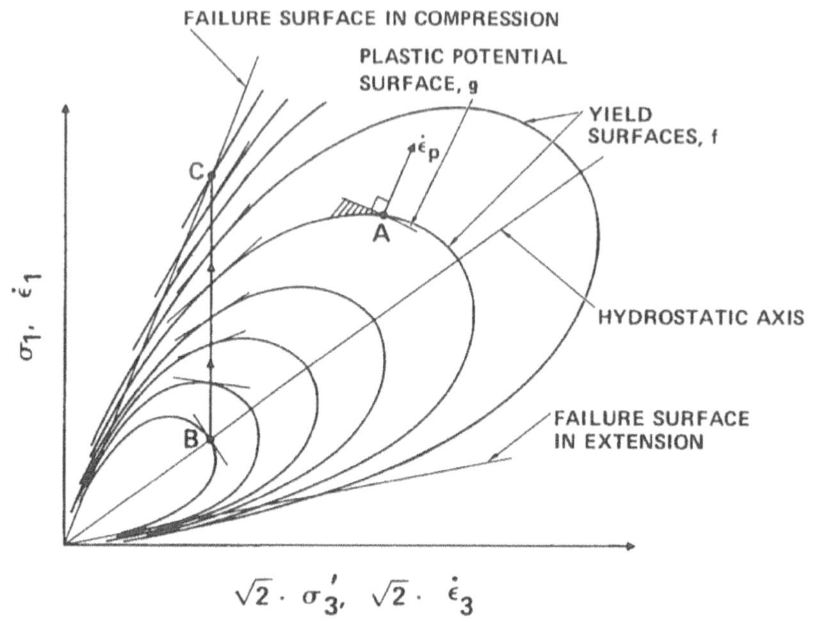

FIG. 1. Pattern of Yield Surfaces for Isotropic Granular Materials. Stress-Path for Conventional Triaxial Compression Test (BC)

surfaces intersect the hydrostatic axis in a perpendicular manner, they bend smoothly backwards towards the origin, and cross the failure surface at sharp angles as indicated in Fig. 1.

Plastic potential surfaces have similar shapes as yield surfaces, but for nonassociated flow the two families of surfaces cross each other. Experimental evidence for frictional materials indicates that the plastic potential surfaces have flatter rounded ends and the apex of the surface is not at the origin, but rather located at a point on the negative pressure axis. A typical plastic potential surface is shown at point A in Fig. 1.

The shaded wedge between the yield surface and the plastic potential surface defines a region in which Inequality 1(*b*) is not fulfilled for a material with nonassociated flow. Since a stress increment vector from point A located inside the wedge and the plastic strain increment vector form an obtuse angle, the scalar product of these two vectors [see Inequality 1(*b*)] is negative. According to Drucker's stability postulate, the sand may exhibit unstable behavior if a stress increment lies in the shaded region. However, as will be seen, the sand is perfectly stable at stress points where the normal to the yield surface points in the outward direction of the hydrostatic axis. For this condition the deviator stress can be safely increased to produce further plastic shear strains (work-hardening). In other words, the sand can sustain higher loads and behave in an inelastic manner without undergoing any instability or collapse.

Potential instability occurs in regions where the yield surface opens up in the outward direction of the hydrostatic axis. This allows plastic strains (loading) to occur while the stresses are decreasing. Here loading occurs inside the failure surface and instability may develop in the form of inability to sustain the current deviator stresses.

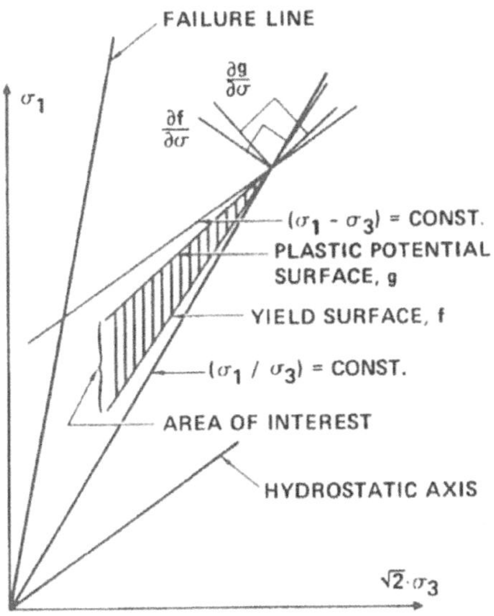

FIG. 2. Wedge-Shaped Region of Stress-Paths with Decreasing Stresses in which Granular Material with Nonassociated Flow may be Unstable during Hardening Inside Failure Surface

Fig. 1 shows the stress-path for a conventional triaxial compression test performed at constant confining pressure. As the specimen is loaded from B to C the inclination of the yield surface changes. At low deviator stresses near point B, the yield surface is inclined towards the outwards direction of the hydrostatic axis. As loading proceeds, the inclination of the yield surface changes gradually and becomes inclined towards the origin as failure is approached at point C. It is in this region of high deviator stresses where the yield surface is inclined towards the origin that instability may occur.

Fig. 2 shows a schematic illustration of the region in which Inequality 1(*b*) is not fulfilled for a material with nonassociated flow. The region is shaped as a wedge between the current yield surface f and the plastic potential surface g corresponding to the current stress point. This wedge shaped region is located within a larger region bounded by lines corresponding to (σ_1/σ_3) = const. and $(\sigma_1 - \sigma_3)$ = const. as indicated on Fig. 2. All stresses, including the stress difference $(\sigma_1 - \sigma_3)$, are decreasing within the wedge between f and g, but the stress ratio (σ_1/σ_3) is increasing in this region. By performing triaxial tests with stress-paths located in this region, experimental evidence regarding the stability of materials with nonassociated flow can be obtained.

It is important to recognize that the material behavior obtained for stress-paths within the shaded wedge in Fig. 2 corresponds to work-hardening with positive plastic work, $dW_p > 0$, and outward motion of the yield surface. Although the stresses in any direction within the wedge are decreasing, failure has not been reached and softening of the material is therefore not occurring. The decreasing stresses along a stress-path within the wedge may be deliberately followed and not imposed by the material. Such a stress-path is not possible in a softening material, because its strength has been exceeded, and it is behaving in an unstable fashion.

Region of Stability

In a previous study (Lade et al. 1987) performed to provide experimental evidence regarding the stability/instability of granular materials, a number of triaxial tests were performed on specimens of dense sand exposed to stress-paths with decreasing stresses and increasing stress ratios. The tests were performed under stress control so that any instability in the material could freely develop. The results of the tests showed that the granular material dilated during shear, following a nonassociated flow rule, and stability was obtained in all cases until the failure surface was reached. Thus, stable behavior was observed in a region in which Drucker's stability postulate was violated. The reliability of the test results, the possible influence of viscous effects on the stability of the material, and the effect of release of elastic energy were investigated experimentally, and discussed in detail. None of these effects could account for the stability of the material.

The occurrence of stable behavior found in the previous study is related to the dilation exhibited by the granular material. The triaxial tests were performed on dense sand at low confining pressures at which the material exhibited the type of behavior which is very characteristic of frictional materials, namely dilation under the test conditions. The tests showed that granular materials appear to be stable in the region of stresses where dilation occurs. This region is located near the failure surface at low to medium confining pressures for materials at a given void ratio. For higher confining pressures the same material undergoes volumetric compression during shear. The unique relation between void ratio and confining pressure for which a granular material dilates or compresses at failure is described by the concept of critical state as defined by Seed and Lee (1967) and by Lee and Seed (1967).

Region of Potential Instability

According to the critical state concept a granular material at a given void ratio will compress during shear at sufficiently high confining pressures. If the question of stability is related to the type of volume change (dilation or compression) occurring in the material, then it should be possible to induce instability inside the failure surface by conducting tests at high confining pressures with stress-paths within the shaded wedge shown in Fig. 2. Under these conditions the specimens will tend to compress during shear.

Undrained triaxial compression tests performed at high confining pressures on saturated granular materials exhibit effective stress-paths whose directions inside the failure surface may fulfill the requirements presented above. Fig. 3 shows the effective stress-paths for four undrained triaxial compression tests on loose Sacramento River Sand performed by Lee (1965). The specimens were initially consolidated isotropically at the effective consolidation pressures, σ'_{3c}, indicated on the figure. The volumetric strain in undrained tests on saturated specimens is zero, but the tendency for compression at high confining pressures results in increasing pore pressures which in turn force the effective stress-path into the desired direction. In the triaxial plane shown in Fig. 3, the stress difference $(\sigma_1 - \sigma_3)$ is proportional to the perpendicular distance from the hydrostatic axis. For the tests at high confining pressures the maximum stress differences occur well within the failure sur-

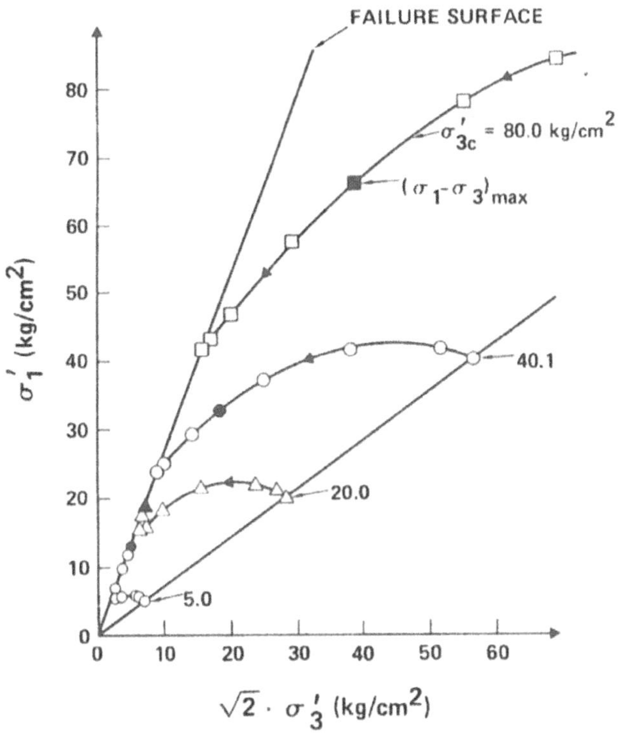

FIG. 3. Effective Stress-Paths for Undrained Triaxial Compression Tests at Four Different Confining Pressures on Loose Sacramento River Sand Shown in Triaxial Plane (1.0 kg/cm^2 = 98 kN/m^2)

face. Thus, true failure does not occur at the maximum stress difference. The region of potential instability is located at and above the solid points indicating the maximum stress differences. This region will be discussed further.

The stress-strain and pore pressure relations for the four undrained tests are shown in Fig. 4. The stress-strain curve for the test at 5.0 kg/cm^2 (490 kN/m^2) exhibits continuously increasing stress difference, although a tendency for leveling off occurs early in the test. The corresponding pore pressure curve shows first increasing and then decreasing pore pressures. This reflects a tendency for compression followed by a tendency for dilation as the failure surface is approached. Since the load carrying ability is continuously increasing until failure at large strains, the specimen appears to be stable.

The test performed at a confining pressure of 20.0 kg/cm^2 exhibits a temporary decline in the load carrying capacity between 1.0 and 4.0% axial strain. This may reflect a temporary instability which is then overcome as further straining occurs. The pore pressure curve exhibits similar behavior as that described above, reflecting the tendencies for compression followed by dilation. The tendency for dilation almost coincides with the recovery and increasing stress difference applied to the specimen.

The two tests with the highest confining pressures of 40.1 and 80.0 kg/cm^2 (3,930 and 7,850 kN/m^2) both show similar behavior. The stress dif-

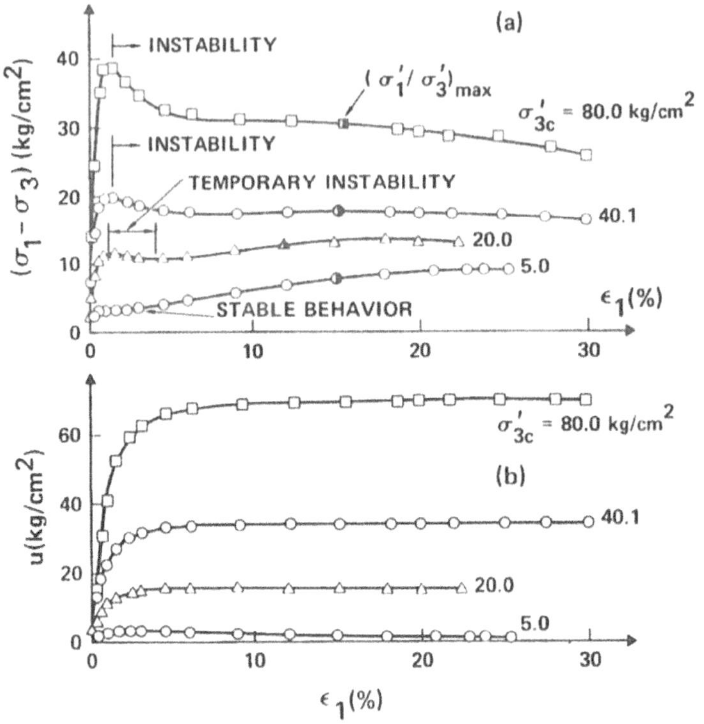

FIG. 4. Undrained Triaxial Compression Tests on Loose Sacramento River Sand (1.0 kg/cm² = 98 kN/m²): (*a*) Stress-Strain Relations; (*b*) Pore Pressure Relations

ference increases to a peak at relatively small axial strains and then decreases at larger strains. The pore pressures increase steadily before reaching essentially constant values at large strains. This reflects tendencies for compression with negligible or no tendency for dilation towards the end of the test. Thus, by increasing the confining pressure to sufficiently high values, the tendency for dilation observed at lower confining pressures can be suppressed.

The effective stress ratios, often taken to indicate failure in soils, occur at axial strains of 12 to 15% in all tests. Thus, the peak stress difference in the tests at high confining pressures are reached much earlier than the stress states corresponding to failure, i.e., the maximum stress difference may not indicate true failure of the granular material. The decline in load carrying capacity may indicate that instability can occur inside the failure surface. However, the tests whose results are shown in Figs. 3 and 4 were all performed under strain control. This method of testing does not allow instabilities of the material to be exposed. To demonstrate instabilities, it is necessary to perform the test under load or stress control.

Nonassociated Flow in Undrained Tests

To study the region of potential instability in stress space and to investigate the occurrence of nonassociated flow in undrained tests, the effective stress-paths are further analyzed below. There are basically three different types of effective stress-paths in undrained tests. These are shown, one at a time, in Fig. 5. Strain increments are superimposed on the stress diagrams to allow analyses of strain increment vector directions and to derive the directions of the plastic potential surfaces *g*. These surfaces are by definition perpendicular to the plastic strain increment vector directions.

For undrained tests the total volumetric strain is zero corresponding to the total strain increment vector being perpendicular to the hydrostatic axis everywhere along the effective stress-path. This is shown in Fig. 5(*a*). This figure also shows that volumetric compression is characterized by a strain increment vector pointing away from the origin, whereas volumetric dilation corresponds to a strain increment vector inclined towards the origin of the stress space.

The stress-strain curves in Fig. 4 show that plastic strains are produced everywhere along the undrained stress-paths. The yield surface must therefore be pushed out, and this requires it to be inclined relative to the effective stress-paths as indicated at each point in Fig. 5.

The direction of the elastic portion of the total strain increment vector in the triaxial plane depends on the value of Poisson's ratio, ν, and the ratio of the stress increments, $K = \dot{\sigma}_1/\dot{\sigma}_3$. The following relations may be deduced from Hooke's law: For $\dot{\sigma}_3 \neq 0$:

$$\frac{\dot{\epsilon}_1}{\sqrt{2}\cdot\dot{\epsilon}_3} = \frac{(K - 2\nu)}{\sqrt{2}\cdot[1 - \nu\cdot(1 + K)]} \quad (2)$$

and for $\dot{\sigma}_3 = 0$:

$$\frac{\dot{\epsilon}_1}{\sqrt{2}\cdot\dot{\epsilon}_3} = -\frac{1}{\sqrt{2}\cdot\nu} \quad (3)$$

These relations give the slopes of the elastic strain increment vectors on the triaxial plane. Since granular materials have $0 \leq \nu \leq 0.5$, these two limits will be used to bracket the directions of the elastic strain increment vectors.

For $\dot{\sigma}_3 \neq 0$ and $\nu = 0$, Eq. 2 produces

$$\frac{\dot{\epsilon}_1}{\sqrt{2}\cdot\dot{\epsilon}_3} = \frac{K}{\sqrt{2}} = \frac{\dot{\sigma}_1}{\sqrt{2}\cdot\dot{\sigma}_3} \quad (4)$$

i.e., the elastic strain increment vector is parallel to the stress increment vector or the stress path on the triaxial plane.

For $\dot{\sigma}_3 = 0$ and $\nu = 0$, Eq. 3 gives

$$\frac{\dot{\epsilon}_1}{\sqrt{2}\cdot\dot{\epsilon}_3} = -\infty \quad (5)$$

i.e., the elastic $\dot{\epsilon}$-vector is again parallel to the $\dot{\sigma}$-vector, both being vertical on the triaxial plane.

For $\nu = 0.5$ both Eqs. 2 and 3 lead to

$$\frac{\dot{\epsilon}_1}{\sqrt{2}\cdot\dot{\epsilon}_3} = -\sqrt{2} \quad (6)$$

i.e., the elastic strain increment vector is perpendicular to the hydrostatic axis corresponding to no elastic volume change.

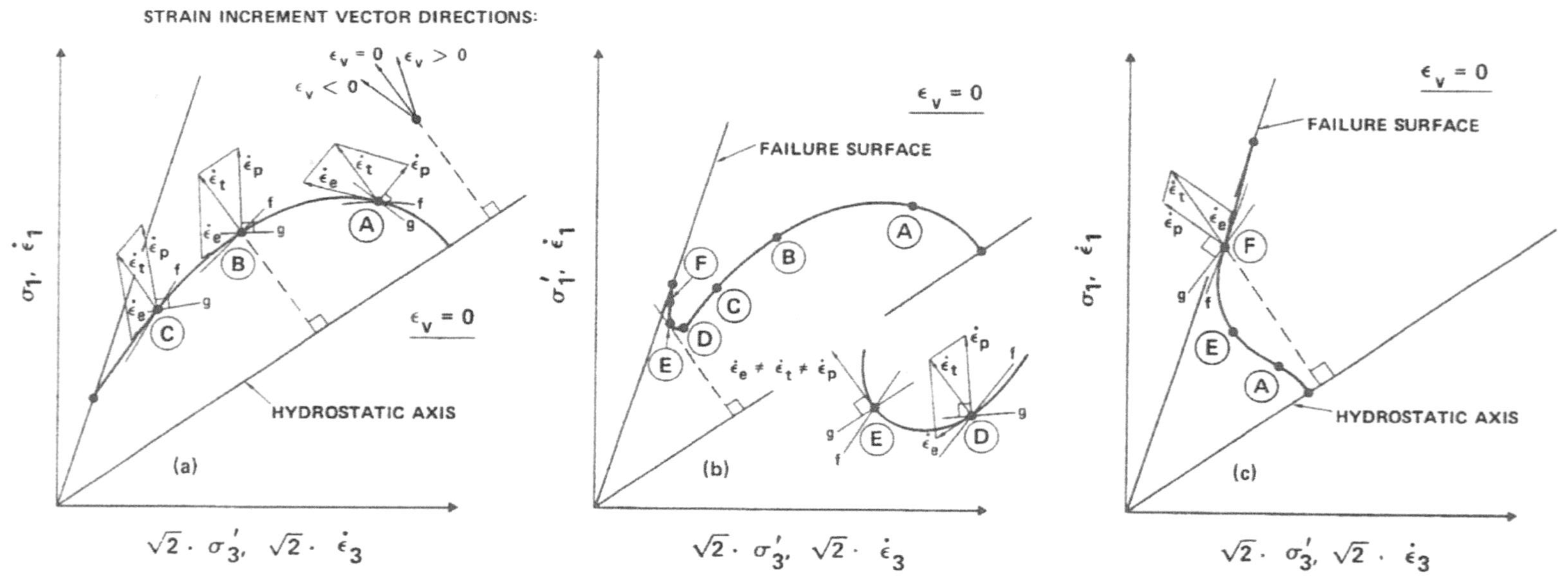

FIG. 5. **Evaluation of Relative Inclinations of Yield (*f*) and Plastic Potential (*g*) Surfaces Along Undrained Effective Stress-Paths: (*a*) Type 1; (*b*) Type 2; and (*c*) Type 3**

For values of Poisson's ratio between zero and 0.5 the elastic strain increment vector will be between the two extreme positions discussed. With no loss of generality, it will be assumed in the following that the elastic strain increment vector is parallel to the stress increment vector or the stress path on the triaxial plane, i.e., corresponding to $\nu = 0$. This simplifies the diagrams in Fig. 5, but does not limit the arguments presented below.

Three points of interest are indicated along the undrained effective stress-path of Type 1 in Fig. 5(*a*). The elastic portion of the total strain increment vector is pointed in the current direction of the effective stress-path. The length and direction of the plastic strain increment vector is obtained by vectorial subtraction of the elastic from the total strain increment vector. It is clear that only the directions of the elastic and the total strain increment vectors are known, and that the derived direction of the plastic strain increment vector depends on the assumed relative magnitudes of the elastic and total strain increments. However, as will be seen, it can be shown that nonassociated flow prevails at points B and C independent of the assumed magnitudes of strains.

At point A the length of the elastic vector is chosen so as to show that nonassociated flow could occur. In fact, in the absence of actual experimental determination of the elastic and plastic strain magnitudes, it is not possible to argue whether the plastic flow is associated or nonassociated, i.e., whether the yield and plastic potential surfaces are identical or not. However, this is of no importance to the study of instability presented here. Whether associated flow (which guarantees stability) or nonassociated flow is observed at point A, stability is obtained everywhere along the effective stress-path from the hydrostatic axis up to point B. Along this portion of the stress-path the load can be maintained constant or increased without any observable instability.

Nonassociated flow is clearly obtained at points B and C. For the limiting case where the elastic strain increments become negligible the plastic strain increment vectors become perpendicular to the hydrostatic axis. Since the yield surfaces must be inclined relative to the effective stress-paths as shown, nonassociated flow is clearly demonstrated to occur at points B and C.

Similar arguments can be made for point D on stress-path Type 2 in Fig. 5(*b*). Point D corresponds to a minimum stress difference along this portion of the effective stress-path. The load carrying capacity increases from point D to failure. At point E the effective stress-path is perpendicular to the hydrostatic axis. The elastic strain increment vector is therefore parallel with the total strain increment vector, and so is the plastic strain increment vector. The plastic potential surface is consequently parallel with the hydrostatic axis. Since the yield surface becomes steeper with increasing stress level, it cannot coincide with the plastic potential surface at point E, and nonassociated flow is demonstrated.

At point F on stress-path Type 3 in Fig. 5(*c*) the relative directions of the elastic and plastic strain increment vectors have been reversed (reversal occurs at point E). Point F is located in the region of plastic dilation. Large plastic strains occur in this region (see stress-strain curve corresponding to $\sigma'_{3c} = 5.0\ \text{kg/cm}^2$ in Fig. 4), and the elastic strains are therefore relatively small. For the limiting case where the elastic strain increments reduce to zero, the plastic strain increment vector becomes perpendicular to the hydrostatic axis and the plastic potential surface becomes parallel with the hydrostic axis. The yield surface is relatively steep, close to failure, as indicated on Fig. 1, and nonassociated flow is therefore obtained. In this region of dilation stability has been demonstrated as discussed.

It is the undrained test with effective stress-path of Type 1 that will be used as a tool to study stability/instability in the following. In this test the effective stress-path from point B to failure (point C being a typical point) is located in the shaded wedge region shown in Fig. 2. The decline in load carrying capacity may indicate that instability can occur inside the failure surface. Stress-controlled tests are required to study this possibility.

Experimental Program and Procedures

Sand Tested

Sacramento River Sand from the same batch previously tested by Lee (1965) was used in the experimental program. This sand is composed of subangular to subrounded particles consisting mostly of feldspar and quartz. The characteristics of this sand are summarized as follows: Mean diameter, 0.21 mm; coefficient of uniformity, 1.5; specific gravity of grains, 2.68; maximum void ratio, 1.03; and minimum void ratio, 0.61.

Effects of membrane penetration on the developed pore pressures are small to negligible in undrained tests when the average diameter of the soil grains is smaller than 0.1–0.2 mm (Frydman et al. 1973; Lade and Hernandez 1977; Martin et al. 1978). Tests were performed on loose specimens with void ratios around 0.84 corresponding to relative densities of 45%.

Specimen Preparation

All triaxial tests were performed on cylindrical specimens with diameter = 1.40 in. (3.56 cm) and height = 3.25 in. (8.26 cm) corresponding to $H/D = 2.32$. Filter stones were employed at each end of the specimens. This setup is similar to that used by Lee (1965).

The loose specimens were prepared by dry deposition of sand through a funnel with essentially zero drop height. The CO_2-method (Lade and Duncan 1973) was employed for saturation, and a nominal back pressure of 30 psi (207 kN/m^2) was applied to the specimens to insure a high degree of saturation. The B-values averaged 0.98 for all tests.

Testing Procedures

All specimens were first isotropically consolidated to 29.9 kg/cm^2 (2,930 kN/m^2). The vertical load was applied by a pneumatic cylinder, thus obtaining stress-controlled loading due to the high compressibility of air.

The vertical load, the confining pressure, the vertical deformation, and the volume change or the pore pressure were measured during each test. Vertical stress differences were calculated using current cross-sectional areas, and all strains were calculated as engineering strains. Corrections were applied to the measured vertical load for uplift forces on the piston. All other corrections to loads, pressures, linear deformations and volumetric strains were found to be negligible.

Effective Stress-Path Tests with Decreasing Stresses

Following application of the isotropic consolidation pressure, the specimens were loaded vertically under drained conditions by increasing the air

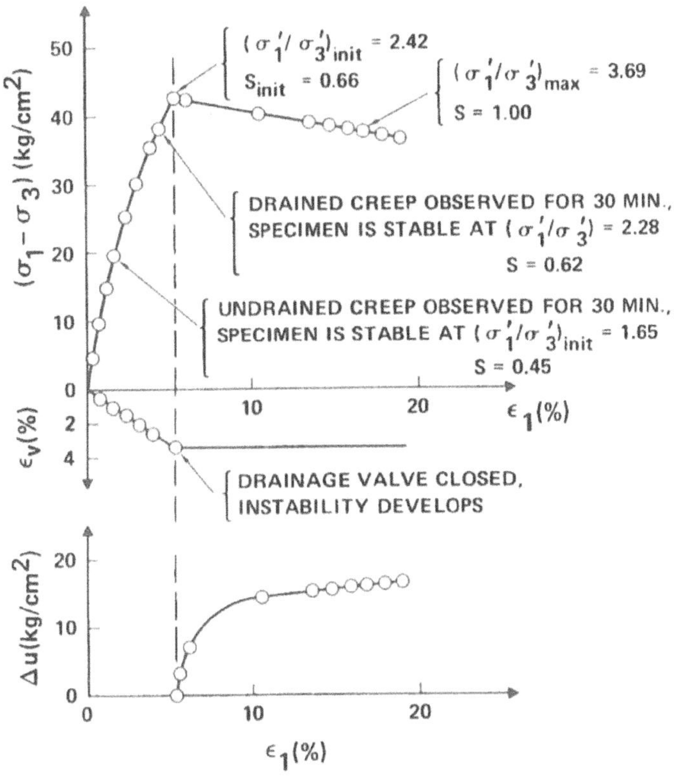

FIG. 6. Stress-Strain, Volume Change, and Pore Pressure Relations in Stress-Controlled Triaxial Compression Test on Loose Sacramento River Sand (1.0 kg/cm² = 98 kN/m²)

pressure in the pneumatic cylinder. Vertical stress differences of $(\sigma_1-\sigma_3)$ = 38.09, 42.72, 47.64, and 51.39 kg/cm² (3,737; 4,191; 4,673; and 5,041 kN/m²) were applied to the specimens. These correspond to effective stress ratios of (σ_1'/σ_3') = 2.28, 2.42, 2.59, and 2.72, whereas the maximum effective stress ratio at failure was $(\sigma_1'/\sigma_3')_{max}$ = 3.69, obtained from the strain-controlled tests discussed in the foregoing as well as from the stress controlled tests presented here. The proximity of these stress states to the failure line may be expressed in terms of the stress level S defined as the ratio of the current to the maximum stress difference at a given confining pressure. Thus, the failure stress state corresponds to a stress level of S = 1.00. The four stress differences reached in the tests correspond to stress levels of 0.62, 0.66, 0.70, and 0.74.

After the desired stress difference was obtained under drained conditions, the drainage valve was closed while the vertical load was maintained constant. The undrained condition imposed on the specimen caused the pore pressure to increase, thus forcing the effective stress-path into the shaded wedge shown in Fig. 2. Unstable behavior of the sand was observed in each test performed for this study.

Fig. 6 shows the results of the test performed with S_{init} = 0.66. These results are typical for all tests. The initial portions of the stress-strain and volume change relations correspond to drained loading. Note that the vol-

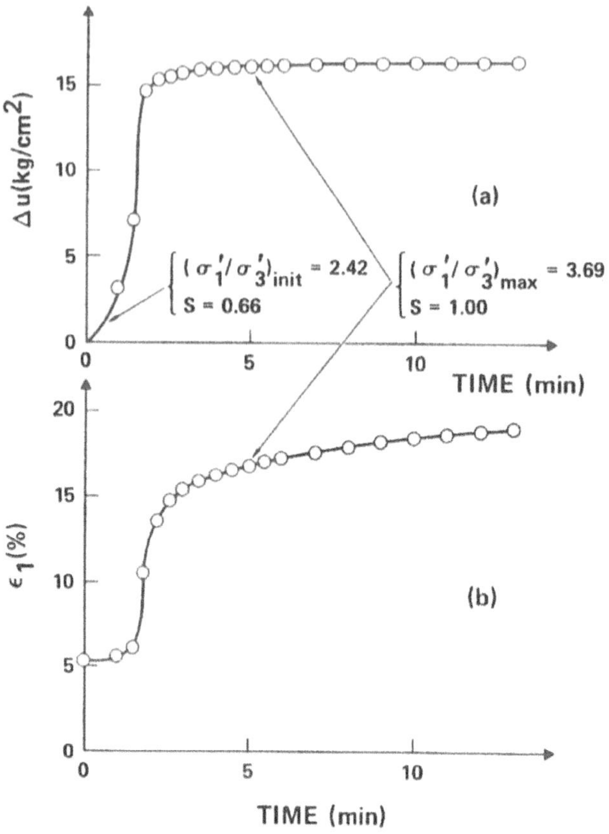

FIG. 7. Relations: (*a*) Pore Pressure; and (*b*) Axial Strain versus Time for Unstable Portion of Stress-Controlled Triaxial Compression Test on Loose Sacramento River Sand (1.0 kg/cm² = 98 kN/m²)

umetric strain is compressive until the drainage valve is closed at a vertical strain of $\epsilon_1 = 5.3\%$. Following closure of the drainage valve the specimen becomes unstable and compresses without any additional vertical load. The effective confining pressure decreases due to increasing pore pressure, and the stress difference decreases due to increasing cross-sectional area of the specimen. However, the decreasing stress difference cannot be sustained by the specimen.

Fig. 7 shows the pore pressure development and the axial strain development versus time after closure of the drainage valve. The pore pressure development slightly precedes the development of axial strain. Although the pore pressure starts increasing immediately upon closure of the drainage valve, the axial strain development initially occurs at a lower rate. The axial strains along this stress-path are very small, because the stresses push the yield surface out very little. If a drained stress-path was followed barely inside the undrained effective stress-path, then the stress state would move inside the yield surface and elastic axial expansion would be recorded due to decreasing stresses. It is therefore understandable why very small compressive strains are recorded along the undrained effective stress-path. This type of behavior was observed in all tests.

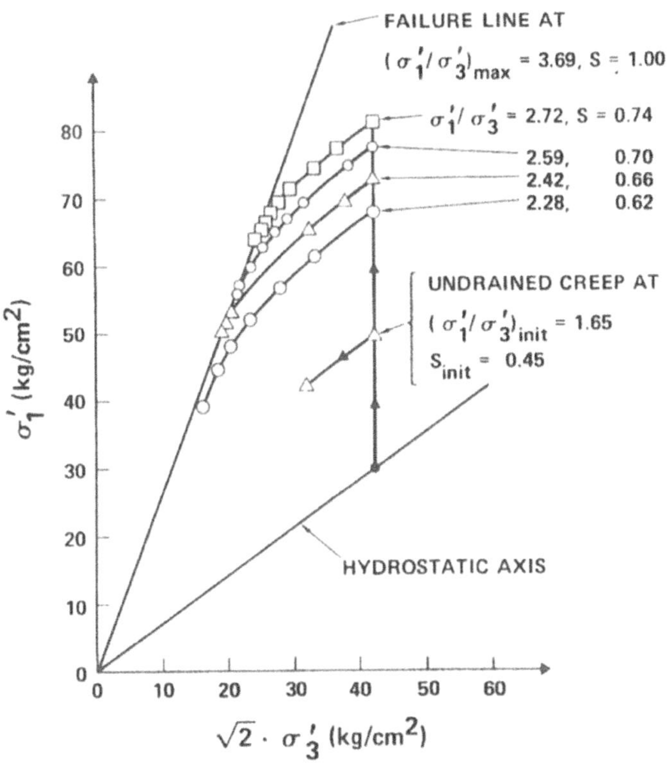

FIG. 8. Effective Stress-Paths for Stress-Controlled Triaxial Compression Tests on Loose Sacramento River Sand (1.0 kg/cm² = 98 kN/m²)

The effective stress-paths for the stress-controlled triaxial compression tests are shown in Fig. 8. The specimens were first loaded at constant confining pressure until the desired vertical loads were reached, corresponding to preselected stress levels. These stress levels were all well below that required to cause failure at the constant confining pressure. Thus, the specimens were perfectly stable at each of these stress levels before the drainage valve was closed. Following closure of the drainage valve, the effective stress-paths were forced into the shaded wedge in Fig. 2 due to the increasing pore pressures. Within this wedge the specimens became unstable and the effective stresses moved to failure with no further change in external load. Due to increasing cross-sectional areas of the specimens, the calculated stress differences, $(\sigma_1 - \sigma_3)$, decreased along these stress-paths. It should be emphasized that the undrained test is only used as a tool to produce effective stress-paths in the shaded wedge within which nonassociated flow has been shown to occur.

Discussion

In the study presented here instability was taken to be the inability to sustain or carry a given load, which includes the ability to sustain small perturbations in the loads. However, the possibility that instability might manifest itself as localization of plastic deformation and development of shear bands or in any other way was not overlooked. The specimens were therefore inspected after the tests (they were enclosed in a steel-walled triaxial apparatus and could not be observed during the tests) at which time they had sustained 15% to 30% axial strain. The instabilities were induced at axial strains of 2.8% to 5.3%. Thus, substantial amounts of strain had occurred after the instabilities were induced. Nevertheless, the specimens had bulged only slightly and there were no visible shear planes. In fact, a uniform distribution of horizontal wrinkles in the membrane was observed after each test, thus revealing a fairly uniform pattern of internal strains in the specimens. Instability did not appear to be associated with localization of plastic deformations and consequent shear banding of the type discussed by Lade (1982), Hettler and Vardoulakis (1984), and Peters et al. (1988). This is consistent with previous observations that shear planes are not observed in soils that compress (Lade 1982).

The yield surfaces at each of the stress points at which the drainage valve was closed and instability developed must have been inclined such as to open up in the outward direction of the hydrostatic axis. This is the only inclincation which would allow plastic strains and instability to occur while the stresses were decreasing. If, on the other hand, the yield surface is inclined such as to close in the outward direction of the hydrostatic axis (as at point A in Fig. 1), then stable behavior should be observed. To test this hypothesis, the drainage valve was closed at an effective stress ratio of $(\sigma_1'/\sigma_3') = 1.65$ corresponding to $S = 0.45$ in the test whose results are shown in Fig. 6. While the external loads were maintained constant, the pore pressure increased some due to a tendency for undrained volumetric creep as indicated in Fig. 8. The axial strains observed over a 30-min period are plotted versus time in Fig. 9. Very little axial strain (0.3%) occurred over this period. This strain was caused by creep in the elastic region within the teardrop shaped yield surface (see Fig. 1). Thus, stability is obtained in the region where the yield surface is closing towards the outward direction of the hydrostatic axis.

To test whether rate/time effects such as viscous flow or creep could cause the instability discussed above, the test whose results are shown in Fig. 6 was stopped at a stress ratio of $(\sigma_1'/\sigma_3') = 2.28$ coresponding to $S = 0.62$. This is precisely the same stress state at which one of the other tests was forced to exhibit instability by closing the drainage valve. In the current test the drainage valve was maintained open, and drained creep was observed for a period of 30 min. The axial creep strains are shown in Fig. 9. It is seen that creep could not be responsible for the large strains observed in the tests discussed above. The axial strain-time relation for the test in which instability was induced is also shown in Fig. 9 for comparison. It is clear that instability causes large strains to occur within relatively short time. The tendency for volumetric creep, however small it may be, causes the pore pressure to increase under undrained conditions, providing the small perturbation which renders the material unstable. However, the large strains discussed here could not be caused by creep or viscous flow, but are the effects of nonassociated flow combined with stress-paths within the shaded wedge in Fig. 2 where plastic volumetric compression occurs.

It should be emphasized that instability is not synonymous with failure. Thus, a failure surface does not go through each of the points at which instability was produced. In fact, these points are well within the failure

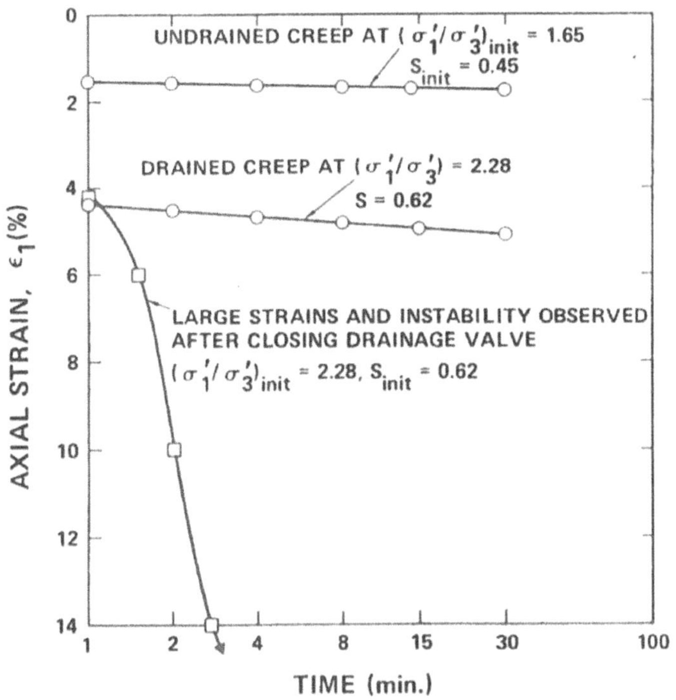

FIG. 9. Axial Strain-Time Relations in Semi-Log Diagram for Undrained Creep, Drained Creep, and Instability Observed in Stress-Controlled Triaxial Compression Tests on Loose Sacramento River Sand

surface as obtained from drained tests. Obviously, each of the specimens could have been loaded to higher stress differences as long as the stress-paths were outside the wedge between the yield and the plastic potential surfaces. Therefore, a failure condition expressed in terms of the maximum stress differences obtained from, e.g., undrained tests initiated at the hydrostatic axis (see Fig. 3) would clearly underestimate the true effective strength of the soil.

Material models which employ associated plastic flow rules must assume the failure surface to go through the point at which instability is produced. This is because stability is guaranteed according to Drucker's stability condition until the failure surface is reached, at which instability ensues. But then stable loading to higher stress differences, which is clearly possible, could not be predicted by such models. Further, since instability as well as stability can be obtained at any point in the region above the line roughly defined by the maximum stress differences in Fig. 3, it is not possible to define any one consistent failure surface to be used in a model which incorporates associated flow.

Stability Conditions for Granular Materials

Based on the arguments and the experimental evidence presented here and in Lade et al. (1987), it is possible to state the conditions for which granular materials exhibit stable behavior. These conditions take the form of additions to Drucker's stability condition [Inequality 1(*b*)], which by itself is sufficient to guarantee stability. However, this is not a necessary condition for stability of granular or frictional materials, i.e., stability may be obtained even when Inequality 1(*b*) is violated. Thus, the tests presented here, together with the work in Lade et al. (1987) indicate that stable material behavior exists when

$$\dot{\sigma}_{ij} \cdot \dot{\epsilon}^{p}_{ij} < 0 \qquad (7)$$

and

$$\frac{\partial f}{\partial \sigma_{ij}} \cdot \boldsymbol{\delta}_{ij} > 0 \qquad (8)$$

where $\boldsymbol{\delta}_{ij}$ = a vector on the hydrostatic axis. Inequality 8 expresses that the yield surface closes in the outward direction of the hydrostatic axis, i.e., stability is obtained along the "front portion" of the teardrop shaped yield surface.

Further, stability is also obtained when Inequality 7 holds and both

$$\frac{\partial g}{\partial \sigma_{ij}} \cdot \boldsymbol{\delta}_{ij} \leq 0 \qquad (9)$$

and

$$\frac{\partial f}{\partial W_p} \geq 0 \qquad (10)$$

Inequality 9 holds when the material exhibits plastic dilation or a tendency for plastic dilation in undrained tests in which the total volumetric strain is zero. Since instability is observed beyond failure, where both Inequalities 7 and 9 are satisfied, it is necessary to add Inequality 10, which expresses that loading and therefore plastic yielding occurs *inside* the failure surface. Thus, stability is obtained for stress-paths within the shaded wedge (see Fig. 2) as long as the plastic volumetric strain show dilation or a tendency for dilation. Experiments to support this stability condition were presented in Lade et al. (1987).

Conditions for Instability of Granular Materials

The conditions for stability stated above may be reversed to produce requirements for instability of granular materials. These may be important and sometimas easier to employ than the stability conditions. Specifically, instability will occur when Inequality 7 is fulfilled together with

$$\frac{\partial f}{\partial W_p} < 0 \qquad (11)$$

which expresses the fact that softening occurs *beyond* failure. This condition for instability is implied in Drucker's stability condition (Inequality 1*b*).

Instability is also obtained *inside* the failure surface when Inequality 7 is combined with

$$\frac{\partial g}{\partial \sigma_{ij}} \cdot \boldsymbol{\delta}_{ij} > 0 \qquad (12)$$

and

$$\frac{\partial f}{\partial \sigma_{ij}} \cdot \delta_{ij} < 0 \quad (13)$$

Inequality 12 holds when the material exhibits plastic volumetric compression or a tendency for plastic compression under undrained conditions where the total volumetric strain is zero. Inequality 13 stipulates that the yield surface must open up in the outwards direction of the hydrostatic axis. Thus, instability is obtained for stress-paths within the shaded wedge (see Fig. 2) when the plastic volumetric strains are compressive or show a tendency for compression under undrained conditions. Again, these results are based on experimental evidence supporting these conditions for instability.

It should be noted that the elastic strains will maintain stability slightly beyond the conditions for instability stated. For example, stability is recovered along the undrained effective stress-path from point D to point E in Fig. 5(*b*) due to elastic expansion. The conditions for instability stated above are fulfilled from D to E, but stability is obtained, because the plastic compression goes to zero at point E and the elastic expansion produces stable behavior beginning at point D. Similar stable behavior is obtained immediately before point B in Fig. 5(*a*) due to elastic expansion. The conditions for instability are fulfilled a little before point B is reached, where the yield surface is parallel with the hydrostatic axis. However, as the elastic strains become negligible, the stated conditions for instability prevail.

Finally, plastic work, W_p, is used as the hardening parameter to indicate loading or hardening in Inequality 10 and softening in Inequality 11. Any other measure of hardening, consistent with the findings presented above, may be used in lieu of plastic work to indicate loading inside the failure surface and softening beyond failure.

Conclusion

The stability condition formulated by Drucker provides a sufficient condition for stability and is satisfied for materials that obey an associated plastic flow rule. Frictional materials exhibit nonassociated plastic flow, and the stability of such materials has often been questioned. To provide experimental evidence regarding the stability of granular materials, two series of triaxial tests were performed under stress control so that any instability in the material could freely develop.

In one series of tests previously reported (Lade et al. 1987), specimens of dense sand were exposed to stress-paths with decreasing effective stresses and increasing stress levels. The results showed that the granular material dilated during shear, it followed a nonassociated flow rule, and stability was obtained in all cases until the failure surface was reached. Thus, stable behavior was observed in a region in which Drucker's stability postulate was violated.

In the second series of tests reported here, specimens of loose sand were also exposed to stress-paths with decreasing effective stresses and increasing stress levels. The results of these tests showed that the granular material tended to compress during shear, it followed a nonassociated flow rule, and instability was obtained in all cases *inside* the failure surface. The possible influence of viscous effects on the instability of the material, and the effect of elastic energy release were investigated experimentally and discussed in detail. None of these effects could account for the instability of the material.

The occurrence of stable and unstable behavior found in these two studies is related to the volume change characteristics of the material. Stable behavior is found when the material dilates, and unstable behavior may occur when the material compresses during shear.

Stability conditions for granular materials exhibiting nonassociated flow are formulated on the basis of the experimental results obtained in the two studies. These conditions take the form of augmentations to Drucker's stability postulate. Conditions for instability of granular materials are also presented. Both types of conditions are expressed in terms of the orientations of the plastic potential surface and the yield surface.

Appendix I. References

Drucker, D. C. (1951). "A more fundamental approach to stress-strain relations." *Proc., First U.S. Nat. Congress Appl. Mech.*, 487–491.

Drucker, D. C. (1956). "On uniqueness in the theory of plasticity." *Quarterly Appl. Math.*, 14, 35–42.

Drucker, D. C. (1959). "A definition of stable inelastic material," *J. Appl. Mech.*, 26, 101–106.

Frydman, S., Zeitlen, J. G., and Alpan, I. (1973). "The membrane effect in triaxial testing of granular soils." *J. Testing and Evaluation*, ASTM, Philadelphia, Pennsylvania, 1(1), 37–41.

Hettler, A., and Vardoulakis, I. (1984). "Behaviour of dry sand tested in a large triaxial apparatus." *Geotechnique*, 34(2), 183–197.

Lade, P. V. (1982). "Localization effects in triaxial tests on sand." *Proc., Symp. on Deformation and Failure of Granular Materials*, International Union of Theoretical and Applied Mechanics, P.A. Vermeer and H. J. Luger, eds., Delft, Holland, Aug., 461–471.

Lade, P. V., and Duncan, J. M. (1973). "Cubical triaxial tests on cohesionless soil." *J. Soil Mech. and Found. Div.*, ASCE, 99(SM10), 793–812.

Lade, P. V., and Hernandez, S. B. (1977). "Membrane penetration effects in undrained tests." *J. Geotech. Engrg. Div.*, ASCE, 103(GT2), 109–125.

Lade, P. V., Nelson, R. B., and Ito, Y. M. (1987). "Nonassociated flow and stability of granular materials." *J. Engrg. Mech.*, ASCE, 113(9), Sept. 1302–1318.

Lee, K. L. (1965). "Triaxial compressive strength of saturated sand under seismic loading conditions." Thesis presented to the University of California, at Berkeley, California, in partial fulfillment of the requirements for the degree of Doctor of Philosophy.

Lee, K. L., and Seed, H. B. (1967). "Drained strength characteristics of sands." *J. Soil Mech. and Found. Div.*, ASCE, 93(SM6), 117–141.

Mandel, J. (1964). "Conditions de stabilite et postulat de Drucker." *Proc. Int. Union of Theoretical and Appl. Mech., Symp. on Rheology and Soil Mechanics*, Grenoble, France, 58–67.

Martin, G. R., Finn, W. D. L., and Seed, H. B. (1978). "Effects of system compliance on liquefaction tests." *J. Geotech. Engrg. Div.*, ASCE, 104(GT4), April, 463–479.

Mroz, Z. (1963). "Nonassociated flow laws in plasticity." *J. de Mechanique*, 2(1), 21–42.

Peters, J. F., Lade, P. V., and Bro, A. (1988). "Shear band formation in triaxial and plane strain tests." *Advanced Triaxial Testing of Soil and Rock*, ASTM XXX, R. T. Donaghe, R. C. Chaney and M. L. Silver, eds., Amer. Soc. Test. Mat.

Seed, H. B., and Lee, K. L. (1967). "Undrained strength characteristics of cohesionless soils." *J. Soil Mech. and Found. Div.*, ASCE, 93(SM6), 333–360.

Chapter VIII
Effects of Non-Associated Plastic Flow in Soils

Preface

Because the author had used non-associated plastic flow in the first and second constitutive models for sands (Lade and Duncan 1975, Lade 1977), several of Drucker's students at various conferences in the 1970s and 1980s pointed out to the author that stability could not be guaranteed, and that was presumably a problem! However, as shown in the articles, this is exactly correct for frictional materials like sands, in which two types of "anomalous" behavior may be encountered: (1) full stability is observed in triaxial compression tests on dense sand that dilate during shearing, even though the plastic flow was clearly non-associated, but shear banding develops in the softening regime in triaxial compression tests, and (2) instability inside the failure surface may lead to liquefaction in loose sand that contracts during shearing.

References

Lade, P.V. and Duncan, J.M. (1975) "Elastoplastic Stress-Strain Theory for Cohesionless Soil," *Journal of the Geotechnical Engineering Division*, ASCE, 101(GT10): 1037-1053.

Lade, P.V. (1977) "Elasto-Plastic Stress-Strain Theory for Cohesionless Soil with Curved Yield Surfaces," *International Journal of Solids and Structures*, 13: 1019-1035.

PERGAMON

International Journal of Solids and Structures 39 (2002) 3337–3357

INTERNATIONAL JOURNAL OF SOLIDS and STRUCTURES

www.elsevier.com/locate/ijsolstr

Instability, shear banding, and failure in granular materials

Poul V. Lade *

Department of Civil Engineering, Aalborg University, Sohngaardsholmsvej 57, 9000 Aalborg, Denmark

Received 13 February 2002

Abstract

Two modes of decrease in load bearing capacity of granular materials are discussed in view of experimental results. Both relate to the fact that frictional materials exhibit nonassociated plastic flow and they undergo considerable volume changes, either contraction or dilation. One mode consists of the instability that may occur in certain regions of stress space and potentially result in liquefaction of the granular material. It is the fact that loading of contracting soil (resulting in large plastic strains) can occur under decreasing stresses that may lead to unstable behavior under undrained conditions. As long as the soil remains drained, it will remain stable in the region of potential instability. The other mode is initiated by localization of plastic strains and subsequent development of shear bands, which in granular materials is followed by a decrease in load bearing capacity. These two modes are mutually exclusive and they occur for different loading and material conditions as discussed here on the basis of experimental observations. © 2002 Elsevier Science Ltd. All rights reserved.

Keywords: Contraction; Dilation; Failure; Granular materials; Instability; Liquefaction; Nonassociated flow; Shear banding

1. Introduction

Several modes of decrease in load bearing capacity, often referred to as failure, have been identified from experiments on granular materials. The strength of granular materials is primarily proportional with the effective confining pressure, and the load bearing capacity therefore decreases with decreasing confining pressure. Reduction in effective confining pressure may be caused by increasing pore pressure and will result in failure. Another well-known mode is the smooth peak failure occurring at constant confining pressure in which the internal structure of the granular material is loaded and deformed sufficiently to cause collapse of load bearing chains of particles. The resulting reduction in load bearing capacity is characterized as material softening, and it involves macroscopically uniform deformations and stresses such that the granular material can be characterized as a continuum.

Two additional modes of decrease in load bearing capacity have been observed. Both relate to the fact that frictional materials exhibit nonassociated plastic flow and they undergo considerable volume changes, either contraction or dilation. One mode consists of the instability that may occur in certain regions of

* Fax: +45-98-142555.
E-mail address: lade@civil.auc.dk (P.V. Lade).

0020-7683/02/$ - see front matter
PII: S0020-7683(02)00157-9

stress space and potentially result in the well-known phenomenon referred to as liquefaction. Another mode is initiated by localization of plastic strains and subsequent development of shear bands, which in granular materials is followed by a decrease in load bearing capacity. These latter two modes are mutually exclusive and they occur for different loading and material conditions as discussed here on the basis of experimental observations.

Results of previously performed series of triaxial compression tests designed to expose the type of behavior displayed by granular materials have shown that materials that exhibit nonassociated flow do not obey the stability postulates by Drucker and Hill. Triaxial tests on fully saturated and partly saturated specimens of sand have been performed under drained and undrained conditions to study the regions of stable and unstable behavior. Granular materials may become unstable *inside* the failure surface if the state of stress is located between the instability line and the failure surface for the material. Thus, instability is not synonymous with failure, although both may lead to catastrophic events such as gross collapse of earth structures. Conventional slope stability methods do not capture the mechanics of instability and subsequent liquefaction that may occur in gently inclined submarine slopes, in tailings dams, and in earth and snow avalanches. It is the fact that loading of a contracting soil (resulting in large plastic strains) can occur under decreasing stresses that leads to unstable behavior under undrained conditions. Loose, fine sands and silts have sufficiently low hydraulic conductivities that small disturbances in load or even small amounts of volumetric creep may temporarily produce undrained conditions in such soils, and instability of the soil mass follows. As long as the soil remains drained, it will remain stable in the region of potential instability.

The occurrence of shear banding in the plastic hardening regime limits the strength of the soil. Shear banding occurs in the softening regime in conventional triaxial compression tests. However, true triaxial tests have been performed to study the effect of shear banding on failure in the full range of the intermediate principal stress. The experiments show that the strength increases as $b = (\sigma_2 - \sigma_3)/(\sigma_1 - \sigma_3)$ increases from 0 to about 0.18, then remains almost constant until b reaches 0.85, and then decreases slightly at $b = 1.0$. Shear banding initiates in the hardening regime for b-values of 0.18–0.85. Thus, peak failure is caused by shear banding in this middle range of b-values, and a smooth, continuous three-dimensional failure surface is therefore not obtained in general for soils. Analysis based on theoretical conditions and experimental results agree on the occurrence of shear banding and its consequent effect on the three-dimensional failure conditions for soil.

2. Instability

Experimental evidence from tests on several types of soils have clearly indicated that the use of conventional associated flow rules in formulation of elasto-plastic constitutive models results in prediction of too large volumetric expansion. To characterize the volume change correctly, it is necessary to employ a nonassociated flow rule. The plastic potential surfaces do therefore not coincide with the yield surfaces, but the two families of surfaces cross each other.

The application of nonassociated flow rules for soil have resulted in questions regarding uniqueness and stability of such materials. The stability postulate for time-independent materials due to Drucker (1951, 1956, 1959) is satisfied provided that associated plastic flow is employed in construction of constitutive models involving convex, plastic yield surfaces. Hill's stability condition (Bishop and Hill, 1951; Hill, 1958) is expressed in terms of total strain increments (elastic and plastic), and it extends the condition for stability a little beyond that due to Drucker. Theoretical considerations have suggested that they are not necessary conditions (Mroz, 1963; Mandel, 1964).

Experiments to study whether or not granular materials are unstable under certain load conditions have been performed. In these studies instability was taken to be the inability of the material to sustain or carry a given load. This includes the inability to sustain small perturbations in the load. Several series of triaxial compression tests designed and performed to expose the type of behavior displayed by granular materials are reviewed. The results of these experiments show that a positive value of the second increment of plastic work is neither a necessary nor a sufficient condition for stability of granular (frictional) materials. Experimental observations of stability and instability are reported, and stability conditions for materials with nonassociated flow are presented.

2.1. Stability postulates

Stability as formulated by Drucker (1951, 1956, 1959) contains two elements as expressed by plastic work increments:

$$dW_p - dW_p^* = (\sigma_{ij} - \sigma_{ij}^*)\dot{\varepsilon}_{ij}^p \geqslant 0 \tag{1}$$

$$d^2W_p = \dot{\sigma}_{ij}\dot{\varepsilon}_{ij} \geqslant 0 \tag{2}$$

in which σ_{ij} is a stress state on the yield surface, σ_{ij}^* is a stress state inside the yield surface, $\dot{\sigma}_{ij}$ is stress increment starting at σ_{ij} causing plastic strain increment $\dot{\varepsilon}_{ij}$.

Eq. (1) is commonly referred to as stability in the large, and it implies that the yield surface must be convex. Eq. (2) is referred to as stability in the small, and it is fulfilled if the plastic strain increment vector is normal to the yield surface. This implies that the plastic potential surface is identical to the yield surface, so that when yielding occurs, the plastic strain increment is normal to the yield surface. Therefore, Drucker's stability postulates imply associated flow rules, and they are applicable for solid metals, which can be reasonably characterized by associated flow rules.

Eq. (2) has often been taken to define stable and unstable behavior. If the condition is satisfied for any stress increment, then the material is unconditionally stable. The material may be conditionally stable, if Eq. (2) is fulfilled for some stress increments and not for others. The material will be completely unstable if Eq. (2) is not satisfied for all possible stress increments. Eq. (2) is always negative for the condition following peak failure, which results in complete instability.

Bishop and Hill (1951) and Hill (1958) extended the condition for stability slightly beyond that proposed by Drucker by expressing it in terms of total strain increments:

$$d^2W = \dot{\sigma}_{ij}\dot{\varepsilon}_{ij}^t = \dot{\sigma}_{ij}(\dot{\varepsilon}_{ij}^e + \dot{\varepsilon}_{ij}^p) = \dot{\sigma}_{ij}\dot{\varepsilon}_{ij}^e + \dot{\sigma}_{ij}\dot{\varepsilon}_{ij}^p \geqslant 0 \tag{3}$$

in which $\dot{\varepsilon}_{ij}^t$ is the total strain increment, and $\dot{\varepsilon}_{ij}^e$ is the elastic strain increment.

Stability postulates developed by Drucker, and by Bishop and Hill provide sufficient conditions to ensure stability. However, theoretical considerations by Mroz (1963), Mandel (1964), and Nemat-Nasser (1983) have suggested that they are not necessary considerations for stability.

Experimental evidence for soils has generally indicated that conventional associated flow rules consistently predict too large volumetric dilation, which is indicated by the condition that the plastic strain increment is not normal to the yield surface. Therefore, proper soil constitutive models need incorporate nonassociated flow rules, in which there is a separate plastic potential surface, which is not coincidental with the yield surface.

2.2. Consequences of nonassociated flow

A typical pattern of yield surfaces for an isotropic soil is shown on the triaxial plane in Fig. 1. In three dimensions these yield surfaces are shaped as asymmetric tear drops. For an isotropic material the yield surfaces intersect the hydrostatic axis at right angles, they bend smoothly backwards towards the origin, and they cross the failure surface at sharp angles.

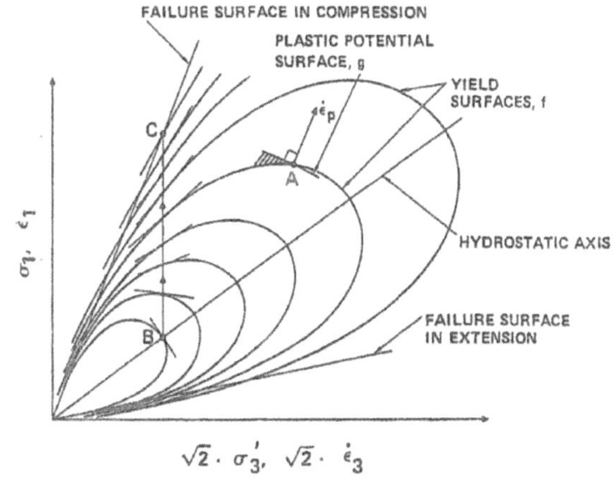

Fig. 1. Pattern of yield surfaces for isotropic soils. Stress path for conventional triaxial compression test (BC).

Plastic potential surfaces for nonassociated flow have similar shapes as yield surfaces, but the two families of surfaces cross each other. Experimental evidence for granular materials indicate that the plastic potential surfaces have more pointed ends and they resemble cigars with asymmetric cross-sections. A typical plastic potential surface is shown at point A in Fig. 1.

The shaded wedge between the yield surface and the plastic potential surface defines a region in which Eq. (2) is not fulfilled, because a stress increment starting at point A and contained in this wedge will produce a negative scalar product. With a negative scalar product Drucker's stability postulate indicates that the material should exhibit unstable behavior. However, drained experiments show that the sand is perfectly stable at stress points where the normal to the yield surface points in the outward direction of the hydrostatic axis. For this condition the deviator stress can be safely increased to produce further plastic shear strains (work-hardening). In other words, the sand can sustain higher loads and behave in an inelastic manner without undergoing any instability or collapse.

Instability may occur when the yield surface opens up in the outward direction of the hydrostatic axis. This allows plastic strains (loading) to occur while the stresses are decreasing. Here loading occurs inside the failure surface and instability may develop in the form of inability to sustain the current deviator stresses.

Fig. 1 shows the stress path for a conventional triaxial compression test performed at constant confining pressure. As the specimen is loaded from B to C the inclination of the yield surface changes. At low deviator stresses near point B, the yield surface is inclined towards the outwards direction of the hydrostatic axis. As loading proceeds, the inclination of the yield surface changes gradually and becomes inclined towards the origin as failure is approached at point C. It is in this region of high deviator stresses where the yield surface is inclined toward the origin that instability may occur. This region is indicated in Fig. 2.

Fig. 2 shows a wedge between the current yield surface f and the plastic potential surface g corresponding to the current stress point. For a dilating material this wedge shaped region is located within a larger region bounded by lines corresponding to (σ_1'/σ_3') = const. and $(\sigma_1 - \sigma_3)$ = const. All stresses, including the stress difference, are decreasing within the wedge between f and g, but the stress ratio (σ_1'/σ_3') is increasing in this region. By performing triaxial tests with stress paths located in this region, experimental evidence regarding the instability of materials with nonassociated flow can be obtained.

It is important to recognize that the material behavior obtained for stress paths within the shaded wedge in Fig. 2 corresponds to work-hardening with positive plastic work, $dW_p > 0$, and outward motion of the

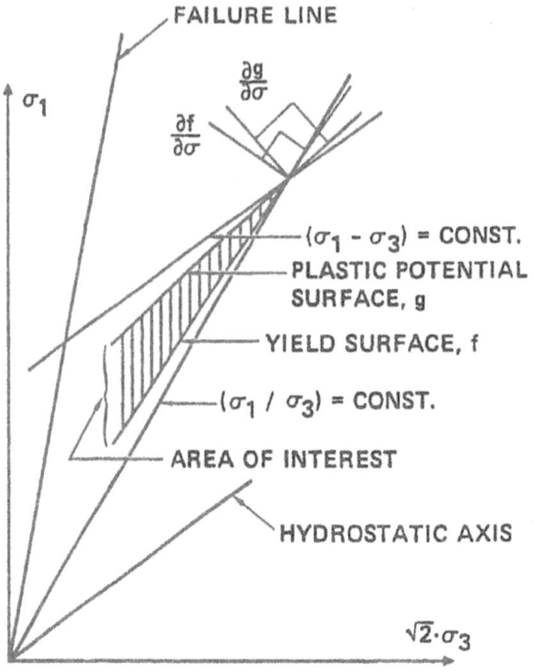

Fig. 2. Wedge-shaped region of stress paths with decreasing stresses in which soils with nonassociated flow may be unstable during hardening inside the failure surface.

yield surface. Although the stresses in any direction within the wedge are decreasing, failure has not been reached and softening of the material is therefore not occurring.

2.3. Stability in drained compression

Several series of triaxial compression tests were performed to study stability and instability in granular materials. The details of these experiments are given by Lade et al. (1987, 1988) and by Lade and Pradel (1990).

To demonstrate that nonassociated plastic flow is obtained for granular materials, the results of a drained triaxial test on fine silica sand are shown in Fig. 3. This test was performed with a stress path inside the failure surface involving primary loading (hardening) with decreasing stress difference $(\sigma_1 - \sigma_3)$ and decreasing confining pressure σ_3. If stresses and strain increments are plotted on the same diagram, as in Fig. 3(b), the direction of the plastic strain increment vector is uniquely determined from the state of stress, and it is independent of the stress path leading to this state of stress as shown by Poorooshasb et al. (1966) and Lade and Duncan (1976). The stress path with decreasing stresses, shown in Fig. 3(b), is so steep as to form an obtuse angle $(\beta - \alpha) = 110°$ with the direction of the plastic strain increment vector. Thus, the condition in Eq. (2) is violated.

The fact that plastic yielding is occurring along the stress path with decreasing stresses is seen from the stress–strain curve in Fig. 3(a). Section BC on the stress path is labeled similarly on the stress–strain curve which indicates large plastic strains between B and C. Since the yield surface is being pushed out, it must be steeper than the stress path direction BC. If it were less steep, so as for example to be perpendicular to the plastic strain increment vector thereby indicating associated flow, then point C would be inside the yield

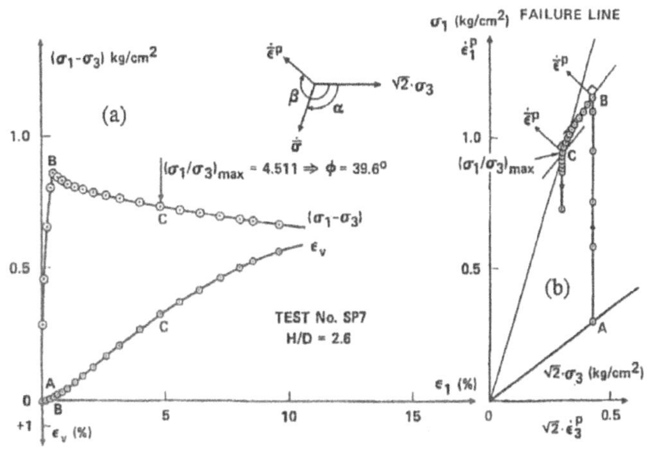

Fig. 3. (a) Stress–strain and volume change curves and (b) stress path and plastic strain increment vectors for test on fine silica sand.

surface passing through point B corresponding to unloading from B to C. This would contradict the large plastic strains shown in Fig. 3(a). Consequently, the yield surface must be steeper than stress path BC, and nonassociated plastic flow is therefore clearly indicated.

In order to expose the potential instabilities in soil behavior that might occur for the stress paths in the prefailure region described earlier (see Fig. 2), it is necessary to perform triaxial tests under stress control. A series of drained tests similar to that shown in Fig. 3 was therefore performed under stress control (Lade et al., 1987). In fact, the test shown in Fig. 3 is one of these tests. Seven experiments were performed. They all exhibited plastic dilation, and they all clearly showed nonassociated flow. When the sand specimens were exposed to stress paths in the region of potential instability, *none was observed*, i.e. run-away instability was not encountered. The tall cylindrical specimens with lubricated ends deformed as perfect right cylinders with no developing shear bands, no bulging or shearing in a nonuniform manner, or any other signs of instability. Little but negligible creep was present in the sand specimens. Thus, stable behavior was observed in the region in which Drucker's stability postulate was violated.

Regardless of the ability to define yield surfaces from experiments, the fact remains that negative values of d^2W_p are obtained from direct experimental measurements. The fact that the specimens remained stable implies that the behavior is not strain softening, i.e. the peak strength has not been exceeded. Therefore, to model this behavior through any type of normality would require a serious departure from plasticity theory.

In order to investigate whether the type of volume change (dilation or contraction) was important for stability of granular soils, a second series of tests was performed on sand that contracted during shear (Lade and Pradel, 1990). Drained tests with stress paths within the shaded wedge shown in Fig. 2 were performed on loose sand that exhibited contraction during shear. Although Drucker's stability condition was violated inside the failure surface, none of the specimens showed any signs of run-away instability.

The results of these two series of triaxial compression tests demonstrated that drained conditions produced stable behavior irrespective of (1) the sign of the second increment of work (2) the sign of the volumetric strain (dilation or contraction), and (3) the sign and direction of the increments in stress components. Thus, under fully controlled conditions such as those prevailing for drained conditions, unstable run-away situations cannot occur. These results therefore clearly show that Drucker's stability postulate is not a necessary condition for stability.

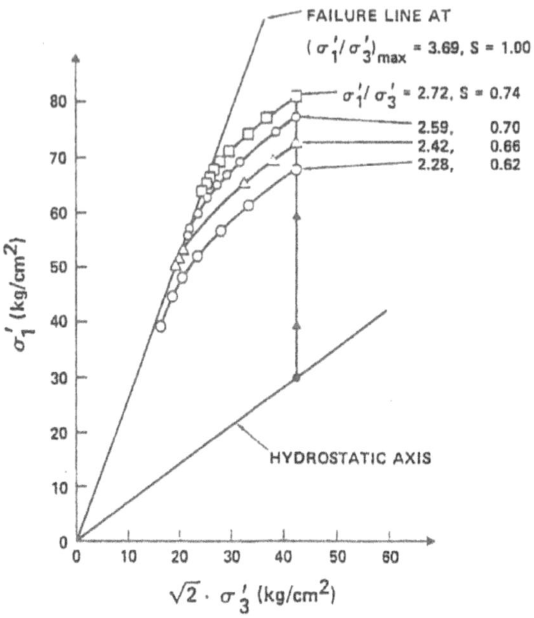

Fig. 4. Effective stress paths for stress-controlled triaxial compression tests on loose Sacramento River sand.

2.4. Instability in undrained compression

On the other hand, fully saturated soils that tend to contract during shear may become unstable *inside* the failure surface and this may lead to liquefaction. A series of stress controlled tests was performed on granular soil that tended to contract during shear (Lade et al., 1988). The specimens were exposed to stress paths within the shaded wedge shown in Fig. 2. Fig. 4 shows the actual stress paths followed in these tests, and Fig. 5 shows the observed stress–strain, volume change, and pore pressure development in one of the tests. In each test the saturated specimen was first loaded under drained conditions to a preselected stress level S (expressed as the ratio of the current to the maximum stress difference at a given confining pressure). The drainage valve was then closed and instability developed in each specimen due to increasing pore pressure, i.e., the specimens could not sustain the applied load. The tendency for volumetric creep, however small it may be, caused the pore pressure to increase under undrained conditions, providing the small perturbation which rendered the material unstable. However, the large strains observed along the unstable stress paths could not be caused by creep or viscous flow as shown by Lade et al. (1988). The effective stress paths shown in Fig. 4 were within the shaded wedge in Fig. 2, the specimens exhibited nonassociated flow and plastic volumetric contraction (although the total volumetric strain was zero), and instability was obtained in all cases in the hardening regime *inside* the failure surface.

In the studies presented here instability was taken to be the inability to sustain or carry a given load, which includes the ability to sustain small perturbations in the loads. However, the possibility that instability might manifest itself as localization of plastic deformation and development of shear planes or in any other way was not overlooked. The specimens were therefore inspected after the tests (they were enclosed in a steel walled triaxial apparatus and could not be observed during the tests) at which time they had sustained 15–30% axial strain. The instabilities were induced at axial strain of 2.8–5.3%. Thus, substantial amounts of strain had occurred after the instabilities were induced. Nevertheless, the specimens

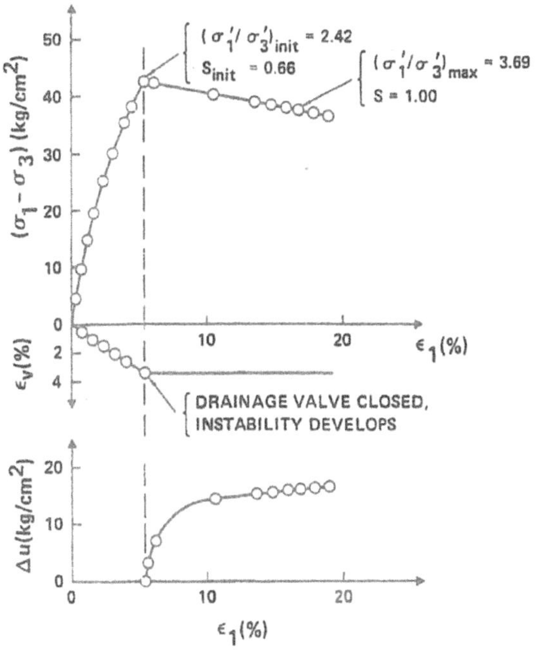

Fig. 5. Stress–strain, volume change, and pore pressure relations in stress-controlled triaxial compression test on loose Sacramento River sand.

had bulged only slightly and there were no visible shear planes. In fact, a uniform distribution of horizontal wrinkles in the membrane was observed after each tests, thus revealing a fairly uniform pattern of internal strains in the specimens. Instability did not appear to be associated with localization of plastic deformations and consequent shear banding of the type discussed by Lade (1982) and Peters et al. (1988). This is consistent with previous observations that shear planes are not observed in soils that contract (Lade, 1982).

2.5. Observations for partly saturated sand

Stress paths with only one particular direction is obtained from undrained tests on fully saturated soils at each particular stress state, as shown in Fig. 4. In order to study the soil behavior for other stress path directions under undrained conditions, a series of tests, similar to those described above, was performed on specimens with decreasing degree of saturation (Lade and Pradel, 1990). By purposely introducing a controlled amount of compressible air into the specimens, the undrained effective stress paths could be pointed in different directions within the shaded wedge in Fig. 2. The specimens in the tests in which the effective stress paths were inside the shaded wedge exhibited unstable behavior. Tests with stress path above the shaded wedge, created by introducing an amount of air in the specimen greater than critical, indicated stable behavior.

The conclusions from the experiments are that sand that tends to contract can become unstable if the degree of saturation is sufficiently high and if the drainage condition is changed from drained to undrained. This may occur in fine sands and silts in which the hydraulic conductivity is relatively low, thus enhancing the undrained condition in the presence of relatively fast loading conditions.

2.6. Instability of dilating sand

In a discussion of the experimental results reported above, Peters (1991) suggested that even dilating sand may become unstable: if a rate of volumetric expansion were imposed on an element of granular material and this rate exceeded the rate of expansion exhibited by the material, then the effective confining pressure would decrease and the element would become unable to sustain the current, applied shear stress. According to the definition, the material element would become unstable. Thus, a rate of volume change (whether positive, zero, or negative) imposed on a material element could cause it to become unstable depending on the rate of volume change exhibited by the soil itself (contraction of dilation). The soil exhibits an inherent rate of volume change at a given point in stress space (as determined from the direction of the plastic strain increment), but small variations in rate of volume changes are associated with the stress path direction. This is because the total volume change consists of plastic and elastic contributions, and the latter is dependent on the stress path direction. Thus, to study this hypothesis experimentally, a series of triaxial compression tests was performed with appropriate test conditions. The results confirm the suggestion that even dilating sand may become unstable (Lade et al., 1993).

Another test was performed differently from the previous experiments. Instead of reducing the effective confining pressure through injection of water and consequent increase in pore pressure, the test was performed in a drained condition and the total confining pressure was reduced. The specimen was unable to hold the load, i.e. it behaved in an unstable manner. Thus, instability of the specimen is conditional on a progressive reduction in effective confining pressure. The similarity between these test results indicates that instability of dilating sand can be produced by decreasing effective confining pressure whether it is caused by increasing pore pressure or decreasing total confining pressure.

2.7. Conditions for stability and instability

The stability of granular materials which exhibit nonassociated flow has been investigated on the basis of experimental observations. Several series of triaxial compression tests designed to expose the type of behavior exhibited by granular materials were performed. The variables in these studies were the sign of the second work increment (positive or negative), the volumetric strain behavior (contraction or dilation), the constraints on the volumetric behavior (free or controlled), and the degree of saturation (fully or partly saturated).

If physical instability is defined as a condition for which the current, applied shear stress cannot be sustained for perturbations in the state of stress, then contractive as well as dilative materials may be considered to be unstable in the region where the yield surface opens up in the outward direction of the hydrostatic axis. In this region, plastic strains can be produced under decreasing stresses. For undrained conditions and contractive material, the instability is self-sustaining and unconditional, i.e. it is not dependent on conditions outside the soil element. For drained conditions, the instability is conditional, i.e. the decrease in load carrying capability depends on the reduction in effective confining pressure. This reduction may occur as a decrease in total confining pressure or as an increase in pore pressure caused by injection of water into the soil element.

Figs. 6 and 7 show schematic diagrams of the conditions for stability and instability of contracting and dilating granular materials. The shear stress is represented by the stress deviator invariant J_2' and the mean normal stress is indicated by the stress invariant I_1. The generic yield surfaces, f, and plastic potential surfaces, g, cross each other at points (A) where the yield surfaces open up in the outward direction of the hydrostatic axis. Stability is obtained when J_2' is constant or increases, whereas instability occurs when J_2' decreases. The signs of the second increment of work d^2W can be negative in the region of stability, whereas d^2W can be positive in the region of instability for dilating material. The stability postulates by Drucker and

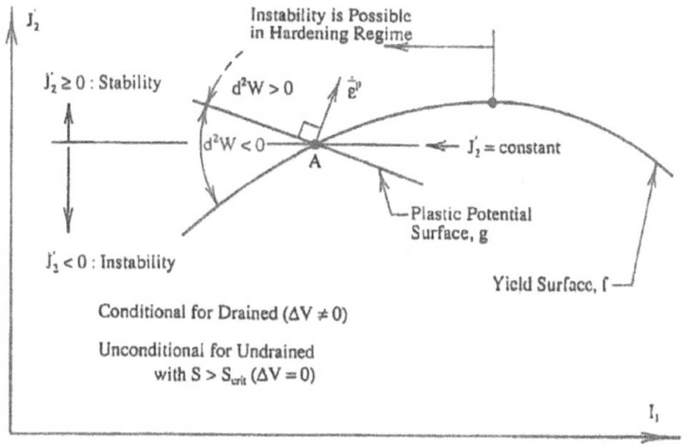

Fig. 6. Schematic diagram of conditions for stability and instability of contracting granular material.

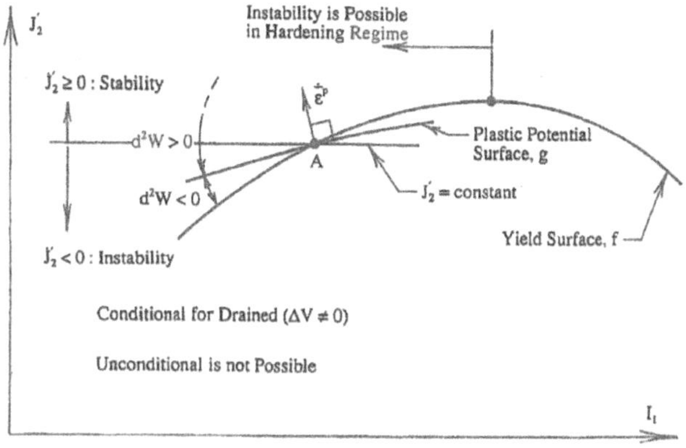

Fig. 7. Schematic diagram of conditions for stability and instability of dilating granular material.

by Hill therefore provide neither necessary nor sufficient conditions for stability of granular (frictional) materials.

2.8. *Region of potential instability*

It is the fact that loading of a contracting soil (resulting in large plastic strains) can occur under decreasing stresses that leads to unstable behavior under undrained conditions. Loose, fine sands and silts have relatively low hydraulic conductivities, and small disturbances in load or even small amounts of volumetric creep may produce undrained conditions in such soils, and instability of the soil mass follows. As long as the soil remains drained, it will remain stable in the region of potential instability.

When the condition of instability is reached, the soil may not be able to sustain the current stress state. This stress state corresponds to the top of the current yield surfaces as shown schematically on the $p' - q$ diagram in Fig. 8. Following this top point the soil can deform plastically under decreasing stresses. The top of the undrained effective stress path, corresponding to $(\sigma_1 - \sigma_3)_{max}$, occurs slightly after but close to the top of the yield surface. The difference is due to the additional stability provided by the elastic strains, as expressed in Eq. (3).

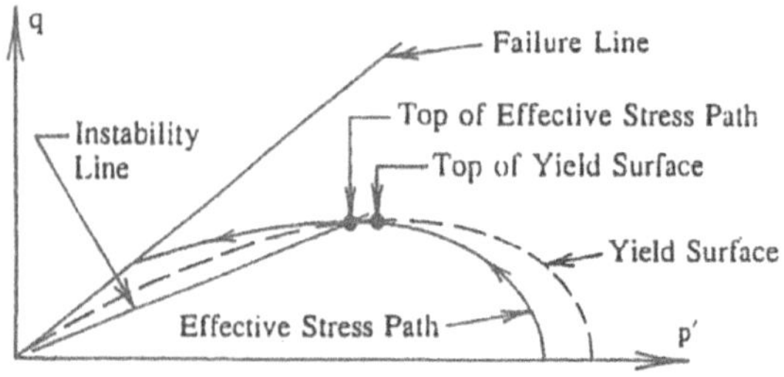

Fig. 8. Location of instability line for loose sand.

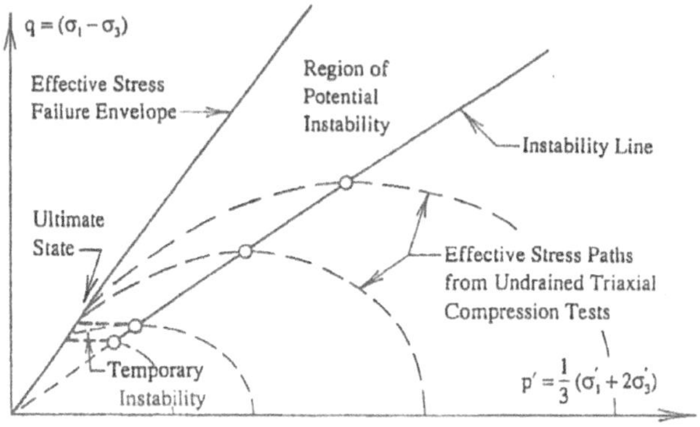

Fig. 9. Schematic diagram of location of instability line in $p' - q$ diagram.

In order for a granular material to become unstable, the state of stress must be located on or above the instability line. Fig. 9 shows a schematic $p' - q$ diagram in which the line connecting the tops of a series of effective stress paths from undrained tests on loose soil provides the lower limit of the region of potential instability. In the region above this instability line the soil can deform plastically under decreasing stresses. Experiments show that this line is straight. Since it goes through the top points of the yield surfaces which evolve from the origin, as shown in Fig. 1, the instability line also intersects the stress origin. Fig. 9 also shows a region of temporary instability which is located in the upper part of the dilating zone. It is a region where instability may initially occur, but conditions allow the soil to dilate after the initial instability, thus causing the soil to become stable again. For very loose soils the region of potential instability reaches down to the origin of the stress diagram.

2.9. Instability versus failure

Instability and failure are two different behavior aspects of soils which exhibit nonassociated flow (Lade, 1989). Although both may lead to catastrophic events, they are not synonymous. However, two criteria for "failure" are commonly used in geotechnical engineering for interpretation of results of triaxial tests on soils: (1) failure occurs when the stress difference reaches a limiting value, $(\sigma_1 - \sigma_3)_{max}$, and (2) failure occurs when the effective principal stress ratio reaches a limiting value, $(\sigma'_1/\sigma'_3)_{max}$. The two conditions are reached simultaneously in drained tests. The confusion as to the definition of failure arises in interpretation of undrained tests on loose sands and sensitive clays in which the pore pressures increase monotonically during shear. For these types of soils the maximum stress difference is reached before the maximum effective stress ratio. The choice of failure criterion varies with the purpose for which the strength is to be used. Typically, $(\sigma_1 - \sigma_3)_{max}$, is employed in total stress analyses, and $(\sigma_1'/\sigma_3')_{max}$ is used in effective stress analyses. These issues have been discussed at length by e.g. Bjerrum and Simons (1960), Seed et al. (1960), and Whitman (1960).

Based on the presentation above, it is clear that the condition of maximum stress difference does not correspond to a true failure condition, i.e. it occurs inside the failure surface. Rather, it corresponds to a condition of minimum stress difference at which instability may develop *inside* the true failure surface. Stress states can be reached above the minimum stress difference condition described by $(\sigma_1 - \sigma_3)_{max}$, and instability can be induced anywhere between the $(\sigma_1 - \sigma_3)_{max}$-condition and the true failure surface described by $(\sigma_1'/\sigma_3')_{max}$. Therefore, a failure condition expressed in terms of the maximum stress difference obtained from e.g. undrained tests initiated at the hydrostatic axis (see Fig. 9) would clearly underestimate the true effective strength of the soil.

Material models which employ associated plastic flow rules must assume the failure surface to go through the point at which instability is produced. This is because stability is guaranteed according to Drucker's and Hill's stability conditions until the failure surface is reached, at which instability ensues. But then stable loading to higher stress differences, which is clearly possible, could not be predicted by such models. Further, since instability as well as stability can be obtained at any point in the region above the line defined by the maximum stress differences in Fig. 9, it is not possible to define any one consistent failure surface to be used in a model which incorporates associated flow.

3. Shear banding

The type of instability discussed above is not the same as that manifesting itself by shear bands. During the performance of the triaxial compression test shown in Fig. 10, the specimen was carefully observed for early detection of developing shear planes. The location on the stress–strain curve at which the first observation of a shear plane was made is indicated on the diagram. Considerable straining beyond the peak failure point may be required before shear bands occur in compression tests. The stress–strain curve clearly shows a drop in strength and the rate of dilation diminishes substantially immediately before the shear plane becomes visible. Thus, the reduced rate of dilation appears to be associated with the occurrence of the shear band and the achievement of critical state conditions within the shear plane in the compression test. Once a shear band has developed fully, the stresses and the volume changes tend to level off, the specimen outside the developing shear band unloads elastically, and the material inside the shear band loosens up to the critical void ratio (Desrues et al., 1996) and rapidly becomes weaker than the remaining major parts of the specimen.

Experimental results with respect to the formation of shear banding under generalized principal stress conditions are available from only few previous studies (e.g. Arthur et al., 1977; Arthur and Dunstan, 1982; Vardoulakis, 1980; Desrues et al., 1985, 1996; Yoshida et al., 1993; Finno et al., 1996, 1997). On the basis of radiographic studies Finno et al. (1996, 1997) pointed to the existence of complex patterns of density fluctuations and early development of strain localizations as precursor to shear banding within plane strain specimens, in which the external deformation apparently continued homogeneously.

Development of shear bands in triaxial compression occurs after peak failure and results in further decrease in strength. This behavior appears to be associated with dilation of the sand. Shear bands are not observed in loose specimens that contract during shear. Loose specimens deform uniformly throughout the entire stress–strain relationship. Fig. 11 shows a critical state diagram for triaxial compression tests on granular materials in which the conditions for stable and unstable behavior as well as for development of shear bands are summarized. Several investigators (Peters et al., 1988, Hettler and Vardoulakis, 1984, Molenkamp, 1985) have shown that the plastic hardening modulus H is required to reach a critical value H_c

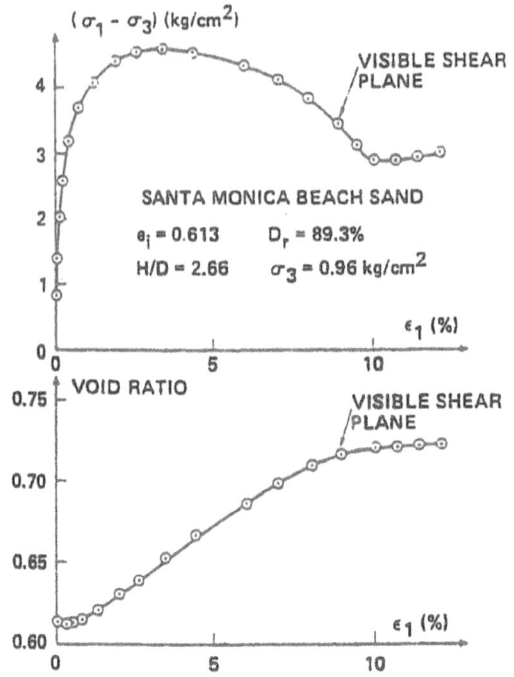

Fig. 10. Stress–strain relation and void ratio change from triaxial compression test exhibiting strain-softening and development of shear plane.

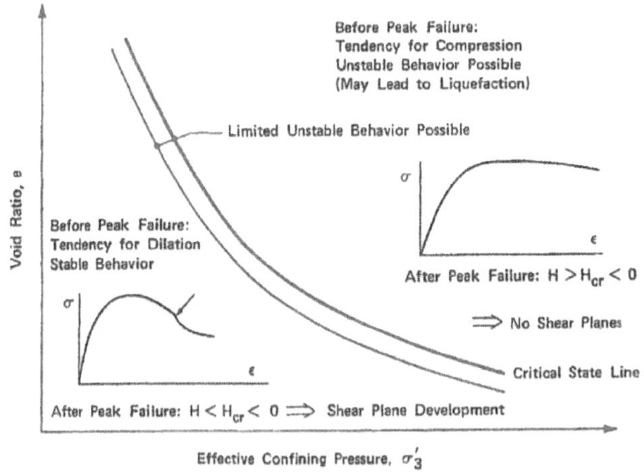

Fig. 11. Critical state diagram for behavior of granular material in triaxial compression. Occurrence of stable and unstable behavior and development of shear planes are indicated.

for development of shear planes. For triaxial compression conditions the critical hardening modulus has to reach a certain negative value before shear planes can develop. This finding is supported by the test results shown in Fig. 10, and it is indicated on Fig. 11.

It appears from these results and from the experimental observations presented above that the type of instability resulting in shear planes is not the same as that discussed in connection with the stability/instability issue presented here. Based on the experimental observations, the two types of instability appear to be mutually exclusive, i.e., shear planes are observed after peak failure in granular materials that dilate, and these materials are perfectly stable for all possible stress path directions before the failure surface has been reached. On the other hand, shear planes are not observed in granular materials that contract during shear, but they can become unstable for certain stress-path directions *inside* the failure surface.

Comparison of stress–strain relations from triaxial and plane strain tests on sand indicates that the three-dimensional stress condition has a significant influence on the formation of shear bands and therefore on the strength of the sand. Peters et al. (1988) found that shear banding can occur in the hardening regime of the stress–strain relationship for sand under plane strain conditions. This experimental finding agrees with the theoretical conclusion by Rudnicki and Rice (1975) for materials with nonassociated plastic flow. On the other hand, shear banding in triaxial compression and extension appears to be a post-peak phenomenon, though it occurs much closer to peak failure in triaxial extension. The fact that peak failure in plane strain tests is associated with the occurrence of shear banding makes it crucial to determine whether failure occurs as a smooth peak or is induced by shear banding over an extended range of intermediate principal stresses.

Experimental studies of shear banding in true triaxial tests have been performed with a modified version of Lade and Duncan's cubical triaxial device (1973). The three-dimensional strength characteristics and the influence of shear banding on failure were studied and the peak strengths are compared with the failure criterion proposed by Lade (1977) and the Mohr–Coulomb failure criterion. Quantitative comparisons are made between the critical conditions for shear band formation obtained from the experiments and from theoretical considerations and expressed by the dimensionless hardening parameter H_c/E immediately prior to the onset of shear banding. Here E is the elastic modulus of the soil.

3.1. Experimental procedures

The cubical triaxial apparatus shown in Fig. 12 and described by Lade and Duncan (1973) was modified for this study. Tests were performed on tall prismatic specimens, 76 mm wide, 76 mm long, and 188 mm high with a height-to-diameter ratio, $H/D = 2.47$ (Wang and Lade, 2001). These tall specimens were employed, because shear bands develop more freely and softening behavior is more pronounced in tall and slender specimens. Extra steel lamellas and balsa wood strips were employed to increase the height of the horizontal loading plates to match the 188 mm tall prismatic specimens. The bars connecting the two horizontal loading plates were still positioned at mid-height, so there was no eccentricity of the horizontal load. Both the horizontal loading plates and the cap and base were provided with lubricated surfaces to avoid development of significant shear stresses between the loading plates and the specimen.

The experiments were conducted on dense, medium dense, and loose Santa Monica Beach sand with void ratios of 0.63, 0.68, and 0.76 corresponding to relative densities of 90%, 74%, and 50%. All specimens were tested with a constant effective confining pressure of $\sigma_3' = 49$ kPa. Twelve drained tests were conducted with constant b-values from 0 to 1.0 for each relative density.

3.2. Experimental results

3.2.1. Stress–strain characteristics

The stress–strain curves obtained from the true triaxial tests on tall prismatic specimens of dense sand are shown in Fig. 13. The arrows on the stress–strain curves indicate the positions of the peak failure points, and the number indicates the b-value. For $b = 0$ (triaxial compression), the stress–strain curve is essentially flat over a relatively large range of strains near failure, and the slope at failure is zero. Considerable

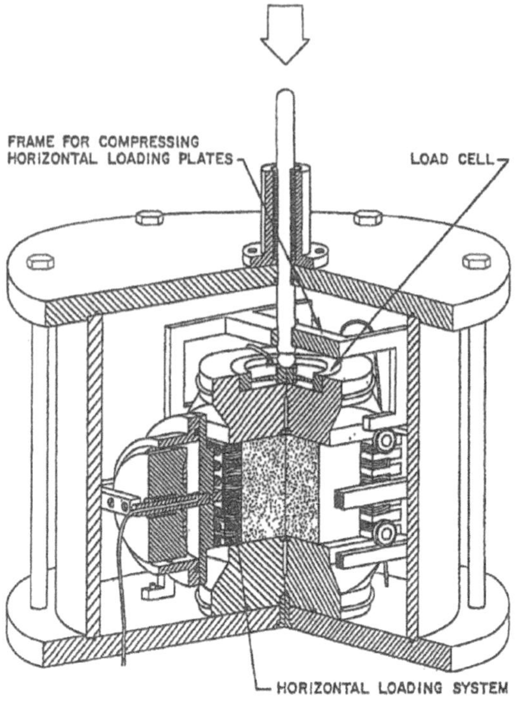

Fig. 12. Cubical triaxial apparatus.

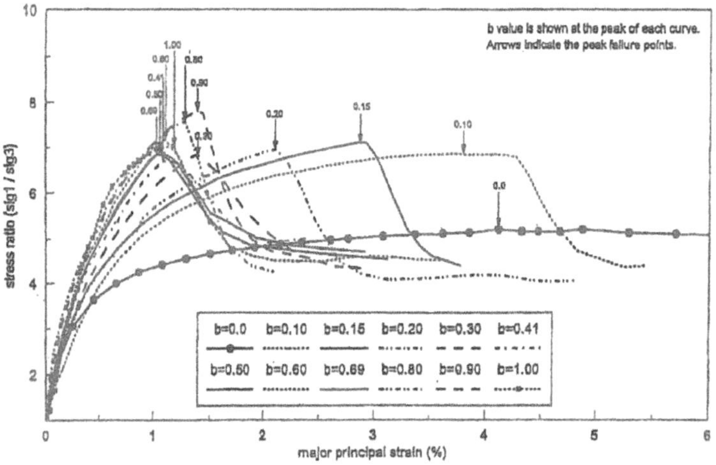

Fig. 13. Comparison of stress ratio–strain characteristics in true triaxial tests on prismatic specimens of dense Santa Monica Beach sand with $\sigma_3' = 49$ kPa.

amounts of softening and strength reduction occurs after failure (not shown in Fig. 13, but indicated in Fig. 10), but shear bands were not observed until a major principal strain of two-and-a-half times the strain-to-peak-failure had been reached.

As b increases from 0 the stress–strain behavior becomes increasingly stiff and the strain-to-failure decreases. The strengths obtained at $b = 0.10$ and 0.15 are considerably higher than that at $b = 0$. The strengths did not drop off after peak failure in these two tests until small amounts of post-peak softening had occurred, as indicated in Fig. 13. After less than 0.5% post-peak straining was obtained, the rate of strength reduction suddenly increased and soon reached its highest value. Shear bands were usually observed when the strength was dropping at the highest rate. After considerable reduction in strength had been obtained, the strength reduction slowed down, the shear bands became well-defined, and the strength reached its residual state.

For the tests performed with b-values in the range of 0.20–0.80, it appears that the strengths drop suddenly at points on the stress–strain curves where the specimens are apparently still being loaded, i.e. the slopes of the curves are positive. There are no smooth peaks on the stress–strain curves obtained in those tests. The peak points appear to represent points of instability occurring before smooth peak failure points can be obtained. Visible shear bands were detected when the strengths were dropping at the highest rates. The residual strengths were reached within 0.7% or less post-peak straining for the tests with b-values in the range of 0.20–0.80.

The peak failure becomes smooth again at $b = 0.90$. However, for the test at $b = 1.00$, the slope of the stress–strain curve prior to failure becomes steeper again and the vertical stress ratio (σ_1'/σ_3') drops abruptly at the peak. The specimen failed in the horizontal direction at $b = 1.00$, and it was traversed by vertical shear planes parallel to the direction of the major principal stress, whereas vertical failure and shear planes parallel to the intermediate principal stress direction always occurred in the tests with $b = 0$–0.90. Failure occurred in the horizontal load-controlled direction at $b = 1.00$, because the specimens were produced by dry pluviation in the vertical direction, and the consequent cross-anisotropic fabric was weaker in the horizontal than in the vertical direction. The peak point and the post-peak stress–strain curve corresponding to the vertical direction, as shown in Fig. 13, cannot represent the real peak strength and the real softening response.

3.2.2. Friction angles

The friction angle is calculated from $\sin\phi = (\sigma_1 - \sigma_3)(\sigma_1 + \sigma_3)$, and Fig. 14 shows that it increases continuously with increasing b-value until $b = 0.15$. It then remains essentially constant as b increases in the

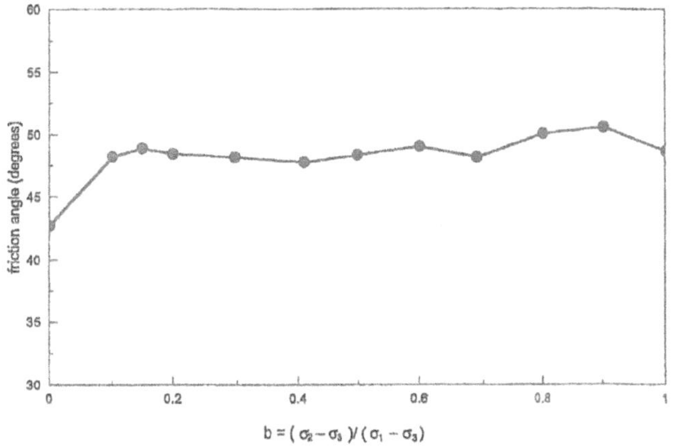

Fig. 14. Variation in friction angle with b-value for true triaxial tests on prismatic specimens of dense Santa Monica Beach sand with $\sigma_3' = 49$ kPa.

range from 0.15 to 0.70. The friction angle increases slightly at $b = 0.90$, and then decreases slightly to $b = 1.0$. The lowest friction angle is obtained at $b = 0$ and the highest friction angle is reached at $b = 0.90$.

3.3. Conditions for shear banding

Bifurcation, i.e. initiation of shear band formation, is the inevitable result of continued shearing under conditions of uniform stresses and uniform strains: at some point, stress and strain states are reached which allow a localized displacement field to satisfy equilibrium, compatibility, boundary conditions, and constitutive relations (Rudnicki and Rice, 1975; Rice, 1976). This analysis procedure has been applied to study the influence of the constitutive model on the predicted shear banding (e.g. Molenkamp, 1985) and to investigate the conditions for shear banding under different three-dimensional stress conditions (e.g. Rudnicki and Rice, 1975; Vermeer, 1982; Peters et al., 1988; Peric et al., 1992).

The results of bifurcation analyses for three-dimensional stress and strain states are a set of critical plastic hardening moduli, H_c, which, when normalized on the elastic Young's modulus, E, vary with the value of b as indicated schematically in Fig. 15. A stress–strain state with a hardening modulus less than the critical hardening modulus can possibly produce shear banding.

The normalized, critical hardening modulus, H_c/E, at the point immediately before onset of shear banding, has been computed for each of the true triaxial tests on tall prismatic specimens of dense, medium dense, and loose sand. In this manner, the experimentally determined conditions for shear band formation under three-dimensional stress conditions can be established, and the pattern may be compared with the that shown in Fig. 15. The analysis required for this computation was presented by Lade and Wang (2001).

Fig. 16 shows the variation with b-value of H_c/E calculated immediately before onset of shear banding. The range of b giving positive H_c/E is about 0.18–0.85. Negative values of H_c/E are obtained for b in the ranges 0.0–0.18 and 0.85–1.0. The calculations for $b = 1.0$ are questionable, because failure in the cross-anisotropic sand specimen occurred in the load-controlled horizontal stress direction and the real peak point and post-peak stress–strain relation are not available. The variation of H_c/E in the mid-range and for high b-values compares well with that indicated by Rudnicki and Rice (1975) and shown in Fig. 15, and the calculated magnitudes compare well with those given by Peters et al. (1988) for medium and high b-values. However, the negative H_c/E-values obtained for small b-values are much smaller than those indicated by these authors. The calculations of critical hardening moduli were found to be rather sensitive to the actually measured slope of the stress–strain curves for low and high b-values, while they were more steady in the

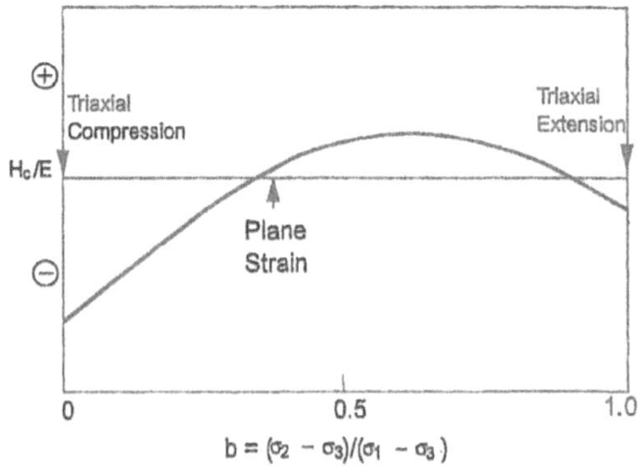

Fig. 15. Schematic diagram of variation of normalized, critical hardening modulus with b according to Rudnicki and Rice (1975) and Rice (1976).

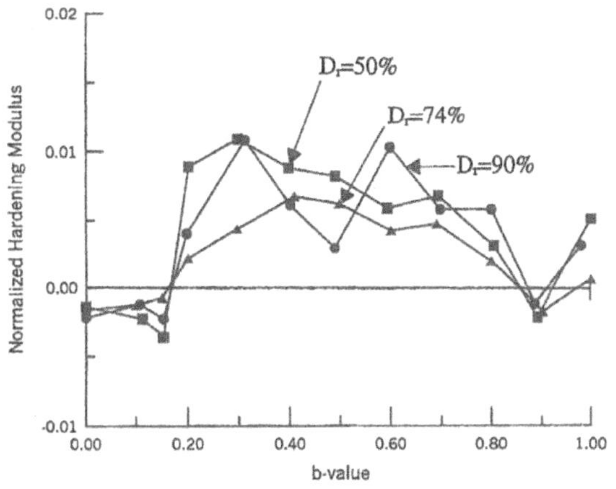

Fig. 16. Variation of normalized, critical hardening modulus, H_c/E with b for true triaxial tests on prismatic specimens of dense, medium dense, and loose Santa Monica Beach sand with $\sigma'_3 = 49$ kPa.

mid-range of b-values where shear banding occurred in the hardening regime. According to the experimental evidence, shear banding causes failure in all experiments with $0.18 < b < 0.85$, although it occurs rather close to smooth peak failure for the tests with high b-values.

It can be concluded that the condition for shear band formation in a test with b in the range of 0.18–0.85 may be fulfilled in the hardening portion of the stress–strain curve, and the maximum strength which the specimen could produce under homogeneous conditions will therefore not be reached. The greater the value of the critical hardening modulus, the earlier the shear banding occurs before smooth peak failure. That is why very limited variations of strength and strain-to-failure were obtained for b in this range. However, for triaxial compression tests and tests at low b-values ($b < 0.18$) and at high b-values ($b > 0.85$), shear banding occurs in the softening regime.

3.4. Comparison of strengths with three-dimensional failure criteria

The strengths from the true triaxial tests on tall prismatic specimens of dense sand are compared in Fig. 17 with those predicted by Lade's failure criterion (1977) at $\sigma'_3 = 49$ kPa. The failure criterion is fitted to the results of triaxial compression tests performed on short, cylindrical specimens, for which a complete set of data for several confining pressures were available. This type of test with height-to-diameter of unity typically produce slightly higher strengths than the tests on tall specimens, as also indicated in Fig. 17. For $b < 0.18$ and $b > 0.85$, the peak strengths are not influenced by shear banding and the experimentally determined strengths compare fairly well with the model predictions. However, this failure criterion does not capture the variation of strength for $0.18 < b < 0.85$. The strengths in this range cannot be fully mobilized due to the influence of shear banding in the hardening regime. This implies that a smooth, continuous three-dimensional failure surface generally is not be obtained for soils.

The failure points from the experiments on dense Santa Monica Beach sand are compared with Lade's failure surface and the Mohr–Coulomb failure surface in the octahedral plane in Fig. 18. It can be seen that while the Mohr–Coulomb criterion always underestimates the three-dimensional strengths produced at b greater than zero, Lade's failure criterion models the strengths with fairly good accuracy, except in the mid-range of b-values in which shear banding occurs in the hardening regime.

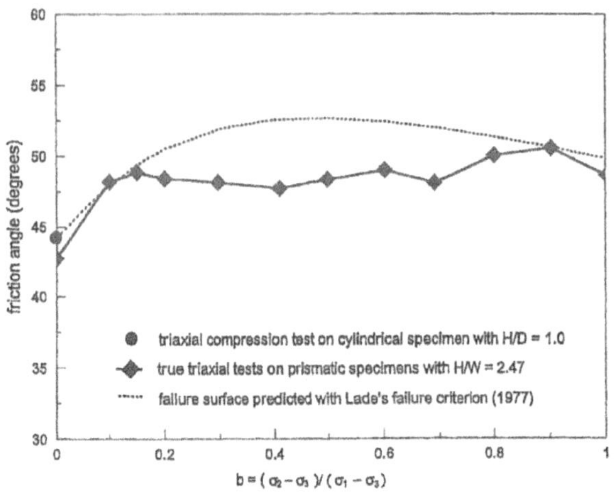

Fig. 17. Comparison Lade's failure surface with measured friction angles for true triaxial tests on prismatic specimens of dense Santa Monica Beach sand with $\sigma_3' = 49$ kPa. Effect of shear banding indicated in middle range of b-values.

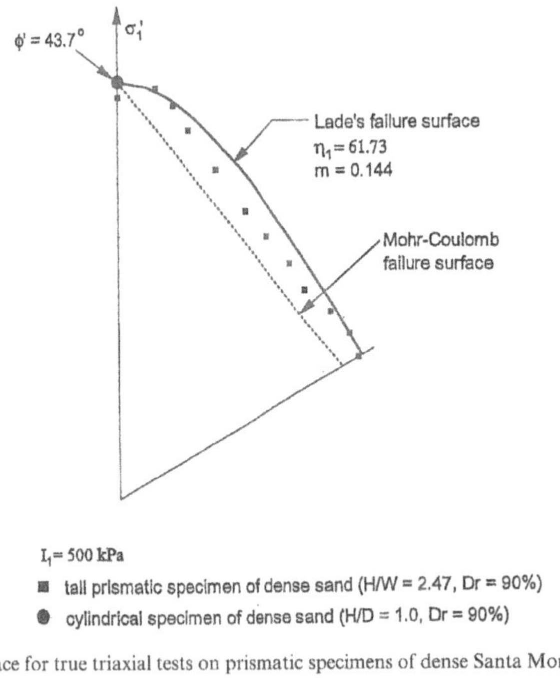

Fig. 18. Experimental failure surface for true triaxial tests on prismatic specimens of dense Santa Monica Beach sand with $\sigma_3' = 49$ kPa compared with the Mohr–Coulomb failure surface and Lade's failure surface.

4. Summary and conclusions

With background in the stability conditions proposed by Drucker and Hill, the stability of granular materials which exhibit nonassociated flow has been investigated on the basis of experimental observations.

Triaxial compression tests designed to expose the type of behavior exhibited by granular materials were performed. The variables in these studies were the sign of the second work increment (positive or negative), the volumetric strain behavior (compression or dilation), the constraints of the volumetric behavior (free or controlled), and the degree of saturation (fully or partly saturated). It is shown that the conditions for stability proposed by Drucker and by Hill are not necessary to guarantee stability of granular (frictional) materials. However, when these criteria are violated, it is possible to create an unstable situation by prescribing a constraint among the kinematic degrees of freedom (e.g. shear and volumetric strain). Conditions for stability are indicated on the basis of experimental observations of the behavior of granular materials in triaxial tests. The occurrence of instability involves small strains, and once it has been initiated it may lead to liquefaction which requires large strains to fully develop.

Shear planes occur in granular materials that dilate. In triaxial compression they develop after peak failure. The type of instability that manifests itself by shear plane development appears to be different from the instability relating to a decrease in load carrying capability occurring *inside* the failure surface for granular materials that contract during shear.

Shear banding, detected through the sudden strength reduction, initiates in the hardening regime of true triaxial tests when $b = (\sigma_2 - \sigma_3)/(\sigma_1 - \sigma_3)$ is in the approximate range of 0.18–0.85. Failure in these tests is considered to be a consequence of shear banding rather than a constitutive response. To verify this, the condition for shear band formation was examined for each test through the computation of the hardening modulus prior to onset of shear banding. For b-values in the ranges from 0 to 0.18 and from 0.85 to 1.0, negative hardening moduli were obtained, implying that conditions for shear band formation are fulfilled in the softening regime. Therefore, failure occurs by smooth peak failure and as a continuum response in these ranges. Positive values of the hardening modulus immediately before onset of shear banding was obtained when b was in the approximate range of 0.18–0.85, indicating that the condition for bifurcation is fulfilled in the hardening portion of stress–strain curve. Thus, peak failure is caused by shear banding in this middle range of b-values, and a smooth, continuous three-dimensional failure surface is therefore not obtained in general for soils.

References

Arthur, J.R.F., Dunstan, T., 1982. Rupture layers in granular media. In: Proceedings, IUTAM Symposium on Deformation and Failure of Granular Materials, Delft. Balkema, Rotterdam, pp. 453–459.

Arthur, J.R.F., Dunstan, T., Al-Ani, Q.A.J.L., Assadi, A., 1977. Plastic deformation and failure in granular media. Geotechnique 27 (1), 53–74.

Bishop, J.F.W., Hill, R., 1951. A theory of the plastic distortion of a polycrystalline aggregate under combined stresses. Philos. J. 42, 414–427.

Bjerrum, L., Simons, N.E., 1960. Comparison of shear strength characteristics of normally consolidated clays. In: Proceedings of the ASCE Research Conference. Boulder, Colorado, pp. 711–726.

Desrues, J., Chambon, R., Mokni, M., Mazerolle, F., 1996. Void ratio evolution inside shear bands in triaxial sand specimens studied by computed tomography. Geotechnique 46 (3), 529–546.

Desrues, J., Lanier, J., Stutz, P., 1985. Localization of the deformation in tests on sand sample. Eng. Fract. Mech. 21 (4), 909–921.

Drucker, D.C., 1951. A more fundamental approach to stress–strain relations. In: Proceedings of the First US Nat. Cong. Appl. Mech., pp. 487–491.

Drucker, D.C., 1956. On uniqueness in the theory of plasticity. Quart. Appl. Math. 14, 35–42.

Drucker, D.C., 1959. A definition of stable inelastic material. J. Appl. Mech. 26, 101–106.

Finno, R.J., Harris, W.W., Mooney, M.A., Viggiani, G., 1996. Strain localization and undrained steady state of sand. J. Geotech. Eng. 122 (6), 462–473.

Finno, R.J., Harris, W.W., Mooney, M.A., Viggiani, G., 1997. Shear bands in plane strain compression of loose sand. Geotechnique 47 (1), 149–165.

Hettler, A., Vardoulakis, I., 1984. Behavior of dry sand tested in a large triaxial apparatus. Geotechnique 34, 183–197.

Hill, R., 1958. A general theory of uniqueness and stability in elastic–plastic solids. J. Mech. Phys. Solids 6, 236–249.

Lade, P.V., 1977. Elasto-plastic stress–strain theory for cohesionless soil with curved yield surfaces. Int. J. Solids Struct. 13, 1019–1035.
Lade, P.V., 1982. Localization effects in triaxial tests on sand. In: Proceedings. IUTAM Symposium on Deformation and Failure of Granular Materials, Delft, pp. 461–471.
Lade, P.V., 1989. Instability and failure of soils with nonassociated flow. In: Proceedings of the Twelvth International Conference on Soil Mech. Found. Eng., vol. 1. Rio de Janeiro, Brazil, pp. 727–730.
Lade, P.V., Bopp, P.A., Peters, J.F., 1993. Instability of dilating sand. In: Mechanics of Materials, vol. 16. Elsevier Science Publishers, pp. 249–264.
Lade, P.V., Duncan, J.M., 1973. Cubical triaxial tests on cohesionless soils. ASCE J. Soil Mech. Found. Div. 99 (SM10), 793–812.
Lade, P.V., Duncan, J.M., 1976. Stress-path dependent behavior of cohesionless soil. J. Geot. Eng. Div., ASCE 102, 51–68.
Lade, P.V., Nelson, R.B., Ito, Y.M., 1987. Nonassociated flow and stability of granular materials. ASCE J. Eng. Mech. 113, 1302–1318.
Lade, P.V., Nelson, R.B., Ito, Y.M., 1988. Instability of granular materials with nonassociated flow. ASCE J. Eng. Mech. 114, 2173–2191.
Lade, P.V., Pradel, D., 1990. Instability and flow of granular materials. I: experimental observations. ASCE J. Eng. Mech. 116, 2532–2550.
Lade, P.V., Wang, Q., 2001. Analysis of shear banding in true triaxial tests on sand. ASCE J. Eng. Mech. 127, 762–768.
Mandel, J., 1964. Conditions de stabilite et postulat de Drucker. In: Proceedings, IUTAM Symposium on Rheology and Soil Mechanics Grenoble, pp. 58–68.
Molenkamp, F., 1985. Comparison of frictional material models with respect to shear band initiation. Geotechnique 35, 127–143.
Mroz, Z., 1963. Nonassociated flow laws in plasticity. J. de Mecanique 2, 21–42.
Nemat-Nasser, S., 1983. On finite plastic flow of crystalline solids and geomaterials. J. Appl. Mech. 105, 1114–1126.
Peric, D., Runesson, K., Sture, S., 1992. Evaluation of plastic bifurcation for plane strain versus axisymmetry. ASCE J. Eng. Mech. 118 (3), 512–524.
Peters, J.F., 1991. Discussion of instability of granular materials with nonassociated flow. ASCE J. Eng. Mech. 117, 934–936.
Peters, J.F., Lade, P.V., Bro, A., 1988. Shear band formation in triaxial and plane strain tests. In: Advanced Triaxial Testing of Soil and Rock, vol. 977. ASTM STP, pp. 604–627.
Poorooshasb, H.B., Holubec, I., Sherbourne, A.N., 1966. Yielding and flow of sand in triaxial compression: Part I. Can. Geot. J. 3, 179–190.
Rice, J.R., 1976. The localization of plastic deformation. In: Koiter, W.T. (Ed.), Proceedings of the Fourth Conf. on Theoret. and Appl. Mech. North-Holland Publishing Co., Amsterdam, pp. 207–229.
Rudnicki, J.W., Rice, J.R., 1975. Conditions for the localization of deformation in pressure-sensitive dilatant materials. J. Mech. Phys. Solids 23, 371–394.
Seed, H.B., Mitchell, J.K., Chan, C.K., 1960. The strength of compacted cohesive soils. In: Proceedings of the ASCE Research Conference. Boulder, Colorado, pp. 877–964.
Vardoulakis, I., 1980. Shear band inclination and shear modulus of sand in biaxial tests. Int. J. Numer. Anal. Meth. Geomech. 4, 103–119.
Vermeer, P.A., 1982. A simple shear-band analysis using compliances. In: IUTAM Conference on Deformation and Failure of Granular Materials, Delft, pp. 493–499.
Wang, Q., Lade, P.V., 2001. Shear banding in true triaxial tests and its effect on failure in sand. ASCE J. Eng. Mech. 127, 754–761.
Whitman, R.V., 1960. Some considerations and data regarding the shear strength of clays. In: Proceedings of the ASCE Research Conference. Boulder, Colorado, pp. 581–614.
Yoshida, T., Tatsuoka, F., Siddiquee, M.S.A., Kamegai, Y., Park, C.-S., 1993. Shear banding in sands observed in plane strain compression. In: Proceedings of the Third International Workshop Localisation and Bifurcation Theory for Soils and Rocks. Grenoble, pp. 165–179.

SOILS AND FOUNDATIONS Vol. 43, No. 4, 161-172, August 2003
Japanese Geotechnical Society

ANALYSIS AND PREDICTION OF SHEAR BANDING UNDER 3*D* CONDITIONS IN GRANULAR MATERIALS

POUL V. LADE[i]

ABSTRACT

A series of true triaxial tests on tall prismatic specimens of sand at three different void ratios form the basis for analyses, prediction and verification of the influence of shear banding on the three-dimensional failure conditions in granular material. The results of the true triaxial tests are reviewed and discussed in view of the shear banding that is observed in these tests. Theoretical conditions for shear banding, which are based on the behavior of the material near the observed shear banding phenomenon, are reviewed. The detailed procedure for shear banding analysis is explained in general and calculations are performed for a single hardening elasto-plastic constitutive model. The predicted shear banding events are compared with those observed in the experiments. It is shown that the predicted stress conditions for shear banding match those observed in the true triaxial tests, and it is concluded that shear banding plays an important role in the three-dimensional strength of granular materials. Thus, peak failure is caused by shear banding in the hardening regime in the approximate range of $b=(\sigma_2-\sigma_3)/(\sigma_1-\sigma_3)$ from 0.18 to 0.85, while it occurs in the softening regime outside this range of b-values. A smooth, continuous 3*D* failure surface is therefore not generally obtained for granular materials.

Key words: bifurcation, constitutive model, failure criterion, granular materials, hardening plasticity, sand, shear banding, strain localization, three-dimensional strength, true triaxial tests (**IGC:** D6)

INTRODUCTION

Analyses of bearing capacity, active and passive pressures on retaining structures, and slope stability assume locations and directions of shear bands in the soil. They are assumed to form the boundaries of the moving soil masses. However, shear bands observed in the soil in experiments and in the field are very often not even close to those assumed for theoretical developments performed for the closed-form analytical solutions. In fact, shear bands may not even develop in real soil (e.g. loose sand) or they may not develop sufficiently to extend all the way to the ground surface at the time of overall peak soil resistance, even in dense sand (e.g. Tatsuoka et al., 1991). Thus, much of the background for analytical expressions is questionable, and furthermore, closed-form solutions exist only for relatively simple configurations, such as axisymmetric and plane strain cases. Many real cases of soil failure and soil-structure interaction involve true 3*D* conditions.

To improve this situation, it may be possible to use advanced and realistic constitutive soil models combined with numerical methods to capture not only the failure condition, but the entire load-deformation relation for the structure in question. However, the conditions for which shear bands occur under general 3*D* circumstances have not been experimentally determined, then compared with theoretical predictions, and predicted in association with a constitutive model. It is, however, important to capture the occurrence of shear bands correctly, because the soil shear strength immediately drops and reaches the residual strength within relatively little displacement. To capture the resulting progressive failure occurring under footings, behind retaining walls and in slope failures depends acutely on a constitutive model with provisions for shear band predictions under 3*D* conditions.

Presented here is a brief review of the results of 3*D* experiments performed on sand in a true triaxial apparatus as well as an analysis of the occurrence of shear banding in these experiments. This is followed by a detailed explanation of the theoretical conditions for shear banding, a brief review of the constitutive model employed for shear band prediction, the calculation procedure leading to prediction of shear banding, and comparisons with the 3*D* experimental results.

SHEAR BANDING IN TRUE TRIAXIAL TESTS ON SAND

Previous Studies

Experimental results with respect to the formation of shear banding under various stress conditions are availa-

[i] Professor, Department of Civil Engineering, Catholic University of America, Washington, DC 20064, USA (Lade@cua.edu).
Manuscript was received for review on July 5, 2002.
Written discussions on this paper should be submitted before March 1, 2004 to the Japanese Geotechnical Society, Sugayama Bldg. 4F, Kanda Awaji-cho 2-23, Chiyoda-ku, Tokyo 101-0063, Japan. Upon request the closing date may be extended one month.

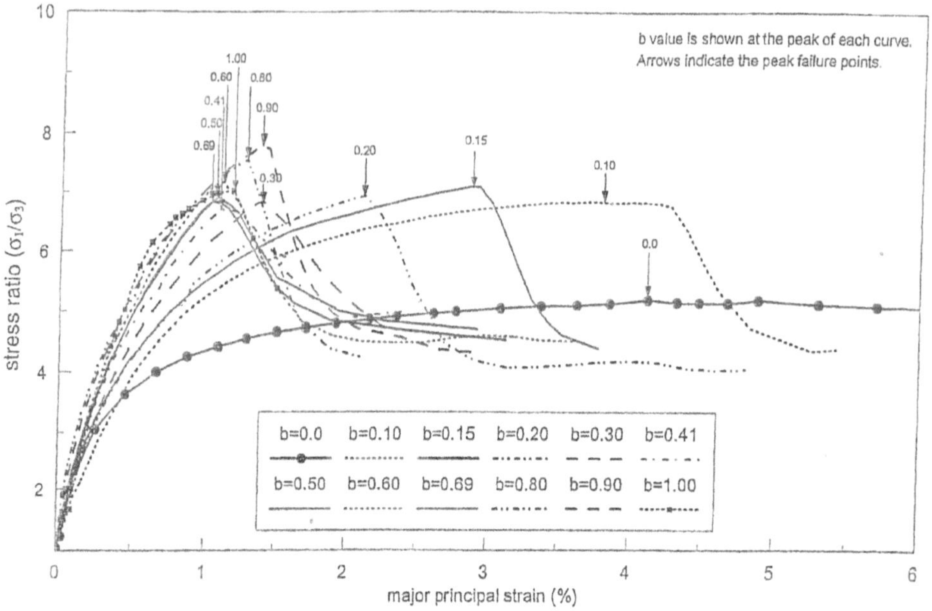

Fig. 1. Comparison of stress-strain characteristics in true triaxial tests on tall prismatic specimens of dense Santa Monica Beach sand with σ_3'

strain curve prior to failure becomes steeper again and the vertical stress ratio (σ_1/σ_3) drops abruptly at the peak. The specimen failed in the horizontal direction at $b=1.00$, and it was traversed by vertical shear planes parallel to the direction of the major principal stress, whereas vertical failure and shear planes parallel to the intermediate principal stress direction always occurred in the tests with $b=0$–0.90. Failure occurred in the horizontal load-controlled direction at $b=1.00$, because the specimens were produced by dry pluviation in the vertical direction, and these cross-anisotropic specimens are weaker in the horizontal direction. The peak point and the post-peak stress-strain curve corresponding to the vertical direction, as shown in Fig. 1, cannot represent the real peak strength and the real softening response.

Plotted in Fig. 2 are the relations between major and intermediate principal strain, ε_1-ε_2, with varying b-value for the true triaxial tests. They form a regular pattern with the variation in b. Expansive ε_2-values were measured in tests conducted with b equal to or less than 0.20, and contractive ε_2-values were obtained for b greater than 0.30. Analysis of the ε_1-ε_2 slopes indicate that plane strain is obtained near $b=0.23$.

Indicated on Fig. 2 are also the points corresponding to the stress peaks, and it is seen that for b-values between 0.20 and 0.80, the peak stress points coincide with the points where the ε_2-strain rates go to zero, i.e. the tangents to the curves become horizontal, except for the test with $b=1.00$ in which failure occurred in the horizontal direction, as explained above. Thus, shear banding and strength reduction correspond with inception of zero strain rate in the intermediate principal direction. Since the specimen becomes weaker and all deformation localizes to the shear band, the stresses reduce and the portions of the specimen outside the shear band unload and slide as rigid blocks along the shear band. Thus, the normal strain in the ε_2-direction, for practical purposes, becomes zero at the inception of shear banding. Note that strains continued to develop past peak failure in the ε_2-direction in the tests with b-values smaller than 0.20 and greater than 0.80, and this lasted until shear banding occurred in the softening regime.

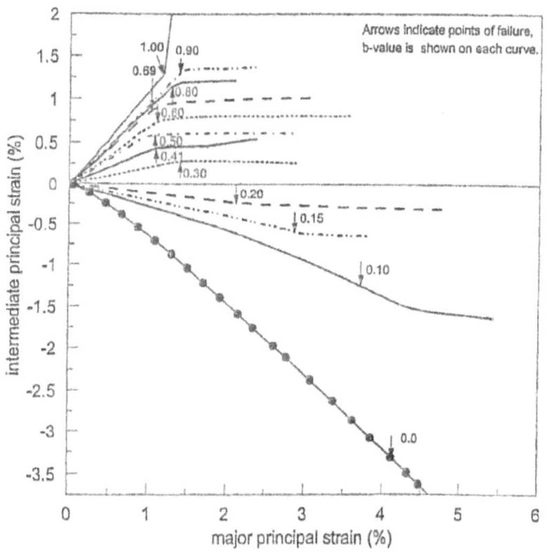

Fig. 2. Relations between major and intermediate principal strains in true triaxial tests on tall prismatic specimens of dense Santa Monica Beach sand with σ_3'

ble from previous investigations (e.g. Arthur et al., 1977; Vardoulakis, 1980; Arthur and Dunstan, 1982; Lade, 1982a; Scarpelli and Wood, 1982; Vardoulakis and Graf, 1982; Desrues et al., 1985, 1996; Tatsuoka et al., 1990; Yoshida et al., 1993; Finno et al., 1996, 1997; Nemat-Nasser and Okada, 2001), while theoretical developments and predictions have been attempted in other studies (e.g. Hansen, 1958; Rudnicki and Rice, 1975; Rice, 1976; Vardoulakis et al., 1978; Vardoulakis, 1980; Vermeer, 1982; Molenkamp, 1985; Muhlhaus and Vardoulakis, 1987; Vardoulakis, 1996a, b).

The stress-strain relations obtained from true triaxial tests commonly show continuously decreasing strain-to-failure, more pointed peak, and more rapid strength reduction after the peak for $b\ (=(\sigma_2-\sigma_3)/(\sigma_1-\sigma_3))$ near and greater than the value at plane strain. This may suggest that the pre-failure shear band occurrence is prevalent over a wide range of b-values. Therefore, failure may be a consequence of the development of shear bands under stress conditions in this range of b-values, and the peak strength may be dependent on the critical condition at which shear banding occurs. True triaxial equipment that produces uniform stresses and strains and covers the full range of variation of the intermediate principal stress is required to establish the influence of shear banding on the three-dimensional stress-strain and strength behavior of sand.

Experimental Procedures

Experimental studies of shear banding in true triaxial tests have been performed with a modified version of Lade and Duncan's cubical triaxial device (1973). The three-dimensional strength characteristics and the influence of shear banding on failure were studied and the peak strengths were compared with the failure criterion proposed by Lade (1977) and the Mohr-Coulomb failure criterion. Quantitative comparisons were made between the critical conditions for shear band formation obtained from experiments and from theoretical considerations and expressed by the dimensionless hardening parameter H_c/E immediately prior to the onset of shear banding.

The cubical triaxial apparatus described by Lade and Duncan (1973) was modified for this study. Tests were performed on tall prismatic specimens, 76 mm wide, 76 mm long, and 188 mm high with a height-to-diameter ratio, $H/D=2.47$ (Wang and Lade, 2001). These tall specimens were employed, because shear bands develop more freely and softening behavior is more pronounced in tall and slender specimens. The experiments were conducted on dense, medium dense, and loose Santa Monica Beach sand with void ratios of 0.63, 0.68, and 0.76 corresponding to relative densities of 90%, 74%, and 50%. The specimens with $D_r=90\%$ and 74% were prepared by pluviating a preweighed amount of dry sand through a No. 30 U.S. sieve and an 8-in.-tall cardboard tube into a prismatic cavity. This cavity was formed by (1) the aluminum base covered with lubricated rubber sheets, and (2) the 0.012-in-thick membrane whose lower end was fixed on the base by O-rings and held by a vacuum on the inner surface of the prismatic specimen mold. The specimens with $D_r=50\%$ were made by placing the preweighed amount of sand inside a funnel and lifting the funnel up such that sand poured out into the specimen cavity with no fall height. After sealing the membrane to the cap, both drainage lines extending from the cap and the base were connected to a vacuum of approximately 20 kPa. The specimen mold was then removed, and the specimen dimensions were measured. The horizontal loading system was then positioned, the triaxial cell was assembled around the specimen, and an initial cell pressure of 49 kPa was applied, while the vacuum was reduced to zero. After saturating the specimens by the CO_2-method, a back pressure of 98 kPa was applied. All specimens were tested with a constant effective confining pressure of $\sigma_3'=49$ kPa. Twelve drained tests were conducted with constant b-values from 0 to 1.0 for each relative density.

Experimental Results

The stress-strain curves obtained from the true triaxial tests on tall prismatic specimens of dense sand are shown in Fig. 1. The arrows on the stress-strain curves indicate the positions of the peak failure points, and the number indicates the b-value. For $b=0.0$ (triaxial compression), the stress-strain curve is essentially flat over a relatively large range of strains near failure, and the slope at failure is zero. Considerable amounts of softening and strength reduction occur after failure (not shown here to maintain reasonable proportions of the diagram), but shear bands were not observed until the major principal strain of two-and-a-half times the strain-to-peak-failure had been reached.

As b increases from zero, the stress-strain behavior becomes increasingly stiff and the strain-to-failure decreases. The strengths obtained at $b=0.10$ and 0.15 are considerably higher than that at $b=0$. The strengths did not drop off after peak failure in these two tests until small amounts of post-peak softening had occurred, as indicated in Fig. 1. After less than 0.5% post-peak straining was obtained, the rate of strength reduction suddenly increased and soon reached its highest rate. After considerable reduction in strength had been experienced, the strength reduction slowed down, the shear bands became well-defined, and the strength reached its residual state.

For the tests performed with b-values in the range of 0.20–0.80, it appears that the strengths drop suddenly at points on the stress-strain curves where the specimens are apparently still being loaded, i.e. the slopes of the curves are positive. There are no smooth peaks on the stress-strain curves observed in those tests. The peak points appear to represent points of instability occurring before smooth peak failure points can be obtained. Visible shear bands were detected when the strengths were dropping at the highest rates. The residual strengths were reached within small amounts of post-peak deformation for the tests with b-values in the range of 0.20–0.80.

The peak failure becomes smooth again at $b=0.90$. However, for the test at $b=1.00$, the slope of the stress-

Variation of Experimental Critical Hardening Modulus

Bifurcation, i.e. initiation of shear band formation in dilating sand, is the result of continued shearing under conditions of uniform stresses and uniform strains. At some point, stress and strain states are reached which allow a localized displacement field to satisfy equilibrium, compatibility, boundary conditions, and constitutive relations (Rudnicki and Rice, 1975; Rice, 1976). This analysis procedure has been applied to study the influence of the constitutive model on the predicted shear banding (e.g. Molenkamp, 1985) and to investigate the conditions for shear banding under different three-dimensional stress conditions (e.g. Rudnicki and Rice, 1975; Vermeer, 1982; Peters et al., 1988; Peric et al., 1992).

The results of bifurcation analyses for three-dimensional stress and strain states are a set of critical plastic hardening moduli, H_c, which, when normalized on the elastic Young's modulus, E, vary with the value of b. The plastic hardening modulus plays the same role for calculation of plastic strain increments as Young's modulus plays for calculation of elastic strain increments. The plastic hardening modulus initiates with some positive value at the hydrostatic axis and decreases towards and becomes zero at smooth peak failure. It becomes negative in the softening regime. The plastic hardening modulus thus reflects the slope of the stress-strain curve. Both the plastic hardening and the elastic moduli have dimensions of stress, and the dimensionless ratio H_c/E, in which E relates to the same stress state as the plastic hardening modulus, expresses the normalized critical plastic hardening modulus. As will be seen below, this quantity emerges automatically, from the bifurcation analyses as a function of dimensionless quantities such as Poisson's ratio, the derivatives of the yield and plastic potential functions, and the direction of the shear band. A stress-strain state with a hardening modulus less than the critical hardening modulus can possibly produce shear banding. Rudnicki and Rice (1975) suggested a range of stress states in which the hardening modulus, critical for bifurcation, is positive, i.e. shear banding may occur during increasing loading before smooth peak failure has been reached, as shown in Fig. 3. According to this schematic diagram, shear banding may occur in the hardening regime in the mid-range of b-values, whereas it may occur for negative critical hardening moduli, i.e. in the softening regime, for b-values at and in the vicinity of triaxial compression and triaxial extension.

The normalized, critical hardening modulus, H_c/E, at the point immediately before onset of shear banding, was computed for each of the true triaxial tests on tall prismatic specimens of dense, medium dense, and loose Santa Monica Beach sand as presented by Wang and Lade (2001). In this manner, the experimentally determined conditions for shear band formation under three-dimensional stress conditions can be established and compared with the those shown in Fig. 3. The analysis required for this computation was presented by Lade and Wang (2001).

Figure 4 shows the variations with b-value of H_c/E

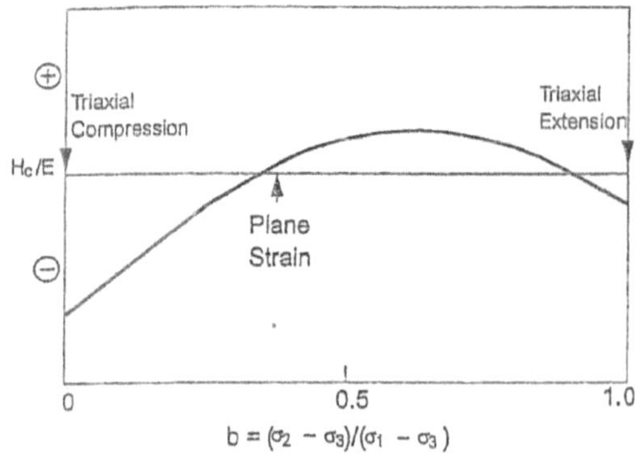

Fig. 3. Schematic diagram of variation of normalized, critical hardening modulus with b according to Rudnicki and Rice (1975) and Rice (1976)

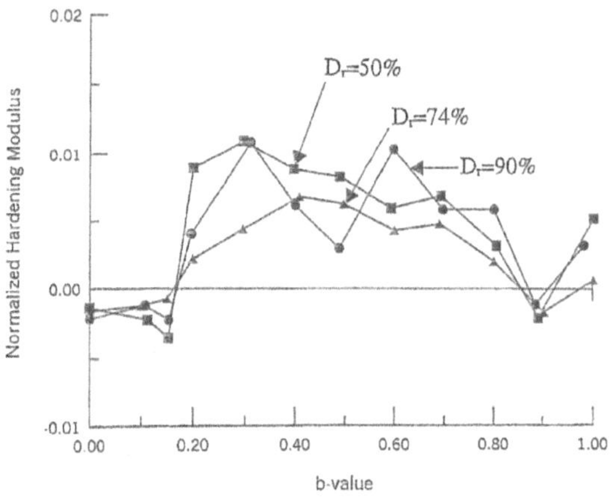

Fig. 4. Variation of normalized, critical hardening modulus, H_c/E, with b for true triaxial tests on tall prismatic specimens of dense Santa Monica Beach sand with σ_3'

calculated immediately before onset of shear banding. The range of b giving positive H_c/E is approximately 0.18–0.85 for all three relative densities. This means that strain localization followed by shear banding occurs in the hardening regime, i.e. before a smooth peak failure is reached. Negative values of H_c/E are obtained for b in the ranges 0.0–0.18 and 0.85–1.0. This means that failure occurs by smooth peak failure for these ranges of b. The calculation for $b = 1.0$ is questionable, because failure in the cross-anisotropic sand specimen occurred in the load-controlled, horizontal stress direction and the real peak point and post-peak stress-strain relation are not available. The variation of H_c/E in the mid-range and for high b-values compares well with that suggested by Rudnicki and Rice (1975) and Rice (1976) and is shown in Fig. 3. However, the negative H_c/E-values obtained for small b-values are much smaller than those indicated by these authors. The calculations of critical hardening moduli were found to be rather sensitive to the actual measured slope of the stress-strain curves for low and high b-

Variation of Experimental Critical Hardening Modulus

Bifurcation, i.e. initiation of shear band formation in dilating sand, is the result of continued shearing under conditions of uniform stresses and uniform strains. At some point, stress and strain states are reached which allow a localized displacement field to satisfy equilibrium, compatibility, boundary conditions, and constitutive relations (Rudnicki and Rice, 1975; Rice, 1976). This analysis procedure has been applied to study the influence of the constitutive model on the predicted shear banding (e.g. Molenkamp, 1985) and to investigate the conditions for shear banding under different three-dimensional stress conditions (e.g. Rudnicki and Rice, 1975; Vermeer, 1982; Peters et al., 1988; Peric et al., 1992).

The results of bifurcation analyses for three-dimensional stress and strain states are a set of critical plastic hardening moduli, H_c, which, when normalized on the elastic Young's modulus, E, vary with the value of b. The plastic hardening modulus plays the same role for calculation of plastic strain increments as Young's modulus plays for calculation of elastic strain increments. The plastic hardening modulus initiates with some positive value at the hydrostatic axis and decreases towards and becomes zero at smooth peak failure. It becomes negative in the softening regime. The plastic hardening modulus thus reflects the slope of the stress-strain curve. Both the plastic hardening and the elastic moduli have dimensions of stress, and the dimensionless ratio H_c/E, in which E relates to the same stress state as the plastic hardening modulus, expresses the normalized critical plastic hardening modulus. As will be seen below, this quantity emerges automatically, from the bifurcation analyses as a function of dimensionless quantities such as Poisson's ratio, the derivatives of the yield and plastic potential functions, and the direction of the shear band. A stress-strain state with a hardening modulus less than the critical hardening modulus can possibly produce shear banding. Rudnicki and Rice (1975) suggested a range of stress states in which the hardening modulus, critical for bifurcation, is positive, i.e. shear banding may occur during increasing loading before smooth peak failure has been reached, as shown in Fig. 3. According to this schematic diagram, shear banding may occur in the hardening regime in the mid-range of b-values, whereas it may occur for negative critical hardening moduli, i.e. in the softening regime, for b-values at and in the vicinity of triaxial compression and triaxial extension.

The normalized, critical hardening modulus, H_c/E, at the point immediately before onset of shear banding, was computed for each of the true triaxial tests on tall prismatic specimens of dense, medium dense, and loose Santa Monica Beach sand as presented by Wang and Lade (2001). In this manner, the experimentally determined conditions for shear band formation under three-dimensional stress conditions can be established and compared with the those shown in Fig. 3. The analysis required for this computation was presented by Lade and Wang (2001).

Figure 4 shows the variations with b-value of H_c/E

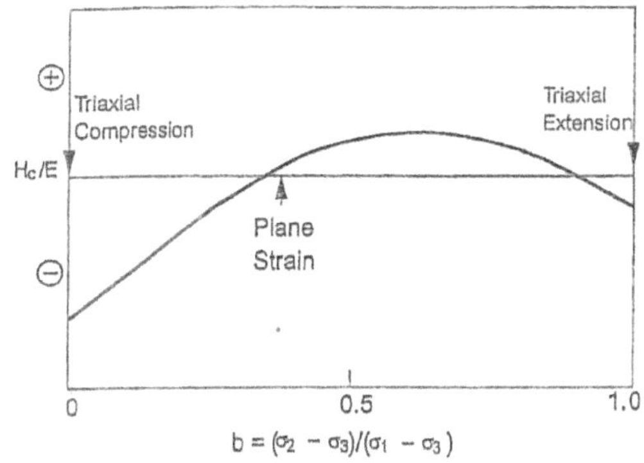

Fig. 3. Schematic diagram of variation of normalized, critical hardening modulus with b according to Rudnicki and Rice (1975) and Rice (1976)

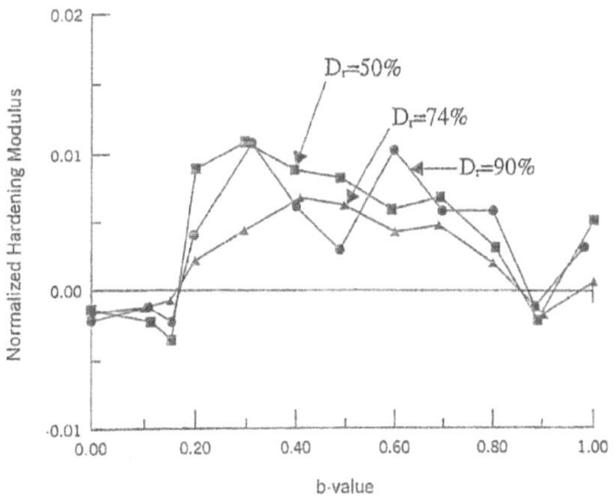

Fig. 4. Variation of normalized, critical hardening modulus, H_c/E, with b for true triaxial tests on tall prismatic specimens of dense Santa Monica Beach sand with σ_3'

calculated immediately before onset of shear banding. The range of b giving positive H_c/E is approximately 0.18–0.85 for all three relative densities. This means that strain localization followed by shear banding occurs in the hardening regime, i.e. before a smooth peak failure is reached. Negative values of H_c/E are obtained for b in the ranges 0.0–0.18 and 0.85–1.0. This means that failure occurs by smooth peak failure for these ranges of b. The calculation for $b = 1.0$ is questionable, because failure in the cross-anisotropic sand specimen occurred in the load-controlled, horizontal stress direction and the real peak point and post-peak stress-strain relation are not available. The variation of H_c/E in the mid-range and for high b-values compares well with that suggested by Rudnicki and Rice (1975) and Rice (1976) and is shown in Fig. 3. However, the negative H_c/E-values obtained for small b-values are much smaller than those indicated by these authors. The calculations of critical hardening moduli were found to be rather sensitive to the actual measured slope of the stress-strain curves for low and high b-

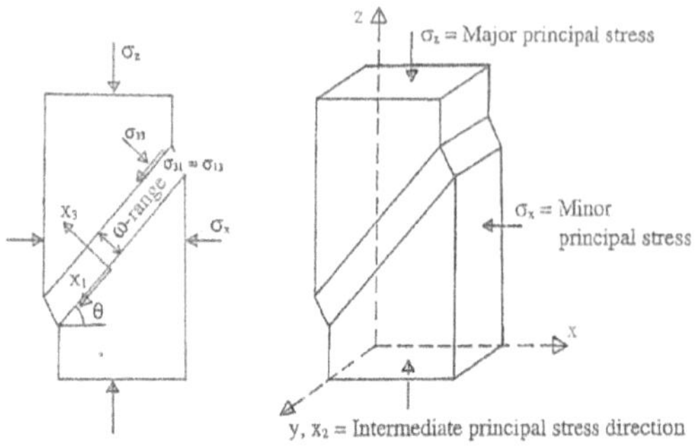

Fig. 6. Two views of tall prismatic specimen with a shear band

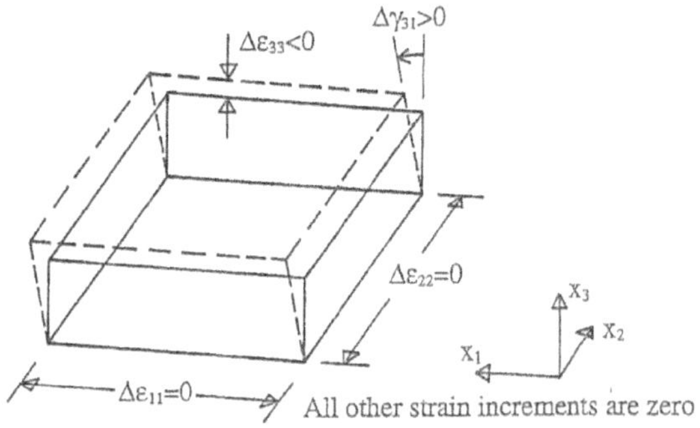

Fig. 7. Overall resulting incremental shear and normal displacement gradient pattern for an element inside the shear band

dicular to the shear band. As indicated above, the direction x_2 is always the direction of the intermediate principal stress σ_{II} (Rudnicki and Rice, 1975; Perrin and Leblond, 1993), and the x_1-x_3-plane therefore corresponds to the plane of the major and minor principal stresses, but these are generally not aligned with these coordinate axes.

The compatibility or kinematic conditions are expressed in terms of the incremental displacement gradients expressed in the local coordinate system shown in Fig. 6. The resulting picture of the incremental shear and normal displacement gradients is shown in Fig. 7.

1. The shear and normal displacement gradients along the shear band (in the x_1-direction) are zero, i.e. no stretching will occur along the shear band:

$$\Delta\varepsilon_{31}=\Delta\frac{\partial u_3}{\partial x_1}=0 \quad (1)$$

$$\Delta\varepsilon_{11}=\Delta\frac{\partial u_1}{\partial x_1}=0 \quad (2)$$

in which Δ indicates the difference between the local field inside and the uniform field outside the shear band.

2. The shear and normal displacement gradients in the plane of the shear band (in the x_2-direction) i.e. in the direction of the intermediate principal stress are zero. Since the x_1-x_3 plane is a principal plane, there are no shear strains or incremental shear strains associated with the x_2-direction. Thus:

$$\Delta\varepsilon_{21}=\Delta\frac{\partial u_2}{\partial x_1}=0 \quad \Delta\varepsilon_{12}=\Delta\frac{\partial u_1}{\partial x_2}=0$$

$$\Delta\varepsilon_{32}=\Delta\frac{\partial u_3}{\partial x_2}=0 \quad \Delta\varepsilon_{23}=\Delta\frac{\partial u_2}{\partial x_3}=0 \quad (3)$$

$$\Delta\varepsilon_{22}=\Delta\varepsilon_{II}=\Delta\frac{\partial u_2}{\partial x_2}=0. \quad (4)$$

3. The shear and normal displacement gradients perpendicular to the shear band (in the x_3-direction) may vary within the shear band, i.e. they are non-zero:

$$\Delta\varepsilon_{13}=\Delta\frac{\partial u_1}{\partial x_3}=g_1(x_3) \quad (5)$$

$$\Delta\varepsilon_{33}=\Delta\frac{\partial u_3}{\partial x_3}=g_2(x_3). \quad (6)$$

Thus, shear straining and dilation occur inside the band, and these strains need not be uniform across the shear band, as indicated by Eqs. (5) and (6), and as observed in laboratory experiments (e.g. Nemat-Nasser and Okada, 2001) and in field observations.

The overall resulting incremental shear and normal displacement gradient pattern for an element inside the shear band are indicated in Fig. 7, in which a uniform incremental shear strain component is indicated.

Equilibrium Conditions

Equilibrium must be maintained as localization occurs, and this requires that no changes in stresses occur in the x_3-direction across the shear band:

$$\Delta\sigma_{3j}=0 \quad \text{for } j=1, 2, 3. \quad (7)$$

With the resulting stress increments shown in Fig. 8, this means that:

1. There is no change in normal stress in the x_3-direction across the shear band:

$$\sigma_{33}(x_3=0)=\sigma_{33}(x_3=\omega) \quad (8)$$

in which the location $x_3=0$ is just outside the shear band, and $x_3=\omega$ corresponds to a location inside the shear band. Therefore:

$$\Delta\sigma_{33}=0. \quad (9)$$

2. There is no change in shear stress in the x_1-x_3-plane across the shear band:

$$\sigma_{31}(x_3=0)=\sigma_{31}(x_3=\omega). \quad (10)$$

Therefore:

$$\Delta\sigma_{31}=\Delta\sigma_{13}=0. \quad (11)$$

3. There is no change in shear stress in the x_2-direction along the shear band:

$$\Delta\sigma_{32}=\Delta\sigma_{23}=0. \quad (12)$$

4. However, there may be variations in normal stresses from outside to inside the shear band in the plane of the shear band (x_1-x_2-plane), i.e. there may be a jump in the normal stress in the x_1-direction across the boundary of the shear band:

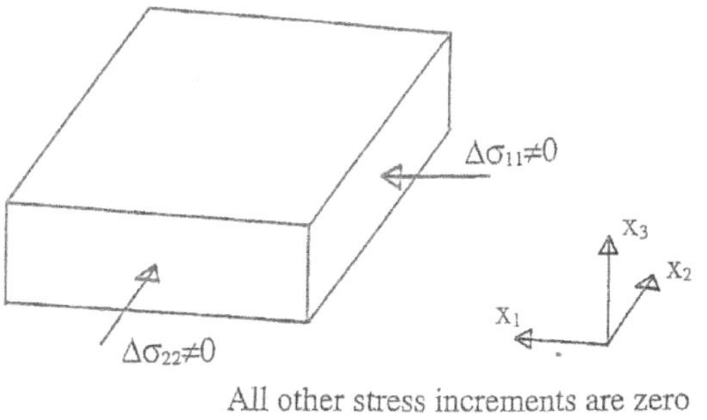

Fig. 8. Diagram of element of shear band with all stress increments indicated

$$\sigma_{11}(x_3=0) \neq \sigma_{11}(x_3=\omega) \tag{13}$$

in which the location $x_3=0$ is just outside the shear band, and $x_3=\omega$ corresponds to a location inside the shear band. Therefore:

$$\Delta\sigma_{11} \neq 0. \tag{14}$$

5. Similarly, there may be a jump in principal stress, σ_{II}, across the boundary of the shear band in the x_2-direction:

$$\sigma_{\mathrm{II}}(x_3=0) \neq \sigma_{\mathrm{II}}(x_3=\omega). \tag{15}$$

Therefore:

$$\Delta\sigma_{\mathrm{II}} \neq 0. \tag{16}$$

6. Because the x_2-direction is a principal stress direction (=intermediate principal stress):

$$\Delta\sigma_{21}=\Delta\sigma_{12}=0. \tag{17}$$

Figure 8 shows an element of the shear band with the non-zero stress increments indicated.

Elasto-Plastic Constitutive Model

The elasto-plastic constitutive model relates the strain increments to the stress increments, which have all been determined above. General expressions for shear banding are first developed, because each model produces its own shear banding conditions. The general expression for an incremental constitutive relationship may be written as follows:

$$\Delta\varepsilon_{ij}=\left(C^{e}_{ijkl}+\frac{1}{H}\cdot Q_{kl}\cdot P_{ij}\right)\cdot\Delta\sigma_{kl} \tag{18}$$

in which

$$\Delta\varepsilon_{ij}=\frac{1}{2}\left(\frac{\partial u_i}{\partial x_j}+\frac{\partial u_j}{\partial x_i}\right) \quad \text{for} \quad i, j=1, 2, 3 \tag{19}$$

are the incremental strain increments determined above, and C^{e}_{ijkl} = matrix of elastic constants in Hooke's law relating incremental elastic strains and stresses.

$$Q_{ij}=\frac{\partial f}{\partial\sigma_{ij}}\Bigg/\sqrt{\left(\frac{\partial f}{\partial\sigma_{ij}}\cdot\frac{\partial f}{\partial\sigma_{ij}}\right)}=n^{f}_{ij} \tag{20}$$

is a unit vector in the direction of the normal to yield surface, and

$$P_{ij}=\frac{\partial g}{\partial\sigma_{ij}}\Bigg/\sqrt{\left(\frac{\partial g}{\partial\sigma_{ij}}\cdot\frac{\partial g}{\partial\sigma_{ij}}\right)}=n^{g}_{ij} \tag{21}$$

is a unit vector in the direction of the normal to the plastic potential, i.e. in the direction of the plastic strain increment vector. In Eq. (18), the dot product of the two unit vectors, Q_{kl} and P_{ij}, represents the length of one unit normal projected on the other.

The remaining factor H in Eq. (18) is the plastic hardening modulus (scalar), which gives magnitude to the plastic strain increments. It plays the same role in hardening plasticity as Young's modulus, E, plays in Hooke's law. The plastic hardening modulus is defined as:

$$H=\frac{\partial g}{\partial\bar{\varepsilon}}\Bigg/\sqrt{\left(\frac{\partial f}{\partial\sigma_{ij}}\cdot\frac{\partial f}{\partial\sigma_{ij}}\right)} \tag{22}$$

in which the term in the denominator represents the length of the normal to the yield surface, g is the plastic potential function, and $d\bar{\varepsilon}$ represents the length of the total plastic strain increment:

$$d\bar{\varepsilon}=\sqrt{d\varepsilon^{p}_{ij}\cdot d\varepsilon^{p}_{ij}}. \tag{23}$$

Determination of the plastic hardening modulus will be demonstrated below for the particular model used in this presentation.

Critical Shear Band Condition

The critical shear band condition is then derived by substitution of the known expressions for the incremental stresses and strains into Eq. (18). This results in the following relations:

$$\begin{bmatrix} \Delta\varepsilon_{11}=0 \\ \Delta\varepsilon_{22}=\Delta\varepsilon_{\mathrm{II}}=0 \\ \Delta\varepsilon_{33} \\ \Delta\varepsilon_{23}=0 \\ \Delta\varepsilon_{13} \\ \Delta\varepsilon_{12}=0 \end{bmatrix} = \begin{bmatrix} C_{1111} & C_{1122} & C & C & C & C \\ C_{2211} & C_{2222} & C & C & C & C \\ C & C & C & C & C & C \\ C & C & C & C & C & C \\ C & C & C & C & C & C \\ C & C & C & C & C & C \end{bmatrix} \begin{bmatrix} \Delta\sigma_{11} \\ \Delta\sigma_{22}=\Delta\sigma_{\mathrm{II}} \\ \Delta\sigma_{33}=0 \\ \Delta\sigma_{23}=0 \\ \Delta\sigma_{13}=0 \\ \Delta\sigma_{12}=0 \end{bmatrix}. \tag{24a-f}$$

The first two equations express that the normal strain is zero in the two perpendicular directions contained in the plane of the shear band. Only the compliance moduli of interest are indicated with indices in Eqs. 24(a) and (b). The first two equations are:

$$\Delta\varepsilon_{11} = C_{1111}\cdot\Delta\sigma_{11} + C_{1122}\cdot\Delta\sigma_{II} = 0 \tag{25}$$

$$\Delta\varepsilon_{II} = C_{2211}\cdot\Delta\sigma_{11} + C_{2222}\cdot\Delta\sigma_{II} = 0. \tag{26}$$

A non-trivial solution for stress increments is only obtained if the determinant of these two simultaneous equations is zero:

$$C_{1122}\cdot C_{2211} - C_{1111}\cdot C_{2222} = 0. \tag{27}$$

This equation is then solved to obtain the hardening modulus, which is critical for shear banding. From the expressions in Eq. (18) for total strain increments:

$$C_{1111} = \frac{1}{E} + \frac{1}{H}\cdot P_{11}\cdot Q_{11} \tag{28}$$

$$C_{2222} = \frac{1}{E} + \frac{1}{H}\cdot P_{22}\cdot Q_{22} \tag{29}$$

$$C_{1122} = -\frac{\nu}{E} + \frac{1}{H}\cdot P_{11}\cdot Q_{22} \tag{30}$$

$$C_{2211} = -\frac{\nu}{E} + \frac{1}{H}\cdot P_{22}\cdot Q_{11}. \tag{31}$$

Substitution of these expressions for the C's into Eq. (27), and reducing and solving for the normalized hardening modulus, H/E, produces:

$$\frac{H}{E} = \frac{-\nu}{1-\nu^2}\cdot(P_{11}\cdot Q_{22} + P_{22}\cdot Q_{11}) - \frac{1}{1-\nu^2}\cdot(P_{11}\cdot Q_{11} + P_{22}\cdot Q_{22}) \tag{32}$$

in which E is young's modulus and ν is Poisson's ratio.

The values of the scalar quantities P_{11} and Q_{11} of the unit vectors P_{ij} (in Eq. (21)) and Q_{ij} (in Eq. (20)) can be related to the principal values by the following transformation:

$$P_{11} = P_n - P_d\cdot\cos 2\theta \tag{33}$$

$$Q_{11} = Q_n - Q_d\cdot\cos 2\theta \tag{34}$$

where

$$P_n = \frac{1}{2}\cdot(P_I + P_{III}) \tag{35}$$

$$P_d = \frac{1}{2}\cdot(P_I - P_{III}) \tag{36}$$

$$Q_n = \frac{1}{2}\cdot(Q_I + Q_{III}) \tag{37}$$

$$Q_d = \frac{1}{2}\cdot(Q_I - Q_{III}). \tag{38}$$

The terms in these equations relate to the values in the x_1-x_3-plane, i.e. the plane of the major and minor principal stresses. The subscripts I and III refer to the values in the directions of the major and minor principal stresses, respectively.

Substitution of the expressions for P_{11} and Q_{11} from Eqs. (33) and (34) into the expression for the normalized plastic hardening modulus H/E in Eq. (32) leads to a quadratic equation in cos 2θ which has real solutions only for $H \le H_c$:

$$\frac{H}{E}\cdot(1-\nu^2) = -[P_{22}\cdot Q_{22} + (P_n - P_d\cdot\cos 2\theta)\times(Q_n - Q_d\cdot\cos 2\theta)] - \nu\cdot[(P_n - P_d\cdot\cos 2\theta)\cdot Q_{22} + P_{22}\cdot(Q_n - Q_d\cdot\cos 2\theta)]. \tag{39}$$

The value of cos $2\theta_c$ that corresponds to H_c is the value that maximizes H. This value can be determined by setting the derivative of H with regard to θ equal to zero. This results in an expression for the critical inclination of the shear band, θ_c, which is determined from:

$$\sin\theta_c = \pm\frac{1}{2}\cdot\sqrt{2 - \left[\frac{(Q_n + \nu\cdot Q_{22})}{Q_d} + \frac{(P_n + \nu\cdot P_{22})}{P_d}\right]}. \tag{40}$$

Note that direction 22 is the direction of the intermediate principal stress, x_2, and therefore $P_{22} = P_{II}$ and $Q_{22} = Q_{II}$.

Calculation Procedure

The calculation procedure for shear band determination is initiated with computation of the unit vectors P_I, P_{II}, P_{III} and Q_I, Q_{II}, Q_{III}. These are then substituted into Eqs. (35)–(38), which in turn are substituted into Eq. (40) for calculation of the critical inclination of the shear band, θ_c. This value is then substituted into Eqs. (33) and (34), and these are in turn inserted into Eq. (32) for determination of the critical hardening modulus, H_c.

As the prediction of the stress-strain relation from the constitutive model proceeds, the plastic hardening modulus, H, is calculated for each step and used to determine the plastic strain increments. The critical hardening modulus, H_c, is also calculated from Eq. (32) for each step, and the two moduli are compared. As long as $H \ge H_c$, the condition for localization of plastic strains and possible shear banding is not critical. However, when this inequality is violated, shear banding may occur in the granular material. Since the strength inside a shear band decreases due to the increasing void ratio, it is assumed that a shear band will, in fact, develop at the time the critical condition for shear banding is reached the first time. This calculated behavior will then be compared with the experimental results.

SINGLE HARDENING ELASTO-PLASTIC CONSTITUTIVE MODEL

It has been shown that prediction of the critical hardening modulus for shear banding depends on the constitutive model employed for predicting the stress-strain behavior of the soil (Molenkamp, 1985). In this presentation, the single hardening constitutive model previously presented in the literature will be used (Kim and Lade, 1988; Lade and Kim, 1988a and b). This elasto-plastic constitutive model with a single yield surface has been developed for the behavior of frictional

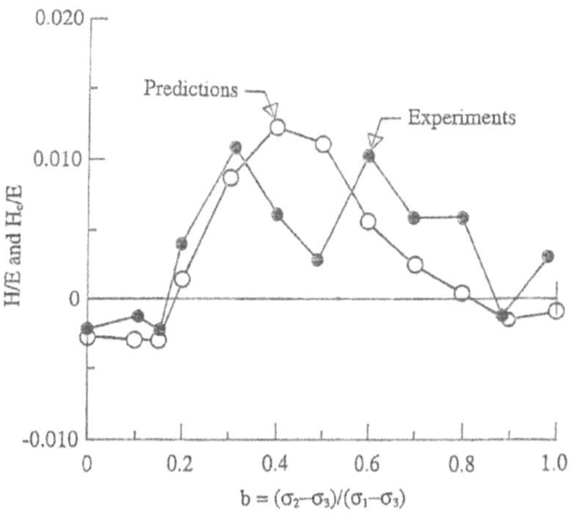

Fig. 9. Comparison of theoretical and experimental variations of normalized critical hardening modulus with *b*-value determined for dense Santa Monica Beach sand

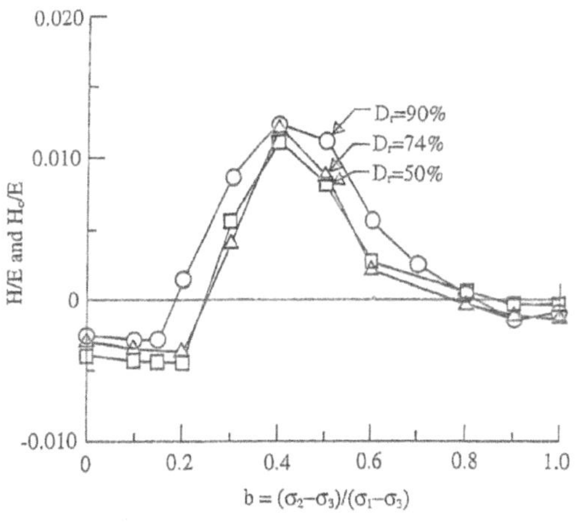

Fig. 10. Comparison of theoretically predicted variations of normalized critical shear modulus with *b*-value determined for three densities of Santa Monica Beach sand

materials such as sand, clay, concrete, and rock, and it has been summarized by Lade (1990) and by Lade and Kim (1995).

Each component of the model has already been developed, presented, and discussed in the literature. In addition to Hooke's law for the elastic behavior (Lade and Nelson, 1987), the framework for the plastic behavior consists of a failure criterion (Lade, 1977, 1982b, 1984; Kim and Lade, 1984), a nonassociated flow rule, a yield criterion that describes contours of equal plastic work, and a work-hardening/softening law. For the purpose of its application to prediction of shear banding in true triaxial tests, the model was further enhanced to capture the smooth peak failure and the softening observed in triaxial compression tests.

COMPARISON OF EXPERIMENTAL AND THEORETICAL SHEAR BANDING

The variation of critical hardening moduli with *b*-value calculated from the theoretical approach is shown in Fig. 9 for dense Santa Monica Beach sand ($e = 0.63$). This diagram indicates that shear banding occurs in the hardening regime for *b*-values in the middle range, and it occurs in the softening regime for low and high values of *b*. The critical hardening moduli obtained from the experiments are also compared with the theoretical values in Fig. 9. More scatter is encountered in the experimentally determined critical hardening moduli, but they generally follow the pattern indicated by the theoretical predictions. Thus, shear banding occurs in the softening regime for *b*-values up to approximately $b = 0.18$. At this point, a sharp break occurs in the variations of both the theoretical and the experimentally obtained hardening moduli, and these become positive, indicating that shear banding now occurs in the hardening regime. The switch to negative critical hardening moduli is less clear, but according to the theory it occurs near 0.77 for the dense sand, while the experiments indicated a value near 0.86. Both the experiments and the calculated stress-strain relations show that shear banding occurs just beyond smooth peak failure in the tests at high *b*-values.

As discussed above, the specimen in the experiment with $b = 1.00$ failed in the horizontal direction, i.e. the direction of the intermediate principal stress, due to the lower strength in the horizontal direction of the cross-anisotropic structure of the vertically pluviated sand specimens. These experiments are therefore not representative of the isotropic sand behavior assumed in the present analyses.

Very similar results were obtained for the other two series of experiments on tall prismatic specimens of Santa Monica Beach sand with void ratios of 0.68 and 0.75. Figure 10 presents a comparison of the predicted critical hardening moduli for all three densities of the Santa Monica Beach sand. The three curves are very similar and they all show an abrupt switch in critical hardening modulus from negative to positive near $b = 0.2$. However, individual switch points are obtained for the three densities. The abrupt switch in critical hardening modulus near $b = 0.2$ was not suggested in the variation indicated by Rudnicki and Rice (1975), as shown in Fig. 3. However, comparison with the variations of critical hardening moduli obtained from the experiments in Fig. 4 indicates similar behavior as predicted from the theory. It is clear that a fair amount of scatter is present in the experimental results, but the overall pattern of occurrence of shear banding under three-dimensional conditions is predicted with good accuracy. The variations of critical hardening moduli shown in Fig. 10 for the three densities are caused by: (1) the difference in and the interaction of the model parameters through the constitutive model and the shear banding analysis, and (2) the round-off errors and very sensitive nature of the predictions for which the two

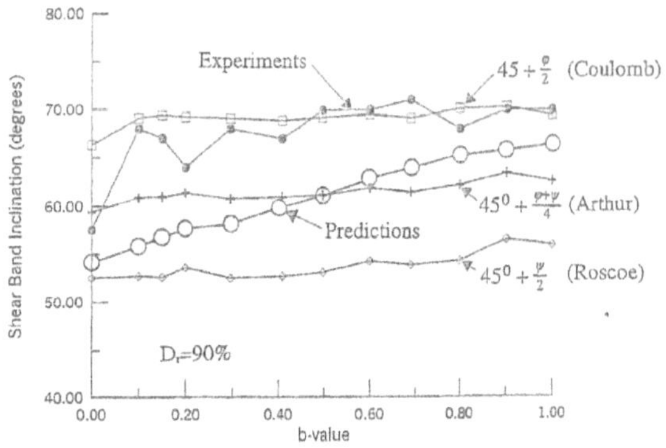

Fig. 11. Experimental shear band inclinations for dense Santa Monica Beach sand and comparisons with shear band calculations and with those proposed by Coulomb, Arthur and Roscoe

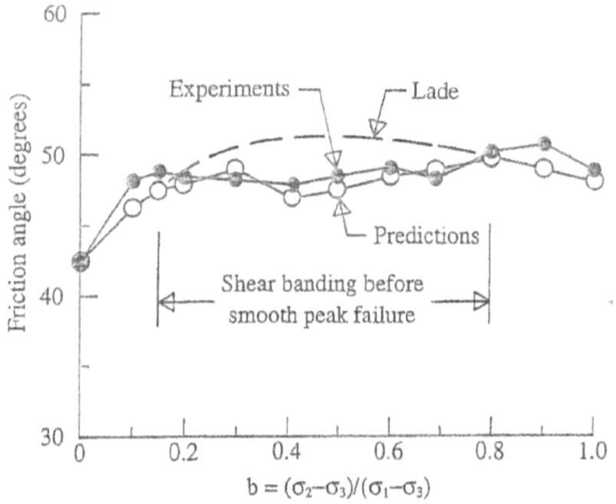

Fig. 12. Comparison of measured and predicted friction angles for true triaxial tests on dense Santa Monica Beach sand. Shear banding controls strength in the middle range of *b*-values

moduli for some *b*-values are rather close to each other over a range of the major principal strain.

CALCULATION OF SHEAR BAND INCLINATIONS

The first step in calculation of critical conditions for shear banding involves determination of the shear band inclination relative to the major principal plane as obtained from Eq. (40). This inclination is then used for determination of the critical hardening modulus from Eq. (32), as explained above. Once the critical condition for shear banding has been determined, the corresponding inclination of the shear band can be compared with the experimentally obtained shear band inclinations. Figure 11 shows comparisons of predicted and observed shear band directions for dense Santa Monica Beach sand. Also shown on this diagram are the directions obtained from the simpler solutions referred to as Coulomb's, Arthur's, and Roscoe's inclinations. It is seen that the inclinations obtained from the present analyses are 5–10 degrees lower than those observed. However, they cross over the Arthur inclination not to far from the plane strain condition, which corresponds to a *b*-value of approximately 0.23. Analyses by e.g. Vardoulakis (1980) and Vermeer (1982) also show results that are closer to the Arthur inclination for plane strain conditions. Based on a theoretical analysis, Vermeer (1990) showed that the actual shear band inclination is very sensitive to boundary conditions such as those imposed by the membrane surrounding the specimen in the experiment. Further discussion of the shear band inclination was presented by Lade and Wang (2001).

VARIATION OF FRICTION ANGLE WITH *b*-VALUE

The effect of shear banding on the variation of the friction angle obtained under three-dimensional conditions can now be investigated. The peak stress point obtained from the calculations of each test is used to calculate the friction angle for the sand, and the resulting variation is shown in Fig. 12. In the ranges of low and high *b*-values where the shear banding occurs in the softening regime, the smooth peak stress is employed for calculation of the friction angle, while the points of shear banding in the middle range of *b*-values correspond to the highest stress point on the stress-strain curves and therefore form the basis for the calculation of the friction angles. Thus, shear banding in the hardening regime over the middle range of *b*-values clearly produces friction angles that are lower than those anticipated by a smooth failure surface, such as that expected from the failure criterion proposed by Lade (1977). However, this failure criterion was a part of the model used to predict the points of shear banding, and these predictions could not have been made with the relative accuracy obtained here unless this criterion had been employed. Thus, the failure criterion, which anticipates the peak failure to be smooth, serves to pull the stress-strain relation up, such that the failure created by shear banding can occur in the hardening regime, i.e. on the way up to smooth peak failure. This failure criterion is therefore only correct for low and high *b*-values, but for the middle range of *b*-values it serves an important role as a target (which is never reached due to shear banding) for the condition of smooth peak failure. Exactly where shear banding occurs simultaneously with smooth peak failure between $b = 0.80$ and $b = 0.90$ was not determined, but it occurs in this interval of *b*-values. The experiments are in agreement with this qualitative and quantitative interpretation of failure in sand under three-dimensional conditions.

Thus, the failure surface for a granular material is not a smooth surface that can be described by a single expression, and smooth peak failure is obtained only for the extreme values of *b*, while shear banding causes failure in the middle range of *b*-values, including plane strain.

CONCLUSIONS

The occurrence of shear banding in true triaxial tests on granular material have been analyzed and predicted theoretically, and both the experimental results and the predictions show that shear banding has an important influence on the shape of the failure surface.

Shear banding has been studied in true triaxial tests on tall prismatic specimens at three different void ratios of a sand, and this study forms the basis for comparison with theoretical predictions of shear banding. The theory of localization and shear banding is presented in detail, and a constitutive model for the sand is reviewed. It is important to employ a realistic constitutive model for the soil, because the model employed influences the predictions of shear banding. Since shear banding occurs near smooth peak failure in the hardening regime or in the softening regime, it is important to employ an accurate description of the transition from hardening to softening in the model.

It is shown that the predicted stress conditions for shear banding match those observed in the true triaxial tests, and it is concluded that shear banding plays an important role in the three-dimensional strength of granular materials. Thus, peak failure is caused by shear banding in the hardening regime in the approximate range of $b=(\sigma_2-\sigma_3)/(\sigma_1-\sigma_3)$ from 0.18 to 0.85, while it occurs in the softening regime outside this range of b-values. A smooth, continuous $3D$ failure surface is therefore not generally obtained for granular materials.

REFERENCES

1) Arthur, J. R. F. and Dunstan, T. (1982): Rupture layers in granular media, *Proc., IUTAM Symp. on Deformation and Failure of Granular Materials*, Delft (eds. by Vermeer, P. A. and Luger, H. J.), Balkema, Rotterdam, 453–459.
2) Arthur, J. R. F., Dunstan, T., Al-Ani, Q. A. J. L. and Assadi, A. (1977): Plastic deformation and failure in granular media, *Géotechnique*, **27** (1), 53–74.
3) Desrues, J., Lanier, J. and Stutz, P. (1985): Localization of the deformation in tests on sand sample, *Engrg. Fracture Mechanics*, **21** (4), 909–921.
4) Desrues, J., Chambon, R., Mokni, M. and Mazerolle, F. (1996): Void ratio evolution inside shear bands in triaxial sand specimens studied by computed tomography, *Géotechnique*, **46** (3), 529–546.
5) Finno, R. J., Harris, W. W., Mooney, M. A. and Viggiani, G. (1996): Strain localization and undrained steady state of sand, *J. of Geotech. Engrg.*, **122** (6), 462–473.
6) Finno, R. J., Harris, W. W., Mooney, M. A. and Viggiani, G. (1997): Shear bands in plane strain compression of loose sand, *Géotechnique*, **47** (1), 149–165.
7) Hansen, B. (1958): Line ruptures regarded as narrow rupture zones—basic equations based on kinematic considerations, *Proc., Conf. on Earth Pressure Problems*, Brussels, 1, 39–48.
8) Kim, M. K. and Lade, P. V. (1988): Single hardening constitutive model for frictional materials, I. Plastic Potential Function, *Computers and Geotechnics*, **5**, 307–324.
9) Lade, P. V. (1977): Elasto-plastic stress-strain theory for cohesionless soil with curved yield surfaces, *Int. J. Solids Struct.*, **13**, 1019–1035.
10) Lade, P. V. (1982a): Localization effects in triaxial tests on sand, *Proc., IUTAM Symp. on Deformation and Failure of Granular Materials*, Delft (eds. by Vermeer, P. A. and Luger, H. J.), Balkema, Rotterdam, 461–471.
11) Lade P. V. (1982b): Three-parameter failure criterion for concrete, *J. Engrg. Mech.*, ASCE, **108** (EM5), 850–863.
12) Lade, P. V. (1984): Failure criterion for frictional materials, *Mechanics of Engineering Materials* (eds. by Desai, C. S. and Gallagher, R. H.), Wiley, New York, 385–402.
13) Lade, P. V. (1990): Single hardening model with application to NC Clay, *J. Geotech. Engrg.*, ASCE, **116** (3), 394–414.
14) Lade, P. V. and Duncan, J. M. (1973): Cubical triaxial tests on cohesionless soils, *J. Soil Mech. Found. Div.*, ASCE, **99** (SM10), 793–812.
15) Lade, P. V. and Kim, M. K. (1988a): Single hardening constitutive model for frictional materials II. Yield criterion and plastic work contours, *Computers and Geotechnics*, **6**, 13–29.
16) Lade, P. V. and Kim, M. K. (1988b): Single hardening constitutive model for frictional materials, III. Comparisons with experimental data, *Computers and Geotechnics*, **6**, 31–48.
17) Lade, P. V. and Kim, M. K. (1995): Single hardening constitutive model for soil, rock and concrete, *Int. J. Solids Struct.*, Pergamon Press, **32**, 1963–1978.
18) Lade, P. V. and Nelson, R. B. (1987): Modeling the elastic behavior of granular materials, *Int. J. for Num. and Analyt. Methods in Geomech.*, **11**, 521–542.
19) Lade, P. V. and Wang, Q. (2001): Analysis of shear banding in true triaxial tests on sand, *J. Engrg. Mech.*, ASCE, **127** (8), 762–768.
20) Molenkamp, F. (1985): Comparison of frictional material models with respect to shear band initiation, *Géotechnique*, **35** (2), 127–143.
21) Muhlhaus, H.-B. and Vardoulakis, I. (1987): The thickness of shear bands in granular materials, *Géotechnique*, **37** (3), 271–283.
22) Nemat-Nasser, S. and Okada, N. (2001): Radiographic and microscopic observation of shear bands in granular materials, *Géotechnique*, **51** (9), 753–765.
23) Peric, D., Runesson, K. and Sture, S. (1992): Evaluation of plastic bifurcation for plane strain versus axisymmetry, *J. Engrg. Mech.*, ASCE, **118** (3), 512–524.
24) Perrin, G. and Leblond, J. B. (1993): Rudnicki and Rice's analysis of strain localization revisited, *J. Appl. Mech.*, ASME, **60**, 842–846.
25) Peters, J. F., Lade, P. V. and Bro, A. (1988): Shear band formation in triaxial and plane strain tests, *Advanced Triaxial Testing of Soil and Rock, ASTM STP 977* (eds. by Donaghe, R. T. et al.), American Society for Testing and Materials, Philadelphia, 604–627.
26) Poorooshasb, H. B., Holubec, I. and Sherbourne, A. N. (1967): Yielding and flow of sand in triaxial compression: Parts II and III, *Can. Geotech. J.*, **4** (4), 376–397.
27) Rice, J. R. (1976): The localization of plastic deformation, *Proc., Fourth Conf. on Theoretical and Applied Mechanics*, (ed. by Koiter, W. T.), North-Holland Publishing Co., Amsterdam, 207–229.
28) Roscoe, K. H. (1970): The influence of strains in soil mechanics, *Géotechnique*, **20** (2), 129–170.
29) Rudnicki, J. W. and Rice, J. R. (1975): Conditions for the localization of deformation in pressure-sensitive dilatant materials, *J. Mech. Phys. Solids*, **23**, 371–394.
30) Scarpelli, G. and Wood, D. M. (1982): Experimental observations of shear band patterns in direct shear tests, *Proc., IUTAM Symp. on Deformation and Failure of Granular Materials*, Delft (eds. by Vermeer, P. A. and Luger, H. J.), Balkema, Rotterdam, 473–484.
31) Tatsuoka, F., Nakamura, S., Huang, C.-C. and Tani, K. (1990): Strength anisotropy and shear band direction in plane strain tests on sand, *Soils and Foundations*, **30** (1), 35–54.
32) Tatsuoka, F., Okahara, M., Tanaka, T., Tani, K., Morimoto, T. and Siddiquee, M. S. A. (1991): Progressive failure and particle size effect in bearing capacity of a footing on sand, *Proc. ASCE Geotech. Engrg. Congress 1991*, Boulder, ASCE GSP 27, 788–802.
33) Vardoulakis, I. (1980): Shear band inclination and shear modulus of sand in biaxial tests, *Int. J. Num. and Analyt. Methods in Geomech.*, **4**, 103–119.
34) Vardoulakis, I. (1996a): Deformation of water-saturated sand: I. uniform undrained deformation and shear banding, *Géotechnique*,

46 (3), 441–456.

35) Vardoulakis, I. (1996b): Deformation of water-saturated sand: I. effect of pore water flow and shear banding, *Géotechnique*, **46** (3), 457–472.

36) Vardoulakis, I., Goldscheider, M. and Gudehus, G. (1978): Formation of shear bands in sand bodies as a bifurcation problem, *Int. J. Num. Analyt. Methods in Geomech.*, **2**, 99–128.

37) Vardoulakis, I. and Graf, B. (1982): Imperfection sensitivity of the biaxial test on dry sand, *Proc., IUTAM Symp. on Deformation and Failure of Granular Materials*, Delft (eds. by Vermeer, P. A. and Luger, H. J.), Balkema, Rotterdam, 485–491.

38) Vermeer, P. A. (1982): A simple shear-band analysis using compliances, *Proc., IUTAM Symp. on Deformation and Failure of Granular Materials*, Delft (eds. by Vermeer, P. A. and Luger, H. J.), Balkema, Rotterdam, 493–499.

39) Vermeer, P. A. (1990): The orientation of shear bands in biaxial tests, *Géotechnique*, **40** (2), 223–236.

40) Wang, Q. and Lade, P. V. (2001): Shear banding in true triaxial tests and its effect on failure in sand, *J. Engrg. Mech.*, ASCE, **127** (8), 754–761.

41) Yoshida, T., Tatsuoka, F., Siddiquee, M. S. A., Kamegai, Y. and Park, C.-S. (1993): Shear banding in sands observed in plane strain compression, *Proc. 3rd Int. Workshop Localisation and Bifurcation Theory for Soils and Rocks*, Grenoble, 165–179.

Computers and Geotechnics 16 (1994) 123–151

0266-352X/94/$7.00

ELSEVIER

INSTABILITY AND LIQUEFACTION OF GRANULAR MATERIALS

P.V. Lade
Department of Civil Engineering
University of California
Los Angeles, California 90024, U.S.A.

ABSTRACT

Stability postulates by Drucker and by Hill are applicable to solid metal. When applied to granular material exhibiting nonassociated flow, these concepts of stability imply that the material may be unstable when exposed to certain stress paths *inside* the failure surface. Series of triaxial tests on fully saturated and partly saturated specimens have been performed under drained and undrained conditions to study the regions of stable and unstable behavior. The stability postulate by Drucker and by Hill do not capture the conditions for stable and unstable behavior of granular materials. The experimentally determined conditions for which granular materials are stable, conditionally unstable, and unconditionally unstable are discussed.

INTRODUCTION

Conditions for well-posed initial and boundary value problems by Drucker [1-3] and by Hill [4-5] are applicable to solid metal for which the shear response does not depend on the hydrostatic stress. Both of these conditions require the second increment of plastic work be positive suggesting a thermodynamic interpretation whereby the conditions for uniqueness coincides with the conditions for material stability. It is now an experimentally established fact that realistic constitutive equations for granular materials, for which the shear response depends strongly on the hydrostatic stress, necessarily violate the sufficient conditions for uniqueness. Further, the violation of the conditions for uniqueness implies a breakdown in the correspondence between uniqueness and stability, for it is possible to find loading combinations that violate the Drucker-Hill criteria yet appear to be stable [6-8]. However, uniqueness is a mathematical concept that describes a characteristic of the solution set of a system of partial differential equations. Stability is a physical description of a mechanical system. While it is reasonable to expect that failure of a system is related to a loss of a unique solution to the equations describing that system, the relationship of the Drucker-Hill criteria to physical stability remains uncertain.

Experiments to study whether or not granular materials are unstable under certain load conditions have been performed. In these studies instability was taken to be the inability of the material to sustain or carry a given load. This includes the inability to sustain small perturbations in the load. Several series of triaxial compression tests designed and performed to expose the type of behavior displayed by granular materials are reviewed. The results of these experiments show that a positive value of the second increment of plastic work is neither a necessary nor a sufficient condition for stability of granular (frictional) materials. Experimental observations of stability and instability are reported, and stability conditions for materials with nonassociated flow are presented.

STABILITY POSTULATES

Experimental evidence from tests on several types of soils have clearly indicated that the use of conventional associated flow rules results in prediction of too large volumetric expansion. To characterize the volume change correctly, it is necessary to employ a nonassociated flow rule. The plastic potential surfaces do therefore not coincide with the yield surfaces, but the two families of surfaces cross each other.

The application of nonassociated plastic flow rules for soil have resulted in questions regarding uniqueness and stability of such materials. The stability postulate for time-independent materials due to Drucker [1-3] is satisfied provided that associated plastic flow is employed in construction of constitutive models involving convex, plastic yield surfaces. Hill's stability condition [4, 5] is expressed in terms of total strain increments (elastic and plastic), and it extends the condition for stability a little beyond that due to Drucker. Theoretical considerations have suggested that they are not necessary conditions [6-8].

As a first approximation, yield and failure conditions are independent of the mean normal stress for solid metals. The yield surfaces are therefore parallel to the hydrostatic axis as shown in the triaxial plane in Fig. 1. Further, solid metals do not change volume during plastic deformation and this leads to normality of plastic strain increment vectors to the yield surfaces, i.e. associated flow is observed for solid metals. Drucker's stability postulate appear to capture the behavior of solid metals with good approximation. According to this postulate, stability requires that the second increment of plastic work is positive or zero:

$$d^2W_p \;=\; \dot{\sigma}_{ij}\;\dot{\varepsilon}^p_{ij} \;\geq\; 0 \tag{1}$$

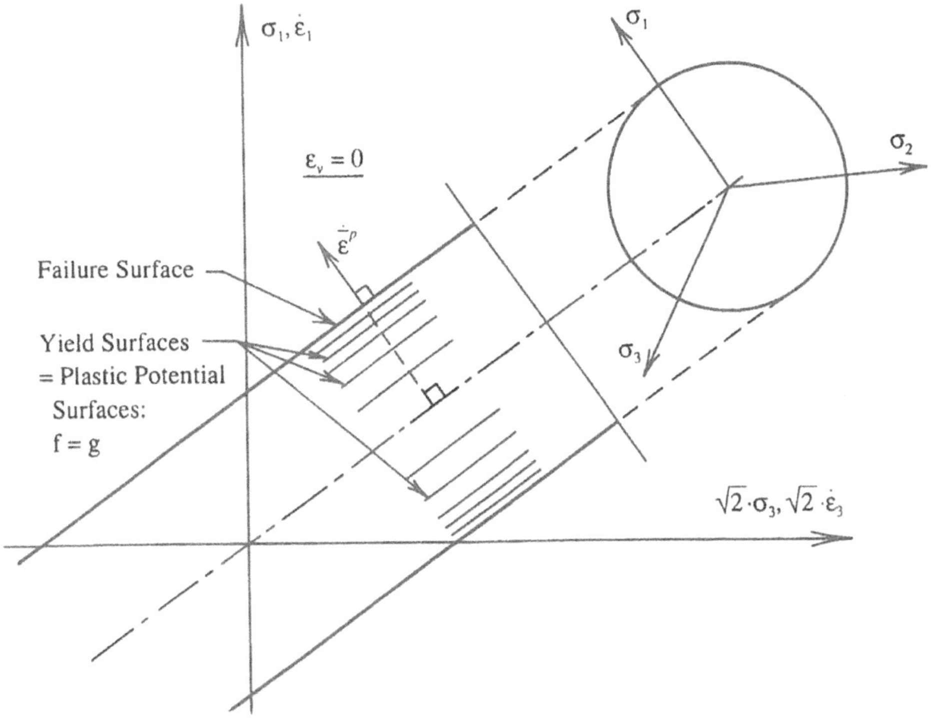

Figure 1. Yield, failure, and plastic potential surfaces for solid metals.

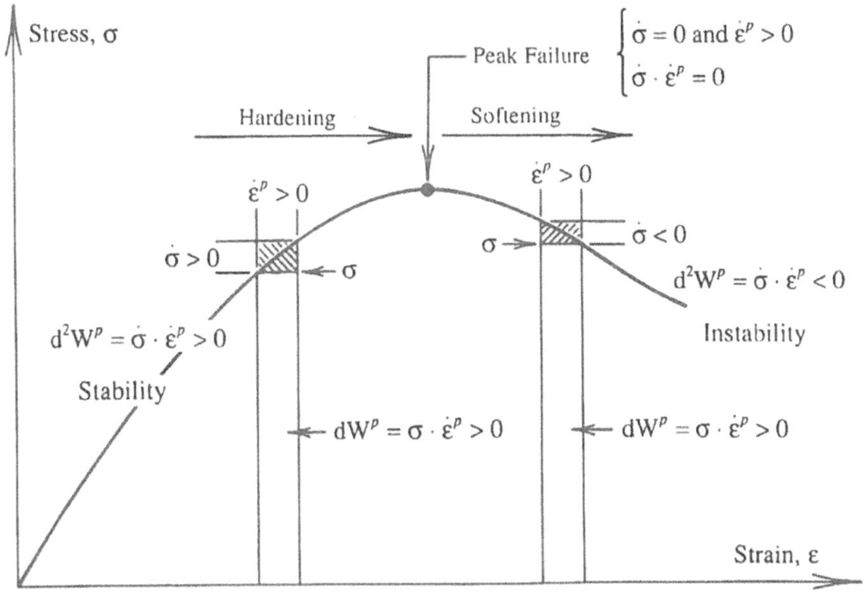

Figure 2. Illustration of Drucker's stability postulate for solid metals.

in which $\dot{\sigma}_{ij}$ = increment of stress and $\dot{\varepsilon}^p_{ij}$ = resulting increment in plastic strain. For metals, positive values of the second increment of plastic work are always associated with the stable, ascending part of the stress-strain relationship, whereas negative values are associated with the unstable, descending part of the stress-strain curve obtained after peak failure, as illustrated in Fig. 2.

According to Hill's condition stability should be maintained as long as

$$\dot{\sigma}_{ij} \ \dot{\varepsilon}^t_{ij} = \dot{\sigma}_{ij} \ (\dot{\varepsilon}^e_{ij} + \dot{\varepsilon}^p_{ij}) = \dot{\sigma}_{ij} \ \dot{\varepsilon}^e_{ij} + \dot{\sigma}_{ij} \ \dot{\varepsilon}^p_{ij} \ \geq \ 0 \tag{2}$$

in which $\dot{\varepsilon}^t_{ij}$ and $\dot{\varepsilon}^e_{ij}$ are the total and elastic strain increments, respectively. Hill's stability condition guarantees stability a little beyond the condition given by Drucker.

CONSEQUENCES OF NONASSOCIATED FLOW

The experimentally determined shape of failure surfaces for granular materials is indicated schematically in Fig. 3. It resembles a pointed bullet with cross sections in octahedral planes which are triangular, monotonically curved surfaces with smoothly rounded corners [9, 10]. The strengths of granular materials are greatly influenced by the magnitude of the mean normal stress. In fact, the strength is essentially proportional to the mean normal stress.

A typical pattern of yield surfaces for an isotropic soil is shown on the triaxial plane in Fig. 4. In three dimensions these yield surfaces are shaped as asymmetric tear drops. For an isotropic material the yield surfaces intersect the hydrostatic axis in a perpendicular manner, they bend smoothly backwards towards the origin, and cross the failure surface at sharp angles as indicated in Fig. 4.

Substantial volume changes (compression and expansion) occur in granular materials during loading, and experiments have clearly shown that nonassociated flow is obtained for such materials.

Plastic potential surfaces have similar shapes as yield surfaces, but for nonassociated flow the two families of surfaces cross each other. Experimental evidence for frictional materials indicates that the plastic potential surfaces have more pointed ends and they resemble cigars with asymmetric cross-sections. A typical plastic potential surface is shown at point A in Fig. 4.

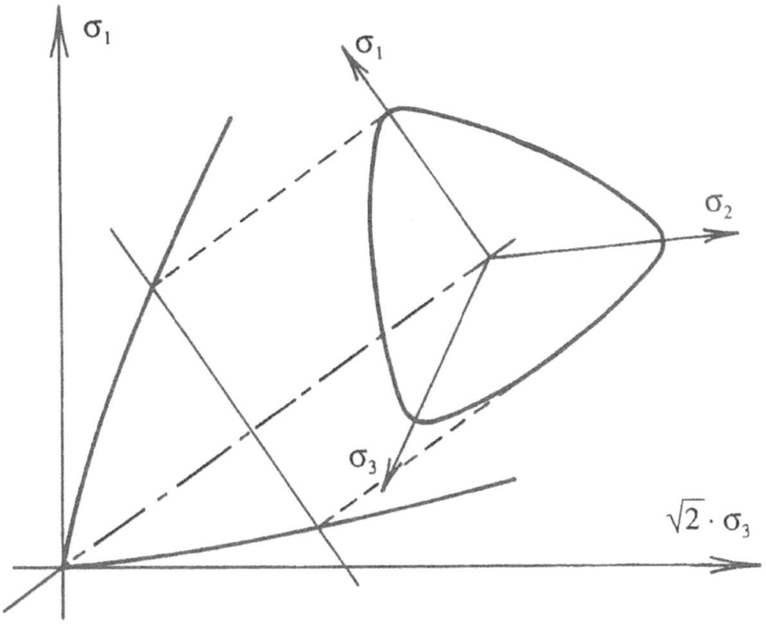

Figure 3. Failure surface for granular materials.

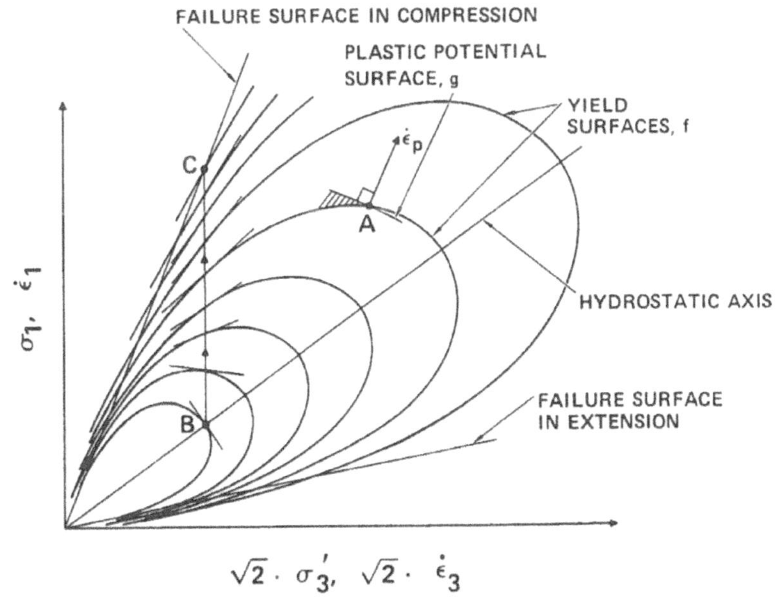

Figure 4. Pattern of yield surfaces for isotropic soils. Stress path for conventional triaxial compression test (BC).

The shaded wedge between the yield surface and the plastic potential surface defines a region in which Inequality (1) is not fulfilled for a material with nonassociated flow. Since a stress increment vector from point A located inside the wedge and the plastic strain increment vector form an obtuse angle, the scalar product of these two vectors (see Inequality (1)) is negative. According to Drucker's stability postulate the sand may exhibit unstable behavior if the stress point lies adjacent to a region (e.g. the shaded region in Fig. 5) into which a stress probe would violate Inequality (1). However, experiments show that the sand is perfectly stable at stress points where the normal to the yield surface points in the outward direction of the hydrostatic axis. For this condition the deviator stress can be safely increased to produce further plastic shear strains (work-hardening). In other words, the sand can sustain higher loads and behave in an inelastic manner without undergoing any instability or collapse.

Potential instability occurs in regions where the yield surface opens up in the outward direction of the hydrostatic axis. This allows plastic strains (loading) to occur while the stresses are decreasing. Here loading occurs inside the failure surface and instability may develop in the form of inability to sustain the current deviator stresses.

Fig. 4 shows the stress path for a conventional triaxial compression test performed at constant confining pressure. As the specimen is loaded from B to C the inclination of the yield surface changes. At low deviator stresses near point B, the yield surface is inclined towards the outwards direction of the hydrostatic axis. As loading proceeds, the inclination of the yield surface changes gradually and becomes inclined towards the origin as failure is approached at point C. It is in this region of high deviator stresses where the yield surface is inclined toward the origin that instability may occur.

Fig. 5 shows a schematic illustration of the region in which Inequality (1) is not fulfilled for a dilating material with nonassociated flow. The region is shaped as a wedge between the current yield surface f and the plastic potential surface g corresponding to the current stress point. This wedge shaped region is located within a larger region bounded by lines corresponding to $(\sigma_1'/\sigma_3') =$ const. and $(\sigma_1 - \sigma_3) =$ const. as indicated on Fig. 5. All stresses, including the stress difference $(\sigma_1 - \sigma_3)$, are decreasing within the wedge between f and g , but the stress ratio (σ_1'/σ_3') is increasing in this region. By performing triaxial tests with stress paths located in this region, experimental evidence regarding the instability of materials with nonassociated flow can be obtained.

It is important to recognize that the material behavior obtained for stress paths within the shaded wedge in Fig. 5 corresponds to work-hardening with positive plastic work, $dW_p > 0$,

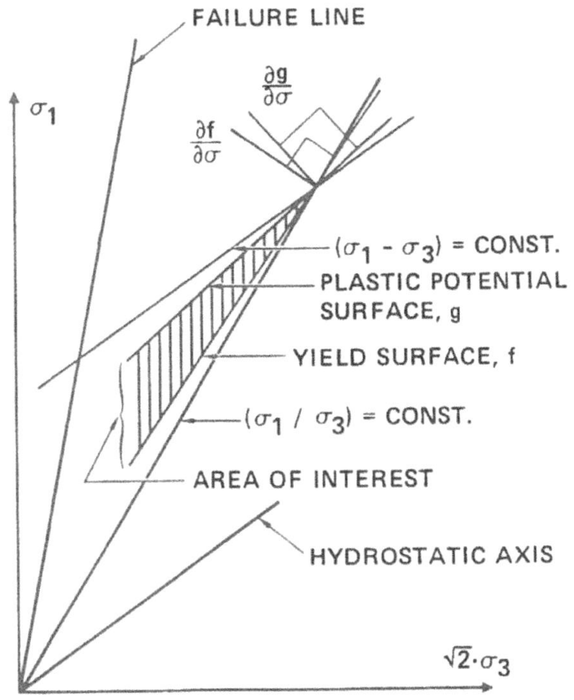

Figure 5. Wedge-shaped region of stress paths with decreasing stresses in which soils with nonassociated flow may be unstable during hardening inside the failure surface.

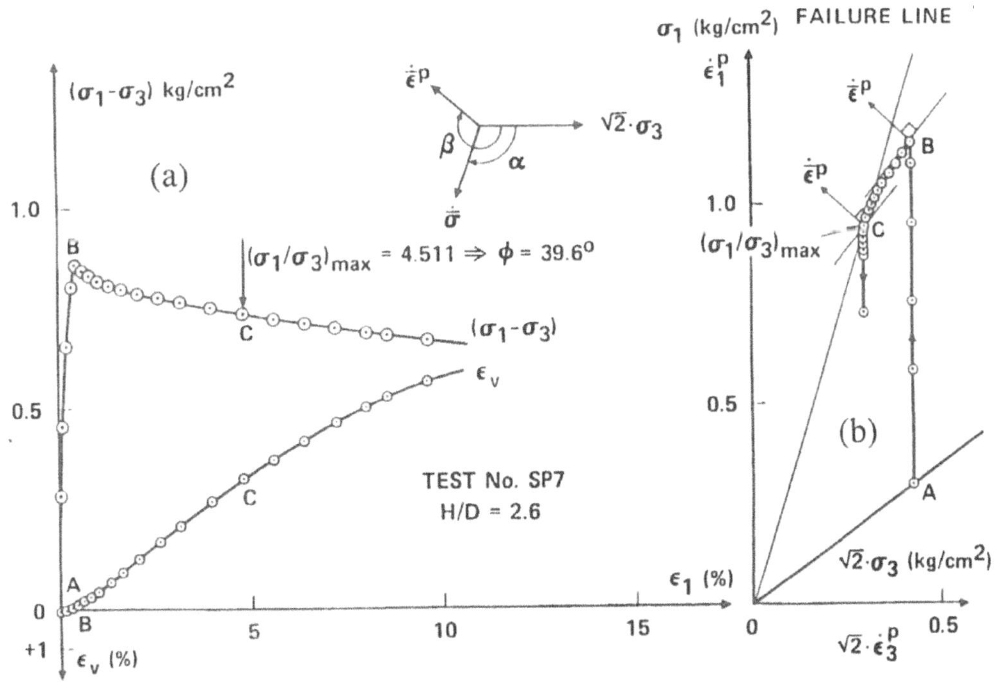

Figure 6. (a) Stress-strain and volume change curves, and (b) stress path and plastic strain increment vectors for test on fine silica sand.

and outward motion of the yield surface. Although the stresses in any direction within the wedge are decreasing, failure has not been reached and softening of the material is therefore not occurring.

NONASSOCIATED FLOW IN GRANULAR SOILS

Drained Conditions

To demonstrate that nonassociated plastic flow is obtained for granular materials, the results of a drained triaxial test on fine silica sand are shown in Fig. 6. This test was performed with a stress path inside the failure surface involving primary loading (hardening) with decreasing stress difference $(\sigma_1 - \sigma_3)$ and decreasing confining pressure σ_3 . If stresses and strain increments are plotted on the same diagram, as in Fig. 6(b), the direction of the plastic strain increment vector is uniquely determined from the state of stress, and it is independent of the stress path leading to this state of stress as shown by Poorooshasb et al. [11] and Lade and Duncan [12]. The stress path with decreasing stresses, shown in Fig. 6(b), is so steep as to form an obtuse angle $(\beta - \alpha = 110°)$ with the direction of the plastic strain increment vector. Thus, the inequality in (1) is violated.

The fact that plastic yielding is occurring along the stress path with decreasing stresses is seen from the stress-strain curve in Fig. 6(a). Section BC on the stress path is labeled similarly on the stress-strain curve which indicates large plastic strains between B and C. Since the yield surface is being pushed out, it must be steeper than the stress path direction BC. If it were less steep, so as for example to be perpendicular to the plastic strain increment vector thereby indicating associated flow, then point C would be inside the yield surface passing through point B corresponding to unloading from B to C. This would contradict the large plastic strains shown in Fig. 6(a). Consequently, the yield surface must be steeper than stress path BC, and nonassociated plastic flow is therefore clearly indicated.

Undrained Conditions

There are basically three different types of effective stress-paths in undrained tests. These are shown, one at a time, in Fig. 7. Strain increments are superimposed on the stress diagrams to allow analyses of strain increment vector directions and to derive the directions of the plastic potential surfaces g. These surfaces are by definition perpendicular to the plastic strain increment vector directions.

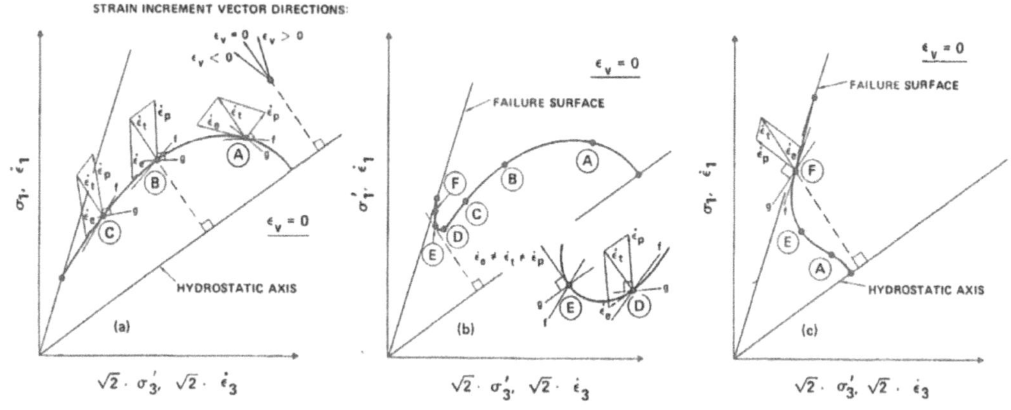

Figure 7. Evaluation of relative inclinations of yield (f) and plastic potential (g) surfaces along undrained effective stress paths, (a) type 1, (b) type 2, and (c) type 3.

For undrained tests the total volumetric strain is zero corresponding to the total strain increment vector being perpendicular to the hydrostatic axis everywhere along the effective stress path. This is shown in Fig. 7(a). This figure also shows that volumetric compression is characterized by a strain increment vector pointing away from the origin, whereas volumetric dilation corresponds to a strain increment vector inclined towards the origin of the stress space.

The stress-strain relations corresponding to the effective stress paths in Fig. 7 show that plastic strains are produced everywhere along the undrained stress paths. The yield surface must therefore be pushed out, and this requires it to be inclined relative to the effective stress path as indicated in Fig. 7.

The direction of the elastic portion of the total strain increment vector in the triaxial plane depends on the value of Poission's ratio, υ. For $\upsilon = 0$ the elastic strain increment vector is parallel to the stress increment vector which is tangential to the stress path on the triaxial plane shown in Fig. 7. For $\upsilon = 0.5$ the elastic strain increment vector is perpendicular to the hydrostatic axis corresponding to no elastic volume change. For values of Poission's ratio between zero and 0.5 the elastic strain increment vector will be between the two extreme positions. This has been discussed in greater detail by Lade et al. [13]. With no loss of generality, it will be assumed in the following that the elastic strain increment vector is parallel to the stress increment vector, i.e., corresponding to $\upsilon = 0$.

Three points of interest are indicated along the undrained effective stress path in Fig. 7(a). The elastic portion of the total strain increment vector is pointed in the current direction of the effective stress path. The length and direction of the plastic strain increment vector is obtained by vectorial subtraction of the elastic from the total strain increment vector. It is clear that only the directions of the elastic and the total strain increment vectors are known, and that the derived direction of the plastic strain increment vector depends on the assumed relative magnitudes of the elastic and total strain increments. However, as will be seen, it can be shown that nonassociated flow prevails at points B and C independent of the assumed magnitudes of strains.

At point A the length of the elastic vector is chosen so as to show that nonassociated flow could occur. In fact, in the absence of actual experimental determination of the elastic and plastic strain magnitudes, it is not possible to argue whether the plastic flow is associated or nonassociated, i.e., whether the yield and plastic potential surfaces are identical or not. However, this is of no importance to the study of instability presented here. Whether associated flow or nonassociated flow is observed at point A, stability is obtained everywhere along the effective stress path from the hydrostatic axis up to point B. Along this portion of the stress path the load can be maintained constant or increased without any observable instability.

Nonassociated flow is clearly obtained at points B and C. For the limiting case where the elastic strain increments become negligible the plastic strain increment vectors become perpendicular to the hydrostatic axis. Since the yield surfaces must be inclined relative to the effective stress paths as shown, nonassociated flow is clearly demonstrated to occur at points B and C. Note that the plastic strain increment vectors in Fig. 7(a) indicate that plastic volumetric compression occurs while the total volumetric strain is zero.

Similar arguments can be made for point D on stress-path type 2 in Fig. 7(b). Point D corresponds to a minimum stress difference along this portion of the effective stress-path. The load carrying capacity increases from point D to failure. At point E the effective stress-path is perpendicular to the hydrostatic axis. The elastic strain increment vector is therefore parallel with the total strain increment vector, and so is the plastic strain increment vector. The plastic potential surface is consequently parallel with the hydrostatic axis. Since the yield surface becomes steeper with increasing stress level, it cannot coincide with the plastic potential surface at point E, and nonassociated flow is demonstrated.

At point F on stress-path type 3 in Fig. 7(c) the relative directions of the elastic and plastic strain increment vectors have been reversed (reversal occurs at point E). Point F is located in the region of plastic dilation. Large plastic strains occur in this region, and the elastic strains are therefore relatively small. For the limiting case where the elastic strain increments reduce to zero, the plastic strain increment vector becomes perpendicular to the hydrostatic axis and the plastic potential surface becomes parallel with the hydrostatic axis. The yield surface is relatively steep close to failure, as indicated don Fig. 4, and nonassociated flow is therefore obtained. Since the load can be increased in the region of dilation, stability is obtained.

BEHAVIOR UNDER DRAINED CONDITIONS

Several series of triaxial compression tests were performed to study stability and instability in granular materials. The details of these experiments are given by Lade et al. [13, 14], and by Lade and Pradel [15].

In order to expose the potential instabilities in soil behavior that might occur for the stress paths in the prefailure region described earlier (see Fig. 5), it is necessary to perform triaxial tests under stress control. A series of drained tests similar to that shown in Fig. 6 was therefore performed under stress control [14]. In fact, the test shown in Fig. 6 is one of these tests. Seven experiments were performed. They all exhibited plastic dilation, and they all clearly showed nonassoicated flow. When the sand specimens were exposed to stress paths in the region of potential instability $(\dot{\sigma}_{ij} \cdot \dot{\varepsilon}^p_{ij} < 0)$, *none was observed*, i.e. run-away instability was not encountered. The tall cylindrical specimens with lubricated ends deformed as perfect right cylinders with no developing shear bands, no bulging or shearing in a nonuniform manner, or any other signs of instability. Little but negligible creep was present in the sand specimens. Thus, stable behavior was observed in the region in which Drucker's stability postulate was violated.

Regardless of the ability to define yield surfaces from experiments, the fact remains that negative values of d^2W_p are obtained from direct experimental measurements. The fact that the specimens remained stable implies that the behavior is not strain softening, i.e. the peak strength has not been exceeded. Therefore, to model this behavior through any type of normality would require a serious departure from plasticity theory.

In order to investigate whether the type of volume change (dilation or compression) was important for stability of granular soils, a second series of tests was performed on sand that compressed during shear [15]. Drained tests with stress paths within the shaded wedge shown in Fig. 5 were performed on loose sand that exhibited compression during shear. Although Drucker's stability condition was violated inside the failure surface, none of the specimens showed any signs of run-away instability.

The results of these two series of triaxial compression tests demonstrated that drained conditions produced stable behavior irrespective of (1) the sign of the second increment of work (2) the sign of the volumetric strain (dilation or compression), and (3) the sign and direction of the increments in stress components. Thus, under fully controlled conditions such as those prevailing for drained conditions, unstable run-away situations cannot occur. These results therefore clearly show that Drucker's stability postulate is not a necessary condition for stability.

BEHAVIOR UNDER UNDRAINED CONDITIONS

Fully Saturated Soil

On the other hand, fully saturated soils that tend to compress during shear may become unstable *inside* the failure surface and this may lead to liquefaction. A series of stress controlled tests was performed on granular soil that tended to compress during shear [13]. The specimens were exposed to stress paths within the shaded wedge shown in Fig. 5. Fig. 8 shows the actual stress paths followed in these tests, and Fig. 9 shows the observed stress-strain, volume change, and pore pressure development in one of the tests. In each test the saturated specimen was first loaded under drained conditions to a preselected stress level S (expressed as the ratio of the current to the maximum stress difference at a given confining pressure). The drainage valve was then closed and instability developed in each specimen due to increasing pore pressure, i.e., the specimens could not sustain the applied load. The tendency for volumetric creep, however small it may be, caused the pore pressure to increase under undrained conditions, providing the small perturbation which rendered the material unstable. However, the large strains observed along the unstable stress paths could not be caused by creep or viscous flow as shown by Lade et al. [13]. The effective stress paths shown in Fig. 8 were within the shaded wedge in Fig. 5, the specimens exhibited nonassociated flow and plastic volumetric compression (although the total volumetric strain was zero), and instability was obtained in all cases in the hardening regime *inside* the failure surface.

In the studies presented here instability was taken to be the inability to sustain or carry a given load, which includes the ability to sustain small perturbations in the loads. However, the possibility that instability might manifest itself as localization of plastic deformation and development of shear planes or in any other way was not overlooked. The specimens were therefore inspected after the tests (they were enclosed in a steel walled triaxial apparatus and could not be observed during the tests) at which time they had sustained 15 to 30% axial strain. The instabilities were induced at axial strain of 2.8 to 5.3%. Thus, substantial amounts of strain had occurred after the instabilities were induced. Nevertheless, the specimens had bulged only slightly and there were no visible shear planes. In fact, a uniform distribution of horizontal wrinkles in the membrane was observed after each tests, thus revealing a fairly uniform pattern

135

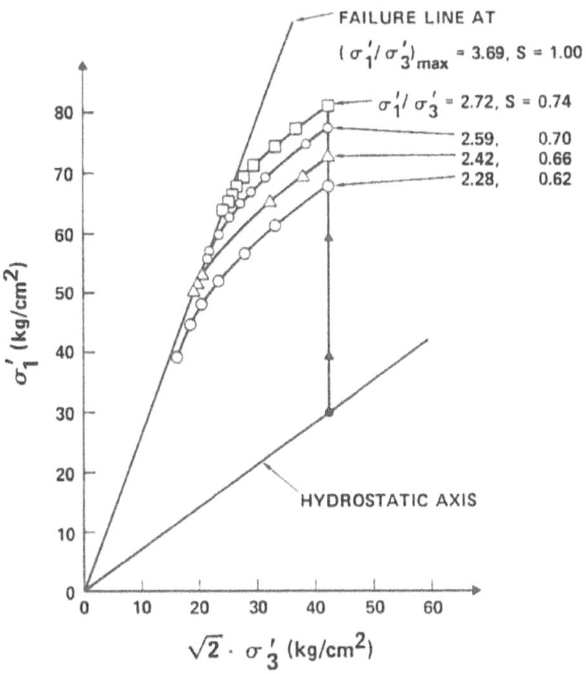

Figure 8. Effective stress paths for stress-controlled triaxial compression tests on loose Sacramento River sand.

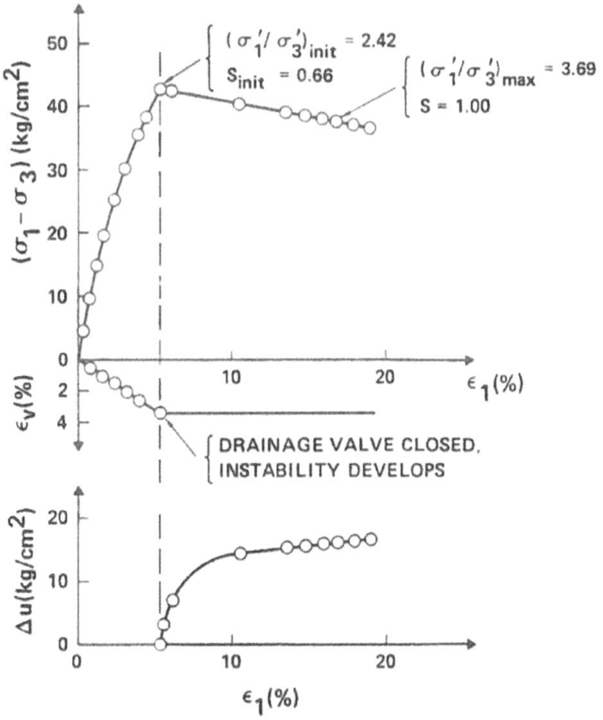

Figure 9. Stress-strain, volume change, and pore pressure relations in stress-controlled triaxial compression test on loose Sacramento River sand.

of internal strains in the specimens. Instability did not appear to be associated with localization of plastic deformations and consequent shear banding of the type discussed by Lade [16] and by Peters et al. [17]. This is consistent with previous observations that shear planes are not observed in soils that compress [16]. The types of instability observed in granular materials are further discussed below.

Partly Saturated Soil

Stress paths with only one particular direction is obtained from undrained tests on fully saturated soils at each particular stress state, as shown in Fig. 8. In order to study the soil behavior for other stress path directions under undrained conditions, a series of tests, similar to those described above, was performed on specimens with decreasing degree of saturation [15]. By purposely introducing a controlled amount of compressible air into the specimens, the undrained effective stress paths could be pointed in different directions within the shaded wedge in Fig. 5. The specimens in both tests, in which the effective stress paths were inside the shaded wedge, exhibited unstable behavior. The stress path in a third test was above the shaded wedge because the amount of air introduced in the specimen was greater than critical. In this test the specimen was perfectly stable.

It is clear from these experiments that a change in drainage conditions for a fully or nearly fully saturated soil can activate unstable behavior. This may occur due to static or dynamic disturbances such as experienced in static flow failures of tailings dams or in liquefaction of granular materials during earthquakes.

INSTABILITY OF DILATING SAND

In a discussion of the experimental results reported above, Peters [18] suggested that even dilating sand may become unstable: If a rate of volumetric expansion were imposed on an element of granular material and this rate exceeded the rate of expansion exhibited by the material, then the effective confining pressure would decrease and the element would become unable to sustain the current, applied shear stress. According to the definition, the material element would become unstable. Thus, a rate of volume change (whether positive, zero, or negative) imposed on a material element could cause it to become unstable depending on the rate of volume change exhibited by the soil itself (compression of dilation). In order to study this hypothesis experimentally, a series of triaxial compression tests was performed with appropriate test conditions. Fig. 10 shows the stress-strain relations, pore pressure development, and effective stress path for test no. FSS-4 on dilating fine silica sand. These results are typical for all tests, and they confirm the suggestion that even dilating sand may become unstable [19].

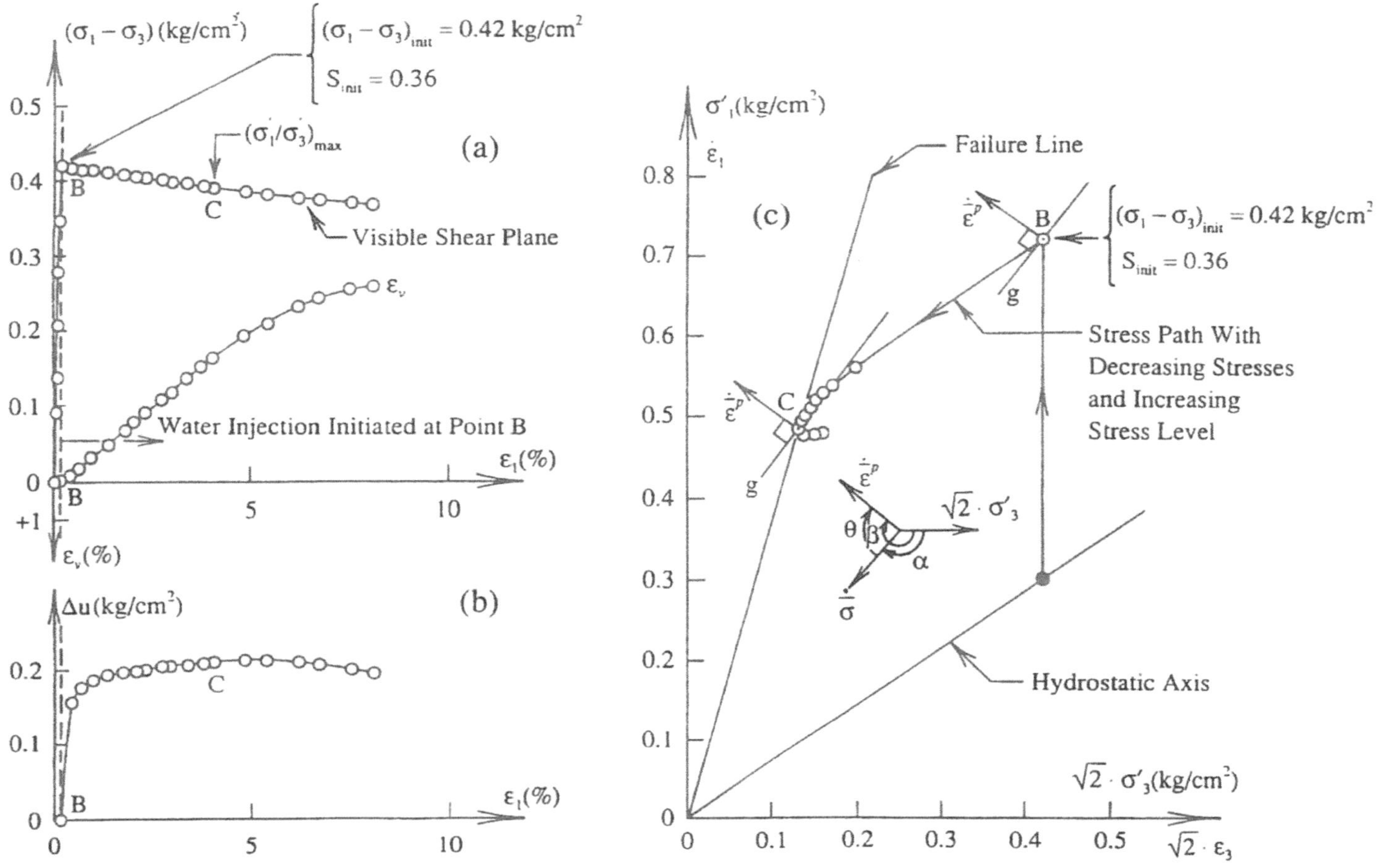

Figure 10. (a) Stress-strain and volume change behavior, (b) pore pressure development, and (c) effective stress-path and plastic strain increment vector directions observed in stress controlled triaxial compression test no. FSS-4 on dilating fine silica sand.

The initial portion of the stress-strain relation in Fig. 10(a) corresponds to drained loading. Note that the volumetric and vertical strains (ε_v and ε_1) are very small until water injection into the specimen is initiated. Following injection of water, the effective confining pressure decreases as a result of increasing pore pressure. In addition, the stress difference decreases since the cross-sectional area of the specimen is increasing. However, even with the lowered stress difference, the specimen cannot sustain the load and large vertical strains are recorded, indicating unstable behavior.

Fig. 10(b) shows the pore pressure development vs. axial strain. It is evident that the pore pressure increases rapidly when water injection is first initiated into the specimen. The jump in pressure levels off once the rate of dilation of the specimen approaches the rate of volumetric expansion imposed by the flow of water. At large axial strains, well beyond peak failure and following development of a shear plane, the pore pressure actually decreases since the rate of specimen dilation has now surpassed the rate of volumetric expansion imposed by the flow rate through the needle valve. At this point negative pore pressure increments begin to develop which decreases the overall pore water pressures. This accounts for the curl at the end of the effective stress path, away from the failure line, as seen in Fig. 10(c).

It is clear from the stress-strain curve that failure cannot be distinguished from observation of the vertical stress difference alone. However, the effective stress ratio reaches a maximum along the stress path indicated in Fig. 10(c). This maximum occurs at the point on the effective stress path that, when connected with the origin, forms the greatest slope of any point on the stress path. This criterion of failure is often used for granular materials and for soils without effective cohesion. Thus, for such materials and for frictional materials in general (soils, rock, concrete), it is important to consider the magnitude of all stresses when conditions of failure are evaluated.

The portion of the stress path of interest in the present study is that from point B to point C in Fig. 10. All stresses, as well as the vertical stress difference, are decreasing along this second branch. The strain increments produced along this branch correspond to dilation of the sand. Superimposed on the triaxial plane in Fig. 10(c) are the strain increment vectors at points B and C. These correspond to essentially plastic strains since the elastic contributions are small compared to the plastic strains, as seen on Fig. 10(a). At failure, the elastic strains are zero since no increments in stress occur near peak failure. These strain increment vectors ($\dot{\varepsilon}^p$), which are almost parallel, are by definition perpendicular to the respective plastic potential surfaces (g) indicated at points B and C in Fig. 10(c).

It is clear that the stress increment vectors, which are tangential to the effective stress path, form both obtuse and acute angles, θ, with the corresponding strain increment directions. Thus, the second increment of plastic work in Inequality (1) is both positive and negative from point B to point C. However, each specimen behaved in an unstable manner, i.e. it was unable to sustain the current stress difference along the entire length of the prefailure stress path BC regardless of the sign of the second increment of plastic work.

One test was performed differently from the previous experiments. Instead of reducing the effective confining pressure through an increase in pore pressure, the test was performed in a drained condition and the total confining pressure was reduced. The specimen was unable to hold the load, i.e. it behaved in an unstable manner. Thus, instability of the specimen is conditional on a progressive reduction in effective confining pressure. The similarity between these test results indicates that instability of dilating sand can be produced by decreasing effective confining pressure whether it is caused by increasing pore pressure or decreasing total confining pressure.

CONDITIONS FOR STABILITY AND INSTABILITY

The stability of granular materials which exhibit nonassociated flow has been investigated on the basis of experimental observations. Several series of triaxial compression tests designed to expose the type of behavior exhibited by granular materials were performed. The variables in these studies were the sign of the second work increment (positive or negative), the volumetric strain behavior (compression or dilation), the constraints on the volumetric behavior (free or controlled), and the degree of saturation (fully or partly saturated).

If physical instability is defined as a condition for which the current, applied shear stress cannot be sustained for perturbations in the state of stress, then compressive as well as dilative materials may be considered to be unstable in the region where the yield surface opens up in the outward direction of the hydrostatic axis. In this region, plastic strains can be produced under decreasing stresses. For undrained conditions and compressive material, the instability is self-sustaining and unconditional, i.e. it is not dependent on conditions outside the soil element. For drained conditions, the instability is conditional, i.e. the decrease in load carrying capability depends on the reduction in effective confining pressure. This reduction may occur as a decrease in total confining pressure or as an increase in pore pressure caused by injection of water into the soil element.

Figures 11 and 12 show schematic diagrams of the conditions for stability and instability of compressing and dilating granular materials. The shear stress is represented by the stress deviator invariant J_2' and the mean normal stress is indicated by the stress invariant I_1. The generic yield surfaces, f, and plastic potential surfaces, g, cross each other at points (A) where the yield surfaces open up in the outward direction of the hydrostatic axis. Stability is obtained when J_2' is constant or increases, whereas instability occurs when J_2' decreases. Experimental observations of conditions for stability and instability of granular materials are summarized in Table. 1.

The signs of the second increment of work d^2W can be negative in the region of stability, whereas d^2W can be positive in the region of instability for dilating material. The stability postulates by Drucker and by Hill therefore provide neither necessary nor sufficient conditions for stability of granular (frictional) materials.

REGION OF POTENTIAL INSTABILITY

It is the fact that loading of a compressible soil (resulting in large plastic strains) can occur under decreasing stresses that leads to unstable behavior under undrained conditions. Loose, fine sands and silts have relatively low permeabilities, and small disturbances in load or even small amounts of volumetric creep may produce undrained conditions in such soils, and instability of the soil mass follows. As long as the soil remains drained, it will remain stable in the region of potential instability.

When the condition of instability is reached, the soil may not be able to sustain the current stress state. This stress state corresponds to the top of the current yield surfaces as shown schematically on the $p' - q$ diagram in Fig. 13. Following this top point the soil can deform plastically under decreasing stresses. The top of the undrained effective stress path, corresponding to $(\sigma_1 - \sigma_3)_{max}$, occurs slightly after but close to the top of the yield surface. The difference is due to the additional stability provided by the elastic strains, as expressed in Inequality (2).

In order for a granular material to become unstable, the state of stress must be located on or above the instability line. Fig. 14 shows a schematic $p' - q$ diagram in which the line connecting the tops of a series of effective stress paths from undrained tests on loose soil provides the lower limit of the region of potential instability. In the region above this instability line the soil can deform plastically under decreasing stresses. Experiments show that this line is straight. Since it goes through the top points of the yield surfaces which evolve from the

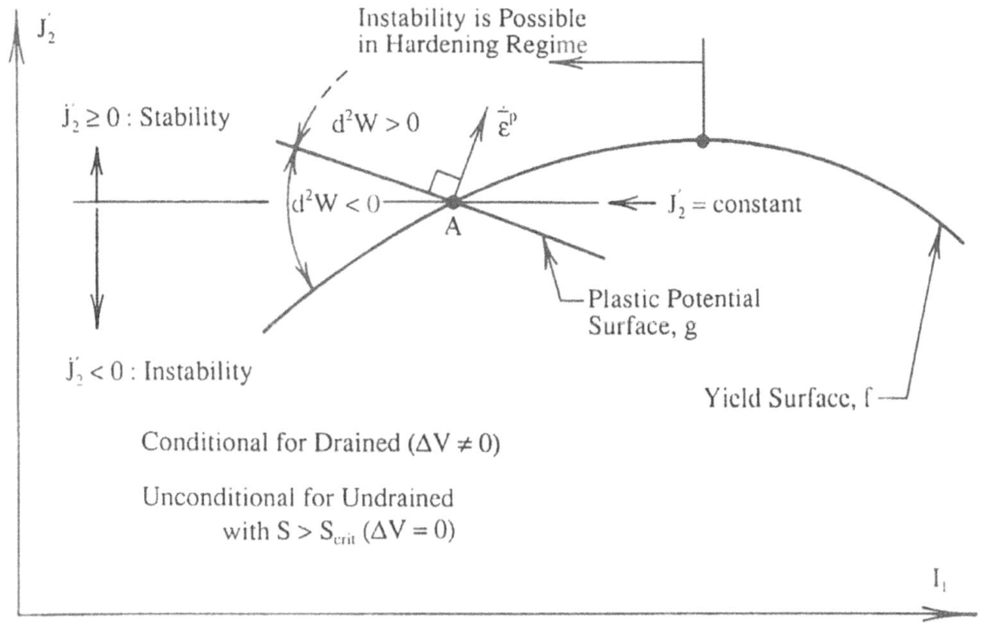

Figure 11. Schematic diagram of conditions for stability and instability of compressing granular material.

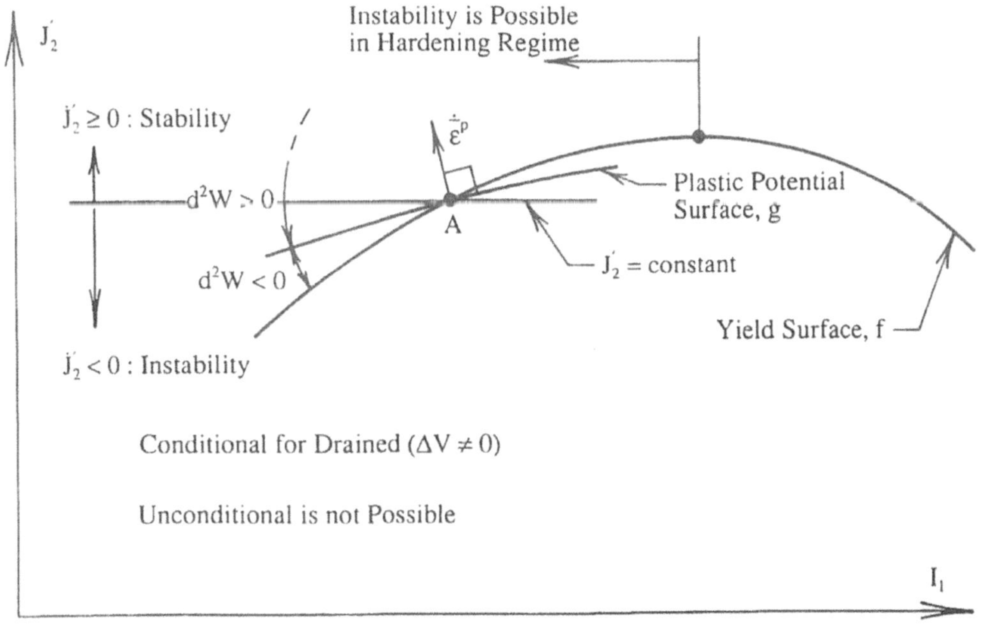

Figure 12. Schematic diagram of conditions for stability and instability of dilating granular material.

Table 1. Experimental Observations of Conditions for Stability, Conditional Instability, and Unconditional Instability of Granular Materials

Soil Volumetric Behavior	Dilation		Compression		
Drainage	Drained $\Delta V \neq 0$	Undrained $\Delta V = 0$	Drained $\Delta V \neq 0$	Undrained, $\Delta V = 0$	
				Degree of Saturation	
$J_2' \geq 0$	Stability	Stability	Stability	$S \geq S_{crit}$	Not Possible
				$S < S_{crit}$	Stability
$J_2' < 0$	Conditional Instability	Not Possible	Conditional Instability	$S \geq S_{crit}$	Unconditional Instability
				$S < S_{crit}$	Not Possible

Note: ΔV refers to fluid or gas crossing the boundaries of the granular material element.

origin, as shown in Fig. 4, the instability line also intersects the stress origin. This location of the instability line is different from that used previously by Sladen et al. [20] and by Lade [21]. However, recent realization of the true origin and nature of the instability line indicates that it goes through the origin of the stress diagram rather than through the ultimate state, indicated on Fig. 14. According to Poorooshasb [22], the steady state and the critical state for a soil are in fact the same, and they are more appropriately referred to as the ultimate state. Fig. 14 also shows a region of temporary instability which is located in the upper part of the dilating zone. It is a region where instability may initially occur, but conditions allow the soil to dilate after the initial instability, thus causing the soil to become stable again. Effective stress paths of type 2 in Fig. 7(b) correspond to this region of temporary instability. For very loose soils the region of potential instability reaches down to the origin of the stress diagram.

INSTABILITY VERSUS FAILURE

Instability and failure are two different behavior aspects of soils which exhibit nonassociated flow [23]. Although both may lead to catastrophic events, they are not synonymous. However, two criteria for "failure" are commonly used in geotechnical engineering for

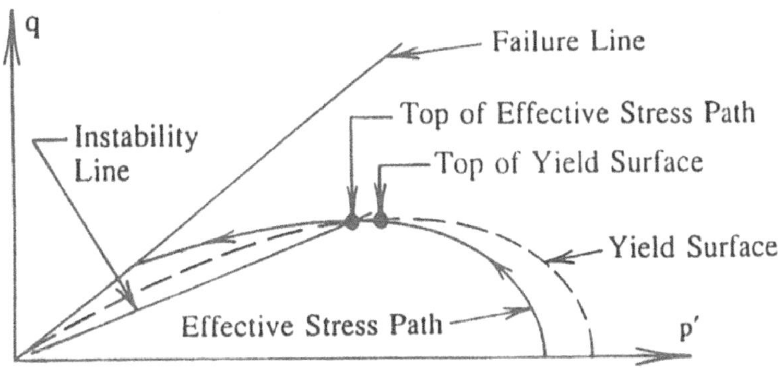

Figure 13. Location of instability line for loose sand.

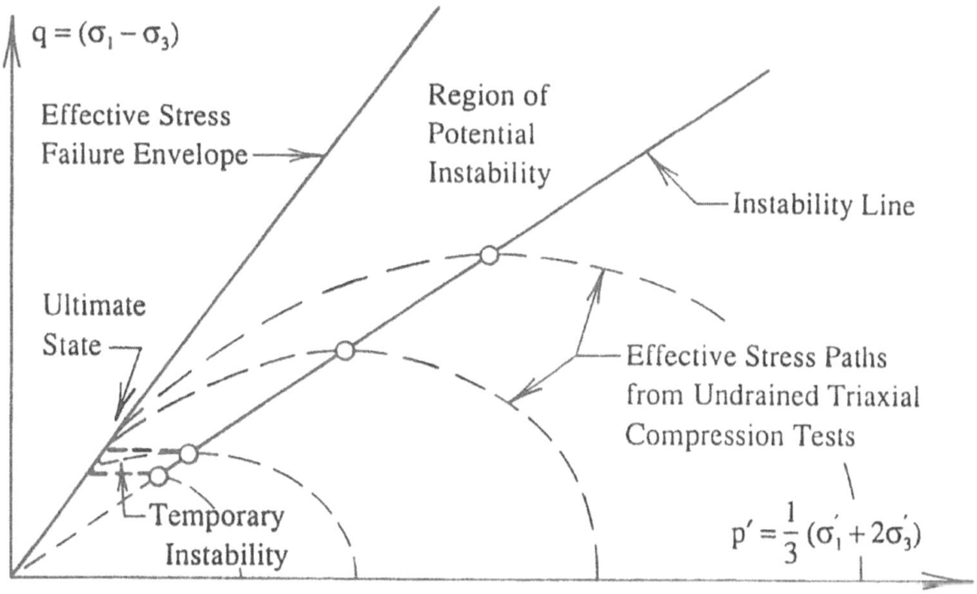

Figure 14. Schematic diagram of location of instability line in $p' - q$ diagram.

interpretation of results of triaxial tests on soils: (1) Failure occurs when the stress difference reaches a limiting value, $(\sigma_1 - \sigma_3)_{max}$, and (2) failure occurs when the effective principal stress ratio reaches a limiting value, $(\sigma_1'/\sigma_3')_{max}$. The two conditions are reached simultaneously in drained tests. The confusion as to the definition of failure arises in interpretation of undrained tests on loose sands and sensitive clays in which the pore pressures increase monotonically during shear. For these types of soils the maximum stress difference is reached before the maximum effective stress ratio. The choice of failure criterion vary with the purpose for which the strength is to be used. Typically, $(\sigma_1 - \sigma_3)_{max}$, is employed in total stress analyses, and $(\sigma_1'/\sigma_3')_{max}$ is used in effective stress analyses. These issues have been discussed at length by e.g. Bjerrum and Simons [24], Seed et al. [25], and Whitman [26].

Based on the presentation above, it is clear that the condition of maximum stress difference does not correspond to a true failure condition, but rather to a condition of minimum stress difference at which instability may develop *inside* the true failure surface. Stress states can be reached above the minimum stress difference condition described by $(\sigma_1 - \sigma_3)_{max}$, and instability can be induced anywhere between the $(\sigma_1 - \sigma_3)_{max}$-condition and the true failure surface described by $(\sigma_1'/\sigma_3')_{max}$. Therefore, a failure condition expressed in terms of the maximum stress difference obtained from e.g. undrained tests initiated at the hydrostatic axis (see Fig. 14) would clearly underestimate the true effective strength of the soil.

Material models which employ associated plastic flow rules must assume the failure surface to go through the point at which instability is produced. This is because stability is guaranteed according to Drucker's and Hill's stability conditions until the failure surface is reached, at which instability ensues. But then stable loading to higher stress differences, which is clearly possible, could not be predicted by such models. Further, since instability as well as stability can be obtained at any point in the region above the line defined by the maximum stress differences in Fig. 14, it is not possible to define any one consistent failure surface to be used in a model which incorporates associated flow.

INSTABILITY AND LIQUEFACTION

Instability and liquefaction occur in the field under static as well as cyclic loading conditions. Initiation of instability requires that the states of stress on some planes in the soil element are located in or are brought into the region of potential instability, shown in Fig. 14. However, the soil will remain perfectly stable as long as it is drained. A trigger mechanism that causes the pore pressure to increase faster than it can dissipate, even for a short moment, is required to cause instability. Thus, the location of the region of potential instability as well as the state of stress in the ground are required for analysis of the potential for instability in the field. For very loose sand the region of potential instability may reach down close to the origin of the stress space. Because sands are perfectly stable above the instability line as long as they are drained, the instability line itself is in principle not a trigger mechanism for instability.

Initiation of instability occurs at very low amounts of strain, whether under static or cyclic conditions. Once the instability has been triggered, it leads to liquefaction at large strains and steady state (or ultimate state) conditions, as shown schematically on Fig. 15. This response may be modified by the effects of creep [29]. However, steady state conditions, achieved at large strains, are not relevant to considerations dealing with initiation of instability. Analyses based on steady state conditions describe the sand behavior at large strains after initiation and

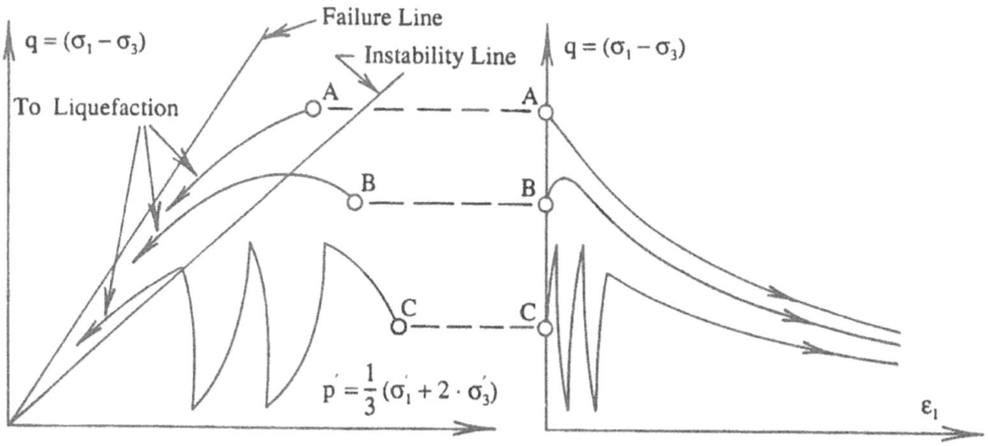

Figure 15. Schematic diagrams showing (a) stress paths and (b) stress-strain relations for initiation of static (A,B) and cyclic (C) instability at small strains and subsequent liquefaction at large strains under undrained conditions in loose sand.

during flow of the liquefied sand.

Because initiation of instability occurs at small strains, the initial grain fabric has an effect on the susceptibility to instability and subsequent liquefaction, as has been observed in experiments [e.g. 30]. Thus, liquefaction, which involves large strains, is always preceded by instability, which occurs at small strains. This explains the relation between grain fabric, which could only influence the sand behavior at small strains, and liquefaction, which occurs at large strains where any effects of the initial grain structure would be negligible.

It is evident from these observations that liquefaction follows upon and necessitates instability, and description of instability requires employment of a nonassociated flow rule in the constitutive model. Nova [31] also concluded that description of static liquefaction within the framework of work-hardening plasticity with a single yield function requires the flow rule to be nonassociated.

COMPARISON OF TYPES OF INSTABILITY

The type of instability discussed above is not the same as that manifesting itself by shear planes or slip bands. Development of slip bands or shear planes in granular materials may be studied experimentally. During the performance of the triaxial compression test shown in Fig. 16, the specimen was carefully observed for early detection of developing shear planes. The location on the stress-strain curve at which the first observation of a shear plane was made is indicated on the diagram. Considerable straining beyond the peak failure point may be required

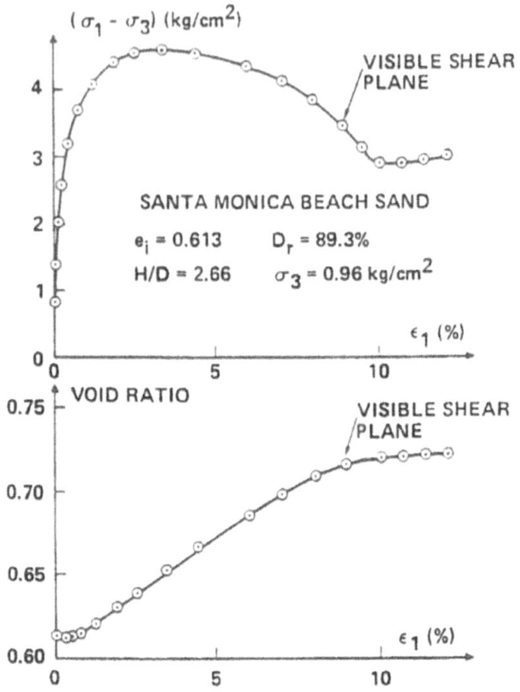

Figure 16. Stress-strain relation and void ratio change from triaxial compression test exhibiting strain-softening and development of shear plane.

before shear planes occur in compression tests. The stress-strain curve clearly shows a drop in strength and the rate of dilation diminish substantially immediately before the shear plane becomes visible. Thus, the reduced rate of dilation appears to be associated with the occurrence of the shear plane and the achievement of critical state conditions within the shear plane in the compression test. Once a shear plane has developed fully, the stresses and the volume changes tend to level off. It is likely that the initiation of a shear plane, i.e., localization of deformation, results in elastic unloading of the specimen outside the developing shear plane which is now becoming rapidly weaker than the remaining major parts of the specimen.

Development of shear planes in triaxial compression ($\sigma_2 = \sigma_3$) occurs after peak failure and results in further decrease in strength. This behavior appears to be associated with dilation of the sand. Shear planes are not observed in loose specimens which compress during shear. Loose specimens deform uniformly throughout the entire stress-strain relationship. Fig. 17 shows a critical state diagram for triaxial compression tests on granular materials in which the conditions for stable and unstable behavior as well as for development of shear planes are summarized. Several investigators [17,27,28] have shown that the hardening modulus H is required to reach a critical value H_{cr} for development of shear planes. For triaxial compression

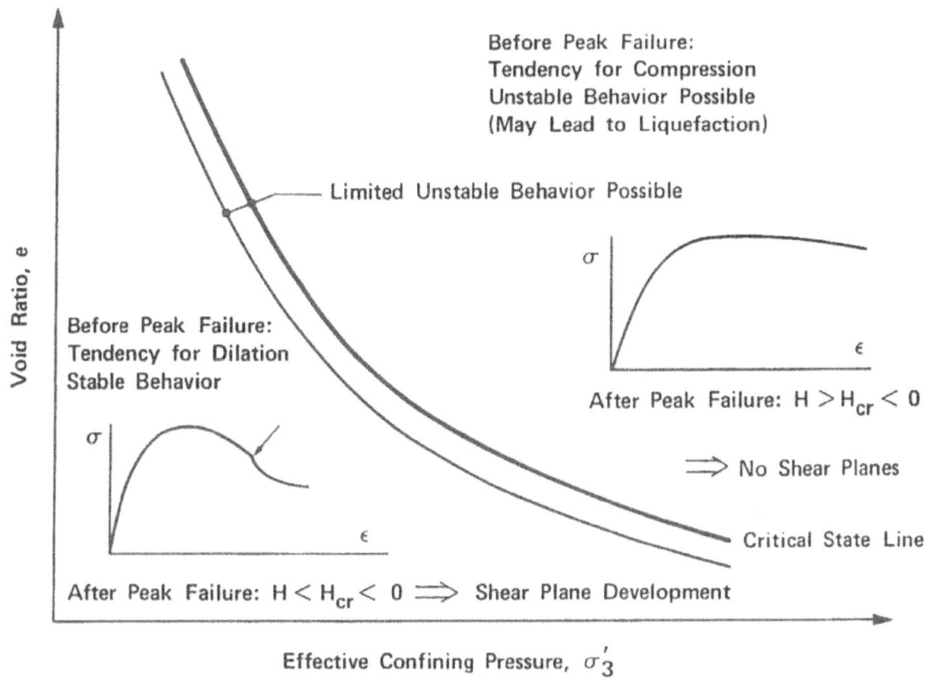

Figure 17. Critical state diagram for behavior of granular material in triaxial compression. Occurrence of stable and unstable behavior and development of shear planes are indicated.

conditions the critical hardening modulus has to reach a certain negative value before shear planes can develop. This finding is supported by the test results shown in Fig. 16, and it is indicated on Fig. 17.

It appears from these results and from the experimental observations presented above that the type of instability resulting in shear planes is not the same as that discussed in connection with the stability/instability issue presented here. Based on the experimental observations, the two types of instability appear to be mutually exclusive, i.e., shear planes are observed after peak failure in granular materials that dilate, and these materials are perfectly stable for all possible stress path directions before the failure surface has been reached. On the other hand, shear planes are not observed in granular materials that compress during shear, but they can become unstable for certain stress-path directions *inside* the failure surface.

CONCLUSIONS

With background in the stability conditions proposed by Drucker and by Hill, the stability of granular materials which exhibit nonassociated flow has been investigated on the basis of experimental observations. Several series of triaxial compression tests designed to expose the type of behavior exhibited by granular materials were performed. The variables in these studies were the sign of the second work increment (positive or negative), the volumetric strain behavior (compression or dilation), the constraints of the volumetric behavior (free or controlled), and the degree of saturation (fully or partly saturated). It is shown that the conditions for stability proposed by Drucker and by Hill are not necessary to guarantee stability of granular (frictional) materials. However, when these criteria are violated, it is possible to create an unstable situation by prescribing a constraint among the kinematic degrees of freedom (e.g. shear and volumetric strain). Conditions for stability are indicated on the basis of experimental observations of the behavior of granular materials in triaxial tests. The occurrence of instability involves small strains, and once it has been initiated it leads to liquefaction which requires large strains to fully develop.

Shear planes occur in granular materials that dilate. In triaxial compression they develop after peak failure. The type of instability that manifests itself by shear plane development appears to be different from the instability relating to a decrease in load carrying capability occurring *inside* the failure surface for granular materials that compress during shear.

Acknowledgment

Part of the study presented here was supported by AFOSR under Grant No. 910117. Grateful appreciation is expressed for this support.

REFERENCES

1. Drucker, D.C.
   A more fundamental approach to stress-strain relations. Proc. 1st U.S. Nat. Cong. Appl. Mech. (1951) 487-491.

2. Drucker, D.C.
   On uniqueness in the theory of plasticity, Quart. Appl. Math., 14, (1956) 35-42.

3. Drucker, D.C.
   A definition of stable inelastic material, J. appl. Mech., 26, (1959) 101-106.

4. Bishop, J.F.W. and Hill, R.
   A theory of the plastic distortion of a polycrystalline aggregate under combined stresses, Phil. Journ., 42 (1951) 414-427.

5. Hill, R.
A general theory of uniqueness and stability in elastic-plastic solids, J. Mech. Phys. Solids, 6 (1958) 236-249.

6. Mroz, Z.
Nonassociated flow laws in plasticity, J. de Mecanique 2 (1963) 21-42.

7. Mandel, J.
Conditions de stabilite et postulat de Drucker, Proc. IUTAM Symp. Rheology and Soil Mechanics, Grenoble (1964) 57-58.

8. Nemat-Nasser, S.
On finite plastic flow of crystalline solids and geomaterials, J. Applied Mechanics, 105 (1983) 1114-1126.

9. Lade, P.V.
Elasto-plastic stress-strain theory for cohesionless soil with curved yield surfaces, Int. J. Solids Structures, Pergamon Press 13 (1977) 1019-1035

10. Lade, P.V.
Prediction of undrained behavior of sand, J. Geotech. Engr. Div., ASCE, 104 (1978) 721-735.

11. Poorooshasb, H.B., Holubec, I. and Sherbourne, A.N.
Yielding and flow of sand in triaxial compression: Part I. Can. Geot. J. 3 (1966) 179-190.

12. Lade, P.V. and Duncan, J.M.
Stress-path dependent behavior of cohesionless soil. J. Geot. Engr. Div., ASCE 102 (1976) 51-68.

13. Lade, P.V., Nelson, R.B. and Ito, Y.M.
Instability of granular materials with nonassociated flow, J. Engr. Mech., ASCE 114 (1988) 2173-2191.

14. Lade, P.V., Nelson, R.B. and Ito, Y.M.
Nonassociated flow and stability of granular materials, J. Engr. Mech, ASCE 113 (1987) 1302-1318.

15. Lade, P.V. and Pradel, D.
Instability and flow of granular materials. I: Experimental observations, J. Engr. Mech., ASCE 116 (1990) 2532-2550

16. Lade, P.V.
Localization effects in triaxial tests on sand. Proc. IUTAM Symp. Deformation and Failure of Granular Materials, Delft (1982) 461-471.

17. Peters, J.F., Lade, P.V. and Bro, A.
Shear band formation in triaxial and plane strain tests, Advanced Triaxial Testing of Soil and Rock ASTM STP 977 (1988) 604-627.

18. Peters, J.F.
Discussion of instability of granular materials with nonassociated flow, J. Engr. Mech., ASCE 117 (1991) 934-936.

19. Lade, P.V., Bopp, P.A. and Peters, J.F.
Instability of dilating sand, Mechanics of Materials, Elsevier Science Publishers, 1993.

20. Sladen, J.A., D'Hollander, R.D. and Krahn, J.
The liquefaction of sands, a collapse surface approach, Can. Geotech. J., 22 (1985) 564-578.

21. Lade, P.V.
Static instability and liquefaction of loose fine sandy slopes, J.Geotech. Engr., ASCE 118 (1992) 51-71.

22. Poorooshasb, H.B.
Description of flow of sand using state parameters, Computers and Geotechnics 8 (1989) 195-218.

23. Lade, P.V.
Instability and failure of soils with nonassociated flow. Proc., 12th Int. Conf. on Soil Mech. Found. Engr., Rio de Janeiro, Brazil 1 (1989) 727-730.

24. Bjerrum, L., and Simons, N.E.
Comparison of shear strength characteristics of normally consolidated clays, Proc. ASCE Research Conf., Boulder, Colorado (1960) 711-726.

25. Seed, H.B., Mitchell, J.K. and Chan, C.K.
The strength of compacted cohesive soils, Proc. ASCE Research Conf., Boulder, Colorado (1960) 877-964.

26. Whitman, R.V.
Some considerations and data regarding the shear strength of clays, Proc. ASCE Research Conf., Boulder, Colorado (1960) 581-614.

27. Hettler, A. and Vardoulakis, I.
Behavior of dry sand tested in a large triaxial apparatus, Geotechnique 34 (1984) 183-197.

28. Molenkamp, F.
Comparison of frictional material models with respect to shear band initiation, Geotechnique 35 (1985) 127-143

29. Lade, P.V.
Influence of creep on static and cyclic instability of granular soils, J. Geotech. Engr. 119 (1993).

30. Mulilis, J.P., Seed, H.B., Chan, C.K., Mitchell, J.K. and Arulanandan, K.
Effects of sample preparation on sand liquefaction, J. Geotech. Engr. Div., ASCE, 103 (1977) 91-108.

31. Nova, R.
A note on sand liquefaction and soil stability, Proc. 3rd Int. Conf. Constitutive Laws for Engineering Materials: Theory and Applications, ASME Press, Tucson, Arizona (1991) 153-156.

Received 12 May 1993; accepted 16 June 1993

Postscript

The following references contain additional information regarding sand liquefaction:

Lade, P.V. (1992) "Static Instability and Liquefaction of Loose Fine Sandy Slopes," *Journal of Geotechnical Engineering*, ASCE, 118(1): pp. 51-71.

Lade, P.V. (1993) "Initiation of static instability in the submarine Nerlerk berm," *Canadian Geotechnical Journal*, 30(5): pp. 895-904.

Lade, P.V. (1994) "Instability and Liquefaction of Granular Materials", *Computers and Geotechnics*, 16: pp. 123-151.

Yamamuro, J.A. and Lade, P.V. (1997) "Static Liquefaction of Very Loose Sands," *Canadian Geotechnical Journal*, 34(6): pp. 905-917.

Lade, P.V. and Yamamuro, J.A. (1997) "Effects of Non-Plastic Fines on Static Liquefaction of Sands," *Canadian Geotechnical Journal*, 34(6): pp. 918-928.

Yamamuro, J.A. and Lade P.V. (1998) "Steady State Concepts and Static Liquefaction of Silty Sands," *Journal of Geotechnical and Geoenvironmental Engineering*, ASCE, 124(9): pp. 868-877.

Lade, P.V. "Instability of granular materials," *Proceedings of the International Workshop on the Physics and Mechanics of Soil Liquefaction*, Baltimore, Maryland, edited by P.V. Lade and J.A. Yamamuro, 10-11 September 1998, Balkema, Rotterdam 1999, pp. 3-16.

Lade, P.V., and Liggio, Carl, Jr. (2014) "Stability and Instability of Granular Materials under Imposed Volume Changes: Experiments and Predictions," *International Journal of Geomechanics*, ASCE, 14(5): 04014020-1-14, ISSN 1532-3641/04014020(14).

Chapter IX

The Single Hardening Model and Its Capabilities

Preface

The developments of the finite element and the finite difference methods in the 1960s and the early 1970s were made for elastic materials, whose material matrices were symmetric. Non-symmetric matrices produced by plasticity theories with non-associated plastic flow could not be handled at that time. There was therefore a desire to formulate constitutive models in terms of the easier elastic behavior. Besides, elasticity is intuitively easier to comprehend.

Thus, early constitutive models were all based on elasticity, whether or not the material to be modeled was entirely elastic or also showed plastic behavior. One prominent elastic model was developed by Duncan and Chang (1970) and known as the Hyperbolic Model. However, elasticity has very limited, realistic application to frictional materials, as also indicated in the closure by Duncan and Chang (1972). This model employed the Mohr-Coulomb criterion in terms of σ_1 and σ_3 to describe failure in these frictional materials. This criterion ignores the influence of σ_2 on failure. At that time (late 1960s) it was not known reliably what the effect of σ_2 was on failure.

Based on true triaxial tests on dense and loose Monterey No. 0 sand, the influence of the intermediate principal stress was studied, and a three-dimensional failure criterion was developed in terms of the first and the third stress invariants, I_1 and I_3 (Lade and Duncan, 1973, 1975):

$$f = I_1/I_3 = \text{const.} \qquad (9.1)$$

This criterion may be depicted in the principal stress space and compared with the Mohr-Coulomb criterion as shown in Chapter VI. It is clear that σ_2 has an important influence on the strengths of cohesionless materials, and this is captured well by the proposed criterion. The yield surface was taken to be expressed by the same criterion with a value of 27 at the hydrostatic axis and increasing values up to failure. The stress-strain behavior was also recorded in these experiments and a model using non-associated plastic flow was developed.

Comments on this first model, especially from Dr. Peter Wroth, who was co-developer of the Cam Clay model (Schofield and Wroth, 1968), led to the following modifications to the proposed model (Lade, 1977):

(1) A separate cap was added, as shown in Fig. 9.2, to account for plastic yielding during isotropic and proportional loading. This model with two yield surfaces was later referred to as a double hardening model.

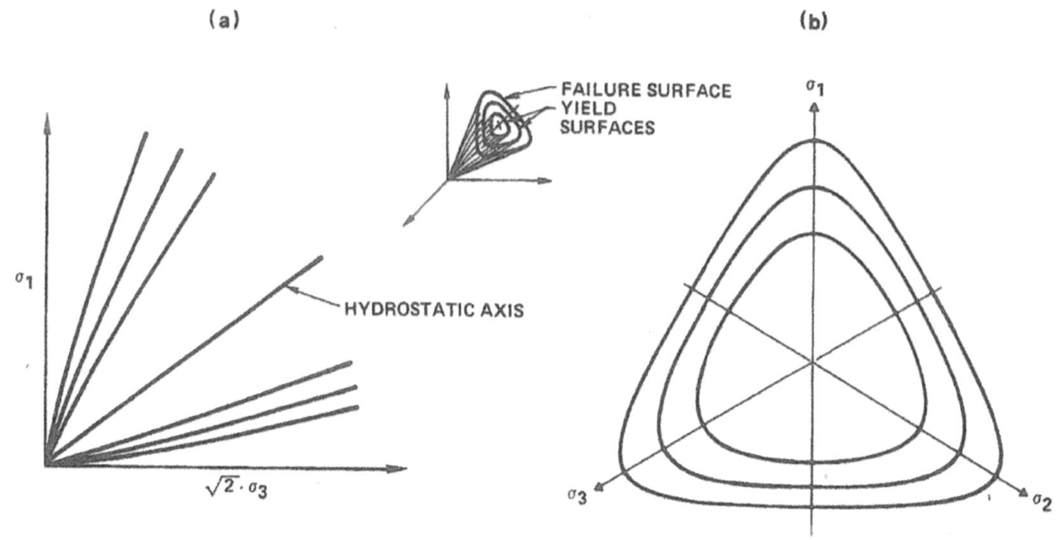

Fig. 9.1. Characteristics of failure and yield surfaces shown in principal stress space. (a) Traces of failure and yield surfaces in triaxial plane, and (b) traces of failure and yield surfaces in octahedral plane.

(2) The curvature of the failure surface in meridian planes of the principal stress space, shown in Figs. 9.1 and 9.2, was included in the formulation of the failure surface as follows:

$$f = (I_1/I_3 - 27)(I_1/p_a)^m = \text{const.} = \eta_1 \text{ at failure} \tag{9.2}$$

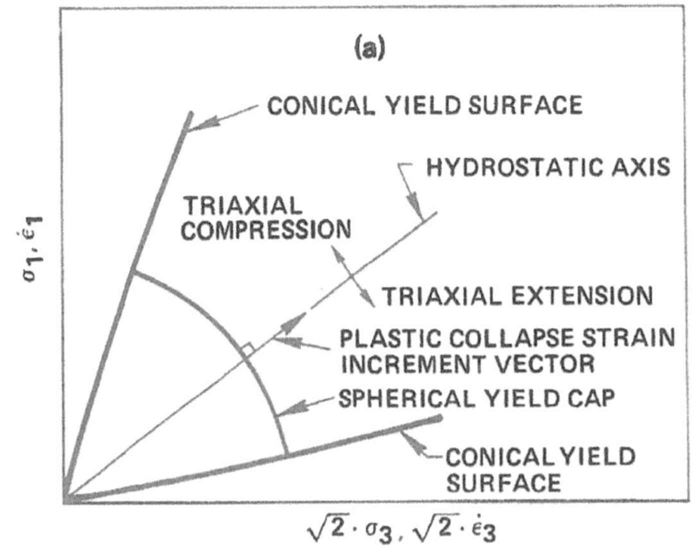

Fig. 9.2. Location of the yield cap relative to conical yield surface shown in triaxial plane.

(3) Material softening following peak failure was added to the formulation.

This model was shown to capture the three-dimensional behavior with good accuracy over a range of confining pressures during drained and undrained loading, unloading and proportional loading. However, this model contains a sharp corner between the two yield surfaces, which is unrealistic relative to experimental results and which complicates separation of the plastic strains into deviatoric and cap yield strains. This was not satisfactory as the basis for further developments.

With the background and experience from the development of these two models and a realization of what was really required for a well-performing constitutive model for frictional materials, a thorough review of the literature was undertaken of the experimentally observed behavior of all types of frictional materials including dense and loose sands, normally and overconsolidated clays, rocks of all types, concrete, cast iron and powder metals. This review served as the basis for development of what is now called the Single Hardening Model for Frictional Materials. The following articles present this development in detail.

References

Duncan, J.M. and Chang, C.Y. (1970) "Nonlinear Analysis of stress and strain in soils," *Journal of the Soil Mechanics and Foundations Division*, ASCE, 96(SM5): 1629-1653.

Duncan, J.M. and Chang, C.Y. (1972) closure to 'Nonlinear Analysis of stress and strain in soils,' *Journal of the Soil Mechanics and Foundations Division*, ASCE, 98(SM5): 495-498.

Lade, P.V. and Duncan, J.M. (1973) "Cubical Triaxial Tests on Cohesionless Soil," *Journal of the Soil Mechanics and Foundations* Division, ASCE, 99(SM10): 793-812.

Lade, P.V. and Duncan, J.M. (1975) "Elastoplastic Stress-Strain Theory for Cohesionless Soil," *Journal of the Geotechnical Engineering Division*, ASCE, 101(GT10): 1037-1053.

Lade, P.V. (1977) "Elasto-Plastic Stress-Strain Theory for Cohesionless Soil with Curved Yield Surfaces," *International Journal of Solids and Structures*, 13: 1019-1035.

Schofield, A. and Wroth, P. (1968) *Critical State Soil Mechanics*, McGraw-Hill.

Computers and Geotechnics **5** (1988) 307–324

SINGLE HARDENING CONSTITUTIVE MODEL FOR FRICTIONAL MATERIALS

I. PLASTIC POTENTIAL FUNCTION

M.K. Kim
Department of Civil Engineering
Yonsei University
Seoul 120, Korea

and

P.V. Lade
Department of Civil Engineering
University of California
Los Angeles, California 90024, U.S.A.

ABSTRACT

Based on review and evaluation of a large number of test data, a general three-dimensional plastic potential function expressed in terms of stress invariants is proposed. Three parameter values are required in this function, but one is redundant with one of the failure surface parameters. The plastic potential surface resembles an asymmetric cigar in the principal stress space. It includes tensile behavior of materials with effective cohesion. Its capabilities are examined by comparisons with experimental data from two- and three-dimensional tests on different frictional materials.

INTRODUCTION

One of the most important components of an elasto-plastic work-hardening constitutive model is the plastic potential function. This is the basis for derivation of the plastic flow rule which establishes the relationship between stresses and plastic strain increments for the material under consideration.

Based on experimental evidence, St. Venant [1] proposed that the principal stress axes coincide with the corresponding principal axes of incremental plastic strain (not with those of total plastic strains) during rotation of stresses. This observation led to the suggestion that the plastic strain increments could be determined from a plastic potential function, g_p [2]:

$$d\varepsilon_{ij}^{p} = d\lambda_p \frac{\partial g_p}{\partial \sigma_{ij}} \tag{1}$$

In this plastic flow rule, $d\lambda_p$ is a positive constant of proportionality. The value of $d\lambda_p$ therefore determines the magnitude of the incremental plastic strain vector.

Computers and Geotechnics 0266-352X/88/$03·50 © 1988 Elsevier Science Publishers Ltd, England. Printed in Great Britain

The plastic potential function is expressed in terms of stresses, and it describes a surface in stress space to which the plastic strain increment vectors are perpendicular, as expressed in Eq. (1). It is usually assumed in classical plasticity theory that the plastic potential function takes the same form as the yield function, i.e. associated flow is employed. However, this assumption is not necessary, and there is clear evidence that the yield surface and the plastic potential surface are not identical for frictional materials, i.e. nonassociated flow is obtained [3-6]. The plastic potential function can be determined independently of any other consideration relating to the constitutive model. A very important premise for Eq. (1) is that the plastic strain increments can in fact be determined from a potential function. This requires that the plastic strain increment vectors are uniquely determined from the state of stress, and is independent of the stress-path leading to this state of stress. This has been demonstrated to be the case by Poorooshasb et al. [3] and by Lade and Duncan [5].

Presented here is a study of the geometrical shape of the plastic potential surface for frictional materials in stress space. The data from many experimental studies of frictional materials have been examined, and the resulting pattern of behavior is presented and discussed with illustrations to support the findings. This study forms the basis for a proposed function to describe a single, isotropic, plastic potential surface and its variation with stress state. The function is expressed in terms of stress invariants, and it is sufficiently flexible to encompass a variety of observed behavior patterns, yet sufficiently structured to allow parameter determination from simple tests.

OBSERVED BEHAVIOR IN THE TRIAXIAL PLANE

To study the plastic potential surface, the plastic strain increments observed in experiments are plotted and superimposed on the stress space. The triaxial plane and the octahedral plane, both from the principal stress space, are used to visualize the directions of the incremental plastic strain vectors.

A strain increment vector that is oriented perpendicular to the hydrostatic axis corresponds to the condition of zero volumetric strain. This is indicated on the insert in Fig. 1. If the volume increases, corresponding to dilation, then the strain increment vector will point towards the origin. On the other hand, if the volume decreases due to compression, then the vector will point in the outward direction of the hydrostatic axis.

Fig. 1(a) shows the directions of plastic strain increment vectors $d\varepsilon^p$ along stress paths from triaxial compression tests performed on Fine Silica Sand. Idealized directions of $d\varepsilon^p$ during isotropic compression are drawn along the hydrostatic axis corresponding to isotropic behavior. Fig. 1(b) shows the results of proportional loading tests on sand as reported by Poorooshasb et al. [3]. From Fig. 1 it can be observed that the plastic volume change is highly compressive at stress states close to the hydrostatic axis, and the rate of change decreases as the stress level, i.e. the proximity to failure, is increased. At high stress levels near failure, the plastic volume change becomes dilative.

Similar observations can be made for materials with effective cohesion. Fig. 2(a) shows results of uniaxial compression ($\sigma_1 > \sigma_2 = \sigma_3 = 0$), biaxial compression ($\sigma_1 = 0$, $\sigma_2 = \sigma_3 > 0$), triaxial compression ($\sigma_1 > \sigma_2 = \sigma_3$, $\sigma_1 + \sigma_2 + \sigma_3 = \text{constant}$), and triaxial extension ($\sigma_1 < \sigma_2 = \sigma_3$, $\sigma_1 + \sigma_2 + \sigma_3 = \text{constant}$) tests on plain concrete reported by Schickert and Winkler [7]. The decreasing rate of compressive volume change with increasing stress level is similar to that of cohesionless soil. It should be noted that the results of triaxial compression and extension tests are not complete, because these tests were reported up to stress states

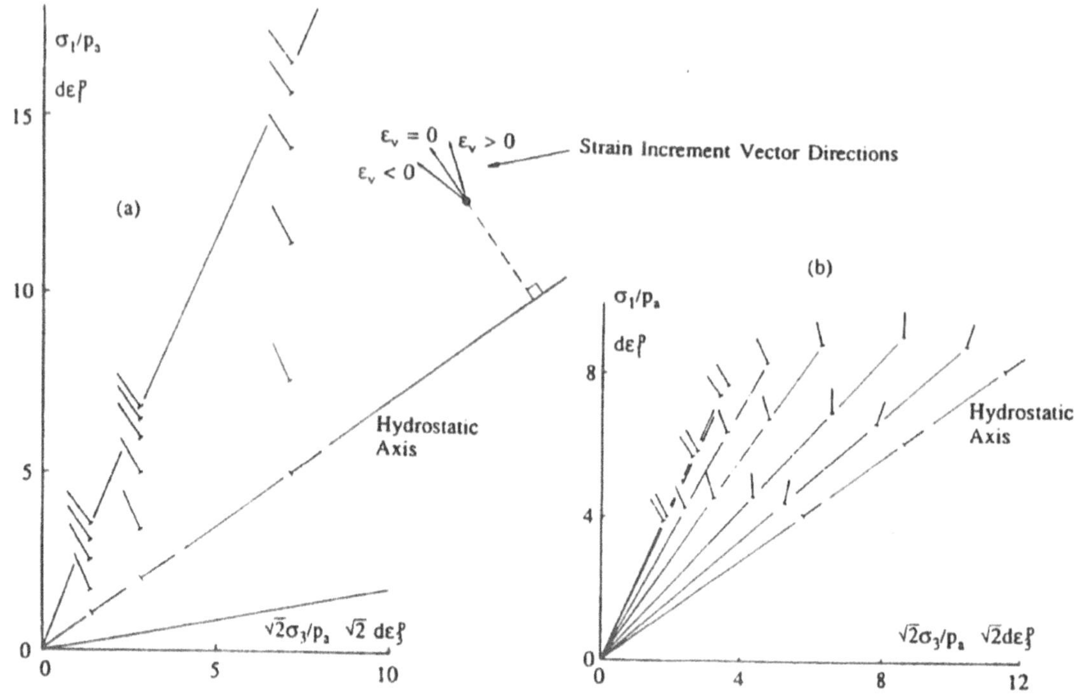

Fig. 1. Directions of Incremental Plastic Strain Vectors in Triaxial Plane for (a) Triaxial Compression Tests on Fine Silica Sand, and (b) Proportional Loading Tests on Sand [3].

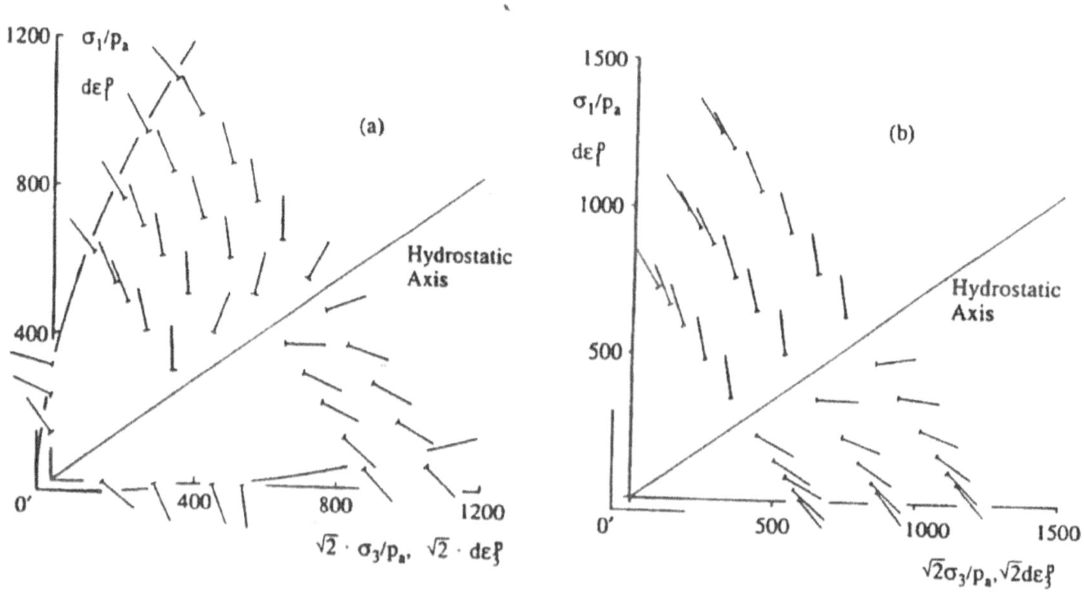

Fig. 2. Directions of Incremental Plastic Strain Vectors in Triaxial Plane for (a) Plain Concrete [7], and (b) Steel Fiber Reinforced Concrete [8].

corresponding to minimum volumes of the specimens. However, if continuous behavior of the material beyond the minimum volume state is assumed, then similar behavior as for sand (i.e. dilation) is expected as indicated on the diagram by the uniaxial and biaxial compression tests. Fig. 2(b) shows the results of triaxial compression and extension tests on steel fibre reinforced concrete performed by Egging and Ko [8]. This set of data is also available up to the minimum volume state, and the type of behavior observed for the plain concrete is repeated for steel fibre reinforced concrete.

The plastic potential surfaces, which are sought in this study, are perpendicular to the plastic strain increment vectors as indicated by the short line at the base of each vector. Based on these short lines and the observations made above, a family of curves may be drawn through the base points in the directions of the short lines. The resulting family of plastic potential surfaces would be shaped as distorted ellipses in the triaxial plane. Whereas the exact shapes of these surfaces seem to depend on the material, they do appear to be of similar type.

OBSERVED BEHAVIOR IN THE OCTAHEDRAL PLANE

Fig. 3 shows the directions of the incremental plastic strain vectors projected on an octahedral plane for the results of cubical triaxial compression tests on Fuji River Sand presented by Yamada and Ishihara [9]. The stress paths for these tests all lie in the same octahedral plane. The specimens were prepared with anisotropic fabric, and the anisotropic deformation of the specimens appears to be dominant at low stress levels (close to the hydrostatic axis). At higher stress levels near failure the effects of the initial anisotropic fabric are small. Similar observations were made by Ochiai and Lade [10] for anisotropically prepared Cambria Sand. They reported that the directions of $d\underline{\varepsilon}^p$ at failure resemble those of isotropically prepared specimens. In Fig. 3 some nonsymmetry in the directions of $d\underline{\varepsilon}^p$ may be noted at low stress levels. This may be due partly to inherent anisotropy and partly to inaccurate elastic parameters used for calculation and subtraction of elastic strain increments.

These observations may be supplemented by those made by Yong and McKeys [11]. They observed that the incremental strain vectors for Kaolinite Clay were almost normal to von Mises circles up to about half of the shear strength on the octahedral plane.

The following conclusions can be drawn regarding the shape of the plastic potential surface on the octahedral plane: (1) At low stress levels, the cross-sections have almost circular shapes, and (2) as the stress level increases towards failure, the plastic potential surfaces gradually change shapes from circular to rounded triangular.

Further inspection reveals that the angularity of the rounded triangle near failure is less pronounced than that of the failure envelope. In order to clarify this point, the Lode angles for stress vectors and for incremental plastic strain vectors are calculated and compared. As indicated on Fig. 4(a), the direction of the stress vector projected on the octahedral plane is measured from the σ_1-axis and indicated by ω_σ. Similarly, ω_ε indicates the angle between the $d\varepsilon_1^p$-axis and the direction of the incremental plastic strain vector projected on the octahedral plane. These angles can be calculated from:

$$\tan\omega_\sigma = \frac{\sqrt{3}(\sigma_2 - \sigma_3)}{2\sigma_1 - \sigma_2 - \sigma_3} \tag{2}$$

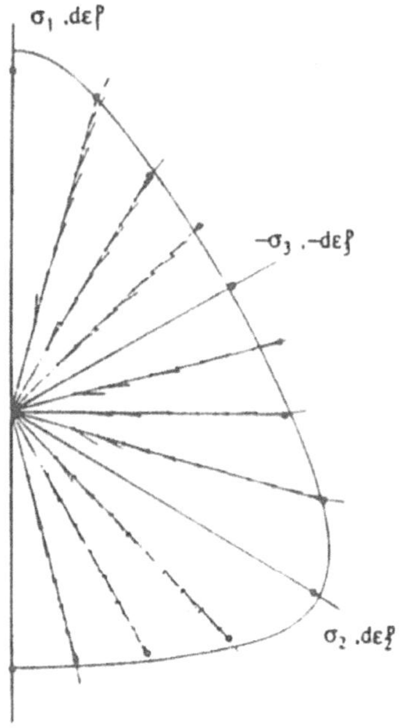

Fig. 3. Directions of Incremental Plastic Strain Vectors Projected on Octahedral Plane for Fuji River Sand [9].

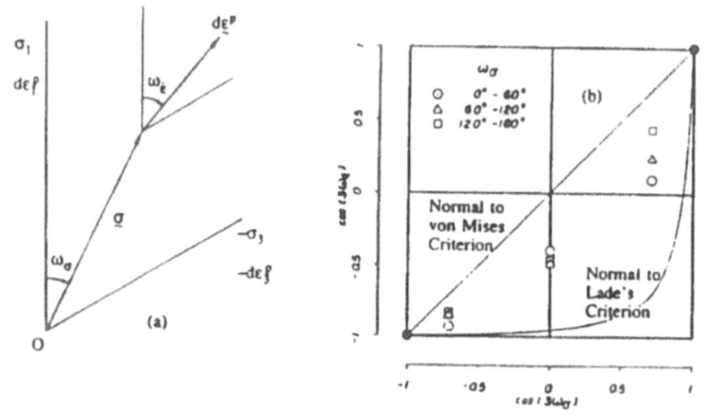

Fig. 4. (a) Definition of Lode Angles for Directions of Stress Vector and Incremental Plastic Strain Vector Projected on Octahedral Plane, and (b) Comparison of Lode Angles at Failure for Fuji River Sand [9] with Normals to Von Mises and Lade's Failure Surfaces.

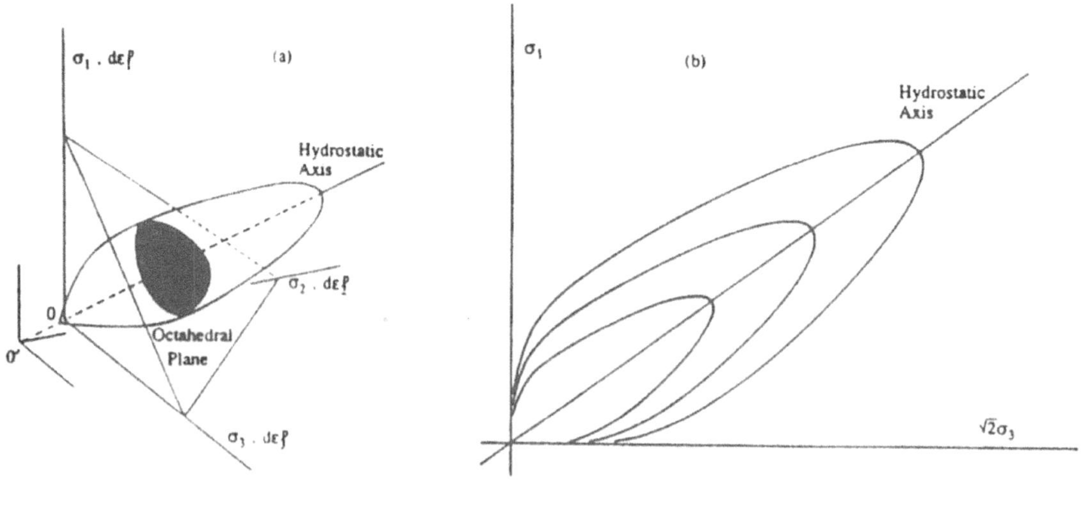

Fig. 5. (a) Schematic Illustration of the Proposed Plastic Potential Surface in Principal Stress Space, and (b) Contours in Triaxial Plane.

312

$$\tan\omega_{\dot\varepsilon} = \frac{\sqrt{3}\,(d\varepsilon_2^p - d\varepsilon_3^p)}{2d\varepsilon_1^p - d\varepsilon_2^p - d\varepsilon_3^p} \qquad (3)$$

Comparison of ω_σ and $\omega_{\dot\varepsilon}$ is enhanced when plotted on a cos $(3\omega_\sigma)$-cos $(3\omega_{\dot\varepsilon})$ diagram. Fig. 4(b) shows such a diagram for failure conditions for Fuji River Sand. The triaxial compression condition ($\omega_\sigma = \omega_{\dot\varepsilon} = 0°$) is located at the bottom left corner, and the triaxial extension condition ($\omega_\sigma = \omega_{\dot\varepsilon} = 60°$) corresponds to the top right corner in the figure. Together with the data points, the analytical curves predicted by the normal directions to the von Mises circle and the normal directions to the failure envelope given by Lade [12] are drawn. the experimental data points are located between the analytical curves of these two criteria. Similar results have been obtained for other frictional materials. The plastic potential surface is therefore smoothly rounded but less angular than the failure surface.

PROPOSED PLASTIC POTENTIAL FUNCTION

Based on the observations presented in the previous sections, a plastic potential may be expressed in the following form:

$$g_p = \left(\psi_1 \cdot \frac{I_1^3}{I_3} - \frac{I_1^2}{I_2} + \psi_2\right) \cdot \left(\frac{I_1}{p_a}\right)^\mu \qquad (4)$$

in which the three stress invariants are give by

$$I_1 = \sigma_1 + \sigma_2 + \sigma_3 = \sigma_x + \sigma_y + \sigma_z \qquad (5a\text{-}b)$$

$$I_2 = -(\sigma_1 \cdot \sigma_2 + \sigma_2 \cdot \sigma_3 + \sigma_3 \cdot \sigma_1) \qquad (6a)$$

$$= \tau_{xy} \cdot \tau_{yx} + \tau_{yz} \cdot \tau_{zy} + \tau_{zx} \cdot \tau_{xz} - (\sigma_x \cdot \sigma_y + \sigma_y \cdot \sigma_z + \sigma_z \cdot \sigma_x) \qquad (6b)$$

$$I_3 = \sigma_1 \cdot \sigma_2 \cdot \sigma_3 \qquad (7a)$$

$$= \sigma_x \cdot \sigma_y \cdot \sigma_z + \tau_{xy} \cdot \tau_{yz} \cdot \tau_{zx} + \tau_{yx} \cdot \tau_{zy} \cdot \tau_{xz} - (\sigma_x \cdot \tau_{yz} \cdot \tau_{zy} + \sigma_y \cdot \tau_{zx} \cdot \tau_{xz} + \sigma_z \cdot \tau_{xy} \cdot \tau_{yx}) \qquad (7b)$$

The parameter ψ_1 in Eq. (4) acts as a weighting factor between the triangular shape (from the I_3 term) and the circular shape (from the I_2 term), the parameter ψ_2 controls the intersection with the hydrostatic axis, and the exponent μ determines the curvature of meridians.

The corresponding plastic potential surface is shaped as an asymmetric cigar with a smoothly rounded triangular cross-section, as shown in Fig. 5(a). The shape of the meridians are shown in the triaxial plane in Fig. 5(b) for Fine Silica Sand. This shape accounts for the rather abrupt change in volume change behavior near the hydrostatic axis observed in experiments and discussed further below. The plastic potential surface is continuous throughout stress space except at the intersection with the hydrostatic axis behind the origin of the space.

Substituting Eq. (4) into the flow equation (1), the components of the palstic strain increments are calculated as:

$$\begin{Bmatrix} d\varepsilon_x^p \\ d\varepsilon_y^p \\ d\varepsilon_z^p \\ d\gamma_{yz}^p \\ d\gamma_{zx}^p \\ d\gamma_{xy}^p \end{Bmatrix} = d\lambda_p \cdot \left(\frac{I_1}{p_a}\right)^{\mu} \begin{Bmatrix} G - (\sigma_y + \sigma_z) \cdot \dfrac{I_1^2}{I_2^2} - \psi_1 \cdot (\sigma_y \cdot \sigma_z - \tau_{yz}^2) \cdot \dfrac{I_1^3}{I_3^2} \\ G - (\sigma_z + \sigma_x) \cdot \dfrac{I_1^2}{I_2^2} - \psi_1 \cdot (\sigma_z \cdot \sigma_x - \tau_{zx}^2) \cdot \dfrac{I_1^3}{I_3^2} \\ G - (\sigma_x + \sigma_y) \cdot \dfrac{I_1^2}{I_2^2} - \psi_1 \cdot (\sigma_x \cdot \sigma_y - \tau_{xy}^2) \cdot \dfrac{I_1^3}{I_3^2} \\ 2 \cdot \dfrac{I_1^2}{I_2^2} \cdot \tau_{yz} - 2 \cdot \psi_1 \cdot (\tau_{xy} \cdot \tau_{zx} - \sigma_x \cdot \tau_{yz}) \cdot \dfrac{I_1^3}{I_3^2} \\ 2 \cdot \dfrac{I_1^2}{I_2^2} \cdot \tau_{zx} - 2 \cdot \psi_1 \cdot (\tau_{xy} \cdot \tau_{yz} - \sigma_y \cdot \tau_{zx}) \cdot \dfrac{I_1^3}{I_3^2} \\ 2 \cdot \dfrac{I_1^2}{I_2^2} \cdot \tau_{xy} - 2 \cdot \psi_1 \cdot (\tau_{yz} \cdot \tau_{zx} - \sigma_z \cdot \tau_{xy}) \cdot \dfrac{I_1^3}{I_3^2} \end{Bmatrix} \qquad \text{(8a-f)}$$

in which

$$G = \psi_1 \ (\mu + 3) \ \frac{I_1^2}{I_3} - (\mu + 2) \ \frac{I_1}{I_2} + \psi_2 \cdot \mu \cdot \frac{1}{I_1} \qquad (9)$$

If only principal stresses are applied to an element, then the plastic strain increment vector is given by:

$$\begin{Bmatrix} d\varepsilon_1^p \\ d\varepsilon_2^p \\ d\varepsilon_3^p \end{Bmatrix} = d\lambda_p \cdot \left(\frac{I_1}{p_a}\right)^{\mu} \begin{Bmatrix} G - \dfrac{I_1^2}{I_2^2} (\sigma_2 + \sigma_3) - \psi_1 \cdot \dfrac{I_1^3}{I_3^2} \sigma_2 \sigma_3 \\ G - \dfrac{I_1^2}{I_2^2} (\sigma_3 + \sigma_1) - \psi_1 \cdot \dfrac{I_1^3}{I_3^2} \sigma_3 \sigma_1 \\ G - \dfrac{I_1^2}{I_2^2} (\sigma_1 + \sigma_2) - \psi_1 \cdot \dfrac{I_1^3}{I_3^2} \sigma_1 \sigma_2 \end{Bmatrix} \qquad \text{(10a-c)}$$

MATERIALS WITH EFFECTIVE COHESION

The pointed apex of the plastic potential surface shown in Fig. 5 is located behind the origin of stress for materials without cohesion. Based on observations for concrete, as exemplified in Fig. 2, it appears that the plastic potential should have its apex at a translated stress origin O′ further behind the real stress origin. If the apex remains at the location for cohesionless soils, then the directions of incremental plastic strain vectors for uniaxial and biaxial compression cannot be derived from the proposed plastic potential. However, the function in Eq. (4) can be extended to materials with effective cohesion by a translation of the principal stress space along the hydrostatic axis from O to O′ in Fig. 5. Thus, a constant stress $a \cdot p_a$ is added to the normal stresses before substitution in Eqs. (5), (6) and (7):

$$\bar{\sigma}_{ij} = \sigma_{ij} + \delta_{ij} \cdot a \cdot p_a \tag{11}$$

in which δ_{ij} is Kronecker's delta ($\delta_{ij} = 1$ for $i = j$, and $\delta_{ij} = 0$ for $i \neq j$). The parameter 'a' is dimensionless and p_a is atmospheric pressure in the same units as σ_{ij}. The value of $a \cdot p_a$ reflects the effect of the tensile strength of the material, and it is determined as shown by Lade [13,14] and by Kim and Lade [15].

The flow rule in Eq. (1) still governs the incremental plastic strains, but the derivatives should be expressed in terms of the translated stresses to maintain mathematical consistency corresponding to a potential function formed in the translated space. The transformed stresses should also be used in the constitutive Eqs. (8), (9) and (10). This is consistent with the notion of cohesion explained in [13-15]. The plastic potential surface in this case has a shape similar to that for cohesionless soil shown in Fig. 5. However, the pointed apex lies at the origin O′ of the translated stress space.

CONDITION OF IRREVERSIBILITY

The condition of irreversibility requires that the plastic work is positive (or zero) whenever a change in plastic strain occurs [16]:

$$dW_p = \sigma_{ij} \cdot d\varepsilon_{ij}^p \geq 0 \tag{12}$$

This requires that the plastic potential surface is convex with regard to the stress origin. This condition may be satisfied when proper parameter values are employed. Since the plastic potential in Eq. 4 is a homogeneous function of the μth degree, Eq. 12 leads to

$$dW_p = \mu \cdot g_p \cdot d\lambda_p \tag{13}$$

Since $d\lambda_p > 0$, the condition of irreversibility requires that

$$\mu \cdot g_p \geq 0 \tag{14}$$

Fig. 6 shows the possible combinations of μ and g_p and the normal directions to the plastic potential. Fig. 6(b) indicates that a negative g_p is not suitable for a plastic potential function. Thus, the only possible combination that satisfies Eq. (14) is that of positive μ and positive g_p.

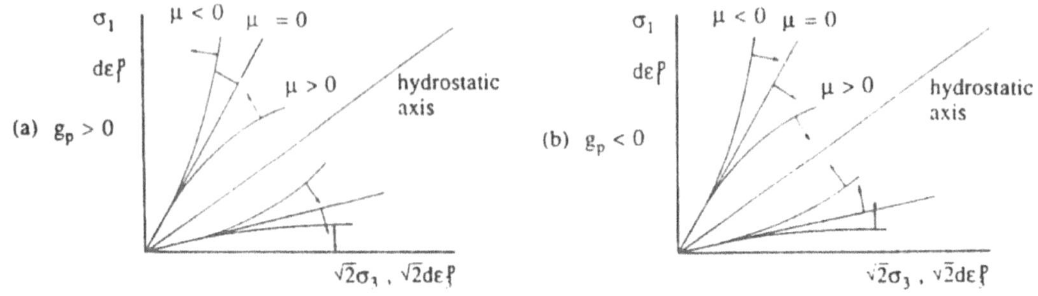

Fig. 6. Combinations of Plastic Potential g_p and Parameter μ in Triaxial Plane. (a) Case of Positive g_p, and (b) Case of Negative g_p.

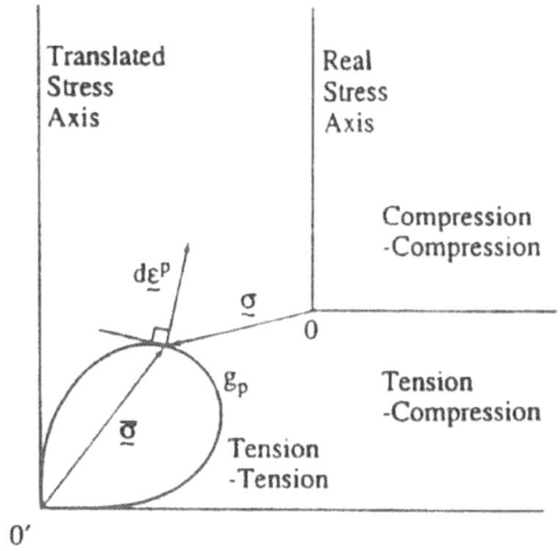

Fig. 7. Relation Between Stress and Incremental Plastic Strain Vectors in Tension-Tension Loading.

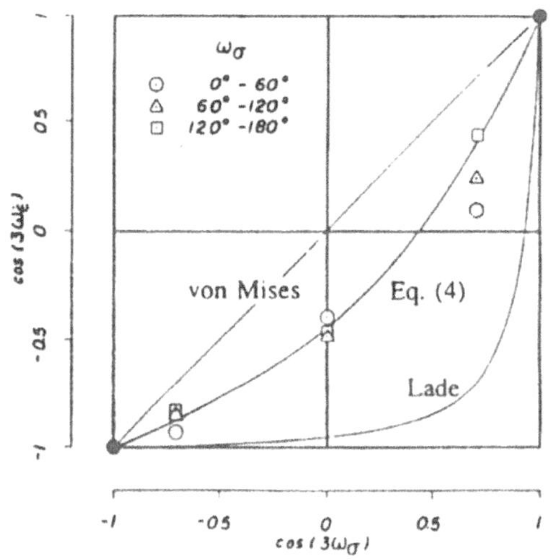

Fig. 8. Comparison of Calculated and Measured Lode Angles for Fuji River Sand (Data from [9]).

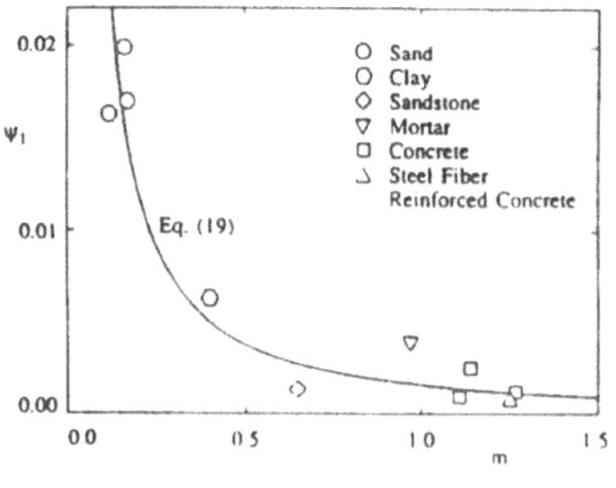

Fig. 9. Relation Between Material Parameters ψ_1 and m for Various Frictional Materials.

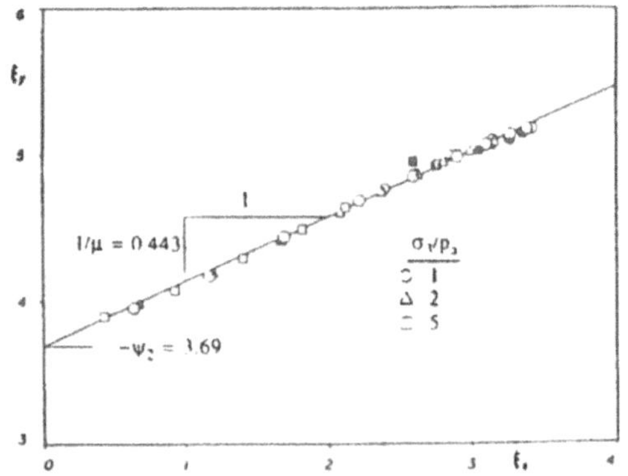

Fig. 10. Determination of Plastic Potential Parameters for Fine Silica Sand.

Since $(I_1^3/I_3) \geq 27$ and $(-I_1^2/I_2) \geq 3$ (where the equal signs hold on the hydrostatic axis) and (I_1/p_a) is always positive, the condition of irreversibility leads to the following requirements:

$$\mu > 0 \tag{15}$$

$$\psi_2 > -(27 \cdot \psi_1 + 3) \tag{16}$$

For materials with effective cohesion, the above conditions hold for the incremental plastic work done by the translated stresses. The real plastic work increment encountered in the tensile zone may be negative as indicated in Fig. 7. The incremental plastic strain vector forms an obtuse angle with the real stress vector thus resulting in a negative plastic work increment, while it forms an acute angle with the translated stress vector. However, this form of the plastic potential in the tensile stress zone never occurs if translation is retained throughout the theory. As will be explained in a companion paper, the stress state in Fig. 7 is treated as unloading and only elastic strains are produced. Thus, the condition of irreversibility is retained for cohesive materials.

PARAMETER DETERMINATION

Parameter ψ_1

As mentioned above, the parameter ψ_1 governs the shape of the plastic potential in octahedral planes. If $\psi_1 = 0$, then g describes a family of circles on the octahedral plane. If ψ_1 approaches infinity, the effect of I_2 vanishes and g_p describes a family of curves like the failure surfaces [12-15].

The incremental plastic strain angle $\omega_{\dot{\epsilon}}$ can be derived by substitution of Eqs. (10) into Eq. (3):

$$\omega_{\dot{\epsilon}} = \arctan \frac{\sqrt{3}\left\{\dfrac{I_1^2}{I_2^2}(\sigma_2 - \sigma_3) + \psi_1 \dfrac{I_1^3}{I_3^2}(\sigma_1\sigma_2 - \sigma_3\sigma_1)\right\}}{\dfrac{I_1^2}{I_2^2}(2\sigma_1 - \sigma_2 - \sigma_3) + \psi_1 \dfrac{I_1^3}{I_3^2}(\sigma_3\sigma_1 + \sigma_1\sigma_2 + \sigma_2\sigma_3)} \tag{17}$$

Note that the parameters μ and ψ_2 are completely eliminated. Solving Eq. (17) for ψ_1 produces:

$$\psi_1 = -\frac{I_3^2}{I_1 I_2^2} \frac{\sqrt{3}(\sigma_2 - \sigma_3) - (2\sigma_1 - \sigma_2 - \sigma_3)\tan\omega_{\dot{\epsilon}}}{\sqrt{3}(\sigma_1\sigma_2 - \sigma_3\sigma_1) - (\sigma_3\sigma_1 + \sigma_1\sigma_2 - 2\sigma_2\sigma_3)\tan\omega_{\dot{\epsilon}}} \tag{18}$$

Table 1 shows calculated values of ψ_1 at failure for Fuji River Sand [9]. The ψ_1-values listed in the right hand column indicate a fair amount of scatter. However, a small variation in $\omega_{\dot{\epsilon}}$ results in a large change in ψ_1. Conversely, a change in ψ_1 does not affect the magnitude of $\omega_{\dot{\epsilon}}$ very much. When the average value of $\psi_1 = 0.00472$ is used, the calculated values of $\omega_{\dot{\epsilon}}$ are not much different from the measured values, as shown in Table 1. This comparison is also

Table 1. Determination of the Parameter ψ_1 for Fuji River Sand (Data from [9]).

Stress angle ω_σ (degrees)	Incremental Plastic Strain angle $\omega_{\dot{\varepsilon}}$ (degrees)		ψ_1
	Test data	Calculated Value	
15	28.1	24.1	0.00834
30	37.7	37.9	0.00451
45	52.7	49.8	0.00109
75	49.4	51.3	0.00269
90	39.6	37.9	0.00636
105	25.3	24.1	0.00561
135	21.2	22.9	0.00326
150	39.1	37.9	0.00583
165	48.8	48.7	0.00480

average ψ_1 = 0.00472

Table 2. Values of Parameters m and ψ_1 for Various Frictional Materials.

Material	m	ψ_1	Reference
Monterey No. 0 Sand			[17]
$D_r = 27\%$	0.12	0.01632	
$D_r = 98\%$	0.16	0.01984	
Cambria Sand	0.017	0.01697	[10]
Grundite Clay	0.4	0.00628	[18]
Mortar	0.969	0.00388	[7]
Sandstone	0.647	0.00130	[19]
Concrete	1.263	0.00122	[7]
Concrete	1.105	0.00078	[20]
Concrete	1.133	0.00250	[21]
Steel Fiber Reinforced Concrete	1.251	0.00067	[8]

shown graphically in Fig. 8. The proposed plastic potential function is seen to be consistent with the observed trend of experimental data.

Similar calculations of ψ_1 were made for various frictional materials for which three-dimensional test results were available. For materials with effective cohesion, calculation of ψ_1 proceed after stress translation according to Eq. (11). Table 2 shows that ψ_1 decreases as the rigidity of the materials increase, and the cross-section of the plastic potential surface changes from a triangular to a rounded shape. The effect of material rigidity was also observed in the characteristics of the failure surfaces. Thus, the curvature of failure surface meridians increases with increasing rigidity. This curvature is modeled by the parameter m in the failure criterion [12-15]. The values of m are listed for the various materials in Table 2, and a diagram of ψ_1 versus m is shown in Fig. 9. The relation between ψ_1 and m can be expressed as a power function:

$$\psi_1 = 0.00155 \cdot m^{-1.27} \tag{19}$$

The solid line in Fig. 9 represents the relation in Eq. (19). Recalling that the incremental plastic strain angle ω_ξ is relatively insensitive to the value of ψ_1, Eq. (19) provides a reasonable approximation for the value of ψ_1. It is therefore not necessary to require results of three-dimensional tests to determine the value of ψ_1.

<u>Parameters ψ_2 and μ</u>

Once the parameter ψ_1 is evaluated, the other parameters, ψ_2 and μ, can be determined using triaxial compression test data. This is done by expressing the incremental plastic strain ratio defined as

$$\nu_p = -\frac{d\varepsilon_3^p}{d\varepsilon_1^p} \tag{20}$$

Substitution of Eqs. (10) for the plastic strain increments under triaxial compression conditions ($\sigma_2 = \sigma_3$) produces the following equation:

$$\xi_y = \frac{1}{\mu}\xi_x - \psi_2 \tag{21}$$

where

$$\xi_x = \frac{1}{1+\nu_p}\left\{\frac{I_1^3}{I_2^2}(\sigma_1+\sigma_3+2\nu_p\sigma_3) + \psi_1 \cdot \frac{I_1^4}{I_3^2}(\sigma_1\sigma_3+\nu_p\sigma_3^2)\right\} - 3\psi_1 \cdot \frac{I_1^3}{I_3} + 2\,\frac{I_1^2}{I_2} \tag{22}$$

and

$$\xi_y = \psi_1 \cdot \frac{I_1^3}{I_3} - \frac{I_1^2}{I_2} \tag{23}$$

Thus, $1/\mu$ and $-\psi_2$ can be determined by linear regression between ξ_x and ξ_y determined from several data points. Fig. 10 is a plot of ξ_x and ξ_y for the triaxial compression tests on Fine Silica Sand. All data points are consistent with Eq. (21). The value of $-\psi_2$ is the intercept value of ξ_y

at $\xi_x = 0$, and the value of $1/\mu$ is the slope of the best fitting straight line.

Table 3 lists the parameter values for plastic potential functions for various materials. Only results of triaxial compression tests (including uniaxial compression tests for materials with effective cohesion) were employed in determination of parameters.

EVALUATION OF PROPOSED PLASTIC POTENTIAL FUNCTION

After the parameters are determined, the incremental plastic strains can be calculated from Eqs. (8), or from Eqs. (10) in the absence of shear stresses. In order to evaluate the plastic potential, the directions of incremental plastic strain vectors have been calculated at the measured stress points for various sets of test data. These calculated directions are compared with the measured directions on the triaxial planes, and the Lode angles on the octahedral plane are compared for three-dimensional test data.

Fig. 11(a) shows that the predicted directions of the vectors are almost identical to the measured directions for Fine Silica Sand. A rather abrupt change in volumetric strain rate near the hydrostatic axis is also predicted by the proposed cigar-shaped plastic potential function shown in Fig. 5.

Fig. 11(b) shows comparisons for uniaxial compression, biaxial compression, triaxial extension, and triaxial compression of concrete [7]. Considering that the parameters were derived from triaxial compression data, the good agreement between predicted and measured directions could be expected for these tests. In biaxial compression tests the measured vectors indicate a rather dilative behavior, and the potential is slow in capturing this pattern. However, the prediction is considered to be within reasonable range of the experimental data in view of the fact that Schickert and Winkler employed different specimen preparation methods and different testing schemes for triaxial and biaxial tests. The observed pattern of vectors correlates well with the potential function for the extension tests shown in Fig. 11(b).

Another comparison is shown for steel fiber reinforced concrete [8] in Fig. 11(c). Here the deviation between measured and predicted directions are significant in the initial loading range of some of the extension tests. It should be noted that the directions of strain increment vectors in this range is governed mainly by the elastic behavior. The elastic modulus for this material, determined from the initial slopes of the stress-strain curves, seems to be too high for this case. However, the influence of plastic strains on the stress-strain behavior in this initial loading region is negligible. Reasonable predictions within the scatter of the experimental data were obtained at higher stress levels.

In order to evaluate the plastic potential under three-dimensional loading conditions, the Lode angles at failure are compared for loose and dense Monterey No. O Sand in Fig. 12. Although the values of the parameter ψ_1 are based on the empirical Eq. (19), the predicted curves simulate the location of the experimental data very well. The density dependent variation of the angles are captured, such that the cross-sections of the plastic potential surfaces for dense sand are more triangular than those for loose sand.

Table 3. Plastic Potential Parameters ψ_1, ψ_2, and μ for Various Frictional Materials.

Material	D_r (%)	ψ_1	ψ_2	μ	Reference
Fine Silica Sand	30	0.0289	-3.69	2.26	
Sacramento River Sand	100	0.0100	-3.09	2.00	[22]
	38	0.0317	-3.72	2.36	[22]
Painted Rock Material	100	0.0113	-3.26	2.82	[23]
	70	0.0159	-3.39	2.72	[23]
Crushed Napa Basalt	100	0.0046	-2.97	2.80	[24]
	70	0.0072	-2.90	2.55	[24]
Monterey No. 0 Sand	98	0.0160	-3.38	2.30	[17]
	27	0.0229	-3.60	2.50	[17]
Edgar Kaolinite Clay		0.0034	-3.02	2.24	
Plain Concrete		0.00115	-2.77	2.51	[7]
		0.00135	-2.92	5.06	[21]
		0.00209	-3.02	4.88	[25]
		0.00119	-2.93	3.45	[20]
Steel Fiber Reinforced Concrete		0.00117	-2.58	2.63	[8]

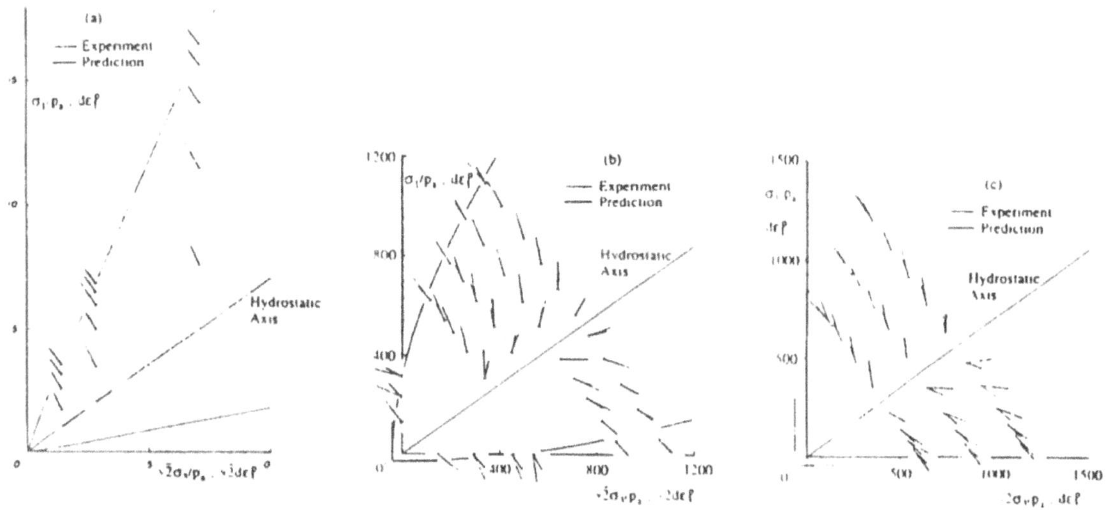

Fig. 11. Comparisons of Predicted and Observed Directions of Incremental Plastic Strain Vectors in Triaxial Plane for (a) Fine Silica Sand, (b) Plain Concrete (Data from [7]), and (c) Steel Fiber Reinforced Concrete (Data from [8]).

Table 3. Plastic Potential Parameters ψ_1, ψ_2, and μ for Various Frictional Materials.

Material	D_r (%)	ψ_1	ψ_2	μ	Reference
Fine Silica Sand	30	0.0289	-3.69	2.26	
Sacramento River Sand	100	0.0100	-3.09	2.00	[22]
	38	0.0317	-3.72	2.36	[22]
Painted Rock Material	100	0.0113	-3.26	2.82	[23]
	70	0.0159	-3.39	2.72	[23]
Crushed Napa Basalt	100	0.0046	-2.97	2.80	[24]
	70	0.0072	-2.90	2.55	[24]
Monterey No. 0 Sand	98	0.0160	-3.38	2.30	[17]
	27	0.0229	-3.60	2.50	[17]
Edgar Kaolinite Clay		0.0034	-3.02	2.24	
Plain Concrete		0.00115	-2.77	2.51	[7]
		0.00135	-2.92	5.06	[21]
		0.00209	-3.02	4.88	[25]
		0.00119	-2.93	3.45	[20]
Steel Fiber Reinforced Concrete		0.00117	-2.58	2.63	[8]

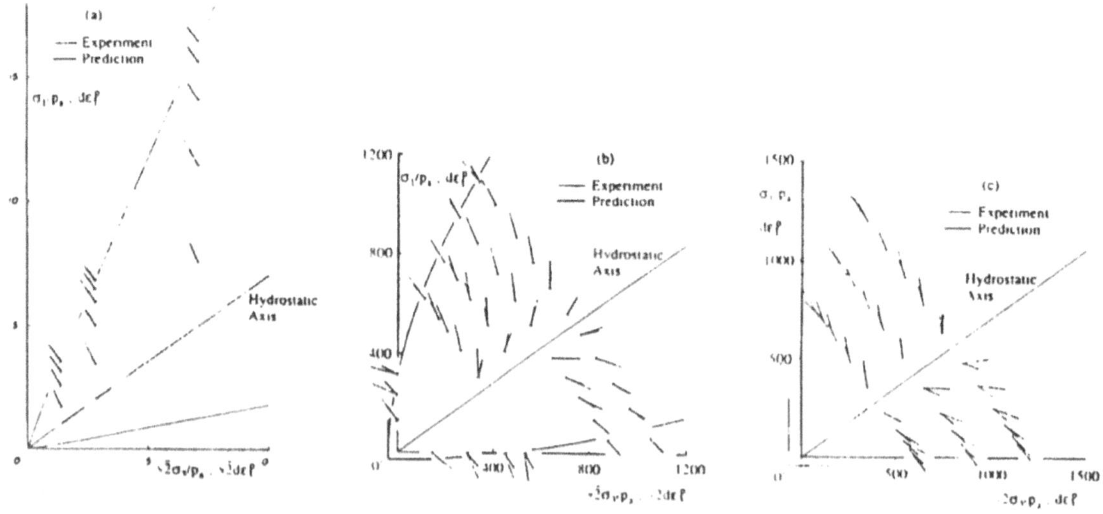

Fig. 11. Comparisons of Predicted and Observed Directions of Incremental Plastic Strain Vectors in Triaxial Plane for (a) Fine Silica Sand, (b) Plain Concrete (Data from [7]), and (c) Steel Fiber Reinforced Concrete (Data from [8]).

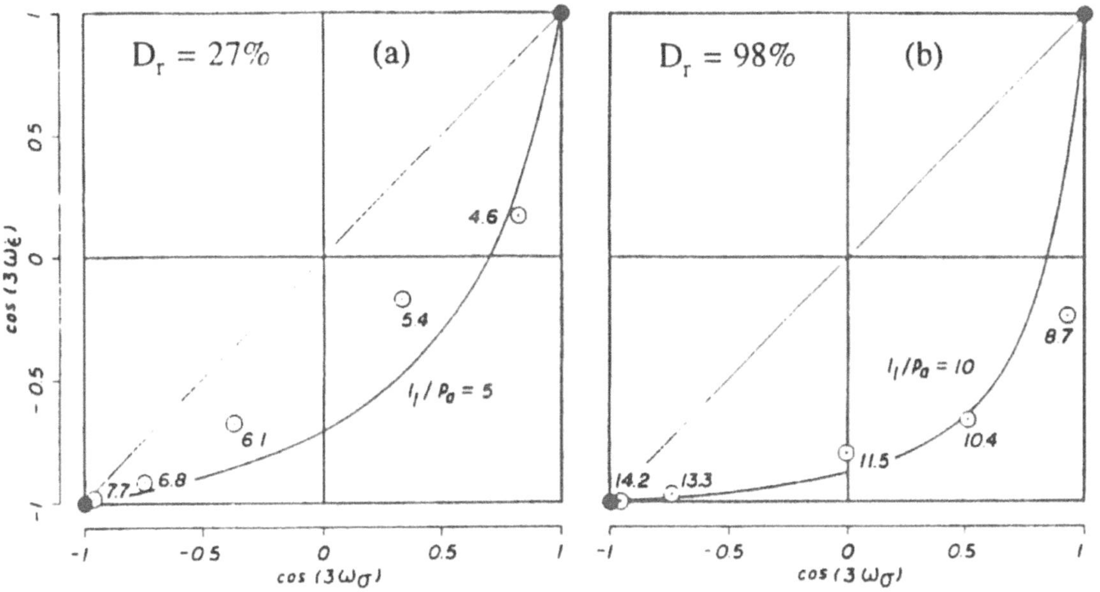

Fig. 12. Comparison of Predicted and Observed Lode Angles for (a) Loose, and (b) Dense Monterey No. 0 Sand (Data from [17]).

CONCLUSION

Based on classical concepts of incremental plasticity and careful review and evaluation of available test data, a general three-dimensional plastic potential function expressed in terms of stress invariants has been proposed. With three parameters (one of which is redundant with one of the failure surface parameters) to be determined, this cigar-shaped plastic potential is unique and continuous in stress space. This makes it applicable as an ingredient of a plasticity model without difficulty. In addition, it includes tensile behavior of cohesive materials without discontinuity or vertices except at the translated stress origin. Applied to general loading in three dimensions, it has the capability of predicting the behavior with good accuracy throughout the loading paths for many different types of frictional materials.

ACKNOWLEDGMENT

The basic developmental work presented here was supported by the National Science Foundation under Grant No. CEE 8211159. Additional developments and computational refinements were supported by Exxon Production Research Company, Houston, Texas, and by California Research and Technology, Chatsworth, California. Grateful appreciation is expressed for their support of this research.

REFERENCES

1. Saint-Venant, B.de,
Memoire sur l'etablissement des equations differentielles des mouvement interieurs operes dans les corps solides ductiles au dela des limites ou l'estasticite pourrait les ramener a leur premier etat.
Comptes Rendus Acad. Sci., Paris, 70 (1870) 473-480.

2. Melan, E.,
Plastizitaet des Raumlichen Kontinuums
Ingr.-Arch., 9 (1938) 116-126.

3. Poorooshasb, H.B., Holubec, I., and Sherbourne, A.N.,
Yielding and Flow of Sand in Triaxial Compression: Part I.
Can. Geot. J., 3 (1966) 179-190.

4. Ko, H.-Y. and Scott, R.F.,
Deformation of Sand in Shear.
J. Soil Mech. Found. Div., ASCE, 93, SM5 (1967) 283-310.

5. Lade, P.V. and Duncan, J.M.,
Stress-Path Dependent Behavior of Cohesionless Soil.
J. Geotech. Eng. Div., ASCE, 102 (1976) 51-68.

6. Lade, P.V., Nelson, R.B. and Ito, Y.M.,
Nonassociated Flow and Stability of Granular Materials.
J. Engr. Mech., ASCE, 113 (1987) 1302-1318.

7. Schickert, G. and Winkler, H.,
Results of Tests Concerning Strength and Strain of Concrete Subjected to Multiaxial Compressive Stresses.
Deutscher Ausschus fur Stahlbeton, 277 Berlin (1977).

8. Egging, D.E. and Ko, H.-Y.,
Constitutive Relations of Randomly Oriented Steel Fiber Reinforced Concrete Under Multiaxial Compressive Loadings.
University of Colorado, Boulder (1981).

9. Yamada, Y. and Ishihara, K.,
Anisotropic Deformation Characteristics of Sand Under Three-Dimensional Stress Conditions.
Soil and Foundations, JSSMFE, 19 (1979) 79-94.

10. Ochiai, H. and Lade, P.V.,
Three-Dimensional Behavior of Sand with Anisotropic Fabric.
J. Geotech. Eng. Div., ASCE, 109 (1983) 1313-1328.

11. Yong, R.N. and McKeys, E.
Yield and Failure of a Clay Under Triaxial Stresses.
J. Soil Mech. Found. Div., ASCE, 97 (1971) 159-176.

12. Lade, P.V.,
Elasto-Plastic Stress-Strain Theory for Cohesionless Soil with Curved Yield Surfaces.
Int. J. Solids Structures, Pergamon Press, 13 (1977) 1019-1035.

13. Lade, P.V.,
Three-Parameter Failure Criterion for Concrete.
J. Engr. Mech. Div., ASCE, 108 (1982) 850-863.

14. Lade, P.V.,
Failure Criterion for Frictional Materials.
Chapt. 20 in Mechanics of Engineering Materials, Ed. by C.S. Desai and R.H. Gallagher, Wiley (1984) 385-402.

15. Kim, M.K., and Lade, P.V.,
Modelling Rock Strength in Three Dimensions.
Int. J. Rock Mech. Min. Sci. & Geomech. Abstr., 21 (1984) 21-33.

16. Prager, W.,
Recent Developments in the Mathematical Theory of Plasticity.
J. Appl. Phys., 20 (1949) 235-241.

17. Lade, P.V.,
The Stress-Strain and Strength Characteristics of Cohesionless Soils.
Ph.D. Dissertation, University of California, Berkeley (1972).

18. Lade, P.V. and Musante, H.M.,
Three-Dimensional Behavior of Remolded Clay.
J. Geotech. Engr. Div., ASCE, 104 (1978) 193-209.

19. Maso, J. and Lerau, J.,
Mechanical Behavior of Darney Sandstone in Biaxial Compression.
Int. J. Rock Mech. Min. Sci. & Geomech. Abstr., 17 (1980) 109-115.

20. Aschl, H., Linse, D. and Stoeckl, S.,
Strength and Stress Strain Behavior of Concrete Under Multiaxial Compression and Tension.
Proc. 2nd Int. Conf. Mechanical Behavior of Materials, Boston Mass. (1976) 102-117.

21. Kupfer, H., Hilsdorf, H.K. and Rusch, H.,
Behavior of Concrete Under Biaxial Stresses.
J. Am. Concr. Inst., 66 (1969) 656-666.

22. Lee, K.L. and Seed, H.B.,
Drained Strength Characteristics of Sands.
J. Soil Mech. Found. Div., ASCE, 93 SM6 (1967) 117-141.

23. Al-Hussaini, M.M.,
Plane Strain and Triaxial Compression Tests on Painted Rock Dam Material.
Tech. Rep. S-71-2, No. 3, USAE, WES, Vicksburg, Miss. (1972).

24. Al-Hussaini, M.M., Drained Plane Strain and Triaxial Compression Tests on Crushed Napa Basalt. Tech. Rep. S-71-2, No. 2, USAE, WES, Vicksburg, Miss. (1971).

25. Tasuji, M.E., Slate, F.O. and Nilson, A.H., Stress-Strain and Fracture of Concrete in Biaxial Loading. J. Am. Concr.Inst., 75 (1978) 306-312.

Received 10 June 1988; revised version received and accepted 2 August 1988

Computers and Geotechnics 6 (1988) 13–29

SINGLE HARDENING CONSTITUTIVE MODEL FOR FRICTIONAL MATERIALS

II. YIELD CRITERION AND PLASTIC WORK CONTOURS

P.V. Lade
Department of Civil Engineering
University of California
Los Angeles, California 90024, U.S.A.

and

M.K. Kim
Department of Civil Engineering
Yonsei University
Seoul 120, Korea

ABSTRACT

A review of numerous experimental data sets for frictional materials has led to a reformulation of the condition for plastic yielding in such materials. A yield function is proposed such that yield surfaces are equivalent to plastic work contours. Two parameters are required in this function. The yield surface resembles an asymmetric tear drop in the principal stress space. It includes tensile behavior of materials with effective cohesion. Work hardening and softening behavior can be simulated through the formulation of plasticity proposed herein. The capabilities of this mathematically consistent formulation are examined by comparisons with experimental data from two- and three-dimensional tests on different frictional materials.

INTRODUCTION

The development of elasto-plastic stress-strain theories for engineering materials requires a criterion to describe the conditions for which plastic yielding occurs. Various criteria expressing combinations of multiaxial stresses that will cause yielding have been proposed in the literature, e.g. [1-12]. These criteria have most often been based on experimental observations, sometimes in combination with assumptions about the type of plastic behavior exhibited by the material (associated or nonassociated flow). Identification of yield points from measured stress-strain curves and combination of these points to form yield surfaces are not always simple. There are two reasons why it may be difficult to identify suitable yield surfaces: (1) Since yielding is a continuous process for frictional materials, there may not be a distinct yield point on a stress-strain curve. (2) In the plasticity theory, yield surfaces must be associated with hardening (and softening) parameters such that the combination uniquely defines the magnitude of incremental plastic strains.

Computers and Geotechnics 0266-352X/88/$03·50

Two methods of identification of yield points and yield surfaces in stress space are reviewed. Plastic yield points for primary loading are then compared to a yield criterion developed to describe contours of plastic work as determined from results of triaxial tests.

OBSERVED YIELD BEHAVIOR FOR SOILS

The existence and shapes of yield surfaces in stress space have been studied often since plasticity theory has been found effective in modeling deformation of frictional materials. Two important studies of yield behavior of soils are reviewed below.

Tatsuoka and Ishihara [13] presented results and analyses of triaxial compression tests on Fuji River Sand at three densities. The different types of stress paths used in their study are summarized in Fig. 1(d). A specimen was first triaxially compressed up to a certain stress level, unloaded at constant confining pressure, then loaded again at a different confining pressure. On a diagram of deviator stress $(\sigma_1-\sigma_3)$ versus shear strain $(\varepsilon_1-\varepsilon_3)$ obtained for reloading, two approximately linear portions were extended near the sudden increase in strains, and a yield point was located at the intersection. Connecting this yield point with the previous maximum stress point, a portion of a yield surface was established. They analyzed an extensive set of test data and found a series of curved yield surfaces as shown in the nondimensional p'–q diagrams in Fig. 1 for (a) loose, (b) medium dense, and (c) dense Fuji River Sand. In these diagrams, p_a is atmospheric pressure in the same units as p' and q. Note that the yield surfaces exhibit varying curvature with stress level, and they do not coincide with the failure envelope. Instead they deviate from and cross the failure envelope, and they are more curved than the failure envelope.

Parry and Nadarajah [14] performed a series of tests on lightly overconsolidated specimens of kaolin. Each specimen was subjected to a different stress path. In a diagram of stress versus volumetric strains, a yield point was located in the same manner as done by Tatsuoka and Ishihara [13]. The resulting yield points were connected to form a cap shaped volumetric yield surface in a triaxial plane as shown in Fig. 2. This cap shaped yield surface is more curved than those shown in Fig. 1, and it cuts across the hydrostatic axis. However, it should be noted that plastic shear strains are not accounted for in this yield locus. Roscoe and Burland [2] performed similar tests examining the plastic shear strains. Their results indicated that significant plastic shear deformation could occur inside the volumetric yield locus.

PLASTIC WORK CONTOURS

As noted in the previous section, a yield criterion in an elasto-plastic model should not only simulate yield surfaces derived under various types of assumptions, but it should also be uniquely related to the hardening and softening characteristics of the materials. An obstacle in achieving these dual goals is that it is not a simple task to investigate yield behavior of the material under various reloading conditions including three-dimensional loading conditions.

An alternative is to examine the contours of the hardening parameters and to compare them with experimental data for yield surfaces. Huang [15] described a study of this type involving contours. In the study presented here the plastic work is used as the hardening parameter.

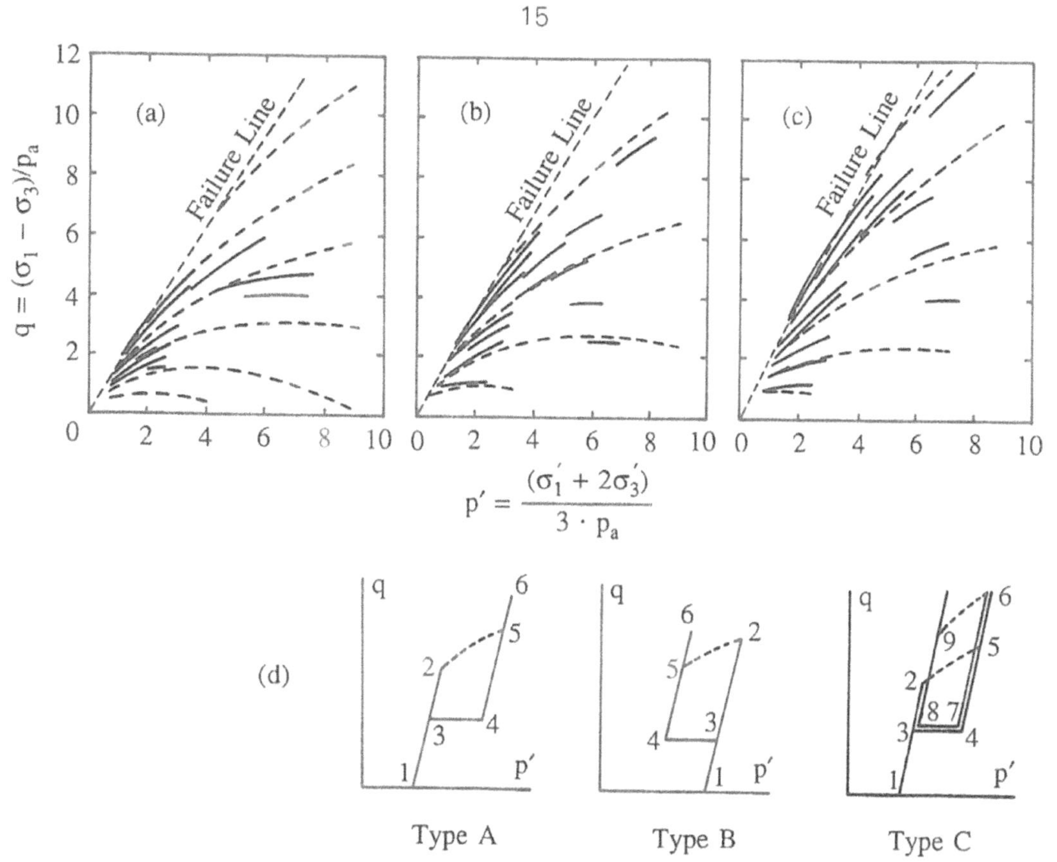

Fig. 1. Observed Yield Loci in p-q Diagram for (a) Loose, (b) Medium Dense, and (c) Dense Fuji River Sand, and (d) Stress-Paths used in Triaxial Tests [13].

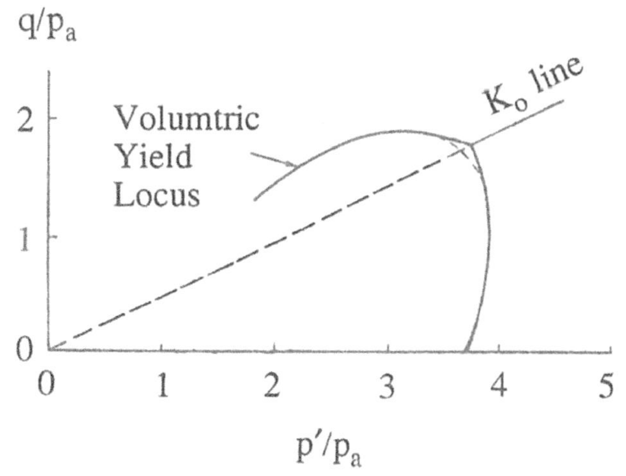

Fig. 2. Observed Volumetric Yield Surface in p-q Diagram for Lightly Overconsolidated Kaolin [14].

Using plastic work contours in the study of yield behavior of a material has several advantages. It does not involve tests with complicated stress-paths and it also avoids difficulties in determination of yield points on stress-strain curves. In comparison, computation of plastic work is straight forward, and the plastic work contours appear to capture yielding in terms of shear strains as well as volumetric strains.

Before further discussion of plastic work contours, it may be advantageous to study the plastic work during isotropic compression tests. An isotropic compression test involves a simple stress path in which the normal stresses are equal in all directions, and the data from this test may serve as a guide in plotting the plastic work contours. Isotropic compression of soil produces partly elastic and partly plastic deformations. The plastic work done by the isotropic pressure increases monotonically as the pressure increases. This plastic work may be modeled by a power function of pressure [4]:

$$W_p = C \cdot p_a \cdot \left(\frac{I_1}{p_a}\right)^p \tag{1}$$

where I_1 is the first invariant of stresses which is three times the isotropic pressure, p_a is atmospheric pressure expressed in the same unit as W_p and I_1, and C and p are dimensionless parameters to be determined. The parameters C and p can be determined by plotting the variation of W_p/p_a and I_1/p_a in a log - log diagram and fitting a straight line, as shown in Fig. 3. The magnitude of C is the value of W_p/p_a at $I_1/p_a = 1$ and the value of p is the slope of the fitted line. Fig. 4 shows the variation of plastic work during an isotropic compression test on Fine Silica Sand (e=0.76) together with the modeling curve from Eq. (1). It can be seen that Eq. (1) models the work-hardening behavior in isotropic compression reasonably well. Eq. (1) is used in the following to determine values of W_p/p_a whenever the isotropic compression condition occurs.

Fig. 5 shows the data for plastic work obtained from three triaxial compression tests on Fine Silica Sand with confining pressures of 1, 2, and 5 kg/cm^2, respectively. For stress states near isotropic compression, i.e. at low stress levels, the plastic work increases slowly with increasing shear stress. As the shear stress increases further, however, the rate of plastic work increases rapidly until it reaches the failure state denoted by shaded circles in the figure. A family of contours can be drawn to connect stress points with the same values of W_p/p_a. It should be noted that the failure surface is not one of these contours. The value of W_p/p_a at failure increases as the confining pressure increases. Consequently, each plastic work contour crosses the failure envelope at a very shallow angle and gradually deviates from it as I_1/p_a increases.

The results of four undrained tests on normally consolidated, remolded Edgar Kaolinite Clay are shown in Fig. 6. A pattern similar to that observed for sand is seen in this figure. However, the plastic work contours enclose areas which are shorter and more rounded. This could be expected, because volume changes and therefore plastic work under isotropic compression are relatively larger for clay than for sand.

Fig. 7 shows the variation of plastic work on an octahedral plane for cubical triaxial tests on Fuji River Sand performed by Yamada and Ishihara [16]. These tests were all performed in the same octahedral plane, i.e. with constant mean normal stress. The specimens were prepared anisotropically, hence, a slight anisotropy can be seen in the figure. The underlying isotropic behavior was discussed by Lewin et al. [17]. However, the data reported by Yamada and Ishihara [16] are adopted without any correction, because only the behavior pattern is of concern at this stage. Qualitatively, the contours have circular shapes at low stress levels and they gradually change into rounded triangular shapes as the stress level increases up to failure. At the

17

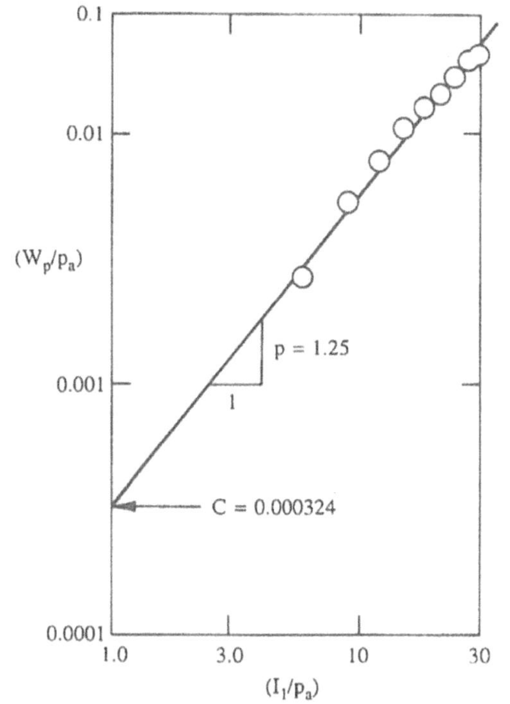

Fig. 3. Determination of Parameters C and p in Hardening Law for Fine Silica Sand.

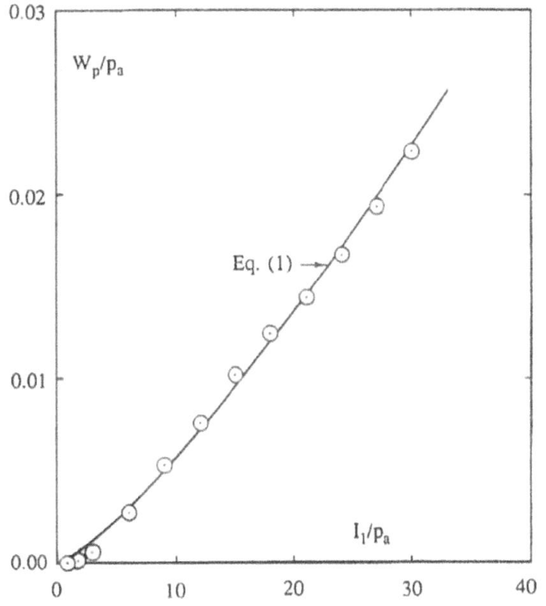

Fig. 4. Comparison of Predicted and Observed Variation of Plastic Work in Isotropic Compression Test on Fine Silica Sand.

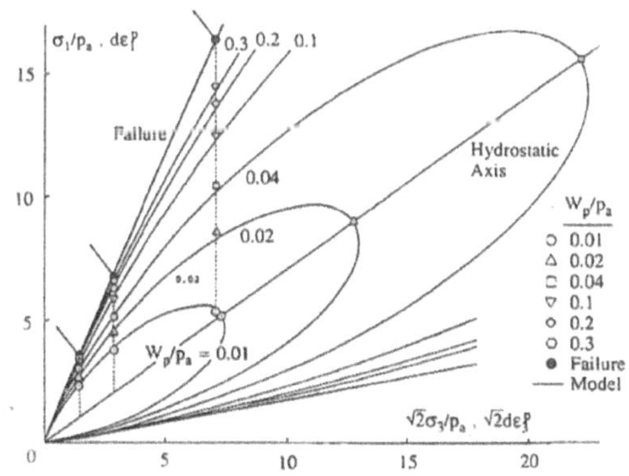

Fig. 5. Experimental Data Points and Contours of Constant Plastic Work Shown on Triaxial Plane for Fine Silica Sand.

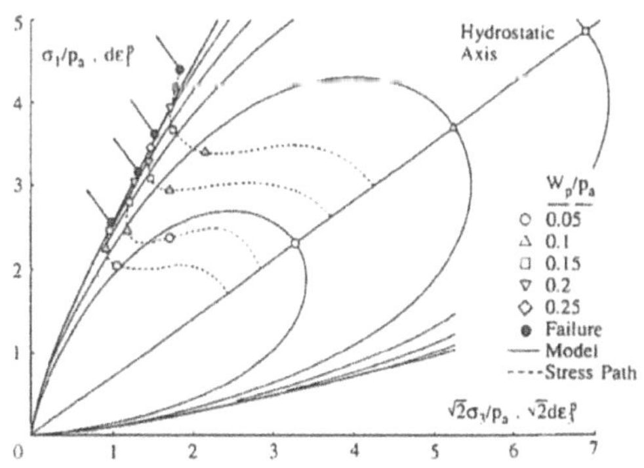

Fig. 6. Experimental Data Points and Contours of Constant Plastic Work Shown on Triaxial Plane for Undrained Tests on Edgar Kaolinite Clay.

18

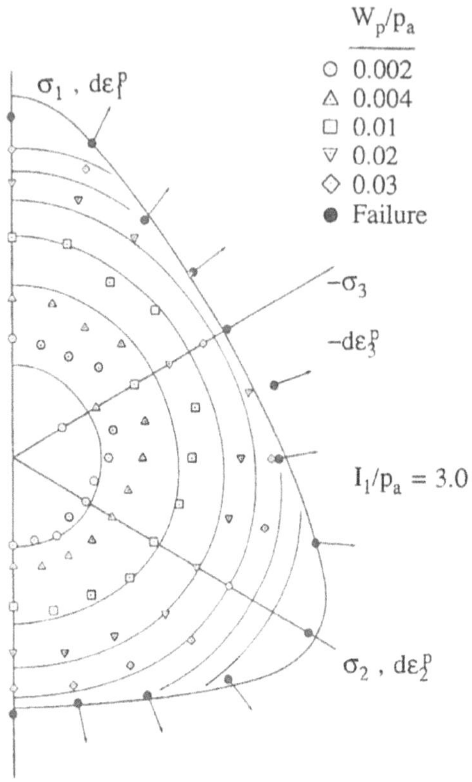

Fig. 7. Experimental Data Points and Contours of Constant Plastic Work Shown on Octahedral Plane for Fuji River Sand (Data from[16]).

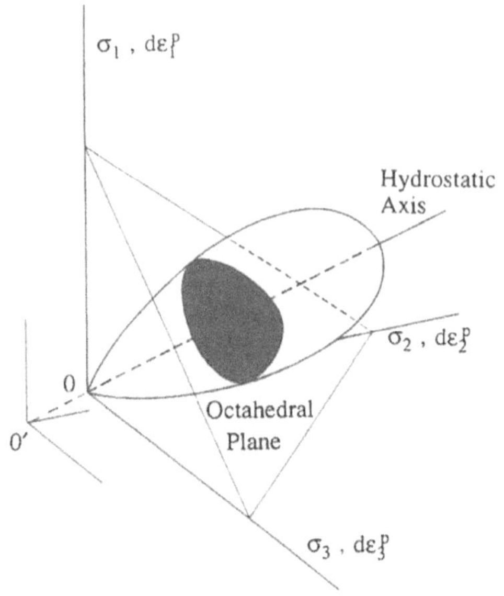

Fig. 8. Schematic Illustration of the Proposed Yield Surface in Principal Stress Space.

Table 1. Calculation of Yield Parameter h for Fine Silica Sand.

Triaxial Compression (B)			Isotropic Compression (A)	h
W_p/p_a	σ_3/p_a	σ_1/p_a	I_1/p_a	
0.313	1.00	3.55	244.2	0.348
0.739	2.00	6.81	485.6	0.341
0.582	2.00	6.74	401.2	0.356
1.401	5.00	16.42	810.3	0.373

Average h = 0.355

failure states, the values of W_p/p_a differ from each other on the same octahedral plane. The plastic work contours have more rounded shapes than the failure envelope.

The shapes of the plastic work contours generally agree with the shapes of the yield surfaces observed in unloading - reloading tests as reviewed in the previous section. The plastic work contours also resemble the material-dependent shapes of the yield surfaces: Contours for clay are shorter and more curved than those for sand.

It is interesting to check the normality of incremental plastic strain vectors to the yield surfaces, or equivalently, to the plastic work contours. The directions of the incremental plastic strain vectors at failure are drawn in Figs. 5 through 7. In Fig. 7 the directions of these vectors appear to coincide with the normal directions to the plastic work contours. However, on the triaxial planes shown in Figs. 5 and 6, the vectors are not normal to the plastic work contours. The directions normal to the yield surfaces at failure correspond to more dilation of the material than measured in experiments. This observation makes it necessary to employ a nonassociated flow rule.

PROPOSED YIELD CRITERION AND HARDENING LAW

Fig. 7 shows that the plastic work contours near failure have more rounded shapes than the failure surface, and that the directions of incremental plastic strain vectors are normal to the plastic work contours. Thus, a possible function to describe the plastic work contours, or a yield function in terms of stress is:

$$f_p' = \left(\psi_1 \cdot \frac{I_1^3}{I_3} - \frac{I_1^2}{I_2}\right) \cdot \left(\frac{I_1}{p_a}\right)^h \cdot e^q \qquad (2)$$

in which ψ_1 was determined in a companion paper dealing with a plastic potential function for frictional material [18]. This parameter controls the relative influence of the two terms within the parentheses, and it is used to model the normality condition on the octahedral plane. The second and third terms are introduced to model the meridional curvature of the yield surface. The value of h is constant and q is variable.

The yield surface is shaped as an asymmetric tear drop with a smoothly rounded triangular cross-section as shown in Fig. 8. The yield surface is continuous throughout stress space except at the origin ($\sigma_1 = \sigma_2 = \sigma_3 = 0$).

A family of yield surfaces described by Eq. (2) is drawn in Fig. 7 for a value of $\psi_1 = 0.00472$ for Fuji River Sand. The contours shown as yield surfaces were drawn for several constant values of f_p' in Eq. (2) and they are not directly related to the values of plastic work. However, the yield function describes the pattern of the plastic work contours reasonably well. The experimental points on Fig. 7 indicate anisotropic behavior as mentioned above, whereas Eq. (2) describes isotropically expanding yield surfaces. This yield function is suitable for an isotropic work-hardening stress-strain model.

The parameter q in Eq. (2) is devised as a function of stress level such that

$q = 0$	during isotropic compression
$0 < q < 1$	during hardening
$q = 1$	at failure stresses

The variation of q in the hardening regime will be investigated later. Once q = 1, a switch to a softening relation is effectuated. This is also discussed later.

For an isotropic compression state the value of f_p' in Eq. (2) is

$$f_p' = (27 \cdot \psi_1 + 3) \cdot \left(\frac{I_1}{p_a}\right)^h \tag{3}$$

Using this expression, the plastic work equation for isotropic compression (Eq. (1)) can be generalized as:

$$W_p = D \cdot p_a \cdot f_p''^{\rho} \tag{4a}$$

or

$$f_p'' = \left(\frac{1}{D}\right)^{1/\rho} \cdot \left(\frac{W_p}{p_a}\right)^{1/\rho} \tag{4b}$$

in which

$$D = \frac{C}{(27\psi_1 + 3)^{\rho}} \tag{5}$$

and

$$\rho = p/h \tag{6}$$

Eq. (4) provides a unique relation between the yield criterion, f_p'', and the plastic work, W_p, and serves as a work-hardening relation for all stress states.

PARAMETER DETERMINATION FOR YIELD CRITERION

Of the parameters that enter into the yield criterion only the constant value of h and the variation of q remain to be determined.

Constant Parameter h

Consider two stress points A and B on the same plastic work contour. Since these points are also located on the same yield surface, then ($f_{pA}' = f_{pB}'$):

$$\left(\psi_1 \cdot \frac{I_{1A}^3}{I_{3A}} - \frac{I_{1A}^2}{I_{2A}}\right) \cdot \left(\frac{I_{1A}}{p_a}\right)^h \cdot e^{q_A} = \left(\psi_1 \cdot \frac{I_{1B}^3}{I_{3B}} - \frac{I_{1B}^2}{I_{2B}}\right) \cdot \left(\frac{I_{1B}}{p_a}\right)^h \cdot e^{q_B} \tag{7}$$

where subscripts A and B denote the stresses at points A and B, respectively. If point A on the hydrostatic axis (where q = 0) and B on the failure envelope (where q = 1) are two points on the same plastic work contour, then

$$(27\psi_1+3)\left[\frac{I_{1A}}{p_a}\right]^h = \left[\psi_1 \cdot \frac{I_{1B}^3}{I_{3B}} - \frac{I_{1B}^2}{I_{2B}}\right]\left[\frac{I_{1B}}{p_a}\right]^h \cdot e \tag{8}$$

and therefore,

$$h = \log \frac{\left[\psi_1 \cdot \frac{I_{1B}^3}{I_{3B}} - \frac{I_{1B}^2}{I_{2B}}\right] \cdot e}{27\psi_1 + 3} \bigg/ \log \frac{I_{1A}}{I_{1B}} \tag{9}$$

in which e is the base of natural logarithms. Table 1 shows the calculation of h for Fine Silica Sand based on four sets of failure stresses (point B) and four corresponding points on the hydrostatic axis (point A). The stresses at the latter points are determined from Eq. (1) based on constant amounts of plastic work along each contour. Thus, the total plastic work at failure (listed in column 1) from a triaxial compression test is used in Eq. (1) to determine the corresponding isotropic state of stress. Table 1 shows that an average value of h = 0.355 is reasonably representative of the individual values determined from Eq. (9).

Variation of q

The value of q varies with stress level S defined as:

$$S = f_n/\eta_1 = \frac{1}{\eta_1} \cdot (I_1^3/I_3 - 27)\left[\frac{I_1}{p_a}\right]^m \tag{10}$$

in which f_n is the function used to express failure for frictional materials [4, 19-21] and η_1 is the value of f_n at failure. The stress level S varies from zero at the hydrostatic axis to unity at the failure surface.

The value of q can now be determined from the test data using Eqs. (2) and (4):

$$q = ln \frac{\left[\frac{W_p}{D \cdot p_a}\right]^{1/\rho}}{\left[\psi_1 \cdot \frac{I_1^3}{I_3} - \frac{I_1^2}{I_2}\right]\left[\frac{I_1}{p_a}\right]^h} \tag{11}$$

Fig. 9 shows the variation of q with stress level S for Fine Silica Sand. Only the portion in the range $0 \le q \le 1$ is of importance, since a different relation between yielding and plastic work is employed in the softening range (discussed blow). In the hardening region the variation may be described by a hyperbolic expression:

$$S = \frac{q}{\alpha + \beta \cdot q} \tag{12}$$

Since the curve passes through (1,1), the value of $\beta = 1 - \alpha$. The best determination of α is

22

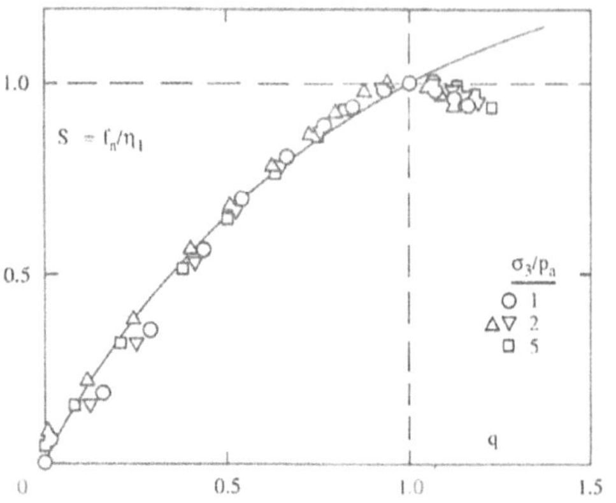

Fig. 9. Observed and Modeled Relation between S and q for Fine Silica Sand.

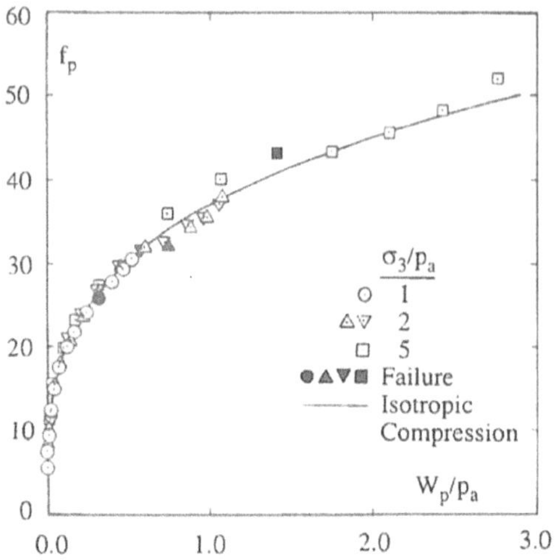

Fig. 10. Comparison of Analytical Hardening Curve Determined from Isotropic Compression Test and Experimental Data from Triaxial Compression Tests on Fine Silica Sand.

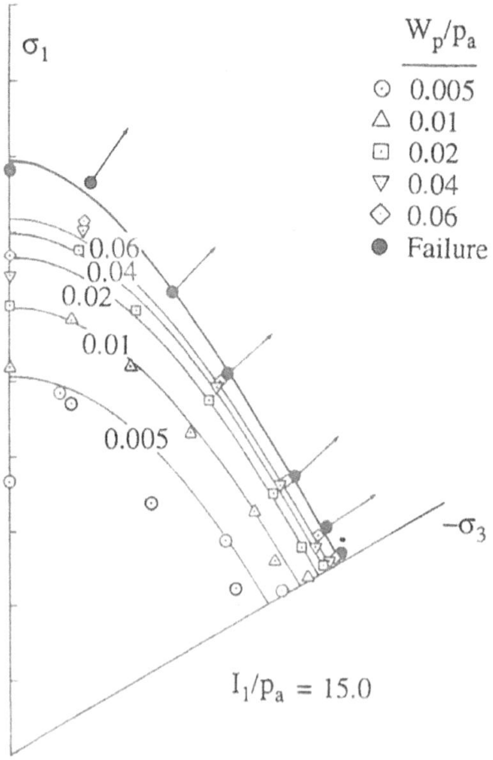

Fig. 11. Comparison of Analytical Contours of Plastic Work with Experimental Data Projected on Octahedral Plane for Dense Monterey No. 0 Sand (Data from [22]).

obtained at a stress level S = 0.80. Substitution in Eq. (12) produces

$$\alpha = \frac{1-S}{S} \cdot \frac{q_s}{1-q_s} = \frac{1}{4} \cdot \frac{q_{80}}{1-q_{80}} \tag{13}$$

For Fine Silica Sand $q_{80} = 0.67$ producing $\alpha = 0.515$. Finally, q is expressed as:

$$q = \frac{\alpha \cdot S}{1-(1-\alpha) \cdot S} \tag{14}$$

With determination of the parameters h and q, the yield function in Eq. (2) is defined and the plastic work during the yielding process can be determined from Eq. (4).

EVALUATION OF PROPOSED YIELD FUNCTION

Four parameters are required to determine the yield-plastic work relation: h and α to define the yield function, and C and p to define the plastic work equation for isotropic compression. The validity and flexibility of the proposed yield function and the plastic work relation can be examined in several independent ways as follows.

The plastic work contours described through the yield function in Eq. (2) have been compared with the experimental points on the triaxial planes in Figs. 5 and 6. The important aspects of the behavior of frictional materials in terms of the plastic work are well captured by the model, including the plastic work during isotropic compression, the slow increase in plastic work at low shear stress levels, and the pressure-dependent amount of plastic work at the peak stress state.

In the development of this model, the basic assumption is that, if yield surfaces are defined properly, the amount of plastic work during yielding may be defined uniquely. This assumption is examined quantitatively in Fig. 10 where the values of f_p' and W_p/p_a calculated from triaxial compression data are compared with those modeled through an isotropic compression test. It is seen that a unique, monotonic relation exists between yield and plastic work, and this relation is modeled through Eq. (4) with f_p' and W_p defined by Eqs. (2) and (1), respectively.

Plastic work under three-dimensional stress states is simulated through the parameter ψ_1 which depends on the rigidity of the material. The three-dimensional behavior is examined in Fig. 11 where experimental and predicted values of plastic work for the tests on Dense Monterey No. 0 Sand ($D_r = 98\%$) [22] are compared on the octahedral plane. Because the stress points do not lie on the same octahedral plane, the data points are projected onto a common octahedral plane ($I_1/p_a = 15$) following the meridians of the proposed yield function (intersection of a yield surface and a constant $b = (\sigma_2-\sigma_3)/(\sigma_1-\sigma_3)$ plane). That is, a projected stress point possesses the same value of f_p' as the original data point. A pattern similar to that in Fig. 7 for Fuji River Sand can be seen. Although relatively higher test values of plastic work appear under triaxial compression conditions, the observed and modeled values of plastic work correspond to each other within a reasonable range. It is worth mentioning that the experimental stress points lie on the octahedral planes whose I_1/p_a values vary from 1.8 to 13.3. Thus, discrepancies between observed and modeled values appear larger on the plane of $I_1/p_a = 15$ than the real discrepancies.

In order to check if the plastic work contours are equivalent to yield surfaces, an independent test was performed on Fine Silica Sand. A specimen was isotropically consolidated up to 10 kg/cm^2 and unloaded isotropically to 1 kg/cm^2. That is, the specimen was prepared with an overconsolidation ratio of 10. Then, triaxial compressive loading was applied up to failure while maintaining the cell pressure at 1 kg/cm^2. The resulting stress-strain curve is compared with that from a normally consolidated specimen in Fig. 12(a). The overconsolidated specimen exhibits a stress range of small deformation which may be considered to be mainly elastic. A yield stress of σ_y = 2.5 kg/cm^2 can be located using a technique similar to that reviewed in the beginning of this paper. This stress state is shown on the triaxial plane in Fig. 12(b) together with the yield surface or plastic work contour corresponding to the initial preconsolidation of 10 kg/cm^2 and described by Eq. (2). The yield stress σ_y lies exactly on the predicted yield surface.

Finally, the yield surfaces reported by Tatsuoka and Ishihara [13] have been simulated using the proposed yield function. The experimental yield surfaces for loose Fuji River Sand are replotted as broken curves in Fig. 13 while the family of curves described by Eq. (2) are plotted as continuous curves. Some disagreement can be observed between experimental and analytical surfaces close to the hydrostatic axis. However, the stress paths adopted by Tatsuoka and Ishihara [13] in this region are sensitive to volumetric strains while more emphasis was placed on the shear strains in their investigation. If yield points were located according to volumetric strains, the yield surfaces would have more curvature and be more perpendicular to the hydrostatic axis. This argument agrees with the observation of Tatsuoka and Ishihara [13] who noted that their yield surfaces are correct as long as yielding is defined in terms of shear deformation. Except for the stress region near the hydrostatic axis, the proposed yield surfaces agree with their observed yield surfaces. Using the plastic work contours as yield surfaces takes into account effects of shear as well as volumetric strains.

SOFTENING BEHAVIOR

The evaluation of the yield criterion was performed in the hardening regime where the soil behavior has been studied experimentally and is reasonably well-known. Much less is known about the soil behavior in the softening regime. However, soils are known to exhibit peak failure and reduction in strength while deforming uniformly under certain conditions including triaxial compression stress states. The actual movement and shape of the yield surface after peak failure have not been studied experimentally in the past.

The simplest possible behavior may be described as isotropic softening following the pattern indicated on Fig. 14. Softening may begin anywhere along the monotonically increasing hardening curve and is initiated when q = 1, which in turn occurs when the stress level reaches unity (S=1) at the failure surface. The softening curve is described by an exponential decay function

$$f_p'' = A \cdot e^{-B \cdot (W_p/p_a)} \tag{15}$$

in which A and B are positive constants. These are devised such that no additional material parameters are required. This seems reasonable in view of the limited knowledge about the actual softening behavior. However, to capture the experimentally observed pattern of greater relative strength decay at lower confining pressures, the slope of the softening curve is set equal to the negative slope of the hardening curve at the point of failure. Since the softening curve also intersects the hardening curve at the peak point, the following expressions for A and B can be derived:

25

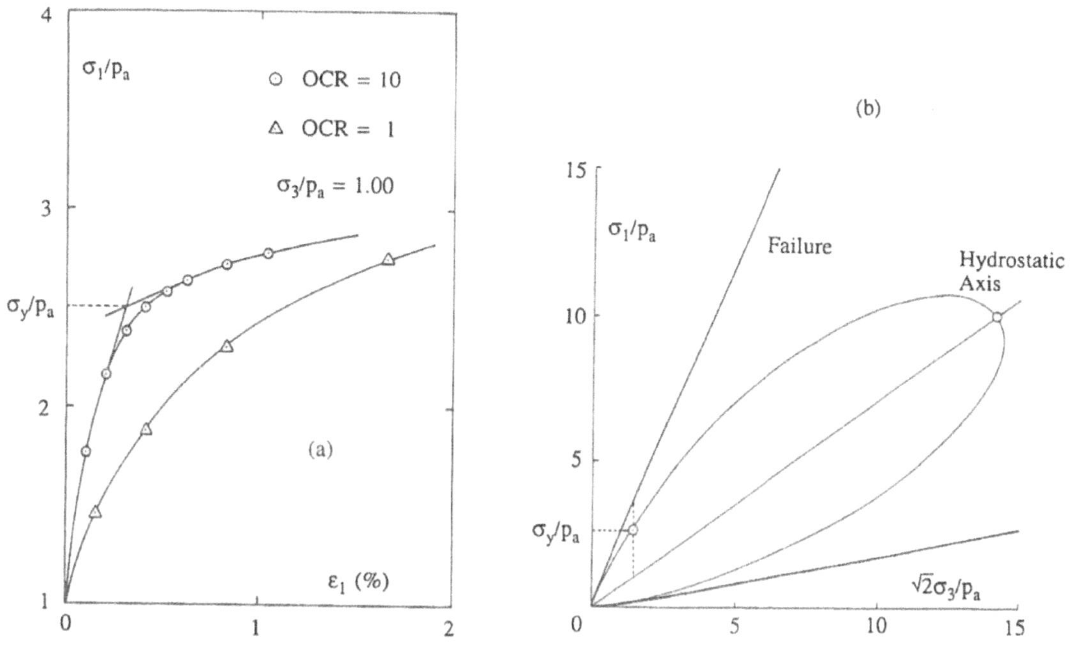

Fig. 12. (a) Stress-Strain Curves from Triaxial Compression Tests on Normally Consolidated and Overconsolidated (OCR = 10) Specimens of Fine Silica Sand, and (b) Comparison of Yield Stress (σ_y/p_a) with Yield Surface Produced by Overconsolidation (OCR = 10) of Fine Silica Sand.

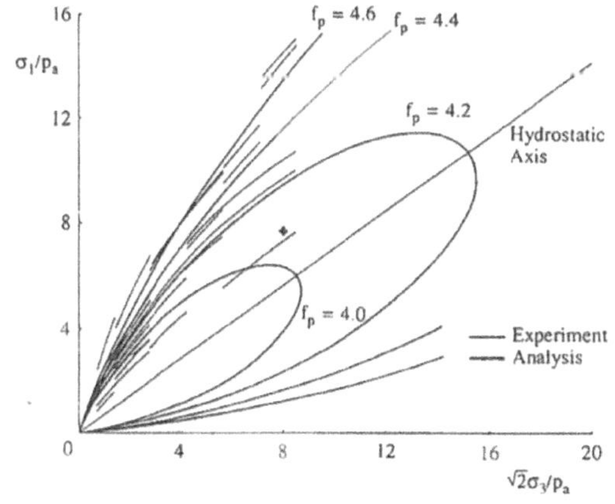

Fig. 13. Comparison of Analytical and Observed Yield Surfaces for Loose Fuji River Sand (Data from [13]).

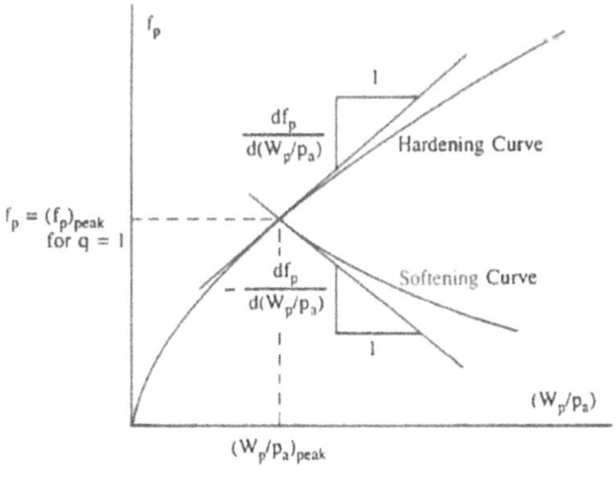

Fig. 14. Modeling of Work-Hardening and -Softening.

$$A = (f_p')_{peak} \cdot e^{B \cdot (W_p/p_a)_{peak}} \tag{16}$$

$$B = \left[\frac{df_p''}{d(W_p/p_a)}\right]_{hard, peak} \cdot \frac{1}{(f_p')_{peak}} \tag{17}$$

The transition from hardening to softening occurs abruptly at the peak failure point. Thus, there are no points with zero slope at the transition anywhere along the hardening curve. The absence of points with zero slope allows greater computational efficiency near peak failure, while the pointed peak is hardly noticeable in actual comparisons with experimental data.

MATERIALS WITH EFFECTIVE COHESION

The proposed yield function and hardening equation may also be employed for materials with effective cohesion. This requires a translation of the principal stress space as demonstrated for the failure criterion [19-21] and for the plastic potential function [18].

If the yield and plastic potential surfaces are treated in the same way as the failure surface, then inconsistencies in the condition of irreversibility may arise. This was discussed in detail in the companion paper [18]. Further inconsistencies in the plastic behavior of materials with effective cohesion are overcome by assuming that there exists an initial yield surface which goes through the origin of the real stress space as shown in Fig. 15. This implies that only elastic strains occur during major portions of tension type tests in which the main parts of the stress-paths are located inside the yield surface. Fig. 16 shows the stress-strain curves obtained from three types of biaxial tension tests on concrete by Kupfer et al. [23]. The strains observed in these tests are almost linear elastic. Small deviations from linear elastic behavior occur prior to the abrupt tensile failure experienced in these tests. Thus, the stresses applied along the stress-paths indicated in Fig. 15 do not produce plastic strains until they are close to failure. The existence of an initial yield surface therefore appears to simulate experimental observations of the type shown in Fig. 16 with good accuracy.

It is therefore assumed that the translated stresses represent the stress components in the development of the constitutive model for frictional materials with effective cohesion. Then, the translated plastic work may be defined as the work done by the translated stresses through the process of plastic deformation:

$$\overline{W}_p = \int \bar{\sigma}_{ij} d\varepsilon_{ij}^p \tag{18}$$

and substitution of the translated stresses produces

$$d\overline{W}_p = (\sigma_{ij} + a \cdot p_a \cdot \delta_{ij})\, d\varepsilon_{ij}^p = dW_p + a \cdot p_a \cdot d\varepsilon_v^p \tag{19}$$

Physically, it has not been shown that such work exists, but this approach provides a convenient tool for mathematical treatment of cohesive materials. Eq. (19) implies that work is done by an imaginary preload '$a \cdot p_a$' through the loading process. Since the plastic work is defined in terms of translated stresses, the yield function and the hardening equation can be generalized by substituting the translated stresses in place of the real stresses. For cohesionless materials for which 'a' vanish, all terms of the theory return to hold conventional physical meaning.

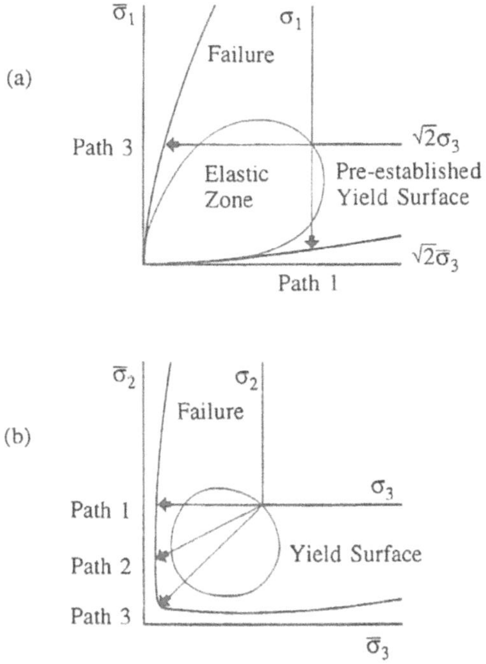

Fig. 15. Biaxial Stress Paths for Concrete Employed by Kupfer et al. [23] and Assumed Initial Yield Surface in (a) Triaxial Plane, and (b) Biaxial Plane.

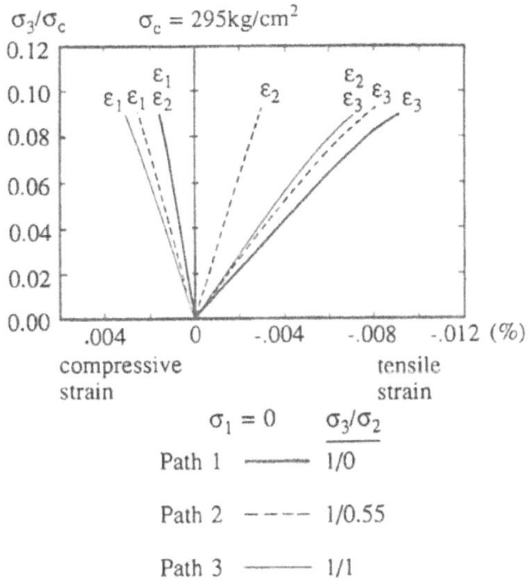

Fig. 16. Stress-Strain Relations Obtained from Biaxial Tension Tests on Concrete by Kupfer et al. [23].

CONCLUSION

The yield behavior of frictional materials has been examined phenomenologically as well as mathematically. A yield function is proposed as a result of the consideration that yield surfaces are equivalent to plastic work contours. Work hardening and softening behavior can be simulated through the formulation of plasticity proposed herein. Various independent checks are made to examine the validity of the formulation. The formulation is also extended to cohesive materials using translation of stresses. Consistency with mathematical work-hardening plasticity is retained.

ACKNOWLEDGMENT

The basic developmental work presented here was supported by the National Science Foundation under Grant No. CEE 8211159. Additional developments and computational refinements were supported by Exxon Production Research Company, Houston, Texas, and by California Research and Technology, Chatsworth, California. Grateful appreciation is expressed for their support of this research.

28

REFERENCES

1. Drucker, D.C., Gibson, R.E. and Henkel, D.J., Soil Mechanics and Work-Hardening Theories of Plasticity. ASCE Transactions, 122 (1957) 338-346.

2. Roscoe, K.H. and Burland, J.B., On the Generalized Stress-Strain Behaviour of 'Wet' Clay. In: Engineering Plasticity, Ed. by J. Heymann and F.A. Leckie, Cambridge University Press, Cambridge (1968) 535-609.

3. DiMaggio, F.L. and Sandler, I.S., Material Model for Granular Soils. J. Engr. Mech. Div., ASCE, 97 (1971) 935-950.

4. Lade, P.V., Elasto-Plastic Stress-Strain Theory for Cohesionless Soil with Curved Yield Surfaces. Int. J. Solids Struct., 13 (1977) 1019-1035.

5. Zienkiewicz, O.C. and Pande, G.N., Some Useful Forms of Isotropic Yield Surfaces for Soil and Rock Mechanics. In: Finite Elements in Geomechanics, Ed. by G. Gudehus, Wiley, New York (1977) 179-190.

6. Prevost, J.-H., Plasticity Theory for Soil Stress-Strain Behavior. J. Engr. Mech. Div., ASCE, 104 (1978) 1177-1194.

7. Dafalias, Y.F. and Herrmann, L.R., A Bounding Surface Soil Plasticity Model. Proc. Int. Symp. Soils under Cyclic and Transient Loading, Balkema, Rotterdam (1980) 335-345.

8. Mroz, Z., Norris, V.A. and Zienkiewicz, O.C., An Anisotropic, Critical State Model for Soils Subject to Cyclic Loading. Geotechnique, 31 (1981) 451-469.

9. Ghaboussi, J. and Momen, H., Modelling and Analysis of Cyclic Behavior of Sands. In: Soil Mechanics – Transient and Cyclic Loads, Ed. by G.N. Pande and O.C. Zienkiewicz, Wiley, London (1982) 313-342.

10. Pande, G.N. and Pietruszak, S., Reflecting Surface Model for Soils. Proc. Int. Symp. Numerical Models in Geomechanics, Balkema, Rotterdam (1982) 50-64.

11. Poorooshasb, H.B. and Pietruszak, S. On Yielding and Flow of Sand; A Generalized Two-Surface Model. Computers and Geotechnics, 1 (1985) 33-58.

12. Frantziskonis, G., Desai, C.S. and Somasundaram, S., Constitutive Model for Nonassociate Behavior. J. Engr. Mech., ASCE, 112 (1986) 932-946.

13. Tatsuoka, F. and Ishihara, K., Yielding of Sand in Triaxial Compression. Soils and Foundations, JSSMFE. 14 (1974) 63-76.

14. Parry, R.H.G. and Nadarajah, V., A Volumetric Yield Locus for Lightly Overconsolidated Clay. Geotechnique, 23 (1973) 450-453.

15. Huang, W., Effect of Work Hardening Rules on the Elasto-Plastic Matrix. Proc. Int. Symp. on Soils under Cyclic and Transient Loading, Balkema, Rotterdam (1980) 277-282.

16. Yamada, Y. and Ishihara, K., Anisotropic Deformation Characteristics of Sand under Three Dimensional Stress Conditions. Soils and Foundations, JSSMFE, 19 (1979) 79-94.

17. Lewin, P.I., Yamada, Y. and Ishihara, K. Correlating Drained and Undrained 3D Tests on Loose Sand. IUTAM Conf. on Deformation and Failure of Granular Materials, Delft (1982) 419-429.

18. Kim, M.K. and Lade, P.V., Single Hardening Constitutive Model for Frictional Materials, I. Plastic Potential Function. Computers and Geotechnics

19. Lade, P.V., Three-Parameter Failure Criterion for Concrete. J. Engr. Mech. Div., ASCE, 108 (1982) 850-863.

20 Lade, P.V., Fiailure Criterion for Frictional Materials. Chapt. 20 in Mechanics of Engineering Material, Ed. by C.S. Desai and R.H. Gallagher, Wiley (1984) 385-402.

21. Kim, M.K. and Lade, P.V., Modelling Rock Strength in Three Dimensions. Int. J. Rock Mech. Min. Sci. & Geomech. Abstr., 21 (1984) 21-33.

22. Lade, P.V., The Stress-Strain and Strength Characteristics of Cohesionless Soils. Ph.D. Dissertation, University of California, Berkeley (1972).

23. Kupfer, H., Hilsdorf, H.K. and Rusch, H., Behavior of Concrete Under Biaxial Stresses. J. Am. Concr. Inst., 66 (1969) 656-666.

Received 10 June 1988; revised version received and accepted 2 August 1988

Computers and Geotechnics 6 (1988) 31–47

SINGLE HARDENING CONSTITUTIVE MODEL FOR FRICTIONAL MATERIALS

III. COMPARISONS WITH EXPERIMENTAL DATA

P.V. Lade
Department of Civil Engineering
University of California
Los Angeles, California 90024,U.S.A.

and

M.K. Kim
Department of Civil Engineering
Yonsei University
Seoul 120, Korea

ABSTRACT

A unified constitutive model for the behavior of frictional materials is described. The model is based on concepts from elasticity and plasticity theories. In addition to Hooke's law for the elastic behavior, the framework for the plastic behavior consists of a failure criterion, a nonassociated flow rule, a yield criterion that describes contours of equal plastic work, and a work-hardening/softening law. The functions that describe these components are all expressed in terms of stress invariants. The model incorporates twelve parameters which can all be determined from simple experiments such as isotropic compression and conventional triaxial compression tests. Validation of the model is achieved by comparison of predicted and measured stress-strain curves for various two- and three-dimensional stress-paths obtained for different types of frictional materials.

INTRODUCTION

Development of constitutive models for the time-independent behavior of frictional materials has most often been based on experimental observations of material behavior and on a framework of elasticity and plasticity theories. The combination of experiment and theory has been employed with varying emphasis depending on access to experimental results, interpretation of physical concepts, and inclination towards mathematical developments. A number of models have therefore resulted with various capabilities relative to observed material behavior. The models contained in references [1-11] are representative of the types and capabilities presented in the literature.

Computers and Geotechnics 0266-352X/88/$03·50

The constitutive model presented here has been developed on the basis of a thorough review and evaluation of data from experiments on frictional materials such as sand, clay, concrete and rock. The framework for the evaluation and subsequent development consists of concepts contained in elasticity and work-hardening plasticity theories.

An important feature of the new model is the application of a single isotopic, yield surface shaped as an asymmetric tear-drop with the pointed apex at the origin of the principal stress space. This yield surface, expressed in terms of stress invariants, describes the locus at which the total plastic work is constant. The total plastic work (due to shear strains as well as volumetric strains) serves as the hardening parameter, and it is used to define the location and shape of the yield surfaces. This concept results in mathematical consistency in the model, and application of a single yield surface produces computational efficiency when used in large computer programs. The nonassociated flow rule is derived from a potential function which describes a three-dimensional surface shaped as a cigar with an asymmetric cross-section.

The model is devised such that the transition from hardening to softening occurs abruptly at the peak failure point. Thus, the transition does not involve any points at which the hardening modulus is zero. The absence of points with zero slope allows greater computational efficiency near peak failure, while the pointed peak is hardly noticeable in actual comparisons with experimental data.

Each component of the model has already been developed, presented, and discussed in the literature [12-19]. Only the main principles and the governing equations for each component will be given here. Values of material parameters are given for different frictional materials, and the stress-strain behavior of various types of laboratory tests is predicted by the constitutive model.

COMPONENTS OF CONSTITUTIVE MODEL

The total strain increments observed in a material when loaded are divided into elastic and plastic components such that

$$d\varepsilon_{ij} = d\varepsilon_{ij}^{e} + d\varepsilon_{ij}^{p} \tag{1}$$

These strains are then calculated separately, the elastic strains by Hooke's law, and the plastic strains by a plastic stress-strain law. Both are expressed in terms of effective stresses.

Below is a brief review of the framework and the components of the constitutive model. In order that the presentation follows a logic developmental sequence, the components are presented in the following order: Elastic behavior, failure criterion, flow rule, yield criterion and work-hardening/softening law.

Elastic Behavior

The elastic strain increments, which are recoverable upon unloading, are calculated from Hooke's law, using a model for the unloading-reloading modulus defined as [12]:

$$E_{ur} = K_{ur} \cdot p_a \cdot \left(\frac{\sigma_3}{p_a}\right)^m \tag{2}$$

The dimensionless, constant values of the modulus number, K_{ur}, and the exponent, n, are determined from triaxial compression tests performed with various values of the confining pressure, σ_3. In Eq. (2) p_a is atmospheric pressure expressed in the same units as E_{ur} and σ_3.

The value of Poisson's ratio has often been found to be close to 0.2 for the elastic parts of unloading-reloading stress-paths. This value is therefore used in the following calculations.

More recently a model for the nonlinear variation of Young's modulus with stress state has been developed [13, 14]. In this model, the value of Poisson's ratio, being limited between zero and one half for most materials, was assumed to be constant. The expression for Young's modulus was then derived from theoretical considerations based on the principle of conservation of energy. According to this derivation, Young's modulus can be expressed in terms of a power function involving nondimensional material constants and stress invariants. Although this more recent model could just as well have been used here, the parameters determined for the plasticity functions presented below were derived on the basis of the elastic strains being calculated using Eq. (2) for the elastic modulus.

Failure Criterion

A general, three-dimensional failure criterion has been developed for soils, concrete, and rock [4, 15-17]. The criterion is expressed in terms of the first and the third stress invariants of the stress tensor as follows:

$$\left(\frac{I_1^3}{I_3} - 27\right)\left(\frac{I_1}{p_a}\right)^m = \eta_1 \tag{3}$$

in which

$$I_1 = \sigma_x + \sigma_y + \sigma_z \tag{4}$$

and

$$I_3 = \sigma_x \cdot \sigma_y \cdot \sigma_z + \tau_{xy} \cdot \tau_{yz} \cdot \tau_{zx} + \tau_{yx} \cdot \tau_{zy} \cdot \tau_{xz} - (\sigma_x \cdot \tau_{yz} \cdot \tau_{zy} + \sigma_y \cdot \tau_{zx} \cdot \tau_{xz} + \sigma_z \cdot \tau_{xy} \cdot \tau_{yx}) \tag{5}$$

The parameters η_1 and m are constant dimensionless numbers.

In principal stress space, the failure surface is shaped like an asymmetric bullet with the pointed apex at the origin of the stress axes, as shown in Fig. 1. The apex angle increases with the value of η_1. The failure surface is always concave towards the hydrostatic axis, and its curvature increases with the value of m. For $m = 0$ the failure surface is straight, and if $m > 1.979$ the failure surface becomes convex towards the hydrostatic axis [17]. Analysis of numerous sets of data for concrete and rock indicates that m-values rarely exceed 1.5 [16]. For constant value of m and increasing η_1-values, the cross-sectional shape in the octahedral plane changes from circular to triangular with smoothly rounded edges in a fashion that conforms to experimental evidence. The shape of these cross-sections do not change with the value of I_1 when $m = 0$. for $m > 0$, the cross-sectional shape of the failure surface changes from triangular to become more circular with increasing value of I_1. Similar changes in cross-sectional shape are observed from experimental studies on soil, concrete, and rock.

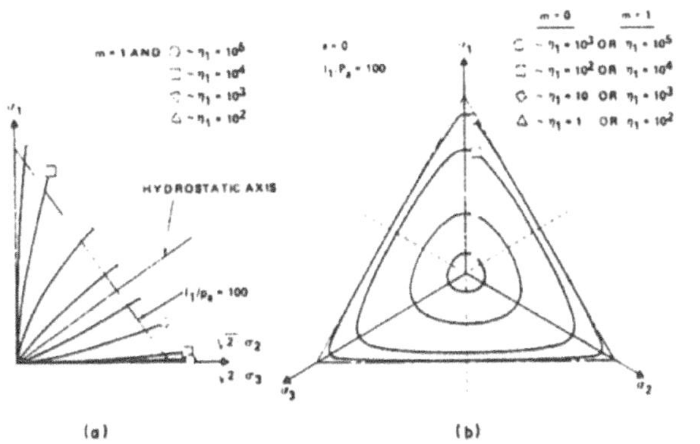

Fig. 1. Characteristics of Failure Surfaces Shown in Principal Stress Space. Traces of Failure Surfaces Shown in (a) Triaxial Plane, and (b) Octahedral Plane.

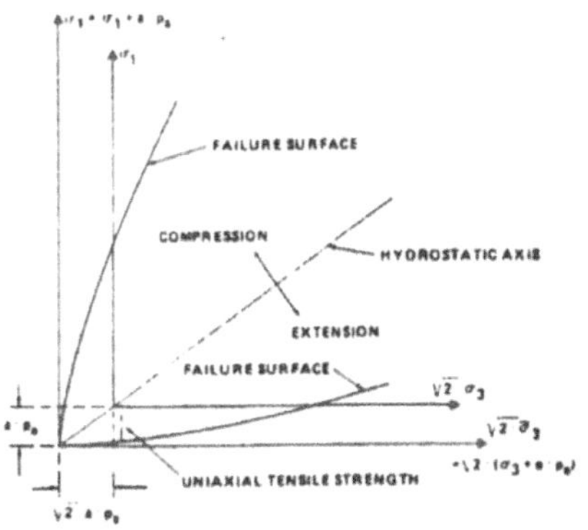

Fig. 2. Translation of Principal Stress Space along Hydrostatic Axis to Include Effect of Tensile Strength in Failure Criterion.

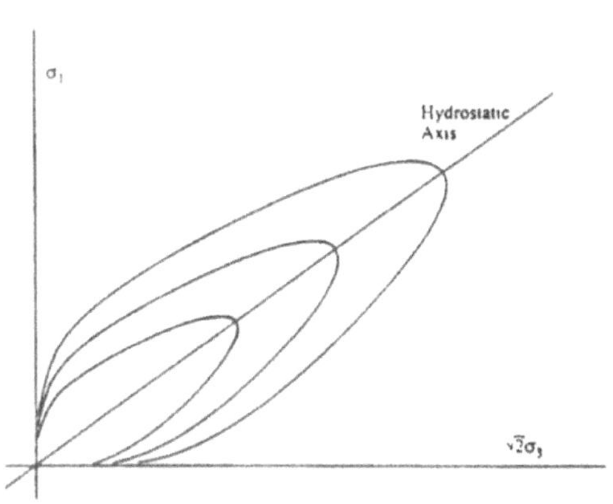

Fig. 3. Plastic Potential Surfaces Shown in Triaxial Plane.

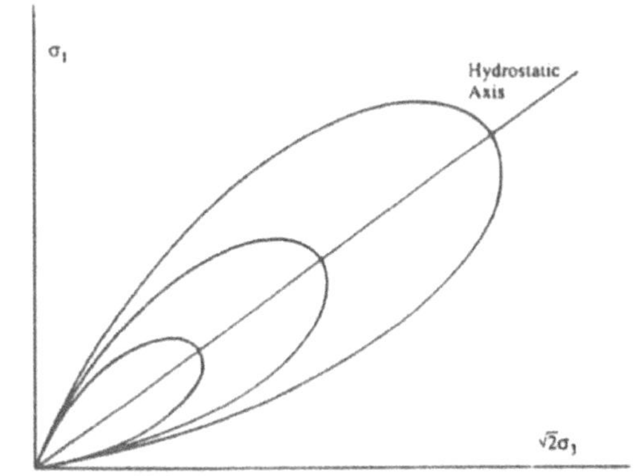

Fig. 4. Yield Surfaces Shown in Triaxial Plane.

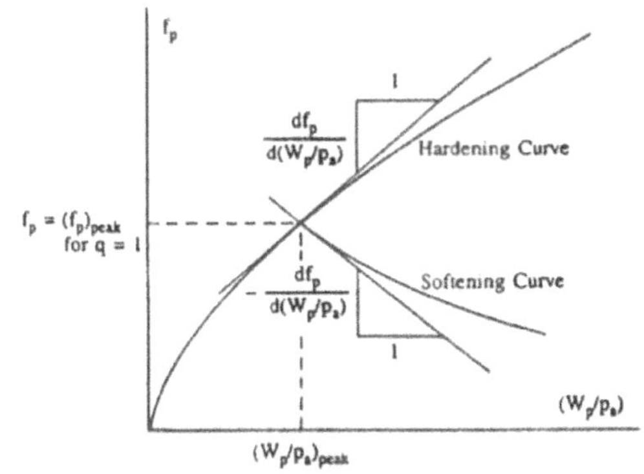

Fig. 5. Modeling of Work-Hardening and Softening.

In order to include the effective cohesion and the tension which can be sustained by concrete and rock, a translation of the principal stress space along the hydrostatic axis is performed as shown in Fig. 2. Thus, a constant stress, $a \cdot p_a$, is added to the normal stresses before substitution in Eq. (3):

$$\bar{\sigma}_{ij} = \sigma_{ij} + \delta_{ij} \cdot a \cdot p_a \tag{6}$$

in which 'a' is a dimensionless parameter, and δ_{ij} is Kronecker's delta (= 1 for i=j, and = 0 for $i \neq j$). The value of $a \cdot p_a$ reflects the effect of the tensile strength of the material. Although the three material parameters describe separate characteristics of the failure surface, they do interact in calculation of, e.g., the uniaxial compressive strength. Thus, an infinite number of combinations of η_1, m, and 'a' could results in the same value of the uniaxial compressive strength.

The three material parameters may be determined from results of simple tests such as triaxial compression tests. Further details on development, description and parameter evaluation for the failure criterion for frictional materials are given in [4,15-17].

Flow Rule

The plastic strain increments are calculated from the flow rule:

$$d\varepsilon_{ij}^{p} = d\lambda_p \cdot \frac{\partial g_p}{\partial \sigma_{ij}} \tag{7}$$

in which g_p is a plastic potential function and $d\lambda_p$ is a proportionality constant. A suitable plastic potential function for frictional materials was developed in a companion paper [18]. This function is different from the yield function and nonassociated flow is consequently obtained. The plastic potential function is written in terms of the three stress invariants of the stress tensor:

$$g_p = \left(\psi_1 \cdot \frac{I_1^3}{I_3} - \frac{I_1^2}{I_2} + \psi_2\right) \cdot \left(\frac{I_1}{p_a}\right)^{\mu} \tag{8}$$

in which I_1 and I_3 are given by Eqs. (4) and (5) and the second stress invariant is defined as:

$$I_2 = \tau_{xy} \cdot \tau_{yx} + \tau_{yz} \cdot \tau_{zy} + \tau_{zx} \cdot \tau_{xz} - (\sigma_x \cdot \sigma_y + \sigma_y \cdot \sigma_z + \sigma_z \cdot \sigma_x) \tag{9}$$

The material parameters ψ_2 and μ are dimensionless constants that may be determined from triaxial compression tests. The parameter ψ_1 is related to the curvature parameter m for the failure criterion as indicated in [18], and it acts as a weighting factor between the triangular shape (from the I_3 term) and the circular shape (from the I_2 term). The parameter ψ_2 controls the intersection with the hydrostatic axis, and the exponent μ determines the curvature of meridians. The corresponding plastic potential surfaces are shown in Fig. 3. They are shaped as asymmetric cigars with smoothly rounded triangular cross-sections similar but not identical to those for the failure surfaces shown in Fig. 1.

The derivatives of g_p with regard to the stresses are

$$\frac{\partial g_p}{\partial \sigma_{ij}} = \left(\frac{I_1}{p_a}\right)^{\mu} \left\{ \begin{array}{l} G - (\sigma_y + \sigma_z) \cdot \dfrac{I_1^2}{I_2^2} - \psi_1 \cdot (\sigma_y \cdot \sigma_z - \tau_{yz}^2) \cdot \dfrac{I_1^3}{I_3^2} \\ G - (\sigma_z + \sigma_x) \cdot \dfrac{I_1^2}{I_2^2} - \psi_1 \cdot (\sigma_z \cdot \sigma_x - \tau_{zx}^2) \cdot \dfrac{I_1^3}{I_3^2} \\ G - (\sigma_x + \sigma_y) \cdot \dfrac{I_1^2}{I_2^2} - \psi_1 \cdot (\sigma_x \cdot \sigma_y - \tau_{xy}^2) \cdot \dfrac{I_1^3}{I_3^2} \\ 2 \cdot \dfrac{I_1^2}{I_2^2} \cdot \tau_{yz} - 2 \cdot \psi_1 \cdot (\tau_{xy} \cdot \tau_{zx} - \sigma_x \cdot \tau_{yz}) \cdot \dfrac{I_1^3}{I_3^2} \\ 2 \cdot \dfrac{I_1^2}{I_2^2} \cdot \tau_{zx} - 2 \cdot \psi_1 \cdot (\tau_{xy} \cdot \tau_{yz} - \sigma_y \cdot \tau_{zx}) \cdot \dfrac{I_1^3}{I_3^2} \\ 2 \cdot \dfrac{I_1^2}{I_2^2} \cdot \tau_{xy} - 2 \cdot \psi_1 \cdot (\tau_{yz} \cdot \tau_{zx} - \sigma_z \cdot \tau_{xy}) \cdot \dfrac{I_1^3}{I_3^2} \end{array} \right\} \tag{10a-f}$$

where

$$G = \psi_1\ (\mu+3)\ \frac{I_1^2}{I_3} - (\mu+2)\ \frac{I_1}{I_2} + \frac{\mu}{I_1}\ \psi_2 \tag{11}$$

These derivatives are used to obtain the plastic strain increments from equation (7).

Yield Criterion and Work Hardening/Softening Law

The yield surfaces are intimately associated with and derived from surfaces of constant plastic work, as explained in a companion paper [19]. The isotropic yield function is expressed as follows:

$$f_p = f_p'(\sigma) - f_p''(W_p) = 0 \tag{12}$$

in which

$$f_p' = \left(\psi_1 \cdot \frac{I_1^3}{I_3} - \frac{I_1^2}{I_2}\right) \cdot \left(\frac{I_1}{p_a}\right)^h \cdot e^q \tag{13}$$

and for *hardening*:

$$f_p'' = \left[\frac{1}{D}\right]^{1/\rho} \cdot \left[\frac{W_p}{p_a}\right]^{1/\rho} \tag{14}$$

The expressions for I_1, I_2, and I_3 are given in Eqs. (4), (9), and (5), respectively. The parameter ψ_1 acts as a weighting factor between the triangular shape (from the I_3 term) and the circular shape (from the I_2 term), as in the expression for the plastic potential (Eq. (8)). The parameter h is constant for a given material and is determined on the basis that the plastic work is constant along a yield surface. The parameter q varies with stress level S from zero at the hydrostatic axis to unity at the failure surface according to the following expression:

$$q = \frac{\alpha \cdot S}{1-(1-\alpha)\cdot S} \tag{15}$$

Determination of the constant parameters h and α is explained in the companion paper [19].

In Eq. (14) the values of D and ρ are constants for a given material. Thus, f_p'' varies with the plastic work only. The value of α depends on p and h, and D is a function of C, ψ_1, and ρ, as explained in [19].

The yield surfaces are shaped as asymmetric tear drops with smoothly rounded triangular cross-sections and traces in the triaxial plane as shown in Fig. 4. As the plastic work increases, the isotropic yield surface inflates until the current stress point reaches the failure surface. The relation between f_p and W_p is described by a monotonically increasing function whose slope decreases with increasing plastic work, as shown in Fig. 5.

For *softening* the yield surface deflates isotropically according to an exponential decay function:

$$f_p'' = A \cdot e^{-B \cdot (W_p/p_a)} \tag{16}$$

in which A and B are positive constants to be determined on the basis of the slope of the hardening curve at the point of peak failure, as indicated in Fig. 5 and explained in the companion paper [19].

Using the expression for the plastic potential in Eq. (8), the relation between plastic work increment and the proportionality constant $d\lambda_p$ in Eq. (7) may be expressed as [18]:

$$d\lambda_p = \frac{dW_p}{\mu \cdot g_p} \tag{17}$$

in which the increment of plastic work can be determined by differentiation of the hardening and softening equations. For *hardening*, Eq. (14) yields:

$$dW_p = D \cdot p_a \cdot \rho \cdot f_p^{\rho-1} \cdot df_p \tag{18}$$

and for *softening* Eq. (16) produces:

$$dW_p = -\left(\frac{1}{B}\right) \cdot p_a \cdot f_p^{-1} \cdot df_p \tag{19}$$

in which df_p is negative during softening.

Combining Eqs. (18) and (19) with Eq. (17) and substituting this and Eqs. (10a-f) into Eq. (7) produces the expression for the incremental plastic strain increments.

MATERIALS WITH EFFECTIVE COHESION

As explained in connection with the failure criterion, it is possible to include the effective cohesion and the tension which can be sustained by concrete and rock. This is done by translating the principal stress space along the hydrostatic axis, as shown in Fig. 2, i.e. by adding a constant stress to the normal stresses, as in Eq. (6), before substitution into the failure criterion in Eq. (3).

A similar technique has been shown to work for the elastic modulus variation [13], the plastic potential [18], the yield criterion and the work-hardening/softening law [19]. The model presented here is therefore fully applicable to materials with effective cohesion such as concrete and rock.

PARAMETER VALUES FOR FRICTIONAL MATERIALS

The required number of parameters to be determined for a given material is twelve:

Elastic Moduli	ν, K_{ur}, n
Failure Criterion	η_1, m, a
Plastic Potential	ψ_2, μ
Yield Criterion	h, α
Hardening Function	C, p

For cohesionless materials the value of a = 0. All other parameters have values different from zero for most frictional materials. Table 1 summarizes the parameter values determined for various types of sand, clay, and concrete. These parameters were employed in the predictions presented below.

PREDICTIONS FOR SOILS AND CONCRETE

In the process of developing the model, individual components of the observed behavior of some frictional materials were compared to the model behavior, and it was shown that the components of the model were able to capture the observed behavior with good accuracy. Here the stress-strain behavior observed in various types of laboratory tests on frictional materials will be predicted. Thus, all model components must work in concert to produce reasonable predictions.

Table 1. Material Parameters for Various Frictional Materials

Material	D_r (%)	Failure Criterion			Elastic Behavior			Plastic Potential		Hardening Fucntion		Yield Function		Reference
		a	m	η_1	K_{ur}	n	ν	ψ_2	μ	C	p	h	α	
Fine Silica Sand	30		0.1	24.7	1170	0.53	0.2	-3.69	2.26	0.324E-03	1.25	0.355	0.515	
Sacramento River Sand	100		0.23	80.	1460	0.47	0.2	-3.09	2.00	0.396E-04	1.82	0.765	0.229	[20]
	38		0.093	28.	890	0.34	0.2	-3.72	2.36	0.127E-03	1.65	0.534	0.794	[20]
Painted Rock Material	100		0.21	101.	891	0.51	0.2	-3.26	2.82	0.351E-03	1.25	0.500	0.313	[21]
	70		0.16	67.	314	0.75	0.2	-3.39	2.72	0.460E-04	1.78	0.698	0.386	[21]
Crushed Napa Basalt	100		0.423	280.	470	0.58	0.2	-2.97	2.80	0.814E-04	1.61	0.546	0.727	[22]
	70		0.3	130.	481	0.36	0.2	-2.90	2.55	0.457E-03	1.39	0.542	0.851	[22]
Monterey No.0 Sand	98		0.16	104.	2300	0.80	0.2	-3.38	2.30	0.269E-04	1.44	0.490	0.896	[23]
	27		0.12	36.	1600	0.86	0.2	-3.60	2.50	0.214E-03	1.26	0.430	0.577	[23]
Edgar Kaolinite Clay			0.54	48.	110	1.23	0.25	-3.02	2.24	0.283E-02	1.48	0.886	0.594	
Plain Concrete		28.1	1.263	367200.	166000	0.	0.20	-2.77	2.51	0.350E-11	3.61	2.150	7.50	[24]
		28.5	1.113	159800.	361800	0.	0.18	-2.92	5.06	0.712E-12	3.80	1.990	0.75	[25]
		29.7	0.79	17370.	230100	0.	0.19	-3.02	4.88	0.934E-12	3.93	3.300	10.0	[26]
		28.2	1.231	336200.	321900	0.	0.14	-2.93	3.45	0.475E-09	2.93	1.666	1.65	[27]
Steel Fiber Reinforced Concrete		45.7	1.251	430100.	188000	0.	0.20	-2.58	2.63	0.252E-08	2.67	2.82	2.50	[28]

Isotropic Compression

Fig. 6 shows predictions of isotropic compression tests on sand and concrete. Loading and unloading is calculated for Fine Silica Sand in Fig. 6(a). Only elastic behavior is predicted for unloading. Both loading and unloading behavior (solid lines) is consistent with the observed behavior for this sand.

The isotropic compression behavior for plane concrete tested by Aschl et al. [27] is shown in Fig. 6(b). Insufficient data was available for this material to determine the elastic modulus variation. A constant elastic modulus was therefore used for prediction of the loading curve. The nonlinear behavior is caused by the plastic compression.

K_o-Compression

An example of sand behavior during K_o-compression is given in Fig. 7. In this test, the lateral strains are zero ($\varepsilon_2 = \varepsilon_3 = 0$) while the axial stress is increased or decreased. Only results for increasing stresses were available for dense Crushed Napa Basalt tested by Al-Hussaini [22]. The predicted behavior (solid lines) is seen to correspond very well with the measured behavior (points) for increasing stresses. The value of K_o increases with increasing stresses and this aspect of the soil behavior is accurately predicted by the model. Tests results were not available for decreasing stresses, but tests by others [29-31] indicate that they follow a pattern shown by the lines denoted "Decreasing Stresses". The directions of stress-path and stress-strain curve indicated by these lines are actually predicted by the model.

Triaxial Compression

Predictions for conventional drained triaxial compression tests on Fine silica Sand were calculated from the model and compared with experimental results in Fig. 8. In these tests, the confining pressure was held constant while the major principal stress was increased to peak failure and beyond. The model predictions capture the observed stress-strain relations accurately. The effects of confining pressure on the axial and the volumetric strains are also captured by the model. The breaks on the stress-strain curves at the peak are relatively shallow and hardly noticeable within the range of scatter of experimental results. Allowing for this shallow break enhances the computational efficiency of the model near failure.

Cubical Triaxial Tests

Particular attention has been paid to the three-dimensional characteristics of the stress-strain relationships. The proposed model can be examined through the predictions of cubical triaxial tests in which the three principal stresses are different form each other. The relative magnitude of the intermediate principal stress is characterized by the value of $b = (\sigma_2 - \sigma_3)/(\sigma_1 - \sigma_3)$, which is zero for triaxial compression and unity for triaxial extension.

The results of cubical triaxial tests on dense Monterey No. 0 Sand are shown in Fig. 9 together with model predictions. All model parameters were determined from triaxial compression tests as well as from an isotropic compression test. The pattern of increasing slope, decreasing strain to failure, and increasing rate of dilation with increasing b-value is captured well by the model. The strength first increases up to b = 0.50 and then decreases to b = 1.00 in both experiments and model predictions.

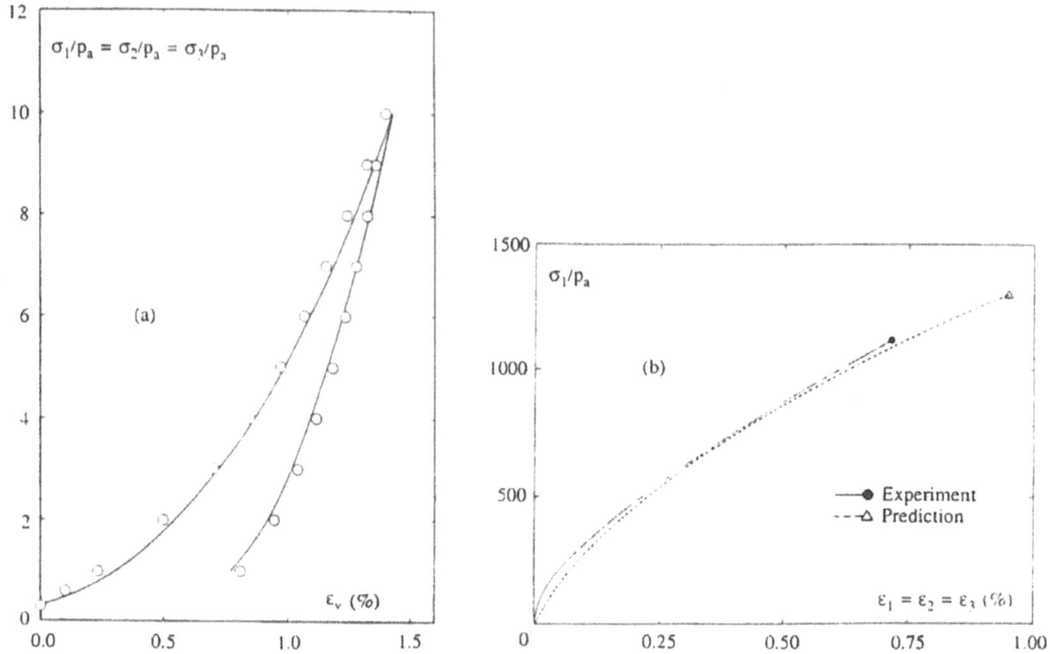

Fig. 6. Comparison of Predicted and Observed Stress-Strain Curves for Isotropic Compression Tests on (a) Fine Silica Sand, and (b) Plain Concrete (Data from [27]).

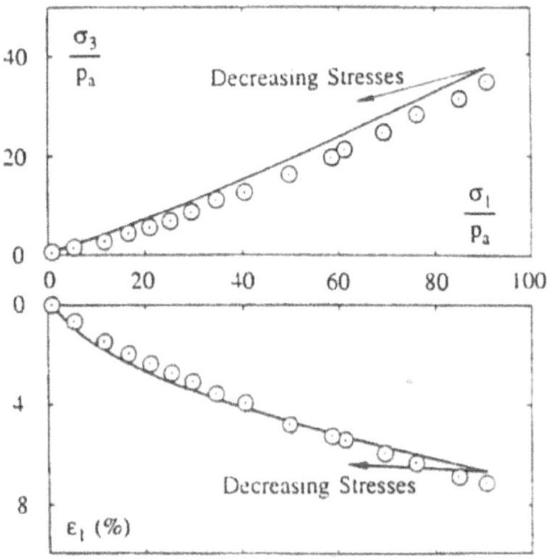

Fig. 7. Comparison of Predicted and Observed Behavior of Dense Crushed Napa Basalt in K_o-Test (Data from [22]).

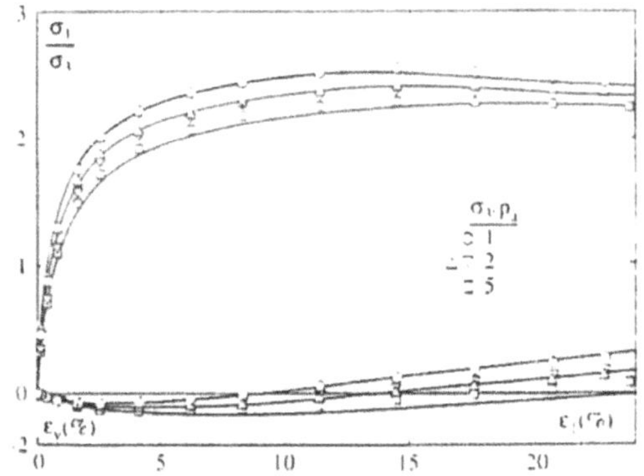

Fig. 8. Comparison of Predicted and Observed Behavior of Fine Silica Sand in Triaxial Compression Tests.

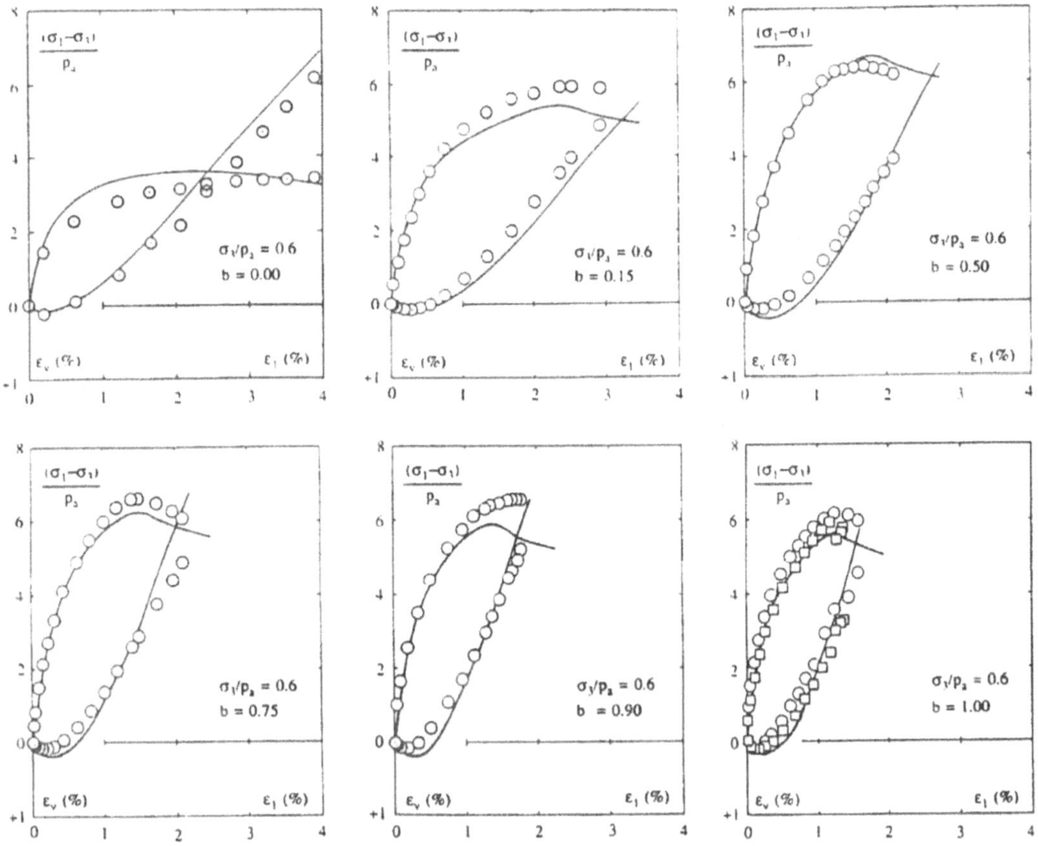

Fig. 9. Comparison of Predicted and Observed Stress-Strain and Volume Change Behavior of Dense Monterey No. 0 Sand in Cubical Triaxial Tests (Data from [23]).

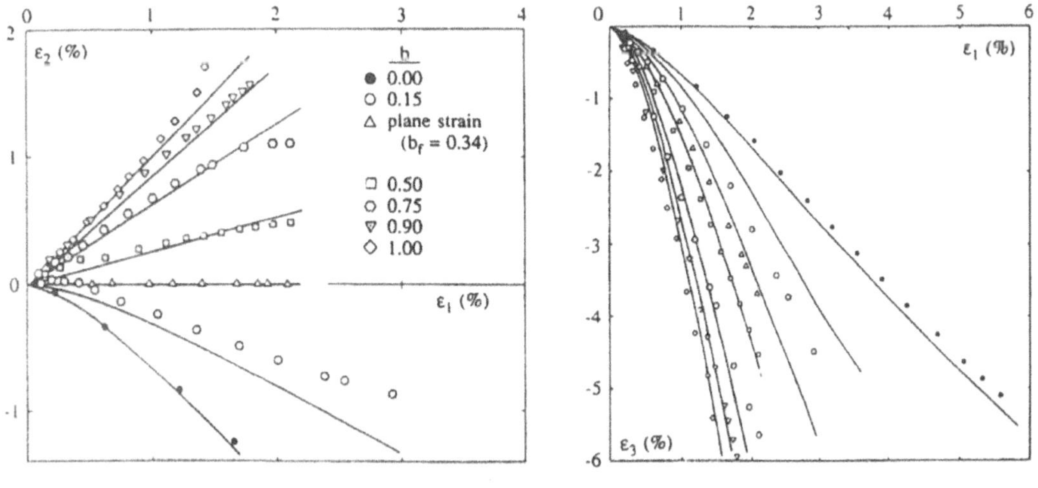

Fig. 10. Comparison of Predicted and Observed Strains in Dense Monterey No. 0 Sand in Cubical Triaxial Tests (Data from [23]).

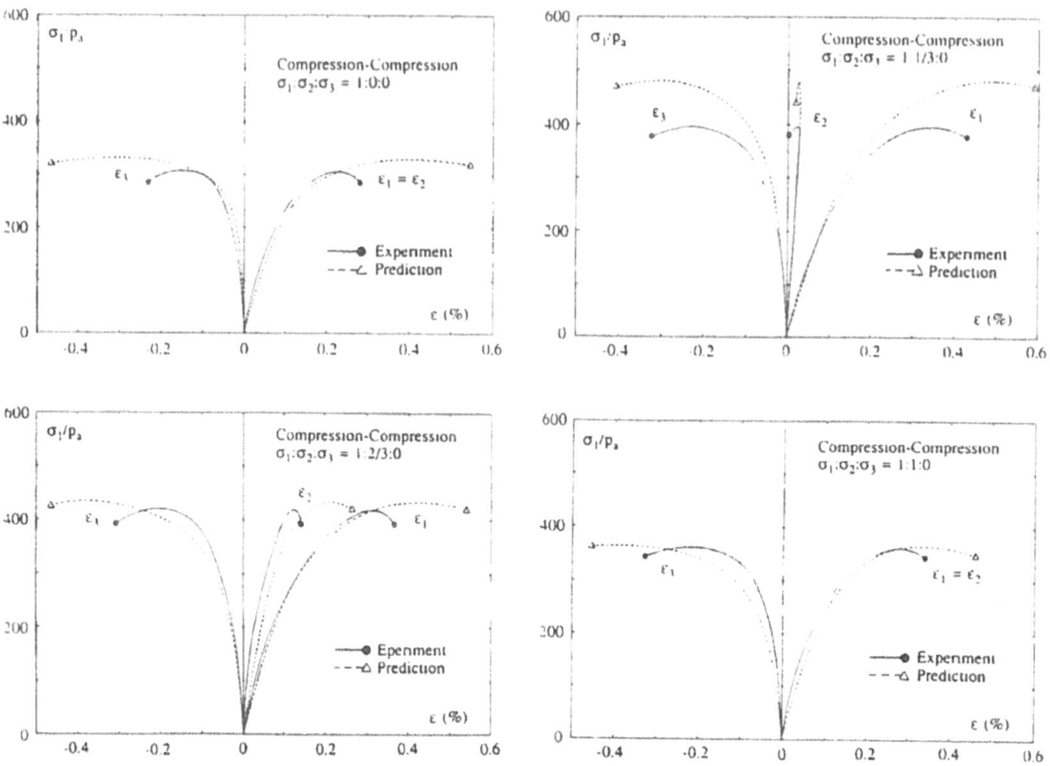

Fig. 11. Comparison of Predicted and Observed Stress-Strain Behavior of Plain Concrete in Biaxial Tests (Data from [27]).

Fig. 10 shows detailed comparisons of the principal strains. the predicted relations between the strains correspond well with the values measured in all the experiments.

Biaxial Tests

The model was also employed to predict the stress-strain behavior of concrete subjected to multiaxial loading. As explained above, for materials with effective cohesion the stress components in all stress functions correspond to translated stresses. The stresses are translated before any calculations are conducted, and the reverse translation is performed before the stresses are obtained for plotting.

The experimental results are compared with predictions for four biaxial tests on plane concrete in Fig. 11. The effect of the intermediate principal stress on the stress-strain curves are captured well in the model predictions. It may be noted that some disagreement appears after the peak stress conditions have been reached. However, measurement of stresses and strains in this stress range is less reliable than in the hardening range, because localized crushing or cracking of the concrete specimen may prevent accurate measurements. Thus, the model may represent the behavior of concrete quite well.

CONCLUSIONS

The behavior of frictional materials is such that a unified approach is possible in describing their failure surfaces as well as their stress-strain relations. A combination of elasticity theory and plasticity theory may provide a reasonable and consistent description of this behavior when appropriate mathematical functions are used in the framework provided by these theories. In addition to Hooke's law, the components in this framework are the failure criterion, the flow rule, the yield criterion and the work-hardening/softening law. Analyses of experimental data indicate that the shapes of the failure, yield, and plastic potential surfaces obtained from different types of frictional materials each have similar characteristics and may be modeled by similar mathematical functions. These functions are all expressible in terms of the three invariants of the stress tensor.

The nonassociated flow rule is derived from a plastic potential whose shape in the principal stress space resembles a cigar with an asymmetric cross-section. Yielding occurs along a single, isotropic yield surface shaped as an asymmetric tear drop. This yield surface describes a contour of constant plastic work. Because plastic yielding is related directly to plastic work contours, the description of the work-hardening/softening law is straight forward. The transition from hardening to softening occurs abruptly at the peak failure point. Thus, no points with a hardening modulus of zero are encountered, and this allows greater computational efficiency near peak failure.

The constitutive model combines all the governing functions studied individually in a mathematically consistent entity. The model requires twelve parameters which can be determined from simple laboratory tests such as isotropic compression and conventional triaxial compression tests.

Validation of the model was achieved by comparison of predicted and experimental stress-strain curves for various two- and three-dimensional stress-paths obtained for different types of frictional materials. Overall acceptable and accurate predictions were produced for soils as well as concrete for which experimental data were available.

ACKNOWLEDGMENT

The basic developmental work presented here was supported by the National Science Foundation under Grant No. CEE 8211159. Additional developments and computational refinements were supported by Exxon Production Research Company, Houston, Texas, and by California Research and Technology, Chatsworth, California. Grateful appreciation is expressed for their support of this research.

REFERENCES

1. Drucker, D.C., Gibson, R.E. and Henkel, D.J., Soil Mechanics and Work-Hardening Theories of Plasticity. ASCE Transactions, 122 (1957) 338-346.

2. Roscoe, K.H. and Burland, J.B.,
On the Generalized Stress-Strain Behaviour of 'Wet' Clay.
In: Engineering Plasticity, Ed. by J. Heymann and F.A. Leckie, Cambridge University Press, Cambridge (1968) 535-609.

3. DiMaggio, F.L. and Sandler, I.S.,
Material Model for Granular Soils.
J.Engr. Mech. Div., ASCE, 97 (1971) 935-950.

4. Lade, P.V.,
Elasto-Plastic Stress-Strain Theory for Cohesionless Soil with Curved Yield Surfaces.
Int. J. Solids Struct., 13 (1977) 1019-1035.

5. Prevost, J.-H.,
Plasticity Theory for Soil Stress-Strain Behavior.
J. Engr. Mech. Div. ASCE, 104 (1978) 1177-1194.

6. Dafalias, Y.F. and Herrmann, L.R.,
A Bounding Surface Soil Plasticity Model.
Proc. Int. Symp. Soils under Cyclic and Transient Loading,
Balkema, Rotterdam (1980) 335-345.

7. Mroz, Z., Norris, V.A. and Zienkiewicz, O.C.,
An Anisotropic, Critical State Model for Soils Subject to Cyclic Loading.
Geotechnique, 31 (1981) 451-469.

8. Ghaboussi, J. and Momen, H.,
Modelling and Analysis of Cyclic Behavior of Sands.
In: Soil Mechanics–Transient and Cyclic Loads, Ed. by G.N. Pande and O.C. Zienkiewicz, Wiley, London (1982) 313-342.

9. Pande, G.N. and Pietruszak, S.,
Reflecting Surface Model for Soils.
Proc. Int. Symp. Numerical Models in Geomechanics,
Balkema, Rotterdam (1982) 50-64.

10. Poorooshasb, H.B. and Pietruszak, S.,
On Yielding and Flow of Sand; A Generalized Two-Surface Model.
Computers and Geotechnics, 1 (1985) 33-58.

11. Frantziskonis, G., Desai, C.S. and Somasundaram, S.,
Constitutive Model for Nonassociative Behavior.
J. Engr. Mech., ASCE, 112 (1986) 932-946.

12. Duncan, J.M. and Chang, C.-Y.,
Nonlinear Analysis of Stress and Strain in Soils.
J. Soil Mech. Found. Div., ASCE, 96 (1970) 1629-1653.

13. Lade, P.V. and Nelson, R.B.,
Modelling the Elastic Behaviour of Granular Materials.
Int. J. Num. Anal. Meth. Geomech., 11 (1987) 521-542.

14. Lade, P.V.,
Model and Parameters for the Elastic Behavior of Soils.
Proc. 6th Int. Conf. Num. Meth. Geomech.,
Balkema, Rotterdam (1988) 359-364.

15. Lade, P.V.,
Three-Parameter Failure Criterion for Concrete.
J. Engr. Mech. Div., ASCE, 108 (1982) 850-863.

16. Lade, P.V.,
Failure Criterion for Frictional Materials.
Chapt. 20 in Mechanics of Engineering Materials, Ed. by C.S. Desai and R.H. Gallagher, Wiley (1984) 385-402.

17. Kim, M.K. and Lade, P.V.,
Modelling Rock Strength in Three Dimensions.
Int. J. Rock Mech. Min. Sci. & Geomech. Abstr., 21 (1984) 21-33.

18. Kim, M.K. and Lade, P.V.,
Single Hardening Constitutive Model for Frictional Materials, I. Plastic Potential Function.
Computers and Geotechnics

19. Lade, P.V. and Kim, M.K.,
Single Hardening Constitutive Model for Frictional Materials, II. Yield Criterion and Plastic Work Contours.
Computers and Geotechnics

20. Lee, K.L. and Seed, H.B.,
Drained Strength Characteristics of Sands.
J. Soil Mech. Found. Div., ASCE, 93, SM6 (1967) 117-141.

21. Al-Hussaini, M.M.,
Plane Strain and Triaxial Compression Tests on Painted Rock Dam Material.
Tech. Rep. S-71-2, No. 3, USAE, WES, Vicksburg, Miss. (1972).

22. Al-Hussaini, M.M.,
Drained Plane Strain and Triaxial Compression Tests on Crushed Napa Basalt.
Tech. Rep. S-71-2, No. 2, USAE, WES, Vicksburg, Miss. (1971).

23. Lade, P.V.,
The Stress-Strain and Strength Characteristics of Cohesionless Soils.
Ph.D. Dissertation, University of California, Berkeley (1972).

24. Schickert, G. and Winkler, H.,
Results of Tests Concerning Strength and Strain of Concrete Subjected to Multiaxial Compressive Stresses.
Deutscher Ausschus für Stahlbeton, 277, Berlin (1977).

25. Kupfer, H., Hilsdorf, H.K. and Rusch, H.,
Behavior of Concrete Under Biaxial Stresses.
J. Am. Concr. Inst., 66 (1969) 656-666.

26. Tasuji, M.E., Slate, F.O. and Nilson, A.H.,
Stress-Strain and Fracture of Concrete in Biaxial Loading.
J. Am. Concr. Inst., 75 (1978) 306-312.

27. Aschl, H., Linse, D. and Stoeckl, S.,
Strength and Stress Strain Behavior of Concrete Under Multiaxial Compression and Tension.
Proc. 2nd Int. Conf. Mechanical Behavior of Materials,
Boston, Mass. (1976) 102-117.

28. Egging, D.E., and Ko, H.-Y.,
Constitutive Relations of Randomly Oriented Steel Fiber Reinforced Concrete Under Multiaxial Compressive Loadings.
University of Colorado, Boulder (1981).

29. Bishop, A.W. and Henkel, D.J.,
The Measurement of Soil Properties in the Triaxial Test,
St. Martin's Press, New York (1962).

30. Hendron, A.J.,
The Behavior of Sand in One-Dimensional Compression.
Ph.D. Dissertation, University of Illinois, Urbana (1963).

31. Brooker, E.W. and Ireland, H.O.,
Earth Pressures at Rest Related to Stress History.
Can. Geotech. J., 2 (1965) 1-15.

Received 10 June 1988; revised version received and accepted 2 August 1988

Postscript

Additional predictions for undrained behavior of dense and loose sands, for normally consolidated clay and for concrete are given in:

Lade, P.V. (1990) "Single Hardening Model with Application to NC Clay," *Journal of Geotechnical Engineering*, ASCE, 116(3): 394-414.

Lade, P.V. and M.K. Kim (1995) "Single Hardening Constitutive Model for Soil, Rock and Concrete," *International Journal of Solids and Structures*, 14: 1963-1978.

Two additional references may be useful for determination of parameter values, and typical values for sands and clays are given as follows:

Lade, P.V. (2005) "Single Hardening Model for soils: Parameter Determination and Typical Values," *ASCE Geotechnical Special Publication No. 128, Soil Constitutive Models: Evaluation, Selection and Calibration*, Edited by J.A. Yamamuro and V.N. Kaliakin, pp. 290-309.

Lade, P.V. (2005) "Calibration of the Single Hardening Constitutive Model for Clays," *Proceedings of the 11th International Conference on Computer Methods and Geomechanics*, Turin, Italy, 10 pages.

Chapter X
Effects of Large Stress Reversals

Preface

Large three-dimensional stress reversals are often encountered in geotechnical engineering projects, and these may result in large plastic strains upon reversal. In the model for primary loading, the yield surface is pushed out and grows in size away from the hydrostatic axis, and large stress reversals result in only elastic strains until the primary yield surface is encountered again. However, experiments show that large stress reversals may result in large plastic strains, and these were studied in triaxial compression tests and true triaxial tests on sand. The Single Hardening model was therefore augmented by rotating the yield surface around the stress origin during stress reversals. The following articles explain the experimental findings and the augmentation of the model to include large, three-dimensional stress reversals. The resulting augmented model is shown to capture this behavior with good accuracy.

Large Stress Reversals in Triaxial Tests on Sand

POUL V. LADE, and SUTJA BOONYACHUT
Associate Professor and former graduate student, respectively,
Mechanics and Structures Department, School of Engineering and Applied Science,
University of California, Los Angeles, California, U.S.A.

SYNOPSIS The behavior of sand in triaxial tests during large stress reversals and large changes in stress involving unloading and reloading has been studied. Tests were performed with various stress-paths, and the results are evaluated in terms of directions of plastic strain increment vectors superimposed on the stress space. Based on results from tests with simple stress-paths, a modeling hypothesis is developed for yield and plastic potential surfaces for plastic strain increments occurring during unloading and reloading conditions. The modeling hypothesis is compared with results from tests with simple and complex stress-paths, and within the scatter of the test results the model is shown to capture the soil behavior with reasonable accuracy. It is shown that the experimental plastic strain increments as well as those obtained from the model depend on the stress-path and the previous stress history. The proposed model for the yield and plastic potential surfaces to be used for unloading and reloading conditions involves a combination of isotropic and kinematic hardening theories.

INTRODUCTION

Successful calculation of deformations and stresses in soil masses due to changes in load using e.g. the finite element method depends to a large extent on a constitutive law for characterization of all aspects of the stress-strain behavior of the soil. Several stress-strain models have been developed for soils based on concepts from elasticity and plasticity theory (see e.g. Drucker et al., 1957; Roscoe and Burland, 1968; Lade and Duncan, 1975; Sandler et al., 1976; Lade, 1977; Prevost, 1978; Dafalias and Herrmann, 1980; Mroz et al., 1981). The model which will be used as reference in the present study is that proposed by Lade (1977).

In the development of this elasto-plastic model, emphasis was placed on characterization of the behavior of cohesionless soil during primary loading. The plastic strains are assumed to obey an isotropic hardening law, thus implying that only elastic strains occur during unloading and reloading until the previously experienced maximum stress level has been exceeded. The behavior of soils during small to moderate stress reversals is essentially elastic, and it can be predicted for many practical purposes by the existing stress-strain model (Lade and Duncan, 1976). However, experimental results indicate that large stress reversals and large changes in stress which involve unloading and/or reloading result in plastic deformations which can be much larger than those predicted from assumed elastic behavior.

The object of the experimental research presented here is to study the basic soil behavior under such unloading and reloading conditions. This is done to form a basis on which an elasto-plastic model may be formulated for general stress conditions including unloading and reloading with arbitrary stress-paths. Results are obtained from triaxial tests in which two of the three principal stresses are equal in magnitude. A testing program involving different stress-paths is employed in a systematic effort to establish the soil behavior of interest. The results of these tests are presented and interpreted in terms of the plastic strain increments superimposed on the triaxial plane, as has been done previously for primary loading conditions (Poorooshasb et al., 1966; Lade and Duncan, 1975). The relative magnitudes of these strain increments are characterized by vector directions in the triaxial plane and interpreted in relation to the stress-path and previous stress history.

The existing elasto-plastic stress-strain model and the principles that led to the development of this model are employed in the analysis of the experimental data. The parts of this model which pertain to the present study are briefly reviewed below.

EXISTING ELASTO-PLASTIC MODEL

For the purpose of modeling the stress-strain behavior of soils by an elasto-plastic theory, the total strain increments, $\{d\varepsilon_{ij}\}$, are divided into an elastic component, $\{d\varepsilon^{e}_{ij}\}$, a plastic collapse component, $\{d\varepsilon^{c}_{ij}\}$, and a plastic expansive component, $\{d\varepsilon^{p}_{ij}\}$, such that:

$$\{d\varepsilon_{ij}\} = \{d\varepsilon^{e}_{ij}\} + \{d\varepsilon^{c}_{ij}\} + \{d\varepsilon^{p}_{ij}\} \qquad (1)$$

These strain components are calculated sepa-

171

Proceedings of the Fourth International Conference on Numerical Methods in Geomechanics, Edmonton, Canada, May 31 – June 4, 1982.

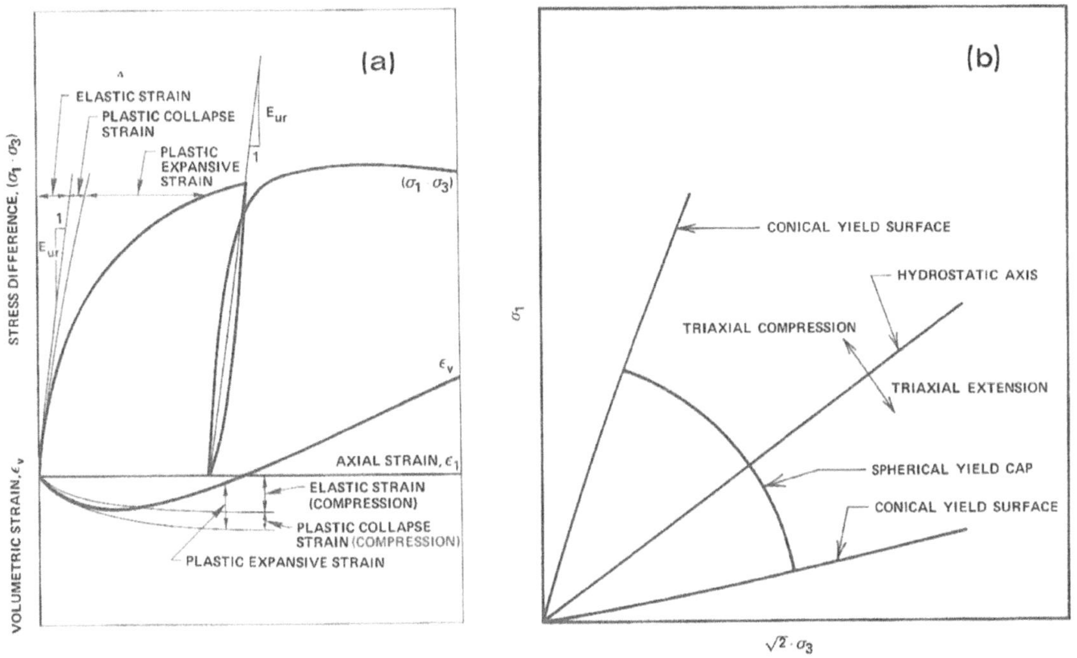

Fig. 1. Schematic Illustrations of (a) Elastic, Plastic Collapse, and Plastic Expansive Strain Components in Drained Triaxial Compression Test, and (b) Conical and Spherical Cap Yield Surfaces in Triaxial Plane.

rately, the elastic strains by Hooke's law, the plastic collapse strains and plastic expansive strains by plastic stress-strain theories that involve, respectively, a cap-type yield surface and a conical yield surface with apex at the origin of the stress space. These two yield surfaces may be activated simultaneously, one at a time, or not at all.

Fig. 1 (a) shows the parts of the total strain that are considered to be elastic, plastic collapse, and plastic expansive components of strain in a drained triaxial compression test. Typical observed variations of stress difference, $(\sigma_1-\sigma_3)$, and volumetric strain, ϵ_v, with axial strain, ϵ_1, are shown on this figure for a test performed with constant value of confining pressure, σ_3. Both elastic and plastic deformations occur from the beginning of loading of a cohesionless soil, the stress-strain relationship is nonlinear, and a decrease in strength may follow peak failure. The volumetric strain is initially compressive and this behavior may be followed by expansion (as shown in Fig. 1 (a)) or by continued compression. The plastic strains are initially smaller than the elastic strains, but at higher values of stress difference the plastic strains dominate the deformation.

Plastic Expansive Strains.-- The yield surfaces for the plastic strain components are indicated on the triaxial plane in Fig. 1 (b). The conical yield surface, which is curved in this plane, is described in terms of the first and the third stress invariants, I_1 and I_3:

$$f_p = (I_1^3/I_3-27)(I_1/p_a)^m \quad (2a)$$

$$f_p = \eta_1 \text{ at failure} \quad (2b)$$

where

$$I_1 = \sigma_x + \sigma_y + \sigma_z \quad (3)$$

and

$$I_3 = \sigma_x\cdot\sigma_y\cdot\sigma_z + \tau_{xy}\cdot\tau_{yz}\cdot\tau_{zx} + \tau_{yx}\cdot\tau_{zy}\cdot\tau_{xz} - (\sigma_x\cdot\tau_{yz}\cdot\tau_{zy} + \sigma_y\cdot\tau_{zx}\cdot\tau_{xz} + \sigma_z\cdot\tau_{xy}\cdot\tau_{yx}) \quad (4)$$

The parameters η_1 and m in Eq. 2 are constants to be determined for specific soils at the desired density, and p_a is atmospheric pressure expressed in the same units as the stresses used

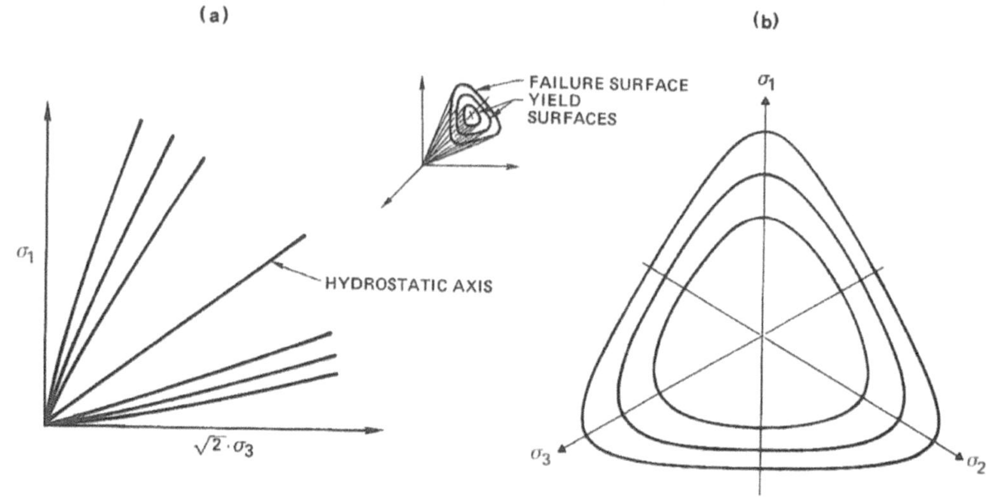

Fig. 2. Characteristics of Failure and Yield Surfaces Shown in Principal Stress Space. Traces of Failure and Yield Surfaces in (a) Triaxial Plane, and in (b) Octahedral Plane.

in this equation. In principal stress space the yield and failure surfaces defined by Eq. 2 are shaped like asymmetric conoids with their pointed apices at the origin of the stress space as shown in Fig. 2 (a). For failure conditions the apex angle increases with the value of η_1, and the curvature of the yield and failure surfaces increases with the value of m. For m=0 these surfaces are straight and the expression becomes identical to that used in a previous theory (Lade and Duncan, 1975). Fig. 2 (b) shows typical cross-sections for given values of I_1^3/I_3 and constant value of I_1. These cross-sections are exactly the same as for the previous theory, and it has been shown that they model the experimentally determined three-dimensional strengths of sands and normally consolidated clay with good accuracy (Lade and Duncan, 1975; Lade and Musante, 1978; Lade, 1978).

The plastic potential function is modeled on the yield function as follows:

$$g_p = I_1^3 - (27+\eta_2\cdot(p_a/I_1)^m)\cdot I_3 \tag{5}$$

where η_2 is constant for given values of f_p and the confining pressure. Thus, the corresponding flow rule is nonassociated, i.e., the plastic expansive strain increment vectors superimposed on the stress space form angles with the conical yield surfaces different from 90° as indicated in Fig. 3 (a).

Plastic Collapse Strains.-- The yield surface corresponding to the plastic collapse strains forms a cap on the open end of the conical yield surface, as shown in Fig. 1 (b). The collapse yield surface is shaped as a sphere with center at the origin of the principal stress space. This yield surface is described in terms of the first and the second stress invariants, I_1 and I_2:

$$f_c = I_1^2+2\cdot I_2 \tag{6}$$

where I_1 is given by Eq. 3 and

$$I_2 = \tau_{xy}\cdot\tau_{yx} + \tau_{yz}\cdot\tau_{zy} + \tau_{zx}\cdot\tau_{xz} - (\sigma_x\cdot\sigma_y + \sigma_y\cdot\sigma_z + \sigma_z\cdot\sigma_x) \tag{7}$$

It should be noted that yielding resulting from outward movement of the cap does not result in eventual failure. Failure is controlled entirely by the conical yield surface.

The plastic potential function is identical to the yield function:

$$g_c = I_1^2 + 2\cdot I_2 \tag{8}$$

Thus, the normality condition used for this surface implies an associated flow rule, and the plastic collapse strain increment vectors are perpendicular to the spherical yield surface, as shown in Fig. 3 (b).

Combination of Plastic Strain Increments.-- The result of a change in stress which involves both yield surfaces is shown in the triaxial plane in Fig. 3 (b). The plastic strain increment vectors are superimposed on the stress space in this diagram. Both plastic collapse and plastic expansive strains are caused by the

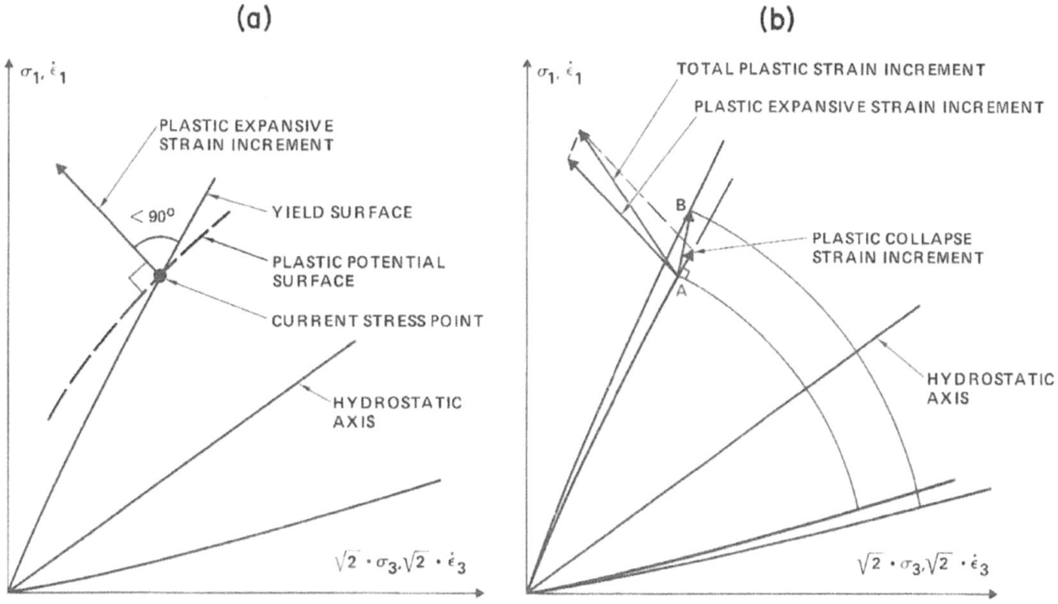

Fig. 3. Schematic Diagrams of (a) Yield and Plastic Potential Surfaces for Plastic Expansive Strains, and (b) Yielding Process with Both Yield Surfaces Activated and Combination of Plastic Strain Increments.

change in stress from point A to point B, because both yield surfaces are active. The magnitudes of the strain increments are indicated by the lengths of the vectors, and the total plastic strain increment is calculated according to Eq. 1 as the vector sum of the two components. The elastic strain components are further added (not shown in Fig. 3 (b)) to obtain the total strain increment for the stress change from A to B.

Note that the conical yield surface expands around the hydrostatic axis when it is activated. This corresponds to isotropic hardening, and it is indicated on Fig. 3 (b) by the movement of the yield surface on the extension side of the hydrostatic axis in response to a stress change on the compression side.

If only one of the two yield surfaces is activated, the other remains in its previous position, and if the stress change is such that none of the yield surfaces are activated, then both remain in their previous positions.

Unloading and Reloading.-- For stress changes involving stress states located on or inside the current yield surfaces, the response of the model in its present form is completely elastic. Soil behavior during small to moderate stress reversals may be predicted with reasonable accuracy, but the model cannot accurately predict soil response during large stress reversals and large changes in stress involving unloading and reloading under general stress conditions. This limitation is illustrated in Fig. 4 (a) which shows the stress-strain behavior for a drained triaxial test in which the specimen is first loaded in compression until the stress level measured by $f_p = 22.4$, then unloaded (decreasing f_p), reloaded in extension (increasing f_p), and finally primary loaded in extension (f_p increasing past $f_p = 22.4$). This stress-path is shown in Fig. 4 (b). Significant plastic yielding occurs already before the hydrostatic state of stress has been reached, and the stress-strain curve continues smoothly as the specimen is reloaded and primary loaded in extension. The fact that plastic yielding starts before the previous maximum stress level has been reached is considered to be the result of the residual stresses left in the material from the initial loading, and it is present whenever a reversal in the stress field occurs. This behavior is referred to as the Bauschinger effect, and it may be modeled if a kinematic hardening law is introduced (Prager, 1959; Mendelson, 1968).

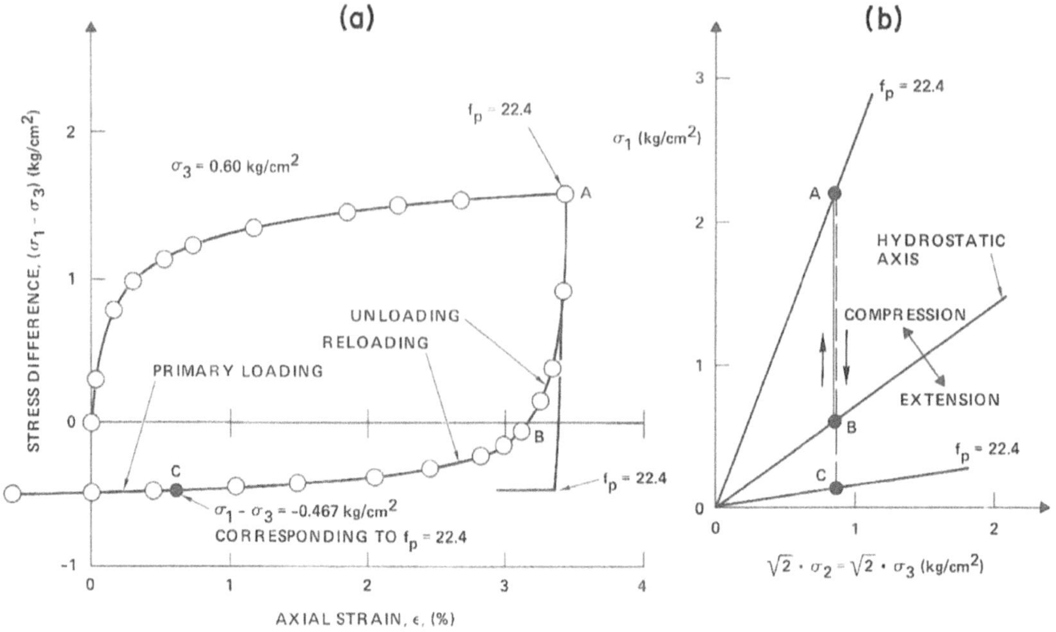

Fig. 4. Triaxial Test on Loose Santa Monica Beach Sand with Unloading, Reloading, and Primary Loading in Extension. (a) Measured Stress-Strain Behavior, and (b) Stress-Path in Triaxial Plane.

EXPERIMENTAL STUDY

Testing Program

Sand Tested.-- All Tests were performed on Santa Monica Beach Sand, which is composed of subangular to subrounded grains consisting mainly of quartz and feldspar. The characteristics of this sand are summarized as follows: Mean diameter, 0.27 mm; coefficient of uniformity, 1.58; specific gravity of grains, 2.66; maximum void ratio, 0.87; and minimum void ratio, 0.58.

The tests presented here were performed on loose specimens in order to obtain relatively large strains, thus facilitating the interpretation of the results. The loose specimens had a void ratio of 0.81 corresponding to a relative density of 20%.

Specimens and Measurements.-- The drained triaxial tests were performed on cylindrical specimens 2.8 in. (71 mm) in diameter and 7.5 in. (190 mm) high. Lubricated ends were used to minimize effects of end restraint. The initially air-dry specimens were saturated using the CO_2-method described by Lade and Duncan (1973). The vertical load, the confining pressure, the vertical deformation, and the volume changes were measured during the tests. Constant or varying confining pressures in the range from 0.3 kg/cm^2 (29.4 kN/m^2) to 6.0 kg/cm^2 (588.6 kN/m^2) were employed in the testing program. The measured vertical deformations were corrected for sand penetration into the lubricated cap and base, and the volumetric deformations were corrected for effects of membrane penetration occurring in tests with varying confining pressure.

Stress-Paths.-- Tests with simple stress paths as well as tests with complex stress-paths were performed. In order to obtain an appreciation of the magnitude and composition (elastic vs. plastic) of the measured strains associated with a given stress-path, each straight unloading or reloading path inside the primary yield surfaces was designed to involve relatively large stress changes.

The tests with simple stress-paths were conducted with constant confining pressure. These tests included conventional triaxial compression tests with several unloading-reloading cycles from increasingly high values of stress difference, tests with compression-extension-compression cycles, and tests with extension-compression-extension-compression cycles. Tests with more complex stress-paths involving varying confining pressure and large unloading and reloading stress changes were also performed to study the pattern of soil behavior in relation to the current stress-path and the previous stress history. The individual stress-paths are presented in greater detail in connection with

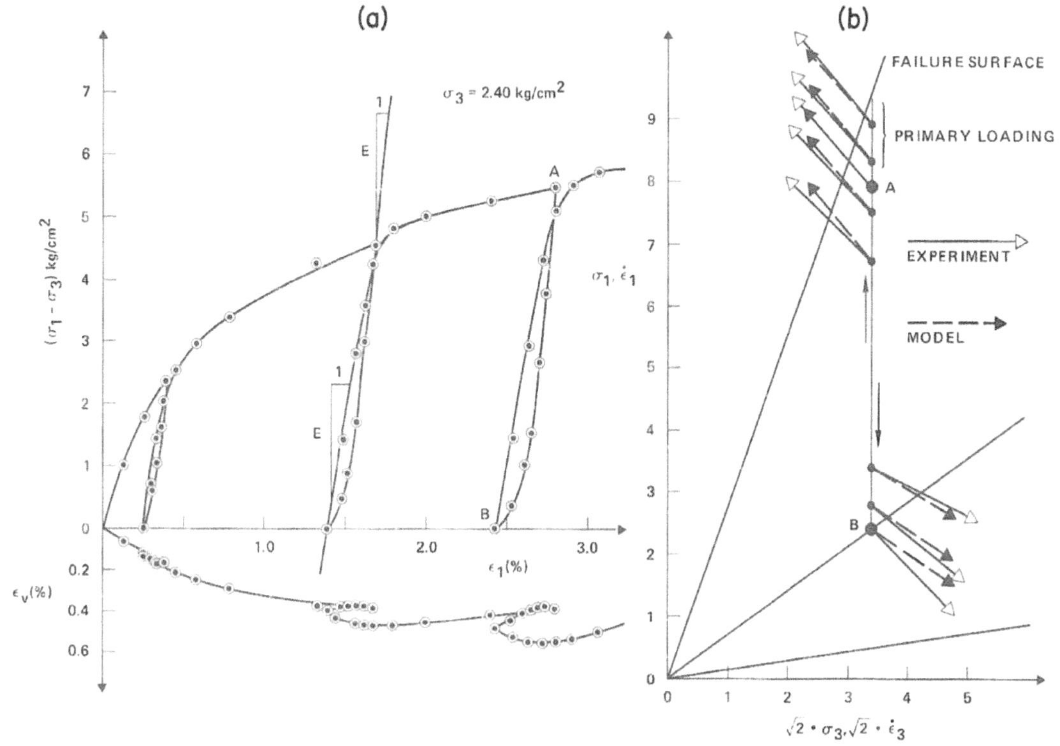

Fig. 5. Triaxial Compression Test with Several Unloading-Reloading Cycles. (a) Measured Stress-Strain and Volume Change Behavior, and (b) Stress-Path for Cycle A-B with Superimposed Plastic Strain Increment Vectors.

the presentation and interpretation of the soil behavior.

Tests with Simple Stress-Paths

Fig. 5 (a) shows the stress-strain and volume change behavior for a simple triaxial compression test with several unloading-reloading cycles. In order to calculate the plastic strain components, it is necessary to determine and subtract the elastic components from the measured strains.

Elastic Strains.-- The strains occurring immediately after each stress reversal are considered to be elastic. Thus, the elastic moduli indicated on Fig. 5 (a) are seen to differ from those previously assumed to represent the elastic behavior (see Fig. 1 (a)). It is possible that a precise description of the variation of the moduli indicated on Fig. 5 (a) may require a complex function involving the stresses, the current void ratio, and possibly the previous stress or strain history (see e.g. Holubec, 1968). However, both theoretical developments and experimental data have previously indicated that the elastic modulus for soil may vary with the mean normal stress (Ko and Scott, 1967; Rowe, 1971). In order to keep the stress-strain model relatively simple, it is proposed to employ this latter approach as expressed in the following equation:

$$E = K \cdot p_a \cdot \left(\frac{I_1}{p_a}\right)^n \tag{9}$$

where I_1 is the first stress invariant given in Eq. 3. The dimensionless constant values of the modulus number, K, and the exponent, n, are determined as indicated in Fig. 6. It may be seen that some scatter in the test data is present, and that the individual modulus values for unloading are somewhat higher than those for reloading. Nevertheless, the solid line in Fig. 6 is used to represent the magnitude and variation of the elastic modulus as expressed in Eq. 9.

The value of Poisson's ratio is assumed to be equal to 0.2 as in the primary loading model.

Plastic Strain Increment Vector Directions.-- After having subtracted the elastic components from the measured strains during the unloading-reloading cycles shown on Fig. 5 (a), the plas-

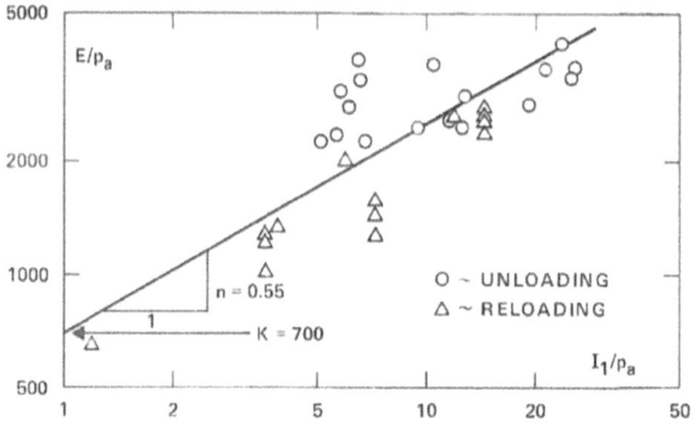

Fig. 6. Variation of Elastic Modulus with the First Stress Invariant, I_1.

tic strain increments may be superimposed on the stress space as indicated in Fig. 5 (b). The directions of the corresponding plastic strain increment vectors depend on the relative magnitudes of the components and not on their actual magnitudes. These vector directions may be used to study yield and plastic potential surfaces suitable for description of the plastic behavior during large stress reversals and large changes in stress involving unloading and reloading.

The observed plastic strain increment vectors indicated on Fig. 5 (b) (solid lines with open arrows) correspond to the third unloading-reloading cycle shown in Fig. 5 (a). Immediately upon unloading the strains are elastic, whereas the plastic strain components increase as the stress difference, $(\sigma_1-\sigma_3)$, approaches zero. Therefore, the plastic strain increment vectors shown on Fig. 5 (b) and on the following diagrams are determined towards the end of each straight section of the stress-paths. Note that although the current stress states are in the region of compression, the strain increment vectors for the unloading branch (A to B) point downwards in a manner similar to those observed for extension tests (Lade and Duncan, 1975). For reloading the vectors point upwards as they do for primary loading in compression. Results similar to those shown in Fig. 5 (b) are obtained for the other unloading-reloading cycles shown in Fig. 5 (a). In fact, analysis of many other tests have shown the same pattern in the directions of the plastic strain increment vectors obtained from simple unloading-reloading stress-paths.

Modeling Hypothesis.-- In order to model the plastic yielding and the flow rule indicated by the experimental data shown in Fig. 5 (b), the principles as well as the yield surfaces and the flow rule used in the existing model for primary loading will be employed together with the concept of kinematic hardening.

Fig. 7 (a) shows the stress-path for a triaxial compression test with the unloading branch B-C. In order to obtain a strain increment vector at point C with direction similar to those indicated in Fig. 5 (b), it is proposed to rotate the coordinate system around the origin O until the isotropic yield surface for primary loading, indicated on Fig. 7 (a), passes through points B and C. The rotated coordinate system is represented by the rotated hydrostatic axis. The value of $f_p=f_\alpha$ corresponding to points B and C in the rotated coordinate system is equal to the value of $f_p=f_\alpha$ in the original coordinate system. The amount of rotation is indicated by the angle between the rotated and the original hydrostatic axes. The yield surface for plastic deformation during unloading at any location of point C always passes through point B.

The direction of the plastic strain increment vector at point C is also obtained from the primary loading model. The direction of this vector at point C' on the primary loading surface corresponding to f_α is rotated to point C. This rotated vector is taken to represent the plastic behavior during unloading at point C.

If the stress-path continues past point C, the value of f_α increases until it reaches f_{pmax} in extension (see Fig. 7 (a)), at which time the primary loading model is used to predict the further soil behavior. The effect of all previous stress history (in this case represented by the stresses at point B) is then erased. Note that only the stresses at reversal (point B) are required in the above procedure. All other information comes from the current stresses (point C) and the existing model for primary loading.

If the soil is reloaded from point C to point D, as indicated on Fig. 7 (b), the same procedure as that described above is used. The coordinate system is rotated around the origin O until the isotropic yield surface for primary loading passes through points C and D. The value of $f_p=f_\alpha$ corresponding to points C (previous point of reversal) and D (current stress point) in the rotated coordinate system is equal to the value of $f_p=f_\alpha$ in the original coordinate system. The yield surface for plastic deformation during reloading at any location of D always passes through the previous point of reversal, C. The direction of the plastic strain increment vector at point D is obtained from the primary loading model at point D' and rotated with the yield surface to point D as shown in Fig. 7 (b).

As further reloading continues past point D, the value of f_α increases until the current stress point coincides with the previous primary yield surface represented by point B. Further loading is described by the primary loading model, and the effect of all previous stress history (represented by the stresses at point C) is then erased.

The advantage of this hypothesis is that the plastic strain increment vectors are described through the existing primary loading model. Only the current stress state and the stress state at the previous point of reversal (and those stress states where other reversals

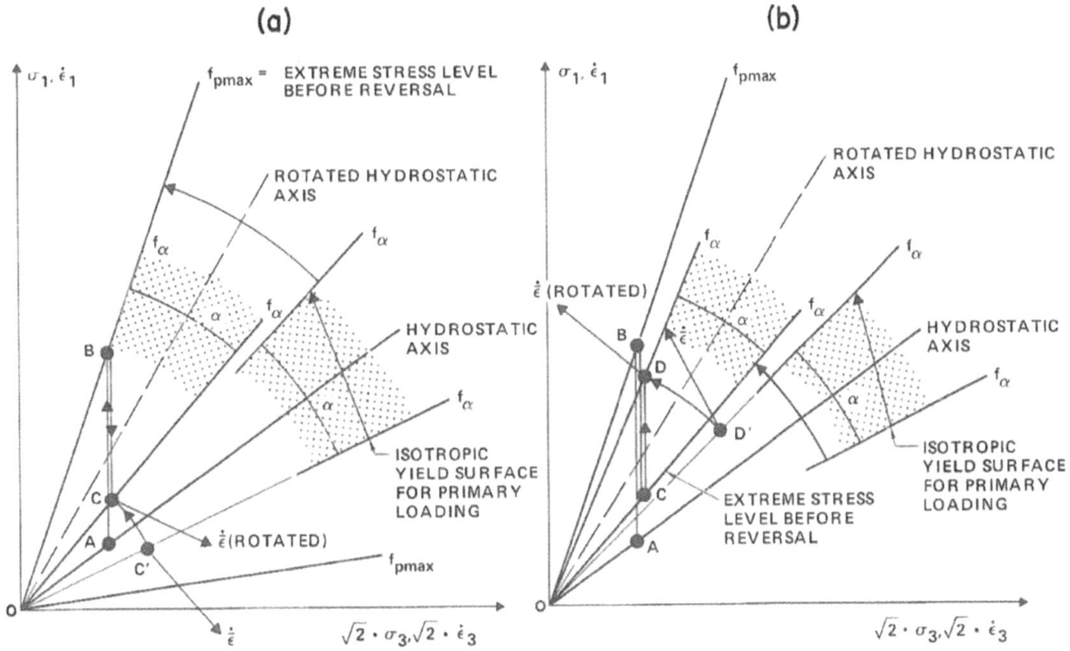

Fig. 7. Modeling Hypothesis for Soil Behavior During Large Stress Reversals involving (a) Unloading, and (b) Reloading.

occurred, if they are outside the current rotating yield surface) are necessary for determination of the current yield surface and the flow rule. All previous stress reversal points inside the current yield surface will not have any effect on the future soil behavior. Thus, it is not necessary to introduce any new soil parameters in order to obtain the yield surfaces and the flow rules for any stress-path involving large stress reversals or large changes in stress which include unloading or reloading.

The concepts presented above involve a combination of isotropic hardening (the primary loading model) and kinematic hardening (the rotating yield and plastic potential surfaces). None of the classical requirements for work-hardening plasticity (such as conditions of continuity, consistency, uniqueness, and irreversibility (Prager, 1949; Mandel, 1964)) have been violated. Furthermore, it is possible to employ the same modeling hypothesis for general three-dimensional stress states with and without rotation of the principal stress axes.

Comparison of Data and Hypothesis.-- The plastic strain increment vectors obtained according to the hypothesis presented above are shown on Fig. 5 (b) (dashed lines with solid arrows). Only the directions of the vectors obtained from the experiments and from the model are important. Their lengths are immaterial to the present considerations.

Fig. 5 (b) shows that the directions of the experimental and the model vectors compare favorably. Due to the relatively small strains encountered during the conventional unloading and reloading cycles (see Fig. 5 (a)), the precision of measurements, and the subtraction of elastic strains, the experimental vector directions can be expected to show considerable scatter. This is demonstrated for the unloading branch in Fig. 5 (b), where the experimental vector directions are on either side of the model vector directions. The latter are generated from a consistent model and they do not show scatter, but they may show systematic deviations to one side of the experimental vector directions. However, it appears that the two sets of vector directions are reasonably close together, and the differences in vector directions shown on Fig. 5 (b) are considered to reflect scatter rather than systematic deviations.

Fig. 8 shows comparisons for simple stress-paths involving primary loading in compression to point B followed by unloading (to A) and reloading in extension to point C. Finally the specimens are unloaded in extension (to A), reloaded in compression to point B and then primary loaded to failure.

The data in Fig. 9 correspond to a simple stress-path initiated with primary loading in extension to point B followed by unloading (to A) and reloading in compression to point C. The specimen is then unloaded again and reloaded in extension, and following a final reversal the

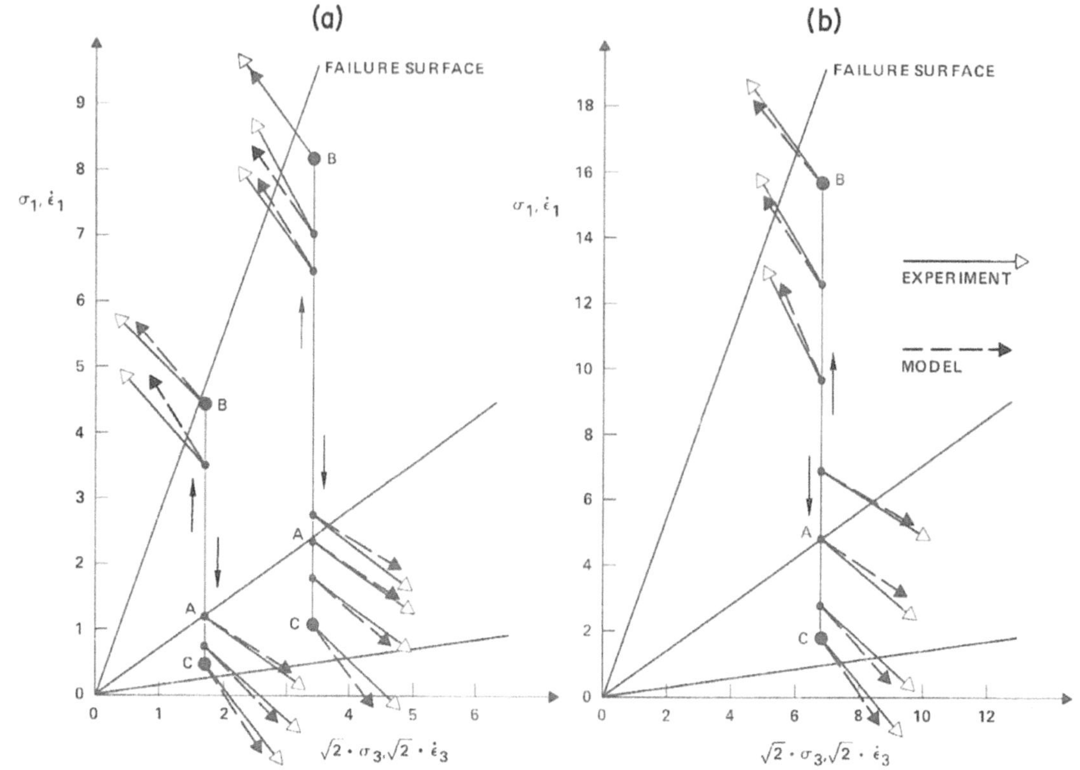

Fig. 8. Simple Stress-Paths with Superimposed Plastic Strain Increment Vectors for Triaxial Tests Performed with Confining Pressures of (a) 1.20 and 2.40 kg/cm^2, and (b) 4.80 kg/cm^2.

specimen is failed in compression. Only the parts which pertain to the present study of unloading and reloading are indicated in Fig. 9.

The comparison of experimental and model vector directions seems to be favorable. It should be noted that the stress changes are larger in these tests which involve states of stress in both compression and extension, and the plastic strains are consequently larger than those shown in Fig. 5. It appears that a slight systematic difference between the experimental and the model vector directions occurs at high stress levels in extension (small values of σ_1). The model predicts slightly higher rates of dilation than those observed. A similar difference was observed in the existing primary loading model for loading in extension (Lade and Duncan, 1975). However, the actual magnitudes of volume change predicted for primary loading stress-paths in extension were comparable with measured volume changes. The slight systematic differences indicated in Figs. 8 and 9 are therefore not expected to have significant consequences.

Tests with Complex Stress-Paths

Figs. 10, 11, and 12 show the results of tests with complex stress-paths. These tests involve varying confining pressures and stress differences, and the stress-paths are designed to explore in principle the plastic soil behavior for various stress-path directions, and to see if the proposed model is able to capture the behavior of the soil for such complex stress-paths.

Immediately following a stress reversal the strains are mainly elastic, whereas the plastic strain components increase as the current stress point is removed from the point of reversal. The experimental plastic strain increment vectors are therefore determined towards the end of each straight section of the stress-paths, where the larger plastic strains provide more reliable evaluations.

The plastic strain increment vectors obtained according to the hypothesis presented above are also shown in Figs. 10, 11, and 12. Comparisons between experimental and model vector directions do in some instances show substantial differences, whereas favorable comparisons are obtained for most cases.

Fig. 10 shows comparisons for a rather complex stress-path. The letters at the stress reversal points indicate the sequences in which the stress points are reached. The effect of previous stress-history can be studied from the results of this test. For example, stress-paths

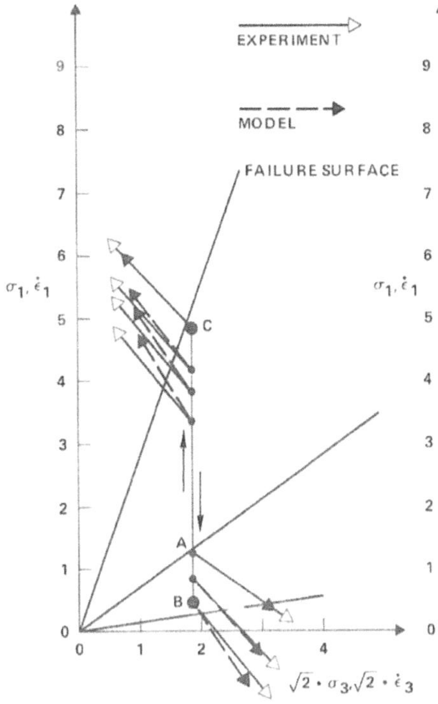

Fig. 9. Simple Stress-Path with Superimposed Plastic Strain Increment Vectors for Triaxial Test with Constant Confining Pressure.

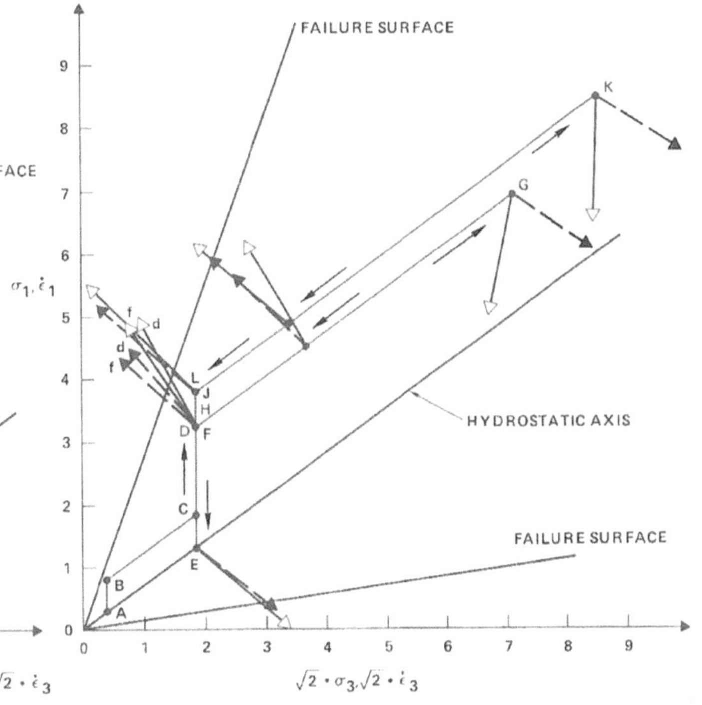

Fig. 10. Complex Stress-Path with Superimposed Plastic Strain Increment Vectors.

C-D and E-F both reach the same stress point (D and F), and both involve reloading, because point B corresponds to a higher stress level than point D (F), as measured by f_p in the original coordinate system. The experimental vector directions are indicated by the letters d and f and they do not coincide. They show the effect of the previous stress-paths, one initiating at point C and the other initiating at point E. Whereas the model vector directions do not coincide with the experimental vector directions, they do show the same pattern in directions as obtained in the tests. Thus, the model is capable of reflecting the effect of previous stress history.

The experimental vector directions shown at points G and K are those corresponding to stress-paths F-G and J-K, respectively. These stress-paths involve plastic collapse strains, and these strains together with the elastic strains have been subtracted from the measured strains. The remaining plastic expansive strains are used on Fig. 10. Clearly, they do not compare favorably with the model vector directions. Plastic strain increment vectors must point away from the interior of a sphere which passes through the current stress point and has its center at the origin of the stress space. This condition is fulfilled by all the model vectors including those at points G and K, but it is not fulfilled by the experimental vectors. The measured strain increment vectors (not shown) do point away from the sphere, but the subtraction of plastic collapse and elastic strains results in the vectors shown on Fig. 10. The measured strains for stress-paths F-G and J-K are very small, and only slightly incorrect measurements or models for the plastic collapse and elastic strains will cause the calculated expansive strains to be in error.

Upon reloading from G to H and from K to L (both with increasing values of f_p in the original coordinate system), experimental and model vector directions are compared at intermediate points. In one case the two vectors point in different directions, whereas they coincide in the other case. This difference reflects the magnitude of scatter that may be expected in cases where small strains are being measured and analyzed.

Fig. 11 shows a stress-path with generally increasing mean normal stress. The comparisons between experimental and model vector directions are very good in all cases. Notably, they com-

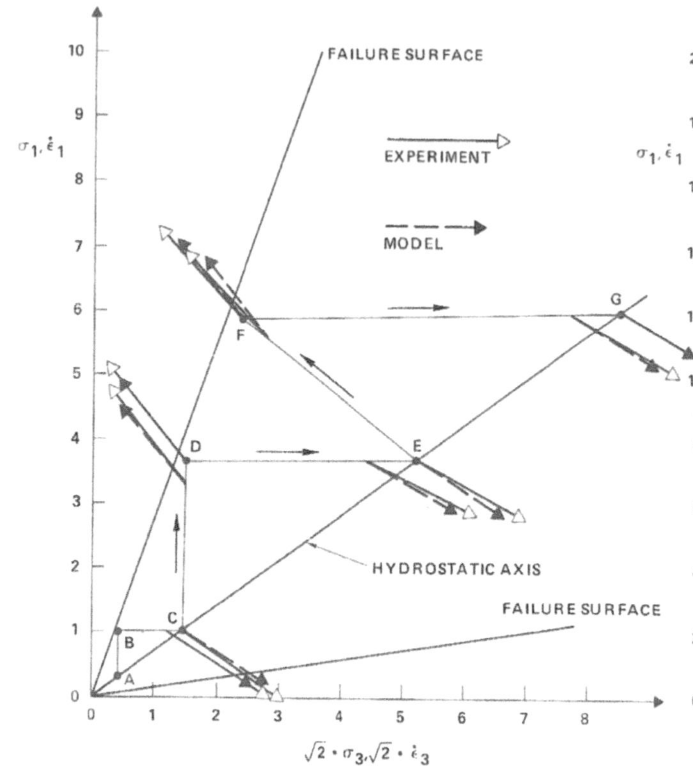

Fig. 11. Complex Stress-Path with Generally Increasing Mean Normal Stress and with Superimposed Plastic Strain Increment Vectors.

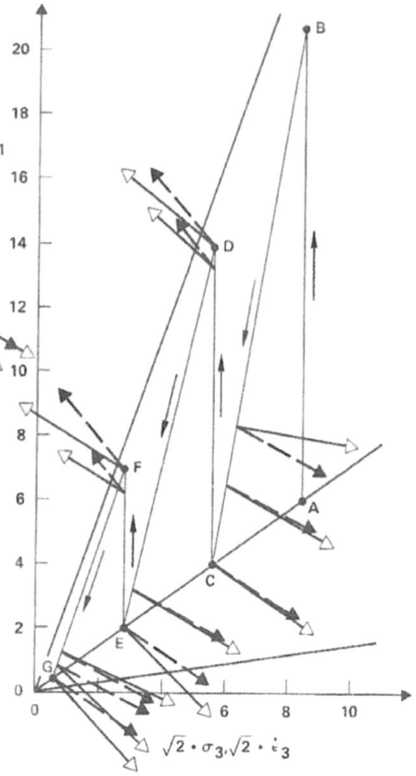

Fig. 12. Complex Stress-Path with Alternating Increases and Decreases in Mean Normal Stress and with Superimposed Plastic Strain Increment Vectors.

pare very favorably at points E and G, which are similar in many respects to points G and K in Fig. 10. Thus, it is likely that the scatter in the test results accounts for at least part of the differences between the experimental and the model vector directions shown in Fig. 10.

Fig. 12 shows a stress-path with alternating increases and decreases in the mean normal stress. Some of the comparisons between experimental and model vector directions are very good, whereas others show some differences. In view of the previous discussions, the magnitudes of these differences appear to be erratic rather than systematic and they are considered to be within the scatter to be expected in tests involving small strains.

Evaluation of Data and Model

In evaluating the experimental soil behavior during large stress reversals and the proposed model for prediction of yield and plastic potential surfaces for these conditions, the following points should be considered:

1) The strains measured during unloading and reloading stress-paths are in many cases relatively small. Although careful measurements were made in the tests performed for this investigation, these measurements do contain errors due to slight variations in void ratios between specimens, corrections for membrane penetration and sand penetration into the lubricated ends, etc. In addition, the quality of the calculated plastic expansive strains employed in the study depends on the models and the soil parameters used to compute the plastic collapse strains and the elastic strains. As shown on Fig. 6, the elastic moduli for unloading and reloading are in some cases different for the same value of I_1. Thus, considering the small strains measured for some stress-paths and the errors associated with the calculation of the plastic expansive strain increment directions, it is clear that some scatter in these directions occurs. As expected, the largest amounts of scatter are associated with the stress-paths producing the smallest strains.

2) The proposed model for the yield and plastic

potential surfaces to be used for large stress reversals involves a combination of isotropic and kinematic hardening. The plastic strain increment vectors are described through the existing model for primary loading. The only additional information necessary consists of the current stress state and the stress state(s) at the previous point(s) of reversal. Thus, no additional soil parameters are required. Furthermore, it is possible to expand the proposed model to general three-dimensional stress conditions with and without rotation of the principal stress axes. In view of the complexity of soil behavior experienced during large stress reversals, the proposed model appears to be relatively simple.

3) Considering the scatter which may be associated with some of the experimental vector directions, it appears that the relatively simple model proposed above captures the soil behavior experienced during large stress reversals with reasonable accuracy.

SUMMARY AND CONCLUSIONS

The behavior of sand in triaxial tests during large stress reversals and large changes in stress involving unloading and reloading has been investigated. The tests were performed with various stress-paths, and the results were evaluated in terms of the directions of the plastic strain increment vectors superimposed on the stress space. Based on results from tests with simple stress-paths, a modeling hypothesis was developed for yield and plastic potential surfaces for plastic strain increments occurring during unloading and reloading conditions. The modeling hypothesis was compared with results from tests with simple and more complex stress-paths, and within the scatter of the test results the model was shown to capture the soil behavior with reasonable accuracy. It was shown that the experimental plastic strain increments as well as those obtained from the model depend on the stress-path and the previous stress history. The proposed model for the yield and plastic potential surfaces to be used for unloading and reloading conditions involves a combination of isotropic and kinematic hardening. The plastic strain increment vectors are described through an existing model for primary loading in addition to the current stress state and the stress state(s) at the previous point(s) of reversal. No additional soil parameters are required.

ACKNOWLEDGMENT

This study was supported by the National Science Foundation under Grant No. ENG-75-05325. Grateful appreciation is expressed for this support.

REFERENCES

Dafalias, Y F & Herrmann, L R (1980). A Bounding Surface Soil Plasticity Model. Soils Under Cyclic and Transient Loading, Balkema, Rotterdam, 335-345.

Drucker, D C, Gibson, R E & Henkel, D J (1957). Soil Mechanics and Work-Hardening Theories of Plasticity. Transactions, ASCE, (122), 338-346.

Holubec, I (1968). Elastic Behavior of Cohesionless Soil. J. Soil Mech. Found. Div., ASCE (94), SM6, pp. 1215-1231.

Ko, H-Y & Scott, R F (1967). Deformation of Sand in Hydrostatic Compression. J. Soil Mech. Found. Div., ASCE (93), SM3, 137-156.

Lade, P V (1977). Elasto-Plastic Stress-Strain Theory for Cohesionless Soil with Curved Yield Surfaces. Int. J. Solids Structures, (13), 1019-1035.

Lade, P V (1978). Prediction of Undrained Behavior of Sand. J. Geot. Eng. Div., ASCE (104), GT6, 721-735.

Lade, P V & Duncan, J M (1973). Cubical Triaxial Tests on Cohesionless Soil. J. Soil Mech. Found. Div., ASCE (99), SM10, 793-812.

Lade, P V & Duncan, J M (1975). Elasto-Plastic Stress-Strain Theory for Cohesionless Soil. J. Geot. Eng. Div., ASCE (101), GT10, 1037-1053.

Lade, P V & Duncan, J M (1976). Stress-Path Dependent Behavior of Cohesionless Soil. J. Geot. Eng. Div., ASCE (102), GT1, 51-68.

Lade, P V & Musante, H M (1978). Three-Dimensional Behavior of Remolded Clay. J. Geot. Eng. Div., ASCE (104), GT2, 193-209.

Mandel, J (1964). Conditions de Stabilité et Postulat de Drucker. Proc. IUTAM, Rheology and Soil Mechanics, Grenoble, 58-67.

Mendelson, A (1968). Plasticity: Theory and Practice. 353 pp., MacMillan, New York.

Mroz, Z, Norris, V A & Zienkiewicz, O C (1981). An Anisotropic, Critical State Model for Soils Subject to Cyclic Loading. Geotechnique (31), 4, 451-469.

Poorooshasb, H B, Holubec, I & Sherborne, A N (1966). Yielding and Flow of Sand in Triaxial Compression: Part I. Can. Geot. J. (3), 4, 179-190.

Prager, W (1949). Recent Developments in the Mathematical Theory of Plasticity. J. Appl. Phys. (20), 4, 235-241.

Prager, W (1959). An Introduction to Plasticity. 148 pp., Addison-Wesley, Reading, MA.

Prevost, J-H (1978). Plasticity Theory for Soil Stress-Strain Behavior. J. Eng. Mech. Div., ASCE (104), EM5, 1177-1194.

Roscoe, K H & Burland, J B (1968). On the Generalized Stress-Strain Behavior of 'Wet' Clay. Engineering Plasticity, Cambridge Univ. Press, 535-609.

Rowe, P W (1971). Theoretical Meaning and Observed Values of Deformation Parameters for Soil. Stress-Strain Behavior of Soils. G T Foulis & Co, Oxfordshire, England, 143-194.

Sandler, I S, DiMaggio, F L & Baladi, G Y (1976). Generalized Models for Geological Materials. J. Geot. Eng. Div., ASCE (102), GT7, 683-699.

ELSEVIER

Computers and Geotechnics, Vol. 21, No. 3, pp. 183–216, 1997

Printed in Great Britain
PII:S0266-352X(97)00022-0 0266-352X/97 $17.00+0.00

Rotational Kinematic Hardening Model for Sand. Part I Concept of Rotating Yield and Plastic Potential Surfaces

Poul V. Lade[a] & Sinan Inel[b]

[a]Department of Civil Engineering, The Johns Hopkins University, Baltimore, Maryland 21218, U.S.A.
[b]Dames & Moore, 911 Wilshire Boulevard, Suite 700, Los Angeles, California 90017, U.S.A.

(Received 28 May 1996; revised version received 2 June 1997; accepted 9 June 1997)

ABSTRACT

Existing kinematic hardening constitutive models are evaluated in the light of experimental studies on sand behavior during large stress reversals. An alternative approach is proposed which incorporates rotation and intersection of yield surfaces to achieve a better fit with experimental data. An existing isotropic hardening constitutive model forms the basic framework to which the rotational kinematic hardening mechanism is added. The behavior of the proposed plastic potential is verified by studying the experimentally obtained directions of plastic strain increment vectors superimposed on the stress space during large stress reversals. The new combined model preserves the behavior of the isotropic hardening model under monotonic loading conditions and the extension from isotropic to kinematic hardening is accomplished in a simple manner which avoids the introduction of complicated rules and new material parameters. © 1997 Elsevier Science Ltd.

INTRODUCTION

The development of suitable stress–strain theories for soils requires comprehensive knowledge about the behavior of soil under various types of loading conditions. However, once realistic and sufficiently accurate stress–strain relationships have been developed, they provide several advantages which include a better understanding of soil behavior, extrapolation to conditions that cannot be produced in laboratory testing and prediction of soil behavior through finite difference or finite element based numerical techniques so that the design of soil structures can be made on a rational basis.

Whereas simple models can be derived to reflect individual phenomena observed in experiments on soils, prediction of true behavior in complex situations requires the employment of comprehensive stress–strain theories of general validity. Many of the typical foundation problems encountered by geotechnical engineers involve stress reversals, rotation of principal stresses and anisotropic behavior. Earthquakes and offshore structures introduce the additional complication of cyclic loading and degradation. A long life-span or sustained loading on a structure may lead to considerations of creep and fatigue. The increasing use and availability of high-speed computers together with the development of readily available programs employing finite difference and finite element techniques to solve static and dynamic continuum problems have allowed and motivated the employment of more sophisticated constitutive laws for the analysis of such problems of increasing complexity.

Elasticity and plasticity theories have been shown to be capable of capturing many aspects of soil behavior. Several constitutive models based on concepts from elastic–plastic theory have been developed for soils [1, 2]. An existing elastic–plastic single-surface isotropic hardening model developed by Kim and Lade [3], and Lade and Kim [4, 5] is used as the basic framework for the present study. In the development of this model, emphasis was placed on characterization of the behavior of frictional materials during monotonic loading. Plastic strains are assumed to obey an isotropic hardening law, thus implying that only elastic strains occur during stress reversals involving unloading and reloading until the previously experienced maximum stress level has been exceeded. This is in accordance with the behavior of soils during small to moderate stress reversals which can essentially be considered as elastic. Yet experimental results indicate that large stress reversals result in plastic deformations which can be much larger than those predicted from assumed elastic behavior. Modeling of the behavior of soils during such large stress reversals is of paramount importance for solving quasi-static foundation engineering problems involving unloading and reloading as well as being a stepping stone towards describing behavior under cyclic loading.

The object of the present study is to evaluate the behavior of cohesionless soils under such unloading and reloading conditions, develop a theoretical basis on which an elasto-plastic model may be formulated for general stress conditions involving unloading and reloading with arbitrary stress paths, and modify the framework of the existing elasto-plastic stress–strain model to predict such behavior.

OUTLINE OF SINGLE SURFACE, ISOTROPIC HARDENING MODEL

The single surface, isotropic hardening model is an elasto-plastic constitutive model for the behavior of frictional materials developed by Kim and Lade

[3], and Lade and Kim [4, 5] on the basis of a thorough review and evaluation of data from experiments on frictional materials such as sand, clay, concrete, and rock.

The total strain increments observed in a material when loaded are divided into elastic and plastic components such that

$$d\varepsilon_{ij} = d\varepsilon_{ij}^{e} + d\varepsilon_{ij}^{p} \tag{1}$$

These strains are then calculated separately, the elastic strains by Hooke's law, and the plastic strains by a plastic stress–strain law. Both are expressed in terms of effective stresses.

Below is a brief review of the framework and the components of the constitutive model. In order that the presentation follows a logical developmental sequence, the components are presented in the following order: elastic behavior, failure criterion, flow rule, yield criterion and work-hardening/softening law.

Elastic behavior

The elastic strain increments are calculated from Hooke's law, using a nonlinear variation of Young's modulus developed by Lade and Nelson [6] from theoretical considerations based on the principle of conservation of energy. The value of Poisson's ratio is assumed to be constant. Young's modulus is expressed in terms of the following power function involving nondimensional constants and stress invariants:

$$E = M \cdot p_a \left[\left(\frac{I_1}{p_a} \right)^2 + R \cdot \frac{J_2'}{p_a^2} \right]^{\lambda} \tag{2}$$

in which

$$R = \frac{1+\nu}{1-2\nu} \tag{3}$$

and the stress invariants are defined as

$$I_1 = \sigma_x + \sigma_y + \sigma_z \tag{4}$$

$$J_2' = \frac{1}{6}[(\sigma_x - \sigma_y)^2 + (\sigma_y - \sigma_z)^2 + (\sigma_{-\sigma_x})^2] + \tau_{xy}^2 + \tau_{yz}^2 + \tau_{zx}^2 \tag{5}$$

The parameters M and λ are constant dimensionless numbers and p_a is atmospheric pressure expressed in the same units as Young's modulus and the stresses.

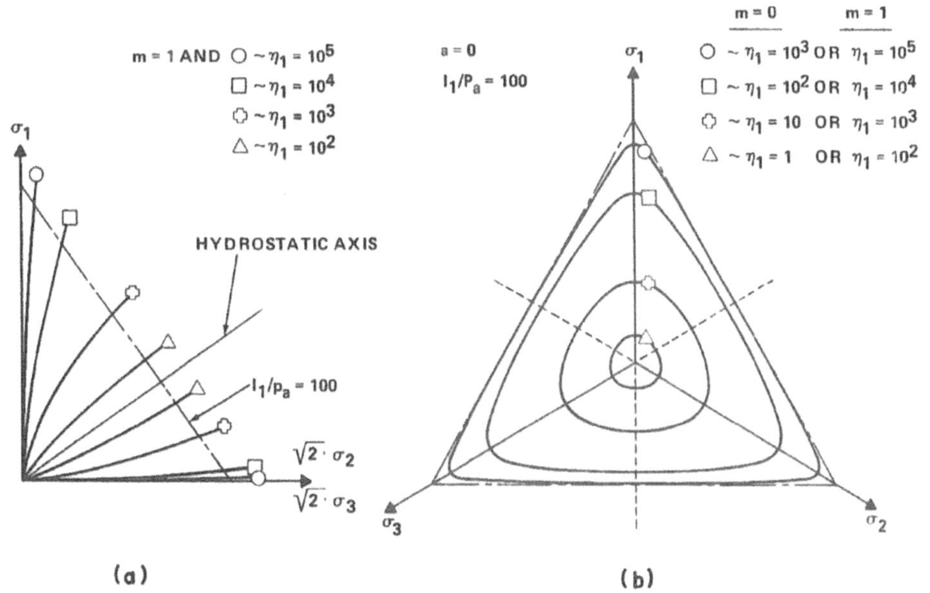

Fig. 1. Cross sections of failure surfaces in (a) triaxial plane and (b) octahedral plane.

Failure criterion

A general, three-dimensional failure criterion developed for soils [7], concrete [8], and rocks [9] is used in the model. The criterion is expressed in terms of stress invariants as follows:

$$\left[\frac{I_1^3}{I_3} - 27\right]\left[\frac{I_1}{p_a}\right]^m = \eta_1 \tag{6}$$

in which

$$\begin{aligned} I_3 = {} & \sigma_x \cdot \sigma_y \cdot \sigma_z + \tau_{xy} \cdot \tau_{yz} \cdot \tau_{zx} + \tau_{yx} \cdot \tau_{zy} \cdot \tau_{xz} \\ & - (\sigma_x \cdot \tau_{yz} \cdot \tau_{zy} + \sigma_y \cdot \tau_{zx} \cdot \tau_{xz} + \sigma_z \cdot \tau_{xy} \cdot \tau_{yx}) \end{aligned} \tag{7}$$

The cross-sectional shape of the failure surface is shown in the triaxial and octahedral planes in Fig. 1. The parameters η_1 and m are constant dimensionless numbers. In order to include the effective cohesion and the tension which can be sustained by concrete and rock, a translation of the principal stress space along the hydrostatic axis is performed by adding a constant stress, $a \cdot p_a$, to the normal stresses before substitution into any of the model functions:

$$\bar{\sigma}_{ij} = \sigma_{ij} + \delta_{ij} \cdot a \cdot p_a \tag{8}$$

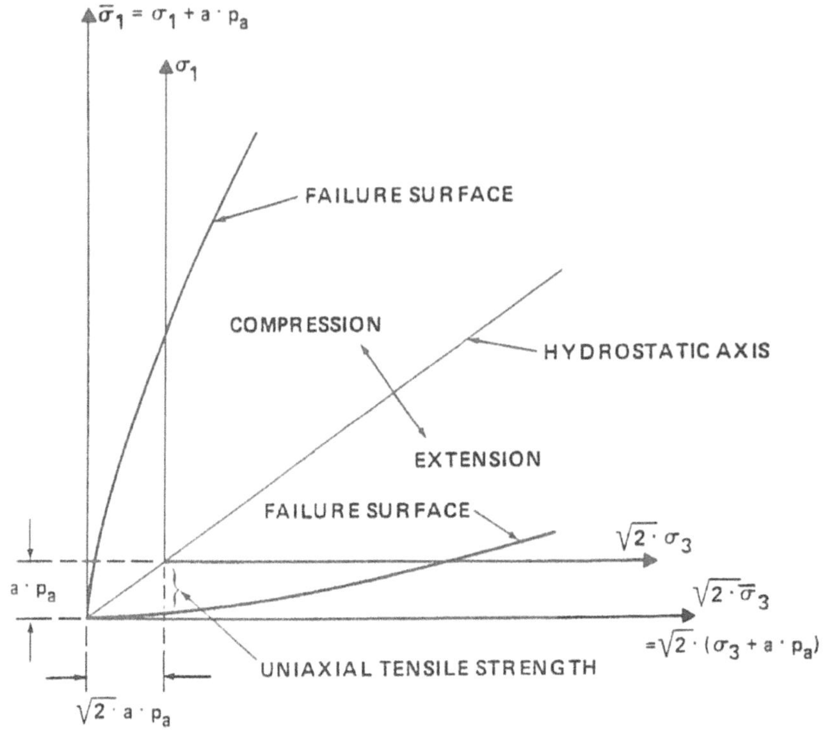

Fig. 2. Translation of principal stress space along hydrostatic axis to include effect of tensile strength in failure criterion.

in which a is a constant dimensionless parameter and δ_{ij} is Kronecker's delta ($=1$ for $i=j$, and $=0$ for $i \neq j$). The corresponding translation of the principal stress space is depicted in Fig. 2.

Flow rule

The plastic strain increments are calculated from the flow rule:

$$d\varepsilon_{ij}^{p} = d\lambda_{\mathrm{p}} \cdot \frac{\partial g_{\mathrm{p}}}{\partial \sigma_{ij}} \tag{9}$$

in which g_{p} is the plastic potential function and $d\lambda_{\mathrm{p}}$ is a proportionality constant. The plastic potential function is different from the yield function leading to nonassociated flow. It is expressed in terms of stress invariants as follows:

$$g_{\mathrm{p}} = \left[\psi_1 \cdot \frac{I_1^3}{I_3} - \frac{I_1^2}{I_2} + \psi_2\right] \cdot \left[\frac{I_1}{p_{\mathrm{a}}}\right]^{\mu} \tag{10}$$

in which

$$I_2 = \tau_{xy} \cdot \tau_{yx} + \tau_{yz} \cdot \tau_{zy} + \tau_{zx} \cdot \tau_{xz} - (\sigma_x \cdot \sigma_y + \sigma_y \cdot \sigma_z + \sigma_z \cdot \sigma_x) \tag{11}$$

The shape of the plastic potential is depicted in the triaxial plane and the principal stress space in Fig. 3. The parameters ψ_1, ψ_2, and μ are constant dimensionless numbers, and a relationship between ψ_1 and m may be expressed in the form of a power function:

$$\psi_1 = 0.00155 \cdot m^{-1.27} \tag{12}$$

The derivatives of the plastic potential function with respect to stresses are used to obtain the plastic strain increments.

Yield criterion and work hardening/softening law

The yield surfaces describe contours of equal total plastic work. The total plastic work (due to shear strains as well as volumetric strains) serves as the hardening parameter, and it is used to define the location and shape of the yield surfaces. The isotropic yield function is defined as:

$$f_{\text{p}} = f'_{\text{p}}(\sigma) - f''_{\text{p}}(W_{\text{p}}) = 0 \tag{13}$$

in which

$$f'_{\text{p}} = \left[\psi_1 \cdot \frac{I_1^3}{I_3} - \frac{I_1^2}{I_2}\right] \cdot \left[\frac{I_1}{p_{\text{a}}}\right]^h \cdot e^q \tag{14}$$

and for hardening,

$$f'_{\text{p}} = \left[\frac{1}{D}\right]^{1/\rho} \cdot \left[\frac{W_{\text{p}}}{p_{\text{a}}}\right]^{1/\rho} \tag{15}$$

The shape of the yield surface is depicted in the triaxial plane and the principal stress space in Fig. 4.

The parameter q varies with stress level S from zero at the hydrostatic axis to unity at the failure surface according to the following expression:

$$q = \frac{a \cdot S}{1 - (1 - a) \cdot S} \tag{16}$$

where S is defined as

$$S = \frac{f_{\text{n}}}{\eta_1} = \frac{1}{\eta_1} \cdot \left[\frac{I_1^3}{I_3} - 27\right] \cdot \left[\frac{I_1}{p_{\text{a}}}\right]^m \tag{17}$$

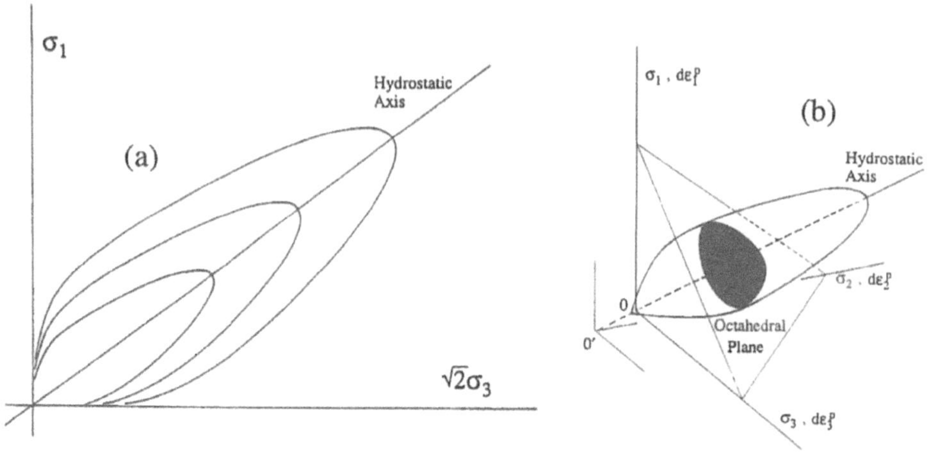

Fig. 3. Plastic potential surfaces shown in (a) triaxial plane and (b) principal stress space.

The values of D and ρ are constants for a given material and are defined as follows:

$$D = \frac{c}{(27\psi_1 + 3)^{\rho}} \tag{18}$$

and

$$\rho = \frac{p}{h} \tag{19}$$

The parameters h, a, C and p are constant dimensionless numbers. For hardening, the isotropic yield surface monotonically inflates until the current stress point reaches the failure surface. For softening, the yield surface deflates isotropically according to an exponential decay function:

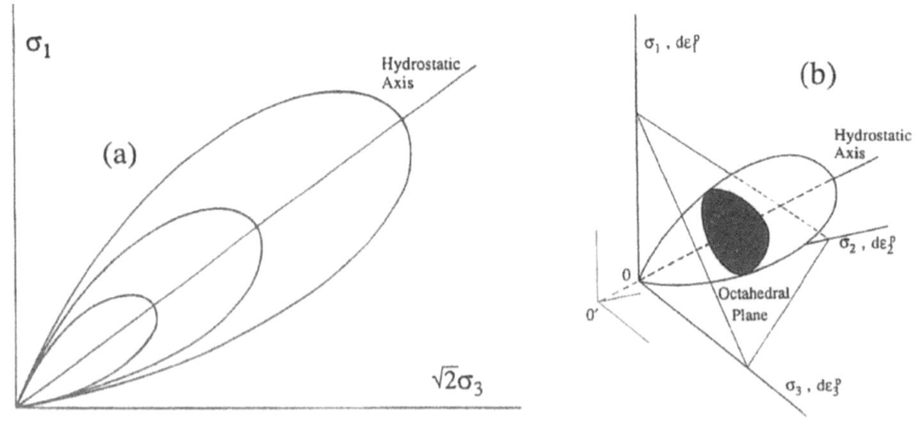

Fig. 4. Yield surfaces shown in (a) triaxial plane, and (b) principal stress space.

$$f''_p = A \cdot e^{-B \cdot (W_p/p_a)} \tag{20}$$

in which A and B are positive constants to be determined on the basis of the slope of the hardening curve at the point of peak failure. The isotropic work hardening and softening functions are shown in Fig. 5.

The required number of parameters to be determined for a given material is 12. For cohesionless materials such as sand, the value of $a=0$, reducing the number of parameters to 11. These parameters can be determined from simple laboratory tests such as isotropic compression and conventional triaxial compression tests. Validation of the model has been achieved by comparison of predicted and experimental stress–strain curves for various two- and three-dimensional stress paths obtained for different types of frictional materials [5, 10–12].

LARGE STRESS REVERSALS IN THE TRIAXIAL PLANE

During small stress reversals, the resulting strains are essentially elastic in nature, and they can be predicted by existing stress–strain relations based on isotropic hardening. Elasto–plastic models based on isotropic hardening assume a yield surface which monotonically expands with loading. Unloading

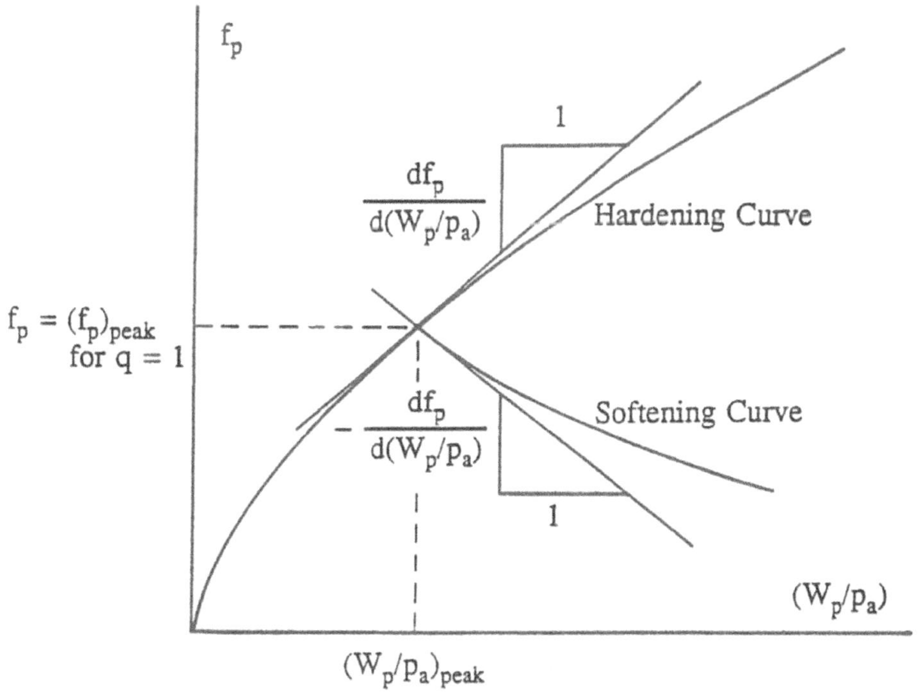

Fig. 5. Modeling of isotropic work hardening and softening.

in any direction from the outermost yield surface which has been reached by the stress path results in elastic strains only. Plastic strains are again produced only when the stress path penetrates the outermost yield surface. According to the isotropic hardening hypothesis, the direction of the plastic strain increment vector is determined by the location of the stress point and is independent of the stress path and the previous loading history.

However, experimental results indicate that large stress reversals result in plastic deformations which are such that the directions of the corresponding plastic strain increment vectors are dependent on the stress path and the previous stress history. Figure 6(a) shows the results of a triaxial test in which the specimen is first loaded in compression up to a stress level measured by $f_p = 22.4$, then unloaded (decreasing f_p), reloaded in extension (increasing f_p) and finally primary loaded in extension (f_p increasing past $f_p = 22.4$). This stress path is shown in the triaxial plane in Fig. 6(b). Significant plastic yielding occurs before the hydrostatic state of stress has been reached, and the stress–strain curve continues smoothly as the specimen is reloaded and primary loaded in extension. This behavior, generally referred to as the Bauschinger effect, is typically modeled within the framework of elastic–plastic theory by introducing a kinematic hardening law.

EXISTING KINEMATIC HARDENING MECHANISMS

The most commonly applied kinematic hardening mechanism has been the translation of an elastic region with the current stress point. Such a mechanism may be visualized by a possibly infinite number of yield surfaces which move in the stress space in a purely translational manner and for which the plastic hardening moduli are defined individually. Such a formulation was proposed independently by Mroz [13] and Iwan [14] with the condition that the surfaces remain tangent to each other as they move. In this manner, yield surfaces are activated and plastic strain is accumulated during initial loading as well as during any arbitrary unloading and reloading stress paths. The nested surface approach has been applied to the behavior of soils in a number of studies [15, 16]. Yet, as noted by Prevost [17] and Norris [18], the necessity of keeping a large number of surfaces in memory in order to predict a fairly complicated stress path with reversals makes the computer implementation of such a model complicated and expensive.

Such considerations were probably the motivation leading to the bounding surface concept introduced independently by Dafalias and Popov [19] and Krieg [20]. Bounding surface plasticity employs an inner surface commonly referred to as the yield or loading surface, and an outer surface commonly

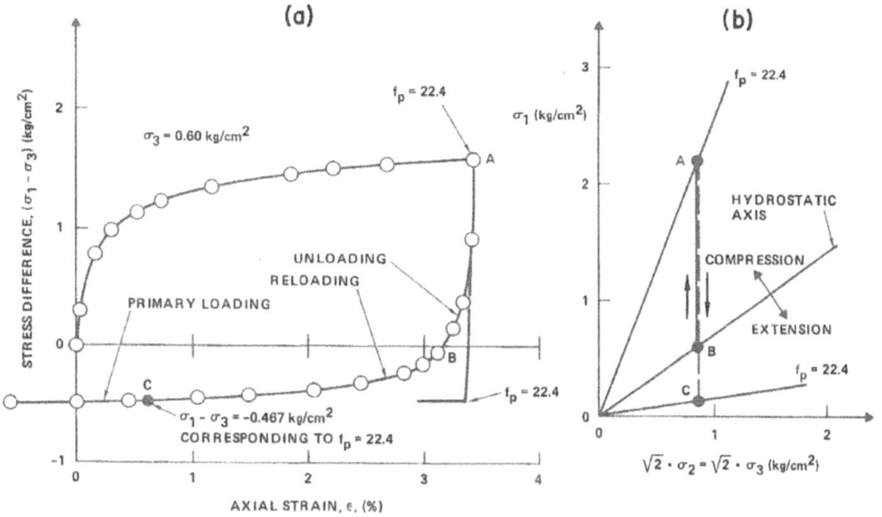

Fig. 6. Triaxial compression–extension test on loose Santa Monica beach sand; (a) measured stress–strain behavior and (b) stress path in triaxial plane.

referred to as the limit or bounding surface. The bounding surface typically obeys some form of isotropic hardening. The hardening of the loading surface usually denoted by a plastic hardening modulus is determined as a function of the distance from the current stress point to a point on the bounding surface which is found by a predefined mapping rule. This rule may change from one model to another. The loading surface encloses an elastic region, although, in some models, this surface shrinks to a point, thus the elastic region vanishes and no explicit surface is postulated within the bounding surface. When the stress state is well within the bounding surface, the plastic hardening modulus is very stiff. On the other hand, as the stress point approaches the bounding surface, the plastic hardening modulus value converges to that of the bounding surface. These conditions enable the material to go through a smooth transition from elastic to pre-reversal plastic state. The bounding surface approach has been applied to the behavior of soils in a number of studies [21–25].

The nested surface approach in its original form defines one extreme of modeling where the whole domain is finally discretized, and the bounding surface approach advocates the other extreme with continuity throughout the domain. Therefore, it is not surprising to find features common to both approaches in many of the elasto-plastic constitutive models developed for soils. On the other hand, the underlying assumption of translating a yield surface is not the only choice for a kinematic hardening mechanism. Due to its simplicity of implementation and its previous use in applications in metal plasticity [26–28], the use of a translational mechanism for kinematic hardening has enjoyed widespread use in soil plasticity and many experimental studies on soil behavior during large stress reversals have been interpreted in such terms [29, 30]. Yet, theoretically, translation of the yield surface allows the possibility that negative incremental plastic work may be generated along stress paths with current stress increments activating a translated yield surface and directing its movement towards the origin of the stress space [31]. This is a physically unacceptable condition which is routinely ignored by most kinematic models involving translation of the yield surface.

Even if this theoretical inconvenience is deemed to have a negligible effect on the actual prediction of material behavior, there is increasing experimental evidence that, for the behavior of soils under large stress reversals, the motion of the yield surface is described in a more accurate and rational manner by rotation rather than translation. Hashiguchi [32, 33] suggested the incorporation of rotational hardening to the model by Sekiguchi and Ohta [34] to predict anisotropic consolidation. Lade [35], Lade and Boonyachut [36] suggested the concept of rotation as an extension of the double hardening model developed by Lade [7]. Ghaboussi and Momen [37], Poorooshasb and Pietruszczak [38] also developed models based on similar rotational concepts. Yet, all these attempts to develop a full model based on a rotational hardening mechanism did not go much beyond the conceptual basis or lacked further development and application. All of these models enforced the tangency condition such that rotating surfaces were not allowed to intersect with each other. The numerical difficulties associated with enforcing the tangency condition for noncircular surfaces typically led to the use of simple conic surfaces which do not conform to the experimentally observed shape of the yield surface. Thus, enforcement of the tangency condition may have been the major obstacle in further successful development of the rotational hardening concept.

PROPOSED ROTATIONAL HARDENING MECHANISM

Attempts in developing plasticity models applicable to the behavior of soils naturally led to investigations of the shape of the yield surfaces. A comprehensive summary of these efforts may be found in Murata *et al.* [39]. The nature of yield surfaces developed during monotonic loading has been well-established [4, 12, 40–43]. The nature of yield surfaces associated with yielding during large stress reversals has been the focus of investigations by Lade and Boonyachut [36], Yamada and Ishihara [44], Tanimoto *et al.* [45], Alawaji *et al.* [29] and Yasufuku *et al.* [46]. Tanimoto *et al.* [45] found that isotropic compression followed by a large stress reversal leads to a yield surface which is nearly symmetrical about the hydrostatic axis, whereas isotropic compression with subsequent shearing followed by a large stress reversal leads to a similarly shaped yield surface which has shrunk and rotated towards the direction of shearing. Connecting data points which indicate the beginning of yielding (as defined by significantly decreasing slope of the stress–strain relation) clearly show a surface formed by the combination of two curves with different curvatures. As shown in Fig. 7, reloading past the prior maximum stress level (established by shearing to point *A*) leads to a curve which may be considered as a portion of the isotropic yield surface, whereas reloading at lower stress levels initiates yielding along a curve which seems to be part of a shrunk and rotated yield surface. Figure 8 shows resulting yield surfaces from the work of Yasufuku *et al.* [46] for anisotropic consolidation in compression and extension. Again, two distinct yielding mechanisms are activated depending on the direction of the stress reversal path, and a combined yield surface oriented in the direction of anisotropic loading is therefore observed.

In general, these experimental observations show that stress reversal paths corresponding to continuing loading in the sense of conventional plasticity tend to initiate a yielding mechanism which is very similar to the isotropic yield surface. On the other hand, stress reversal paths corresponding to continuing unloading and eventual reloading within the isotropic yield surface indicate the activation of another yield mechanism rotated in the direction of the previous monotonic loading history. Thus, a new hybrid yield

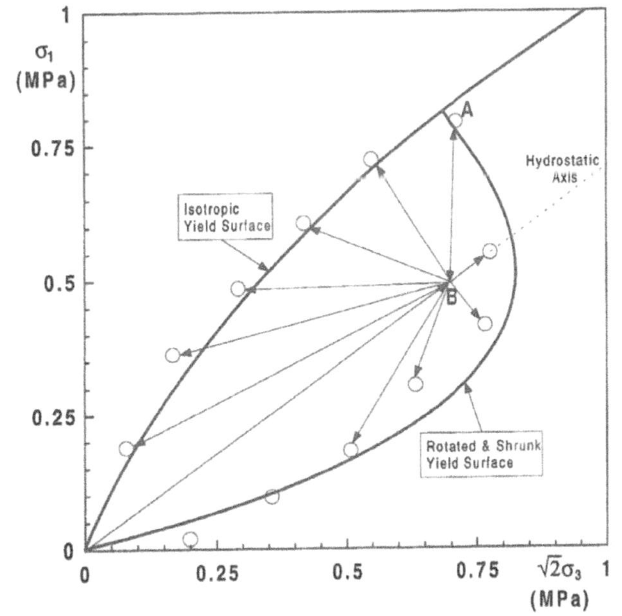

Fig. 7. Yield loci observed from exploratory stress paths on sand specimens following pre-shearing to point A and stress reversal to point B (after Tanimoto *et al.* [45]).

surface may be considered to be active during reversal of loading which may be viewed as a combination of the isotropic yield surface and a new rotated kinematic yield surface. These observations are consistent with kinematic behavior in the triaxial plane as discussed by Lade and Boonyachut [36] and

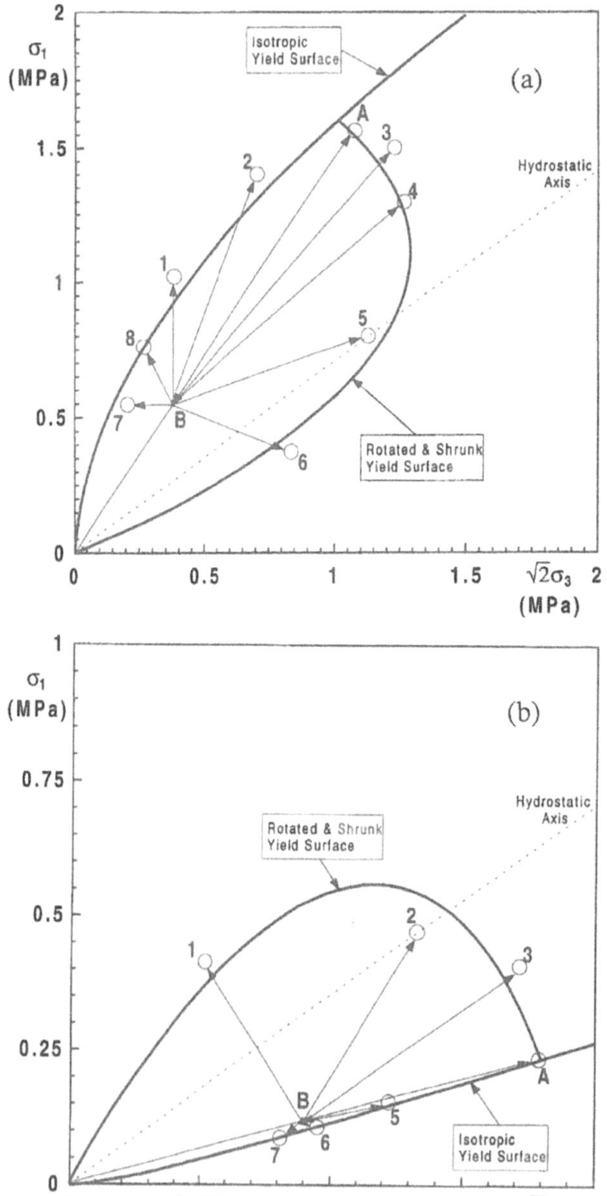

Fig. 8. Yield loci observed from exploratory stress paths on sand specimens following preshearing to point A and stress reversal to point B in (a) compression and (b) extension (after Yasufuku *et al.* [46]).

in principal stress space as discussed by Inel [47]. Observed translation of the yield surface in the octahedral plane in either distorted form as by Yamada and Ishihara [44] or in undistorted bubble form by Alawaji *et al.* [29] may also be explained within the framework of a rotating kinematic surface since rotation in the principal stress space will result in translational motion of the corresponding cross-section in the octahedral plane with a relative distortion depending on material characteristics, size of the rotating surface and the amount of rotation. The interactive interpretation of a rotating yield surface in the triaxial plane and its corresponding cross-section in the octahedral plane is shown in Fig. 9.

The concept of a hybrid yield surface defined as the intersection of the isotropic yield surface and a rotating kinematic yield surface necessarily refutes the requirement of a tangency condition. The intersection of yield surfaces has been regularly employed in double hardening models in which corresponding contributions of the separate yield surfaces have to be calculated individually. In the present concept of kinematic yielding, such calculations are avoided since it is only pertinent to know which yielding mechanism is active at the instant in question. Since both mechanisms cannot be active at the same moment, a uniqueness problem does not exist. It is recognized that minor discontinuities may be introduced for certain sudden

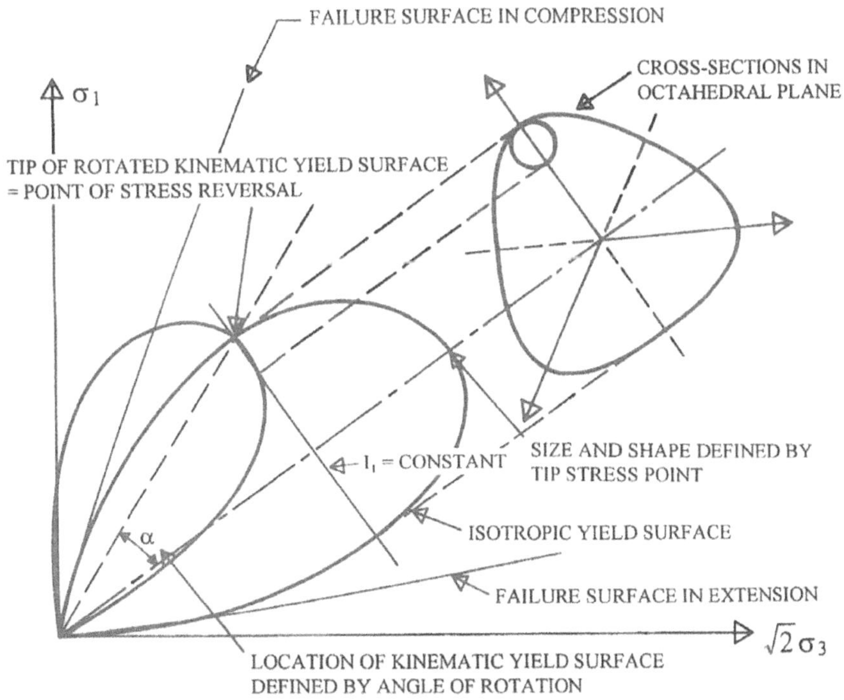

Fig. 9. Rotation of kinematic yield surface in triaxial plane and traces of isotropic and rotated kinematic yield surfaces in octahedral plane.

changes in the stress path direction in the close vicinity of the intersecting point of the two surfaces, yet such discontinuities are considered to be negligible compared to the overall behavior of the soil and further experimental work may be used to develop rules for a smooth transition under such situations if deemed necessary.

EVOLUTION OF THE KINEMATIC YIELD SURFACE

The existing single hardening constitutive model allows the location and size of both the yield surface and plastic potential to be described by a single scalar value. This scalar value may be chosen as the hydrostatic stress value at which the surface in question intersects the hydrostatic axis. In the following, this point of intersection will be referred to as the tip of the surface. Such a description is sufficient as long as isotropic hardening is valid and the surfaces expand uniformly with no rotation involved.

In the case of rotation, the orientation of each surface must be specified as well. The rotation of each surface may be defined relative to the hydrostatic axis. Since the hydrostatic axis uniquely defines the orientation of a surface, a pseudo hydrostatic axis may be employed to keep track of a rotating surface. The only variable required to describe this rotation in the triaxial plane is the angle between the original and the pseudo hydrostatic axes as shown in Fig. 9.

It is assumed that isotropic hardening is valid as long as the yield surface is expanding under primary monotonic loading. Upon any stress reversal leading to unloading within the yield surface, kinematic hardening rules will be employed. The isotropic yield surface will act as a limit surface which bounds all activity concerning stress reversals. Any stress reversal which continues in a fixed direction long enough to reach the isotropic yield surface will activate isotropic hardening and wipe out all past history effects due to kinematic hardening. The failure surface and the elastic behavior of the soil is assumed to be unaffected by stress reversals. Failure and subsequent softening will be initiated when the stress level reaches the failure surface regardless of the state of loading.

The kinematic hardening process will be described in terms of a secondary yield surface which is rotated around the origin. As shown in Fig. 9, on stress reversal from primary loading, the pseudo hydrostatic axis defining the position of the kinematic surface passes through the stress reversal point, and the tip of the rotated kinematic surface is at the stress reversal point. In this manner, both the orientation and size of the initial kinematic surface is determined by the stress reversal point. This choice follows immediately from consideration of soil behavior during isotropic loading along the hydrostatic axis followed by unloading. Experimental results show mainly elastic behavior for any unloading stress path following isotropic compression. Such behavior may be captured only if coincident kinematic and isotropic yield surfaces are formed for this stress path, which implies that the pseudo hydrostatic axis must pass through the stress reversal point and the tip of the initial kinematic yield surface must be at the stress reversal point.

For continuity, it is desirable that the kinematic surface will rotate and expand in such a way that it will merge precisely into the isotropic yield surface when primary loading is initiated again. Figure 10 shows that such continuity may be accomplished if the pseudo hydrostatic axis always rotates towards the original hydrostatic axis, and simultaneously, the kinematic surface expands such that its tip moves towards the tip of the isotropic yield surface. The condition of continuity may be satisfied in a simple manner by enforcing the tip of the kinematic surface to move on a straight line connecting the initial stress reversal point to the tip of the isotropic surface.

The region between the kinematic and isotropic surfaces defines an elastic domain which evolves as the kinematic surface rotates and expands in accordance with the stress reversal path. Figure 10 indicates that the kinematic surface may be determined at any time by requiring that (1) it passes through the current stress point, C, and (2) that its tip, D, lies on a straight line connecting the initial stress reversal point, B, with the isotropic yield surface tip, A. When the current stress point, C, reaches the isotropic surface at any point, both kinematic and isotropic tips, D and A, will be coincident,

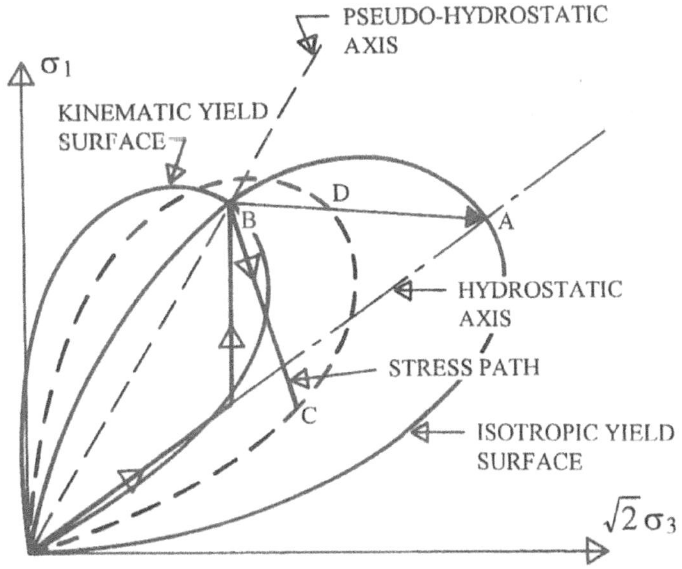

Fig. 10. Growth of kinematic yield surface (in magnitude and orientation) towards isotropic yield surface upon stress reversal at point B.

that is the kinematic surface will have merged into the isotropic surface. The proposed rotational mechanism is depicted in Fig. 10 for a conventional triaxial compression stress path followed by stress reversal at point B.

The stress reversal path within the isotropic yield surface may consist of any combination of minor reversals forming corners or cycles of unloading and reloading. For each such reversal, a progression of new kinematic surfaces must develop. This may easily be accomplished by forming consecutive kinematic surfaces according to the same logic used for the initial stress reversal. Eventually, such an approach will create a system of kinematic surfaces which are nested in the sense that each must grow towards its previously created counterpart. Since it is not desirable to keep in memory a large number of surfaces, and experimental observations show that soil tends to forget previous yielding mechanisms after sufficient yielding is activated by the latest surface, it was found to be appropriate to remember only one surface prior to the current kinematic surface. This surface will be referred to as the memory surface. Therefore, upon a new stress reversal at point C within the isotropic yield surface, a new kinematic yield surface is created as shown in Fig. 11. The previous kinematic yield surface is now referred to as the memory surface, with its tip at point D. The new kinematic surface will evolve towards the isotropic yield surface along the bilinear path, C–D–A, shown in Fig. 11. If a new kinematic surface is created, due to another stress reversal within the yield surface, as shown in Figs 12 and 13, the previous kinematic surface becomes the memory surface and the previous memory surface is erased.

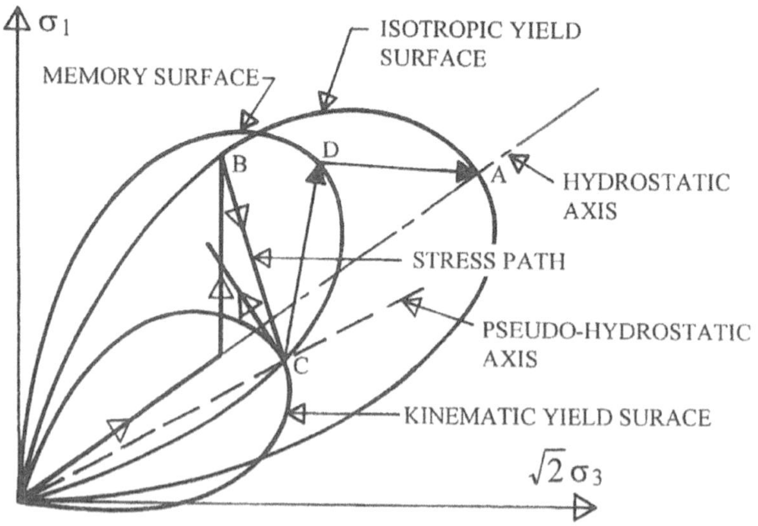

Fig. 11. Creation of new kinematic yield surface due to second stress reversal at point C; old kinematic yield surface becomes memory surface.

The kinematic surface is related to the memory surface in the same manner as the memory surface is related to the isotropic yield surface. The kinematic yield surface evolves in accordance with the stress reversal path and it eventually merges with the memory surface. In turn, the memory surface is activated and it subsequently evolves and merges with the isotropic yield surface.

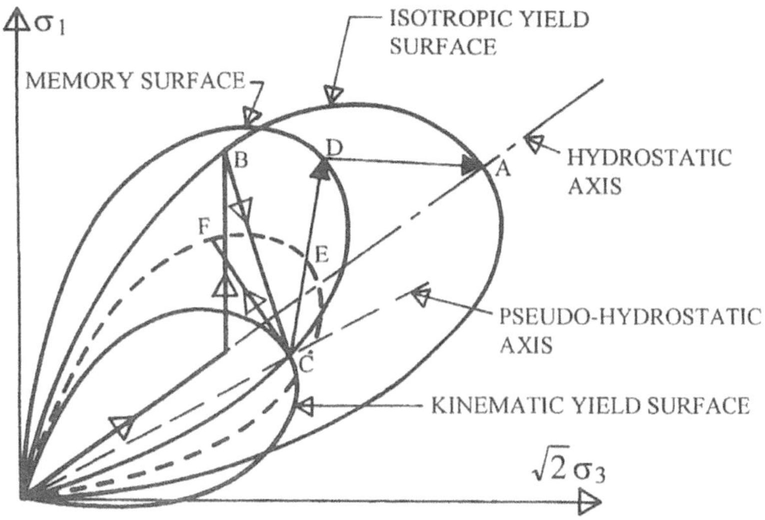

Fig. 12. Growth of new kinematic yield surface (in magnitude and orientation) towards memory surface due to second stress reversal at point C.

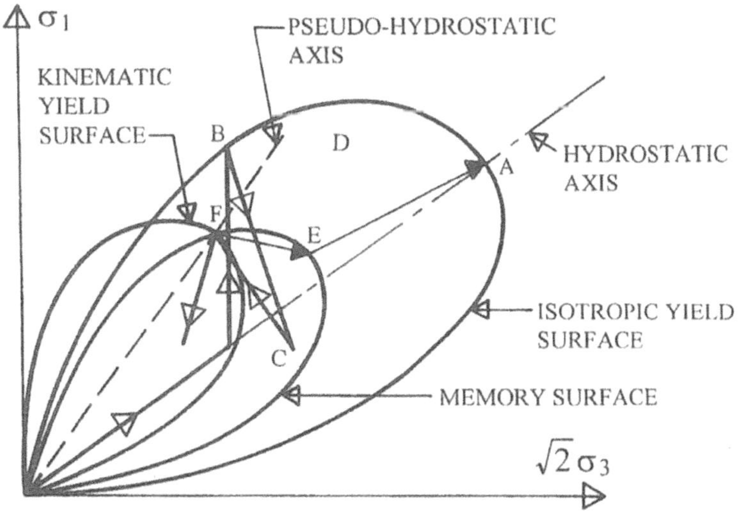

Fig. 13. Creation of yet another kinematic yield surface due to third stress reversal at point F. Previously active kinematic yield surface becomes memory surface. Old memory surface at point D fades away.

If yet another stress reversal occurs at point F in Fig. 13, then the previously active yield surface becomes the memory surface, and the old memory surface at point D fades away. The new kinematic surface created by the stress reversal at point F will evolve towards point E, the tip of the previously active kinematic surface, which has now become the memory surface. Thus, the greatest number of active yield surfaces is three. The yield surfaces further in the past fade away and any memory of their existence need not be retained. This is supported by experimental results, as will be demonstrated later.

The behavior observed from experiments and modeled with the three levels of yield surfaces, as explained above, will be further investigated and formulated mathematically in the following section.

MOTION OF PLASTIC POTENTIAL

The prediction of plastic strains for soil loaded along a given stress path may be separated into two parts: determination of plastic strain increment (1) directions and (2) magnitudes. The plastic strain increment directions are determined by the shape and position of the plastic potential, and the plastic strain increment magnitudes are determined by a work hardening rule. In the present paper, the behavior of the plastic potential is formulated in terms of the new rotational kinematic hardening mechanism. A testing program involving various stress paths in the triaxial plane is employed to establish the validity of the proposed rotational mechanism. The results of these tests are presented and interpreted in terms of plastic strain increments superimposed on the triaxial plane, as has been done previously for primary loading conditions [10, 40, 48] and stress reversals [35, 36, 47, 49]. The relative magnitudes of these strain increments are characterized by vector directions in the triaxial plane, and the accuracy of the plastic potential during large stress reversals is then assessed by comparing predicted and measured plastic strain increment directions. The extension of the isotropic work-hardening law to kinematic hardening conditions and an assessment of the accuracy of predictions of the plastic strain increment magnitudes will be made in a companion paper [50].

In the single hardening model, plastic strain increment directions are determined by the normal to the plastic potential surface at the current stress point. The plastic potential surface is attached to the isotropic yield surface which expands monotonically as the stress point moves in the stress space. In the proposed kinematic hardening model, the behavior of sand during large stress reversals is described by a kinematic yield surface which rotates around the origin and expands or shrinks as the stress point moves inside the isotropic yield surface. It is proposed that the plastic potential surface is again attached to the kinematic yield surface and moves along and expands or shrinks with this surface. This is in accordance with the experimentally observed behavior discussed by Lade and Boonyachut [36] in the triaxial plane and Lade [35] in the principal stress space. In their work, they approximated the plastic potential as a cone with smoothly rounded triangular cross section passing through the stress reversal point and the current stress point. Once the size of the cone was determined, the plastic strain increment vector direction was determined relative to the hydrostatic axis and then the cone was rotated back so that the correct orientation of the plastic strain increment vector relative to the rotated pseudo hydrostatic axis could be found. The concept was found to be quite successful in predicting the direction of the plastic strain increment vectors in the triaxial plane. Similar to following the rotation of a cylindrical cone by its axis of symmetry, the plastic potential of the proposed kinematic hardening model can be followed using a pseudo rotating hydrostatic axis. This procedure is depicted in Fig. 14 which shows the plastic potential surface and the corresponding plastic strain increment direction in the actual and in the rotated stress spaces.

In order to determine the rotation of the yield surface or the plastic potential relative to the original hydrostatic axis, an equation governing the motion of both surfaces must be established. Such an equation may be

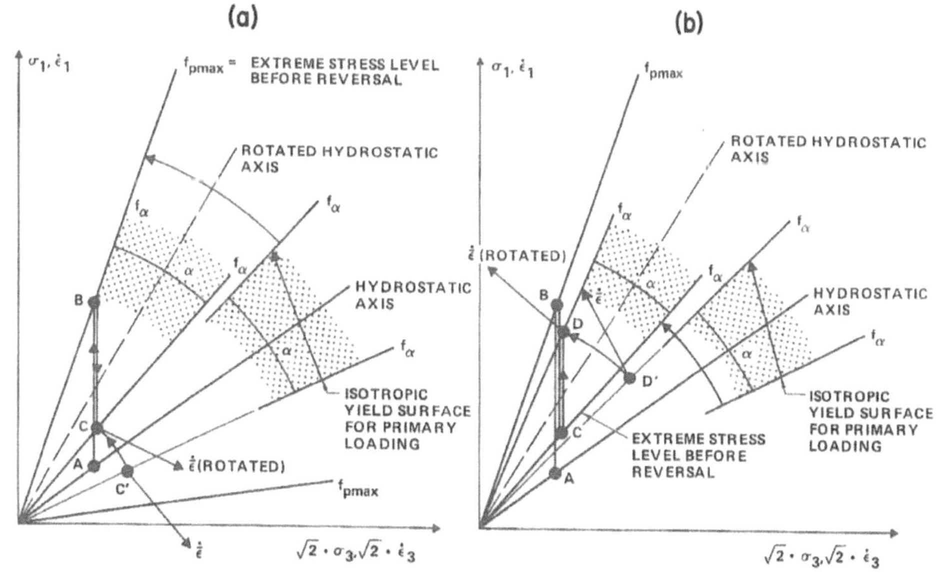

Fig. 14. Modeling hypothesis involving rotation of plastic potential and plastic strain increment vector in triaxial plane proposed by Lade and Boonyachut [36] for soil behavior during large stress reversals.

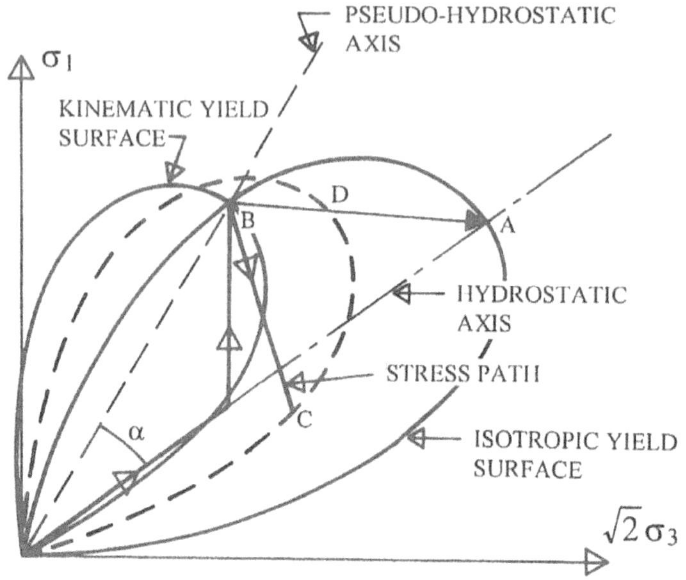

Fig. 15. Proposed modeling hypothesis consisting of rotation and growth of kinematic yield surface towards the isotropic yield surface upon stress reversal at point B.

derived from the specific rules and constraints proposed above, which introduced the rotational kinematic hardening mechanism. Once the governing equation for the rotational mechanism is determined, predictions of plastic strain increment directions can be made in the rotated stress space in a manner similar to that employed for the isotropic single hardening model. Measured and predicted plastic strain increment directions may then be compared to assess the accuracy of the proposed behavior for the plastic potential.

ANALYSIS OF ROTATION IN THE TRIAXIAL PLANE

The kinematic mechanism in the triaxial plane consists of plane rotation around the stress origin as defined by an angle of rotation, a, between the true hydrostatic axis and a rotated pseudo hydrostatic axis indicating the orientation of the kinematic surface, as shown in Fig. 15. In order to develop an analytical formulation for the kinematic mechanism, it is necessary to deal with stresses in the rotated coordinate system. Rotated stress values indicated by a star may be calculated as follows:

$$\begin{bmatrix} \sigma_1^* \\ \sqrt{2}\sigma_3^* \end{bmatrix} = \begin{bmatrix} \cos a & \sin a \\ -\sin a & \cos a \end{bmatrix} \begin{bmatrix} \sigma_1 \\ \sqrt{2}\sigma_3 \end{bmatrix} \tag{21}$$

The angle of rotation is found by enforcing two constraints: the first constraint is that the current stress point, σ_c, must lie on the active kinematic yield surface, and therefore:

$$f'_p(\overline{\sigma}^*_C) = f''_p[(W^*_p)C] \tag{22}$$

The second constraint is that the kinematic yield surface must evolve in such a manner that its tip moves on a straight line connecting the initial stress reversal point to the tip of the maximum past isotropic yield surface, as shown in Fig. 15:

$$\begin{bmatrix}\sigma_{1D}\\ \sigma_{3D}\end{bmatrix} = \begin{bmatrix}\sigma_{1B}\\ \sigma_{3B}\end{bmatrix} + \kappa \cdot \begin{bmatrix}\sigma_{1A} - \sigma_{1B}\\ \sigma_{3A} - \sigma_{3B}\end{bmatrix} \tag{23}$$

where the subscripts A, B and D correspond to the tips of the isotropic yield surface, the initial kinematic yield surface and the currently active kinematic yield surface, respectively. The scalar ratio, κ, describes the displacement of stress point D in terms of the distance from point B to A and thus has a value between 0 and 1.

The initial angle of rotation defining the first point of stress reversal, a_{AB}, is determined from the dot product of the radial position vectors to points A and B:

$$\vec{\sigma}_A \cdot \vec{\sigma}_B = \| \vec{\sigma}_A \| \cdot \| \vec{\sigma}_B \| \cdot \cos a_{AB} \tag{24}$$

and similarly, the angle of rotation, a, is defined by the dot product of the radial position vectors to points A and D:

$$\cos a = \frac{\vec{\sigma}_A \cdot \vec{\sigma}_D}{\| \vec{\sigma}_A \| \cdot \| \vec{\sigma}_D \|} = \frac{\sigma_{1A} \cdot \sigma_{1D} + 2 \cdot \sigma_{3A} \cdot \sigma_{3D}}{\sqrt{(\sigma_{1A}^2 + 2 \cdot \sigma_{3A}^2) \cdot (\sigma_{1D}^2 + 2 \cdot \sigma_{3D}^2)}} \tag{25}$$

These relations can be combined with Eqn (23) to solve for the angle of rotation, a, in terms of the scalar ratio, κ, as follows:

$$\cos a = \frac{(\xi - \cos a_{AB}) \cdot \kappa + \cos a_{AB}}{\sqrt{(1 - 2\xi \cos a_{AB} - \xi^2) \cdot \kappa^2 + 2\xi \cdot \kappa \cdot \cos a_{AB} + \xi^2}} \tag{26a}$$

in which

$$\xi = \frac{\| \vec{\sigma}_B \|}{\| \vec{\sigma}_A \|} \tag{26b}$$

If the stress values corresponding to the tip of the yield surface are known, then the expression for the yield surface, given by Eqn (14), may be used to quantify f''_p, the right hand side of Eqn (22). Using the subscript D to indicate the stress point at the tip of the yield surface, this expression may be simplified:

$$f''_p[W^*_p)C] = (27 \cdot \psi_1 + 3) \cdot \left[\frac{I_{1D}}{p_a}\right]^h \tag{27}$$

Therefore, using the current stress point, C, and the stress point at the tip of the yield surface, D, and employing Eqn (14) for both sides of the yield function, the following relation is obtained:

$$\left[\psi_1 \cdot \frac{I^{*3}_{1C}}{I^*_{3C}} - \frac{I^{*2}_{1C}}{I^*_{2C}}\right] \cdot \left[\frac{I^*_{1C}}{p_a}\right]^h \cdot e^{q*} = (27 \cdot \psi_1 + 3) \cdot \left[\frac{I_{1D}}{p_a}\right]^h \tag{28}$$

From Eqn (23), stress point D, and therefore the right hand side of Eqn (29), may be expressed in terms of κ and stress points A and B. Using Eqn (21), the left hand side of Eqn (28) may be expressed in terms of the current stress point in the unrotated stress space and the angle of rotation, a. Further, the relation given in Eqn (26a) is substituted for the angle of rotation, a. Finally, Eqn (22) is rewritten as the following nonlinear equation:

$$f'_p(\vec{\sigma}_A, \vec{\sigma}_B, \vec{\sigma}_C, \kappa) = f''_p(\vec{\sigma}_A, \vec{\sigma}_B, \kappa) \tag{29}$$

Equation (29) forms the governing equation defining the motion of the kinematic yield surface and the attached plastic potential surface in stress space. At any load step, the only unknown in this equation is the displacement ratio, κ. Due to the transcendental nature of the equation, the unknown value of κ has to be determined numerically. It is preferred to solve the equation in terms of κ rather than a since the iterative solution procedure is simplified by the knowledge that the value of κ ranges between 0 and 1, and that it monotonically increases unless a new stress reversal occurs, in which case the value of κ for the new kinematic yield surface is initialized to 0. Typically, a few iterations are sufficient to find the new value of κ for continuous portions of the stress path. The corresponding computational burden is therefore small. Once the value of κ is known, the corresponding rotation and new position of the kinematic yield surface may be calculated in a straight forward manner.

EXPERIMENTAL STUDY

The behavior of sand during large stress reversals in triaxial tests on cylindrical specimens was studied by Boonyachut [49]. The testing program consisted of:

(1) a series of conventional triaxial compression tests with constant confining pressure involving cycles of unloading to and reloading from hydrostatic stress conditions,
(2) a series of triaxial compression tests with different stress paths involving large stress reversals,
(3) a series of triaxial tests with constant confining pressure involving unloading and reloading from compression to extension and vice versa.

The sand used in the testing program was obtained from Santa Monica Beach, California. Details on the properties of the sand, its preparation for testing and the testing procedure are given by Boonyachut [49] and by Lade and Boonyachut [36]. All tests were isotropically consolidated drained tests. The tests presented here were performed on loose specimens in order to obtain relatively large strains, thus facilitating the interpretation of the results. The loose specimens had a void ratio of 0.81 corresponding to a relative density of 20%.

The accuracy of the proposed plastic potential was evaluated by comparing predicted and measured behavior using the results of this experimental study.

EVALUATION OF PROPOSED PLASTIC POTENTIAL

After subtracting the elastic components from the measured strains, the plastic strain increments may be superimposed on the stress space. The directions of the corresponding plastic strain increment vectors depend on the relative magnitudes of the components and not on their actual magnitudes. These vector directions may be used to study yield and plastic potential surfaces suitable for the description of plastic behavior during large stress reversals.

The plastic strain increment vectors obtained according to the proposed rotational hypothesis are presented in Figs 16–21. Experimental data are indicated with solid arrows. Only the directions of the vectors obtained from the experiments and the model are relevant. Their lengths are immaterial to the present considerations.

Figure 16 shows predicted and measured plastic strain increment directions for a series of simple triaxial compression tests with several unloading–reloading cycles. Immediately following a stress reversal, the strains are mainly elastic, whereas the plastic strain components increase as the current stress point moves away from the point of stress reversal. The experimental plastic strain increment vectors are therefore determined towards the end of

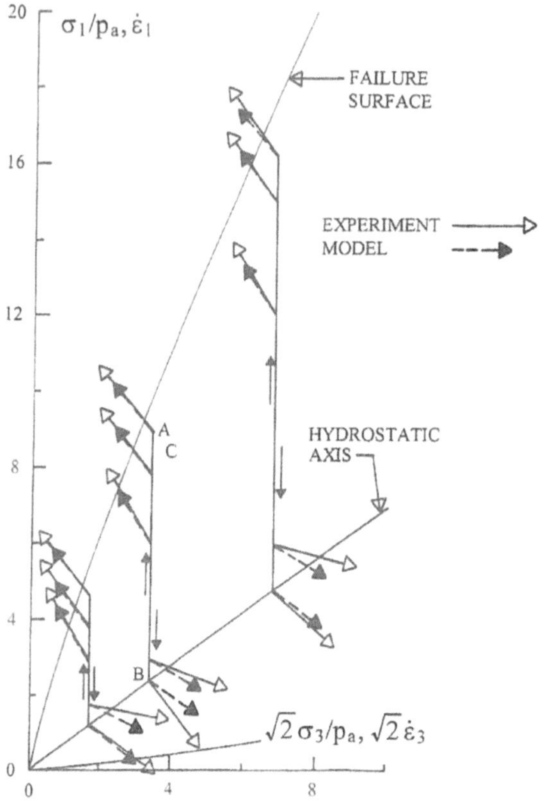

Fig. 16. Stress paths with superimposed plastic strain increment vectors in triaxial plane for unloading and reloading in conventional triaxial compression tests on loose Santa Monica beach sand.

each straight section of the stress paths, where the larger plastic strains provide more reliable evaluations. Note that although some of the stress states are in the region of compression, the strain increment directions for unloading point downwards similar to those observed for extension tests [48]. For reloading, the vectors point upwards as they do for primary loading in compression.

The directions of the experimental and predicted vectors compare favorably. Although careful measurements were made in the tests performed for this investigation, these measurements do contain errors due to slight variations in void ratios between specimens, corrections for membrane penetration and sand penetration into the lubricated ends, etc. In addition, the quality of the calculated plastic strains employed in the study depends on the model and soil parameters used to compute the elastic strains. Therefore, the experimental vector directions can be expected to show considerable scatter, especially when the strains are small. This is demonstrated in Fig. 3, where the experimental vector directions are on either side of the predicted vector

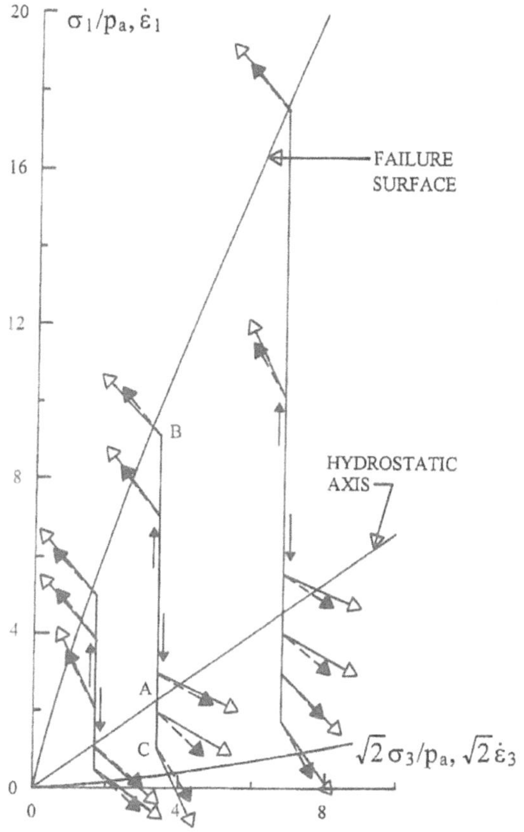

Fig. 17. Simple stress paths with superimposed plastic strain increment vectors in triaxial plane for triaxial tests on loose Santa Monica beach sand.

directions. The vector directions predicted by the model are consistent and they do not show scatter, but they may show systematic deviations to one side of the experimental vector directions. However, it appears that the two sets of vector directions are reasonably close together, and the differences in vector directions shown in Fig. 3 are considered to reflect scatter rather than systematic deviations.

Figure 17 shows comparisons for simple stress paths involving primary loading in compression followed by unloading and reloading in extension. The specimens are then unloaded in extension, reloaded in compression and finally primary loaded to failure.

The data in Fig. 18 correspond to a simple stress path initiated with primary loading in extension followed by unloading and reloading in compression. The specimen is then unloaded again and reloaded in extension and following a final reversal, the specimen is failed in compression. The comparison of experimental and predicted vector directions seem to be quite favorable.

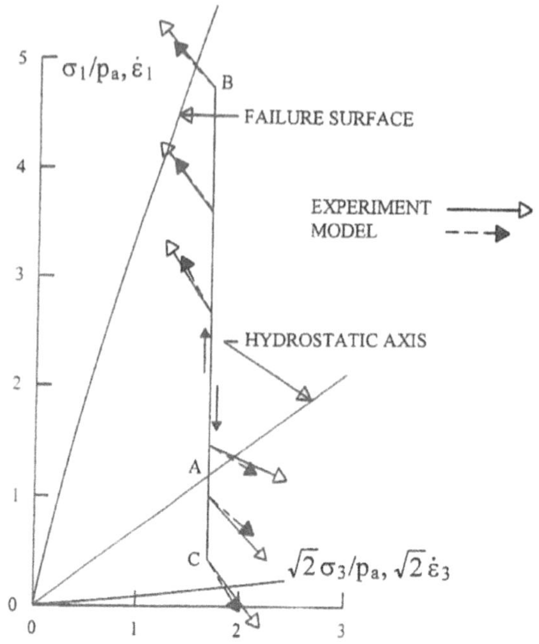

Fig. 18. Simple stress paths with superimposed plastic strain increment vectors in triaxial plane for triaxial tests on loose Santa Monica beach sand.

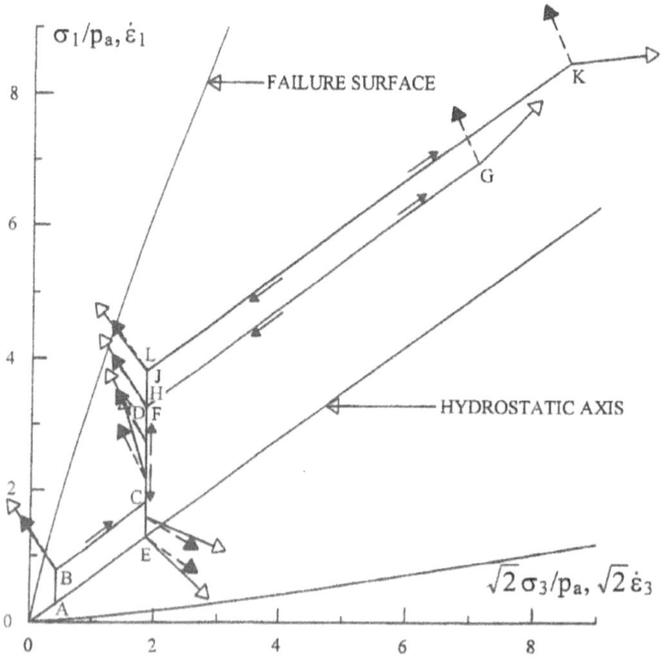

Fig. 19. Complex stress path with superimposed plastic strain increment vectors in triaxial plane for triaxial tests on loose Santa Monica beach sand.

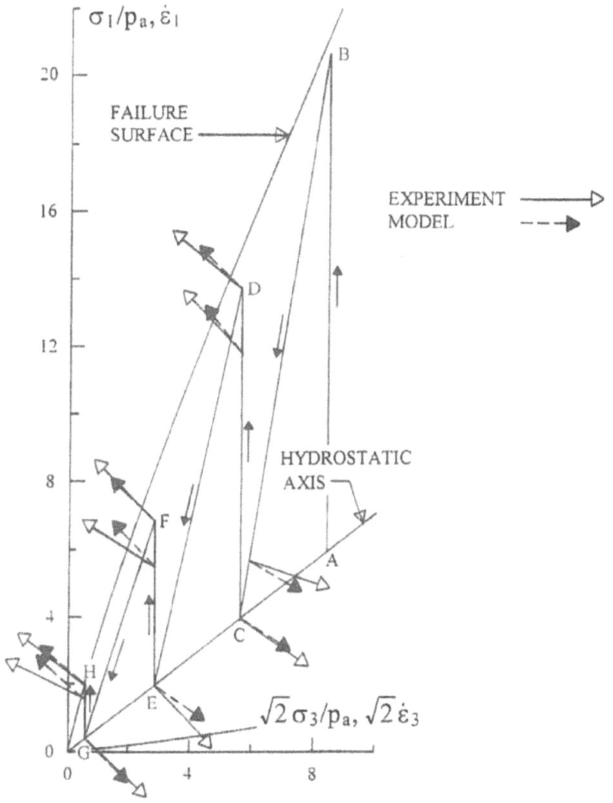

Fig. 20. Complex stress path with generally decreasing mean normal stress and with superimposed plastic strain increment vectors in triaxial plane for triaxial test on loose Santa Monica beach sand.

Figures 19–21 show the results of tests with complex stress paths. These tests involve varying confining pressures, and the stress paths are designed to explore plastic soil behavior in various directions and to see if the proposed model is able to capture the behavior of soil for such complex stress paths. Comparisons between experimental and predicted vector directions do in some instances show substantial differences, yet favorable comparisons are obtained for most cases.

Figure 19 shows comparisons for a rather complex stress path. The letters at the stress reversal points indicate the sequences in which the stress points are reached. The effect of previous stress history can be studied from the results of this test. For example, stress paths C–D and E–F both reach the same stress point (D and F), and both involve reloading, because point B corresponds to a higher stress level than point D (F), as measured by f_p in the original coordinate system. The predicted vector directions compare very favorably with the experimental vector directions, and they show little to no effect of the previous stress paths.

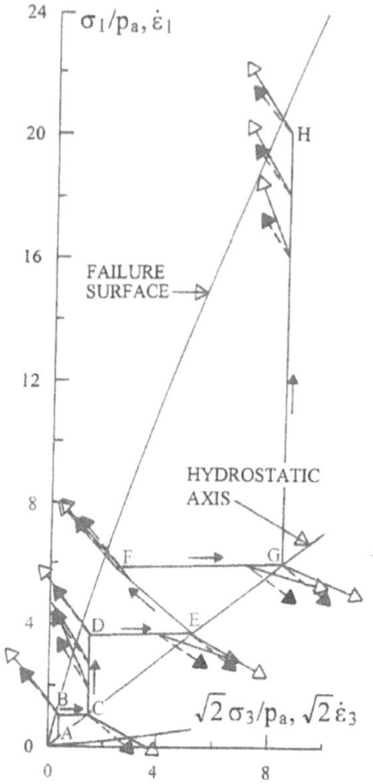

Fig. 21. Complex stress path with generally increasing mean normal stress and with superimposed plastic strain increment vectors in triaxial plane for triaxial test on loose Santa Monica beach sand.

The experimental vector directions shown at points G and K are those corresponding to stress paths F–G and J–K, respectively. Clearly, they do not compare favorably with the predicted vector directions. The measured strains for stress paths F–G and J–K are very small, and only slightly incorrect elastic strains will cause the calculated plastic strains to be in error.

Figure 20 shows a stress path with alternating increases and decreases in mean normal stress, and Fig. 21 shows a stress path with generally increasing mean normal stress. The comparisons between experimental and model vector directions are very good in all cases. Notably, they compare favorably at points E and G, which are very similar in many respects to points G and K in Fig. 19. Thus, it is likely that the scatter in the test results accounts for at least part of the differences between the experimental and model vector directions at points G and K shown in Fig. 19.

The difference between measured and predicted directions of plastic strain increments are plotted against measured plastic strain increment magnitudes in Fig. 22 for the data previously shown in Fig. 16 for the intermediate

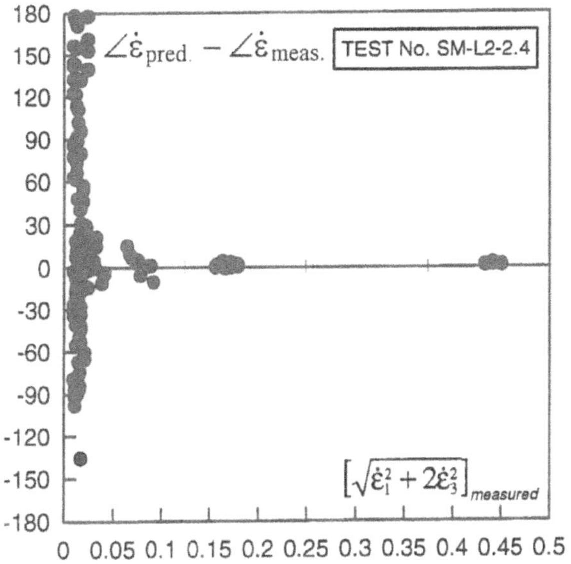

Fig. 22. Diagram of differences between predicted and measured plastic strain increment vector directions for triaxial test with simple stress path.

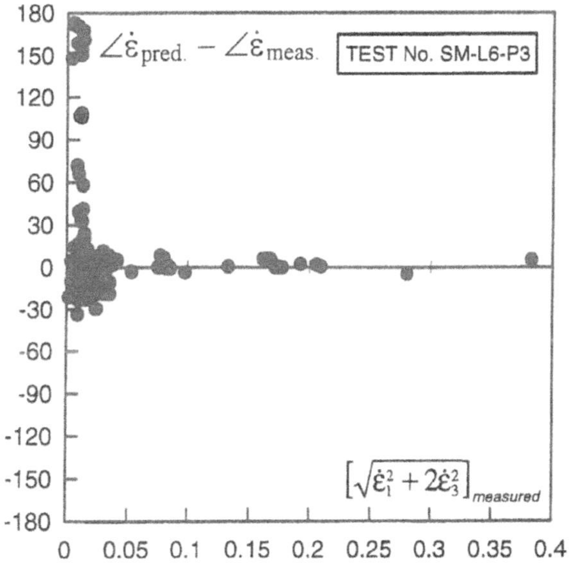

Fig. 23. Diagram of differences between predicted and measured plastic strain increment vector directions for triaxial test with complex stress path.

confining pressure. A similar evaluation for the test shown in Fig. 21 is presented in Fig. 23. Considering the small strains measured for some stress paths, the inevitable variations from one test to another, the corrections due to membrane penetration, and the errors associated with the calculation of plastic strain increment vector directions, it is clear that some scatter in the plastic strain increment directions occurs. As expected, the largest amounts of scatter are associated with the stress paths producing the smallest strains. Nevertheless, the relationships depicted in the diagrams in Figs 22 and 23 clearly show that there is a strong correlation between the accuracy of the model predictions and the magnitude of measured strains. Thus, the prior conclusion that scatter in the test results is at least partially responsible for some of the deviations between measured and predicted responses is supported by the trends in Figs 22 and 23.

CONCLUSIONS

Current approaches to the formulation of kinematic hardening constitutive models based on the theory of plasticity have been reviewed and evaluated in light of experimental studies on sand behavior during large stress reversals. Based on test results, an alternative kinematic hardening mechanism is proposed which incorporates rotation and intersection of yield surfaces to achieve a more consistent and physically rational fit with behavior observed in both the triaxial and octahedral planes. The existing single hardening model [3–5] forms the basic framework to which the rotational kinematic mechanism is added.

Predictions of plastic strain increment vector directions from the proposed rotational kinematic model were compared with experimental results from triaxial compression and extension tests in the triaxial plane. Within the scatter of the test results, the model is shown to capture the behavior of soil during large stress reversals with reasonable accuracy.

The proposed model for the yield and plastic potential surfaces involves a combination of isotropic hardening based on an existing single hardening model and kinematic hardening based on a new rotational mechanism. The only additional information necessary to predict plastic strain increment directions during large stress reversals consists of the current stress state and the stress states defining the configuration of the memory and kinematic yield surfaces. Thus, no additional soil parameters are required. The expansion of the proposed model to general three-dimensional stress conditions is straightforward and will be presented in future papers. In view of the apparent complexity of soil behavior during large stress reversals, the proposed model appears to be relatively simple.

The new combined model preserves the behavior of the isotropic hardening model under monotonic loading conditions and the extension from isotropic to kinematic hardening is accomplished without introducing complex rules and new material parameters.

ACKNOWLEDGEMENTS

The research presented here was performed in the Department of Civil Engineering at University of California, Los Angeles, with support from the National Science Foundation under Grant No. MSS 9119272/MSS 9396271. S.I. was supported partly by the Kenneth L. Lee Memorial Fellowship and partly by the Department of Civil Engineering at UCLA. Grateful appreciation is expressed for this support.

REFERENCES

1. Chen, W. F. and Baladi, G. Y., *Soil Plasticity: Theory and Implementation.* Elsevier, Amsterdam, 1985.
2. Desai, C. S. and Siriwardane, H. J., *Constitutive Laws for Engineering Materials with Emphasis on Geologic Materials.* Prentice-Hall, New Jersey, 1984.
3. Kim, M. K. and Lade, P. V., Single hardening constitutive model for frictional materials, I. plastic potential function. *Computers and Geotechnics*, 1988, **5**, 307–324.
4. Lade, P. V. and Kim, M. K., Single hardening constitutive model for frictional materials, II. yield criterion and plastic work contours. *Computers and Geotechnics*, 1988, **6**, 13–30.
5. Lade, P. V. and Kim, M. K., Single hardening constitutive model for frictional materials, III. comparisons with experimental data. *Computers and Geotechnics*, 1988, **6**, 31–48.
6. Lade, P. V. and Nelson, R. B., Modeling the elastic behavior of granular materials. *International Journal of Numerical and Analytical Methods in Geomechanics*, 1987, **11**, 521–542.
7. Lade, P. V., Elasto–plastic stress–strain theory for cohesionless soil with curved yield surfaces. *International Journal of Solids and Structures*, 1977, **13**, 1019–1035.
8. Lade, P. V., Three-parameter failure criterion for concrete. *Journal of Engineering Mechanics, ASCE*, 1982, **108**(EM5), 850–863.
9. Kim, M. K. and Lade, P. V., Modelling rock strength in three dimensions. *International Journal of Rock Mechanics and Mining Science Geomechanical Abstract*, 1984, **s21**, 21–33.
10. Kim, M. K., A study of constitutive models for frictional materials. Ph.D. dissertation, University of California, Los Angeles, 1984.
11. Lade, P. V., Single-hardening model with application to NC clay. *Journal of Geotechnical Engineering, ASCE*, 1990, **116**(GT3), 394–414.
12. Lade, P. V. and Kim, M. K., Single hardening constitutive model for soil, rock and concrete. *International Journal of Solids and Structures*, 1995, **32**(14), 1963–1978.
13. Mroz, Z., On the description of anisotropic work hardening. *Journal of Mechanics and Physics of Solids*, 1967, **15**, 163–175.
14. Iwan, W. D., On a class of models for the yielding behavior of continuous and composite systems. *Journal of Applied Mechanics*, 1967, **34**, 612–617.
15. Mroz, Z., Norris, V. A. and Zienkiewicz, O. C., Application of an anisotropic hardening model in the analysis of elastoplastic deformation of soils. *Geotechnique*, 1979, **29**, 1–34.

16. Prevost, J. H., Plasticity theory for soil stress–strain behavior. *Journal of Engineering and Mechanics Divisions, ASCE*, 1978, **104**(EM5), 1177–1194.
17. Prevost, J. H., Two-surface versus multi-surface plasticity theories: a critical assessment. *International Journal of Numerical and Analytical Methods in Geomechanics*, 1982, **6**, 323–338.
18. Norris, V. A., Numerical modeling of soil response to cyclic loading using 'stress reversal surfaces' In *Proceedings of the International Symposium on Numerical Models in Geomechanics*, ed. R. Dungar, G. N. Pande and J. A. Studer. Zurich, 1982, pp. 50–64.
19. Dafalias, Y. F. and Popov, E. P., A model for nonlinearly hardening materials for complex loading. *Acta Mechanica*, 1975, **21**, 173–192.
20. Krieg, R. D., A practical two-surface plasticity theory. *Journal of Applied Mechanics*, 1975, **42**, 641–646.
21. Aboim, C. A. and Roth, W. H., Bounding-surface-plasticity theory applied to cyclic loading of sand. In *Proceedings of the International Symposium on Numerical Models in Geomechanics*, ed. R. Dungar, G. N. Pande and J. A. Studer. Zurich, Switzerland, 1982, pp. 65–72.
22. Bardet, J. P., Application of bounding surface plasticity to cyclic sand behavior. In *Proceedings of the 2nd International Conference Soil Dynamics & Earthquake Engineering*, ed. C. A. Brebbia and A. M. Abdel-Ghaffar. Springer-Verlag, Berlin, 1985, pp. 3–15.
23. Bardet, J. P., Hypoplastic model for sands. *Journal of Engineering Mechanics, ASCE*, 1990, **116**(EM9), 1973–1994.
24. Dafalias, Y. F. and Herrmann, L. R., Bounding surface formulation of soil plasticity. In *Soil Mechanics—Transient Cyclic Loads*, ed. G. N. Pande and O. C. Zienkiewicz. John Wiley and Sons, New York. 1982, pp. 253–282.
25. Wang, Z., Dafalias, Y. F. and Shen, C. K., Bounding surface hypoplasticity model for sand. *Journal of Engineering Mechanics, ASCE*, 1990, **116**(EM5), 983–1001.
26. Prager, W., A new method of analyzing stresses and strains in work-hardening plastic solids. *Journal of Applied Mechanics*, 1956, **78**, 493–496.
27. Ziegler, H., A modification of Prager's hardening rule. *Quarterly in Applied Mathematics*, 1959, **17**.
28. Phillips, A. and Weng, G. J., An analytical study of an experimentally verified hardening law. *Journal of Applied Mechanics*, 1975, **42**, 375–378.
29. Alawaji, H., Alawi, M., Ko, H.-Y., Sture, S., Peters, J. F. and Wood, D. M., Experimental observations of anisotropy in some stress controlled tests on dry sand. In *Yielding, Damage, and Failure of Anisotropic Solids*, ed. J. P. Boehler. Elsevier, Amsterdam, 1990, pp. 251–264.
30. Hirai, H., Modelling of cyclic behavior of sand with combined hardening. *Soils and Foundations*, 1987, **27**, 1–11.
31. Mosaddad, B. and Powell, G. H., *Computational Models for Cyclic Plasticity, Rate Dependence, and Creep in Finite Element Analysis.* Earthquake Engineering Research Center, University of California, Berkeley, 82/26, 1982.
32. Hashiguchi, K., An expression of anisotropy in a plastic constitutive equation of soils (Discussion on H. Sekiguchi and H. Ohta's Paper). Constitutive Equations of Soils. Specialty Session 9, IX ICSMFE, Tokyo, 1977, pp. 302–305.
33. Hashiguchi, K., Constitutive equations of granular media with an anisotropic hardening rule. Proceedings of the 3rd International Conference on Numerical Methods in Geomechamics. Aachen, 1979, pp. 435–439.

34. Sekiguchi, H. and Ohta, H., Induced anisotropy and time dependency in clays. Constitutive Equations of Soils. Specialty Session 9, IX ICSMFE, Tokyo, 1977, pp. 163–175.
35. Lade, P. V., Three-dimensional stress–strain behavior and modeling of soils. *Serie Grundbau*, **4**, Ruhr-Universitat, Bochum, 1979.
36. Lade, P. V. and Boonyachut, S., Large stress reversals in triaxial tests on sand. Proceedings of the 4th International Conference on Numerical Methods in Geomechanics. Edmonton, 1982, pp. 171–181.
37. Ghaboussi, J. and Momen, H., Plasticity model for cyclic behavior of sands. Proceedings of the 3rd International Conference on Numerical Methods in Geomechanics. Aachen, 1979, pp. 423–434.
38. Poorooshasb, H. B. and Pietruszczak, S., On yielding and flow of sand; a generalized two-surface model. *Computers and Geotechnics*, 1985, **1**, 33–58.
39. Murata, H., Miura, N., Hyodo, M. and Yasufuku, N., Experimental study on yielding of sand, mechanics of granular materials. Report of ISSMFE Committee on mechanics of granular materials, Rio de Janeiro, 1989.
40. Poorooshasb, H. B., Holubec, I. and Sherbourne, A. N., Yielding and flow of sand in triaxial compression: Part I. *Canadian Geotechnical Journal*, 1996, **3**, 179–190.
41. Roscoe, K. H. and Burland, J. B., On the generalized stress–strain behavior of 'wet' clay, In *Engineering Plasticity*, ed. J. Heymann and F. A. Leckie. Cambridge University Press, Cambridge, 1968, 535–609.
42. Parry, R. H. G. and Nadarajah, V., A volumetric yield locus for lightly overconsolidated clay. *Geotechnique*, 1973, **23**, 450–453.
43. Tatsuoka, F. and Ishihara, K., Yielding of sand in triaxial compression. *Soils and Foundations*, 1974, **14**, 63–76.
44. Yamada, Y. and Ishihara, K., Yielding of loose sand in three dimensional stress conditions. *Soils and Foundations*, 1982, **22**(3), 15–31.
45. Tanimoto, K., Tanaka, Y. and Kagotani, N., The effect of preshearing on yield locus of sandy soil. Proceedings of the 8th Asian Regional Conference on Soil Mechanics and Foundations Engineering. 1987, Vol. 1, pp. 105–108.
46. Yasufuku, N., Murata, H. and Hyodo, M., Yield characteristics of anisotropically consolidated sand under low and high stresses. *Soils and Foundations*, 1991, **31**, 95–109.
47. Inel, S., Kinematic hardening model for sand behavior during large stress reversals. Ph. D. dissertation, University of California, Los Angeles, 1992.
48. Lade, P. V. and Duncan, J. M., Elasto–plastic stress–strain theory for cohesionless soil. *Journal of Geotechnical Engineering, ASCE*, 1975, **101**(GT10), 1037–1053.
49. Boonyachut, S., Experimental study of the behavior of cohesionless soil during large stress reversals. Ph.D. dissertation, University of California, Los Angeles, 1977.
50. Inel, S. and Lade, P. V., Rotational kinematic hardening model for sand II. characteristic work hardening law and predictions. *Computers and Geotechnics*, 1997, **21**(3), 217–234.

ELSEVIER

Computers and Geotechnics, Vol. 21, No. 3, pp. 217–234, 1997

Printed in Great Britain
PII:S0266-352X(97)00023-2 0266-352X/97 $17.00+0.00

Rotational Kinematic Hardening Model for Sand. Part II Characteristic Work Hardening Law and Predictions

Sinan Inel[a] & Poul V. Lade[b]*

[a]Dames & Moore, 911 Wilshire Boulevard, Suite 700, Los Angeles, California 90017, U.S.A.
[b]Department of Civil Engineering, The Johns Hopkins University, Baltimore, Maryland 21218, U.S.A.

(Received 28 May 1996; revised version received 2 June 1997; accepted 9 June 1997)

ABSTRACT

A new kinematic hardening mechanism has been proposed which incorporates rotation and intersection of yield surfaces to achieve a more consistent and physically rational fit with experimentally observed soil behavior during large stress reversals. An existing elasto–plastic model with isotropic hardening is used as the basic framework to which the rotational kinematic hardening mechanism has been added. The existing, isotropic work hardening law is shown here to apply to isotropic as well as kinematic behavior of sand. The new combined model preserves the behavior of the isotropic hardening model under monotonic loading conditions, and the extension from isotropic to rotational kinematic hardening is accomplished without introducing new material parameters. The capability of the proposed model is examined by comparing predictions with experimental data for simple and complex stress paths involving large stress reversals in the triaxial plane. © 1997 Elsevier Science Ltd.

INTRODUCTION

A new kinematic hardening mechanism has been proposed by Lade and Inel [1], which incorporates rotation and intersection of yield surfaces. Analyses of data from triaxial tests performed at different confining pressures indicate that plastic yielding during unloading and reloading, and the directions of the plastic strain increment vectors tend to form a pattern of behavior, which may be described by an elasto-plastic constitutive model, involving kinematic

*To whom correspondence should be addressed.

hardening with rotational, rather than translational, evolution of the yield surfaces. An existing isotropic hardening model [2–4] is used as the basic framework to which the rotational kinematic hardening mechanism has been added. The new combined model preserves the behavior of the isotropic hardening model under monotonic loading conditions, and the extension from isotropic to kinematic hardening is accomplished in a simple manner, which avoids the introduction of complex rules and new material parameters.

The prediction of plastic strains for soil loaded along a given stress path may be separated into two parts: plastic strain increment *directions* and *magnitudes*. The plastic strain increment directions are determined from the shape and position of the plastic potential, and the plastic strain increment magnitudes are determined from a work hardening rule. In Lade and Inel [1], the behavior of the plastic potential was formulated in terms of the new rotational kinematic hardening mechanism. Comparisons of predicted and measured plastic strain increment directions showed that the proposed model was capable of capturing the behavior of sand during large stress reversals in the triaxial plane. In this paper, the complete stress–strain behavior of sand during large stress reversals will be used to evaluate the proposed model in terms of both plastic strain increment directions and magnitudes. For this purpose, the isotropic work-hardening law is extended to cover kinematic hardening conditions. In the process, it is suggested that the proposed work hardening law is a material characteristic which defines the accumulation of plastic work and the evolution of yield surfaces along any stress path. An assessment of the accuracy of the proposed model is made using results from a testing program which includes simple and complex stress paths involving unloading and reloading behavior in the triaxial plane.

WORK-HARDENING AS A MATERIAL CHARACTERISTIC

In the present approach to constitutive modeling of sand behavior within the framework of plasticity theory, the yield surface is defined as a contour of constant plastic work as measured from the origin of the stress space. The existing correlation between plastic work and yield surfaces is based on experimentally observed behavior of frictional materials. The isotropic work-hardening rule is described by an empirically derived power function which maps plastic work values to stress states and vice versa. Thus, as a stress path is traversed so that yielding occurs, the yield surface expands and plastic work accumulates at a positive rate governed by this function. Similarly, at any stress state, the accumulated plastic work also defines a unique yield surface in stress space.

The isotropic work-hardening law is extended to cover kinematic work-hardening by proposing that the already developed empirical relation between a yield surface and its corresponding plastic work state is universally valid for both isotropic and kinematic hardening. Thus, the work-hardening relation is invariant in the sense that, given any rotation of the coordinate system in stress space, a yield surface can uniquely be mapped into a plastic work value and vice versa using the same empirically observed relation. Consequently, as the kinematic yield surface rotates and expands during a large stress reversal, a unique plastic work value may be computed at each step of its evolution. The procedure is schematically depicted in Figs 1 and 2. As shown in Fig. 1, the plastic strain increment direction is calculated using the plastic potential in the rotated frame of reference. Figure 2 depicts the calculation of plastic strain increment magnitude using the increment of plastic work generated by the expansion of the yield surface. As shown schematically in the figure, the plastic strain increment magnitude is proportional to the difference in size of the initial and current kinematic yield surfaces rotated back to the hydrostatic axis. Note that plastic strains can now be calculated both during primary loading and stress reversals without introducing additional soil parameters.

Computation of incremental plastic work may be described in terms of the isotropic work-hardening curve. This curve, shown schematically in Fig. 3, may now be recognized as defining a characteristic work-hardening law which is unique for a given type of soil. When a stress reversal occurs, the

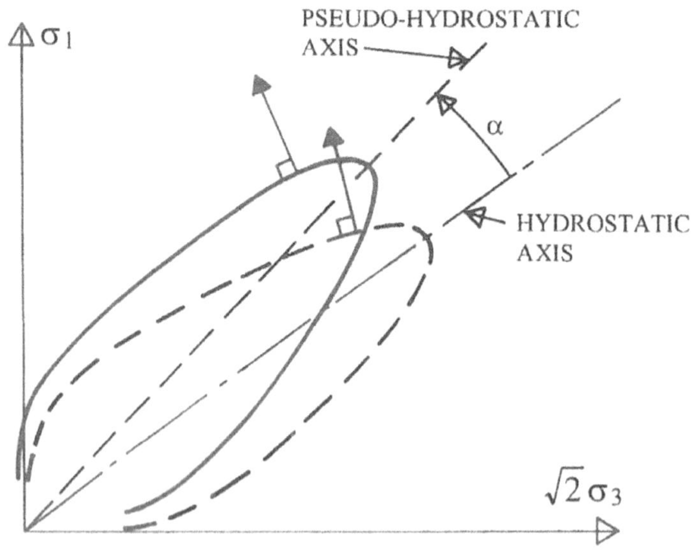

Fig. 1. Rotation of plastic potential and plastic strain increment vector in triaxial plane as proposed modeling hypothesis.

new rotated kinematic yield surface is constructed as described in Lade and Inel [1]. By definition, the kinematic yield surface is always smaller than the isotropic yield surface, and therefore, it will correspond to a unique point, B, on the work-hardening curve, which is located at a lower value in comparison to point A corresponding to the isotropic yield surface. As the stress reversal path eventually moves towards and reaches the isotropic yield surface, the kinematic surface will evolve into the isotropic surface. As indicated by point C in Fig. 3, its evolution will follow the work-hardening curve from point B to point A. When the kinematic surface merges with the isotropic surface, point A is reached again on the characteristic work-hardening curve, and work-hardening due to primary loading is commenced in a continuous manner.

For any load step, the increment of plastic work is determined by the slope of the work-hardening curve which may be referred to as the plastic work-hardening modulus. Immediately after a stress reversal, the plastic modulus is quite high, and, as loading continues, the modulus decreases, thus generating increasing amounts of plastic work and strain. In this manner, the value of the overall modulus is smoothly transformed from initially elastic to fully plastic behavior in a manner similar to that employed in the bounding surface theory [5] without resorting to complicated rules of mapping.

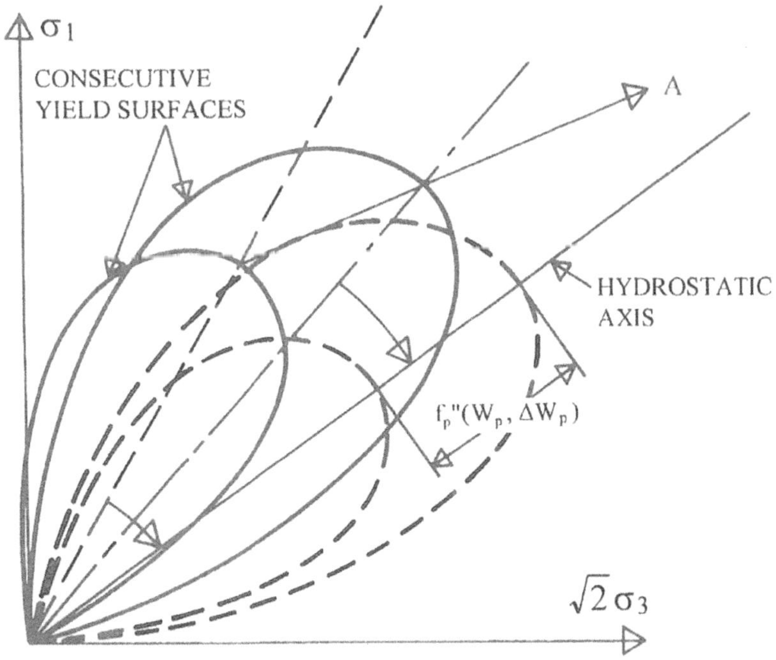

Fig. 2. Determination of magnitude of plastic strain increment using increment of plastic work generated by expansion of kinematic yield surface rotated to position of isotropic yield surface.

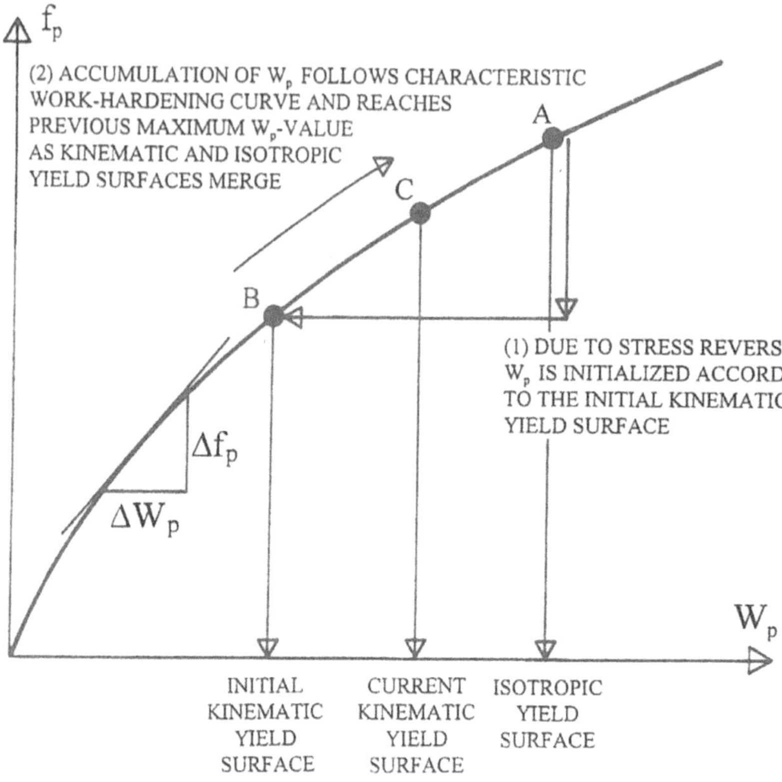

Fig. 3. Procedure for computation of incremental plastic work during kinematic hardening using isotropic hardening relationship.

IMPLEMENTATION INTO SINGLE HARDENING MODEL

In an isotropic hardening model, yielding is determined by checking the sign of the partial derivative of the yield function with respect to stress. There are essentially three possibilities for the partial derivative of the yield function for an isotropic hardening model: elastic behavior for a negative sign, elasto-plastic behavior for a positive sign and neutral loading, which is interpreted as elastic behavior, for a value of zero.

In a kinematic hardening model, a non-positive value would only indicate that a stress reversal has taken place relative to the currently active isotropic or kinematic yield surface, but this will not suffice to determine the elastic or plastic nature of soil behavior. An algorithmic procedure must be used to determine which of the seven possible modes of action will be taken to determine the active yielding condition. A flowchart depicting this decision structure is given in Fig. 4. Following the flowchart, the process may be described as follows:

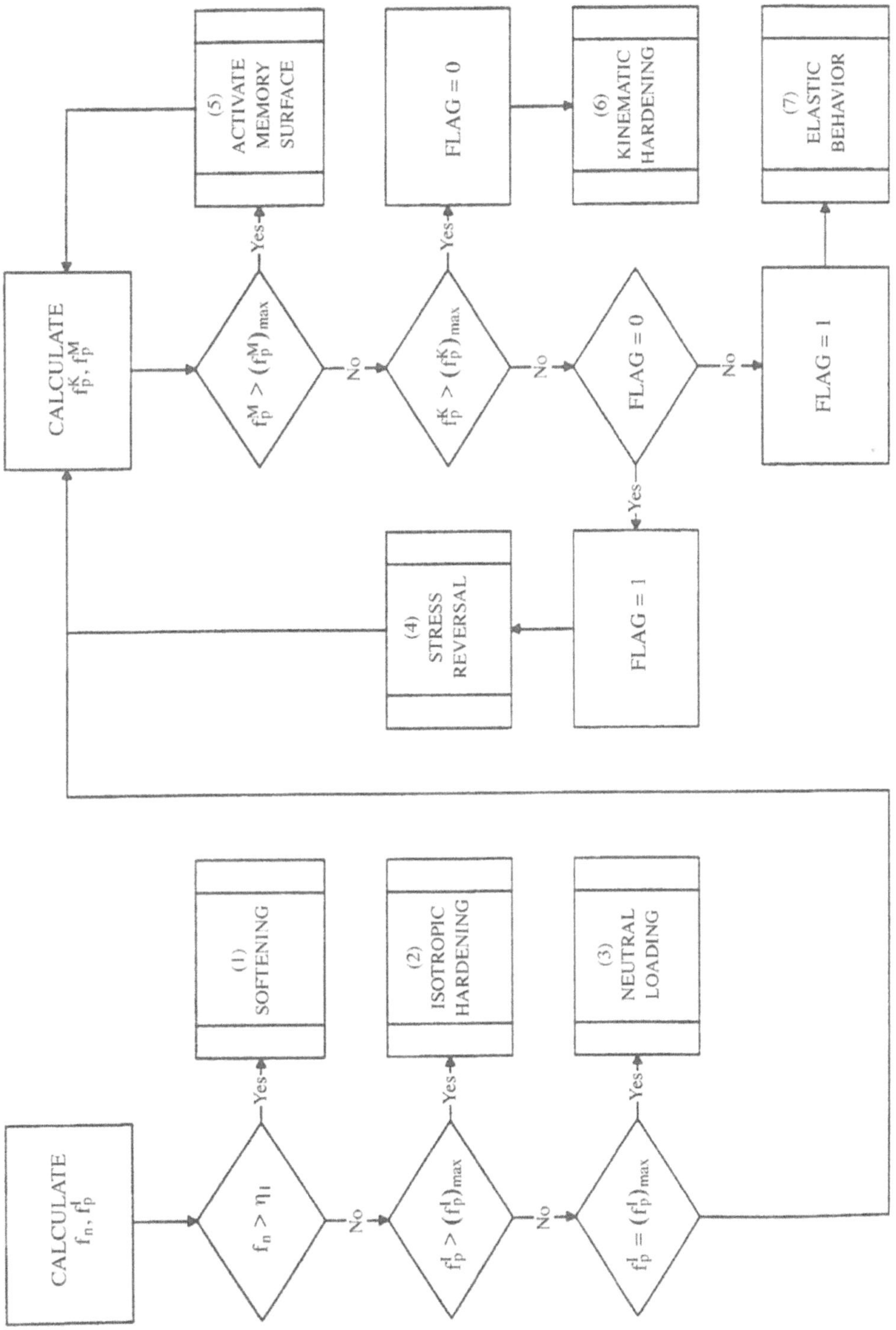

Fig. 4. Flowchart of decision processes to determine loading mode in soil element.

(1) The material is softening if the stress state steps outside the failure surface.
(2) Isotropic hardening will apply if the isotropic yield surface is being penetrated (primary loading).
(3) Neutral loading will apply if the stress path is moving along the contour of the isotropic yield surface.
(4) If none of the above conditions are met, the stress path is within the isotropic yield surface. In the case of initial stress reversal from primary loading, a flag is turned on to indicate stress reversal, and a kinematic surface is activated. The same flag is turned on for the case of stress reversal from the currently active kinematic yield surface so that a new kinematic yield surface is formed. The currently active kinematic yield surface is then stored as a memory surface.
(5) If the stress path penetrates the existing memory surface, then the memory surface is activated as the kinematic yield surface. Kinematic hardening applies as the merged kinematic/memory surface follows a path of evolution towards the isotropic yield surface.
(6) Kinematic hardening will apply if the stress reversal path penetrates the current kinematic yield surface. The kinematic yield surface will evolve along a path towards the memory surface. The flag indicating stress reversal is turned off.
(7) The material will behave elastically if the stress reversal path does not penetrate the currently active kinematic yield surface. The flag indicating stress reversal is not turned off until yielding commences.

The computational procedures involved in isotropic hardening, elastic behavior and softening are identical to that of the single hardening model. The transitional operations for stress reversal or memory surface activation simply involve noting the position of the new kinematic yield surface and/or the exchange of the stored position of a memory surface with that of the active kinematic yield surface. Evolution of the kinematic yield surface is determined by calculating the resulting rotation and expansion of the current kinematic yield surface in relation to the given stress increment. Once the new position of the kinematic yield surface is established, a plastic work increment may be calculated. The calculated plastic work increment is used together with the rotated plastic potential to compute new plastic strain increments in a manner similar to that of the isotropic hardening model.

EVALUATION OF MODEL ASSUMPTIONS

The basic assumptions made in developing the proposed model may be summarized as follows:

(1) The present formulation is an extension of the single-surface isotropic hardening model which has been verified by comparisons with experimental data covering a wide range of frictional materials and two- and three-dimensional stress and strain conditions [4, 6, 7].
(2) The proposed kinematic mechanism based on rotation of the yield surface and the plastic potential is strongly supported by experimental observations [1]. In addition to experimental data obtained during large stress reversals and results of stress probing to locate and define the shape of anisotropic yield surfaces, the rotational behavior of the plastic potential has been verified by comparison of measured and predicted plastic strain increment directions [1, 8–10].
(3) The initial orientation and size of the kinematic yield surface has been defined by specifying the tip of the initial kinematic yield surface to be coincident with the initial stress reversal point. This assumption is consistent with the experimentally observed elastic behavior of soils during isotropic compression and subsequent unloading.
(4) The accumulation of plastic work is recognized to be a unique characteristic of soil type. Plastic work may be accumulated by employing the same characteristic work-hardening law for both isotropic and kinematic hardening. Thus, the underlying theoretical framework for the isotropic hardening model may be generalized to all types of loading in a consistent manner.
(5) It is assumed that the kinematic yield surface evolves into the past isotropic yield surface following a linear path connecting the tips of both surfaces.
(6) Further stress reversals within the isotropic yield surface lead to the formation of new kinematic yield surfaces which are related to the past kinematic surface, now called the memory surface, in the same manner that the initial kinematic surface was related to the isotropic yield surface. Prior kinematic surfaces are erased from memory so that a maximum of three surfaces, the isotropic yield, kinematic yield and memory surfaces, are stored in memory for any given step along the stress reversal path. As the stress reversal path moves towards the past maximum stress level, the active kinematic surface merges into the memory surface, which, in turn, merges into the isotropic surface.

The evolution of the kinematic surface along a linear path (5) and the existence of a single memory surface (6) are the only assumptions that have been made for the sake of simplicity rather than based on analysis of experimentally observed behavior. Yet, these assumptions are shown in the following to be consistent with experimentally observed behavior when approached from a viewpoint based on the unique relation between yield surface and plastic work.

THE ROLE OF MEMORY

Knowing the stress path and corresponding experimentally observed plastic strains, the plastic work done along this stress path may easily be calculated. Using the characteristic work-hardening rule, a yield surface can be determined for each stress point. For the monotonic loading portion of the stress path, the determined yield surface will correspond to the isotropic yield surface. For the portion of the stress path involving stress reversals, the size of the corresponding kinematic yield surface can be calculated by keeping track of the number of stress reversals, and initializing accumulated plastic work in accordance with the initial point of stress reversal for each branch. The problem is how to determine the orientation of the yield surface.

This can be accomplished by a numerical search routine which will rotate the calculated kinematic yield surface, as shown in Fig. 5, until it lies on the corresponding current stress point. In this manner, the exact location and size of the kinematic yield surface can be determined for each stress point. Note that the only assumption made in the process is rotation of the yield surface and re-initialization of accumulated plastic work. Both assumptions are based on features of the proposed model which are supported by experimental observations. No *a priori* assumptions have been made on the existence of a memory surface or how the yield surface expands as it rotates. Therefore, the above method of construction of yield surfaces from experimental data may be used to reach a conclusion regarding the validity of assumptions (5), evolution of the kinematic surface along a linear path, and (6), the existence of a single memory surface.

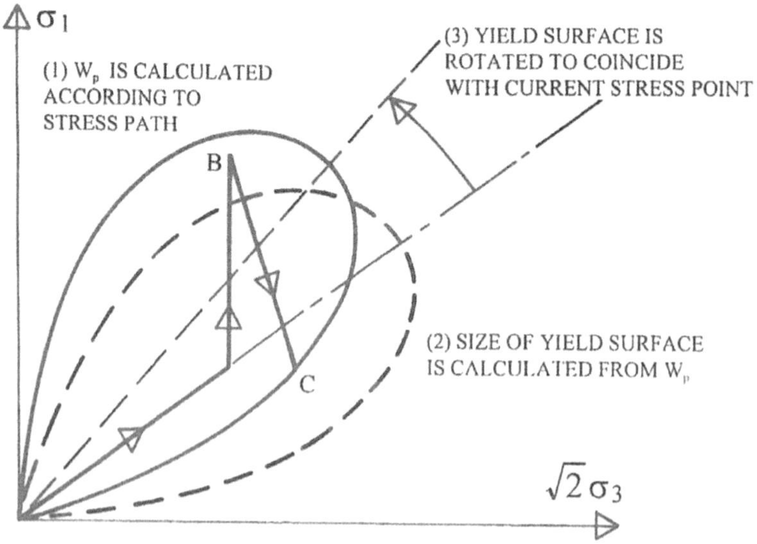

Fig. 5. Rotation of calculated kinematic yield surface until coincidence with current stress point.

The above procedure was applied to a drained triaxial compression–extension test and the results are shown in Fig. 6. The size and orientation of calculated isotropic or kinematic yield surfaces is represented by circles indicating the stress point corresponding to the tip of that surface. This presentation of data clearly shows any deviation from a linear path of growth, as well as indicating possible distinctive trajectories of growth which would indicate multiple memory surfaces.

Referring to Fig. 6, the results show a pattern of points which closely follow a straight line of growth. The points lying on the hydrostatic axis may be ignored in the present context since they indicate the tips of expanding isotropic yield surfaces. Two separate trajectories of growth may be observed from the figure. The upper trajectory shows a kinematic yield surface growing along a linear path from the initial stress reversal point, B, toward the tip of the isotropic yield surface, A. The other trajectory traces the stress reversal from loading in extension at point C. Upon stress reversal at point C, a new kinematic yield surface is formed which evolves along a linear path, not towards the tip of the isotropic yield surface, but towards the tip of the previous kinematic yield surface, now the memory surface. Due to the simplicity of the analyzed stress reversal path, it is possible to clearly observe the existence of this memory surface.

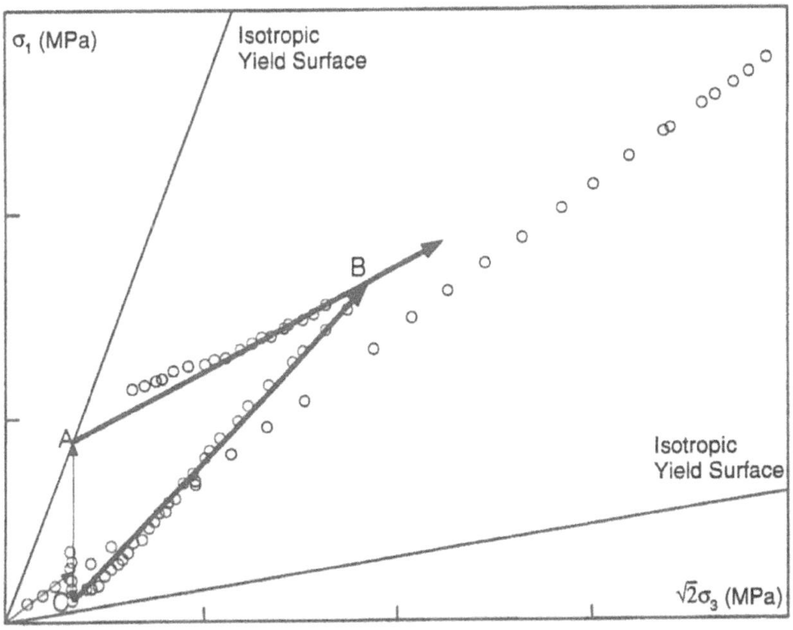

Fig. 6. Comparison of measured (points) and predicted (solid line) evolution of kinematic yield surface as represented by tip of tear-drop shaped yield surface in triaxial compression–extension test on loose Santa Monica beach sand.

The results shown in Fig. 6 are also representative of other tests involving simple cycles of unloading and reloading. No conclusions could be made based on more complicated stress paths due to the large scatter in computed trajectories. This is probably due to experimental uncertainties discussed by Lade and Inel [1], or may possibly be due to an increasing nonlinearity with continuing arbitrary stress reversals. Yet, for the simple case of a cycle of unloading and reloading, the model assumptions seem to closely follow actual soil behavior. This partial verification of the assumptions involved in the proposed model is very encouraging since it shows that the basic framework of the model is consistent within itself, and it is able to closely predict the actual behavior of soil in a rational manner.

COMPARISON OF EXPERIMENTAL DATA AND PREDICTIONS

The stress–strain behavior of sand observed in conventional triaxial laboratory tests were predicted to evaluate the proposed model. By comparing measured and predicted stress–strain curves, the model is tested in all aspects and conclusions may be made on its validity. The testing program performed by Boonyachut [8] consisting of triaxial tests with large stress reversals on loose and dense cylindrical specimens of Santa Monica beach sand was used for comparison. The same study was employed in showing the validity of the proposed plastic potential behavior in Lade and Inel [1].

The required number of model parameters to be determined for a given material is 12. For cohesionless materials, such as sand, the value of $a = 0$. Table 1 summarizes the model parameter values determined for loose and dense Santa Monica beach sand. These parameters were employed in the predictions presented below.

TABLE 1
Material properties for Santa Monica beach sand

Material parameter	*Loose Santa Monica beach sand*	*Dense Santa Monica beach sand*
M	600	1270
λ	0.27	0.23
ν	0.26	0.15
a	0.00	0.00
m	0.107	0.23
η_1	32.6	102.3
C	0.000204	0.000094
p	1.51	1.50
ψ_2	−3.65	−3.19
μ	2.10	2.12
h	0.60	0.62
α	0.79	0.68

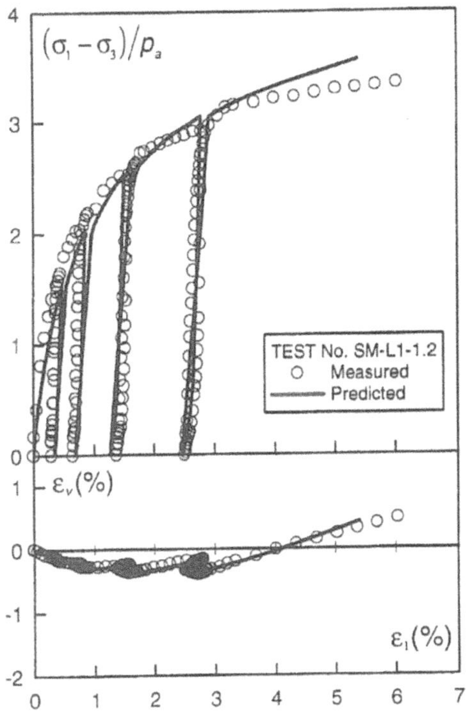

Fig. 7. Comparison of measured and predicted stress–strain and volume change relations for triaxial compression test on loose Santa Monica beach sand.

Figure 7 shows the prediction of a triaxial compression test at a constant confining pressure of 117.7 kPa (1.2 kg cm^{-2}) involving four consecutive cycles of unloading to and reloading from the hydrostatic stress state. The model prediction captures the observed stress–strain behavior during each cycle. Figure 8 shows the prediction of an identical test performed at a higher confining pressure of 235.4 kPa (2.4 kg cm^{-2}). As shown in the diagram, the predicted and observed stress–strain behavior match and the effect of confining pressure is captured. Note also that the volumetric strain behavior during stress reversals is captured quite accurately.

In both figures, sudden jumps in the plastic hardening modulus value are indicated by the sharp bends observed as a cycle of unloading–reloading ends and primary loading is continued. These bends are due to the stress reversal path penetrating the isotropic yield surface before the memory surface is able to merge with the isotropic yield surface. They may be an indicator of the expansion of the isotropic yield surface due to plastic work generated along the stress reversal path. Further testing is required to reach a conclusion on this matter. In the meantime, these bends do not affect the quality of the predictions, and they do not warrant a solution which would further complicate the model and increase the amount of numerical computation.

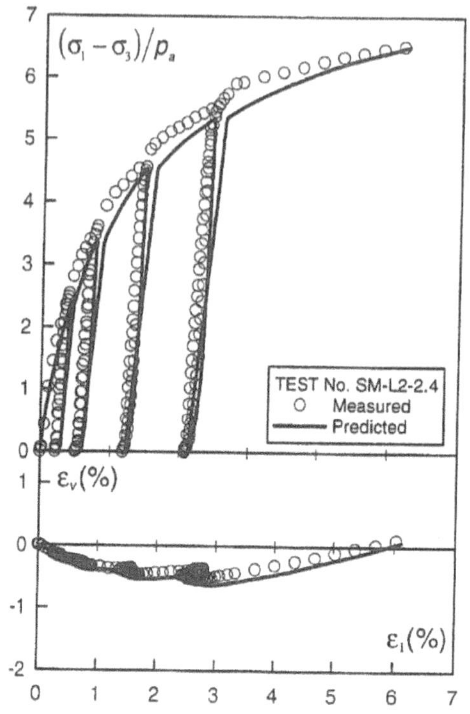

Fig. 8. Comparison of measured and predicted stress–strain and volume change relations for triaxial compression test on loose Santa Monica beach sand.

Figures 9 and 10 show the prediction of triaxial compression–extension tests at constant confining pressures of 117.7 and 235.4 kPa (1.2 and 2.4 kg cm^{-2}), respectively. Both cycles of unloading and reloading are captured quite accurately. The volumetric strain behavior is slightly off in magnitude, yet the qualitative behavior of the stress–strain relation is followed closely. Closer observation shows that the deviation in the magnitude of volumetric strains is accumulated in the initial primary loading portion of the test. If shifted, the predicted volumetric strain curve would fit the measured volumetric strain behavior with reasonable accuracy. Therefore, this deviation in magnitude may indicate some inconsistency in the measured stress–strain response or in the employed soil parameters rather than an inadequacy in the model.

Figure 11 shows the prediction of a triaxial test with primary loading into extension followed by a cycle of unloading and reloading in compression and extension and then primary loading to failure in compression. The axial strain behavior of the soil is captured very accurately. The predicted volumetric behavior is still slightly off in magnitude, but as before, the initial loading in extension accounts for the difference between measured and predicted behavior.

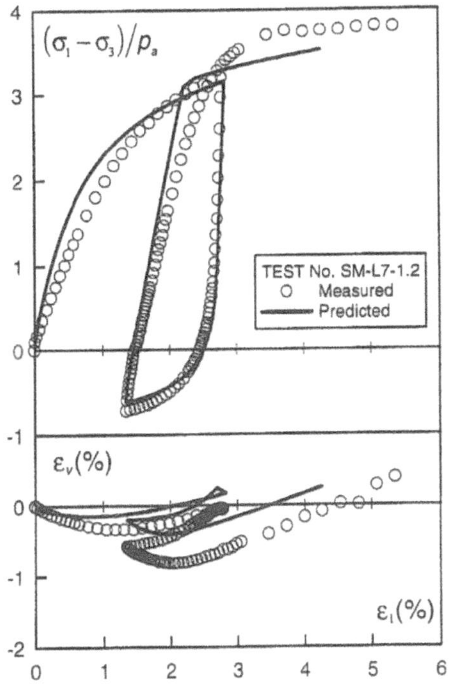

Fig. 9. Comparison of measured and predicted stress–strain and volume change relations for triaxial compression–extension test on loose Santa Monica beach sand.

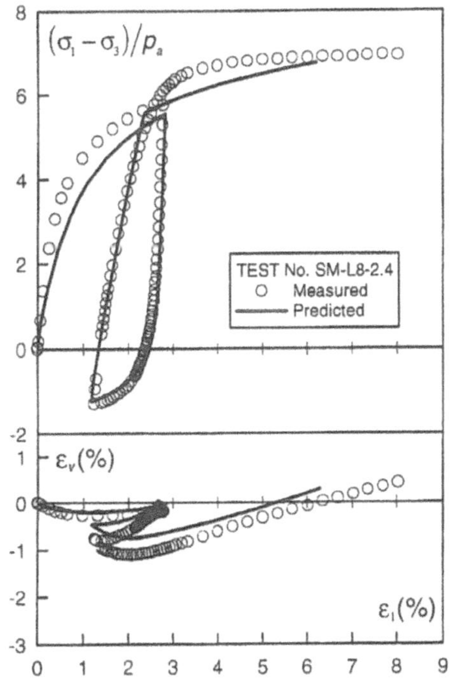

Fig. 10. Comparison of measured and predicted stress–strain and volume change relations for triaxial compression–extension test on loose Santa Monica beach sand.

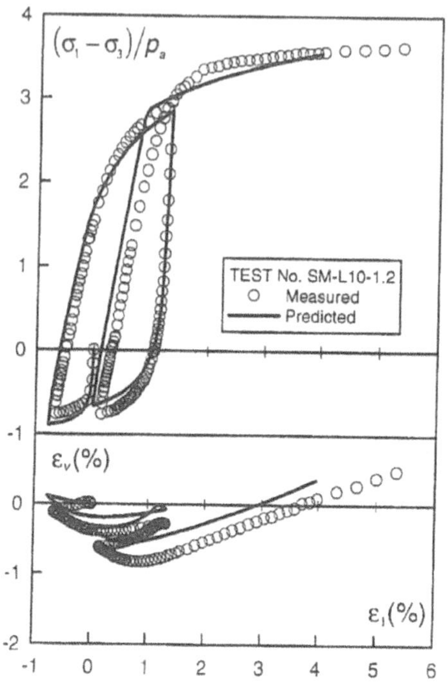

Fig. 11. Comparison of measured and predicted stress–strain and volume change relations for triaxial extension–compression–extension test on loose Santa Monica beach sand.

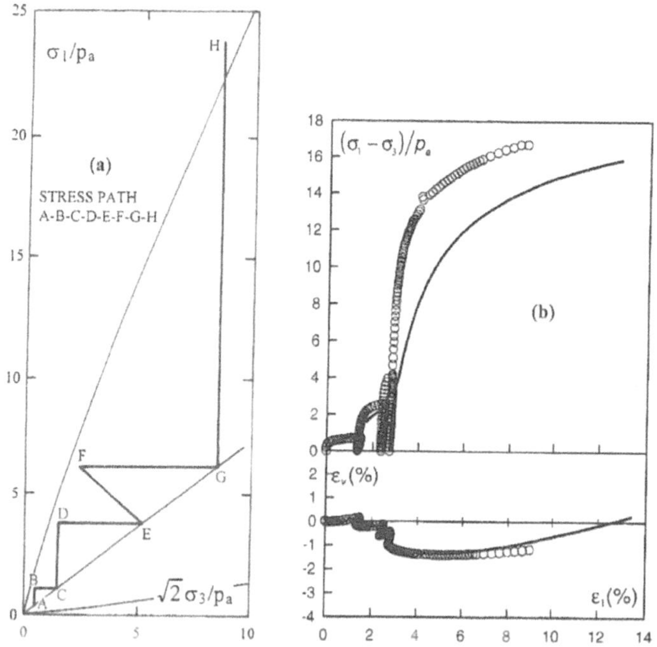

Fig. 12. (a) Complex stress path with generally increasing mean normal stress and (b) comparison of measured and predicted stress–strain and volume change relations for triaxial test on loose Santa Monica beach sand.

Figures 12–14 show predictions of triaxial tests involving more complicated stress paths with varying confining pressure. Figure 12(a) shows a stress path with generally increasing mean normal stress. The measured and predicted stress–strain behavior is shown in Fig. 12(b). The model prediction closely matches actual soil behavior along this complicated stress path. The prediction of volumetric strain behavior is almost perfect, and the prediction of stress–strain behavior deviates only as the soil is primary loaded to failure. Figure 13(a) shows a stress path with generally decreasing mean normal stress. The prediction, shown in Fig. 13(b), closely matches the measured stress–strain behavior. Figure 14(a) shows a stress path with generally increasing mean normal stress followed by unloading and reloading along a path of constant deviatoric stress. Figure 14(b) shows the prediction of the stress–strain curve which is slightly off in magnitude, yet captures the overall qualitative behavior with reasonable accuracy. The deviation between measured and predicted strain magnitudes seems to arise during the loading portion from point C to point D.

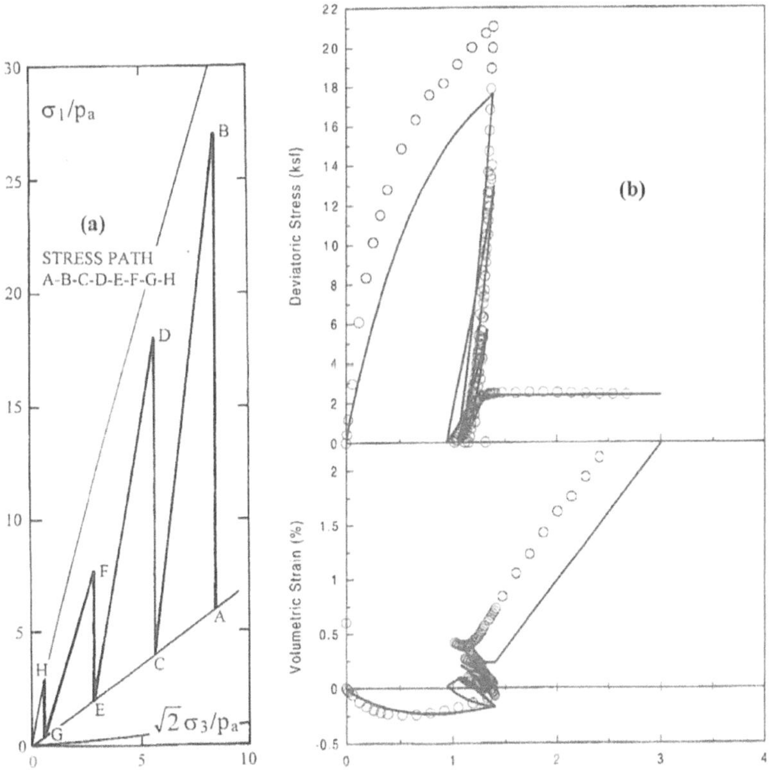

Fig. 13. (a) Complex stress path with generally decreasing mean normal stress and (b) comparison of measured and predicted stress–strain and volume change relations for triaxial test on dense Santa Monica beach sand.

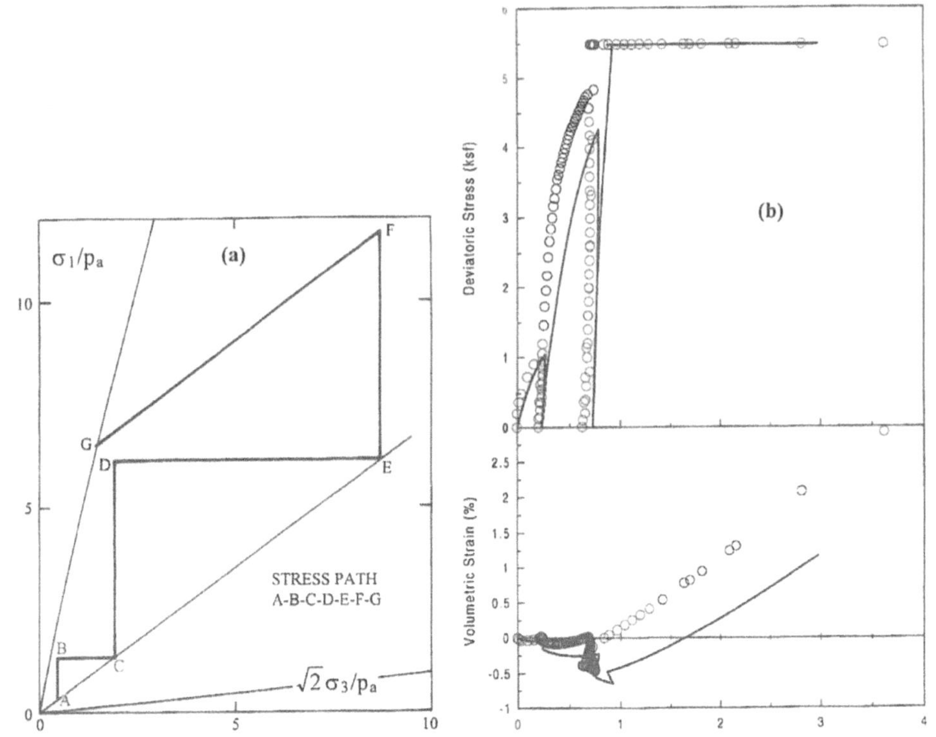

Fig. 14. (a) Complex stress path and (b) comparison of measured and predicted stress–strain and volume change relations for triaxial test on dense Santa Monica beach sand.

Overall, the predictions from the rotational kinematic hardening model match the experimental behavior with good accuracy. When deviations occur, they appear to be caused by inconsistencies or scatter in the experimental results. The model seems to capture the stress–strain behavior in principle for stress paths in the triaxial plane.

CONCLUSIONS

A new kinematic hardening mechanism has been proposed which incorporates rotation and intersection of the yield surfaces to achieve a more consistent and physically rational fit with experimentally observed behavior. The existing single hardening constitutive model [2–4] is used as the basic framework to which the rotational kinematic hardening mechanism has been added. The new combined model preserves the behavior of the isotropic hardening model under monotonic loading conditions, and the extension from isotropic to kinematic hardening is accomplished without introducing new material parameters. The characteristic work hardening law applies to isotropic as well as kinematic behavior of sand. The capability of the new model is examined by comparing predictions with experimental data for various stress paths involving large stress reversals in the triaxial plane. Within the scatter of test results, the proposed model is shown to capture the behavior of sand with reasonable accuracy.

ACKNOWLEDGEMENTS

The research presented here was performed in the Department of Civil Engineering at University of California, Los Angeles, with support from the National Science Foundation under Grant No. MSS 9119272/MSS 9396271. S.I. was supported partly by the Kenneth L. Lee Memorial Fellowship and partly by the Department of Civil Engineering at UCLA. Grateful appreciation is expressed for this support.

REFERENCES

1. Lade, P. V. and Inel, S., Rotational kinematic hardening model for sand I. concept of rotating yield and plastic potential surfaces. *Computers and Geotechnics*, 1997, **21**(3), 183–216.
2. Kim, M. K. and Lade, P. V., Single hardening constitutive model for frictional materials I. plastic potential function. *Computers and Geotechnics*, 1988, **5**(4), 307–324.
3. Lade, P. V. and Kim, M. K., Single hardening constitutive model for frictional materials II. yield criterion and plastic work contours. *Computers and Geotechnics*, 1988, **6**(1), 13–30.
4. Lade, P. V. and Kim, M. K., Single hardening constitutive model for frictional materials, III. comparisons with experimental data. *Computers and Geotechnics*, 1988, **6**(1), 31–48.
5. Dafalias, Y. F. and Popov, E. P., A model for nonlinearly hardening materials for complex loading. *Acta Mechanica*, 1975, **21**, 173–192.
6. Lade, P. V., Single-hardening model with application to NC clay. *Journal of Geotechnical Engineering, ASCE*, 1990, **116**(GT3), 394–414.
7. Lade, P. V. and Kim, M. K., Single hardening constitutive model for soil, rock and concrete. *International Journal of Solids and Structures*, 1995, **32**(14), 1963–1978.
8. Boonyachut, S., Experimental study of the behavior of cohesionless soil during large stress reversals. Ph.D. dissertation, University of California, Los Angeles, 1977.
9. Lade, P. V. and Boonyachut, S., Large stress reversals in triaxial tests on sand. Proceedings of the 4th International Conference on Numerical Methods in Geomechanics. Edmonton, 1982, pp. 171–181.
10. Inel, S., Kinematic hardening model for sand behavior during large stress reversals. Ph.D. dissertation, University of California, Los Angeles, 1992.

Computers and Geotechnics 36 (2009) 1285–1297

Contents lists available at ScienceDirect

Computers and Geotechnics

journal homepage: www.elsevier.com/locate/compgeo

Kinematic hardening predictions of large stress-reversals in 3-D test on loose sand

Poul V. Lade [a,*], Suresh K. Gutta [b], Jerry A. Yamamuro [c]

[a] Department of Civil Engineering, The Catholic University of America, Washington, DC 20064, USA
[b] American Geotechnical and Environmental Services, Inc., Canonsburg, PA 15317, USA
[c] School of Civil and Construction Engineering, Oregon State University, Corvallis, OR 97331, USA

ARTICLE INFO

Article history:
Received 11 October 2008
Received in revised form 12 May 2009
Accepted 14 May 2009
Available online 27 June 2009

Keywords:
Constitutive model
Elasto-plastic behavior
Rotational kinematic hardening
Sand
True triaxial tests
Stress reversal

ABSTRACT

A rotational kinematic hardening constitutive model with the capability of predicting the behavior of soil during 3-D stress-reversals has been developed. An existing elasto-plastic constitutive model, the Single Hardening Model, utilizing isotropic hardening serves as the basic framework in these formulations. To this framework is added the capability of handling cross-anisotropic behavior as well as the kinematic hardening mechanism to capture inherent anisotropy of the sand in addition to the large stress-reversals. The model involves thirteen parameters, which can be determined from simple laboratory experiments, such as isotropic compression, drained triaxial compression and triaxial extension tests. The results from a series of true triaxial tests with large stress-reversals performed on loose cross-anisotropic Santa Monica Beach sand are employed for comparison with predictions.

1. Introduction

A rotational kinematic hardening mechanism in principal stress space has been proposed and added to an existing constitutive model, the Single Hardening Model, by Lade and Inel [14] and its capabilities to capture the stress–strain and volume change behavior during large stress-reversals was demonstrated by Inel and Lade [6]. This rotational kinematic hardening mechanism was then expanded to the principal stress space by Lade et al. [17]. It was realized that the cross-anisotropy exhibited by most soils, including sand deposits created in laboratory specimens, plays an important role, and the Single Hardening Model was therefore modified to include effects of cross-anisotropy by Abelev at al. [1]. These two mechanisms were then combined [17] and the capabilities of the combined model were presented by Gutta et al. [5] by comparing predictions with experimental results from true triaxial tests on cross-anisotropic, medium dense Santa Monica Beach sand.

Presented here is a review of the combined model together with the predictions of true triaxial tests with large stress-reversals performed on loose Santa Monica Beach sand by Boonyachut [4]. The experiments are presented alongside the model predictions that include the effects of rotational kinematic mechanism and cross-anisotropic hardening.

* Corresponding author.
E-mail address: Lade@cua.edu (P.V. Lade).

0266-352X/$ - see front matter © 2009 Elsevier Ltd. All rights reserved.
doi:10.1016/j.compgeo.2009.05.014

2. Review of combined constitutive model

The constitutive model used here is a combination of two models individually developed with capabilities to handle: (1) large 3-D stress-reversals, and (2) inherent cross-anisotropic behavior. Both models have at the core the Single Hardening Model that was developed for frictional materials such as sand, clay, rock and concrete [7,15,16].

2.1. Rotational kinematic hardening model in triaxial plane

Experimental evidence [12] indicates that the effects of large stress-reversals may be predicted by rotating the plastic potential and the yield surfaces around the stress origin of the principal stress space. The essence of the resulting rotational kinematic hardening model [14,6] is presented in the triaxial plane in Fig. 1. In addition to rotation around the origin, the kinematic yield surface intersects the isotropic yield surface. During primary loading the isotropic yield surface is active. When a stress reversal occurs, the pseudo-hydrostatic axis defining the position of the rotated kinematic surface passes through the stress reversal point. In this manner, both the orientation and the size of the kinematic surface are determined by the stress reversal point.

The kinematic surface rotates in such a manner that it will merge into the isotropic yield surface when primary loading is encountered again. As shown in Fig. 1, this is accomplished if the pseudo-hydrostatic axis always rotates towards the true hydrostatic axis, and the kinematic surface expands such that its tip

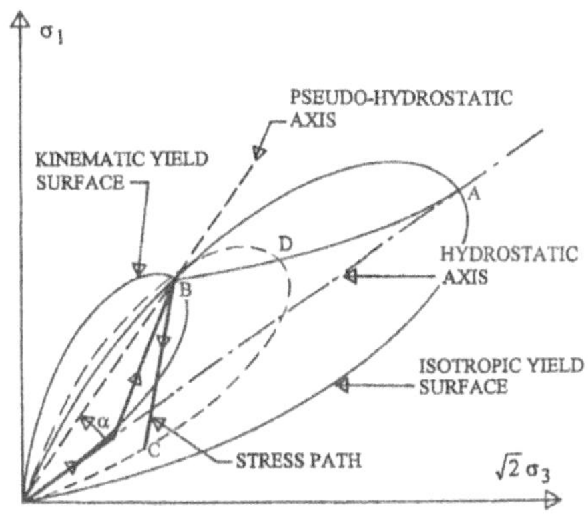

Fig. 1. Growth of rotational kinematic yield surface in triaxial plane (in magnitude and orientation) towards isotropic yield surface upon stress reversal at point *B*.

moves along a parabolic line towards the tip of the isotropic surface. The region between the kinematic and the isotropic surfaces defines an elastic domain that evolves as the kinematic surface rotates and expands in accordance with the stress reversal path. Thus, the model involves a combination of isotropic and kinematic hardening theories such that is preserves the behavior of the isotropic hardening model under monotonic loading conditions.

If a second stress reversal occurs, then a new kinematic surface is initiated with the same evolutionary behavior as explained above, and the previous kinematic surface becomes the memory surface. If still another stress reversal takes place, then another kinematic surface is created, the previous kinematic surface becomes the memory surface, and the very first kinematic surface is forgotten. Experiments indicate that the sand remembers only the past two yield surfaces. Anything further in the past no longer leaves an impression with the sand.

2.2. Extension of rotational kinematic hardening model to principal stress space

The rotational kinematic hardening model was further expanded to the principal stress space by Lade et al. [17], as reviewed below. The model does not require any more parameters than the Single Hardening Model. The effects of large stress-reversals are modeled by mathematical formulations as well as the remembrance (i.e. storage) of the tip points of the isotropic yield surface, the kinematic yield surface, and the memory surface.

The extension of the model is done is such a way that the essence of the rotational kinematic hardening mechanism developed with reference to the triaxial plane is preserved in the principal stress space, with one exception, as indicated in Fig. 2 and discussed below. In addition to rotation around the origin, the kinematic yield surface intersects the isotropic yield surface. During primary loading the isotropic teardrop-shaped yield surface is active. When a stress reversal occurs, the pseudo-hydrostatic axis defining the position of the rotated kinematic surface passes through the stress reversal point. In this manner, both the orientation and the size of the kinematic surface are determined by the stress reversal point. The region between the kinematic and isotropic surfaces defines an elastic domain which evolves as the kinematic surface rotates and expands in accordance with the stress reversal path.

The kinematic surface rotates and grows in such a manner that it will merge into the isotropic yield surface when primary loading is encountered again. This pattern is accomplished if the pseudo-hydrostatic axis always rotates towards the true hydrostatic axis, and the kinematic surface expands such that its tip, at point *B* in Fig. 2, moves towards the tip of the isotropic surface, point *A*. Thus, the rotation of the kinematic surface is defined to occur about a unit vector, $\vec{n}$, through the origin and normal to the plane containing the real and the pseudo hydrostatic axes. This is depicted in Fig. 2 by rotational degrees of freedom shown as angles corresponding to in-plane, out-of-plane, and axial rotations.

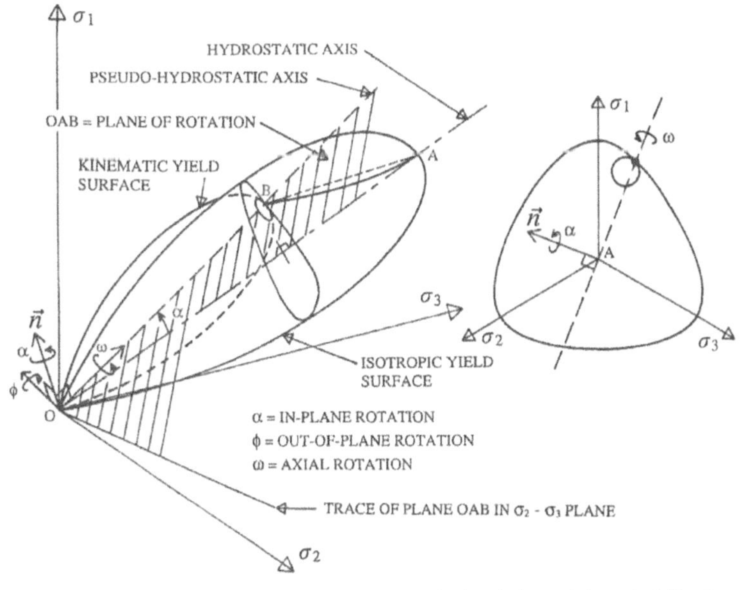

Fig. 2. Growth of rotational kinematic yield surface in principal stress space (in magnitude and orientation) towards isotropic yield surface upon stress reversal at point *B*.

In accordance with the above discussion, the out-of-plane angle will always be zero. The diagram also shows that the kinematic surface would be free to rotate about its own axis resulting in an additional axial rotation which must be taken into account. It is considered appropriate to restrict this degree of freedom of the kinematic surface, since it introduces a level of complexity that is not justified by the experimental results. Such a rotation would also introduce probable numerical difficulties during implementation as well as difficulties in determining uniqueness of the solutions. Thus, it was simply decided to fix the kinematic surface against rotation about its own axis.

Once the type of rotation in the principal stress space has been decided, all other features of the kinematic mechanism from the triaxial plane will remain unchanged in the principal stress space, with one exception. In the model developed for the triaxial plane, the kinematic yield surface and the plastic potential move in the principal stress space obeying a linear evolution from the initial stress reversal point B, to the maximum stress point A, of the isotropic surface or of the memory surface. However, further analyzes showed that the movement of point B along a parabolic path in plane OAB (shown by solid line in Fig. 1 for the triaxial plane and in Fig. 2 for the principal stress space) rather than a straight line (shown by dashed line in Fig. 2) to point A produced a better representation of the real soil behavior during large stress-reversals.

Failure characteristics, softening regime, and elastic behavior of the soil is unaffected by the kinematic mechanism. The elastic domain defined by the intersection of the kinematic and isotropic yield surfaces is now a spatial region in the principal stress space, and the plane in which rotation takes place is one of an infinite number of planes containing the hydrostatic axis. Kinematic yielding may take place due to 3-D stress paths moving within or out of the current plane of rotation.

2.3. Formulation of rotation in principal stress space

The formulation of rotation in the principal stress space is a generalization of the formulation in the triaxial plane. Thus, rotation can occur in any plane containing the hydrostatic axis. In the principal stress space the rotated stress values, indicated by a star, may be calculated as follows:

$$\begin{bmatrix}\sigma_1^*\\ \sigma_2^*\\ \sigma_3^*\end{bmatrix} = \begin{bmatrix} n_1^2(1-\cos\alpha)+\cos\alpha & n_1n_2(1-\cos\alpha)+n_3\sin\alpha & n_1n_3(1-\cos\alpha)-n_2\sin\alpha \\ n_1n_2(1-\cos\alpha)-n_3\sin\alpha & n_2^2(1-\cos\alpha)+\cos\alpha & n_2n_3(1-\cos\alpha)+n_1\sin\alpha \\ n_1n_3(1-\cos\alpha)+n_2\sin\alpha & n_2n_3(1-\cos\alpha)-n_1\sin\alpha & n_3^2(1-\cos\alpha)+\cos\alpha \end{bmatrix} \begin{bmatrix}\sigma_1\\ \sigma_2\\ \sigma_3\end{bmatrix} \tag{1}$$

in which α is positive if anti clockwise, and

$$\bar{n} = (n_1, n_2, n_3) = \frac{\vec{\sigma}_A \times \vec{\sigma}_B}{|\vec{\sigma}_A||\vec{\sigma}_B|\sin\alpha_{AB}} = \frac{(\sigma_3-\sigma_2, \sigma_1-\sigma_3, \sigma_2-\sigma_1)}{\sqrt{(\sigma_3-\sigma_2)^2+(\sigma_1-\sigma_3)^2+(\sigma_2-\sigma_1)^2}} \tag{2}$$

is a unit normal vector defining the plane in which the kinematic surface rotates.

The angle of rotation must be found by enforcing that the motion of the tip of the kinematic surface, point D, moves on a parabolic path from point B to point A, as shown in Fig. 3, together with the condition that the current stress point must be located on the kinematic surface. Points B and A are located at the start of the stress reversal and at the tip of the isotropic yield surface, respectively. A 2-D X–Y coordinate system is placed in plane OAB such that the X-axis lies along the hydrostatic axis, and the origin coincides with the origin of the principal stress space.

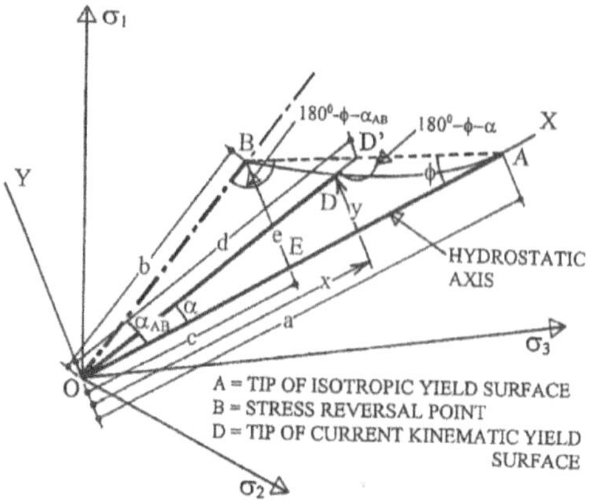

Fig. 3. Diagram showing the parabolic path of the tip of the kinematic yield surface.

The initial angle of rotation defining the first point of stress reversal, α_{AB}, is determined from the dot product of the radial position vectors to points A and B in the principal stress space according to:

$$\vec{\sigma}_A \cdot \vec{\sigma}_B = \|\vec{\sigma}_A\| \cdot \|\vec{\sigma}_B\| \cdot \cos\alpha_{AB} \tag{3}$$

The various lengths of the sides in the triangles shown in Fig. 3 are used in the expressions for the parabolic path, and they are as follows:

$$OA = a = \sqrt{\sigma_{1A}^2 + \sigma_{2A}^2 + \sigma_{3A}^2} \tag{4}$$

$$OB = b = \sqrt{\sigma_{1B}^2 + \sigma_{2B}^2 + \sigma_{3B}^2} \tag{5}$$

$$OE = c = b \cdot \cos\alpha_{AB} \tag{6}$$

$$OD' = d = a \cdot \frac{\sin\phi}{\sin(\phi+\alpha)} \tag{7}$$

$$BE = e = b \cdot \sin\alpha_{AB} \tag{8}$$

The parabola may then be expressed in the X–Y coordinate system as follows:

$$y = \frac{e}{(a-c)^2} \cdot (x-a)^2 \tag{9}$$

Point D in the X–Y coordinate system is represented by (x,y), where x is expressed as:

$$x = c + t \cdot (a-c) \tag{10}$$

The variable t in Eq. (10) varies from 0 to 1 along the X-axis and is used as a computational aid to represent the movement of point D from B to A. Substitution of Eq. (10) into Eq. (9) produces:

$$y = e \cdot (1-t)^2 \tag{11}$$

With known values of x and y, the length of OD, i.e. the length of the vector from the origin to stress point D, can be expressed as:

$$\|\vec{\sigma}_D\| = \sqrt{x^2 + y^2} = \sqrt{(t \cdot a + (1-t) \cdot c)^2 + e^2 \cdot (1-t)^4} \tag{12}$$

and the rotation angle α can be calculated from:

$$\tan\alpha = \frac{y}{x} = \frac{e \cdot (1-t)^2}{c + t \cdot (a-c)} \tag{13}$$

The coordinates of point D' in Fig. 3 may be expressed as:

$$\vec{\sigma}_{D'} = \vec{\sigma}_B + \vec{\sigma}_{BD'} = s \cdot \vec{\sigma}_A + (1-s) \cdot \vec{\sigma}_B \tag{14}$$

in which the variable s varies from 0 to 1 depending on the position of D'. The length of OD', i.e. the length of the vector from the origin to stress point D', can be expressed as:

$$\|\vec{\sigma}_{D'}\| = d = \sqrt{(s \cdot \vec{\sigma}_A + (1-s) \cdot \vec{\sigma}_B)^2} \tag{15}$$

Substitution of expressions from Eqs. (4) and (5) into Eq. (15) and squaring the expression results in:

$$(a^2 + b^2 - 2ac) \cdot s^2 - 2(b^2 - ac) \cdot s + (b^2 + d^2) = 0 \tag{16}$$

which is a quadratic equation in s, in which only d (expressed in Eq. (7)) depends on the variable t through the expression for tan α in Eq. (13). The solution to s is therefore a function of t, and this may be substituted into Eq. (14) to produce the coordinates for point D'.

Finally, the coordinates for stress point D may be obtained from:

$$\vec{\sigma}_D = (\sigma_{1D}, \sigma_{2D}, \sigma_{3D}) = \vec{\sigma}_{D'} \cdot \frac{\|\vec{\sigma}_D\|}{\|\vec{\sigma}_{D'}\|} \tag{17}$$

The variable t is embedded in these coordinates.

2.4. Calculation procedure

The calculation procedure is very similar to that used for the triaxial plane, but with the variable being t. It is recalled that two constraints are enforced to find the angle of rotation. The first constraint is that the current stress point, σ_C, must lie on the active kinematic yield surface, and therefore the following equation must be fulfilled:

$$f_p'(\vec{\sigma}_C^*) = f_p''[(W_p^*)_C] \tag{18}$$

The second constraint is that the kinematic yield surface must evolve in such a manner that its tip moves on a parabolic curve, as developed above. Thus, the stresses for the tip, point D, are given by Eq. (17), which contains the variable t. Therefore, using the current stress point C and the stress point D at the tip of the yield surface, and employing the expression for the yield function:

$$f_p' = \left[\psi_1 \cdot \frac{I_1^3}{I_3} - \frac{I_1^2}{I_2}\right] \cdot \left(\frac{I_1}{p_a}\right)^h \cdot e^q \tag{19}$$

for both sides of the yield function, the following equation must be fulfilled:

$$\left[\psi_1 \cdot \frac{I_{1C}^{*3}}{I_{3C}^*} - \frac{I_{1C}^{*2}}{I_{2C}^*}\right]\left(\frac{I_{1C}^*}{p_a}\right)^h \cdot e^q = (27 \cdot \psi_1 + 3) \cdot \left(\frac{I_{1D}}{p_a}\right)^h \tag{20}$$

in which the asterisk (*) refers to the rotated stresses.

From Eq. (17), stress point D, and therefore the right hand side of Eq. (20), may be expressed in terms of t and the stress points A and B. Using Eq. (1), the left hand side of Eq. (20) may be expressed in terms of the current stress point in the unrotated stress space and the angle α from Eq. (13). Finally, Eq. (18) is rewritten as the following nonlinear equation:

$$f_p'(\vec{\sigma}_A, \vec{\sigma}_B, \vec{\sigma}_C, t) = f_p''(\vec{\sigma}_A, \vec{\sigma}_B, t) \tag{21}$$

This equation forms the governing equation defining the motion of the kinematic yield surface in the principal stress space. At any load step, the only unknown in this equation is the variable t.

As in the triaxial plane, the isotropic plastic potential surface is attached to the yield surface, and it expands with the yield surface as the stress point moves in the stress space. The failure surface remains stationary during rotation of the yield and plastic potential surfaces. Thus, the model involves a combination of isotropic and kinematic hardening theories such that it preserves the behavior of the isotropic hardening model under monotonic loading conditions. The extension from isotropic to rotational kinematic hardening in the principal stress space is accomplished entirely within the mathematical framework and without introducing new material parameters.

2.5. Cross-anisotropic hardening model

The isotropic failure surface proposed by Lade [8] does not fully capture the failure surface for cross-anisotropic sands, as observed in true triaxial tests [2,11]. They found that a failure surface rotated in the triaxial plane around the stress origin of the principal stress space provided a better fit with experimental data [3], as shown in Fig. 4a and b. Similarly, to predict the stress–strain and volume change behavior of cross-anisotropic sand, it was necessary to rotate the plastic potential and the yield surfaces around the stress origin. While the latter two surfaces rotate together, the failure surface rotates separately. Therefore, two additional parameters are included with the isotropic model to describe the rotations required to model the cross-anisotropic behavior. The modified model and its capabilities are described by Abelev et al. [1]. It should be

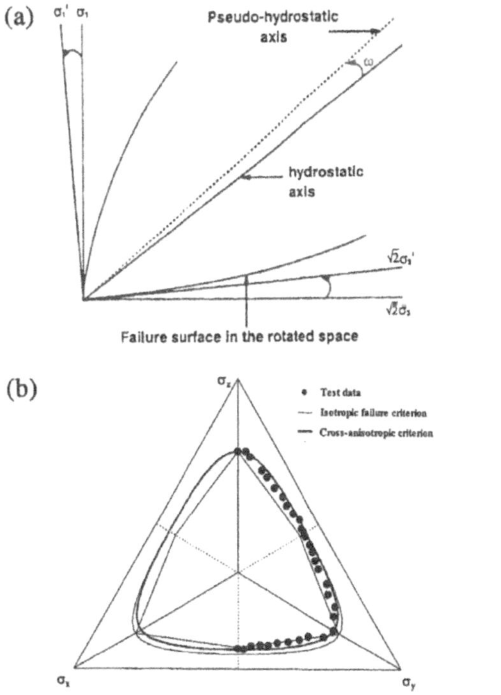

Fig. 4. (a) Rotation of principal stress coordinate system around the stress origin in the triaxial plane to capture the cross-anisotropic failure surface observed in true triaxial tests on dense Santa Monica Beach sand, and shown in (b) octahedral plane.

mentioned that the employment of a rotational mechanism in the principal stress space to model the effects of cross-anisotropy works well for cases in which stress rotation in the field does not occur, but this approach cannot handle stress rotation in the physical space.

The two mechanisms of rotational kinematic hardening and cross-anisotropic hardening were then combined and the capabilities of the combined model were presented by Gutta et al. [5] by comparing predictions with experimental results from true triaxial tests performed on cross-anisotropic, medium dense Santa Monica Beach sand by Yamamuro and Lade [18].

3. Experimental results

Boonyachut [4] performed drained, true triaxial tests to study the behavior of sand during large 3-D stress-reversals on loose Santa Monica Beach sand. The results of some of these tests have previously been reviewed and interpreted as presented by Lade and Boonyachut [12]. The true triaxial tests performed by Boonyachut [4] tests were performed in a manner similar to that explained by Yamamuro and Lade [18]. The experiments were performed in a true triaxial apparatus [13,9] on cubical specimens with side length of 76 mm. Loose specimens were prepared from Santa Monica Beach sand by deposition through a funnel that was lifted up slowly thereby leaving the sand in a loose condition inside the specimen mold. Void ratios of 0.809 corresponding to a relative density of 20.3% were obtained by this method. All tests were performed at a constant confining pressure, σ_3 = 117.5 kPa.

4. Model parameters

The results of three drained triaxial compression tests and an isotropic compression test conducted on loose Santa Monica Beach sand [4] are utilized to calculate the model parameters. Since no extension test was available for this sand and density, the friction angle in triaxial extension is assumed to be equal to the value in the corresponding compression test. Isotropic soils typically exhibit significantly greater friction angles in extension over that in compression. It was assumed that cross-anisotropy reduced the friction angle in extension to that exhibited in compression. This is often found to be the case for sand deposits [10]. The strength in extension is necessary in order to calculate the rotation of the failure surface, ω, for the cross-anisotropic deposits as well as the parameters, η_1 and m. The model parameters for the loose Santa Monica Beach sand are listed in Table 1. They were determined as explained by Abelev et al. [1].

Table 1
Summary of parameter values for loose cross-anisotropic Santa Monica Beach sand.

Model component	Parameter	Value
Elastic parameter	Modulus number, M	600
	Exponent, λ	0.27
	Poisson's Ratio, ν	0.26
Failure criterion	Intercept, η_1	62
	Exponent, m	0.383
Plastic potential	Intercept, ψ_2	−3.1
	Exponent, μ	2.27
Yield criterion	Exponent, h	0.58
	Constant, α	0.62
Hardening function	Intercept, C	0.00019
	Exponent, p	1.5
Rotation angles for		
Failure surface	Angle, ω	2.55°
Yield and plastic potential surfaces	Angle, χ_o	3.0°

5. Prediction procedures

In the following, the stress paths from the true triaxial tests are plotted on a plane of constant σ_3 = 117.5 kPa in which all stresses occur and they are projected on an octahedral plane corresponding to I_1 = 490 kPa. This value of I_1 is an average value for all stresses in the experiments.

The model predictions of the true triaxial tests are performed such that the predicted major principal stress difference relationship (σ_1–σ_3 vs ε_1) for each stress path segment is plotted up to the magnitude of the major principal strain, ε_1, as measured in the experiments. Thus, the history of each of the three principal stresses was specified and predictions were made until the predicted value of ε_1 reached the experimental value. At this point, the subsequent branch of the stress path was initiated. Thus, the comparison between the experimental and predicted behavior of soil is made for the same levels of ε_1 for each stress path segment. The differences in the stresses that cause these equal strains then can be analyzed using this technique. Therefore, the length of each stress path segment is extended or shortened to produce the experimental ε_1-value corresponding to that segment. The stress paths of the true triaxial tests with superimposed plastic strain increment vector directions projected onto the octahedral and constant σ_3-planes are plotted for each test. The changes in the direction of experimental stress paths (stress-reversals) are indicated as A, B, C, etc., whereas significant points of interest in the model predictions are indicated as I, II, III, etc.

6. Prediction of true triaxial tests with large 3-D stress-reversals

The true triaxial tests and their predictions are presented in sequence of increasing complexity of the 3-D stress paths.

6.1. Test SMC-L6

Fig. 5a and b shows the stress path for test SMC-L6, together with the experimental and predicted plastic strain increment vector directions. The major principal stress difference relationship (σ_1–σ_3 vs ε_1) and the volumetric strain behavior are shown in Fig. 6a and b. The intermediate principal stress difference relationship (σ_2–σ_3 vs ε_1) and the intermediate principal strain relationship, ε_2 vs ε_1, are shown in Fig. 7a and b.

The stress path for SMC-L6 in Fig. 5 shows that the specimen is initially loaded along a true triaxial extension path, between points *A* and *B*. The initial cross-anisotropic yield surface at point *A* makes an angle (χ_o) of 3.0° with the hydrostatic axis. The cross-anisotropic yield surface expands and rotates back towards the hydrostatic axis during loading along segment A–B. The rotation angle of the yield surface (χ) decreases from 3.0° at point *A* to 0.38° at point *B*. The stress path reverses at point *B* resulting in a kinematic yield surface forming with its tip located at *B*. Path B–C lies within the elastic domain of the model. Hence, pure elastic strains are produced during path B–C. The stress path changes at point *C* and follows a path of increasing σ_1 with constant σ_2 and σ_3. A new kinematic yield surface is not formed at point *C* since the stress reversal occurs within the elastic domain. Only elastic strains are produced along path C–D until it intersects and activates the kinematic yield surface at point I. This surface expands and rotates such that the tip of the current kinematic yield surface lies on a parabola joining the tip of the original cross-anisotropic yield surface (associated with point *B*) and the stress reversal point *B*. Plastic strains are produced during kinematic hardening. The evolution of the kinematic yield surface continues during loading along path I-D until the stress path reaches point II where the original cross-anisotropic yield surface is located. The kinematic yield surface

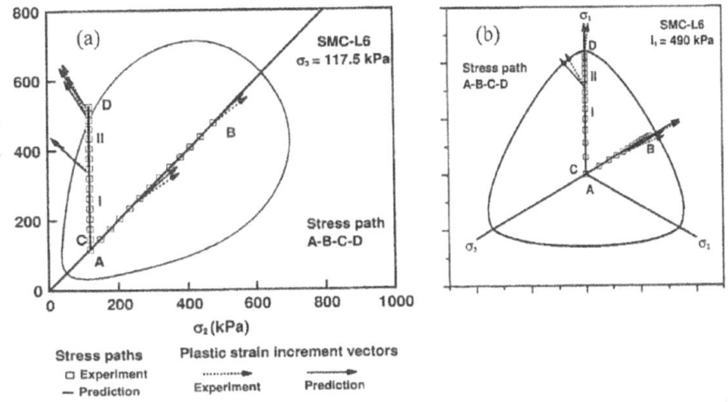

Fig. 5. Stress path with superimposed plastic strain increment vectors from true triaxial test no. SMC-L6 on loose, cross-anisotropic Santa Monica Beach sand (a) shown in plane with σ_3' = constant, and (b) projected on octahedral plane.

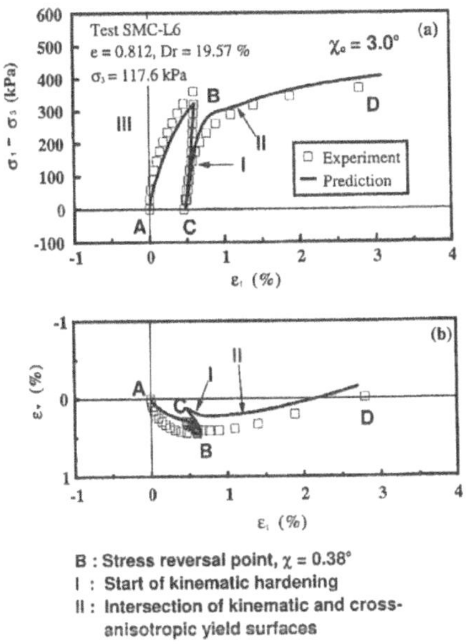

Fig. 6. Experimental test data of SMC-L6 and kinematic hardening predictions performed with cross-anisotropic model; (a) $(\sigma_1 - \sigma_3)$ vs ε_1, and (b) ε_v vs ε_1.

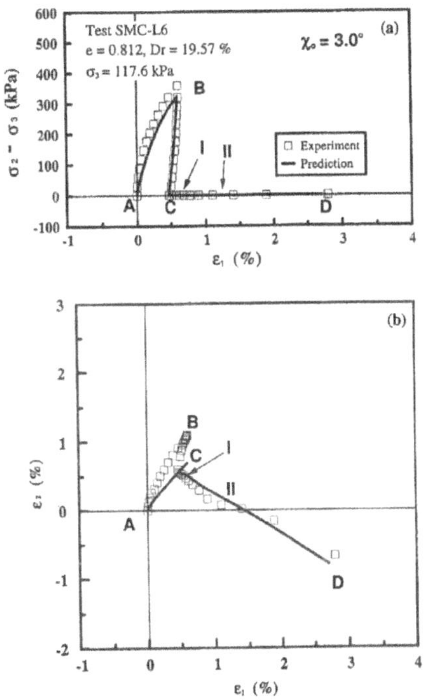

Fig. 7. Experimental test data of SMC-L6 and kinematic hardening predictions performed with cross-anisotropic model; (a) $(\sigma_2 - \sigma_3)$ vs ε_1, and (b) ε_2 vs ε_1.

merges with the cross-anisotropic yield surface at point II. Once the cross-anisotropic yield surface is activated, the history of stress-reversals is assumed to be erased from the memory of the soil.

Fig. 5 also shows that the experimental and predicted plastic strain increment vector directions follow each other closely along the entire 3-D stress path.

Fig. 6a shows that the predicted major principal stress difference relationship, $(\sigma_1 - \sigma_3$ vs $\varepsilon_1)$, follows the experimental behavior very closely for the entire stress path. Fig. 6b shows that the predicted volumetric strains are slightly less contractive throughout the stress path A–B–C–D. Close inspection indicates that this deviation is accumulated during the initial loading portion (A–B) of the test. The strains associated with the loading path A–B are produced by the cross-anisotropic version of the Single Hardening Model, not the kinematic hardening model. If the experimental and predicted curves were combined at point B, where the stress reversal occurred, the predicted volumetric strain curve would fit very well with the measured volumetric strain behavior.

The prediction follows very closely with the variation of the experimental intermediate principal stress difference relationship, $(\sigma_2 - \sigma_3$ vs $\varepsilon_1)$, as shown in Fig. 7a. Fig. 7b indicates that the intermediate principal strain is under-predicted during stress path A–B, but follows closely for the remainder of the stress path.

P.V. Lade et al. / Computers and Geotechnics 36 (2009) 1285–1297

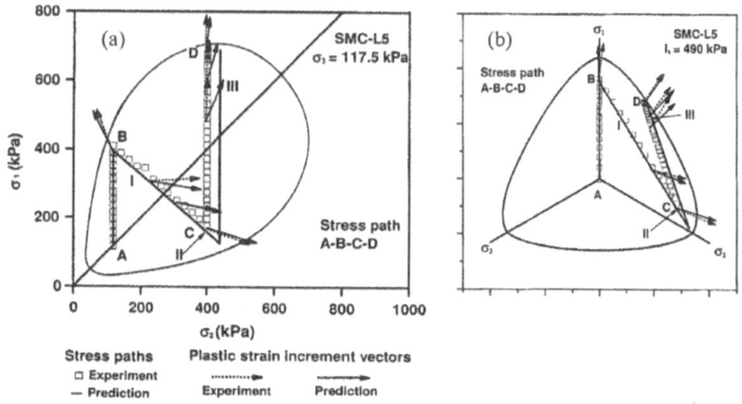

Fig. 8. Stress path with superimposed plastic strain increment vectors from true triaxial test no. SMC-L5 on loose, cross-anisotropic Santa Monica Beach sand (a) shown in plane with σ_3' = constant, and (b) projected on octahedral plane.

6.2. Test SMC-L5

Fig. 8a and b shows the stress path for test SMC-L5 plotted on a constant σ_3-plane and an octahedral plane, respectively. The experimental and predicted behavior of SMC-L5 is shown in Figs. 9 and 10.

The stress path in Fig. 8 shows that the specimen is initially loaded along path A–B in triaxial compression. The stress path changes direction at point *B*. The rotation angle of yield surface decreases from its initial value of 3.0° at point *A* to 0.59° at *B*. A kinematic yield surface is formed with its tip at point *B*. The specimen is unloaded along segment B–C, as shown in Fig. 8a and b. Pure elastic strains are developed until the stress path intersects the kinematic yield surface at point I. The kinematic yield surface then evolves with loading along segment I-C and it consequently produces plastic strains. The kinematic yield surface merges with the cross-anisotropic yield surface at point II, which is associated with the stress reversal point *B*. The stress path direction changes at point *C* and follows the path of increasing values of σ_1 while σ_2 and σ_3 remain constant. A new kinematic yield surface develops at point *C*, which then expands with loading along path C–D. This stress path segment moves in a direction that does not pass through the purely elastic domain. Therefore, plastic strains are produced immediately upon the stress reversal. The stress path intersects the cross-anisotropic yield surface at point III which was associated with point *C*. The cross-anisotropic yield surface then expands and rotates with loading along path III-D.

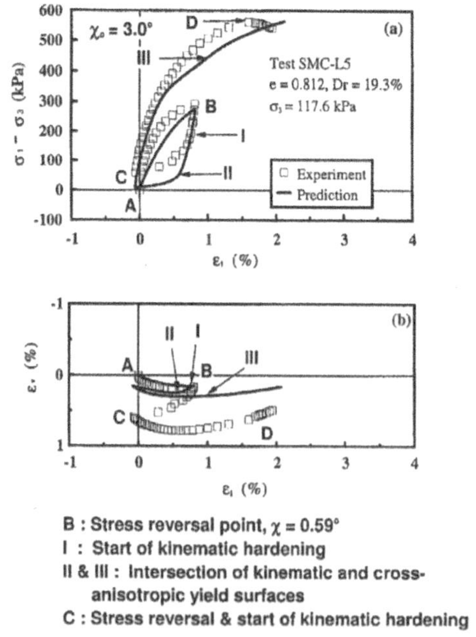

Fig. 9. Experimental test data of SMC-L5 and kinematic hardening predictions performed with cross-anisotropic model: (a) (σ_1–σ_3) vs ε_1 and (b) ε_v vs ε_1.

The predicted plastic strain increment vector directions follow the experimental results very closely during path A–B–C, as shown in Fig. 8a and b. The predictions deviate some following point III along path C–D, where the experimental intermediate principal strains were observed to be constant. However, the deviations tend to decrease with increased loading along III-D. Predicted and experimental plastic strain increment vectors are very close to each other at point *D*.

The predicted major principal stress difference relationship, (σ_1–σ_3 vs ε_1), deviates somewhat from the experimental behavior during the triaxial compression stress path A–B, as shown in Fig. 9a. However, the overall predicted (σ_1–σ_3) vs ε_1 curve fits the measured behavior with reasonable accuracy. The experimental curve is slightly stiffer than the prediction after the stress-reversals. This may be due to the large amount of densification occurring in the specimen during segment B–C. Fig. 9b indicates that the volumetric strain relationship, (ε_v vs ε_1), deviates from experimental results in magnitude, yet the qualitative trends in behavior are observed. The predictions exhibit much less volumetric contraction than the experiments during stress path B–C.

Fig. 10a indicates that the prediction of the intermediate principal stress difference relationship, (σ_2–σ_3 vs ε_1), coincides with the experimental behavior reasonably well. Fig. 10b shows that the predicted intermediate principal strain relationship, (ε_2 vs ε_1), fits closely with the measured results during path A–B. However, it deviates from the experimental behavior for the remainder of the stress path. Insufficient ε_2 is produced along segment B–C and an excessive amount is predicted along path C–D. The experimental intermediate principal strain, ε_2, remains approximately constant

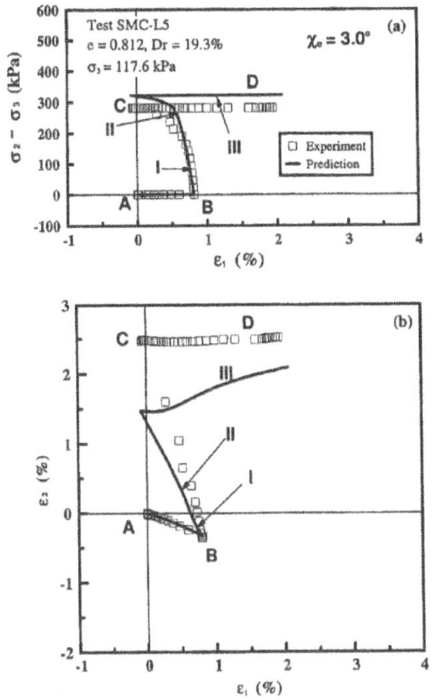

Fig. 10. Experimental test data of SMC-L5 and kinematic hardening predictions performed with cross-anisotropic model: (a) (σ_2–σ_3) vs ε_1 and (b) ε_2 vs ε_1.

with a constant value of σ_2 along segment C–D, even after the original yield surface is rejoined. This experimental behavior was also observed in the other true triaxial tests involving large stress-reversals on medium dense Santa Monica Beach sand [5].

6.3. Test SMC-L7

Fig. 11a and b shows the stress path of test SMC-L7 plotted on constant the σ_3-plane and the octahedral plane, respectively. The experimental and predicted behavior of SMC-L7 is shown in Figs. 12 and 13.

The stress path in Fig. 11a and b shows that the specimen is initially loaded along segment A–B in triaxial compression. The stress path changes direction at point *B*. The rotation angle of the cross-anisotropic yield surface decreases from 3.0° at point *A* to 0.56° at *B*. A kinematic yield surface is formed with its tip located at point *B* and it evolves during loading along segment B–C. The stress path segment B–C moves in a direction that does not pass through the purely elastic domain. Therefore, plastic strains are produced immediately. The stress path again changes at point *C* and a new kinematic yield surface is formed. The previous kinematic yield surface becomes the memory surface. Specimen is unloaded at constant σ_2 and σ_3 during segment C–D, as shown in Fig. 11a and b. Only elastic strains are predicted along segment C–D until point I is reached, whereupon the stress path intersects the kinematic yield surface. The kinematic yield surface expands and rotates such that its tip moves on a parabola joining the tip of the memory surface and stress reversal point *C*. Plastic strains are subsequently predicted due to the activation of kinematic hardening from point I to II, where the kinematic yield surface merges with the memory surface. Then with further loading along path II-D, the kinematic yield surface expands and rotates until the stress path intersects the cross-anisotropic yield surface at point III. The cross-anisotropic yield surface expands with loading along III-D. The stress path reverses at point *D*. The kinematic hardening model continues to develop in a similar manner as previously discussed. The stress path segment D–E moves in a direction that does not pass through the purely elastic domain. Therefore, plastic strains are produced immediately. Finally, the kinematic yield surface merges with the cross-anisotropic yield surface at point IV. At this point strains are computed based on the cross-anisotropic Single Hardening Model.

The predicted plastic strain increment vector directions follow closely with the measured directions for the whole stress path except during segment B–C, as indicated in Fig. 11a and b. This deviation explains the reason the model under-predicts the intermediate principal strains during path B–C as shown in Fig. 13b below.

Fig. 12a shows that the predicted major principal stress difference relationship (σ_1–σ_3 vs ε_1) follows reasonably closely with the experimental behavior during stress path A–C. However, the prediction deviates from the experiment during path C–D. The start of kinematic hardening, point I, appears to lie at a lower value of (σ_1–σ_3) in the model prediction than in the experiment. This may imply that the true yield surface of the soil may not be shaped

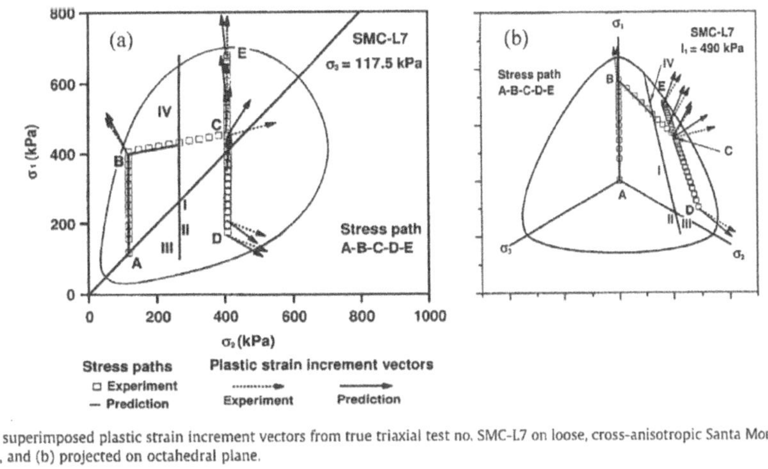

Fig. 11. Stress path with superimposed plastic strain increment vectors from true triaxial test no. SMC-L7 on loose, cross-anisotropic Santa Monica Beach sand (a) shown in plane with σ_3' = constant, and (b) projected on octahedral plane.

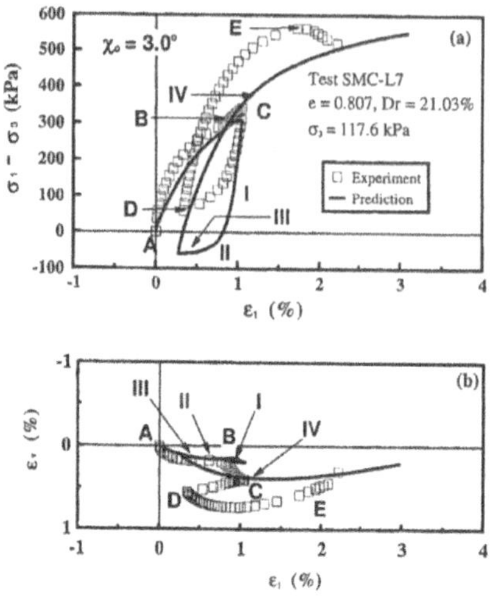

B : Stress reversal point & start of kinematic hardening (χ = 0.56°)
C : Stress reversal
I : Start of kinematic hardening
II : Intersection of kinematic and memory surfaces
II & IV : Intersection of kinematic and cross-anisotropic yield surfaces
D : Stress reversal & start of kinematic hardening

Fig. 12. Experimental test data of SMC-L7 and kinematic hardening predictions performed with cross-anisotropic model: (a) (σ_1–σ_3) vs ε_1 and (b) ε_v vs ε_1.

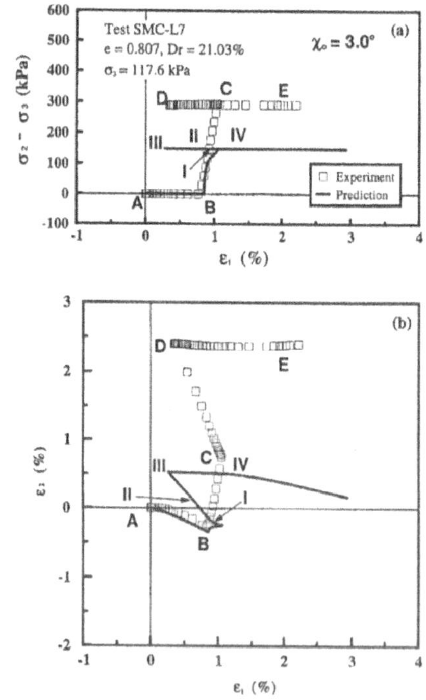

Fig. 13. Experimental test data of SMC-L7 and kinematic hardening predictions performed with cross-anisotropic model: (a) (σ_2–σ_3) vs ε_1 and (b) ε_2 vs ε_1.

the same along the three principal stress axes. The experimental (σ_1–σ_3) vs ε_1 curve tends to be stiffer than the model prediction along stress path D–E. This effect may be the effect of soil densification during segment B–C.

Fig. 12b shows that the predicted volumetric strain relationship (ε_v vs ε_1) deviates from the experimental behavior. The model does not capture the large amount of volumetric contraction in path B–D. However, the model predicts the significant trends in the behavior during the stress reversal.

The predicted intermediate principal stress difference relationship (σ_2–σ_3 vs ε_1) follows reasonably closely with the experimental trends as shown in Fig. 13a. The intermediate principal strains (ε_2) appear to be significantly under-predicted on segment B–D, but reasonably well predicted on paths A–B and D–E, as shown in Fig. 13b. The experimental behavior along segment D–E shown in Fig. 13b indicates that the intermediate principal strains, ε_2, remain approximately constant with loading at constant values of σ_2 and σ_3. The model also appears to predict the nearly constant ε_2 during kinematic loading D-IV, and slightly decreasing values of ε_2 during segment IV-E.

6.4. Test SMC-L9

Fig. 14a and b shows the stress path of test SMC-L9 plotted on the constant σ_3-plane and the octahedral plane, respectively. The experimental and predicted behavior of SMC-L9 is shown in Figs. 15 and 16.

The stress path for SMC-L9 in Fig. 14a and b shows that the specimen is initially loaded along a true triaxial extension path A–B. Consequently, the rotation angle of the cross-anisotropic yield surface (χ) decreases from 3.0° at point *A* to 0.29° at *B*. The stress path direction changes at point *B* and therefore a kinematic yield surface is formed with its tip located at B. Along segment B–C, σ_2 decreases and σ_1 increases as shown in Fig. 14a and b. Only elastic strains are produced by the model along path B–C until stress point I is reached, where the stress path intersects the kinematic yield surface. Plastic strains are subsequently produced due to activation of kinematic hardening from point I to C. The stress path direction changes again at point *C*. Consequently, a new kinematic yield surface is formed at point *C* and the previous kinematic yield surface becomes the memory surface. The specimen is unloaded at constant σ_2 and σ_3 during segment C–D. Only elastic strains are produced until point II, where the stress path intersects the kinematic yield surface. The kinematic yield surface expands and rotates toward the memory surface until the stress path reverses at point *D*, where a new kinematic yield surface is formed. Consequently, the previous memory surface is eliminated and the previous kinematic yield surface becomes the current memory surface. The specimen is reloaded from point *D* at constant σ_2 and σ_3 as shown in Fig. 14a and b. The stress path moves in a direction that does not pass through the elastic domain and therefore, plastic strains are produced immediately. The kinematic yield surface then merges with the memory surface at point III. The kinematic surface expands and rotates towards the cross-anisotropic yield surface that is associated with the initial stress reversal point *B*. The stress path joins the cross-anisotropic yield surface at point IV.

The predicted plastic strain increment vector directions follow closely with those of the measured vectors, as shown in Fig. 14a and b.

Fig. 15a shows that the predicted major principal stress difference relationship (σ_1–σ_3 vs ε_1) fits very closely with the experimental behavior. The predicted volumetric strain behavior (ε_v vs ε_1) is slightly more dilatant than the experimental results as shown in Fig. 15b. This deviation is associated with the initial loading path A–B where the strains are calculated by the cross-anisotropic Single Hardening Model, not the kinematic hardening model. If the

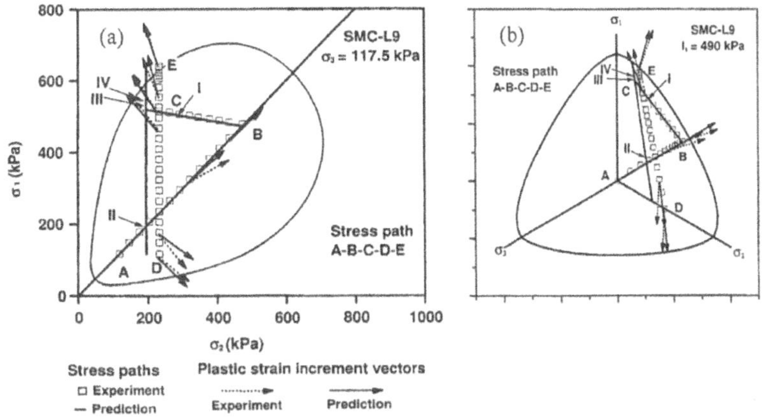

Fig. 14. Stress path with superimposed plastic strain increment vectors from true triaxial test no. SMC-L9 on loose, cross-anisotropic Santa Monica Beach sand (a) shown in plane with σ'_3 = constant, and (b) projected on octahedral plane.

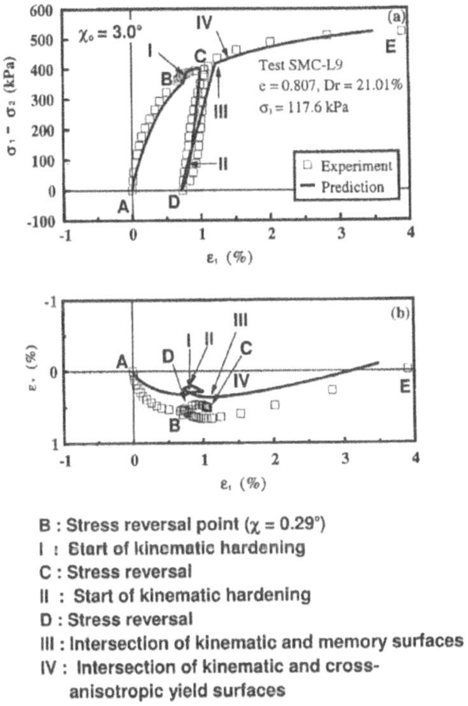

Fig. 15. Experimental test data of SMC-L9 and kinematic hardening predictions performed with cross-anisotropic model: (a) (σ_1–σ_3) vs ε_1 and (b) ε_v vs ε_1.

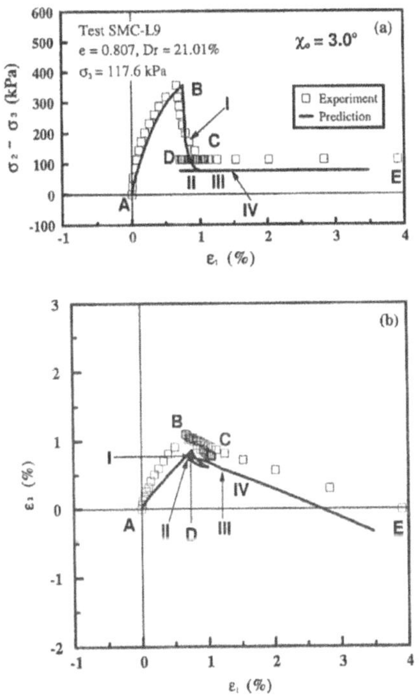

Fig. 16. Experimental test data of SMC-L9 and kinematic hardening predictions performed with cross-anisotropic model: (a) (σ_2–σ_3) vs ε_1 and (b) ε_2 vs ε_1.

experimental and predicted curves were combined at the initial stress reversal point *B*, the predicted volumetric strain curve would fit the experimental data with excellent accuracy.

The predicted intermediate principal stress difference relationship (σ_2–σ_3 vs ε_1) fits closely with the experimental behavior as shown in Fig. 16a. The predicted intermediate principal strain relationship (ε_2 vs ε_1) slightly deviates from the experimental results as shown in Fig. 16b. However, the deviation is mainly associated with the initial loading path A–B. The kinematic hardening model appears to correctly capture trends observed in the experimental results during the load reversals as shown in Fig. 16b.

6.5. Test SMC-L8

Fig. 17a and b shows the stress path of test SMC-L8 plotted on the constant σ_3-plane and the octahedral plane, respectively. The experimental and predicted behavior of SMC-L8 is shown in Figs. 18 and 19.

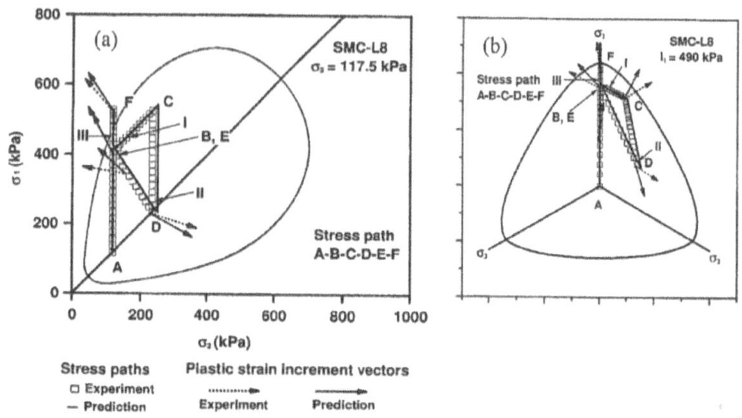

Fig. 17. Stress path with superimposed plastic strain increment vectors from true triaxial test no. SMC-L8 on loose, cross-anisotropic Santa Monica Beach sand (a) shown in plane with σ_3' = constant, and (b) projected on octahedral plane.

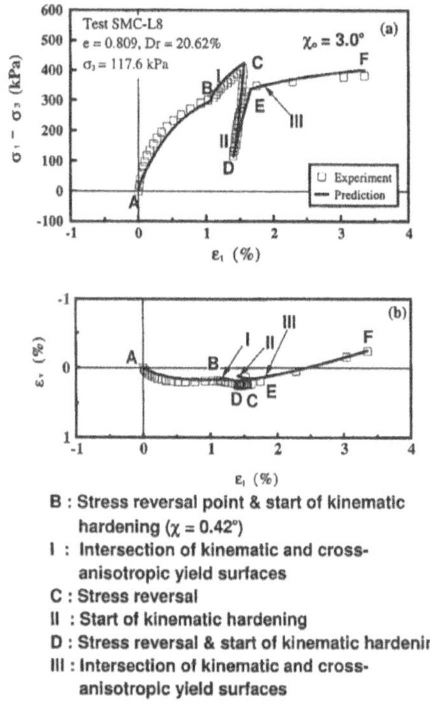

Fig. 18. Experimental test data of SMC-L8 and kinematic hardening predictions performed with cross-anisotropic model: (a) $(\sigma_1 - \sigma_3)$ vs ε_1 and (b) ε_v vs ε_1.

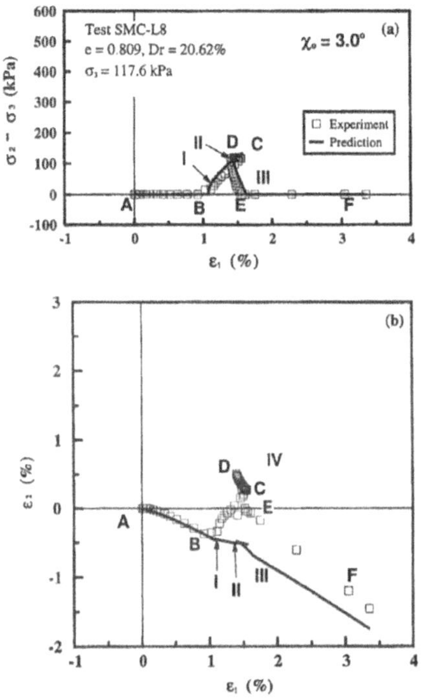

Fig. 19. Experimental test data of SMC-L8 and kinematic hardening predictions performed with cross-anisotropic model: (a) $(\sigma_2 - \sigma_3)$ vs ε_1 and (b) ε_2 vs ε_1.

The stress path for SMC-L8 in Fig. 17a and b shows that the specimen is initially loaded along triaxial compression path A–B. The rotation angle at point *B* is 0.42°, whereas it was 3.0° at point *A*. The stress path changes direction at point *B* and hence, a kinematic yield surface is formed with its tip located at B. Along segment B–C, σ_1 and σ_2 increase. The path moves in a direction that does not pass through the elastic domain and therefore, plastic strains are produced immediately. The kinematic yield surface merges with the cross-anisotropic yield surface at point I. The cross-anisotropic yield surface expands and rotates towards the hydrostatic axis with loading along I-C. The stress path reverses at point *C* and a new kinematic yield surface is developed with its tip located at C. The specimen is unloaded at constant values of σ_2 and σ_3 along segment C–D. Only elastic strains are produced along path C–D until it intersects the kinematic yield surface at point II. Plastic strains are the produced due to the activation of kinematic hardening during path II-D. The stress path then reverses at point *D*. A new kinematic yield surface is developed with its tip located at point *D*, and the previous kinematic surface now becomes the memory surface. The specimen is loaded towards triaxial compression with increasing σ_1 and decreasing σ_2. The path D–E moves in a direction that does not pass through the elastic domain and therefore, plastic strains are produced immediately. The kinematic surface expands and rotates towards the memory surface during loading path D–E. The stress path direction changes at point *E*, but the new segment is directed outside of the kinematic yield surface. Hence, a new kinematic yield surface is not formed at point *E* and the current kinematic surface still continues to expand and rotate towards the memory surface with loading along path E–F. The path intersects the cross-anisotropic yield surface at point III (associated with the stress reversal at point *C*) before joining the memory surface. The plastic strains for the remainder of the stress path III-F are calculated by the cross-anisotropic Single Hardening Model.

Fig. 17a and b indicates that predicted plastic strain increment vector directions slightly deviate from the measured directions during the stress-reversals. This is the underlying reason why the model under-predicts the intermediate principal strains during stress reversal path B–C, as discussed below. In the remaining sections, predicted directions follow the experimental results with better accuracy.

Fig. 18a indicates that the predicted major principal stress difference relationship (σ_1–σ_3 vs ε_1) follows very closely with the experimental behavior. The predicted volumetric strain behavior (ε_v vs ε_1) also fits very well with the experimental data, as shown in Fig. 18b.

Fig. 19a indicates that the predicted intermediate principal stress difference relationship (σ_2–σ_3 vs ε_1) follows closely with the experimental behavior. However, the predicted intermediate principal strain relationship (ε_2 vs ε_1) deviates from the observed experimental behavior during stress path B–C, as shown in Fig. 19b. It appears that the model greatly under-predicts the values of ε_2 during this section of the stress path. Model predictions essentially follow the experimental results during segments A–B and C–F.

7. Discussion of model predictions

The predictions of the true triaxial tests with 3-D stress-reversals on loose cross-anisotropic Santa Monica Beach sand were presented above using a combined model that takes into account rotational kinematic hardening and effects of inherent cross-anisotropic behavior. Overall, the model essentially captures the soil behavior during primary loading and large stress-reversals under three dimensional conditions. This is done without introducing more parameters for the rotational kinematic hardening, but by handling the model enhancement within the mathematical framework. Nevertheless, a few, relatively small shortcomings are observed in the comparison between the experimental results and the predictions. Improvements on the current model require more research and detailed analysis in the future.

Tests SMC-L5 and SMC-L7 showed that the model predictions exhibit a less stiff major principal stress difference relationship (σ_1–σ_3 vs ε_1) after the stress reversal has taken place than observed from the experimental results. This effect may be due to some densification of the soil during stress-reversals. This relatively small densification effect can be observed in the volumetric strain relationship for each test. However, this small densification should not have any significant effect on the predictions of the Single Hardening Model. Some of the differences between experimental results and predictions may be due to experimental scatter and deviations from the true behavior. Previous predictions have shown that such soil behavior can be predicted satisfactorily. Overall, the trends in the measured behavior are reproduced by the model predictions.

During tests SMC-L5, SMC-L7, and SMC-L8, the magnitudes of ε_2 were under-predicted in the stress path segments beginning at point *B* relative to the measured strains in this direction. As an example, SMC-L7 starts the first stress reversal at point *B* in triaxial compression, as shown in Fig. 11a. In the following segments, B–C and C–D, there are significant strains in the ε_2 direction, as shown in Fig. 13b, while the predictions show smaller amounts of strains in the ε_2 direction. This aspect was also observed in other true triaxial tests involving large stress-reversals on medium dense Santa Monica Beach sand [5]. However, the overall trend in predictions is correct.

The volumetric strain along the initial stress paths are predicted as slightly less contractive in all of the tests. This is a product of the model parameters in the basic Single Hardening Model. In addition, in all but test SMC-L8, the predicted volumetric strains during the first stress reversal are more dilatant than the experiments indicate. Thus, it appears that part of the problem with the predicted strains during the large stress-reversals is due to the less than perfect parameter determination for the primary loading conditions.

Test SMC-L5 showed that after the first stress reversal at point *B*, the yield surface of the soil was rejoined before that predicted by the model. Similar behavior was also observed after the reversal at point *C* in test SMC-L7. This may be due to the shape of the true yield surface. The true yield surface may be distorted, such that when established in the σ_1-direction, it may be reached at a lower stress level in the σ_2-direction. This speculation is somewhat verified in these tests by the observation that once the yield surface is established in the σ_2-direction, the model appears to more accurately capture its location when the stress path moves back towards the σ_1-direction (e.g., SMC-L6). These observations would tend to indicate that the yield surface is distorted and not completely symmetrical (isotropic) about the hydrostatic axis. However, detailed improvements require more research and comprehensive analysis in the future.

Overall, the predictions based on the kinematic hardening model with the cross-anisotropic Single Hardening Model follow the experimental behavior with reasonable accuracy, even for the relative more complex stress paths of tests SMC-L9 and SMC-L8. It should be mentioned again that the employment of a rotational mechanism in the principal stress space to model the effects of cross-anisotropy works well for cases without stress rotation in the field, but this approach cannot handle stress rotation in the physical space.

8. Conclusions

Verification of a kinematic hardening model with a cross-anisotropic formulation is presented using comparisons of predictions with true triaxial test data. A rotational kinematic hardening mechanism has been proposed for 3-D large stress-reversals in sands [17]. An existing elasto-plastic, non-associated plastic flow, isotropic Single Hardening Model serves as the basic framework for this model. To enable better predictions, a cross-anisotropic model was utilized in addition to the proposed kinematic hardening model. The cross-anisotropic Single Hardening Model proposed by Abelev et al. [1] and Gutta et al. [5] is utilized for this purpose. The cross-anisotropic model uses the concept of kinematic hardening and rotation of the principal stress coordinate system such that the anisotropic behavior of soils is reduced to a pseudo-isotropic state. The cross-anisotropic model with the kinematic hardening mechanism incorporates thirteen parameters that can be determined from three drained triaxial compression tests, an isotropic compression test and an extension test. The formulation and parameter evaluation procedure was discussed by Abelev et al. [1].

Verification of the kinematic hardening model with the cross-anisotropic formulation is presented using comparisons of predictions with true triaxial test data. The true triaxial tests incorporating large 3-D stress-reversals were performed on loose cross-anisotropic Santa Monica Beach sand [4]. The model is shown to capture the overall trends of the observed behavior during large 3-D stress-reversals with reasonably good accuracy within the scatter of the test results.

Acknowledgments

The research presented here was performed, in part, in the Department of Civil Engineering at The Johns Hopkins University with support from the National Science Foundation under Grant Nos. MSS 9119272/MSS 9396271. The research was finalized in the Department of Civil Engineering at University of Delaware under Grant No. CMS 0096341. Grateful appreciation is expressed for this support.

References

[1] Abelev AV, Gutta SK, Lade PV, Yamamuro JA. Modeling cross anisotropy in granular materials. ASCE J Eng Mech 2007;133(8):919–32.

[2] Abelev AV, Lade PV. Effects of cross-anisotropy on three-dimensional behavior of sand. Part I: Stress-strain behavior and shear banding. ASCE J Eng Mech. 2003;129(2):160–6.

[3] Abelev AV, Lade PV. Characterization of failure in cross-anisotropic soils. ASCE J Eng Mech 2004;130(5):599–606.

[4] Boonyachut S. Experimental study of the behavior of cohesionless soil during stress reversals. Ph.D. thesis. University of California: Los Angeles; 1976.

[5] Gutta SK, Yamamuro JA, Lade PV. Predictions of large stress reversals in true triaxial tests on cross-anisotropic sand. Int J Numer Anal Meth Geomech 2009;33(8):1013–37.

[6] Inel S, Lade PV. Rotational kinematic hardening model for sand. Part II. Characteristic work hardening law and predictions. Comput Geotech 1997;21(3):217–34.

[7] Kim MK, Lade PV. Single hardening constitutive model for frictional materials. I. Plastic potential function. Comput Geotech 1988;5:307–24.

[8] Lade PV. Elasto-plastic stress-strain theory for cohesionless soil with curved yield surfaces. Int J Solid Struct 1977;13:1019–35.

[9] Lade PV. Cubical triaxial apparatus for soil testing. ASTM Geotech Test J 1978;1(2):93–101.

[10] Lade PV. Assessment of test data for selection of 3-D failure criterion for sand. Int J Numer Anal Meth Geomech 2006;30(4):307–33.

[11] Lade PV, Abelev AV. Effects of cross-anisotropy on three-dimensional behavior of sand. Part II: Volume change behavior and failure. ASCE J Eng Mech 2003;129(2):167–74.

[12] Lade PV, Boonyachut S. Large stress reversals in triaxial tests on sand. In: Eisenstein Z, editor. Proceedings of the 4th international conference on numerical methods in geomechanics. Edmonton, Canada; 1982: 1. p. 171–82.

[13] Lade PV, Duncan JM. Cubical triaxial tests on cohesionless soil. ASCE J Soil Mech Found Div 1973;99(SM10):793–812.

[14] Lade PV, Inel S. Rotational kinematic hardening model for sand. Part I. Concept of rotating yield and plastic potential surfaces. Comput Geotech 1997;21(3):183–216.

[15] Lade PV, Kim MK. Single hardening constitutive model for frictional materials. II. Yield criterion and plastic work contours. Comput Geotech 1988;6:13–29.

[16] Lade PV, Kim MK. Single hardening constitutive model for frictional materials. III. Comparison with experimental data. Comput Geotech 1988;6:31–47.

[17] Lade PV, Yamamuro JA, Gutta SK. Rotational kinematic hardening model for three-dimensional stress reversals in sand. Int J Numer Anal Meth Geomech 2009;33(7):967–91.

[18] Yamamuro JA, Lade PV. Large stress reversals in true triaxial tests on cross-anisotropic sand. Int J Numer Anal Meth Geomech 2009;33(7):953–65.

Chapter XI
Effects of Stress Rotation

Preface

Important to the development of constitutive models is that a model has to be able to accommodate stress rotation correctly. To this end, the use of stress invariants is very useful as shown below. Experimental evidence shows that during stress rotation, such as in twisting of thin-walled metal tubes, the observed principal strain increments initially coincide in direction with that of the stress increments. This is predicted correctly by elasticity theory. But when the metal tubes are twisted further into the plastic range, it is found that the principal directions of total stress coincide with the principal directions of incremental plastic strain (Taylor and Quinney, 1931), as shown in the diagram in Fig. 11.1.

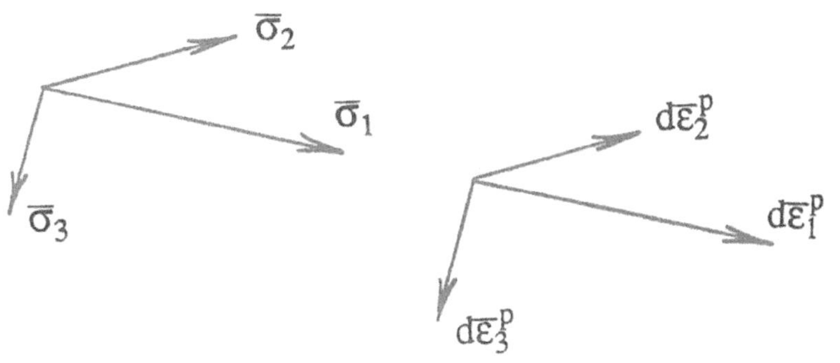

Fig. 11.1. Coincidence of total stress direction and incremental strain directions in physical space.

Stress rotation also occurs in geotechnical structures, such as, e.g.

(a) Under footings during loading
(b) In slopes during excavations
(c) In simple shear and torsion shear of hollow cylinder experiments

The equation expressing the plastic strain increments in isotropic materials in terms of the stresses is given as follows (see Chapter II):

$$d\varepsilon_{ij}^{p} = d\lambda \cdot \frac{\partial g}{\partial \sigma_{ij}}$$

where $d\lambda$ = proportionality constant
g = plastic potential function
$g = f$ for associated flow
$g \neq f$ for nonassociated flow

$$\left\{\frac{\partial g}{\partial \sigma_{ij}}\right\} = \text{normal to plastic potential surface}$$

For an isotropic material the plastic potential function may be expressed in terms of stress invariants: $g = g(\sigma_{ij}) = G(I_1, I_2, I_3)$ in which I_1, I_2, and I_3 are stress invariants of the stress tensor.

It will be shown that for an isotropic material Saint Venant's experimental observation is contained in the basic equations for plastic strain increments:

$$d\varepsilon_{ij}^{p} = d\lambda \cdot \frac{\partial G}{\partial \sigma_{ij}} \tag{A.1}$$

Using the chain rule of differentiation the following expression is obtained:

$$d\varepsilon_{ij}^{p} = d\lambda \cdot \left(\frac{\partial G}{\partial I_1} \cdot \frac{\partial I_1}{\partial \sigma_{ij}} + \frac{\partial G}{\partial I_2} \cdot \frac{\partial I_2}{\partial \sigma_{ij}} + \frac{\partial G}{\partial I_3} \cdot \frac{\partial I_3}{\partial \sigma_{ij}}\right) \tag{A.2}$$

Now, the plastic strain increments $d\varepsilon_{ij}^{p}$ will have the same principal axes as σ_{ij} if it can be demonstrated that all plastic shear strain increments vanish, $d\varepsilon_{ij}^{p} = 0$ for $i \neq j$, whenever all shear stresses vanish, $\sigma_{ij} = 0$ for $i \neq j$. To demonstrate this, it will be shown that the derivatives of I_1, I_2 and I_3 with respect to the shear stresses σ_{ij} $(i \neq j)$ are equal to zero when all shear stresses $\sigma_{ij} = 0 (i \neq j)$

$$I_1 = \sigma_{11} + \sigma_{22} + \sigma_{33} \tag{A.3}$$

$$\left(\frac{\partial I_1}{\partial \sigma_{ij}}\right)_{i \neq j} = 0 \tag{A.4}$$

$$I_2 = \tau_{12} \cdot \tau_{21} + \tau_{23} \cdot \tau_{32} + \tau_{31} \cdot \tau_{13} - (\sigma_1 \cdot \sigma_2 + \sigma_2 \cdot \sigma_3 + \sigma_3 \cdot \sigma_1) \tag{A.5}$$

$$\left(\frac{\partial I_2}{\partial \sigma_{ij}}\right)_{i \neq j} = \sigma_{ji} = 0 \quad \text{for } \sigma_{ij} = \sigma_{ji} = 0 \tag{A.6}$$

$$I_3 = \sigma_{11} \cdot \sigma_{22} \cdot \sigma_{33} + \tau_{12} \cdot \tau_{23} \cdot \tau_{31} + \tau_{21} \cdot \tau_{32} \cdot \tau_{13} - (\sigma_{11} \cdot \tau_{23} \cdot \tau_{32} + \sigma_{22} \cdot \tau_{31} \cdot \tau_{13} + \sigma_{33} \cdot \tau_{12} \cdot \tau_{21}) \tag{A.7}$$

$$\left(\frac{\partial I_3}{\partial \sigma_{ij}}\right)_{i \neq j} = \sigma_{jk} \cdot \sigma_{ki} - \sigma_{kk} \cdot \sigma_{ji} = 0 \quad \text{for } \sigma_{ij} = \sigma_{ji} = 0 \tag{A.8}$$

Thus, the derivatives of the stress invariants with respect to the shear stresses are equal to zero when the shear stresses are zero.

Substitution of the derivatives in Eqs. (A.4), (A.6) and (A.8) into Eq. (A.2) produces:

$$\mathrm{d}\varepsilon_{ij}^{\mathrm{p}} = \mathrm{d}\lambda \cdot \left(\frac{\partial G}{\partial I_1} \cdot 0 + \frac{\partial G}{\partial I_2} \cdot 0 + \frac{\partial G}{\partial I_3} \cdot 0\right) = 0 \quad \text{for } \sigma_{ij} = \sigma_{ji} = 0 \tag{A.9}$$

Therefore, $\mathrm{d}\varepsilon_{ij}^{\mathrm{p}} = 0$ for $i \neq j$, i.e., when all shear stresses are zero, and the principal stress axes coincide with the principal axes of plastic strain increments in isotropic plastic materials. Consequently, the basic equations for plastic strain increments contain the experimental observations for isotropic materials (Lade, Nam and Hong, 2009).

This is also found for the behavior of sands, as demonstrated in the following article.

References:

Lade, P.V., Nam, J. and Hong, W.P. (2009) "Interpretation of strains in torsion shear tests," *Computers and Geotechnics*, 36(1-2): 211-225.

Taylor, G.I. and Quinney, H. (1931) "The plastic distortion of metals," *Phil. Trans. Roy. Soc.*, A230: 323-362.

Strain Increment and Stress Directions in Torsion Shear Tests

By Won Pyo Hong[1] and Poul V. Lade,[2] Members, ASCE

Abstract: The directions of the major principal strain increment, stress, and stress increment during rotation of the principal stress axes at any stress level are studied for K_0-consolidated clay using a torsion shear apparatus with individual control of the vertical normal stress, the confining pressure, and the shear stress on hollow cylinder specimens under undrained and drained conditions. The torsion shear tests are performed along predetermined stress-paths, which are chosen to cover the full range of rotation of principal stress axes from 0° to 90° relative to vertical. The test results indicate that the major principal strain increment directions coincide with the major principal stress directions at failure. The directions of major principal strain increment coincide with the directions of major principal stress increment at low stress levels and with the directions of major principal stress at higher stress levels. This indicates that the behavior of clay gradually changes from elastic to plastic as the stress level is increased. Elasto-plastic theory is therefore suitable for modeling the behavior of clay during rotation of principal stress axes.

Introduction

The behavior of soils during reorientation of principal stresses in the field should be understood for correct modeling of their stress-strain behavior. Rotation of principal stress axes can only be accomplished in the laboratory by applying a combination of normal stresses and shear stresses to the surface of a soil specimen. A torsion shear apparatus, in which there is individual control of vertical normal stress, confining pressure, and shear stress applied to a hollow cylinder specimen, was employed in this study. Rotation of principal stress axes due to construction on or in soils can be reproduced in the torsion shear apparatus by various combinations of vertical loading and torque.

One of the most important pieces of information required for modeling the stress-strain behavior of soils concerns the direction of the principal strain increment relative to the directions of principal stress and stress increment during rotation of stress axes at any stress level. For the purpose of investigating the relations among the directions of strain increment, stress, and stress increment as shearing progresses, undrained and drained torsion shear tests were performed on hollow cylinder specimens according to predetermined stress-paths. The stress-paths were chosen to cover the full range of rotation of principal stress axes from 0° to 90° with vertical. Hollow cylinder specimens were prepared by consolidating Edgar plastic kaolinite clay under K_0-conditions before shearing, thus introducing stress states in the clay specimens similar to those found in field deposits.

[1]Assoc. Prof., Dept. of Civ. Engrg., Coll. of Engrg., Chung-Ang Univ., Seoul, 156-756, Korea.

[2]Prof., Dept. of Civ. Engrg., School of Engrg. and Appl. Sci., Univ. of Calif., Los Angeles, CA 90024.

Note. Discussion open until March 1, 1990. To extend the closing date one month, a written request must be filed with the ASCE Manager of Journals. The manuscript for this paper was submitted for review and possible publication on January 17, 1989. This paper is part of the *Journal of Geotechnical Engineering*, Vol. 115, No. 10, October, 1989. ©ASCE, ISSN 0733-9410/89/0010-1388/$1.00 + $.15 per page. Paper No. 23934.

Previous Studies

To rotate the directions of principal stresses in soil specimens during testing, a combination of normal and shear stresses must be applied to the surface of the specimen. A review of the advantages and limitations of equipment employed for this purpose was presented by Saada and Townsend (1981).

Various forms of ring shear apparatus have been developed for measurement of residual strength of soils (Bishop et al. 1971). As the ring shear test proceeds, however, the directions of principal stresses are uncertain and the strains are not distributed uniformly in the specimen. This type of apparatus can therefore not be used for the purpose of investigating effects of reorientation of principal stresses.

Reorientation of principal stresses can be achieved in simple shear devices (Bjerrum and Landva 1966; Roscoe 1953, 1970; Roscoe et al. 1967) and in torsion shear equipment (Bojanowski 1970; Broms and Casparian 1965; Hight et al. 1983; Hong and Lade 1988; Ishibashi and Sherif 1974; Lade 1975, 1976, 1981; Lomize et al. 1969; Saada and Baah 1967; Saada and Bianchini 1975; Saada and Ou 1973). The simple shear device offers the advantage of testing soils under plane strain conditions. Roscoe and his coworkers (Roscoe 1970; Roscoe et al. 1967) have investigated the relation between the directions of strain increment and stress in sand during shear in simple shear devices. However, these apparatuses also have some disadvantages, which limit their use for more elaborate investigations (Lade 1975). The initial state of stress before shearing cannot be arbitrarily imposed. Only K_0-stress states can be imposed on the specimens before shearing. Further, the distributions of normal and shear stresses within the specimen are nonuniform (Roscoe et al. 1967; Wright et al. 1967). The evolution of the horizontal normal stress during shear is generally unknown (Saada and Townsend 1981).

Torsion shear equipment in which a hollow cylinder specimen is confined horizontally between membranes has also been employed to a limited extent during the last two decades. This equipment offers the advantage that no shear stresses are induced on the vertical surfaces of the specimen, complementary shear stresses are automatically applied to the specimen, and large and fairly uniform shear strains can be produced. Furthermore, individual control of the vertical normal stress, the confining pressure, and the shear stress makes it possible to impose various initial states of stress before the specimen is sheared (Lade 1981). However, this type of equipment also has limitations in terms of nonuniformity of stress and strain distributions, especially in specimens with inappropriate dimensions (Saada and Townsend 1981; Wright et al. 1978). In a torsion shear test, the hollow cylinder specimen should be exposed to a plane stress state (when the inside and outside pressures are equal). In order to obtain this state, the stresses and strains should be distributed uniformly in the specimen (Saada and Townsend 1981). But nonuniformities can be minimized in specimens with appropriate dimensions (Lade 1981). Thus: (1) The ratio of wall thickness and specimen diameter should be small; and (2) the height-to-diameter ratio should be high enough to avoid severe effects of end restraint, and it should be small enough to avoid buckling of the hollow cylinder specimen. With careful attention to the geometry of the specimen, the torsion shear apparatus is considered to be well suited for investigating the influence of reorientation of principal stresses. Saada and his coworkers (Saada and Baah 1967; Saada and Bianchini 1975; Saada and Ou 1973; Saada and Townsend 1981; Saada and Zamani 1973) have used the torsion shear apparatus to determine the stress-strain behavior of anisotropic clay. Recently, Hight and his coworkers (Hight et al. 1983; Symes et al. 1984) built a torsion shear apparatus and studied the effects of initial anisotropy of sand in a limited number of tests. Tatsuoka et al. (1986) have also studied failure and deformation of sand.

Torsion Shear Apparatus

Fig. 1 shows a three-dimensional drawing of a torsion shear apparatus designed and constructed to permit individual control of confining pressure, vertical deviator stress, and shear stress applied to a hollow cylinder specimen.

The hollow cylinder specimen employed here has height of 25 cm, average diameter of 20 cm, and wall thickness of 2 cm. Based on previous studies (Lade 1981) these dimensions were sufficient to neglect nonuniformities in stresses and strains in the specimens due to boundary effects. The specimen is confined between cap and base rings and inside and outside rubber membranes with thicknesses of 0.03 to 0.04 cm. The inside membrane was formed directly on the specimen, while the outside prefabricated membrane was placed using a membrane stretcher. In order to form the inside membrane, the specimen was first sprayed with diluted rubber cement to seal it off. The inside surface was then sprayed or painted with several layers of fluid latex rubber. The rubber was allowed to dry between layers. In order to transfer shear stresses from the cap and base rings to the specimen and to avoid slippage at their interfaces, full friction surfaces were provided on the rings. These consisted of a layer of sand grains glued to the rings by epoxy.

The entire setup is contained in a pressure cell, and the integrated hydraulic loading system is located below the table on which the apparatus sits. The same confining pressure is applied to the inside and outside surfaces of the specimen through the cell water that surrounds the specimen. The design and operational principles of this apparatus are given in Lade (1981).

For the purpose of performing torsion shear tests on clay specimens, some modifications were made to the vertical loading cylinder and the torsion shear loading cylinders. Upward loading capability was added to the vertical loading cylinder, and the strokes of the torsion shear loading cylinders were increased so that larger torsional displacements could be imposed on the specimen. The maximum strains that can be achieved in this apparatus for a 25-cm tall specimen are $\pm\epsilon_z = 25\%$ for vertical strains and $\gamma_{z\theta} = \pm 40\%$ for torsion shear strains.

The vertical load and the torque can be applied independently to the specimen through the center shaft and the cap plate, which is attached to the cap ring, as shown in Fig. 1. Both the vertical load and the torque can be either stress- or strain-controlled. The resulting deviator stress, with the cell pressure, provides for a vertical normal stress either larger or smaller than the confining pressure. Two sets of two interconnected pressure cylinders supply horizontal load in either a clockwise or a counterclockwise direction to the torsion arm connected to the center shaft.

During the torsion shear tests, the vertical load, the torque, the cell pressure, the vertical deformation, the thickness change of the hollow cylinder

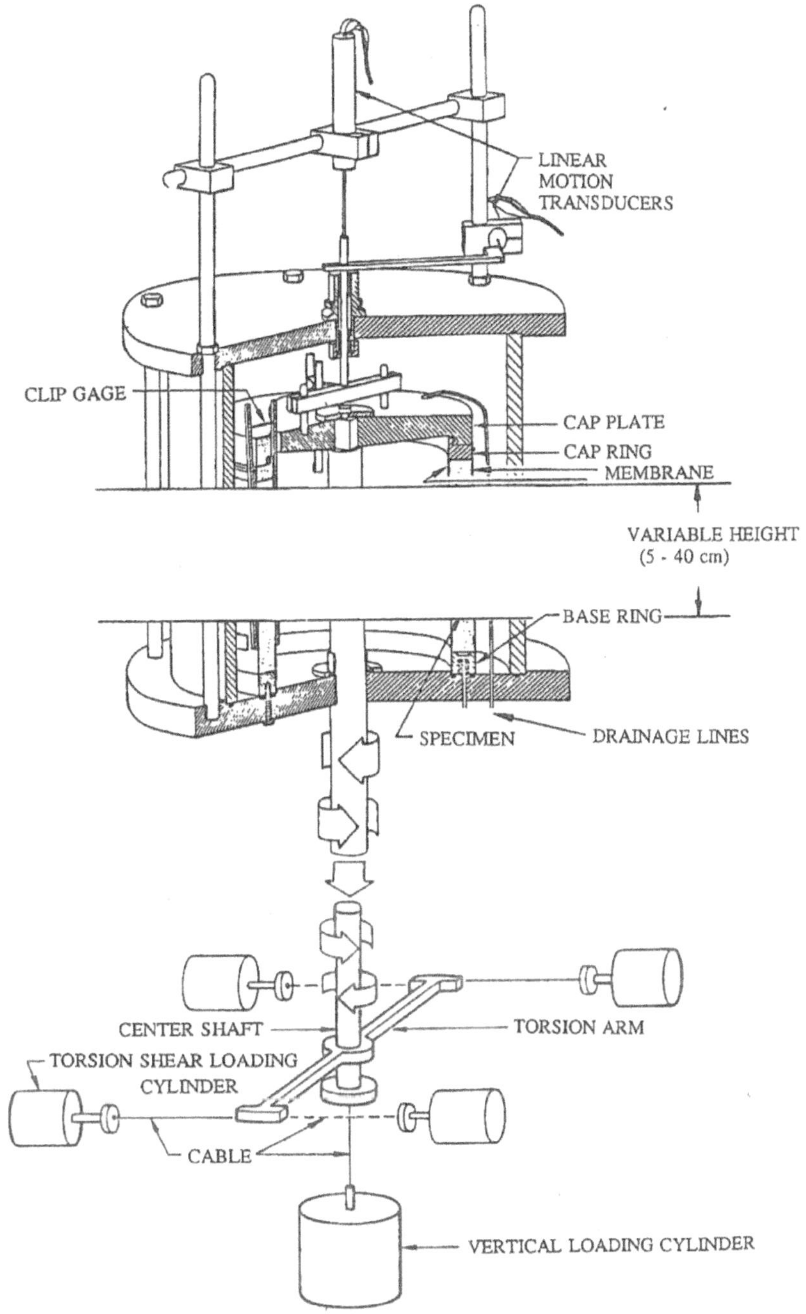

FIG. 1. Torsion Shear Apparatus

wall, and the shear deformation were measured in addition to the pore pressure in undrained tests or the volumetric strain in drained tests.

Clay Tested

All tests were performed on Edgar plastic kaolinite (EPK) clay whose specific gravity was 2.62. The particle size distribution indicated that the clay consisted of 40% silt and 60% clay-size particles. Atterberg limit tests produced $LL = 60$ and $PL = 30$. The activity was 0.5.

Specimen Preparation

Remolded specimens consolidated under K_0-stress conditions were prepared from a slurry of EPK clay mixed at a water content of twice the liquid limit. The clay slurry was poured into a special consolidometer and consolidated at an effective vertical pressure of 2.00 kg/cm^2 (196 kPa). The special consolidometer was built to make hollow cylinder specimens for torsion shear tests. The consolidometer has a height of approximately 60 cm, an outer diameter of 26 cm, and an inner diameter of 16.5 cm. Two drainage plates consisting of porous plastic were placed on the top and bottom of the clay slurry. In addition, long drainage strips made of porous plastic were attached along the outside wall to form a spiral from top to bottom, so it could be compressed easily as the clay slurry consolidated. After consolidation, the oversized hollow cylindrical clay sample was trimmed to its final dimensions in a specially designed trimming device with a rotating table and common trimming tools.

The specimen was then installed in the torsion shear apparatus and again consolidated under K_0-stress conditions by maintaining the volumetric strain equal to the vertical strain. Slotted filter paper drains were installed on the outside face of the specimen. The maximum drainage path was therefore equal to the thickness of the specimen wall (2 cm). During K_0-consolidation, a cell pressure of 4.00 kg/cm^2 (392 kPa) and a back pressure of 2.00 kg/cm^2 (196 kPa) were applied. The vertical stress difference was increased gradually to maintain $\epsilon_v = \epsilon_1$ throughout the entire K_0-consolidation. The average value of K_0 for all tests was 0.55. The pore pressure parameter B ($=\Delta u/\Delta\sigma_{cell}$) measured after the consolidation stage indicated that the specimens were fully saturated.

Testing Program

Fifteen undrained and two drained torsion shear tests were performed on hollow cylinder specimens consolidated under K_0-stress conditions. Each test was conducted with constant confining pressure, σ_r, according to predetermined stress-paths shown in Fig. 2. The diagrams in this figure indicate the relations between the vertical stress difference, $(\sigma_z - \sigma_\theta)$, and the torsion shear stress, $\tau_{z\theta}$. The stress-paths are shown on two diagrams and different symbols are used for the sole purpose of clarity. Some stress-paths appear on both diagrams for purposes of comparison. In order to follow these stress-paths to failure and beyond, either the vertical stress difference or the torsion shear stress was applied under strain-control, and the other stress was stress-controlled. The vertical stress difference could be either increased or decreased. Strain rates of approximately 0.005%/min for undrained tests and 0.001%/min for drained tests were employed for both vertical strain and shear strain in the strain-controlled directions.

All shear tests were started after K_0-consolidation in the torsion shear apparatus. The stress-paths projected on the plane of $\sigma_r' = 1.00$ kg/cm^2 are shown in Fig. 2. Projection of effective stress-paths is performed by mul-

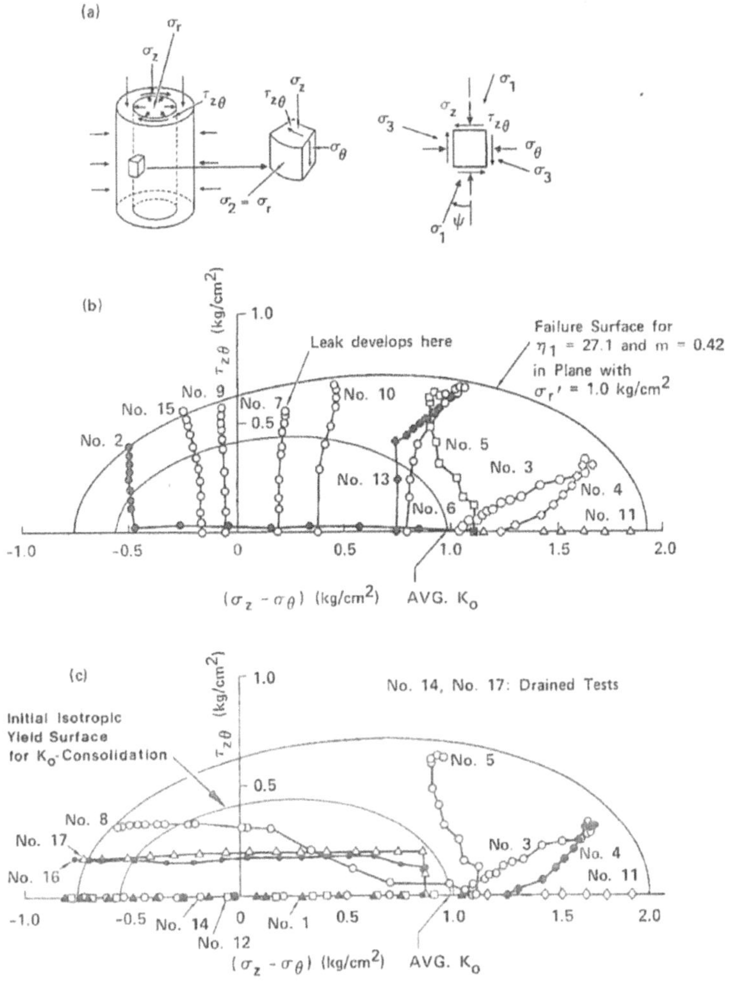

FIG. 2. (*a*) Stress in Hollow Cylinder Specimen; (*b*) and (*c*) Stress-Paths in Torsion Shear Tests in Plane with $\sigma_r' = 1.0$ kg/cm^2 (98 kPa)

tiplying the vertical stress difference and the applied shear stress by the ratio of $\sigma_r' = 1.00$ kg/cm^2 and the measured value of σ_r' (in the same stress units) at each stress point. Test 11 was performed as a triaxial compression test by increasing the vertical load under strain-control. Tests 3, 4, 5, 6, and 13 were performed by increasing the torque under strain-control with only small changes in the vertical load. Tests 2, 7, 9, 10, and 15 were performed by adjusting the vertical stress difference under stress-control to the predetermined values and then increasing the torque under strain-control. A leak developed in test 7 and it could not be continued, as indicated in Fig. 2(*b*). Tests 8, 16, and 17 were performed by increasing the torsion shear stress under stress-control and then decreasing the vertical load under strain-control. Tests 1, 12, and 14 were performed by decreasing the vertical load under strain control to produce triaxial extension tests. Tests 14 and 17 were conducted under drained conditions, whereas all other tests were performed under undrained conditions.

Relations between Strains

The shear strains, $\epsilon_{z\theta}$ ($=\gamma_{z\theta}/2$), are plotted versus the normal strain differences, $(\epsilon_z - \epsilon_\theta)/2$, in Fig. 3 for the specimens in tests 2, 3, 10, and 16, whose stress-paths are shown in Fig. 2. The points corresponding to failure according to the maximum effective stress ratio are indicated with a black solid point on each curve, and the values of $b = (\sigma_2 - \sigma_3)/(\sigma_1 - \sigma_3)$ at failure are indicated for each test in Fig. 3. The relations between the shear strains and the normal strain differences are almost linear, except at the beginning of the tests. The almost linear relations are more obvious at strains close to failure, where the behavior of the clay is plastic in nature.

Even though essentially constant vertical loads were supplied in tests 2 and 10, as shown by the stress-paths in Fig. 2, Fig. 3 indicates that normal strain differences occurred in addition to the shear strains caused by the torque. This "coupling effect" was more pronounced in test 2 than in test 10. A similar coupling effect between shear strains and axial stresses was observed in test 16, in which the axial stress was decreased while the shear stress was held constant.

Fig. 3 shows that the normal strain differences are expansive for values of $b[=(\sigma_2 - \sigma_3)/(\sigma_1 - \sigma_3)]$ larger than approximately 0.34 and compressive for smaller values of b. A given increment in b has a greater effect on the relation between the shear strains and the normal strain differences at small b-values than at high b-values.

Strain Increment and Stress Directions

The general pattern of directions of major principal strain increments at failure is of importance in considerations regarding the applicability of isotropic plasticity theory and the general framework required for modeling of

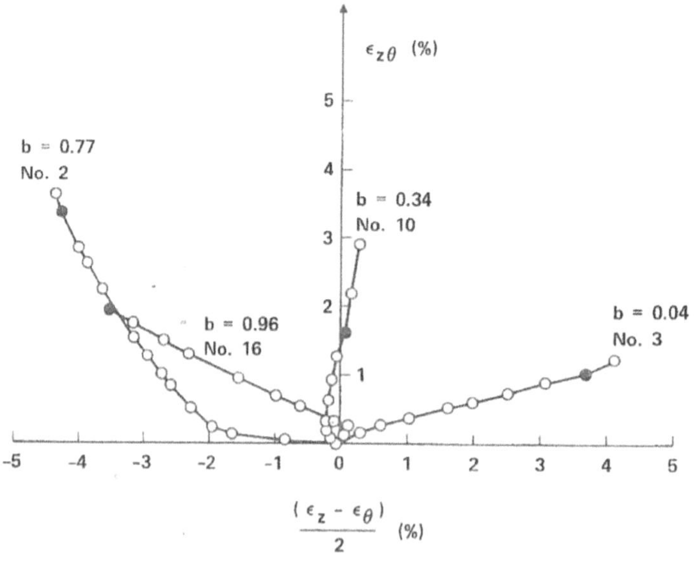

FIG. 3. Relations between Normal Strain Differences and Shear Strains in Four Torsion Shear Tests

clay with initial K_0-stress states. In order to investigate the soil behavior at failure, it is necessary to relate the directions of major principal strain increments to the directions of major principal stress in the physical space.

According to plasticity theory, the principal axes of plastic strain increments coincide in direction with the principal axes of stress during reorientation of stresses in isotropic materials, whereas the principal axes of elastic strain increments coincide in direction with the principal axes of stress increments according to isotropic elasticity theory.

Triaxial compression tests performed in a companion study on vertical and horizontal specimens of K_0-consolidated EPK clay produced stress-strain relations and undrained strengths that indicated negligible influence of orientation. The laboratory-prepared clay was therefore taken to behave in an isotropic manner, although it had initially been K_0-consolidated. More detailed studies of the clay behavior might have indicated some anisotropic features, but these were not likely to be as pronounced as those found in natural clays.

Natural clays have been produced by deposition of individual clay particles or clusters of particles, each allowed to settle and come to equilibrium with preferred orientation (most likely horizontal), and consolidation under K_0-conditions over long time periods, thus allowing the majority of creep to occur under current in situ stresses. In comparison, the slurry of EPK clay mixed at two times the liquid limit may not have provided the individual particles with sufficient space to allow extensive alignment in a preferred direction. Further, sufficient time was allowed for the slurry to consolidate, but the amount of time allowed for creep after the end of primary consolidation was small. The EPK clay prepared for the present study, therefore, exhibited little anisotropy as a result of the initial K_0-consolidation. The experimental results of the torsion shear tests are consequently evaluated in view of isotropic elastic and plastic behavior.

Mohr Circles

The directions of major principal stress, major principal stress increment, and major principal strain increment during primary loading are shown in Fig. 4, where Ψ = angle between σ_1 and vertical, X = angle between $\Delta\sigma_1$ (=maximum of increments in $\tau_{z\theta}$, σ_z, and σ_θ) and vertical, ξ = angle between $\Delta\epsilon_1$ (=maximum of increments in $\epsilon_{z\theta}$, ϵ_z, and ϵ_θ) and vertical, respectively. The relationships from Fig. 4 are as follows:

$$\tan 2\Psi = \frac{2 \cdot \tau_{z\theta}}{\sigma_z - \sigma_\theta} \qquad (1)$$

$$\tan 2X = \frac{2 \cdot \Delta\tau_{z\theta}}{\Delta\sigma_z - \Delta\sigma_\theta} \qquad (2)$$

$$\tan 2\xi = \frac{2 \cdot \Delta\epsilon_{z\theta}}{\Delta\epsilon_z - \Delta\epsilon_\theta} \qquad (3)$$

Therefore, isotropic elastic behavior implies that X = ξ, and isotropic plastic behavior requires that $\Psi = \xi$.

Directions of Major Principal Strain Increments at Failure

Fig. 5 shows a diagram of $(\sigma_z - \sigma_\theta)/2$ versus $\tau_{z\theta}$ superimposed on a diagram of $(\Delta\epsilon_z - \Delta\epsilon_\theta)/2$ versus $\Delta\epsilon_{z\theta}$. The inclinations of stress directions should

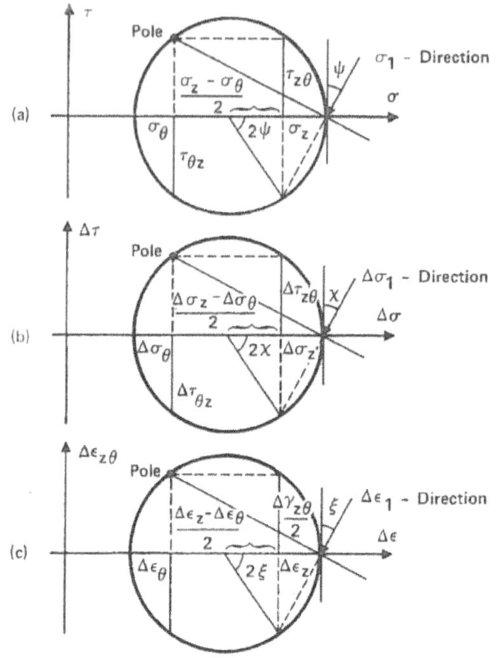

FIG. 4. Directions of Major Principal: (*a*) Stress; (*b*) Stress Increments; and (*c*) Strain Increment in Torsion Shear Tests

coincide with the inclination of strain increment directions for isotropic plastic behavior at failure, i.e.

$$\frac{2 \cdot \tau_{z\theta}}{\sigma_z - \sigma_\theta} = \frac{2 \cdot \epsilon_{z\theta}}{\Delta\epsilon_z - \Delta\epsilon_\theta} \quad \text{(4)}$$

where $2 \cdot \tau_{z\theta}/(\sigma_z - \sigma_\theta)$ = the inclination of the stress direction; and $2 \cdot \Delta\epsilon_{z\theta}/(\Delta\epsilon_z - \Delta\epsilon_\theta)$ = the inclination of the strain increment direction. In Fig. 5 the

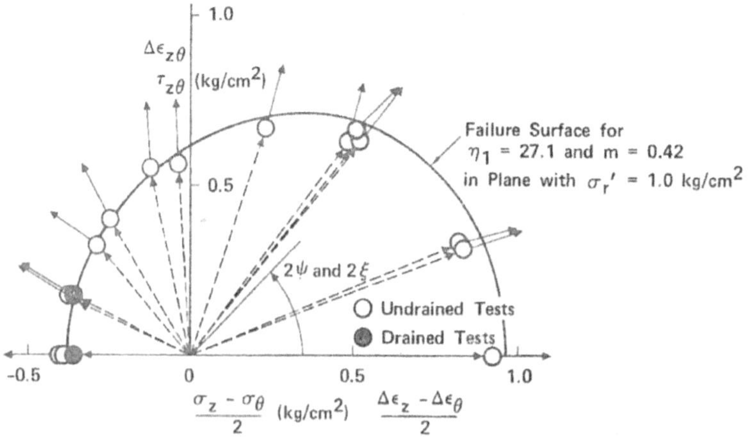

FIG. 5. Directions of Major Principal Stress and Strain Increments at Failure (Note that Double Angles are Shown)

1396

major principal stress direction is represented by a line connecting the stress point at failure and the origin. The direction of the major principal strain increment is shown initiating from the stress point at failure. All strain increments are plastic at failure, because the changes in stress at failure are negligible, thus resulting in negligible elastic strain increments.

It should be noted that the stress states at failure in undrained tests most often do not correspond to the same value of effective confining stress, σ_r'. However, in order to indicate that failure was achieved in the tests under consideration, it may be desirable to plot the data points on the same effective confining stress plane. By projection of a given stress point along the curved failure envelope using a technique involving Lade's failure criterion (Lade 1984), it is possible to obtain the adjusted stress states. The failure stress points and the failure surface are shown in Fig. 5 on the plane corresponding to $\sigma_r' = 1.00\ \text{kg/cm}^2$ (98 kPa). This failure surface is given by the following failure criterion proposed for isotropic soils (Lade 1984):

$$\eta_1 = \left(\frac{I_1^3}{I_3} - 27\right)\left(\frac{I_1}{p_a}\right)^m \quad \text{(5)}$$

in which I_1 and I_3 = the first and the third stress invariants which for the conditions in the torsion tests are defined as follows:

$$I_1 = \sigma_1 + \sigma_2 + \sigma_3 = \sigma_z + \sigma_r + \sigma_\theta \quad \text{(6)}$$

$$I_3 = \sigma_1 \cdot \sigma_2 \cdot \sigma_3 = \sigma_z \cdot \sigma_r \cdot \sigma_\theta - \sigma_r \cdot \tau_{z\theta} \cdot \tau_{\theta z} \quad \text{(7)}$$

and p_a = atmospheric pressure expressed in the same units as the stresses. The material parameters, η_1 and m, were determined to be 27.1 and 0.42, respectively, for Edgar plastic kaolinite (Hong and Lade 1989).

Only one half of the failure surface is shown in Fig. 5, because of symmetry in material behavior. The failure envelope is not a circle, but it has a unique shape that is symmetrical about the horizontal axis. It may be seen that the failure envelope represents the failure points obtained from the torsion shear tests with reasonable accuracy.

Fig. 5 shows that the major principal strain increment directions at failure for practical purposes coincide with the major principal stress directions. Only a few directions deviate slightly from each other. Because double angles are shown in Fig. 5, the deviations represent twice the deviations in physical space. The magnitudes of these deviations are so small that they could even be due to experimental inaccuracies. It appears that the effect of the initial K_0-stress state applied to the clay specimens to a large extent has been eliminated at failure, which occurred at relatively large strains in the torsion shear tests. Thus, any cross-anisotropy that might have developed in response to the initial K_0-consolidation did not affect the results at failure. Plasticity theory for isotropic materials can therefore be applied to modeling of K_0-consolidated clay at failure in torsion shear tests.

Effects of Rotation of Principal Stress Axes

A more detailed evaluation of the clay behavior in four representative tests is shown in Fig. 6. The variations of the angles between vertical and the directions of major principal stress Ψ, major principal stress increment X, and major principal strain increment ξ are plotted versus shear strain, $\gamma_{z\theta}$, for tests 2, 3, 10, and 16. The angles Ψ, X, and ξ were calculated from

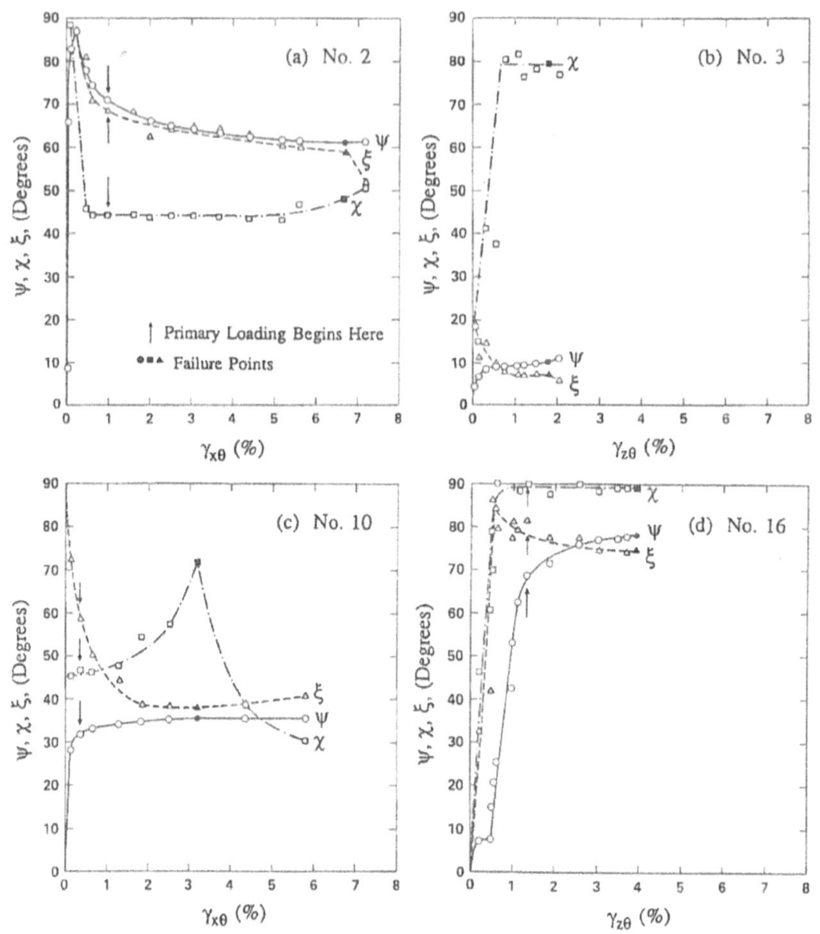

FIG. 6. Variations of Angles Ψ, X, and ξ versus Engineering Shear Strain in Four Torsion Shear Tests

Eqs. 1, 2, and 3, respectively. The arrows in the diagrams indicate the initiating points of primary loading. These points are indicated in Fig. 2, where the stress-paths cross the initial isotropic yield surface corresponding to K_0-consolidation. Thus, smaller shear strains than those shown at these points in Fig. 6 represent stress states of unloding or reloading from the K_0-stress states. The points corresponding to failure are also indicated on each curve.

Fig. 6 shows that the direction of major principal strain increment essentially coincide with the major principal stress increment direction at small shear strains or in the early stages of the tests. This indicates that the clay behaves as an isotropic elastic material at the beginning of the tests. As the stress level increased towards failure, the major principal strain increment direction approached and essentially coincided with the major principal stress direction. This is typical of isotropic plastic behavior.

Following K_0-consolidation, the stress-path in Fig. 2(*b*) shows that the specimen in test 2 was first unloaded vertically into the region of negative deviator stresses, then reloaded and finally primary-loaded by application of shear stresses. Fig. 6(*a*) shows that plastic behavior (i.e., $\Psi = \xi$) was obtained slightly before the isotropic yield surface was crossed.

In test 3 [see stress-path in Fig. 2(*c*)] the specimen was primary-loaded directly from the K_0-stress state, and in this test there was essentially no elastic behavior, as indicated in Fig. 6(*b*).

Fig. 2(*b*) shows that the specimen in test 10 was first unloaded vertically, then reloaded and primary-loaded in shear. Fig. 6(*c*) indicates that a gradual transition from elastic to plastic behavior at higher stress levels was obtained in this test.

Finally, the stress-path for test 16 shown in Fig. 2(*c*) indicates that a shear stress was applied to the specimen immediately after K_0-consolidation, after which the vertical load was decreased to failure in the extension re~ion. Most of this stress-path was located inside the isotropic yield surface, and elastic behavior is clearly indicated in Fig. 6(*d*) for the beginning of this test. A gradual transition occurred around the point where the stresses exceed the isotropic yield surface, and plastic behavior is obtained in the remaining portion of the test.

These observations from the torsion shear tests on Edgar plastic kaolinite indicate that the clay behaved as an elasto-plastic material. Thus, the applicability of isotropic elastoplastic theory for modeling the behavior of clay during rotation of the principal stress axes in torsion shear tests has been demonstrated.

Results from simple shear tests on sand reported by Roscoe and his coworkers (Roscoe 1970; Roscoe et al. 1967) showed behavior similar to that obtained in the present study on clay.

Conclusions

Undrained and drained torsion shear tests on hollow cylinder specimens of Edgar plastic kaolinite clay have been performed along various stress-paths for the purpose of investigating the relations among the directions of major prinicipal stress, stress increment, and strain increment during rotation of principal stress axes.

The clay prepared by K_0-consolidation of a slurry was found to exhibit little anisotropy. The effects of any initial cross-anisotropy caused by K_0-consolidation in the torsion shear tests were, to a large extent, eliminated at failure, which occurred at relatively large strains. The major principal strain increment directions coincided with the major principal stress directions at failure. Thus, isotropic plasticity theory may be applied for modeling the behavior of this clay at failure.

The experimental results also indicated that the directions of major principal strain increment essentially coincided with the directions of major principal stress increment at small stress levels inside the isotropic yield surface. This is typical of elastic behavior. A gradual transition from elastic to plastic behavior was observed as the stress level was increased, and the directions of major principal strain increment approached the directions of major principal stress at higher stress levels. These observations from the torsion shear tests show that the clay behaved as an elasto-plastic material. Isotropic elastoplastic theory may therefore be suitable for modeling the behavior of K_0-consolidated EPK clay during rotation of the principal stress axes.

The study presented here provides the basic behavior of a laboratory-prepared, isotropic clay during rotation of principal stresses. The results may therefore serve as a basis for comparison and evaluation of the behavior of natural clays that exhibit cross-anisotropic characteristics.

Acknowledgment

A portion of the study presented here was supported by the National Science Foundation under Grant No. ENG-78-20239. The study was continued while the first writer was on leave at the University of California, Los Angeles, Calif. Support for the first writer was provided by the Korea Science and Engineering Foundation, Korea. Grateful appreciation is expressed to both foundations for their support of this study.

Appendix. References

Bishop, A. W., et al. (1971). "A new ring shear apparatus and its application to the measurement of residual strength." *Geotechnique,* 21(4), 273–328.
Bjerrum, L., and Landva, A. (1966). "Direct simple-shear tests on a Norwegian quick clay." *Geotechnique,* 16(1), 1–20.
Bojanowski, W. (1970). "The study of the influence of intermediate principal stress on the strength of granular material." *Proc., 2nd Seminar on Soil Mech. and Found. Engrg.*, Lodz, Poland, 41–58.
Broms, B. B., and Casbarian, A. O. (1965). "Effects of rotation of the principal stress axes and of the intermediate principal stress on shear strength." *Proc., 6th Int. Conf. on Soil Mech. and Found. Engrg.*, Montreal, Canada, I, 179–183.
Hight, D. W., Gens, A., and Symes, M. J. (1983). "The development of a new hollow cylinder apparatus for investigating the effects of principal stress rotation in soils." *Geotechnique,* 33(4), 355–383.
Hong, W. P., and Lade, P. V. (1989). "Elasto-plastic behavior of K_0-consolidated clay in torsion shear tests." *Soils and Found.*, Tokyo, Japan, 29(2), 57–70.
Ishibashi, I., and Sherif, M. A. (1974). "Soil liquefaction by torsional simple shear device." *J. Geotech. Engrg. Div.*, ASCE, 100(8), 871–888.
Lade, P. V. (1975). "Torsion shear tests on cohesionless soil." *Proc., 5th Panamerican Conf. on Soil Mech. and Found. Engrg.*, Buenos Aires, Argentina, I, 117–127.
Lade, P. V. (1976). "Interpretation of torsion shear tests on sand." *Proc., 2nd Int. Conf. on Numerical Methods in Geomech.*, Blacksburg, Va., I, 381–389.
Lade, P. V. (1981). "Torsion shear apparatus for soil testing." *Laboratory Shear Strength of Soil,* ASTM STP 740, R. N. Yong and F. C. Townsend, eds., American Society for Testing and Materials, Philadelphia, Pa., 145–163.
Lade, P. V. (1984). "Failure criterion for frictional materials." *Mechanics of Engineering Materials,* C. S. Desai and R. H. Gallagher, eds., John Wiley and Sons, New York, N.Y., 385–402.
Lomize, G. M., et al. (1969). "Study on deformation and strength of soils under three-dimensional state of stress." *Proc., 7th Int. Conf. on Soil Mech. and Found. Engrg.*, International Society for Soil Mechanics and Foundation Engineering, I, 257–265.
Roscoe, K. H. (1953). "An apparatus for the application of simple shear to soil samples." *Proc., 3rd Int. Conf. on Soil Mechanics and Foundation Engineering,* Zurich, Switzerland, 186–191.
Roscoe, K. H. (1970). "Some results of tests in the latest model of the simple shear apparatus and new biaxial apparatus." *Veröffentlichungen des Institutes für Bodenmechanik and Felsmechanik der Universität Fridericiana,* Karlsruhe, W. Germany, Heft 44, 62–94.
Roscoe, K. H., Bassett, R. H., and Cole, E. R. I. (1967). "Principal axes observed during simple shear of a sand." *Proc., Geotech. Conf.*, Oslo, Norway, I, 231–237.
Saada, A. S., and Baah, A. K. (1967). "Deformation and failure of a cross aniso-

tropic clay under combined stresses."*Proc., 3rd Panamerican Conf. on Soil Mech. and Found. Engrg.*, Caracas, Venezuela, I, 67–88.

Saada, A. S., and Bianchini, G. F. (1975). "Strength of one-dimensionally consolidated clays." *J. Geotech. Div.*, ASCE, 101(11), 1151–1164.

Saada, A. S., and Ou, C.-D. (1973). "Stress-strain relations and failure of anisotropic clays." *J. Soil Mech. and Found. Div.*, ASCE, 99(12), 1091–1111.

Saada, A. S., and Townsend, F. C. (1981). "State-of-the-art: Laboratory strength testing of soils." *Laboratory Shear Strength of Soil*, ASTM STP 740, R. N. Yong and F. C. Townsend, eds., American Society for Testing and Materials, Philadelphia, Pa., 7–77.

Saada, A. S., and Zamani, K. K. (1973). "The mechanical behavior of cross-anisotropic clays." *Proc., 7th Int. Conf. on Soil Mech. and Found. Engrg.*, International Society for Soil Mechanics and Foundation Engineering, 1, 351–359.

Symes, M. J. P. R., Gens, A., and Hight, D. W. (1984). "Undrained anisotropy and principal stress rotation in saturated sand." *Geotechnique*, 34(1), 11–27.

Tatsuoka, F., et al. (1986). "Failure and deformation of sand in torsion shear." *Soils and Foundations*, Tokyo, Japan, 26(4), 79–97.

Wright, D. K., Gilbert, P. A., and Saada, A. S. (1978). "Shear devices for determining dynamic soil properties." *Proc., ASCE Specialty Conf. on Earthquake Engrg. and Soil Dynamics*, Pasadena, Calif., 2, 1056–1075.

International Journal of Solids and Structures xxx (2013) xxx–xxx

Contents lists available at ScienceDirect

International Journal of Solids and Structures

journal homepage: www.elsevier.com/locate/ijsolstr

Non-coaxiality of strain increment and stress directions in cross-anisotropic sand

Nina M. Rodriguez [a], Poul V. Lade [b,*]

[a] Department of the Navy, Naval Sea Systems Command 05C, Washington Navy Yard, Washington, DC 20376, USA
[b] Department of Civil Engineering, The Catholic University of America, Washington, DC 20064, USA

ARTICLE INFO

Article history:
Received 4 June 2013
Received in revised form 25 November 2013
Available online xxxx

Keywords:
Cross-anisotropy
Non-coaxiality
Sand
Three-dimensional
Torsion shear

ABSTRACT

An experimental program was carried out in a recently developed torsion shear apparatus to study the non-coaxiality of strain increment and stress directions in cross-anisotropic deposits of Fine Nevada sand. Forty-four drained torsion shear tests were performed at constant mean confining stress, σ_m, constant intermediate principal stress ratios, as indicated by $b = (\sigma_2 - \sigma_3)/(\sigma_1 - \sigma_3)$, and constant principal stress directions, α. The experiments were performed on large hollow cylinder specimens deposited by dry pluviation and tested in an automated torsion shear apparatus. The specimens had height of 40 cm, and average diameter of 20 cm, and wall thickness of 2 cm. The stress-strain behavior of Fine Nevada sand is presented for discrete combinations of constant principal stress direction, α, and intermediate principal stress. The effects of these two variables on the non-coaxiality are presented. The experiments show that the directions of the strain increments do not in general coincide with the directions of stresses, and there is a switch from one to the other side between the two quantities.

1. Introduction

The issue of the direction of the major principal strain increment relative to the direction of the major principal stress has interest for modeling of these quantities using elasticity and plasticity theories. For cross-anisotropic materials these directions are not expected to coincide. A detailed experimental investigation was performed for cross-anisotropic sand deposits in a hollow cylinder torsion shear apparatus. The notation for the physical quantities used here is given in Table 1. A series of torsion shear experiments was performed on large hollow cylinder specimens of Fine Nevada sand with constant principal stress directions relative to vertical, α, varying between 0° and 90° and with the intermediate principal stress, σ_2, varying from σ_3 to σ_1 as indicated by $b = (\sigma_2 - \sigma_3)/(\sigma_1 - \sigma_3)$. The Fine Nevada sand was deposited by dry pluviation, thus producing a sand fabric with horizontal bedding planes and cross-anisotropic characteristics. The various stress conditions were achieved by varying the pressures inside and outside the hollow cylinder specimen relative to the shear stress and the vertical deviator stress according to a pre-calculated pattern. All stresses and all strains were determined from careful measurements so that analysis of the soil behavior could be made reliably. The soil behavior was determined for a pattern of combinations of α varying with increments of 22.5° from 0° to 90° and b varying with increments of 0.25 from 0.0 to 1.0. Thus, 25 test locations were established, but many tests were repeated to study the consistency of the results. The measured stress–strain behavior is presented for 15 tests in which the major principal stress was not aligned with or perpendicular to the bedding planes. The results show that strain increment directions are generally not coinciding with the major principal stress directions. When the major principal stress forms angles greater than approximately 45° with the bedding planes, the major principal strain increment directions are closer to the bedding plane direction. However, a switch occurs such that when the major principal stress direction forms more shallow angles below 45° with the bedding planes, the direction of the major principal strain increment is steeper than the direction of the major principal stress. All tests except those with $b = 0.0$ resulted in shear bands, after which strains are not uniform anymore.

2. Previous studies

A significant aspect of soil behavior, which may be modeled by hardening plasticity theory, is that relating to coincidence in physical space of principal strain increments with principal stresses and the influence of anisotropy on this behavior. The stress–strain behavior of sands measured in torsion shear tests has previously been analyzed in terms of the directions of major principal strain increment and major principal stress during

* Corresponding author. Tel.: +1 202 319 6942; fax: +1 202 319 6677.
E-mail addresses: nina.m.rodriguez1@navy.mil (N.M. Rodriguez), Lade@cua.edu (P.V. Lade).

0020-7683/$ - see front matter © 2013 Elsevier Ltd. All rights reserved.
http://dx.doi.org/10.1016/j.ijsolstr.2013.12.003

Please cite this article in press as: Rodriguez, N.M., Lade, P.V. Non-coaxiality of strain increment and stress directions in cross-anisotropic sand. Int. J. Solids Struct. (2013), http://dx.doi.org/10.1016/j.ijsolstr.2013.12.003

Table 1
Notation for physical quantities.

Symbol	Physical quantity
b	$=(\sigma_2-\sigma_3)/(\sigma_1-\sigma_3)$ = relative magnitude of the intermediate principal stress
D_{50}	Mean diameter of sand grains
e	Void ratio
F_v	Vertical load on hollow cylinder specimen
h	Height of hollow cylinder specimen
LVDT	Linear variable differential transducer
p_0, p_i	Outside and inside pressures on hollow cylinder specimen
R	σ_1/σ_3
R_0, R_i	Outside and inside radii of hollow cylinder specimen
T	Torque applied to hollow cylinder specimen
α	Inclination of major principal stress to vertical
β_R	Non-uniformity index $=(R_{max}-R_{min})/R_{avg}$
$\gamma_{z\theta}$	$=2\cdot\varepsilon_{z\theta}$ = engineering shear strain
Δ	$=(\xi-\alpha)$ = angle between directions of major principal strain increment and major principal stress
ΔI_{vol}	Volume change of inner cell
ΔV	Volume change of specimen
$\Delta\theta$	Twist angle of hollow cylinder specimen
ε_z, ε_θ, ε_r	Vertical, tangential and radial normal strains
ε_1	Major principal strain
$\varepsilon_{z\theta}$	Shear strain
ξ	Inclination of major principal strain increment to vertical
σ_z, σ_θ, σ_r	Effective vertical, tangential, and radial stresses on hollow cylinder specimen
σ_1, σ_2, σ_3	Effective major, intermediate, and minor principal stresses
σ_m	Effective mean normal stress
$\tau_{z\theta}$	Torsional shear stress

rotation of principal stress axes in physical space (e.g. Lade 1975, 1976; Hight et al., 1983; Ishihara and Towhata, 1983; Tatsuoka et al., 1986; Symes et al., 1984, 1988; Pradhan et al., 1988; Vaid et al., 1990; Vaid and Sayao, 1995; Saada et al., 1999; Zdravkovic and Jardine, 2000; Sivathayalan and Vaid, 2002; Chaudhary and Kuwano, 2003; Yang et al., 2007; Lade et al., 2008, 2009). Similar testing and analyses have also been performed on clay (e.g. Broms and Casbarian, 1965; Saada and Baah, 1967; Hicher and Lade, 1987; Saada et al., 1994; Frydman et al., 1995; Hong and Lade, 1989a and b; Lade and Kirkgard, 2000; Nishimura et al., 2007). Lade et al., (2009) showed that modeling the observed behavior by an isotropic, plastic hardening soil implies that the two sets of directions coincide during stress rotation, but loading along the principal axes of an anisotropic material will also result in coincidence of principal stress axes with principal strain increment axes. The influence of cross-anisotropy, as is most often found in naturally deposited soils, may not be present in slope talus, sand dunes, or for fluvial or marine sands that have been mobile. However, the effect of cross-anisotropy on the soil behavior has drawn most attention and has been at the root of the use of torsion shear tests with stress rotation in many studies exemplified by those listed above.

3. Torsion shear tests on hollow cylinder specimens

Torsion shear tests performed on hollow cylinder specimens are conducted in an apparatus as shown in Fig. 1. These tests are suitable for investigating the effects of principal stress directions on the behavior of soil, because in these tests the bedding planes remain horizontal and the principal stresses can be rotated relative to the bedding planes. In addition, there are only minor effects of imposed non-uniform stress conditions if the specimen is sufficiently tall (Lade, 1981). The state of stress in the hollow cylinder specimen is shown in Fig. 2. In these experiments the direction of the major principal stress may be changed relative to the cross-anisotropic deposit created by dry pluviation. This allows studying the direction of major principal strain increment relative

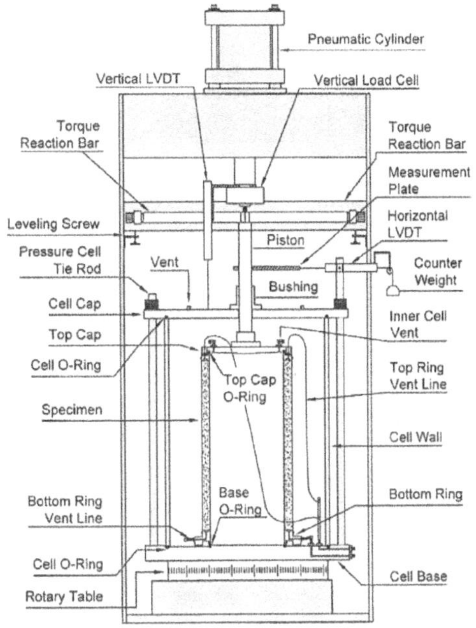

Fig. 1. Schematic drawing of torsion shear apparatus.

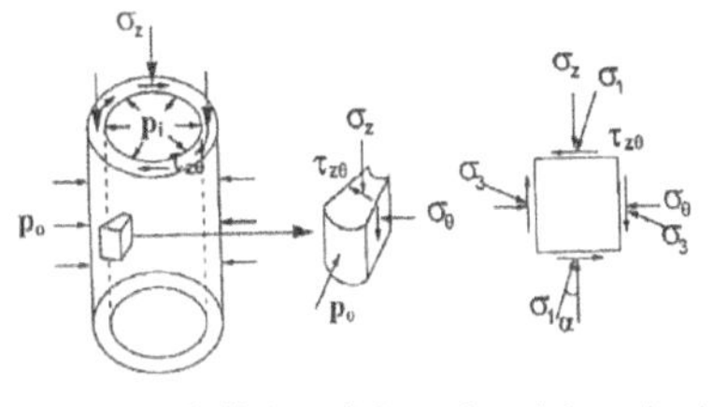

Fig. 2. Stress state in wall of hollow cylinder specimen during torsion shear test.

to the direction of the major principal stress in cross-anisotropic sand deposits. The strength results of such torsion shear tests on dense, Fine Nevada sand were presented by Lade et al. (2013), and the non-coaxiality of strain increment and stress directions are presented and analyzed here.

3.1. Sand tested

All torsion shear tests were performed on Fine Nevada sand, which is composed of subangular to subrounded grains consisting mainly of quartz (98%). The properties of this sand are as follows: Mean diameter, $D_{50}=0.23$ mm; coefficient of uniformity, 2.08; coefficient of curvature, 1.05; specific gravity, 2.65; maximum void ratio, 0.771; and minimum void ratio, 0.507.

3.2. Preparation of hollow cylinder specimens

The boundaries of the hollow cylinder specimen consisted of custom molded inner and outer latex rubber membranes attached to stainless steel end rings. Hollow cylinder specimens with hori-

Please cite this article in press as: Rodriguez, N.M., Lade, P.V. Non-coaxiality of strain increment and stress directions in cross-anisotropic sand. Int. J. Solids Struct. (2013), http://dx.doi.org/10.1016/j.ijsolstr.2013.12.003

zontal bedding planes were prepared using the pluviation and saturation techniques described by Lade et al. (2013). All specimens had inner and outer diameters of 18.0 cm and 22.0 cm, respectively, and the wall thickness was therefore 2.0 cm. The height of the hollow cylinder was 40 cm. These dimensions are in agreement with those recommended by Sayao and Vaid (1991). They proposed that the wall thickness should be in the order of 2.0–2.6 cm to include a sufficiently large number of sand grains across the wall to ensure uniform sand density and to minimize the significance of membrane penetration effects on the measured volume changes. They also suggested that a satisfactory geometry of the hollow cylinder specimen with regard to non-uniformity and experimental control would be obtained if R_i/R_o were within 0.65 to 0.82. The present hollow cylinder has $R_i/R_o = 0.82$. To reduce the influence of radial friction, the height was recommended be such that $H/2R_o = 1.8$–2.2. The present apparatus employs $H/2R_o = 40/22$ cm = 1.82. However, actual experiments on sand show that a height of 25 cm is sufficient to obtain acceptable results (Lade, 1981; Lade et al., 2008). Thus, the specimen dimensions are favorable to producing satisfactory experimental results.

Molds or forming jackets were used to hold the inner and outer latex rubber membranes while pluviating the sand. Two factors affect the void ratio when pouring the sand: The drop height and the rate of sand pluviation. In order to ensure the same void ratio for each specimen, the sand was poured into a funnel with a small tube inside the mouth of the funnel ensuring a constant flow rate of sand. It was determined empirically that a drop height of 35 cm at the employed flow rate would create the desired void ratio. As the sand was deposited, the funnel was carefully raised to ensure a drop height of 35 cm and even bedding planes in the assembled mold. A void ratio, $e = 0.53$ was targeted for each specimen. This corresponds to a relative density of 91% for the Fine Nevada sand.

Fig. 1 shows that the base plate of the torsion shear apparatus was rigidly attached to a rotary table, whose rotation was driven by a gear motor with constant rate and resisted by a piston protruding through the top plate. The hollow cylinder specimen was sitting between a base ring and a top ring which in turn were attached to the base plate and the piston. Thus, the hollow cylinder specimen was exposed to the torque between the base plate and the piston.

In torsion shear tests with the same inside and outside pressures, the value of $b = (\sigma_2 - \sigma_3)/(\sigma_1 - \sigma_3)$ is tied to the inclination, α, of the major principal stress such that $b = \sin^2\alpha$. In the torsion shear apparatus employed here different pressures could be applied to the inner and outer cells, thus making it possible to separate the major principal stress inclination, α, from the value of b. The torque was measured at the piston and it was used to calculate and control the vertical load and the inside and outside pressures in such a way as to follow a prescribed stress path. The pressures and forces in the torsion shear apparatus were computer controlled. The necessary equations were installed in the computer and for each prescribed increment in torque the vertical load and the inner and outside pressures were updated and applied. Since deformations were also measured, it was possible to calculate the actual pressures and shear stresses and apply them in real time. Further details regarding the specimen construction and saturation of the inner cell are given by Lade et al. (2013).

3.3. Calculation of stresses and strains

Representative average stresses and strains for thin-walled cylinders were determined from the following expressions:

Vertical stress:

$$\sigma_z = \frac{F_v}{\pi(R_0^2 - R_i^2)} + \frac{R_0^2 \cdot p_0 - R_i^2 \cdot p_i}{R_0^2 - R_i^2} \quad (1)$$

in which F_v is the vertical load, p_0 and p_i are the outside and the inside pressures, and R_0 and R_i are the outside and the inside radii of the hollow cylinder.

Radial stress:

$$\sigma_r = \frac{R_0 \cdot p_0 + R_i \cdot p_i}{R_0 + R_i} \quad (2)$$

Tangential stress:

$$\sigma_\theta = \frac{R_0 \cdot p_0 - R_i \cdot p_i}{R_i - R_0} \quad (3)$$

Shear stress:

$$\tau_{z\theta} = \frac{3 \cdot T}{2\pi(R_0^3 - R_i^3)} \quad (4)$$

in which T is the torque applied to twist the hollow cylinder.

Corrections were applied to the data after the tests were performed due to vertical piston uplift, membrane strength, etc.

Vertical strain:

$$\varepsilon_z = \frac{\Delta h}{h_0} \quad (5)$$

in which h_0 is the initial height and Δh is the change in height of the hollow cylinder specimen.Radial strain:

$$\varepsilon_r = -\frac{\Delta R_0 - \Delta R_i}{R_0 - R_i} \quad (6)$$

Tangential strain:

$$\varepsilon_\theta = -\frac{\Delta R_0 + \Delta R_i}{R_0 + R_i} \quad (7)$$

in which ΔR_0 and ΔR_i are the changes in outside and inside radii determined from the following expressions (see Lade et al., 2009):

$$\Delta R_0 = \sqrt{\frac{\pi(R_0^2 h_0) + \Delta I_{vol} + \Delta V}{\pi h}} - R_0 \quad (8)$$

$$\Delta R_i = \sqrt{\frac{\pi(R_i^2 h_0) + \Delta I_{vol}}{\pi h}} - R_i \quad (9)$$

in which ΔI_{vol} and ΔV are the changes in volumes of the inner cell and the specimen, respectively.

Shear strain:

$$\varepsilon_{z\theta} = \frac{\Delta\theta\left(R_0^3 - R_i^3\right)}{3h\left(R_0^2 - R_i^2\right)} \quad (10)$$

in which:

$$\Delta\theta = \frac{\Delta H_{LVDT}}{r_{measurementplate}} \quad (11)$$

where ΔH_{LVDT} is the recorded change in the horizontal LVDT reading and $r_{measurementplate}$ is the distance from the center to the radio wire cord of the pie-shaped measurement plate.

It should be pointed out that no unloading and reloading cycles were performed to study yield surfaces, both primary and kinematic, which require resolution of strains to better than 0.1%, such as done by e.g. Shibuya et al. (2003) in hollow cylinder tests and by Kuwano and Jardine (2007) in triaxial tests.

3.4. Stress non-uniformity

To study stress concentrations across the cylinder wall, Sayao and Vaid (1991) assumed the wall to be thick compared with the cylinder radius ($t/R > 0.1$) and the influence of the material's consti-

Please cite this article in press as: Rodriguez, N.M., Lade, P.V. Non-coaxiality of strain increment and stress directions in cross-anisotropic sand. Int. J. Solids Struct. (2013), http://dx.doi.org/10.1016/j.ijsolstr.2013.12.003

tutive behavior on the stresses therefore becomes more important. The vertical stress is not affected by the material behavior and is obtained from equilibrium alone, as given in Eq. (1).

Sayao and Vaid (1991) produced the following expressions for the horizontal normal stresses and the shear stress in the hollow cylinder:

$$\sigma_r = \frac{R_0^2 \cdot p_0 - R_i^2 \cdot p_i}{R_0^2 - R_i^2} - \frac{2(p_0 - p_i) \cdot R_0^2 \cdot R_i^2 \cdot \ell n(R_0/R_i)}{R_0^2 - R_i^2} \tag{12}$$

$$\sigma_\theta = \frac{R_0^2 \cdot p_0 - R_i^2 \cdot p_i}{R_0^2 - R_i^2} + \frac{2(p_0 - p_i) \cdot R_0^2 \cdot R_i^2 \cdot \ell n(R_0/R_i)}{R_0^2 - R_i^2} \tag{13}$$

$$\tau_{z\theta} = \frac{4 \cdot T \cdot (R_0^3 - R_i^3)}{3\pi(R_0^4 - R_i^4)(R_0^2 - R_i^2)} \tag{14}$$

Using these expressions the principal stresses can be determined everywhere across the wall and for any inclination of the major principal stress relative to vertical. Sayao and Vaid (1991) then devised an index of stress non-uniformity in terms of the principal stress ratio, $R = \sigma_1/\sigma_3$, as follows:

$$\beta_R = \frac{R_{max} - R_{min}}{R_{avg}} \tag{15}$$

As an example, Fig. 3 shows the variation of the stress non-uniformity index across the wall in the hollow cylinder specimen (inner radius = 5.0 cm, outer radius = 7.5 cm) employed by Sayao and Vaid (1991) for an average stress ratio of $R = 3.0$. Other dimensions will produce different but similar stress non-uniformity indices. It is seen from the diagram in Fig. 3 that the index is zero for conventional triaxial compression ($b = 0.0$) with a vertical major principal stress ($\alpha = 0°$), while the conventional triaxial extension conditions ($b = 1.0$) and a vertical major principal stress ($\alpha = 0°$) exhibits the highest stress concentration index. For horizontal major principal stresses ($\alpha = 90°$), the highest index is obtained for triaxial compression ($b = 0.0$) and the lowest index is produced for triaxial extension ($b = 1.0$). The effects of these stress non-uniformities in the torsional hollow cylinder specimens were studied and reported by Lade and Rodriguez (2013).

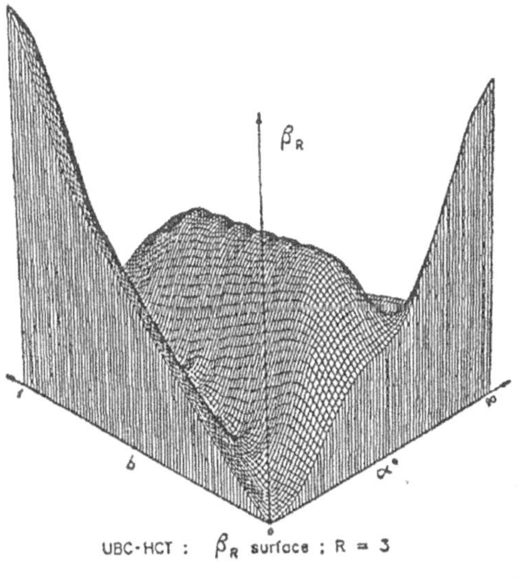

Fig. 3. Variation of stress non-uniformity index across the wall of hollow cylinder specimens with dimensions employed in torsion shear tests performed at University of British Columbia by Sayao and Vaid (1991).

3.5. Instrumentation and control of experiments

The physical quantities of loads, pressures and deformations were measured by load cells, pressure transducers, LVDTs, and volume change devices. Fig. 1 shows that a load cell was placed between the vertical piston and an air pressure cylinder located at the top of the rigid loading frame. To measure the shear stresses, a crossbar assembly was rigidly attached to the piston and two torque arms with load cells were connected to a back plate of the rigid frame. Springs were placed between the torque load cells and the rigid body frame to allow for compression and extension of the arms. A vertical LVDT was fastened to the piston to measure the axial deformation of the specimen. The shear deformation was measured at a constant distance from the center of the piston by a horizontal LVDT attached to the top plate of the chamber. The volume changes of the specimen and the inner cell were measured and allowed determination of the normal strains in the horizontal directions.

A main panel board applied pressures and measured the volume changes of the specimen during shearing. The inner and outer pressures and the back pressure applied to the hollow cylinder specimen were measured by pressure transducers and displayed on pressure gages. These pressures could be applied both manually by a pressure regulator and automatically by the computer when the LabView program was turned on. An on/off switch valve allowed for the switch between automatic and manual control. The constant back pressure was applied with a manual regulator and indicated on a pressure gage. The vertical load could also be applied both manually and automatically to the specimen. Lower and upper air pressure lines were connected to the vertical air pressure cylinder allowing for both compression and extension tests to be performed. Differential pressure transducers were employed to measure the volume changes of the inner cell and the specimen, and together with the LVDT measurements, they provided sufficient information for calculation of the current strains in the hollow cylinder specimen. This allowed calculation of simultaneous stresses and strains, given by the formulas in Section 3.3, for accurate control of the experiment.

The internal and external pressure transducers, inner cell and specimen volume change transducers, vertical load cell, the two torque load cells and horizontal LVDT and vertical LVDT were all connected to a data acquisition unit and displayed in LabView. This computer program reads the electrical signals from the measurement devices and through calibration inputs in the program, the signals were converted to pressures, loads and distances. These were used to compare with the intended quantities and then used for control of b-values, α-values and mean normal stress σ_m, as initially input to the program. Once a set of measurements was acquired by the LabView program, the internal and external pressures and the vertical load applied to the hollow cylinder specimen were changed to ensure the designated b-value and α-value were applied to the specimen while keeping the mean normal stress constant. The applied pressures and loads were recorded, as well as the inner volume change, specimen volume change, horizontal LVDT and vertical LVDT readings. These readings provided the necessary data to study the behavior of the sand under different stress paths.

3.6. Experimental procedure and program

A series of 44 drained tests was performed in this experimental program to study the behavior of dense, Fine Nevada sand while

Please cite this article in press as: Rodriguez, N.M., Lade, P.V. Non-coaxiality of strain increment and stress directions in cross-anisotropic sand. Int. J. Solids Struct. (2013), http://dx.doi.org/10.1016/j.ijsolstr.2013.12.003

keeping b, α, and σ_m constant. The specimens were saturated by the CO_2-method (Lade and Duncan, 1973). Skempton's B-value was determined and all specimens showed B-values above 0.94, which indicates sufficient degrees of saturation for drained tests on dense sand. Once ready to begin testing, the specimens underwent isotropic consolidation in increments in effective confining pressure of 6.9 kPa from 48.3 kPa to 101 kPa. This initial isotropic consolidation may have affected the anisotropy of the specimens with the result that they may have become slightly more isotropic in behavior, as observed by Lade and Abelev (2005) for triaxial specimens and by Shibuya et al. (2003) for torsion shear hollow cylinder specimens.

Shearing began for all specimens at the initial isotropic effective confining pressure of 101 kPa. Depending on the stress path indicated for the particular test, the vertical force was either in compression or tension for constant mean stress, b-value and α-value. Note that the vertical stress was always in compression, even when the vertical force applied through the central shaft was in tension, because it was counteracted by the confining pressures. The inner and outer confining pressures either increased or decreased from the initial value of 101 kPa. All efforts were made to stay as close to the targeted b- and α-values as possible during testing.

To establish the conditions for all combinations of intermediate principal stress, as expressed by $b = (\sigma_2 - \sigma_3)/(\sigma_1 - \sigma_3)$, and σ_1-directions from $\alpha = 0°$ to $\alpha = 90°$, experiments were performed at each of the 25 intersection points of $b = 0.0$, 0.25, 0.50, 0.75, and 1.00 and $\alpha = 0°$, 22.5°, 45.0°, 67.5°, and 90.0°. For each of these experiments the stress–strain and volume change and strength behavior were determined.

These experiments were performed while maintaining the mean normal stress constant at 101 kPa. The tests were conducted by rotating the rotary table thereby increasing the torque and then adjusting all other quantities to the torque to follow the desired stress path. Each experiment would typically take 2–3 h. However, effects of loading rate are negligible at low stresses because only very small amounts of crushing (which is responsible for time effects in granular materials) occur at low stresses.

Tests without stress rotation and with $\alpha = 0°$ and $\alpha = 90°$ do not require torque, and they were stress controlled tests. For these tests, the major principal stress and the major principal strain increments coincide in vertical or horizontal directions. The results of these tests were compared with true triaxial tests by Lade and Rodriguez (2013) and these results will not be discussed further here.

4. Stress–strain and volume change behavior

The stress–strain and volume change behavior are shown in Figs. 4–6 for torsion shear tests with $\alpha = 22.5°$, 45°, and 67.5° and for all b-values from 0 to 1. The results are presented as stress ratio, σ_1/σ_3, versus the major principal strain, ε_1. These values are calculated from the measured normal and shear stresses and normal and shear strains. The results in Figs. 4(a), 5(a), and 6(a) clearly indicate that the initial slopes of the $\sigma_1/\sigma_3 - \varepsilon_1$ relations increase, the strains-to-failure decrease, and the strengths, as indicated by the maximum values of σ_1/σ_3 (or the friction angles), increase with increasing b-value, except for $b = 1$, where these quantities decrease. The arrows indicate the points of failure. The volume change curves in Figs. 4(b), 5(b), and 6(b) show that the rate of dilation, defined as $d\varepsilon_v/d\varepsilon_1$, increases with increasing b-values. Thus, with respect to these quantities the variation in behavior with b-values is very similar to those observed in true triaxial tests on sands (Lade and Duncan, 1973; Wang and Lade, 2001; Abelev and Lade, 2003).

The strength results of the torsion shear tests on Fine Nevada sand were presented and discussed by Lade et al. (2013).

5. Analyses of strain increment and stress directions

The directions of total strain increments and stress directions for the tests with $\alpha = 22.5°$, 45°, and 67.5° and for all b-values from 0 to 1 were analyzed using Mohr circles for strain increments and stresses throughout each test. Total strain increments were analyzed, i.e. the strain increments consisted of both elastic and plastic strain increments. In previous publications (e.g. Lade et al., 2009) the directions of major principal strain increments in physical space were superimposed on diagrams of directions of major principal effective stress so the progressive alignment (or lack thereof) could be seen. To show more directly the deviations between these two directions in this presentation, the directions are shown as angles (α = major principal effective stress direction relative to vertical and ξ = major principal strain increment direction relative to vertical)) versus engineering (torsional) shear strain ($\gamma_{z\theta} = 2 \cdot \varepsilon_{z\theta}$). As is seen in Figs. 7–11, the initial response which is dominated by elastic strains creates an initial deviation of major strain increment direction (open circles) from the major stress direction (open squares) due to the cross-anisotropic behavior of the sand. As more stress is applied, the specimen moves into the plastic range of the stress–strain curve, and the major strain increment directions tends towards the major principal stress directions. The largest deviation is seen in the tests with $\alpha = 22.5°$. Since there are no external shear strains in the tests with $\alpha = 0°$ and $\alpha = 90°$, their directions coincide with the directions of the major principal stress and they are not plotted on the diagrams. Points are plotted up to failure represented by shear banding in all test except those with $b = 0.0$. Strains after failure are not considered as they are highly non-uniform due to development of shear bands. Not all tests followed the desired stress paths in terms of the α- and b-values, but representative results of tests that had the correct stress path have been presented in Figs. 7–11.

As seen in Fig. 7 for tests with $b = 0$, test # TS 3 starts out having a significantly higher ξ-direction than α-direction, which is reasonably constant at 22.5°. As the specimen is sheared, the ξ-direction approaches the α-direction. The ξ-direction becomes constant near engineering shear strains of $\gamma_{z\theta} = 1.5\%$, and the deviation from the α-direction remains constant at approximately 6°. In test # TS 14 with $\alpha = 67.5°$, a small deviation develops between the two directions until about $\gamma_{z\theta} = 0.5\%$ after which the ξ-direction asymptotically approaches the α-direction. In test # TS 31 the two directions start out at 45°, but the ξ-direction begins to deviate from the α-direction at around $\gamma_{z\theta} = 0.5\%$ and the two directions remain apart at around 3° for the rest of the test.

Similar to the results in Fig. 7, tests with higher b-values in Figs. 8–11 show domination of the elastic behavior at the beginning of the tests. While smaller plastic strains occur from the beginning of shearing, the strains include larger plastic components as failure is approached, and the major principal strain increment directions tend to approach the major principal stress directions, and the experiments with $\alpha = 67.5°$ show slightly lower ξ-directions. Although not large, the relatively largest variation occurs at $\alpha = 22.5°$. At $\alpha = 45°$ the major principal strain increment directions are closest to the principal stress directions throughout the entire experiments on the hollow cylinder specimens. Tests with $b = 1.0$, shown in Fig. 11, exhibit the largest variation near the beginning of the test, where the elastic strains dominate.

Fig. 12 summarizes the deviations of principal strain increment directions and principal stress directions at failure for all torsion shear experiments. On the average these deviations vary between $\Delta = \xi - \alpha = +3.6°$ at $\alpha = 22.5°$ over $\Delta = \xi - \alpha = +1.1°$ at $\alpha = 45°$ to $\Delta = \xi - \alpha = -5.5°$ at $\alpha = 67.5°$. While there is some scatter in the results for each of the principal stress inclinations, i.e. the standard deviations on the Δ-values are ±2.3°, ±1.4°, and ±3.1°, respectively,

Please cite this article in press as: Rodriguez, N.M., Lade, P.V. Non-coaxiality of strain increment and stress directions in cross-anisotropic sand. Int. J. Solids Struct. (2013), http://dx.doi.org/10.1016/j.ijsolstr.2013.12.003

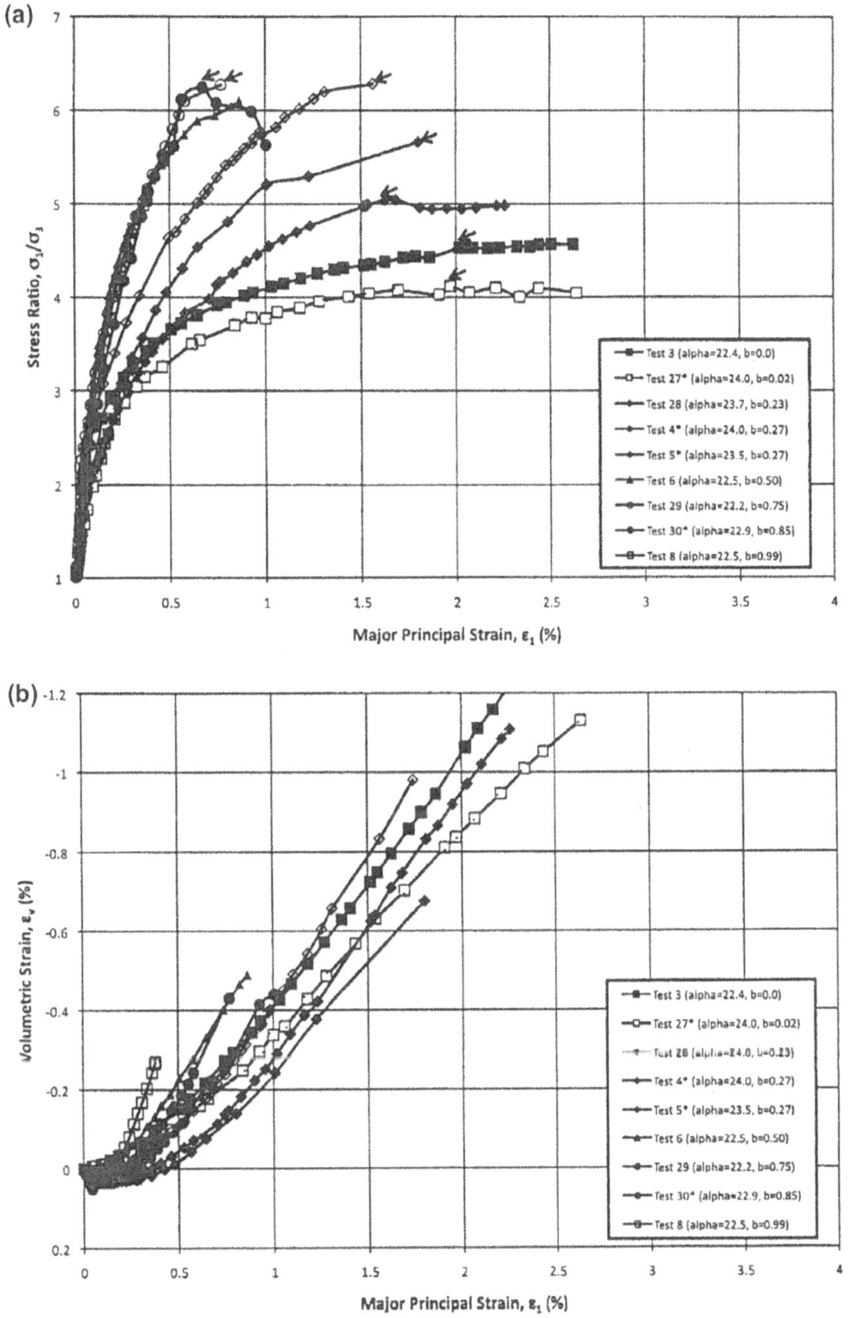

Fig. 4. (a) Stress–strain and (b) volume change relations for torsion shear test with $\alpha = 22.5°$ on hollow cylinder specimens of dense Nevada sand.

Please cite this article in press as: Rodriguez, N.M., Lade, P.V. Non-coaxiality of strain increment and stress directions in cross-anisotropic sand. Int. J. Solids Struct. (2013), http://dx.doi.org/10.1016/j.ijsolstr.2013.12.003

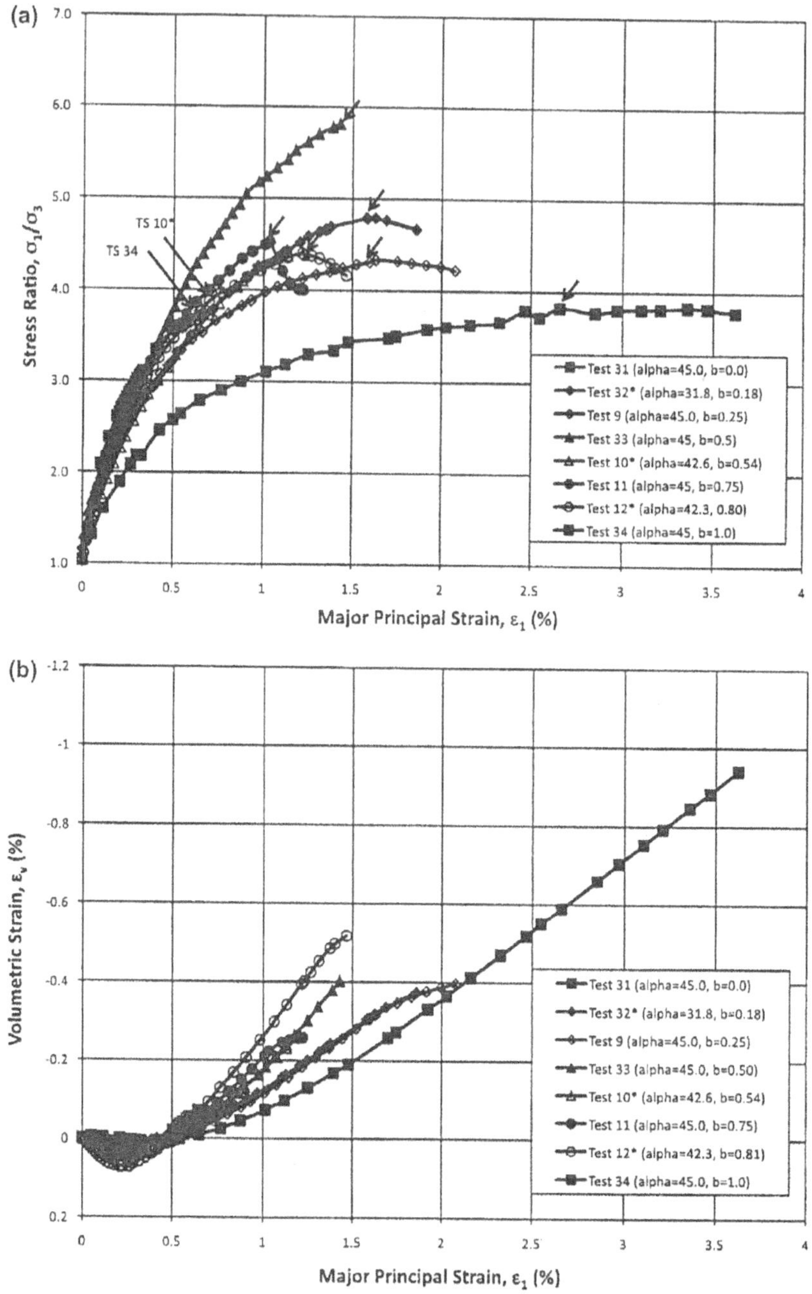

Fig. 5. (a) Stress–strain and (b) volume change relations for torsion shear test with $\alpha = 45°$ on hollow cylinder specimens of dense Nevada sand.

Please cite this article in press as: Rodriguez, N.M., Lade, P.V. Non-coaxiality of strain increment and stress directions in cross-anisotropic sand. Int. J. Solids Struct. (2013), http://dx.doi.org/10.1016/j.ijsolstr.2013.12.003

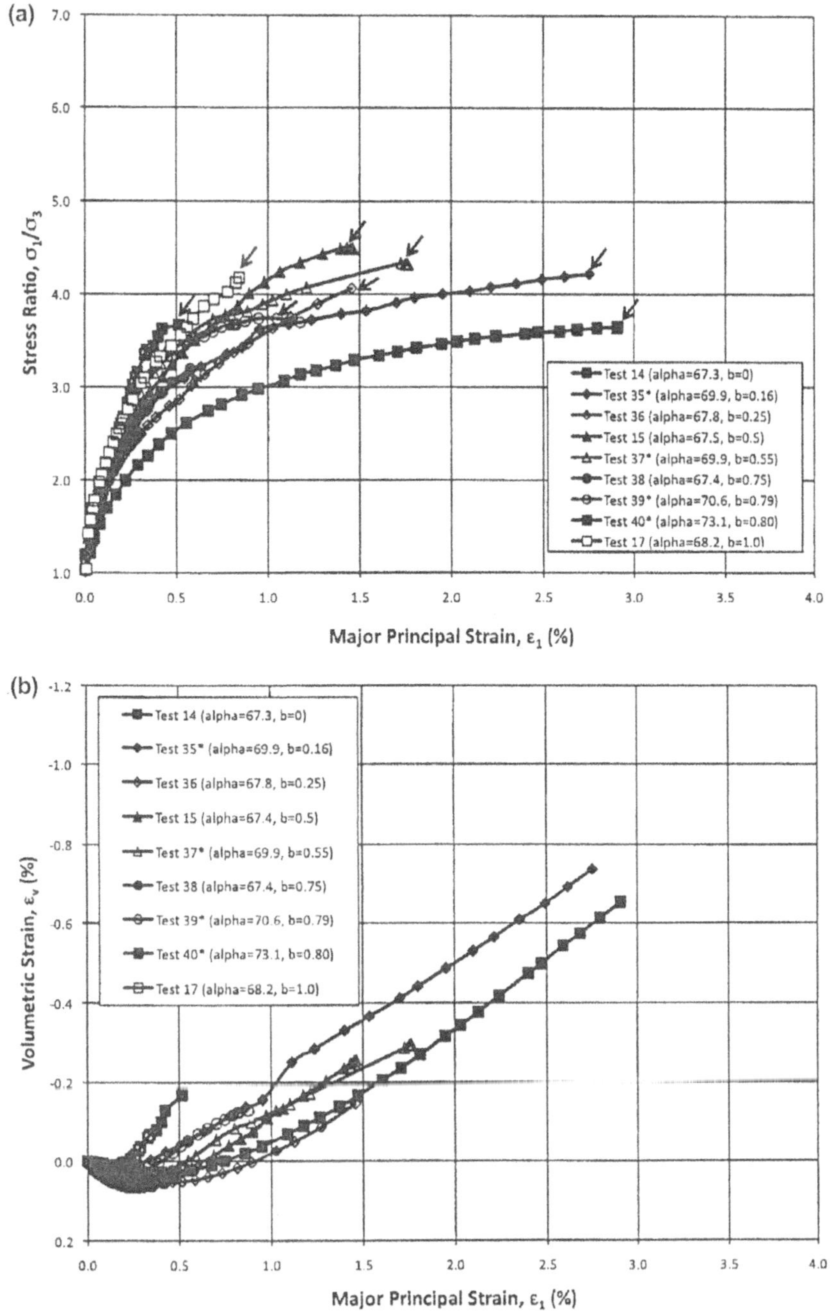

Fig. 6. (a) Stress–strain and (b) volume change relations for torsion shear test with $\alpha = 67.5°$ on hollow cylinder specimens of dense Nevada sand.

Please cite this article in press as: Rodriguez, N.M., Lade, P.V. Non-coaxiality of strain increment and stress directions in cross-anisotropic sand. Int. J. Solids Struct. (2013), http://dx.doi.org/10.1016/j.ijsolstr.2013.12.003

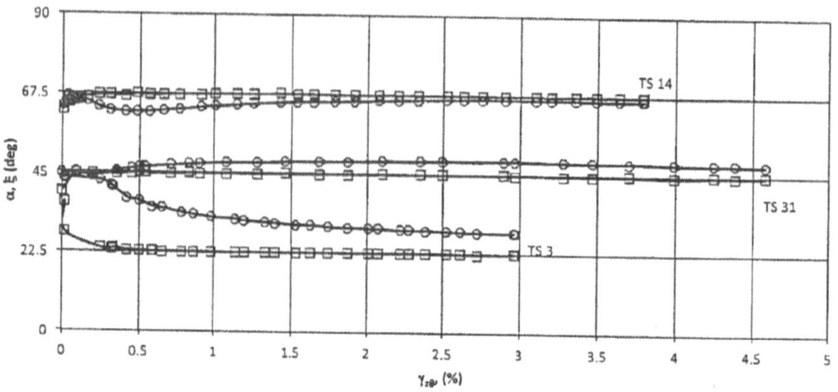

Fig. 7. Directions of principal strain increments (open circle) and stresses (open square) versus engineering shear strain for torsion shear tests with $b = 0.0$ on dense Nevada sand.

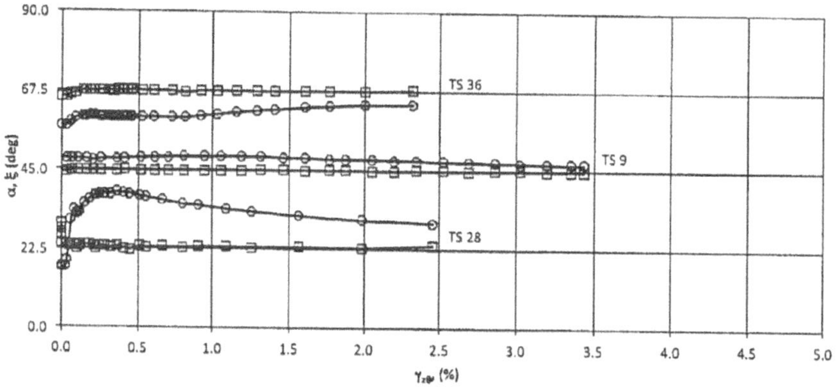

Fig. 8. Directions of principal strain increments (open circle) and stresses (open square) versus engineering shear strain for torsion shear tests with $b = 0.25$ on dense Nevada sand.

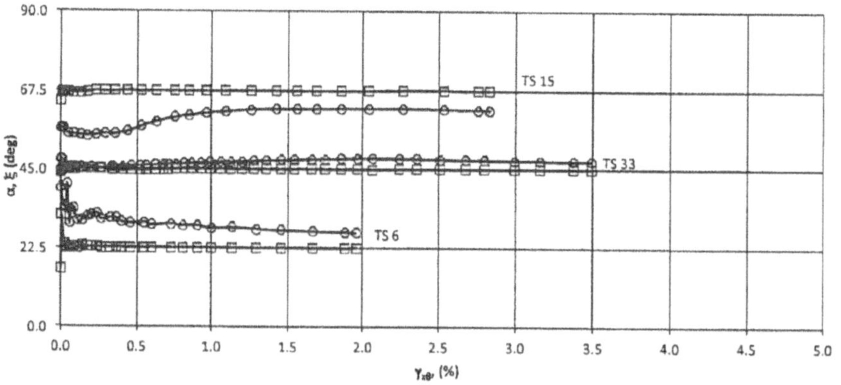

Fig. 9. Directions of principal strain increments (open circle) and stresses (open square) versus engineering shear strain for torsion shear tests with $b = 0.50$ on dense Nevada sand.

Please cite this article in press as: Rodriguez, N.M., Lade, P.V. Non-coaxiality of strain increment and stress directions in cross-anisotropic sand. Int. J. Solids Struct. (2013), http://dx.doi.org/10.1016/j.ijsolstr.2013.12.003

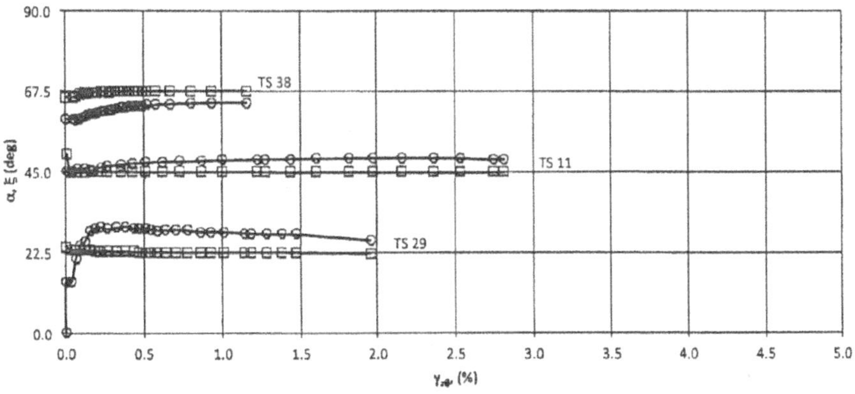

Fig. 10. Directions of principal strain increments (open circle) and stresses (open square) versus engineering shear strain for torsion shear tests with $b = 0.75$ on dense Nevada sand.

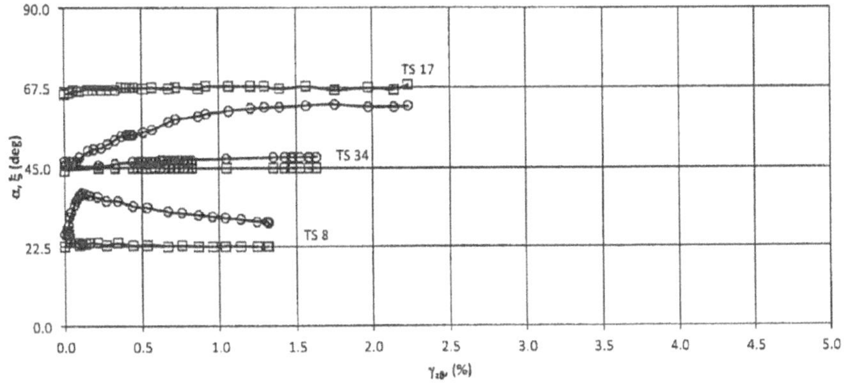

Fig. 11. Directions of principal strain increments (open circle) and stresses (open square) versus engineering shear strain for torsion shear tests with $b = 1.0$ on dense Nevada sand.

there does not appear to be any systematic variation with the value of $b = (\sigma_2 - \sigma_3)/(\sigma_1 - \sigma_3)$.

These deviations are caused by scatter similar to e.g. the scatter in strain-to-failure observed for repeated tests in triaxial compression. It is not caused by inaccuracy of the measurements, because the same instrumentation was used for all the experiments, and the errors due to experimental accuracy are small. Thus, it was estimated that the error on the α-values were in the order of 0.3°, and the error on the ξ-values were of similar magnitude. These are insufficient to invalidate the measurements and the conclusions presented here.

It is interesting to note that the major strain increment directions for $\alpha = 22.5°$, and $\alpha = 45°$ are slightly above the stress directions. For $\alpha = 67.5°$, the major principal strain increment directions are slightly below the major principal stress directions. This is due to the cross-anisotropic nature of the specimen in which the bedding planes are horizontal. At $\alpha = 22.5°$, the horizontal bedding planes cause the strain increment direction to become more horizontal and therefore show a slightly higher inclination than the σ_1-direction. At $\alpha = 45°$, this also occurs but to a lesser extent. However, at $\alpha = 67.5°$, the strain increment direction moves over to the other side of the principal stress direction. This pattern is observed for all b-values. Fig. 13 presents a schematic diagram showing the bedding planes of a specimen and the general pattern of directions of stress directions and strain increment directions for all b-values.

In nature, where the initial stress state is not isotropic but at a stress level closer to failure, the initial portions of the relations shown in Figs. 7–11 are not present. At higher stress levels (proximity to failure) the directions of major principal strain increments are controlled by the plastic strain components and the major principal strain increment directions are likely to be closer to the directions of major principal stress, but in general the two directions are not expected to coincide, even at failure. The deviations between the two directions are expected to follow the pattern indicated and discussed above.

It is clear from the experiments that the directions of the strain increment vectors do not coincide generally with the directions of the stress, i.e. the ξ- and the α-directions are generally non-coaxial. The pattern of behavior observed in these torsion shear experiments, which were performed with constant stress directions indicated by α, has to be modeled by a constitutive model that involves cross-anisotropic plastic hardening behavior.

Please cite this article in press as: Rodriguez, N.M., Lade, P.V. Non-coaxiality of strain increment and stress directions in cross-anisotropic sand. Int. J. Solids Struct. (2013). http://dx.doi.org/10.1016/j.ijsolstr.2013.12.003

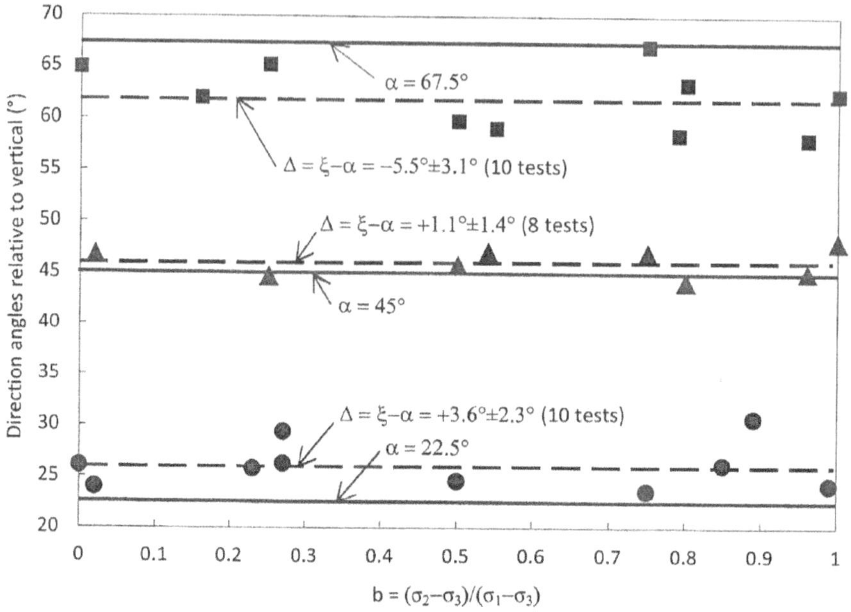

Fig. 12. Diagram of the deviations between major principal strain increment directions (ξ) and major principal stress directions (α) relative to vertical at failure.

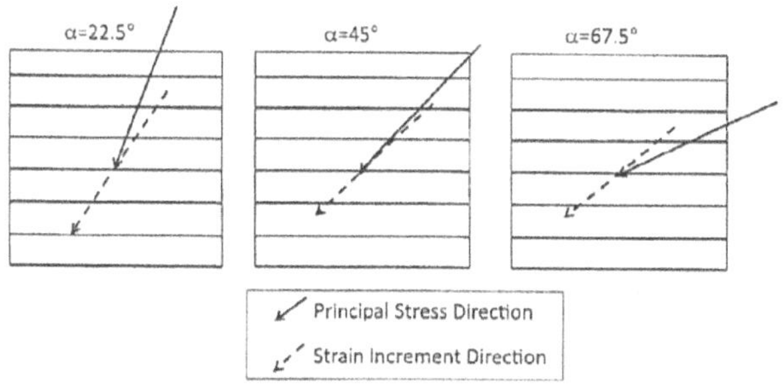

Fig. 13. Schematic patterns of principal strain increment directions and principal stress directions observed in torsion shear tests on dense Nevada sand.

6. Conclusion

The results of carefully performed torsions shear tests on hollow cylinder specimens of dense Nevada sand have been interpreted in terms of the directions in physical space of the major principal strain increments relative to the major principal stresses, which were maintained in constant directions during each test. The two directions were found to be generally non-coaxial, even at failure, which was manifested as development of shear bands. The average deviations between the two directions varied in a systematic manner between +3.6° and −5.5°, but a fair amount of scatter was observed in the experiments. To model this behavior, a constitutive model is required that includes effects of cross-anisotropy.

In a previous study (Lade et al., 2009) it was concluded that the directions of major principal strain increments essentially coincide with the directions of the major principal stress at the condition of failure. In that study the sand behaved as an isotropic material at failure, a surprising finding in view of the fact that the same torsion shear experiments showed considerable influence of cross-anisotropy on shear band orientations and on the shape of the failure surface. In those tests the stress paths were quite general with large variations in directions of major principal stress, and there might have been a tendency to oversimplify the conclusions from that study. In comparison, the present study employed constant directions of the major principal stress throughout the tests and the major principal strain increment directions therefore had a larger range of strains to establish themselves.

Please cite this article in press as: Rodriguez, N.M., Lade, P.V. Non-coaxiality of strain increment and stress directions in cross-anisotropic sand. Int. J. Solids Struct. (2013), http://dx.doi.org/10.1016/j.ijsolstr.2013.12.003

Acknowledgments

The research presented here was performed with support from the National Science Foundation under Grant No. CMMI-0757827. Grateful appreciation is expressed for this support.

References

Abelev, A.V., Lade, P.V., 2003. Effects of cross-anisotropy on three dimensional behavior of sand. I: stress–strain behavior and shear banding. ASCE J. Eng. Mech 129 (2), 160–166.

Broms, B.B., Casbarian, A.O., 1965. Effects of rotation of the principal stress axes and of the intermediate principal stress on shear Strength. In: Proc. 6th Int. Conf. Soil Mech. Found. Eng., Montreal, Canada, vol. I, pp. 179–183.

Chaudhary, S.K., Kuwano, J., 2003. Anisotropic multiple yielding of dense Toyoura sand in p'-constant shear plane. Soils Found. 43 (4), 59–69.

Frydman, S., Talesnick, M., Puzrin, A., 1995. Colinearity of stresses, strains, and strain increments during shearing of soft clay. ASCE J. Geotech. Eng. 121 (2), 174–184.

Hicher, P.-Y., Lade, P.V., 1987. Rotation of principal directions in K_0-consolidated clay. ASCE J. Geotech. Eng. 113 (7), 774–788.

Hight, D.W., Gens, A., Symes, M.J., 1983. The development of a new hollow cylinder apparatus for investigating the effects of principal stress rotation in soils. Geotechnique 33 (4), 355–383.

Hong, W.P., Lade, P.V., 1989a. Elasto-plastic behavior of K_0-consolidated clay in torsion shear test. Soils Found. 29 (2), 57–70.

Hong, W.P., Lade, P.V., 1989b. Strain increment and stress directions in torsion shear tests. ASCE J. Geotech. Eng. Div. 115 (10), 1388–1401.

Ishihara, K., Towhata, I., 1983. Sand response to cyclic rotation of principal stress direction as induced by wave loads. Soils Found. 33 (4), 11–26.

Kuwano, R., Jardine, R.J., 2007. A triaxial investigation of kinematic yielding in sand. Geotechnique 57 (7), 563–579.

Lade, P.V., 1975. Torsion shear tests on cohesionless soil. In: Proc. 5th Panamerican Conf. Soil Mech. Found. Eng., Buenos Aires, Argentina. pp. 117–127, I.

Lade, P.V., 1976. Interpretation of torsion shear tests on sand. In: Proc. 2nd Int. Conf. Numerical Meth. Geomech., Blacksburg, VA, vol. I, pp. 381–389.

Lade, P.V., 1981. Torsion shear apparatus for soil testing. In: Yong, R.N., Townsend, F.C. (Eds.), Laboratory Shear Strength of Soil, ASTM STP 740. ASTM, Philadelphia, Pa., pp. 145–163.

Lade, P.V., Abelev, A.V., 2005. Characterization of cross-anisotropic soil deposits from isotropic compression tests. Soils Found. 45 (5), 89–102.

Lade, P.V., Duncan, J.M., 1973. Cubical triaxial tests on cohesionless soil. ASCE J. Soil Mech. Found. Div. 99 (SM10), 793–812.

Lade, P.V., Kirkgard, M.M., 2000. Effects of stress rotation on cross-anisotropic behavior of natural K_0-consolidated soft clay. Soils Found. 40 (6), 93–105.

Lade, P.V., Nam, J., Hong, W.P., 2008. Shear banding and cross-anisotropic behavior observed in laboratory sand tests with stress rotation. Can. Geotech. J. 45 (1), 74–84.

Lade, P.V., Nam, J., Hong, W.P., 2009. Interpretation of strains in torsion shear tests. Comput. Geotech. 36 (1–2), 211–225.

Lade, P.V., Rodriguez, N.M., Van Dyck, E.J., 2013. Effects of principal stress directions on 3D failure conditions in cross-anisotropic sand. ASCE J. Geotech. Geoenvironm. Eng., in press.

Lade, P.V., Rodriguez, N.M., 2013. Comparison of true triaxial and hollow cylinder tests on cross-anisotropic sand specimens. ASTM Geotech. Test. J., submitted for publication.

Nishimura, S., Minh, N.A., Jardine, R.J., 2007. Shear strength anisotropy of natural London clay. Geotechnique 57 (1), 49–62.

Pradhan, B.S., Tatsuoka, F., Horii, N., 1988. Simple shear testing on sand in a torsional shear apparatus. Soils Found. 28 (2), 95–112.

Saada, A.S., Baah, A.K., 1967. Deformation and failure of a cross anisotropic clay under combined stresses. In: Proc. 3rd Panamerican Conf. Soil Mech. Found. Eng., I. Caracas, Venezuela, pp. 67–88.

Saada, A.S., Bianchini, G.F., Liang, L., 1994. Cracks, bifurcation, and shear bands propagation in saturated clays. Geotechnique 44 (1), 35–64.

Saada, A.S., Liang, L., Figueroa, J.L., Cope, C.T., 1999. Bifurcation and shear band propagation in sands. Geotechnique 49 (3), 367–385.

Sayao, A., Vaid, Y.P., 1991. A critical assessment of stress nonuniformities in hollow cylinder test specimens. Soils Found. 31 (1), 60–72.

Shibuya, S., Hight, D.W., Jardine, R.J., 2003. Local boundary surfaces of a loose sand dependent on consolidation path. Soils Found. 43 (3), 85–93.

Sivathayalan, S., Vaid, Y.P., 2002. Influence of generalized initial state and principal stress rotation on the undrained response of sands. Can. Geotech. J. 39 (1), 63–76.

Symes, M.J.P.R., Gens, A., Hight, D.W., 1984. Undrained anisotropy and principal stress rotation in saturated sand. Geotechnique 34 (1), 11–27.

Symes, M.J.P.R., Gens, A., Hight, D.W., 1988. Drained principal stress rotation in saturated sand. Geotechnique 38 (1), 59–81.

Tatsuoka, F., Sonoda, S., Hara, K., Fukushima, S., Pradhan, T.B.S., 1986. Failure and deformation of sand in torsional shear. Soils Found. 26 (4), 79–97.

Vaid, Y.P., Sayao, A., 1995. Proportional loading behaviour of sand under multiaxial stresses. Soils Found. 35 (3), 23–29.

Vaid, Y.P., Sayao, A.S.F., Hou, E., Negussey, D., 1990. Generalized stress path dependent soil behaviour with a new hollow cylinder torsional apparatus. Can. Geotech. J. 27 (5), 601–616.

Wang, Q., Lade, P.V., 2001. Shear banding in true triaxial tests and its effect on failure in sand. ASCE J. Eng. Mech. 127 (8), 754–761.

Yang, Z.X., Li, X.S., Yang, J., 2007. Undrained anisotropy and rotational shear in granular materials. Geotechnique 57 (4), 371–384.

Zdravkovic, L., Jardine, R.J., 2000. Undrained anisotropy of K_0-consolidated silt. Can. Geotech. J. 37 (1), 178–200.

Please cite this article in press as: Rodriguez, N.M., Lade, P.V. Non-coaxiality of strain increment and stress directions in cross-anisotropic sand. Int. J. Solids Struct. (2013), http://dx.doi.org/10.1016/j.ijsolstr.2013.12.003

Chapter XII
Incrementalization and Implementation of Single Hardening Model

Preface

The plasticity model reviewed above is expressed as strain increments as functions of total stresses. However, it is necessary to develop expressions for relations between strain increments and stress increments, which are used in calculation procedures in large finite element or finite difference computer programs. This incrementalization procedure and the following implementation procedure are shown in the following two articles for the Single Hardening model.

INTERNATIONAL JOURNAL FOR NUMERICAL AND ANALYTICAL METHODS IN GEOMECHANICS
Int. J. Numer. Anal. Meth. Geomech., 2002; **26**:647–659 (DOI: 10.1002/nag.216)

Incrementalization of a single hardening constitutive model for frictional materials

P. V. Lade[1,*,†] and K. P. Jakobsen[2]

[1] *Department of Civil Engineering, Aalborg University, Sohngaardsholmsvej 57, 9000 Aalborg, Denmark*
[2] *DHI—Water and Environment, Agern Allé 11, 2970 Hørsholm, Denmark*

SUMMARY

The governing equations for an elasto-plastic constitutive model for frictional materials such as soil, rock, and concrete are presented, and the incremental form is indicated in preparation for implementation of the model in a user-defined module for finite element calculations. This isotropic, work-hardening and -softening model employs a single yield surface, it incorporates non-associated plastic flow, and its capability of capturing the behaviour of different types of frictional materials under various three-dimensional conditions has been demonstrated by comparison with measured behaviour, as presented in the literature. The incrementalization procedure is indicated and the resulting equations for the single hardening model are presented together with parameters for a dense sand. Following the implementation of the model, these parameters are used for evaluation of different integration schemes as presented in a companion paper by Jakobsen and Lade (*Int. J. Numer. Anal. Meth. Geomech.* 2002; **26**:661). Copyright © 2002 John Wiley & Sons, Ltd.

KEY WORDS: incrementalization procedure; elasto-plastic constitutive model; frictional materials; yield surface; non-associated flow

1. INTRODUCTION

An elasto-plastic hardening constitutive model with a single yield surface has been developed on the basis of a thorough review and evaluation of data from experiments on frictional materials such as sand, clay, concrete and rock [1–3]. The numerical implementation of this model into a finite element program, such as ABAQUS [4], through a user-defined material module, requires the elasto-plastic model on incremental form. The components of the constitutive model which involve non-associated plastic flow are reviewed, and the incremental form of the model is presented. This form of the model is then used in the numerical implementation presented and evaluated in a companion paper [5].

The model employs a single, isotropic yield surface shaped as an asymmetric tear-drop with the pointed apex at the origin of the principal stress space. This yield surface, expressed in terms of stress invariants, describes the locus at which the total plastic work is constant. The total

*Correspondance to: P. V. Lade, Department of Civil Engineering, Aalborg University, Sohngaardsholmsvej 57, 9000 Aalborg, Denmark
† E-mail: lade@civil.auc.dk

Published online 15 April 2002

Received 22 April 2001
Revised 15 November 2001

plastic work (due to shear strains as well as volumetric strains) serves as the hardening parameter, and it is used to define the location and shape of the yield surfaces. The use of contours of constant plastic work (or any other measure of hardening) as yield surfaces results in mathematical consistency in the model, because the measure of yielding and the measure of hardening are uniquely related through one monotonic function. In addition, application of a single yield surface produces computational efficiency when used in large computer programs. The non-associated flow rule is derived from a potential function which describes a three-dimensional surface shaped as a cigar with an asymmetric cross-section.

The model is devised such that the transition from hardening to softening occurs abruptly at the peak failure point. Thus, the transition does not involve any points at which the hardening modulus is zero, but the rate of softening may be controlled and the pointed peak is hardly noticeable in actual comparisons with experimental data.

The main principles of the model and the governing equations for each component are reviewed below. The incremental form of the constitutive model is presented as detailed mathematical expressions required for numerical implementation in a finite element program. The model is then implemented in a user-defined module and employed in evaluation of the efficiency of several integration schemes.

2. SINGLE HARDENING STRESS–STRAIN MODEL

The total strain increments observed in a material when loaded are divided into elastic and plastic components such that

$$d\varepsilon = d\varepsilon^{e} + d\varepsilon^{p} \tag{1}$$

These strains are then calculated separately, the elastic strains by Hooke's law, and the plastic strains by a plastic stress–strain law. Both are expressed in terms of effective stresses.

Below, is a brief review of the framework and the components of the constitutive model. In order that the presentation follows a logic developmental sequence, the components are presented in the following order: Elastic behaviour, failure criterion, plastic potential and flow rule, yield criterion and work-hardening/softening law.

2.1. Elastic behaviour

The elastic strain increments, which are recoverable upon unloading, are calculated from Hooke's law, using a recently developed model for the non-linear variation of Young's modulus with stress state [6]. The value of Poisson's ratio, ν, being limited between zero and one-half for most materials, is assumed to be constant. The expression for Young's modulus was derived from theoretical considerations based on the principle of conservation of energy. According to this derivation, Young's modulus E can be expressed in terms of a power law involving non-dimensional material constants and stress functions as follows:

$$E = Mp_a\left[\left(\frac{I_1}{p_a}\right)^2 + 6\left(\frac{1+\nu}{1-2\nu}\right)\frac{J_2'}{p_a}\right]^{\lambda} \tag{2}$$

in which I_1 is the first invariant of the stress tensor, and J_2' is the second invariant of the deviatoric stress tensor, given as follows:

$$I_1 = \sigma_{11} + \sigma_{22} + \sigma_{33} \tag{3}$$

$$J_2' = \frac{1}{6}[(\sigma_{11} - \sigma_{22})^2 + (\sigma_{22} - \sigma_{33})^2 + (\sigma_{33} - \sigma_{11})^2] + \sigma_{12}^2 + \sigma_{23}^2 + \sigma_{31}^2 \tag{4}$$

The parameter p_a is the atmospheric pressure expressed in the same units as E, I_1 and $\sqrt{J_2'}$, and the modulus number M and the exponent λ are constant, dimensionless numbers. The three material parameter v, M, and λ may be obtained from simple tests such as triaxial compression tests. The model can be used for materials with effective cohesion.

2.2. *Failure criterion*

A general, three-dimensional failure criterion has been developed for soils, concrete, and rock [7–11]. The criterion is expressed in terms of the first and third stress invariants of the stress tensor:

$$f_n = \left(\frac{I_1^3}{I_3} - 27\right)\left(\frac{I_1}{p_a}\right)^m \tag{5a}$$

$$f_n = \eta_1 \text{ at failure} \tag{5b}$$

in which I_1 is given in (3) and

$$I_3 = \sigma_{11}\sigma_{22}\sigma_{33} + \sigma_{12}\sigma_{23}\sigma_{31} + \sigma_{21}\sigma_{32}\sigma_{13} - (\sigma_{11}\sigma_{23}\sigma_{32} + \sigma_{22}\sigma_{31}\sigma_{13} + \sigma_{33}\sigma_{12}\sigma_{21}) \tag{6}$$

The parameters η_1 and m are constant dimensionless numbers.

Figure 1 shows that in principal stress space, the failure criterion is shaped like an asymmetric bullet with the pointed apex at the origin of the stress axes, and the cross-sectional shape in the octahedral plane is triangular with smoothly rounded edges in a fashion that conforms to experimental evidence.

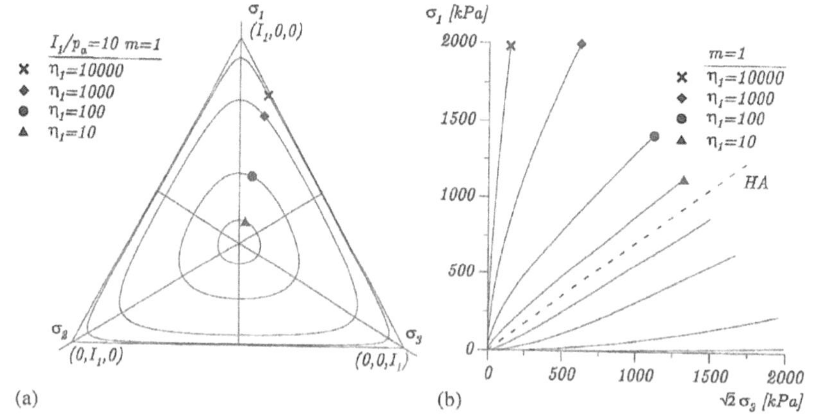

Figure 1. Characteristics of failure criterion in principal stress space: traces shown in (a) octahedral plane, and (b) triaxial plane.

In order to include the effective cohesion and the tension which can be sustained by concrete and rock, a translation of the principal stress space along the hydrostatic axis is performed [8–11]. Thus, a constant stress, $a p_a$, is added to the normal stresses before substitution in (5)

$$\sigma = \sigma + \delta a p_a \tag{7}$$

in which 'a' is a dimensionless parameter, and δ is Kronecker's delta. The value of $a p_a$ reflects the effect of the tensile strength of the material. The three material parameters, η_1, m, and a, may be determined from results of simple tests such as triaxial compression tests.

2.3. *Plastic potential and flow rule*

The plastic strain increments are calculated from the flow rule

$$d\varepsilon^p = d\lambda_p \frac{\partial g_p}{\partial \sigma} \tag{8}$$

in which g_p is a plastic potential function and $d\lambda_p$ is a scalar factor of proportionality. A suitable plastic potential function for frictional materials was developed and presented by Kim and Lade [1]. This function is different from the yield function and non-associated flow is consequently obtained. The plastic potential function is written in terms of the three invariants of the stress tensor:

$$g_p = \left(\psi_1 \frac{I_1^3}{I_3} - \frac{I_1^2}{I_2} + \psi_2\right)\left(\frac{I_1}{p_a}\right)^{\mu} \tag{9}$$

in which I_1 and I_3 are given in (3) and (6) and the second stress invariant is defined as

$$I_2 = \sigma_{12}\sigma_{21} + \sigma_{23}\sigma_{32} + \sigma_{31}\sigma_{13} - (\sigma_{11}\sigma_{22} + \sigma_{22}\sigma_{33} + \sigma_{33}\sigma_{11}) \tag{10}$$

The material parameters ψ_2 and μ are dimensionless constants that may be determined from triaxial compression tests. The parameter ψ_1 is related to the curvature parameter m of the failure criterion as follows:

$$\psi_1 = 0.00155 m^{-1.27} \tag{11}$$

The parameter ψ_1 acts as a weighting factor between the triangular shape (from the I_3 term) and the circular shape (from the I_2 term). The parameter ψ_2 controls the intersection with the hydrostatic axis, and the exponent μ determines the curvature of meridians. The corresponding plastic potential surfaces are shown in Figure 2. They are shaped as asymmetric cigars with smoothly rounded triangular cross-sections similar but not identical to those for the failure surfaces.

2.4. *Yield criterion and work hardening/softening relations*

The yield surfaces are intimately associated with and derived from surfaces of constant plastic work, as explained by Lade and Kim [2]. The isotropic yield function is expressed as follows:

$$f_p = f_p'(\sigma) - f_p''(W_p) = 0 \tag{12}$$

in which

$$f_p' = \left(\psi_1 \frac{I_1^3}{I_3} - \frac{I_1^2}{I_2}\right)\left(\frac{I_1}{p_a}\right)^{h} e^{q} \tag{13}$$

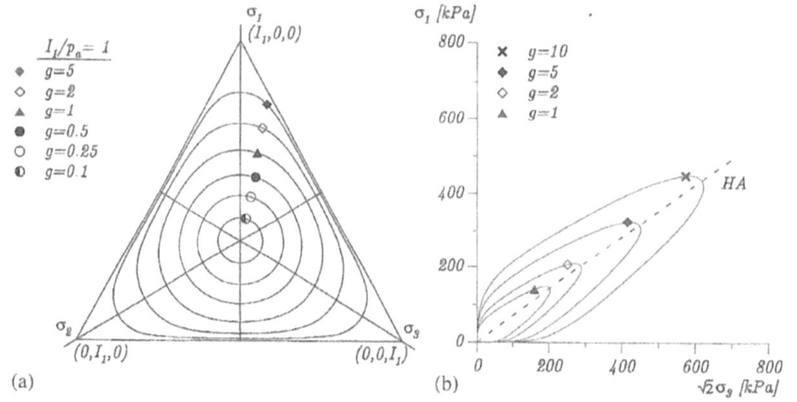

(a) (b)

Figure 2. Characteristics of the plastic potential function in principal stress space: traces shown in (a) octahedral plane, and (b) triaxial plane.

where h is constant and q varies from zero at the hydrostatic axis to unity at the failure surface. The expressions for I_1, I_2, and I_3 in (13) are given in (3), (10), and (6), respectively. The parameter ψ_1 acts as a weighting factor between the triangular shape (from the I_3 term) and the circular shape (from the I_2 term), as in the expression for the plastic potential (9). The constant parameter h is determined on the basis that the plastic work is constant along a yield surface.

The value of q varies with stress level S defined as

$$S = \frac{f_n}{\eta_1} = \frac{1}{\eta_1}\left(\frac{I_1^3}{I_3} - 27\right)\left(\frac{I_1}{p_a}\right)^m \tag{14}$$

in which f_n is the expression for the failure criterion in (5a), and η_1 is the value of f_n at failure, given in (5b). The stress level S varies from zero at the hydrostatic axis to unity at the failure surface, and the variation of q with S is expressed as

$$q = \frac{\alpha S}{1 - (1 - \alpha)S} \tag{15}$$

in which α is constant.

For *hardening*, the yield surface inflates isotropically with plastic work according to

$$f_p'' = \left(\frac{1}{D}\right)^{1/\rho}\left(\frac{W_p}{p_a}\right)^{1/\rho} \tag{16}$$

In (16) the values of ρ and D are constants for a given material. Thus, f_p'' varies with the plastic work only. The values of D and ρ are given by

$$D = \frac{C}{(27\psi_1 + 3)^\rho} \tag{17}$$

and

$$\rho = \frac{p}{h} \tag{18}$$

The parameters C and p in (17) are used to model the plastic work during isotropic compression:

$$W_{\mathrm{p}} = C p_{\mathrm{a}} \left(\frac{I_1}{p_{\mathrm{a}}} \right)^{\mathrm{p}} \tag{19}$$

The yield surfaces are shaped as asymmetric tear drops with smoothly rounded triangular cross-sections and traces in the triaxial plane as shown in Figure 3. As the plastic work increases, the isotropic yield surface inflates until the current stress point reaches the failure surface. The relation between f_{p}'' and W_{p} is described by a monotonically increasing function whose slope decreases with increasing plastic work, as shown in Figure 4.

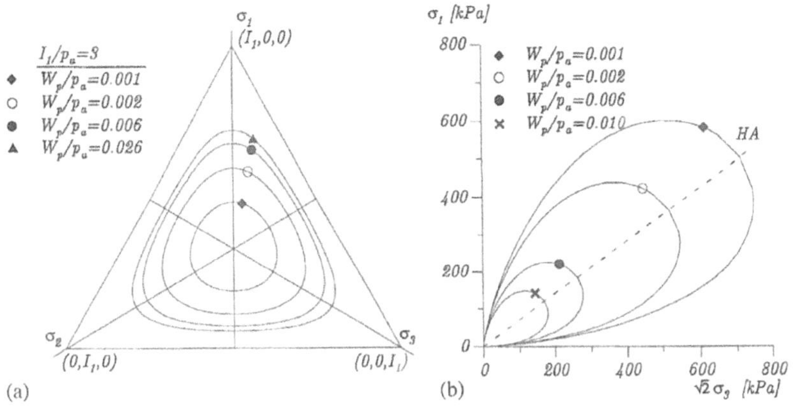

Figure 3. Characteristics of yield function in principal stress space: traces shown in (a) octahedral plane, and (b) triaxial plane.

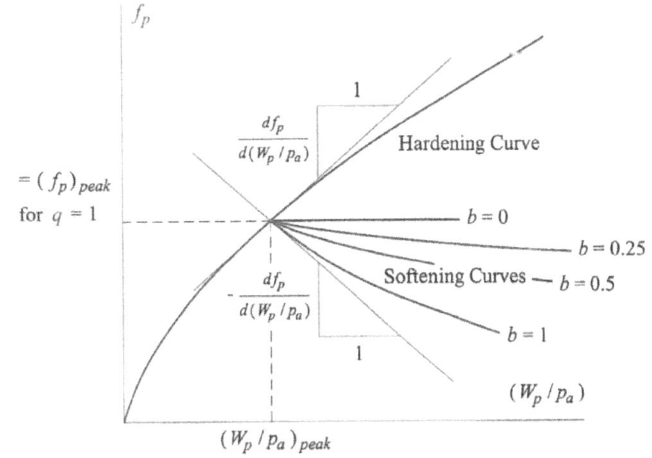

Figure 4. Modelling of work hardening and softening.

For *softening*, the yield surface deflates isotropically according to an exponential decay function:

$$f_p'' = A e^{-BW_p/p_a} \tag{20}$$

in which A and B are positive constants to be determined on the basis of the slope of the hardening curve at the point of peak failure, $S = 1$, as indicated in Figure 4. Thus

$$A = [f_p'' e^{BW_p/p_a}]_{S=1} \tag{21}$$

and

$$B = \left[b \frac{d f_p''}{d(\frac{W_p}{p_a})} \frac{1}{f_p''} \right]_{S=1} \tag{22}$$

in which both the size of the yield surface f_p'' and the derivative $d f_p''/d(W_p/p_a)$ are obtained from the hardening curve at peak failure, indicated by $S = 1$. The value of $d f_p''$ is negative during softening. The parameter b is greater than or equal to zero, where the lower limit corresponds to that of a perfect plastic material.

Using the expression for the plastic potential in (9), the relation between plastic work increment and the scalar factor of proportionality $d\lambda_p$ in (8) may be expressed as

$$d\lambda_p = \frac{dW_p}{\mu g_p} \tag{23}$$

in which the increment of plastic work can be determined by differentiation of the hardening and softening equations.

Combining (21) and (22) with (23) and substituting this into (8) produces the expression for the incremental plastic strain increments.

2.5. *Materials with effective cohesion*

As explained in connection with the failure criterion, it is possible to include the effective cohesion and the tension which can be sustained by concrete and rock. This is done by translating the principal stress space along the hydrostatic axis, i.e. by adding a constant stress to the normal stress, as in (7), before substitution into the failure criterion in (5).

A similar technique has been shown to work for the elastic modulus variation [6], the plastic potential [1], the yield criterion and the work-hardening/softening law [2]. Inconsistencies in the plastic behaviour of materials with effective cohesion are overcome by assuming that there exists an initial yield surface which goes through the origin of the real stress space, as shown in Figure 5. This implies that only elastic strains occur during major portions of tension type tests in which the main parts of the stress paths are located inside the yield surface. Thus, the stresses applied along the stress paths indicated in Figure 5 do not produce plastic strains until they are close to failure. The existence of an initial yield surface appears to simulate experimental observations with good accuracy.

The technique of performing all calculations in the translated stress space provides a convenient tool for mathematical treatment of cohesive materials. Following all the necessary calculations, the stresses are again modified to hold conventional physical meaning. For cohesionless materials for which 'a' is zero, the calculations are performed in the original stress space.

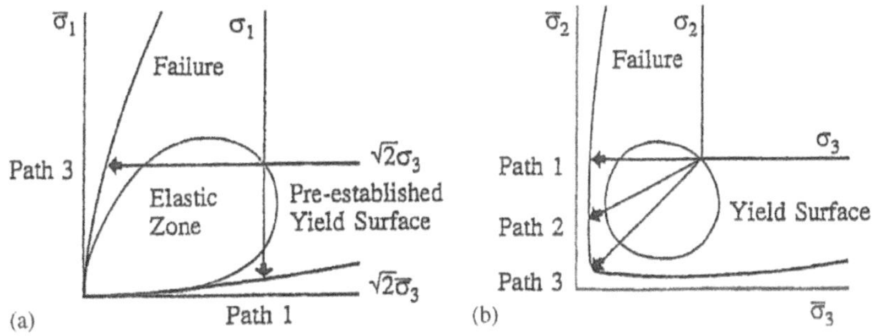

Figure 5. Biaxial stress paths for concrete employed by Kupfer *et al.* [12] and assumed initial yield surface in (a) triaxial plane, and (b) biaxial plane.

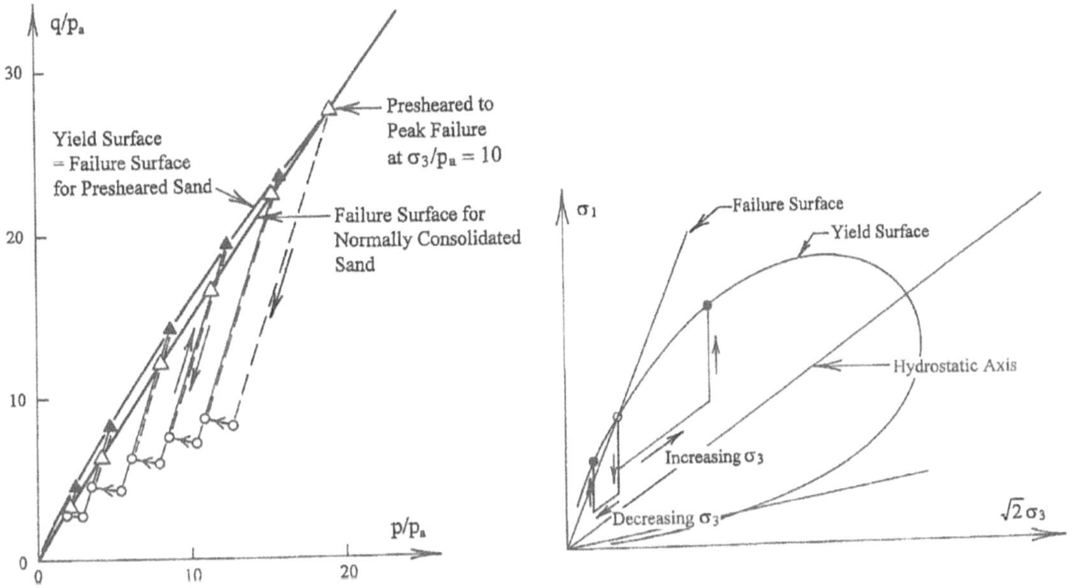

Figure 6. Effects of preshearing and overconsolidation on failure surface for soils.

2.6. *Effects of preshearing and overconsolidation*

An experimental investigation of the behaviour of sand near failure was undertaken by Lade and Prabucki [13] to study the shape, location and movement of the plastic yield surface in the hardening regime and the softening regime near peak failure. This investigation confirmed that the yield surface, defined as a contour of constant plastic work as measured from the origin of stress, captures the behaviour of soil with good accuracy in both the hardening and in the softening regime. It was found that preshearing to peak failure produced effects similar to overconsolidation observed in clays in the region of lower confining pressures, as shown in Figure 6. In this region, the yield surface was found to move out beyond the failure surface for

Table I. Material parameters for dense Eastern Scheldt sand used with the single hardening model

Model component	Parameter names	Parameter values			Equation
Elastic properties	M, λ, ν	458.45	0.4142	0.20	2
Failure criterion	m, η_1, a	0.2879	70.19	0	5, 7
Plastic potential	ψ_2, μ	−3.1540	2.0611		9
Yield criterion	h, α	0.5525	0.6166		13, 15
Hardening/softening law	C, p, b	1.2748×10^{-4}	1.6078	0.5	19, 22

normally consolidated sand, i.e. the sand became stronger. The observed pattern of yielding is captured with good accuracy by the yield criterion employed in the single hardening constitutive model.

2.7. *Summary*

The governing functions of the single hardening model have been presented and the material parameters identified. The material parameters depend on the specific material and must be calibrated to results of isotropic and triaxial compression tests as outlined by Lade [7], Kim and Lade [1], Lade and Kim [2,3]. The components of the constitutive model and the corresponding material parameters are listed in Table I with values indicated for dense Eastern Scheldt sand [14]. These material parameters are employed in the companion paper [5] for an evaluation of the integration schemes used in a user-defined material module employed in a finite element programme.

3. INCREMENTAL FORM OF CONSTITUTIVE MODEL

3.1. *Elasto-plastic stiffness matrix for single hardening model*

Application of an elasto-plastic model in a finite element procedure requires an incremental form in which the stress increments are expressed in terms of the total strain increments, as follows:

$$d\sigma = \mathbf{C}^{ep}\, d\varepsilon = (\mathbf{C}^{e} - \mathbf{C}^{p})\, d\varepsilon \tag{24}$$

in which $\mathbf{C}^{ep}$ is the elasto-plastic stiffness matrix. This is obtained through an incrementalization procedure as described by, e.g. Zienkiewicz and Taylor [15], Smith and Griffiths [16], Chen and Mizuno [17], Lade and Nelson [18]. The incrementalization results in

$$d\sigma = \left[\mathbf{C}^{e} - \frac{\mathbf{C}^{e}(\frac{\partial g_p}{\partial \sigma})(\frac{\partial f_p}{\partial \sigma})^{T}\mathbf{C}^{e}}{(\frac{\partial f_p}{\partial \sigma})^{T}\mathbf{C}^{e}(\frac{\partial g_p}{\partial \sigma}) + H}\right] d\varepsilon \tag{25}$$

The term in the square brackets is the general expression for the incremental stiffness matrix, $\mathbf{C}^{ep}$, for an elasto-plastic model with a single yield surface.

If the stress point is in the elastic domain, then the plastic stiffness, $\mathbf{C}^{p}$ in (24), vanishes and $\mathbf{C}^{ep}$ and $\mathbf{C}^{e}$ become identical. It is noted that $\mathbf{C}^{p}$ and therefore $\mathbf{C}^{ep}$ is asymmetric for non-associated plastic flow.

It is possible to extract the proportionality factor $d\lambda_p$ from (25)

$$d\lambda_p = \frac{\mathbf{C}^e(\frac{\partial f_p}{\partial \sigma})^T}{(\frac{\partial f_p}{\partial \sigma})^T \mathbf{C}^e (\frac{\partial g_p}{\partial \sigma}) + H} d\varepsilon \tag{26}$$

in which $d\lambda_p$ is expressed in terms of the total strain increment, to be used in the integration scheme presented in the companion paper [5].

3.2. *Elastic stiffness matrix*

In (25) $\mathbf{C}^e$ is the elastic stiffness matrix defined as

$$\mathbf{C}^e = \frac{E}{(1+v)(1-2v)} \times \begin{bmatrix} 1-v & v & v & 0 & 0 & 0 \\ v & 1-v & v & 0 & 0 & 0 \\ v & v & 1-v & 0 & 0 & 0 \\ 0 & 0 & 0 & \frac{(1-2v)}{2} & 0 & 0 \\ 0 & 0 & 0 & 0 & \frac{(1-2v)}{2} & 0 \\ 0 & 0 & 0 & 0 & 0 & \frac{(1-2v)}{2} \end{bmatrix} \tag{27}$$

where Young's modulus E is given in (2) and Poisson's ratio v is constant.

3.3. *Derivatives of the plastic potential*

The plastic potential function, g_p, is given in (9). Since the plastic potential function is expressed in terms of stress invariants, the derivatives of g_p with regard to stresses may be obtained through the chain rule:

$$\frac{\partial g_p}{\partial \sigma} = \frac{\partial g_p}{\partial I_1}\frac{\partial I_1}{\partial \sigma} + \frac{\partial g_p}{\partial I_2}\frac{\partial I_2}{\partial \sigma} + \frac{\partial g_p}{\partial I_3}\frac{\partial I_3}{\partial \sigma} \tag{28}$$

in which

$$\frac{\partial g_p}{\partial I_1} = \left(\psi_1(\mu+3)\frac{I_1^2}{I_3} - (\mu+2)\frac{I_1}{I_2} + \frac{\mu\psi_2}{I_1}\right)\left(\frac{I_1}{p_a}\right)^{\mu} \tag{29}$$

$$\frac{\partial g_p}{\partial I_2} = \frac{I_1^2}{I_2^2}\left(\frac{I_1}{p_a}\right)^{\mu} \tag{30}$$

$$\frac{\partial g_p}{\partial I_3} = -\psi_1\frac{I_1^3}{I_3^2}\left(\frac{I_1}{p_a}\right)^{\mu} \tag{31}$$

and the derivatives of the stress invariants with respect to stresses are

$$\frac{\partial I_1}{\partial \sigma} = \begin{bmatrix} 1 \\ 1 \\ 1 \\ 0 \\ 0 \\ 0 \end{bmatrix} \tag{32}$$

$$\frac{\partial I_2}{\partial \sigma} = \begin{bmatrix} -(\sigma_{22} + \sigma_{33}) \\ -(\sigma_{33} + \sigma_{11}) \\ -(\sigma_{11} + \sigma_{22}) \\ 2\sigma_{23} \\ 2\sigma_{31} \\ 2\sigma_{12} \end{bmatrix} \tag{33}$$

$$\frac{\partial I_3}{\partial \sigma} = \begin{bmatrix} \sigma_{22}\sigma_{33} - \sigma_{23}^2 \\ \sigma_{33}\sigma_{11} - \sigma_{31}^2 \\ \sigma_{11}\sigma_{22} - \sigma_{12}^2 \\ 2(\sigma_{12}\sigma_{31} - \sigma_{11}\sigma_{23}) \\ 2(\sigma_{23}\sigma_{12} - \sigma_{22}\sigma_{31}) \\ 2(\sigma_{31}\sigma_{23} - \sigma_{33}\sigma_{12}) \end{bmatrix} \tag{34}$$

3.4. *Derivatives of the yield function*

The yield function, f'_p, is given in (13). Using the chain rule, the derivatives of the yield function can be written as

$$\frac{\partial f'_p}{\partial \sigma} = \frac{\partial f'_p}{\partial I_1}\frac{\partial I_1}{\partial \sigma} + \frac{\partial f'_p}{\partial I_2}\frac{\partial I_2}{\partial \sigma} + \frac{\partial f'_p}{\partial I_3}\frac{\partial I_3}{\partial \sigma} \tag{35}$$

in which the derivatives of the stress invariants are given in (32)–(34). The derivatives of f'_p with respect to the stress invariants are

$$\frac{\partial f'_p}{\partial I_1} = \left(\frac{3+h}{I_1} + \frac{\partial q}{\partial I_1}\right) f'_p + \frac{I_1}{I_2}\left(\frac{I_1}{p_a}\right)^h e^q \tag{36}$$

$$\frac{\partial f'_p}{\partial I_2} = \frac{I_1^2}{I_2^2}\left(\frac{I_1}{p_a}\right)^h e^q \tag{37}$$

$$\frac{\partial f'_p}{\partial I_3} = f'_p \frac{\partial q}{\partial I_3} - \psi_1 \frac{I_1^3}{I_3^2}\left(\frac{I_1}{p_a}\right)^h e^q \tag{38}$$

It is recalled that the exponent q varies with the actual stress level as defined by (14) and (15). The derivatives of q with respect to the stress invariants are

$$\frac{\partial q}{\partial I_1} = \frac{\alpha}{\eta_1(1-(1-\alpha)S)^2}\left(\frac{mS\eta_1}{I_1} + \frac{3I_1^2}{I_3}\left(\frac{I_1}{p_a}\right)^m\right) \tag{39}$$

$$\frac{\partial q}{\partial I_3} = -\frac{\alpha}{\eta_1(1-(1-\alpha)S)^2}\frac{I_1^3}{I_3^2}\left(\frac{I_1}{p_a}\right)^m \tag{40}$$

3.5. *Hardening modulus*

The last term in the denominator in (25) represents the hardening modulus:

$$H = -\frac{\partial f''_p}{\partial W_p}\sigma\frac{\partial g_p}{\partial \sigma} = -\frac{\partial f''_p}{\partial W_p}\mu g \tag{41}$$

Since the function g_p in (9) is homogeneous and of the order μ, Euler's theorem for homogeneous functions (see, e.g. The International Dictionary of Applied Mathematics, 1960 [19]) produces the last expression for the hardening modulus in (41). The derivative of the yield function with respect to the plastic work depends on the stress history. The derivatives are determined for the hardening and the softening relations in (16) and (20) as follows:

$$\textit{Hardening}: \quad \frac{\partial f_p}{\partial W_p} = \frac{\partial f''_p}{\partial W_p} = \frac{1}{\rho(Dp_a)^{1/\rho}} W_p^{\frac{1}{\rho}-1} \tag{42}$$

$$\textit{Softening}: \quad \frac{\partial f_p}{\partial W_p} = \frac{\partial f''_p}{\partial W_p} = -\frac{AB}{p_a} e^{-BW_p/p_a} \tag{43}$$

The above expressions represent the incremental form of the single hardening model. They form the basic expressions used for the implementation in a user-defined material module presented in a companion paper [5].

4. SUMMARY

An elasto-plastic constitutive model developed for frictional materials has been reviewed, and the incremental version of the model has been given in the form of mathematical expressions to be implemented in a user-defined material module for a finite element program. This three-dimensional model for isotropic materials is formulated in terms of stress invariants, it involves hardening and softening, non-associated plastic flow, and it handles volume changes such a contraction and dilation as observed in experiments on frictional material. The model is implemented in a material subroutine and used in a finite element program, as described in a companion paper [5]. Several integration schemes are investigated and evaluated for this elasto-plastic model.

REFERENCES

1. Kim MK, Lade PV. Single hardening constitutive model for frictional materials. I. plastic potential function. *Computers and Geotechnics* 1988; **5**:307–324.
2. Lade PV, Kim MK. Single hardening constitutive model for frictional materials. II. yield criterion and plastic work contours. *Computers and Geotechnics* 1988a; **6**:13–29.
3. Lade PV, Kim MK. Single hardening constitutive model for frictional materials. III. comparisons with experimental data. *Computers and Geotechnics* 1988b; **6**:30–47.
4. ABAQUS *version 5.5—manuals*, Hibbit, Karlson and Sorensen Inc, 1995.
5. Jakobsen KP, Lade PV. Implementation algorithm for a single hardening constitutive model for frictional materials. *International Journal for Numerical and Analytical Methods in Geomechanics* 2002; **26**:661.
6. Lade PV, Nelson RB. Modelling the elastic behavior of granular materials. *International Journal for Numerical and Analytical Methods in Geomechanics* 1987; **11**:521–542.
7. Lade PV. Elasto-plastic stress–strain theory for cohesionless soil with curved yield surfaces. *International Journal of Solids and Structures* 1977; **13**:1019–1035.
8. Lade PV. Three-parameter failure criterion for concrete. *ASCE Journal of Engineering Mechanics Division* 1982; **108**:850–863.
9. Lade PV. Failure criterion for frictional materials. In *Mechanics of Engineering Materials*, Desai CS, Gallagher RH (eds). Wiley: New York, 1984; 385–402.
10. Lade PV. Rock Strength Criteria: The Theories and the Evidence, Vol. I, Chapter 11. In *Comprehensive Rock Engineering, Principles, Practice and Projects*, Hudson JA, et al. (eds). Pergamon Press: Oxford, 1993; 255–284.
11. Kim MK, Lade PV. Modelling rock strength in three dimensions. *International Journal of Rock Mechanics and Mining Sciences & Geomechanics Abstracts* 1984; **21**:21–33.
12. Kupfer H, Hilsdorf HK, Rusch H. Behavior of concrete under biaxial stresses. *Journal of the American Concrete Institute* 1969; **66**:656–666.
13. Lade PV, Prabucki M-J. Softening and preshearing effects in sand. *Soils and Foundations* 1995; **35**(4): 93–104.
14. Jakobsen KP, Praastrup U. Drained triaxial tests on Eastern Scheldt sand. *Aalborg University Geotechnical Engineering Papers*, 1998; ISSN 1398-6465 R9822.
15. Zienkiewicz OC, Taylor RL. *The Finite Element Method—Vol. 2: Solid and Fluid Dynamics and Non-linearity*. McGraw-Hill: London, UK, 1991.
16. Smith IM, Griffiths DV. *Programming the Finite Element Method (2nd Edition)*. Wiley: New York, 1988.
17. Chen WF, Mizuno E. *Non-linear Analysis in Soil Mechanics*. Elsevier: New York, 1990.
18. Lade PV, Nelson RB. Incrementalization procedure for elasto-plastic constitutive model with multiple, intersecting yield surfaces. *International Journal for Numerical and Analytical Methods in Geomechanics* 1984; **8**:311–323.
19. *The International Dictionary of Applied Mathematics*. Van Nostrand Company, Inc: Princeton, NJ, 1960.

INTERNATIONAL JOURNAL FOR NUMERICAL AND ANALYTICAL METHODS IN GEOMECHANICS
Int. J. Numer. Anal. Meth. Geomech., 2002; **26**:661–681 (DOI: 10.1002/nag.217)

Implementation algorithm for a single hardening constitutive model for frictional materials

K. P. Jakobsen[1,*,†] and P. V. Lade[2]

[1] *DHI—Water and Environment, Agern Allé 11, 2970 Horsholm, Denmark*
[2] *Department of Civil Engineering, Aalborg University, Sohngaardsholmsvej 57, 9000 Aalborg, Denmark*

SUMMARY

An advanced elasto-plastic constitutive model for frictional materials, whose incremental version is presented in a companion paper (*Int. J. Numer. Anal. Meth. Geomech.*, 2002; **26**:647), is implemented in a user-defined material module. The general calculation strategy inside this module is presented and discussed, including the initial intersection of the yield surface and the techniques for updating of stresses and hardening modulus. Several integration schemes are implemented in the module and their capabilities in relation to the advanced, three-dimensional constitutive model are evaluated. The forward Euler, modified Euler, and Runge–Kutta–Dormand–Prince integration schemes are explained in detail, compared, and evaluated in view of error tolerances and computational efficiency. Copyright © 2002 John Wiley & Sons, Ltd.

KEY WORDS: implementation algorithm; elasto-plastic constitutive model; frictional materials; yield surface; non-associated flow

1. INTRODUCTION

Implementation of an advanced elasto-plastic constitutive model in a finite element program involves formulation of an incremental version of the model before actual coding of the model for computer application can be performed. A single hardening model for frictional materials and its incremental form are presented in a companion paper by Lade and Jakobsen [1]. The present paper contains the implementation procedures as well as an evaluation of how well several available integration schemes perform for this advanced elasto-plastic model with non-associated flow. Such integration schemes have been developed and used with relatively simple constitutive models and with some advanced models, and their qualities are studied here for the single hardening model.

The employment of more advanced and accurate elasto-plastic constitutive models in finite element codes is occasionally omitted due to a substantial increase in the computational costs. Analyses based on more accurate modelling of the true material behaviour can obviously not be obtained without a loss of efficiency. However, by a critical assessment of the used integration

*Correspondance to: K. P. Jakobsen, DHI—Water and Environment, Agern Allé 11, 2970 Hørsholm, Denmark
†E-mail: kpj@dhi.dk

Published online 15 April 2002

Received 22 April 2001
Revised 15 November 2001

schemes it is possible to limit the additional computational costs. The problem of selection of suitable integration schemes is two-fold as the analysis implies the use of integration schemes for the computation of updated stresses and hardening parameters by an integration of the constitutive relation, and an integration for establishment of the overall equilibrium of the system. In the latter case, the rate of convergence and therefore the efficiency rely on the determination of the material tangent stiffness (see e.g. References [2,3]). The use of the 'consistent' tangent stiffness may preserve the quadratic convergence inherited from the Newton–Raphson method. However, this implies that the integration scheme used for the overall equilibrium is of the Newton–Raphson type [3, p. 178]. The method of updating stresses and hardening parameters has on the other hand a direct impact on both accuracy and efficiency.

The solution of non-linear finite element problems involving the mechanical behaviour of materials usually consists of a series of load steps, each involving iterations to establish equilibrium between internal and external forces at the new load level. This global iteration process is entirely handled by the finite element program, e.g. Reference [4]. However, the evaluation of the internal forces and displacements, used in the global iterations, is dependent on the applied stress–strain relation. The role of the user material module, which includes the single hardening model, is therefore two-fold. It must provide (1) an update of stresses for evaluation of internal forces, and (2) a material stiffness for establishment of the global stiffness matrix used for equilibrium iterations and estimation of the corresponding displacement field.

Since ABAQUS uses a Gauss integration scheme to establish the tangent stiffness matrix for each element, it is only necessary to consider a single material point within an element. Whenever a new estimate of updated stresses and material stiffness is needed, the user material module is called once for each Gauss point. The determination of the updated quantities will essentially depend on the imposed strain increment, but as the material behaviour is of the path-dependent type, knowledge of the stress and strain history is also required.

The input to each step consists of the material parameters, the current stresses, the current maximum plastic work, and the new strain increments. The subroutine then produces the six new stresses, the plastic work corresponding to the new stress state as well as the updated tangent stiffness matrix, the so-called Jacobian matrix. The following section explains the overall calculation strategy.

2. CALCULATION STRATEGY

The calculation strategy is illustrated schematically in Figure 1 and a flow chart is shown in Figure 2. Based on the current stresses (σ_0, point A in Figure 1), the maximum plastic work, $W_{p,0}$, and the new strain increments ($\Delta\varepsilon$), stress point B is calculated assuming the strain increment to be entirely elastic (step 1 in Figure 2). The stresses at point B are used for calculation of f', being the value of the yield function corresponding to the current state of stress, and compared with the current value of the yield function f'' calculated from the current maximum plastic work (step 2). The expressions for f' and f'' are defined in the companion paper [1], but the subscript p has been omitted here. If the difference is less than or equal to zero $[f = (f' - f'') \leqslant 0]$, then the new stress point (B) is located inside the yield surface and the strain increment is truly elastic. The calculation for this increment have already been performed and the Jacobian matrix is elastic (step 3a).

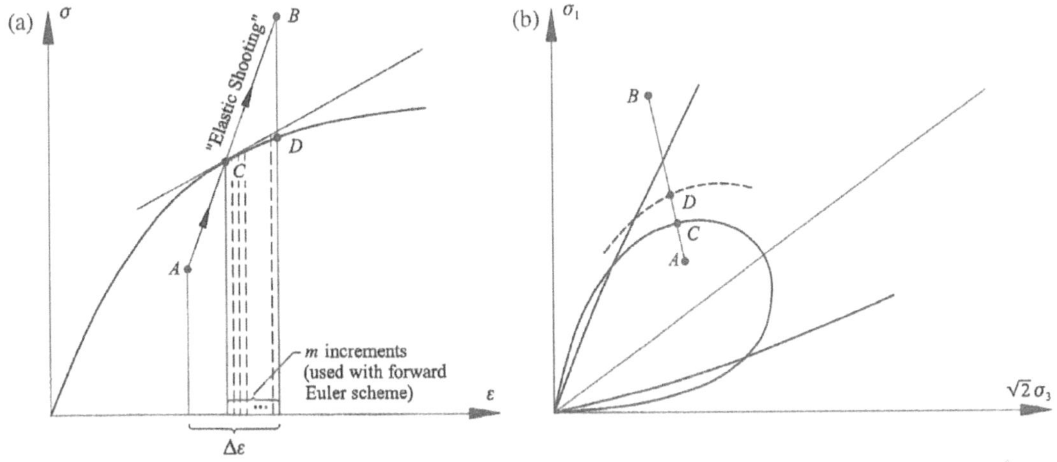

Figure 1. Schematic illustration of (a) stress–strain relation, and (b) triaxial plane with stress states used in explanation of calculation strategy.

If on the other hand $f = (f' - f'') > 0$, then point B is outside the current yield surface and a portion of the strain increment is plastic. In this case it is determined whether the current stress state (point A) is located inside or on the current yield surface. The stresses at point A are used for calculation of f' and compared with the previously calculated f'' (step 3b). If $f = (f' - f'') = 0$, then point A is on the yield surface and coincides with point C in Figure 1, and $\alpha = 0$ (step 4a).

If $f = (f' - f'') < 0$ (step 3b), then point A is inside the yield surface, as shown schematically in Figure 1. In this case it is necessary to determine the portion of the stress-path from A to B that is purely elastic (AC) by determining the ratio $\alpha = AC/AB$ (step 4b). For explicit integration schemes it is required that point C be found, and the technique for finding point C is further discussed below (Section 3).

Having calculated the elastic strains from A to C, the remaining portion of the total strain increment (calculation in step 5) is used to update the stresses and the hardening parameter (steps 6–8). Three integration schemes are implemented: (1) forward Euler scheme with subincrementation, (2) modified Euler scheme with error control, and (3) Runge–Kutta–Dormand–Prince scheme with error control. These techniques are discussed in detail in Sections 4–8. At the end of either one of these schemes, the user material module is exited with the stresses, the Jacobian matrix, and the plastic work at point D (step 8).

3. INITIAL INTERSECTION WITH YIELD SURFACE

If the stress point changes from an elastic to an elasto-plastic state, as it occurs for presheared or overconsolidated materials, then it is necessary to determine the portion of the stress increment that causes purely elastic deformations, as indicated in Figure 1. The technique for finding point C is illustrated in Figure 3, indicated in step 4b in Figure 2, and described mathematically below.

The initial state of stress at point A lies within the yield surface corresponding to:

$$f(\sigma_A, W_{p,A}) = f_A < 0 \tag{1}$$

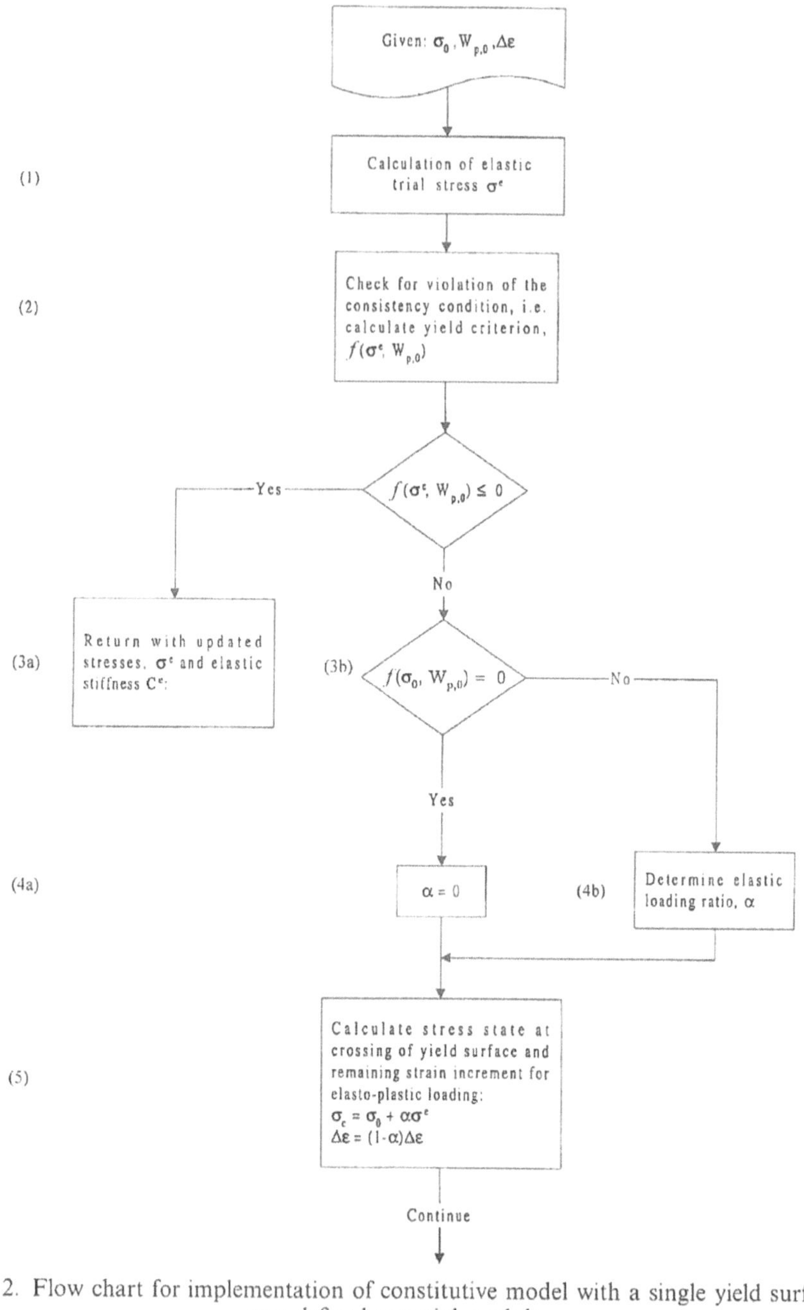

Figure 2. Flow chart for implementation of constitutive model with a single yield surface in user-defined material module.

The 'elastic shooting' from A to B produces an elastic trial stress increment calculated from Hooke's law:

$$\Delta\sigma^{e} = \mathbf{C}^{e}(\sigma_{A})\Delta\varepsilon \tag{2}$$

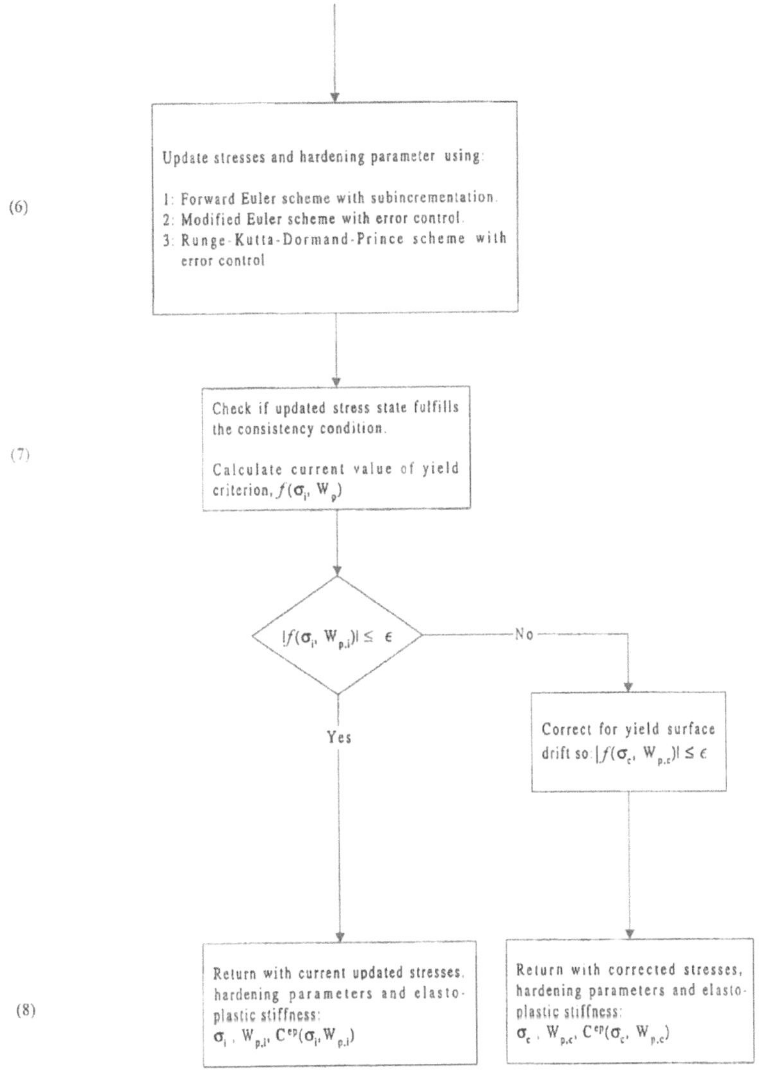

Figure 2. (*Continued*)

Thus, the elastic non-linearity is accounted for in the implementation. Hence, Young's modulus and the elastic stiffness matrix are determined on the basis of (2) and (27) of the companion paper. If the stress point changes from an elastic state (point A) to an elasto-plastic state (point B), then the trial stresses at B may violate the yield criterion:

$$f(\sigma_A + \Delta\sigma^e, W_{p,C}) = f(\sigma_B, W_{p,C}) = f_B > 0 \tag{3}$$

It is therefore necessary to determine a scalar α $(0<\alpha<1)$ corresponding to the portion of the stress increment that lies within the yield surface so that the stress state at point C fulfills the

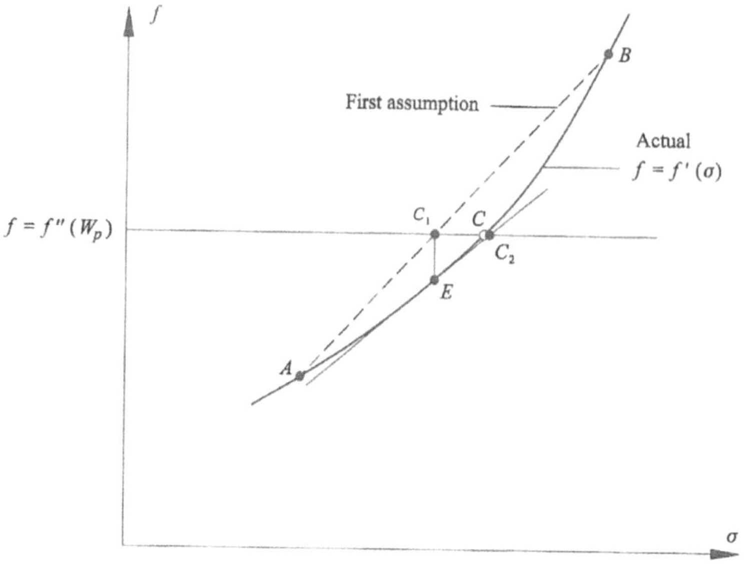

Figure 3. Schematic illustration of technique for finding intersection with yield surface.

yield criterion:

$$f(\sigma_A + \alpha\Delta\sigma^e, W_{p,C}) = f(\sigma_C, W_{p,C}) = f_C = 0 \tag{4}$$

Explicit expressions for the scalar α can be derived only for simple types of yield functions. A first estimate may be determined by a simple linear interpolation in f [5,6]

$$\alpha_0 = -\frac{f_A}{f_B - f_A} \tag{5}$$

The yield function is, however, highly non-linear and the scalar estimate determined by (5) will generally not satisfy the yield criterion:

$$f(\sigma_A + \alpha_0\Delta\sigma^e, W_{p,C}) = f(\sigma_E, W_{p,C}) = f_E \neq 0 \tag{6}$$

A more accurate estimate for α may be obtained by a Taylor series expansion around point E:

$$\alpha = \alpha_0 - \frac{f_E}{(\partial f/\partial(\sigma_A + \alpha_0\Delta\sigma^e)^T\Delta\sigma^e} \tag{7}$$

This approach for the calculation of the stresses on the yield surface at point C will be accurate for small strain increments only. In order to avoid any initial yield surface drift in the integration schemes and enhance the prediction capabilities of the stress–strain relations it is advisable to apply an iterative scheme [5]. Using the Newton–Raphson technique, for example, the stresses and α are updated as outlined in Figure 4. The iterative procedure is started by assuming the initial stresses at point A and using α_0 from (5). The procedure is terminated when the stress norm $\|\sigma_i - \sigma_{i-1}\|/\|\sigma_{i-1}\|$ is less than a specified tolerance. The resulting elastic stress increments $\alpha\Delta\sigma^e$ correspond to an elastic strain increment $\alpha\Delta\varepsilon$ and the strain increment used in the integration of the elasto-plastic stress–strain relation equals $(1 - \alpha)\Delta\varepsilon$.

Initial state $\sigma_0, W_{p,0}$

If $f(\sigma_0, W_{p,0}) > 0$

$$\alpha_0 = -\frac{f_A}{f_B - f_A}$$

Iterations $i = 1, 2, \ldots\ldots, i_{\max}$

$$\sigma_i = \sigma_{i-1} + \alpha_{i-1}\Delta\sigma^e$$

$$\Delta\alpha_i = -\frac{f(\sigma_i, W_p)}{(\frac{\partial f}{\partial \sigma_i})^T \Delta\sigma^e}$$

$$\alpha_i = \alpha_{i-1} + \Delta\alpha_i$$

Stop iteration when $\|\sigma_i - \sigma_{i-1}\| / \|\sigma_{i-1}\| \leq \epsilon$

Final state $\sigma_c = \sigma_i, \alpha_c = \alpha_i$

Figure 4. Algorithm with improved procedure for finding the initial intersection with the yield surface.

3.1. *Handling effects of preshearing and overconsolidation on the material strength*

The numerical problems that occur for presheared or overconsolidated materials are generally handled by applying the procedure outlined in the previous section. However, in special cases the current yield surface, which indicates the previously experienced highest load level, extends beyond the failure surface and actually strengthens the material, as indicated in Figure 6 of the companion paper. Thus, the material should not reach a state of failure until the yield surface is reached. This is accomplished numerically by calculating the intersection of the yield surface, as presented above, and then performing a check for failure. The stress level relative to failure, as defined by (14) in the companion paper, will in such cases exceed unity, but is for consistency set equal to unity.

3.2. *Back-scaling of elastic stresses*

The single hardening model is developed for frictional materials, and only stress combinations in the pressure octant of the principal stress space are allowed, even for materials with effective cohesion, as explained in Section 2.5 of the companion paper. This implies that if the elastic trial stresses correspond to one (or more) negative or zero principal stresses, then they must be scaled back into the pressure octant, in order to facilitate the calculation of the elasto-plastic stress–strain response. The problem is most likely to occur for large strain increments and near failure in the extension region.

4. UPDATING STRESSES AND HARDENING PARAMETER

The requirement of an accurate integration of the constitutive law is traditionally accomplished by use of a backward Euler or an explicit subincrementation scheme (see e.g. References [5,7,8]).

Both types of schemes have a number of advantages and disadvantages. The backward Euler scheme is an elastic predictor and plastic corrector type of method and is attractive because it does not require the initial intersection of the yield surface to be computed if the stress point passes from an elastic to an elasto-plastic state. The plastic correction is obtained by solving a small system of non-linear equations [3] by an iterative procedure, which ensures that the consistency condition may be satisfied within a specified tolerance. The approach, however, has some major disadvantages. The establishment of the system of equations becomes laborious for more advanced models and convergence is not necessarily guaranteed. The backward Euler scheme has apparently only been used and documented for simple constitutive models having associated flow rules (e.g. von Mises, Tresca, Mohr–Coulomb and Cam–Clay). Actually, Abbo and Sloan [8] make comparisons between the backward Euler and explicit integration schemes for a non-associated Mohr–Coulomb model. However, the handling of the non-associated flow rule is not described at all and the conclusions are merely based on a table containing the final results. Furthermore many soil models have a separate failure criterion and the numerical handling may be obstructed near failure as the elastic trial stress intersects the failure criterion. Alternatively, integration schemes of the forward Euler and Runge–Kutta types, which use explicit formulations, can be used. These types of schemes have the disadvantage that the initial intersection with the yield surface must be computed if a stress point passes from an elastic to an elasto-plastic state and they do not necessarily ensure that the consistency condition is fulfilled. The validity of these types of methods can, however, be enhanced and the disadvantages are balanced by their robustness.

As already explained, the forward Euler integration scheme is only accurate for small strain increments. By subdividing the strain increment into a fixed number of subincrements, as done in Section 5, the accuracy will be improved. However, this approach turns out to be computationally expensive [5]. Hence, two other schemes for explicit integration have been investigated and implemented. These include a refined subincremental version of the forward Euler scheme with active error control and an enhanced Runge–Kutta scheme. These methods are applicable to non-linear stress–strain relations in general, but the stress–strain relations given by the single hardening model are employed here. The paper is concluded by an example where the capabilities of the different methods are studied.

5. FORWARD EULER INTEGRATION SCHEME WITH SUBINCREMENTATION

Following the determination of the intersection with the yield surface, as described above, the subsequent objective is to determine the updated state of stress and hardening parameter as a total strain increment is imposed. A commonly used integration scheme for elasto-plastic stress–strain relations is the forward Euler scheme. In the following it is presumed that the initial state of stress is located on the yield surface and that the next stress increment causes elasto-plastic strains.

5.1. Procedure

In the forward Euler scheme the stresses are updated by replacing the infinitesimal elasto-plastic stress–strain relation in (25) of the companion paper by a finite incremental relation:

$$\Delta\sigma = \mathbf{C}^{\mathrm{ep}}(\sigma_0, W_{\mathrm{p},0})\Delta\varepsilon \tag{8}$$

where the constitutive matrix is evaluated at the initial state of stress. As the elasto-plastic constitutive matrix depends on the stress and strain history, this linear approximation is only accurate for very small strain increments. The method may be refined by a piece wise linear integration where the strain increment is subdivided into smaller subincrements [3,5,6,9,10]:

$$\delta\varepsilon = \Delta T \Delta\varepsilon = \frac{\Delta\varepsilon}{m} \tag{9}$$

where ΔT is a dimensionless time step of fixed size and the finite stress increment is determined as the sum of the m subincrements, $\delta\sigma_i$, each evaluated as a forward Euler step:

$$\delta\sigma_i = \mathbf{C}^{\mathrm{ep}}(\sigma_0 + \Delta\sigma_{i-1}, W_{\mathrm{p},0} + \Delta W_{\mathrm{p},i-1})\delta\varepsilon \tag{10}$$

$$\delta W_{\mathrm{p},i} = \delta\lambda_{\mathrm{p}}(\sigma_0 + \Delta\sigma_{i-1}, W_{\mathrm{p},0} + \Delta W_{\mathrm{p},i-1}, \delta\varepsilon)(\sigma_0 + \Delta\sigma_{i-1})\frac{\partial g}{\partial(\sigma_0 + \Delta\sigma_{i-1})} \tag{11}$$

where

$$\Delta\sigma_{i-1} = \sum_{j=1}^{i-1} \delta\sigma_j \tag{12}$$

$$\Delta W_{\mathrm{p},i-1} = \sum_{j=1}^{i-1} \delta W_{\mathrm{p},j} \tag{13}$$

The given change to the hardening parameter is valid for work hardening and is computed by use of (26) and (41) of the companion paper, with derivatives evaluated at the stress state $(\sigma_0 + \Delta\sigma_{i-1})$. As illustrated in Figure 5, the forward Euler scheme has the advantage of being straight forward and easy to implement compared to other schemes.

However, the yield criterion is not necessarily fulfilled, and the stress increments tend to drift away from the yield surface, i.e. $f(\sigma_0 + \Delta\sigma, W_{\mathrm{p},0} + \Delta W_{\mathrm{p}}) \neq 0$. Even though the subdivision may

Initial state $\sigma_0, W_{p,0}, \Delta\varepsilon, \delta\varepsilon = \frac{\Delta\varepsilon}{m}$

Strain subincrements $i = 1, 2, \ldots\ldots, m$

$$\delta\sigma_i = C^{ep}(\sigma_{i-1}, W_{p,i-1})\delta\varepsilon$$

$$\delta W_{p,i} = \delta\lambda_p(\sigma_{i-1}, W_{p,i-1}, \delta\varepsilon)\sigma_{i-1}\frac{\partial g}{\partial\sigma_{i-1}}$$

$$\sigma_i = \sigma_{i-1} + \delta\sigma_i$$

$$W_{p,i} = W_{p,i-1} + \delta W_{p,i}$$

Stop strain subincrementation when $i = m$

Final state $\sigma_i, W_{p,i} C^{ep}(\sigma_i, W_{p,i})$

Figure 5. Algorithm for subincremental forward Euler scheme for evaluation of new stress state, plastic work, and Jacobian matrix.

reduce the yield surface drift, the procedure may lead to unacceptable results as the error accumulates during subsequent load steps (e.g. References [3,5,11]). Moreover, the subincremental form has the disadvantage that it uses subincrements of equal size. This turns out to be computationally inefficient as the number of subincrements must be determined by trial-and-error so that the maximum error is within some close tolerance.

5.2. Correction for yield surface drift

When using explicit integration schemes such as the forward Euler scheme for updating of stresses, the stress state predicted at the end of the elasto-plastic load increment may not lie on the current yield surface and the consistency condition is violated. The error will essentially depend on the size of the strain increment and number of subdivisions, but as the error is cumulative it is important to ensure that the stresses are corrected back to the yield surface during each increment of loading.

A method proposed by Potts and Gens [11] accounts for the changes in elastic strains that accompany any stress correction. The problem of yield surface drift and correction is illustrated schematically in Figure 6, where the material is subjected to loading that causes elasto-plastic deformation. The initial stress state is initially located on the yield surface $f(\sigma_A, W_{p,A}) = 0$. Further loading involves elasto-plastic deformation and a change in stresses as determined by the integration scheme. This new state σ_B will, due to the tendency for yield surface drift, not necessarily be located on the new yield surface $f(\sigma, W_{p,B}) = 0$, and the objective is therefore to correct the stresses so that the yield criterion is fulfilled. During the correction process the total strain increment will remain constant, and this implies that any elastic strain change must be balanced by an equal and opposite change in plastic strain. These changes will affect the stresses and the hardening parameter, and the new updated elasto-plastic stress state is denoted σ_C and $W_{p,C}$.

The requirement of an unaltered total strain increment throughout the correction process can be formulated as

$$d\varepsilon^p = -d\varepsilon^e = -\mathbf{C}^{e^{-1}}(\sigma_C - \sigma_B) \tag{14}$$

The plastic strain increment is also proportional to the gradient of the plastic potential:

$$d\varepsilon^p = \beta \frac{\partial g}{\partial \sigma} \tag{15}$$

where β is a scalar quantity. The corrected stress state is obtained by substituting (15) into (14) and solving for σ_C:

$$\sigma_C = \sigma_B - \beta \mathbf{C}^e \frac{\partial g}{\partial \sigma_B} \tag{16}$$

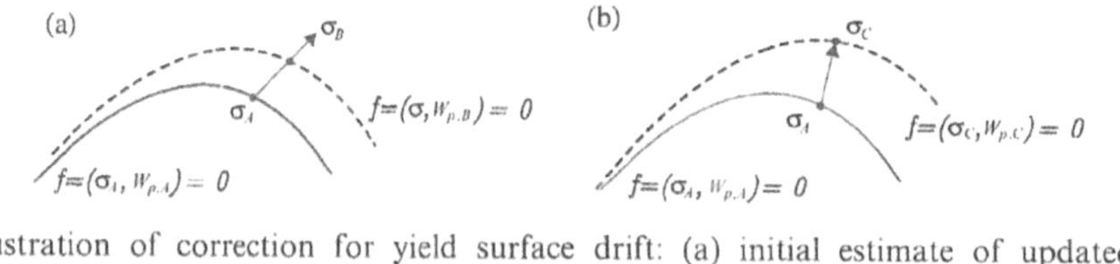

Figure 6. Illustration of correction for yield surface drift: (a) initial estimate of updated hardening parameter and stresses causing yield surface drift, and (b) corrected values of hardening parameter and stresses located on the yield surface.

The change in plastic strains will also affect the hardening parameter:

$$W_{p,C} = W_{p,B} + \Delta W_P \tag{17}$$

where ΔW_P for work-hardening is given by

$$\Delta W_P = \beta \left(\frac{\partial g}{\partial \sigma_B}\right)^T \sigma_B \tag{18}$$

The corrected stress state must satisfy the yield criterion:

$$f(\sigma_C, W_{p,C}) = f\left(\sigma_B - \beta \mathbf{C}^e \frac{\partial g}{\partial \sigma_B}, W_{p,B} + \beta\left(\frac{\partial g}{\partial \sigma_B}\right)^T \sigma_B\right) = 0 \tag{19}$$

A first order estimate of the scalar β can be obtained by a Taylor series expansion around σ_B:

$$\beta = \frac{f(\sigma_B, W_{p,B})}{(\frac{\partial f}{\partial \sigma_B})^T \mathbf{C}^e (\frac{\partial g}{\partial \sigma_B}) - (\frac{\partial f}{\partial W_{p,B}})(\frac{\partial g}{\partial \sigma_B})^T \sigma_B} \tag{20}$$

The above procedure may be sufficiently accurate if the load increment is small. However, it must in general be checked that the corrected stress state fulfils the yield criterion $f(\sigma_C, W_{p,C}) = 0$ to some close tolerance. Otherwise, an iterative procedure as outlined in Figure 7 must be applied. The iterative procedure is started by assuming $\sigma_0 = \sigma_B$ and $W_{p,0} = W_{p,B}$.

Initial state $\sigma_0, W_{p,0}$

If $|f(\sigma_0, W_{p,0})| > \epsilon$

Iterations $i = 1, 2, \ldots, i_{max}$

$$\beta_{i-1} = \frac{f(\sigma_{i-1}, W_{p,i-1})}{(\frac{\partial f}{\partial \sigma_{i-1}}) C^e (\frac{\partial g}{\partial \sigma_{i-1}}) - (\frac{\partial f}{\partial W_{p,i-1}})(\frac{\partial g}{\partial \sigma_{i-1}})^T \sigma_{i-1}}$$

$$\sigma_i = \sigma_{i-1} - \beta_{i-1} C^e \frac{\partial g}{\partial \sigma_{i-1}}$$

$$\Delta W_{p,i-1} = \beta_{i-1} \sigma_{i-1} \frac{\partial g}{\partial \sigma_{i-1}}$$

$$W_{p,i} = W_{p,i-1} + \Delta W_{p,i-1}$$

Stop iteration when $|f(\sigma_i, W_{p,i})| \leq \epsilon$

Final state $\sigma_c = \sigma_i, W_{p,c} = W_{p,i}$

Figure 7. Algorithm for correction of yield surface drift.

6. MODIFIED FORWARD EULER SCHEME WITH ERROR CONTROL

In order to reduce the yield surface drift and computational costs of the forward Euler scheme, a modified Euler scheme with active control can be used [5,12]. Instead of using a fixed number of subincrements of equal size, the size of the subincrements is varied throughout the integration process. Hence, the size of each subincrement is determined so that the new stress state fulfils the yield criterion within some small tolerance and only the absolutely necessary number of subdivisions are applied.

The modified scheme uses a pair of first- and second-order Euler formulas to estimate the error produced by the standard forward Euler scheme at the end of a strain increment, $\delta\varepsilon = \Delta T \Delta\varepsilon$. The first estimate of the updated stresses and the work hardening parameter at the end of the strain increment is given by

$$\sigma = \sigma_0 + \delta\sigma^{\mathrm{I}} \tag{21}$$

$$W_{\mathrm{p}} = W_{\mathrm{p},0} + \delta W_{\mathrm{p}}^{\mathrm{I}} \tag{22}$$

where

$$\delta\sigma^{\mathrm{I}} = \mathbf{C}^{\mathrm{ep}}(\sigma_0, W_{\mathrm{p},0})\delta\varepsilon \tag{23}$$

$$\delta W_{\mathrm{p}}^{\mathrm{I}} = \delta\lambda_{\mathrm{p}}(\sigma_0, W_{\mathrm{p},0}, \delta\varepsilon)\sigma_0 \frac{\partial g}{\partial \sigma_0} \tag{24}$$

A more accurate estimate of the updated stress state may be found by

$$\hat{\sigma} = \sigma_0 + \tfrac{1}{2}(\delta\sigma^{\mathrm{I}} + \delta\sigma^{\mathrm{II}}) \tag{25}$$

$$\hat{W}_{\mathrm{p}} = W_{\mathrm{p},0} + \tfrac{1}{2}(\delta W_{\mathrm{p}}^{\mathrm{I}} + \delta W_{\mathrm{p}}^{\mathrm{II}}) \tag{26}$$

where

$$\delta\sigma^{\mathrm{II}} = \mathbf{C}^{\mathrm{ep}}(\sigma_0 + \delta\sigma^{\mathrm{I}}, W_{\mathrm{p},0} + \delta W_{\mathrm{p}}^{\mathrm{I}})\delta\varepsilon \tag{27}$$

$$\delta W_{\mathrm{p}}^{\mathrm{II}} = \delta\lambda_{\mathrm{p}}(\sigma_0 + \delta\sigma^{\mathrm{I}}, W_{\mathrm{p},0} + \delta W_{\mathrm{p}}^{\mathrm{I}}, \delta\varepsilon)(\sigma_0 + \delta\sigma^{\mathrm{I}}) \frac{\partial g}{\partial(\sigma_0 + \delta\sigma^{\mathrm{I}})} \tag{28}$$

The method is seen to employ two evaluations of the elasto-plastic constitutive matrix in each subincrement. The difference between the stress states given by (21) and (25) can be seen as an estimate of the local error in σ:

$$\delta\sigma^{\mathrm{I-II}} \approx \hat{\sigma} - \sigma = \tfrac{1}{2}(\delta\sigma^{\mathrm{II}} - \delta\sigma^{\mathrm{I}}) \tag{29}$$

This error estimate serves as a guide for selecting the size of the next time step, ΔT, when integrating over the total strain increment $\Delta\varepsilon$. Thus, the relative error for a subincrement is defined by the norm:

$$\xi = \frac{\|\delta\sigma^{\mathrm{I-II}}\|}{\|\hat{\sigma}\|} \tag{30}$$

and the size of each step is continually adjusted until ξ is less than some specified tolerance, ϵ.

The integration is started by choosing a value of the dimensionless time step ΔT and computing $\delta\varepsilon, \sigma, W_{\mathrm{p}}, \hat{\sigma}, \hat{W}_{\mathrm{p}}, \delta\sigma^{\mathrm{I-II}}$ and ξ using (21)–(30). If $\xi \leqslant \epsilon$, then the new updated stresses and hardening parameters are taken as $\hat{\sigma}$ and $\hat{W}_{\mathrm{p}}$. Otherwise it is necessary to reduce ΔT and repeat the calculation. The size of the next dimensionless time step is generally given by local extrapolation:

$$\Delta T = q\Delta T \tag{31}$$

where

$$q = 0.9\left(\frac{\epsilon}{\xi}\right)^{1/2} \tag{32}$$

The exponent $\frac{1}{2}$ relates to the local truncation error $O(\Delta T^2)$ of the first-order formulae, whereas the factor of 0.9 is introduced to reduce the number of subincrements that are likely to be rejected during the integration process. The size of the new increment is furthermore constrained as q must lie within the interval:

$$0.01 \leqslant q \leqslant 2 \tag{33}$$

The modified Euler scheme with active error control is summarized in Figure 8.

7. RUNGE–KUTTA–DORMAND–PRINCE SCHEME WITH ERROR CONTROL

Variations of the Runge–Kutta scheme are widely used for integration purposes. The classical Runge–Kutta scheme uses a fourth-order integration scheme with only one strain increment for stress updating. Several higher order schemes with subincrementation have been proposed (e.g. References [12–15]). The Runge–Kutta scheme modified by Dormand and Prince [15] is used in the following. The scheme uses a pair of fourth- and fifth-order formulas to estimate the new updated stresses and work hardening parameter. The coefficients in these formulas have been chosen to control and minimize the error, and to estimate the new updated stresses and work hardening parameter as accurately as possible:

$$\sigma = \sigma_0 + \frac{31}{540}\delta\sigma^{\mathrm{I}} + \frac{190}{297}\delta\sigma^{\mathrm{III}} - \frac{145}{108}\delta\sigma^{\mathrm{IV}} + \frac{351}{220}\delta\sigma^{\mathrm{V}} + \frac{1}{20}\delta\sigma^{\mathrm{VI}} \tag{34}$$

$$W_{\mathrm{p}} = W_{\mathrm{p},0} + \frac{31}{540}\delta W_{\mathrm{p}}^{\mathrm{I}} + \frac{190}{297}\delta W_{\mathrm{p}}^{\mathrm{III}} - \frac{145}{108}\delta W_{\mathrm{p}}^{\mathrm{IV}} + \frac{351}{220}\delta W_{\mathrm{p}}^{\mathrm{V}} + \frac{1}{20}\delta W_{\mathrm{p}}^{\mathrm{VI}} \tag{35}$$

$$\hat{\sigma} = \sigma_0 + \frac{19}{216}\delta\sigma^{\mathrm{I}} + \frac{1000}{2079}\delta\sigma^{\mathrm{III}} - \frac{125}{216}\delta\sigma^{\mathrm{IV}} + \frac{81}{88}\delta\sigma^{\mathrm{V}} + \frac{5}{56}\delta\sigma^{\mathrm{VI}} \tag{36}$$

$$\hat{W}_{\mathrm{p}} = W_{\mathrm{p},0} + \frac{19}{216}\delta W_{\mathrm{p}}^{\mathrm{I}} + \frac{1000}{2079}\delta W_{\mathrm{p}}^{\mathrm{III}} - \frac{125}{216}\delta W_{\mathrm{p}}^{\mathrm{IV}} + \frac{81}{88}\delta W_{\mathrm{p}}^{\mathrm{V}} + \frac{5}{56}\delta W_{\mathrm{p}}^{\mathrm{VI}} \tag{37}$$

where

$$\delta\sigma^{\mathrm{I}} = \mathbf{C}^{\mathrm{ep}}(\sigma_0, W_{\mathrm{p},0})\delta\varepsilon \tag{38}$$

$$\delta W_{\mathrm{p}}^{\mathrm{I}} = \delta\lambda_{\mathrm{p}}(\sigma_0, W_{\mathrm{p},0}, \delta\varepsilon)\sigma_0\frac{\partial g}{\partial\sigma_0} \tag{39}$$

$$\delta\sigma^{\mathrm{II}} = \mathbf{C}^{\mathrm{ep}}(\sigma^{\mathrm{I}}, W_{\mathrm{p}}^{\mathrm{I}})\delta\varepsilon \tag{40}$$

$$\delta W_{\mathrm{p}}^{\mathrm{II}} = \delta\lambda_{\mathrm{p}}(\sigma^{\mathrm{I}}, W_{\mathrm{p}}^{\mathrm{I}}, \delta\varepsilon)\sigma^{\mathrm{I}}\frac{\partial g}{\partial\sigma^{\mathrm{I}}} \tag{41}$$

Initial state $\sigma_0, W_{p,0}, \Delta\varepsilon, \delta\varepsilon_0 = \Delta\varepsilon, q = 1$

Strain increment $i = 1, 2, \ldots, n$

$$\delta\varepsilon_i = \min\left[q\delta\varepsilon_{i-1}, 1 - \Sigma_{j=1}^{i-1}\delta\varepsilon_j\right]$$

do

$$\delta\sigma_i^I = \boldsymbol{C}^{ep}(\sigma_{i-1}, W_{p,i-1})\delta\varepsilon_i$$

$$\delta W_{p,i}^I = \delta\lambda_p(\sigma_{i-1}, W_{p,i-1}, \delta\varepsilon_i)\sigma_{i-1}\frac{\partial g}{\partial\sigma}$$

$$\delta\sigma_i^{II} = \boldsymbol{C}^{ep}(\sigma_{i-1} + \delta\sigma_i^I, W_{p,i-1} + \delta W_{p,i}^I)\delta\varepsilon_i$$

$$\delta W_{p,i}^{II} = \delta\lambda_p(\sigma_{i-1} + \delta\sigma_i^I, W_{p,i-1} + \delta W_{p,i}^I, \delta\varepsilon_i)(\sigma_{i-1} + \delta\sigma_i^I)\frac{\partial g}{\partial\sigma}$$

$$\delta\sigma_i^{I-II} \approx \tfrac{1}{2}(\delta\sigma_i^{II} - \delta\sigma_i^I)$$

$$\hat{\sigma}_i = \sigma_{i-1} + \tfrac{1}{2}(\delta\sigma_i^I + \delta\sigma_i^{II})$$

$$\hat{W}_{p,i} = W_{p,i-1} + \tfrac{1}{2}(\delta W_{p,i}^I + \delta W_{p,i}^{II})$$

$$\xi_i = \frac{\|\delta\sigma_i^{I-II}\|}{\|\hat{\sigma}_i\|}$$

$$\textit{If } \xi_i > \in \textit{ then } q = \max\left[0.9(\tfrac{\in}{\xi})^{\frac{1}{2}}, 0.01\right]; \delta\varepsilon_i = q\delta\varepsilon_i$$

Until $\xi_i \leq \in$

$$q = \min\left[0.9(\tfrac{\in}{\xi})^{\frac{1}{2}}, 2\right]$$

Stop strain incrementation when $\Sigma_{i=1}^{n}\delta\varepsilon_i = \Delta\varepsilon$

Final state $\hat{\sigma}_i, \hat{W}_{p,i}, \boldsymbol{C}^{ep}(\hat{\sigma}_i, \hat{W}_{p,i})$

Figure 8. Algorithm for modified Euler scheme with active error control.

in which

$$\sigma^{\mathrm{I}} = \sigma_0 + \tfrac{1}{5}\delta\sigma^{\mathrm{I}}$$

$$W_{\mathrm{p}}^{\mathrm{I}} = W_{\mathrm{p},0} + \tfrac{1}{5}\delta W_{\mathrm{p}}^{\mathrm{I}}$$

and further

$$\delta\sigma^{\mathrm{III}} = \mathbf{C}^{\mathrm{ep}}(\sigma^{\mathrm{II}}, W_{\mathrm{p}}^{\mathrm{II}})\delta\varepsilon \tag{42}$$

$$\delta W_{\mathrm{p}}^{\mathrm{III}} = \delta\lambda_{\mathrm{p}}(\sigma^{\mathrm{II}}, W_{\mathrm{p}}^{\mathrm{II}}, \delta\varepsilon)\sigma^{\mathrm{II}}\frac{\partial g}{\partial\sigma^{\mathrm{II}}} \tag{43}$$

in which

$$\sigma^{\text{II}} = \sigma_0 + \frac{3}{40}\delta\sigma^{\text{I}} + \frac{9}{40}\delta\sigma^{\text{II}}$$

$$W_{\text{p}}^{\text{II}} = W_{\text{p},0} + \frac{3}{40}\delta W_{\text{p}}^{\text{I}} + \frac{9}{40}\delta W_{\text{p}}^{\text{II}}$$

and further

$$\delta\sigma^{\text{IV}} = \mathbf{C}^{\text{ep}}(\sigma^{\text{III}}, W_{\text{p}}^{\text{III}})\delta\varepsilon \tag{44}$$

$$\delta W_{\text{p}}^{\text{IV}} = \delta\lambda_{\text{p}}(\sigma^{\text{III}}, W_{\text{p}}^{\text{III}}, \delta\varepsilon)\delta^{\text{III}}\frac{\partial g}{\partial\sigma^{\text{III}}} \tag{45}$$

in which

$$\sigma^{\text{III}} = \sigma_0 + \frac{3}{10}\delta\sigma^{\text{I}} - \frac{9}{10}\delta\sigma^{\text{II}} + \frac{6}{5}\delta\sigma^{\text{III}}$$

$$W_{\text{p}}^{\text{III}} = W_{\text{p},0} + \frac{3}{10}\delta W_{\text{p}}^{\text{I}} - \frac{9}{10}\delta W_{\text{p}}^{\text{II}} + \frac{6}{5}\delta W_{\text{p}}^{\text{III}}$$

and further

$$\delta\sigma^{\text{V}} = \mathbf{C}^{\text{ep}}(\sigma^{\text{IV}}, W_{\text{p}}^{\text{IV}})\delta\varepsilon \tag{46}$$

$$\delta W_{\text{p}}^{\text{V}} = \delta\lambda_{\text{p}}(\sigma^{\text{IV}}, W_{\text{p}}^{\text{IV}}, \delta\varepsilon)\delta^{\text{IV}}\frac{\partial g}{\partial\sigma^{\text{IV}}} \tag{47}$$

in which

$$\sigma^{\text{IV}} = \sigma_0 + \frac{226}{729}\delta\sigma^{\text{I}} - \frac{25}{27}\delta\sigma^{\text{II}} + \frac{880}{729}\delta\sigma^{\text{III}} + \frac{55}{729}\delta\sigma^{\text{IV}}$$

$$W_{\text{p}}^{\text{IV}} = W_{\text{p},0} + \frac{226}{729}\delta W_{\text{p}}^{\text{I}} - \frac{25}{27}\delta W_{\text{p}}^{\text{II}} + \frac{880}{729}\delta W_{\text{p}}^{\text{III}} + \frac{55}{729}\delta W_{\text{p}}^{\text{IV}}$$

and further

$$\delta\sigma^{\text{VI}} = \mathbf{C}^{\text{ep}}(\sigma^{\text{V}}, W_{\text{p}}^{\text{V}})\delta\varepsilon \tag{48}$$

$$\delta W_{\text{p}}^{\text{VI}} = \delta\lambda_{\text{p}}(\sigma^{\text{V}}, W_{\text{p}}^{\text{V}}, \delta\varepsilon)\sigma^{\text{V}}\frac{\partial g}{\partial\sigma^{\text{V}}} \tag{49}$$

in which

$$\sigma^{\text{V}} = \sigma_0 - \frac{181}{270}\delta\sigma^{\text{I}} + \frac{5}{2}\delta\sigma^{\text{II}} - \frac{266}{297}\delta\sigma^{\text{III}} - \frac{91}{27}\delta\sigma^{\text{IV}} + \frac{189}{55}\delta\sigma^{\text{V}}$$

$$W_{\text{p}}^{\text{V}} = W_{\text{p},0} - \frac{181}{270}\delta W_{\text{p}}^{\text{I}} + \frac{5}{2}\delta W_{\text{p}}^{\text{II}} - \frac{266}{297}\delta W_{\text{p}}^{\text{III}} - \frac{91}{27}\delta W_{\text{p}}^{\text{IV}} + \frac{189}{55}\delta W_{\text{p}}^{\text{V}}$$

Even though the integration process requires six evaluations of the elasto-plastic constitutive matrix, the scheme rapidly becomes competitive with the modified Euler scheme as the error tolerance is tightened. As for the modified Euler scheme, the estimated relative error can be expressed as

$$\xi = \frac{\|\hat{\sigma} - \sigma\|}{\|\hat{\sigma}\|} \tag{50}$$

As the local truncation error in the fourth-order formulae is $O(\Delta T^5)$ the exponent in (32) is replaced by $\frac{1}{5}$ and the factor that controls the size of the next dimensionless time step is instead given by:

$$q = 0.9\left(\frac{\epsilon}{\xi}\right)^{1/5} \tag{51}$$

But the constraints in (33) still apply. The Runge–Kutta–Dormand–Prince scheme with active error control is summarized in Figure 9.

8. COMPARISON OF INTEGRATION SCHEMES

The effectiveness of the integration schemes presented in Sections 5–7 is examined in the following. For this purpose a single point of Eastern Scheldt sand [16] with the parameters listed in Table I of the companion paper is undergoing triaxial compression under constant volume. The triaxial compression as started from an anisotropic stress state of $\sigma_0^{\mathrm{I}} = [450\ 400\ 400\ 0\ 0\ 0]$, with all stresses in kPa. The resulting stress path is somewhat simple, because it does not involve rotation of principal axes, but even a relatively small strain increment will nevertheless lead to considerable change in principal stresses. The simulation is conducted by imposing a number of strain increments of equal size $\Delta\varepsilon^{\mathrm{T}} = 10^{-4} \times [5\ \ -2.5\ \ -2.5\ 0\ 0\ 0]$ and using the three integration schemes for determination of the corresponding change in stresses. The resulting stress path and development in hardening parameter, shown in Figure 10, is composed of 40 sequential strain increments of equal size. All the shown simulations are performed without correction for yield surface drift.

Figure 10 shows a comparison of the calculated effective stress paths for the undrained triaxial compression of the sand element, for which the parameter values are listed in Table I in the companion paper. The modified Euler and Runge–Kutta–Dormand–Prince integration schemes yield similar results, whereas the forward Euler integration scheme produces visibly different results for decreasing values of σ_3. This deviation becomes less distinct on the linear part of the undrained stress path, but the forward Euler scheme generally overestimates the final stress state considerably. The applicability of the forward Euler scheme, however, improves as the number of subincrements in the integration process are increased. Hence, the discrepancy is more or less reduced by a factor of two as the number of subincrements are doubled. Similar observations hold for the development in the hardening parameter W_{p}, as indicated in Figure 10c.

The observations above illustrate the effect of integration method for rate independent material behaviour. A reduction of the size of the strain increments, $\Delta\varepsilon$, may:

1. Change the appearance of the stress path for all methods and lead to results that are in better agreement with the 'correct' solution.
2. Reduce the discrepancy between the integration schemes.
3. Decrease the computational efficiency.

Initial state $\sigma_0, W_{p,0}, \Delta\varepsilon, \delta\varepsilon_0 = \Delta\varepsilon, q = 1$

Strain increments $i = 1, 2, \ldots\ldots, n$

$\delta\varepsilon_i = \min\left[q\delta\varepsilon_{i-1}, 1 - \Sigma_{j=1}^{i-1}\delta\varepsilon_j\right]$

do

Stress increments $m = I, II, \ldots\ldots, VI$

$$\delta\sigma_i^m = C^{ep}(\sigma_i^{m-1}, W_{p,i}^{m-1})\delta\varepsilon_i$$

$$\delta W_{p,i}^m = \delta\lambda_p(\sigma_i^{m-1}, W_{p,i}^{m-1}, \delta\varepsilon_i)\sigma_i^{m-1}\frac{\partial g}{\partial\sigma_i^{m-1}}$$

Stop stress incrementation when $j = VI$

$$\sigma_i = \sigma_{i-1} + \tfrac{31}{540}\delta\sigma_i^I + \tfrac{190}{297}\delta\sigma_i^{III} - \tfrac{145}{108}\delta\sigma_i^{IV} + \tfrac{351}{220}\delta\sigma_i^V + \tfrac{1}{20}\delta\sigma_i^{VI}$$

$$W_{p,i} = W_{p,i-1} + \tfrac{31}{540}\delta W_{p,i}^I + \tfrac{190}{297}\delta W_{p,i}^{III} - \tfrac{145}{108}\delta W_{p,i}^{IV} + \tfrac{351}{220}\delta W_{p,i}^V + \tfrac{1}{20}\delta W_{p,i}^{VI}$$

$$\hat{\sigma}_i = \sigma_{i-1} + \tfrac{19}{216}\delta\sigma_i^I + \tfrac{1000}{2079}\delta\sigma_i^{III} - \tfrac{125}{216}\delta\sigma_i^{IV} + \tfrac{81}{88}\delta\sigma_i^V + \tfrac{5}{56}\delta\sigma_i^{VI}$$

$$\hat{W}_{p,i} = W_{p,i-1} + \tfrac{19}{216}\delta W_{p,i}^I + \tfrac{1000}{2079}\delta W_{p,i}^{III} - \tfrac{125}{216}\delta W_{p,i}^{IV} + \tfrac{81}{88}\delta W_{p,i}^V + \tfrac{5}{56}\delta W_{p,i}^{VI}$$

$$\xi_i = \frac{\|\hat{\sigma}_i - \sigma_i\|}{\|\hat{\sigma}_i\|}$$

If $\xi_i > \in$ *then* $q = \max\left[0.9(\frac{\in}{\xi})^{\frac{1}{5}}, 0.01\right]; \delta\varepsilon_i = q\delta\varepsilon_i$

Until $\xi_i \leq \in$

$q = \min\left[0.9(\frac{\in}{\xi})^{\frac{1}{5}}, 2\right]$

Stop strain incrementation when $\Sigma_{i=1}^n \delta\varepsilon_i = \Delta\varepsilon$

Final state $\hat{\sigma}_i, \hat{W}_{p,i}, C^{ep}(\hat{\sigma}_i, \hat{W}_{p,i})$

Figure 9. Algorithm for Runge–Kutta–Dormand–Prince scheme with error control.

The choice of the strain increment will therefore essentially depend on the required accuracy of the global solution, and the purpose of the integration schemes is merely to provide an accurate stress update for a given strain increment.

As already seen, it is possible to obtain similar results using the three different schemes, but another consideration involve their computational efficiencies. The computational cost is strongly related to the evaluation of the elasto-plastic constitutive matrix, and as described in

Table I. Comparison of results of integration of the single hardening model using three different schemes.

Method	E_{max}/ϵ			E_{avg}/ϵ		
	$\epsilon=10^{-3}$	$\epsilon=10^{-4}$	$\epsilon=10^{-5}$	$\epsilon=10^{-3}$	$\epsilon=10^{-4}$	$\epsilon=10^{-5}$
FE	1.16	1.16	1.16	0.99	1.00	1.00
	(0.30)	(0.34)	(0.31)	(0.22)	(0.22)	(0.23)
ME	0.26	0.41	0.45	0.15	0.22	0.24
	(0.41)	(0.44)	(0.45)	(0.22)	(0.25)	(0.26)
RKDP	0.04	0.04	0.13	0.02	0.03	0.10
	(0.04)	(0.04)	(0.11)	(0.01)	(0.01)	(0.08)

Method	Maximum number of subincrements			Relative CPU time		
	$\epsilon=10^{-3}$	$\epsilon=10^{-4}$	$\epsilon=10^{-5}$	$\epsilon=10^{-3}$	$\epsilon=10^{-4}$	$\epsilon=10^{-5}$
FE	72	710	7091	9.8	102.9	924.0
ME	10	29	89	0.8	1.6	4.0
RKDP	1	2	3	1.0	1.1	1.1

Notes: Values denoted by ( ) indicate errors in stresses after correction for yield surface drift. Forward Euler (FE). Modified Euler (ME), Runge–Kutta–Dormand–Prince (RKDP)

Sections 5–7, the forward Euler, the modified Euler, and the Runge–Kutta–Dormand–Prince schemes require one, two and six evaluations per subincrement, respectively. The computational costs are evaluated by performing a number of runs, in which the updated stresses must lie within some tolerance, ϵ, of a reference state. Since no analytical solution is available for integrating the relations of the single hardening model exactly, the reference stresses are computed by using the Runge–Kutta–Dormand–Prince scheme with 250 subincrements of equal size. The error in each strain increment for the different schemes can then be expressed as

$$E_i = \frac{\|\sigma_i - \sigma_{ref}\|}{\|\sigma_{ref}\|} \tag{52}$$

and the average and maximum errors along the stress path are given by:

$$E_{avg} = \frac{1}{40}\sum_{i=1}^{40} E_i \tag{53}$$

$$E_{max} = \max E_i \quad \text{for } 1 \leqslant i \leqslant 40 \tag{54}$$

The maximum number of subincrements within a single strain increment and the accumulated relative CPU time required for fulfilment of a given error tolerance along the stress path in Figure 10 are listed in Table I. The simulations have been performed with and without correction for yield surface drift. The CPU time is a measure of the computational cost due to the application of the user defined material module. For the forward Euler scheme the number of subincrements is adjusted until the average error equals the specified tolerance. The modified Euler and Runge–Kutta–Dormand–Prince schemes use the specified tolerance for adjustment of the size of the subincrements. The simulations have been performed both with and without correction for yield surface drift.

As indicated in Table I, the simulations reveal that the two higher order schemes have normalized errors that are substantially less than unity and are largely unaffected by correction

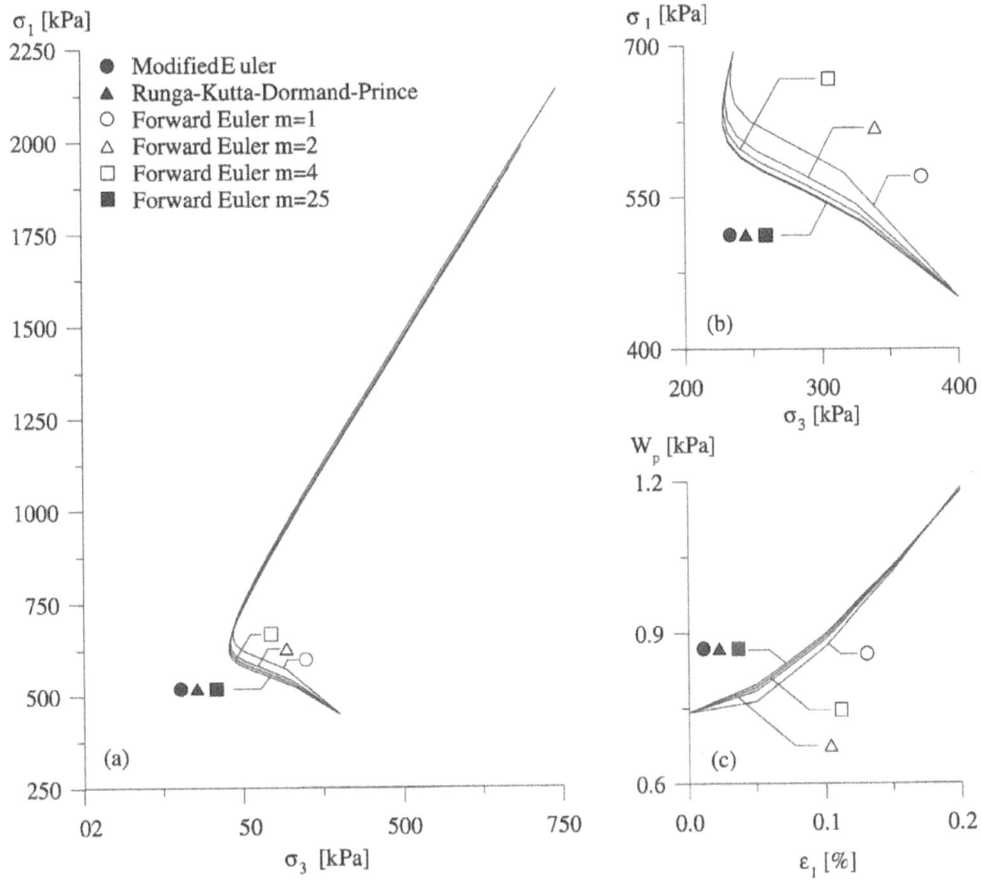

Figure 10. Comparison of various integration schemes used for determination of constant volume stress path: (a) stress path in $\sigma_3 - \sigma_1$ diagram, (b) segment of stress path, and (c) development of the hardening parameter W_p with the major principal strain.

for yield surface drift. However, the correction for yield surface drift becomes very important when using the forward Euler scheme, because it reduces the maximum and average errors in stresses by at least 70%. Moreover, if the subdivision of the strain increment is insufficient the correction for yield surface drift tends to diminish the overestimation of the final stress state shown in Figure 10.

As expected the forward Euler scheme shows a more or less proportional growth in the maximum number of subincrements and relative CPU time with the tightening of the error tolerance. For the modified Euler scheme the maximum number of subincrements and relative CPU time grows slightly with the reduction of the error tolerance, whereas the Runge–Kutta–Dormand–Prince scheme is barely affected. For all the schemes the first few increments (where the stress path is changing direction) are decisive for the maximum number of subincrements needed for fulfilment of a given error tolerance. Thus, for the two higher order schemes only a few of the imposed strain increments are subject to subincrementalization.

The Runge–Kutta–Dormand–Prince integration scheme is in general found to be superior to both the modified Euler and the forward Euler schemes in terms of both accuracy and computational costs as the error tolerance is tightened. Even the modified Euler scheme is only competitive for large error tolerances.

9. CONCLUSIONS

The procedures involved in implementation of a single hardening constitutive model for frictional materials have been employed in creation of a user-defined material module based on the incremental form of the model presented in a companion paper by Lade and Jakobsen [1]. The general calculation strategy inside this module is presented and discussed, including the initial intersection of the yield surface, the techniques for updating of stresses and hardening modulus, and correction for a possible yield surface drift. Several integration schemes are implemented in the module and their capabilities in relation to the three-dimensional constitutive model are evaluated. The forward Euler, modified Euler, and Runge–Kutta–Dormand–Prince integration schemes have been compared in view of error tolerances and computational efficiency. The modified Euler and the Runge–Kutta–Dormand–Prince schemes, which include active error control, posses the advantage of allowing greater strain increments to be imposed as subincrementalization automatically is performed if needed for fulfilment of a specific error tolerance. The traditional forward Euler scheme is, due to the principle of a fixed number of subincrements, not competitive with the higher order schemes, because too many subincrements are needed in order to obtain the required accuracy. However, the capability of the forward Euler scheme can be improved by adopting an iterative scheme correcting for a possible yield surface drift.

The present study shows that the Runge–Kutta–Dormand–Prince scheme is superior in terms of both accuracy and computational costs, and it is conclusively the most advantageous integration scheme for this advanced constitutive model.

REFERENCES

1. Lade PV, Jakobsen KP. Incrementalization of a single hardening constitutive model for frictional materials. *International Journal for Numerical and Analytical Methods in Geomechanics* 2002; **26**:647.
2. Simo JC, Taylor RL. Consistent tangent operator for rate-independent elastoplasticity. *Computer Methods in Applied Mechanics and Engineering* 1985; **48**:101–118.
3. Crisfield MA. *Non-linear Finite Element Analysis of Solids and Structures*. Wiley: Chichester, U.K. 1991.
4. ABAQUS *version 5.5—manuals*. Hibbit, Karlson and Sorensen Inc., 1995.
5. Sloan SW. Substepping schemes for the numerical integration of elasto-plastic stress–strain relations. *International Journal for Numerical Methods in Engineering* 1987; **24**:893–911.
6. Chen WF, Mizuno E. *Non-linear Analysis in Soil Mechanics*. Elsevier: New York, 1990.
7. Ortiz M, Simo JC. An analysis of a new class of integration algorithms for elastoplastic constitutive relations. *International Journal for Numerical Methods in Engineering* 1986; **23**:353–366.
8. Abbo AJ, Sloan SW. Backward Euler and subincrementation schemes in computational plasticity. In *Computational Plasticity*, Valliapan et al. (eds). Balkema: Rotterdam, 1993.
9. Smith IM, Griffiths DV. *Programming the Finite Element Method* (2nd edn). Wiley: New York, 1988.
10. Zienkiewicz OC, Taylor RL. *The Finite Element Method—Volume 2: Solid and Fluid Dynamics and Non-linearity*. McGraw-Hill: London, U.K. 1991.
11. Potts DM, Gens A. A critical assessment of methods of correcting for drift from the yield surface in elasto-plastic finite element analysis. *International Journal for Numerical and Analytical Methods in Geomechanics* 1985; **9**:149–159.
12. Sloan SW, Booker JR. Integration of Tresca and Mohr–Coulomb constitutive relations in plane strain elastoplasticity. *International Journal for Numerical Methods in Engineering* 1992; **33**:163–196.

13. England R. Error estimate for Runge–Kutta type solutions to systems of ordinary differential equations. *Computer Journal* 1969; **12**:166–170.
14. Fehlberg E. Klassiche Runge–Kutta formeln vierter und niedrigerer ordnung mit schrittweiten-kontrole und ihre anwendungen auf wärmleitungs-probleme. *Computing* 1970; **6**:61–71.
15. Dormand JR, Prince PJ. A family of embedded Runge–Kutta formulae. *Journal of Computer Applied Mathematics* 1980; **6**:19–26.
16. Jakobsen KP, Praastrup U. Drained triaxial tests on Eastern Scheldt sand. *Aalborg University Geotechnical Engineering Papers*, 1998; ISSN 1398-6465 R9822.

Chapter XIII
Time Effects

Preface

Time effects, which relate to crushing in sands, play important roles both in experimentation and in constitutive modeling. The original development of the Single Hardening model was based on experiments on quartz sand, which show very little crushing at lower confining pressures, so the development was free of these time effects. However, the time effects showed up in other experiments in which crushing were clearly important. This also affected the determination of material parameters for other sands. The time effects such as creep and stress relaxation are explained in the following, and their modeling can be handled by simply modifying the plastic work by addition of a multiplier, as shown in the second paper.

SOILS AND FOUNDATIONS Vol. 50, No. 5, 573–583, Oct. 2010
Japanese Geotechnical Society

STATIC FATIGUE CONTROLS PARTICLE CRUSHING AND TIME EFFECTS IN GRANULAR MATERIALS

Poul V. Lade[i] and Hamid Karimpour[ii]

ABSTRACT

Based on observations from constant strain rate experiments and from creep and stress relaxation experiments initiated at different stress levels it is found that sand exhibits patterns of time effects different from those observed in clays. It appears that time effects in sand may be associated with crushing of particles, and a mechanistic picture of time effects in granular materials is constructed in which time effects depend on interparticle friction, grain crushing and grain rearrangement. This mechanistic picture is based on measured behavior in drained triaxial compression tests on three different sands in which strain rate effects are observed as small to negligible. While creep and relaxation are caused by the same underlying phenomenon, it appears that results of creep tests cannot be obtained from results of relaxations tests, and vice versa. The phenomenon of static fatigue of individual particles seems to be at the root of time effects in sand. A review of previous studies of static fatigue is presented. Triaxial tests on a beach sand incorporating creep and stress relaxation are followed by grain size analysis to prove that grain crushing relate to the observed time effects. Additional triaxial tests are presented in which the effect of water is demonstrated in support of the static fatigue mechanism. Load-controlled tests on individual sand particles in the form of spherical glass beads (quartz) were performed by maintaining constant loads lower than the short term crushing loads. As do rock and concrete specimens in triaxial compression, the glass beads show effects of time to crushing.

Key words: creep, sand, static fatigue, stress relaxation, time dependence, triaxial compression (**IGC**: D6)

INTRODUCTION

Time dependent behavior of granular materials is quite different from the viscous behavior observed in clays. The effects of strain-rate, creep and stress relaxation of clays follow a classic pattern of viscous behavior observed for most materials. For such materials the stiffness and strength increase with increasing strain rate and phenomena such as creep, relaxation and strain rate effects are governed by the same basic mechanism. This behavior is denoted as "isotach" behavior, i.e. there is a unique stress-strain-strain rate relation for a given clay. For such materials, the creep properties may be obtained from, say, a triaxial compression test and used for prediction of stress relaxation in another experiment.

For granular materials the effects of strain rate or loading rate are observed to be small to negligible, and while creep and stress relaxation are caused by the same phenomenon, namely grain crushing followed by grain rearrangement, the prediction of one phenomenon can apparently not be accomplished on the basis of the other. Such behavior is referred to as "nonisotach" behavior. The role of grain crushing is explained, and the transfer of forces through the grain structure is different in creep and relaxation tests. While each phenomenon follows similar patterns, the strains produced after one day of creep does not produce a result that correlates with the changes in stress due to one day of stress relaxation.

To throw further light on the effects of time on the behavior of sand, and as background for development of constitutive models incorporating time effects, presented here is a study of strain rate effects, creep, and stress relaxation in crushed coral sand. A mechanistic picture is introduced in which static fatigue of individual grains account for the time effects observed in experimental studies. Additional triaxial tests on a beach sand incorporating creep and stress relaxation are followed by grain size analyses to prove that grain crushing relate to the observed time effects. Further, a triaxial test is presented in which the effect of introducing water into an initially dry specimen is presented in support of the static fatigue mechanism.

PREVIOUS STUDIES

Comprehensive reviews of time dependent behavior of soils and models for characterization of this behavior have recently been presented in the literature (Augustesen

i) Professor, Department of Civil Engineering, The Catholic University of America, Washington, D.C., U.S.A. (LADE@cua.edu).
ii) Graduate Student, ditto.
The manuscript was received on August 24, 2009 for publication in the Special Issue on "IS-Kyoto International Symposium on Prediction and Simulation Methods for Geohazard Mitigation" published in Vol. 50, No. 4.
Written discussions on this paper should be submitted before May 1, 2011 to the Japanese Geotechnical Society, 4-38-2, Sengoku, Bunkyo-ku, Tokyo 112-0011, Japan. Upon request the closing date may be extended one month.

et al., 2004; Liingaard et al., 2004). The essence of these reviews is that clay and sand behave differently with respect to time. They show that strain rate has important influence on the stress-strain behavior of clay, while widely different strain rates produce essentially the same stress-strain relation for sand, as seen in experiments presented by e.g., Tatsuoka et al. (2002), Kuwano and Jardine (2002), and Kiyota and Tatsuoka (2006). Changes in strain rate have permanent effects in clay, where switches from one to another stress-strain curve occur in response to changes in strain rate. Only temporary changes occur in the stress-strain relations for sand, as also observed by the authors listed above.

Observed behavior shows (Augustesen et al., 2004) that the phenomena of creep and stress relaxation are also different in clay and sand. For clay, creep and relaxation properties can be obtained from constant rate of strain tests, and vice versa, as shown by e.g., Leroueil and Marques (1996). The fact that creep, relaxation and strain rate effects can be modeled by the same basic viscous mechanism is referred to as "isotach" behavior.

An investigation of strain rate effects in dense Cambria sand under drained and undrained conditions at high pressures performed by Yamamuro and Lade (1993) showed no significant rate effects on the stress-strain relations. The Hostun and Toyoura sands tested by Matsushita et al. (1999) exhibited noticeable amounts of creep and relaxation but no strain rate effects. This led to one of the main conclusions: The phenomena of creep and relaxation cannot be predicted based on results of constant rate of strain tests. This is because the changes in stress-strain relations due to changes in strain rate are temporary. This behavior of sand is labeled "nonisotach" behavior. Note that Tatsuoka et al. (2008) have proposed additional sub-classifications for different types of nonisotach behavior of granular materials.

To provide a more comprehensive background for development of an appropriate constitutive model for time effects in sand, which shows nonisotach behavior, experiments have been performed to study time effects in sand.

EXPERIMENTAL STUDIES OF SANDS

The initial studies of time-dependent behavior of sand were conducted in a conventional triaxial apparatus. Modifications to this equipment were made to improve its capability to carry out long-term tests with steady stresses and accurate measurements at a constant temperature. Mechanical equipment with negligible drift in applied pressures and loads and measurement systems without zero drift or devices in which the zero position could be verified during experiments were employed for all testing. The triaxial equipment, the loading systems, the deformation measurement systems, and the temperature control were explained by Lade and Liu (1998).

The sand tested was crushed coral sand. The gradation consisted of grain sizes between the No. 30 and No. 140 U.S. sieves (0.60 to 0.106 mm) with a nearly straight line gradation between these two sizes. The maximum and minimum void ratios were 1.22 and 0.70. The specific gravity of sand grains was 2.88. The tests on crushed coral sand were performed on specimens with a relative density of 60% corresponding to a void ratio of 0.91. The experiments on crushed coral sand have been presented by Lade et al. (2009, 2010), and a few results relating to the present issue are given below.

Additional triaxial tests have been performed on Virginia Beach sand at high confining pressures to study particle crushing and effects of water, as implied in the phenomenon of static fatigue. The portion contained between the No. 20 and No. 40 U.S. sieves (0.850 mm to 0.425 mm) was used in these tests so that crushing would be easily visible and result in identifiable changes in the gradation curve. The sand consists of subangular to subrounded quartz grains with mean diameter = 0.638 mm; coefficient of uniformity = 1.40; specific gravity = 2.65; and maximum and minimum void ratios = 0.759 and 0.532, respectively. Tests were performed on Virginia Beach sand at a relative density of 100%.

EXPERIMENTS ON CRUSHED CORAL SAND

Strain Rate Effects

Triaxial compression tests were performed on crushed coral sand with an effective confining pressure of 200 kPa and with five different, constant axial strain rates varying from 0.00665%/min to 1.70%/min, corresponding to a

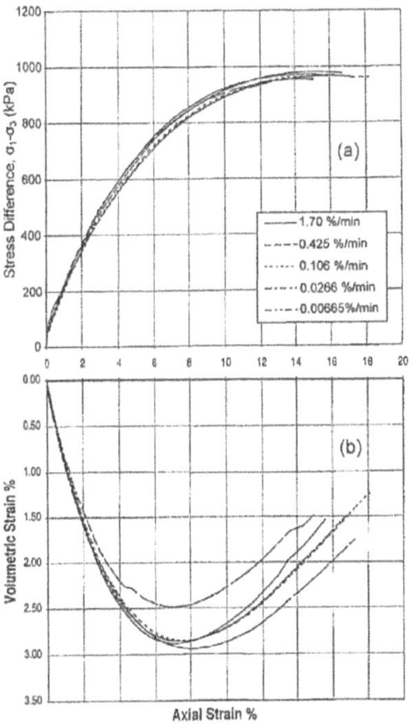

Fig. 1. Comparison of (a) stress-strain and (b) volume change relations for drained triaxial compression tests on crushed coral sand performed with five different strain rates

256-fold increase in strain rate. The results of these tests are shown in Figs. 1(a) and (b). They indicate that the influence of strain rate on the characteristics of the stress-strain and volume change curves is small and within the scatter of such results. Several experiments were performed with each of the five strain rates and the curves shown in Fig. 1 are those deviating most from each other. Note that the scale on the volumetric strain axis is approximately 5 times larger than the axial strain scale. Thus, deviations in volumetric strains are accentuated to demonstrate the magnitude of scatter in the worst case. Therefore, the experimental results confirm that the slopes of the curves, the volumetric strains, as well as the strengths are very little affected by the strain rate. Similar results for sand have been found by Yamamuro and Lade (1993) and by Matsushita et al. (1999). This departure from classic time-dependent behavior, according to which the stiffness and the strength increase with increasing strain rate, is significant, because it indicates that it may not be possible to employ conventional viscous type models to capture the time-dependent behavior of sands. Such models have been successfully used to characterize a number of other materials, including soils such as clays (*see* Liingaard et al. (2004) for comprehensive review of time effect models).

Creep Experiments

Conventional creep experiments were performed after the specimen had been sheared under load control corresponding to the middle strain rate of 0.106%/min. The behavior obtained under load control is slightly different from that obtained under deformation control, as discussed in detail by Lade et al. (2009). Once the desired deviator stresses of 500, 700, and 900 kPa had been reached, the specimen was allowed to creep for approximately one day (= 1440 min). After the creep stage, the deviator stress was again increased sufficiently to join the virgin or primary stress-strain curve before another creep test was initiated.

Figure 2(a) shows the stress-strain and volume change curves, and superimposed on these diagrams are the results of corresponding load controlled experiment. As creep proceeds at a given stress, the plastic yield surface moves out to higher stresses. This may be seen from the fact that further loading first produces what appears to be elastic reloading.

The volume change curves corresponding to creep, shown in Fig. 2(b), do not follow the reference curve, unlike the previous experiments on Antelope Valley sand presented by Lade (2007). This means that the potential for inelastic creep strains cannot be taken to be the same as the potential for plastic strains, as was the case for the Antelope Valley sand. The potential for inelastic strains for crushed coral sand must be inclined such that the creep volumetric strains are more contractive than those obtained from the plastic potential at the same stress point.

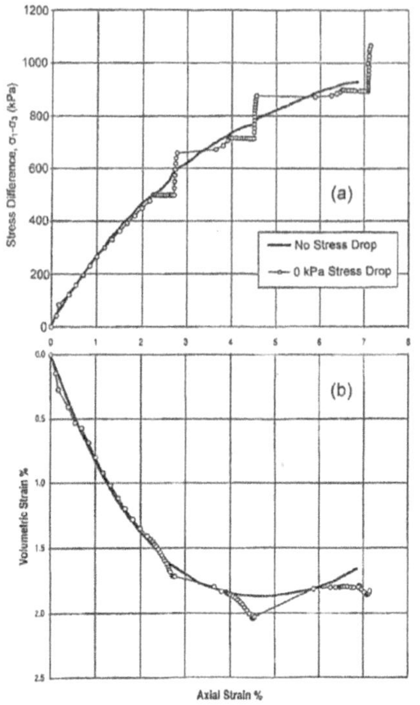

Fig. 2. **Comparison of results from load controlled tests with and without creep stages: (a) stress-strain and (b) volume change relations from drained triaxial compression test on crushed coral sand with creep stages of 1440 min at stress differences of 500, 700, and 900 kPa**

Stress Relaxation Experiments

Stress relaxation experiments were performed after primary loading with a strain rate of 0.106%/min. Once the desired stress differences of 500, 700, and 900 kPa had been reached, the axial deformation was held constant to observe the stress difference relax. Relaxation periods of 1,000 min (i.e., a little less than one day) were employed in all but a few experiments in which longer relaxation periods were used. After the stress relaxation stage, the stress difference was again increased sufficiently to join the virgin or primary stress-strain curve before another relaxation test was initiated.

Figure 3(a) shows the stress-strain relations obtained from the basic experiment performed with stress relaxation at the three desired stress differences and at a stress point beyond peak failure. Reloading after stress relaxation exhibits structuration effects similar to those observed after periods of creep, i.e., a temporary increase in the deviator strength beyond that corresponding to the primary loading curve. In all cases, the stress-strain curve appears to unite with or become the primary stress-strain relation well before the next stress relaxation point is reached.

Figure 3(b) shows that volumetric contraction of the specimen is associated with relaxation of the stress difference, whether or not the specimen is contracting or dilat-

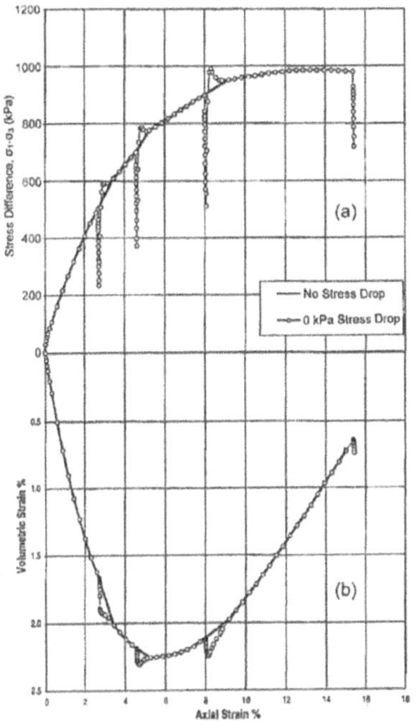

Fig. 3. Comparison with results from deformation controlled test with and without stress relaxation stages: (a) stress-strain and (b) volume change relations from drained triaxial compression test on crushed coral sand with stress relaxation stages of 1,000 min at stress differences of 500, 700, and 900 kPa

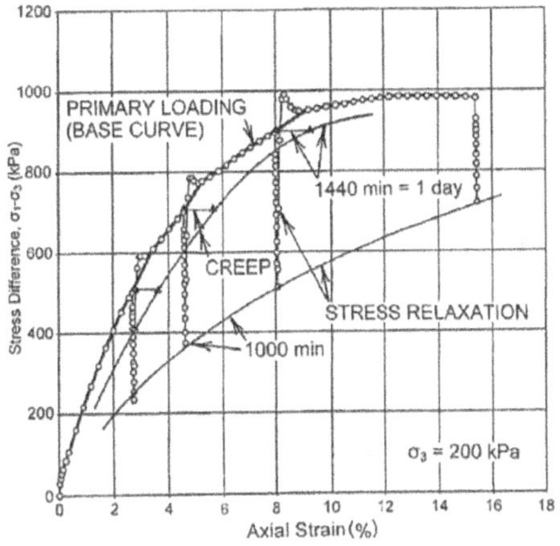

Fig. 4. Comparison of creep and stress relaxation experiments performed in two triaxial compression tests on crushed coral sand

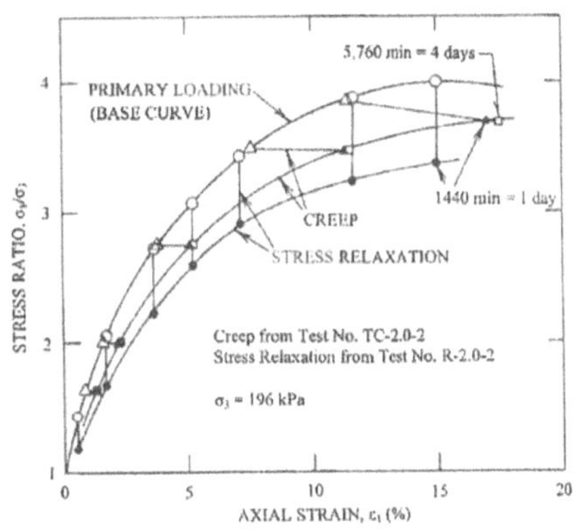

Fig. 5. Comparison of creep and stress relaxation experiments performed in two triaxial compression tests on Antelope Valley sand

ing during primary loading. These periods of volumetric contraction produce offsets in the volume change curve and like the stress-strain relation, the volume changes also seem to recover to the basic volume change curve upon reloading.

Comparison of Creep and Stress Relaxation

The stress relaxation may be compared with the creep observations by plotting the points of initiation and the end points after a certain amount of time on the same diagram. To overcome the small differences in the primary stress-strain curves from the two experiments in Figs. 2 and 3, the stress-strain curve shown in Fig. 3(a) is used as the base curve from which creep and stress relaxation are initiated.

The comparison of stress relaxation and creep after one day is shown in Fig. 4 in which the points of initiation of creep have been located on this base curve, while the end points obtained after approximately 1 day of creep are shown relative to the initiation points. The data from both types of tests are very consistent, and they show how much the axial strains change due to creep and how much the axial stresses change due to relaxation, respectively. It is clear that the amount of creep and the amount of relaxation resulting after 1 day define curves that are located at quite different positions.

Figure 5 shows a comparison of creep and stress relaxation effects after one day plotted from a common stress-strain curve for comparable experiments performed to study time effect on Antelope Valley sand (Lade, 2007). As for the crushed coral sand, the experiments on Antelope Valley sand showed that strain rate effects are negligible, and Fig. 5 indicates that the observed stress relaxation behavior does not correspond with the measured creep behavior.

To avoid interference with the creep process during the experiments, the deviator loads were constant and not adjusted according to the cross-sectional area to achieve constant deviator stresses. Thus, both Figs. 4 and 5 show branches of creep with slightly decreasing deviator stresses.

EXPERIMENTS ON VIRGINIA BEACH SAND

Comparison of Creep and Stress Relaxation

It is proposed herein that particle crushing is at the root of time effects in granular materials through the

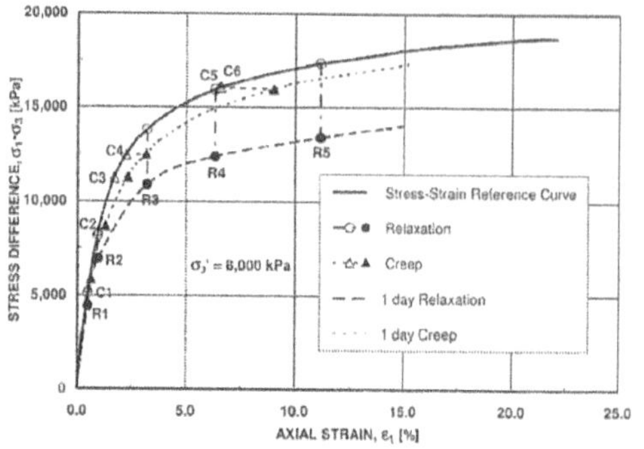

Fig. 6. Comparison of creep and stress relaxation experiments performed in triaxial compression tests on Virginia Beach sand

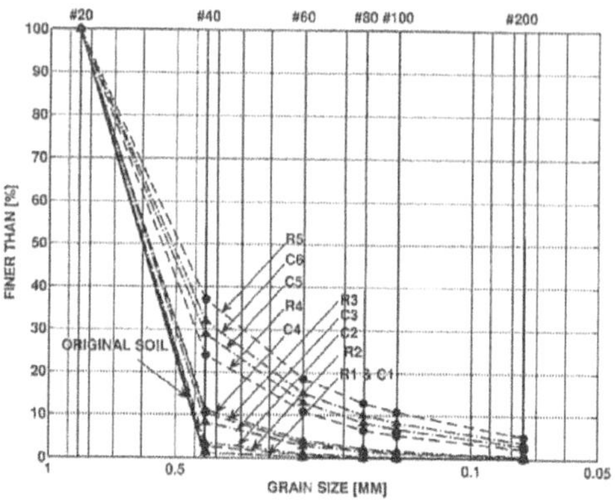

Fig. 7. Grain size distribution curves obtained after creep and stress relaxation experiments performed in triaxial compression tests on Virginia Beach sand

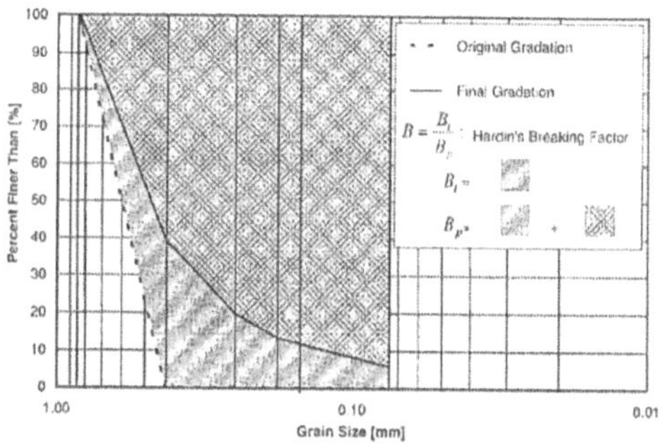

Fig. 8. Definition of particle breakage factor proposed by Hardin (1985)

phenomenon of static fatigue. However, sieve analysis of the tested crushed coral sand and Antelope Valley sand were not performed, because such sieve analyses could not be trusted as explained below. Therefore, a study of the much stronger Virginia Beach sand, which consists of subangular to subrounded quartz particles, was conducted and sieve analyses were performed to prove that particle crushing relate directly to the observed time effects. The details of the equipment, specimen preparation and testing procedures are given by Karimpour and Lade (2010). The experiments were performed in a high pressure triaxial cell with an effective confining pressure of 8,000 kPa at which pressure significant crushing occurs in the Virginia Beach sand.

Similar to observations for crushed coral sand (*see* Fig. 1) and Antelope Valley sand, experiments on Virginia Beach sand also indicated that strain rate effects on stress-strain and strength behavior are small (Karimpour and Lade, 2010). Figure 6 shows a comparison of creep and stress relaxation effects in Virginia Beach sand. Each of these experiments were performed by shearing a freshly deposited specimens at a constant strain rate of 0.0416% /min up to the point at which the specimens were allowed to creep or to stress relax for one day. The results are plotted from a common stress-strain curve and the initial or final points of each creep and stress relaxation test are indicated by an identification number. This allowed the grain size distribution to be determined at the end of each creep and stress relaxation test.

Relations between Particle Crushing, Energy Input and Time Effects

The grain size distribution curves are shown in Fig. 7 for each of the experiments shown in Fig. 6. It is clear that considerable particle crushing occurs in the tests, and some of the observed crushing relates to creep, as explained in the following. It should be noted that negligible amounts of particle breakage occurs during stress relaxation for which negligible strains occur (Karimpour and Lade, 2010) and the solid circles in Fig. 7 therefore correspond to the breakage that occurred during shearing up to the point at which the stress relaxation is initiated.

Figure 7 shows that the amounts of particle breakage are related to the locations along the stress-strain curve at which the creep and stress relaxation are terminated. To quantify this relationship, the particle breakage is represented by Hardin's particle breakage factor (Hardin, 1985) and the locations at which creep and stress relaxation are terminated is represented by the energy input per unit volume of the specimen.

Hardin's particle breakage factor is defined as:

$$B=\frac{B_p}{B_t} \tag{1}$$

in which B_t is the total breakage represented by the area between the original and the final gradation curves, as shown in Fig. 8, and B_p is the breakage potential represented by the area over the original grain size curve and limited to U.S. sieve No. 200.

The energy input per unit volume is calculated from

$$E=\Sigma\sigma_c'\cdot\Delta\varepsilon_v+\Sigma(\sigma_1-\sigma_3)\cdot\Delta\varepsilon_a \tag{2}$$

in which σ_c' is the effective confining pressure, $\Delta\varepsilon_v$ is an increment in volumetric strain, $(\sigma_1-\sigma_3)$ is the deviator stress, and $\Delta\varepsilon_a$ is an increment in axial strain.

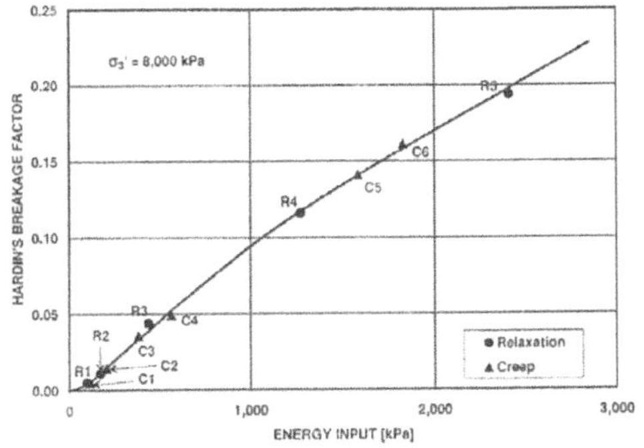

Fig. 9. Relation between Hardin's particle breakage factor and input energy per unit volume of triaxial compression specimens of Virginia Beach sand performed with a confining pressure of 8,000 kPa

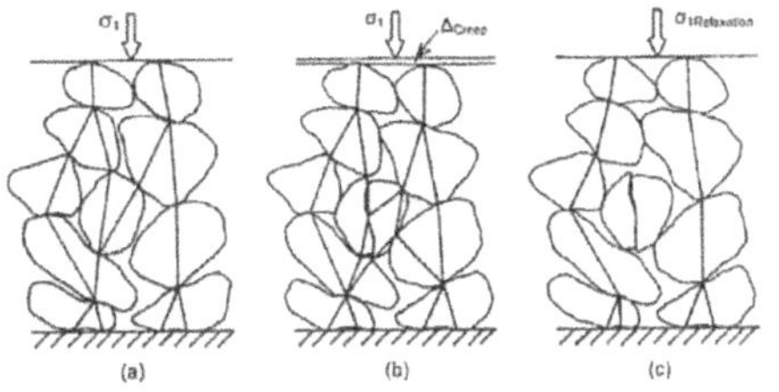

Fig. 10. (a) Initial force chains in particle structure and effects of grain crushing in (b) creep test and in (c) stress relaxation test

Figure 9 shows the relation between Hardin's particle breakage and the energy input per unit volume for the experiments shown in Fig. 6. A unique and almost linear relation is observed between the amount of particle breakage and the energy input per unit volume to the specimens, which were all sheared at an effective confining pressure of 8,000 kPa. While stress relaxation occurs with no strains and therefore relates to zero energy input, a portion of the energy input relates to the time-dependent phenomenon of creep. It is therefore evident that time effects are related to the crushing of particles. This will be further elaborated below, where the phenomenon of static fatigue is reviewed.

Figure 6 indicates that the observed stress relaxation behavior does not correspond with the measured creep behavior. It is concluded that none of the three sands exhibits classic viscous effects. Note also that the amounts of disagreement between the stress-strain relations after one day of creep or stress relaxation are quite different for the three sands. The fact that the same basic mechanism can account for creep, stress relaxation, and rate dependency and can serve as basis for prediction of one from the other, as is the case for clays, indicates that the material complies with the "correspondence principle" according to Sheahan and Kaliakin (1999). The experiments presented here showed noticeable amounts of creep and relaxation but no strain rate effects. Further, the stress relaxation and the creep responses do not appear to follow the correspondence principle, i.e., two different stress-strain relations are obtained after 1 day, as indicated for three different sands in Figs. 4, 5 and 6. Thus, it appears that the phenomena of creep, stress relaxation and strain rate effects in sand cannot be predicted from the same type of test using a viscous type model.

PROPOSED MECHANISTIC PICTURE OF TIME EFFECTS IN SANDS

Particle breakage has often been observed to be associated with time effects in granular materials, and a mechanistic picture of time effects may be constructed on the basis of this phenomenon. Particle breakage may not occur and time effects are negligible in granular materials at very low stresses. Time effects become significant with increasing confining pressure and increasing stress difference, as has been observed in several studies (*see* Yamamuro and Lade, 1993). These studies also noted the association between particle crushing and its occurrence with time.

Figure 10(a) shows an assembly of grains that have been loaded up to a given stress difference and either creep or stress relaxation occurs from this point. The diagram shows the force chains down through the assembly. The grain in the middle fractures in the beginning of either of these two types of time effects. The responses of the grain assembly are quite different for the two phenomena. Figure 10(b) shows what happens during creep in which the vertical stress is held constant. The assembly adjusts its structure to carry the vertical stress. This requires adjustment of the grains and it results in some vertical deformation and new force chains are created to match the externally applied stress. The redundancy in the grain structure allows new force chains to be created and engage other grains that may break. But slowly the amount of breakage will reduce and the creep will slow down with time, just as observed in the experiments.

Figure 10(c) shows what happens in the stress relaxation experiment. After the grain has broken, the grain structure is not able to carry the vertical stress, but since the assembly is prevented from vertical deformation, the stress relaxes. New force chains are created around the broken grain which does not carry any load. It is the small amount of grain movement in the creep tests that allows new contacts to be created and forces to be carried through the grain skeleton. Without this adjustment and consequent deformation to achieve the adjustment, the grain structure is able to transmit only a reduced load and stress relaxation is the consequence. It can also be seen that if a lower limit to the relaxed stress exists, then it depends on the grain strength rather than its frictional properties. Thus, the amount of difference between the creep and stress relaxation for the three sands shown in Figs. 4, 5 and 6 may be explained in terms of and relates primarily to different grain strengths.

It may be seen that a relation between creep and stress relaxation does not exist, because the two explanations do not allow a transition from one to the other phenomenon. Thus, Figs. 4, 5, and 6 show that the amounts of creep and the amounts of stress relaxation after one day do not converge towards the same curves. In fact, the two sets of curves are quite different and one cannot be obtained from the other. This nonisotach behavior is quite different from that of the isotach, viscous behavior exhibited by clays.

With only friction (and slippage when the frictional resistance is overcome) and particle breakage (when the strengths of the particles are overcome) as basic behavior constituents, how can the observed time effects be explained for granular materials? Experiments on rock specimens have clearly shown that their strengths are strongly dependent on time. Crushing of single sand particles indicates that they behave similar to rock specimens in the sense that their crushing strengths are time-dependent. This phenomenon is referred to as static fatigue or delayed fracture.

STATIC FATIGUE

Static fatigue is a phenomenon that leads to fracture and crushing of individual soil particles. Brittle fracture of materials such as quartz, feldspar, concrete and rock are similar to those of glasses and other ceramics, all brittle materials, in which fracture occurs due to time-dependent crack propagation and with negligible deformation prior to fracture. Sustained static loading of a soil particle may eventually lead to fracture at stresses considerably smaller than the short-term strength (Lawn, 1993; Lemaitre and Chaboche, 1994; Suresh, 1998; Callister, 2005).

Internal Microcracks

Rock materials, from which most soil particles derive, contain a distribution of flaws or microcracks caused by their formation history. In igneous rocks, residual stresses are created at grain boundaries and at interfaces between grains with dissimilar thermal expansion coefficients and with dissimilar elastic properties, and they produce a distribution of microcracking during cooling from the high formation temperatures of the magma. Similar microcracks are formed in metamorphic rocks as the exposure of the parent material to high pressures and temperatures is reduced as they emerge and are exposed to mechanical and chemical weathering at the Earth's surface. Sedimentary rocks are full of microcracks between the grains and their boundaries with cementing agents. In addition, pores and gas bubbles may be included in any type of rock. These pre-existing flaws serve as potential sites of nucleation and further propagation of major cracks.

As a result, the strengths of rock specimens vary with size as described by Weibull (1951), who observed that the strength of brittle solids is a statistical phenomenon caused by a pre-existing distribution of flaws in specimens with the same shape:

$$\sigma_{1ult} = \sigma_{2ult} \cdot \left(\frac{V_2}{V_1} \right)^{1/m} \tag{1}$$

in which σ_{1ult} is the strength of a particle with volume V_1 and σ_{2ult} is the strength a particle with similar shape and volume V_2. The value of the exponent m is in the order of 6–18 for different types of coal (Jaeger and Cook, 1969). Thus, larger particles are weaker than smaller particles, because larger particles are more likely to contain critically oriented microcracks and are therefore exposed to fracture at lower stresses than smaller particles. Considerable scatter occurs in the fracture of brittle solids and statistical treatment of experimental results is necessary to achieve sensible results.

Single particle crushing has been studied in a number of experimental investigations performed in recent years (e.g., McDowell and Bolton, 1998; Nakata et al., 1999, 2001a, b; Tang et al., 2001; Kou et al., 2001; Bolton et al., 2008). In a study of the fracture strength of spherical sodium lime glass beads with diameters of 2, 3, and 6 mm, the expression in Eq. (1) was found to describe the strength variation with sphere volume very well, and the value of the exponent m was determined to be $m=3$ (Lade and Karimpour 2010).

Surface Cracks

In addition to the internal microscopic flaws, surface cracks may be produced near particle contacts (Suresh, 1998) due to (1) stationary normal loading, which produces elastic indentation and tensile stresses parallel to the surface outside the actual contact, (2) partial slip and complete sliding at the particle contacts, which results in non-uniform shear stress distribution and tearing parallel to the surface, and (3) rolling of particles, which results in varying normal and shear stresses underneath the contact point. Thus, surface cracks may be produced as explained by contact mechanics and tensile fracture of the particles while the soil mass is compressed and sheared.

The surfaces of granular soil particles are also known to be rough such that contacts between particles occur at extremely small asperities. At the atomic level the particle surfaces are very uneven and full of rough features at which stress concentrations may produce additional cracks.

Effects of Blasting

Rockfill is produced by blasting of solid rock and this process results in very angular particles with a large number of surface cracks created by the stress waves, especially by the tensile stress through following the initial compressive wave from the blast. Surface cracks play an important role in static fatigue, because the environment, especially the presence of humidity or water, in which the soil is present, has great influence on the speed of crack propagation, as explained below.

Loading of Brittle Solids

As a brittle solid is loaded in shear, it goes through

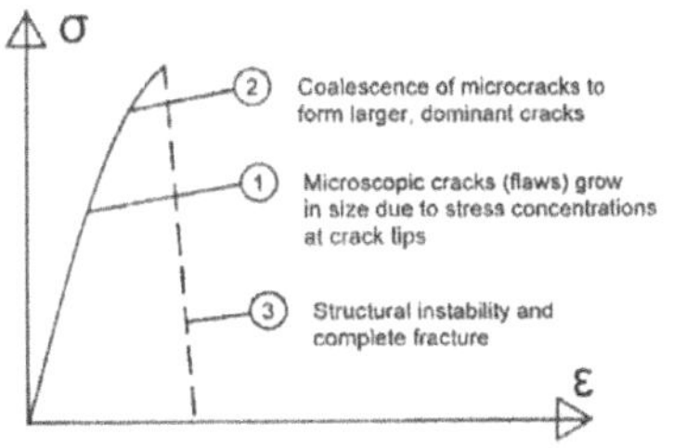

Fig. 11 **Three-stage fracture process in soil particle loaded in compression (after Van Mier, 2009)**

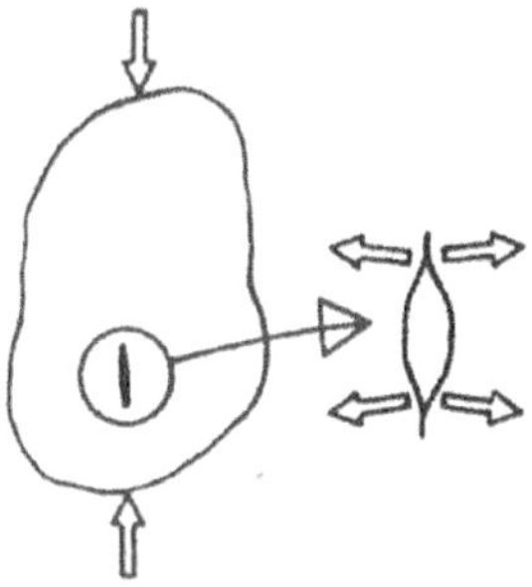

Fig. 12. **Fracture process in soil particle involving tensile stresses near crack tips**

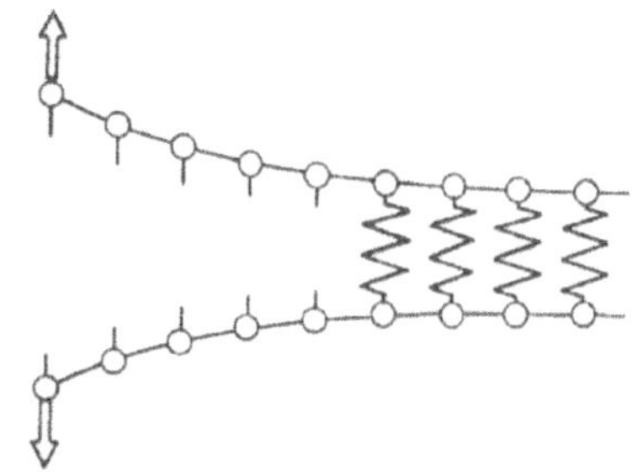

Fig. 13. **One-dimensional model of sharp crack with nonlinear crack-tip bond (after Lawn, 1993)**

several stages, indicated in Fig. 11 (after van Mier, 2009). During the first stage the microscopic flaws grow in size due to stress concentrations at the highly stressed regions around the crack tips, as shown schematically in Fig. 12. During the second stage the microcracks begin to coalesce to form larger cracks which eventually lead to catastrophic fracture of the specimen. During this second stage, which is basically stable in the sense that additional load can be added, the propagation of the larger cracks occurs at an accelerating rate with very little accompanying plastic deformation. The third stage consists of structural instability and complete fracture of the specimen. Static fatigue refers to the crack growth that occurs under sustained loading and that leads to the catastrophic fracture in the third stage. The crack growth occurs at the sharp ends of the cracks due to stress corrosion, and it is primarily the strength of the atomic bonds at the crack tips that controls the fracture resistance.

Speed of Fracture Propagation

The speed with which fractures propagate in brittle solids and therefore the progression of static fatigue is strongly influenced by mechanical and environmental factors. Clearly the state of stress is of primary importance. At higher tensile stresses the microcracks begin to grow at a faster rate, and the time to complete fracture diminishes. Brittle fracture test results can exhibit considerable amounts of scatter, but Charles (1958) has expressed a relation between tensile strength of glass specimens and time-to-fracture as follows:

$$\sigma_{1ult} = \sigma_{2ult} \cdot \left(\frac{t_2}{t_1}\right)^{1/n} \tag{2}$$

in which σ_{1ult} is the strength of a particle with time-to-fracture t_1 and σ_{2ult} is the strength of a particle with time-to-fracture t_2. Typical values of the exponent n are reported as 16 for glass (Charles, 1958), 98 for Carrara marble and 8 for Pennant sandstone (Cruden, 1974).

Lade and Karimpour (2010) loaded sodium lime glass beads of three different sizes and sustained the load at certain percentages of the short-term strengths to study the time-dependent static fatigue. They found the scatter of test results to be so pronounced that it was not possible to characterize the time effects for single particles by the expression in Eq. (2). Nevertheless, static fatigue was clearly observed in the sense that fracture occurred after some time (from seconds to days) when the beads were loaded up to the vicinity of their short term strengths. The time-dependent strengths of non-spherical single grains with a variety of shapes are likely to be even more difficult to predict. However, when integrated over a large number of particles, as in a triaxial specimen which contains millions of particles, the time-dependent behavior is very systematic and conforms to a regular pattern.

Environmental Effects

The speed of crack growth in the second stage is also decisively influenced by the environment to which the surface cracks are exposed. It is the rupture of atomic bonds at the crack tips that constitute the principal mode of failure. Figure 13 shows a model of a sharp crack at the atomic level. The bonds between the atoms act as nonlinear elastic springs that snap when the atoms are sufficiently separated. In addition, it has been found that the presence of water or water vapor has profound influence on the chemical reactions that lead to weakening and tensile failure near the crack tips. Thus, in glass (quartz) the water molecules may react with the silica atoms at the crack tips exposed to tensile stresses with the effect of unzipping the atomic bonds, as exemplified in the diagram in Fig. 14. The chemical reactions occurring at the crack tip are schematically indicated in Fig. 15(a) for glass and water and 15(b) for glass and water in an alkaline solution (Michalske and Freiman, 1981, 1983;

Gong and Du, 2000). This in turn sharpens the crack tips and increases the crack propagation speeds. Therefore, the brittle material appears to be more brittle with accelerating crack propagation speeds in the presence of water. Thus, the pH in the humid environment as well as the presence of other chemical components near the crack tips may serve to increase the speed of stress corrosion (Wiederhorn, 1975). Oldecop and Alonso (2007) investigated the effect of water on the time-dependent behavior of rockfill and they found that the settlement of rockfill dams with time could be understood in terms of the water-enhanced static fatigue of the rockfill.

Effects of Water Introduction in Dry Granular Material

Because brittle material appears to be more brittle with accelerating crack propagation speeds in the presence of water, an experiment was performed on an initially dry sand to study the effect of introducing water into an already highly stressed granular material. A triaxial specimen was prepared of oven-dry Virginia Beach sand, isotropically confined at 8,000 kPa, and percolated with gaseous CO_2 before shearing under strain control to a deviator stress near 14,000 kPa as shown in Fig. 16. Water was then introduced from the base to saturate the specimen. It is clear that introduction of water causes fracture and partial collapse of the particle structure during the upwards migration of the water front in the specimen. This experiment confirms that particles are more brittle in the presence of water, and together with the relations between particle crushing, energy input, and the observed time effects, this helps confirm that the mechanism of static fatigue controls time effects in granular materials.

Overall Effect of Static Fatigue

The overall stress-strain-time behavior of specimens or particles of brittle solids that exhibit static fatigue is indicated in Fig. 17 (Rüsch, 1960). As the stress on the specimen is held constant below the short-term fracture stress, the time to fracture increases with decreasing stress until fracture does not occur at all at some limiting stress,

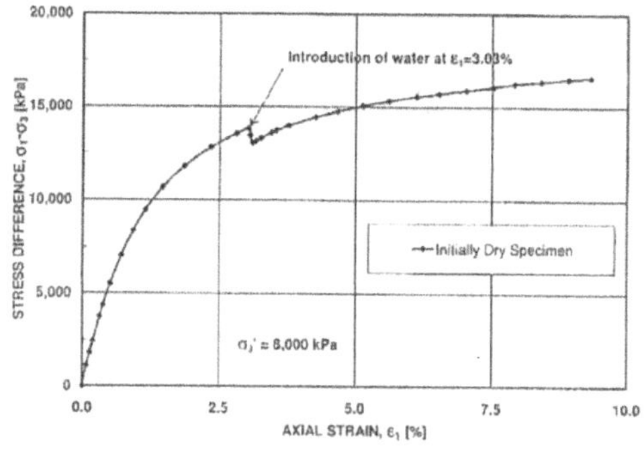

Fig. 16. Stress-strain curve from triaxial compression test on oven-dry Virginia Beach sand with introduction of water to show partial collapse of soil structure due to local water-induced grain fractures

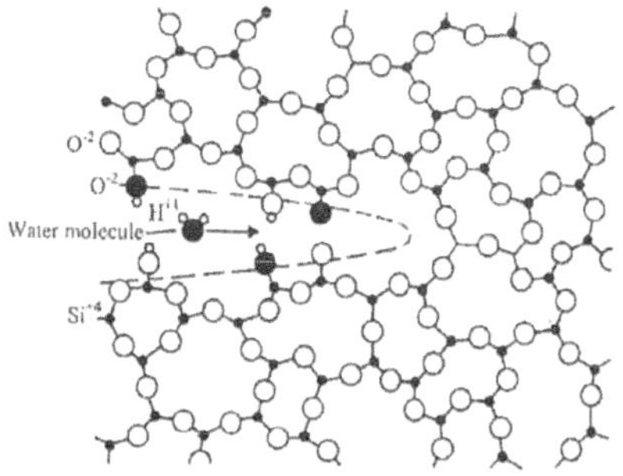

Fig. 14. Two-dimensional model of water-induced bond rupture in silica glass (after Lawn, 1993)

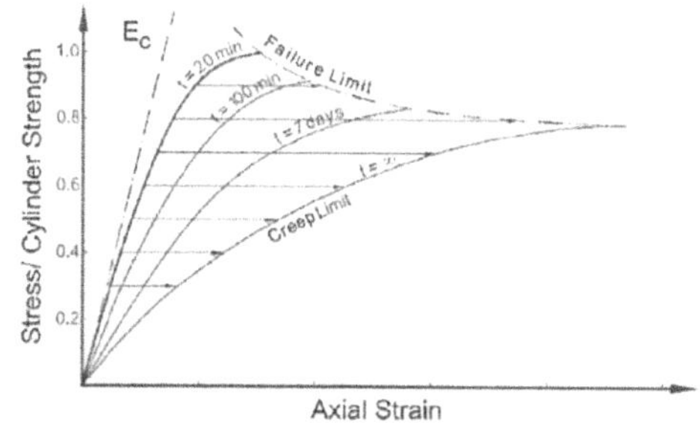

Fig. 17. Effect of time on strength of concrete specimens (after Rusch, 1960)

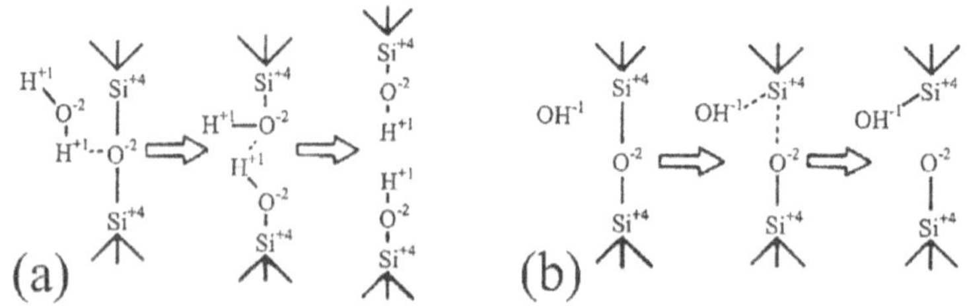

Fig. 15. Schematic representation of chemical reactions at crack-tip between (a) water and quartz, and (b) hydroxyl and quartz in alkaline solution (after Michalske and Freiman, 1981; Gong and Du, 2000)

which may be characterized as a percentage of the short-term fracture strength.

The resulting overall stress-strain-time behavior of a triaxial specimen consists of the integrated response of millions of particles in the granular material in which force chains are established, meander, and become stronger with increasing load. Compared with the response of individual particles, which may show considerable variation and scatter, the responses of the triaxial specimens exhibit very little scatter in their stress-strain-time relations, as seen in the studies of crushed coral sand, Antelope Valley sand and Virginia Beach sand discussed here.

EFFECT OF PARTICLE STRENGTH

While the strengths of the particles in the three types of sand whose results are shown in Figs. 4, 5, and 6 were not explicitly measured, an impression of the crushability of the particles was obtained from the sand that came out of the specimens after testing. For the purpose of the present study, the crushed coral sand was initially sieved and the sand used for testing was composed to create an initial straight grain size curve with the original intent of possibly sieving the soil after testing. However, what was retrieved after testing was a paste that was not amiable to sieve analysis. Thus, the crushed coral sand consisted of very soft and weak grains.

The Antelope Valley sand, whose results are shown in Fig. 5, was a relatively friable sand obtained from a streambed in the desert north of Los Angeles. While crushing was evident from the grain sizes that came out of the specimens after testing, the sand was so friable that reliable grain size analyses could not be obtained from shaking a stack of sieves, which could easily cause additional particle crushing. However, individual grains were visible after testing and they were likely stronger that the crushed coral sand. Both these studies were conducted at effective confining pressures near 200 kPa.

The third study was performed on Virginia Beach sand, consisting mainly of quartz. This study was carried out with effective confining pressures of 8,000 kPa at which considerable amounts of crushing occur. It is clear that this sand had the strongest particles of the three sands, because considerable stress magnitudes were required to crush this sand.

Comparing the differences between the one day creep curves and the one day stress relaxation curves for the three sands in Figs. 4, 5, and 6, it appears that these differences may depend primarily on the grain crushability. While other factors such as particle size distribution, particle shape and mineral composition may also play a role, the pattern is not discernable on the basis of the three studies presented here, because both the particle strengths and the stress magnitudes varied in the experiments. However, it appears that the particle strength is related to the nonisotach behavior observed for granular materials. Additional studies are required to determine any behavior pattern.

CONCLUSIONS

Observations from experiments show that strain rate effects are negligible for sands, unlike for clays in which strain rate effects are significant. Further, the observed stress relaxation behavior of granular materials were not in ''correspondence'' with the measured creep behavior for three different sands. Triaxial experiments with branches of creep at different deviator stresses formed a consistent stress-strain relation after one day of creep, and similar tests with stress relaxation also formed a consistent stress-strain relation after one day of relaxation. However, the two relations were quite different with the relation after one day of stress relaxation showing much larger strains at the same deviator stresses. These differences appear to be related to the particle strength, but no pattern is discernable from the existing experiments. It is concluded that sands do not exhibit classic viscous effects, and their behavior is indicated as ''nonisotach'', while the typical viscous behavior of clay is termed ''isotach''. Thus, there are significant differences in the time-dependent behavior patterns of sands and clays.

A mechanistic picture of time effects in sands is proposed in which interparticle friction, grain crushing, and grain rearrangement play the key roles. Grain crushing is a time dependent phenomenon known as static fatigue and this accounts for the time dependency observed in granular materials. The characteristics of the phenomenon of static fatigue are reviewed and the experiments on crushed coral sand, Antelope Valley sand and Virginia Beach sand confirmed that the phenomenon of static fatigue can explain the behavior patterns observed in granular materials.

REFERENCES

1) Augustesen, A., Liingaard, M. and Lade, P. V. (2004): Evaluation of time dependent behavior of soils, *Int. J. Geomechanics*, ASCE **4**(3), 137–156.
2) Bolton, M. D., Nakata, Y. and Cheng, Y. P. (2008): Micro- and macro-mechanical behaviour of DEM crushable materials, *Geotechnique*, **58**(6), 471–480.
3) Callister, W. D. Jr. (2005): *Fundamentals of Materials Science and Engineering—An Integrated Approach*, 2nd Edition, Wiley.
4) Charles, R. J. (1958): Static fatigue of glass, *J. Applied Physics*, **29**, 1549.
5) Cruden, D. M. (1974): The static fatigue of brittle rock under uniaxial compression, *Int. J. Rock Mechanics and Mining Sciences & Geomechanics Abstracts*, **11**, 67–73.
6) Gong, J. and Du, C. (2000): Environmental effects on the room-temperature static fatigue behavior of polycrystalline mullite, *Material Science and Engineering*, **A283**, 76–81.
7) Jaeger, J. C. and Cook, N. G. W. (1969): *Fundamentals of Rock Mechanics*, Methuen & Co. Ltd., London.
8) Karimpour, H. and Lade, P. V. (2010): Time effects relate to crushing in sand, *J. Geotecnical and Geoenvironmental Engineering*, ASCE, **136**(9), 1209–1219.
9) Kiyota, T. and Tatsuoka, F. (2006): Viscous property of loose sand in triaxial compression, extension and cyclic loading, *Soils and Foundations*, **46**(5), 665–684.
10) Kou, S. Q., Liu, H. Y., Lindqvist, P.-A., Tang, C. A. and Xu, X. H. (2001): Numerical investigation of particle breakage as applied to mechanical crushing—Part II: Interparticle breakage, *Int. J.*

Rock Mechanics & Min. Sci., **38**, 1163–1172.

11) Kuwano, R. and Jardine, R. (2002): On measuring creep behaviour in granular materials through triaxial testing, *Canadian Geotechnical Journal*, **39**(5), 1061–1074.
12) Lade, P. V. (2007): Experimental study and analysis of creep and stress relaxation in granular materials, *GeoDenver*, ASCE, Denver, Colorado, CD-ROM.
13) Lade, P. V. and Liu, C.-T. (1998): Experimental study of drained creep behavior of sand. *J. Engineering. Mechanics*, ASCE, **124**(8), 912–920.
14) Lade, P. V. and Karimpour, H. (2010): Static fatigue produces time effects in granular materials, *GeoFlorida*, ASCE, West Palm Beach, Florida, CD-ROM.
15) Lade, P. V., Liggio, Jr., C. D. and Nam, J. (2009): Strain rate, creep, and stress drop-creep experiments on crushed coral sand, *J. Geotechnical and Geoenvironmental Engineering*, ASCE, **135**(7), 941–953.
16) Lade, P. V., Nam, J. and Liggio, Jr., C. D. (2010): Effects of particle crushing in stress drop-relaxation experiments on crushed coral sand, *J. Geotechnical and Geoenvironmental Engineering*, ASCE, **136**(3), 500–509.
17) Lawn, B. (1993): *Fracture of Brittle Solids*, 2nd edition, Cambridge University Press.
18) Lemaitre, J. and Chaboche, J.-L. (1994): *Mechanics of Solid Materials*, Cambridge University Press.
19) Leroueil, S. and Marques, M. E. S. (1996): State of the Art: Importance of train rate and temperature effects in geotechnical engineering, *Measuring and Modeling Time Dependent Soil Behavior*, Geotech. Spec. Publ. No. 61, eds. by Sheahan, T. C. and Kaliakin, V. N., ASCE, Reston, VA, 1–60.
20) Liingaard, M., Augustesen, A. and Lade, P. V. (2004): Characterization of models for time-dependent behavior of soils, *Int. J. Geomechanics*, ASCE, **4**(3), 157–177.
21) Matsushita, M., Tatsuoka, F. Koseki, J., Cazacliu, B., Benedetto, H. and Yasin, S. J. M. (1999): Time effects on the pre-peak deformation properties of sands, *Pre-failure Deformation Characteristics of Geomaterials*, eds. by Jamiolkowski, Lancelotta, and Lo Presti, Balkema, Rotterdam, 681–689.
22) McDowell, G. R. and Bolton, M. D. (1998): On the micromechanics of crushable aggregates, *Geotechnique*, **48**(5), 667–679.
23) Michalske, T. A. and Freiman, S. W. (1981): A molecular interpretation of stress corrosion in silica, *Nature*, **295**, 511.
24) Michalske, T. A. and Freiman, S. W. (1983): A molecular mechanism for stress corrosion in vitreous silica, *J. American Ceramics Society*, **66**, 284–288.
25) Nakata, Y., Hyde, A. F. L., Hyodo, M. and Murata, H. (1999): A probabilistic approach to sand particle crushing in the triaxial test, *Geotechnique*, **49**(5), 567–583.
26) Nakata, Y., Hyodo, M., Hyde, A. F. L., Kato, Y. and Murata, H. (2001a): Microscopic particle crushing of sand subjected to high pressure one-dimensional compression, *Soils and Foundations*, **41**(1), 69–82.
27) Nakata, Y., Kato, Y., Hyodo, M., Hyde, A. F. L. and Murata, H. (2001b): One-dimensional compression behaviour of uniformly graded sand related to single particle crushing strength, *Soils and Foundations*, **41**(2), 39–51.
28) Oldecop, L. A. and Alonso, E. E. (2007): Theoretical investigation of time-dependent behaviour of rockfill, *Geotechnique*, **57**(3), 289–301.
29) Rüsch, H. (1960): Researches toward a general flexural theory for structural concrete, *J. American Concrete Institute*, **57**(1), 1–28.
30) Sheahan, T. C. and Kaliakin, V. N. (1999): Microstructural considerations and validity of the correspondence principle for cohesive soils, *Engineering Mechanics, Proc. 13th Conference*, eds. by N. Jones and R. Ghanem, ASCE, Baltimore, MD, USA.
31) Suresh, S. (1998): *Fatigue of Materials*, 2nd edition, Cambridge University Press.
32) Tatsuoka, F., Shihara, M., Di Benedetto, H. and Kuwano, R. (2002): Time-dependent shear deformation characteristics of geomaterials and their simulation, *Soils and Foundations*, **42**(2), 103–129.
33) Tatsuoka, F., Di Benedetto, H., Enomoto, T., Kawabe, S. and Kongkitkul, W. (2008): Various viscosity types of geomaterials in shear and their mathematical expression, *Soils and Foundations*, **48**(1), 41–60.
34) Tang, C. A., Xu, X. H., Kou, S. Q., Lindqvist, P.-A. and Liu, H. Y. (2001): Numerical investigation of particle breakage as applied to mechanical crushing—Part I: Single-particle breakage, *Int. J. Rock Mechanics & Min. Sci.*, **38**, 1147–1162.
35) Van Mier, J. G. M. (2009): Mode II fracture localization in concrete loaded in compression, *J. Engineering Mechanics*, ASCE, **135**(1), 1–8.
36) Weibull, W. (1951): A statistical distribution function of wide applicability, *J. Applied Mechanics*, **18**, 293–297.
37) Wiederhorn, S. M. (1975): Crack growth as an interpretation of static fatigue, *J. Non-Crystalline Solids*, **19**, 169–181.
38) Yamamuro, J. A. and Lade, P. V. (1993): Effects of strain rate on instability of granular soils, *Geotechnical Testing Journal*, **16**(3), 304–313.

Computer Methods and Advances in Geomechanics, Desai et al. (eds) © 2001 Balkema, Rotterdam, ISBN 90 5809 183 X

Modeling creep behavior of granular materials

Poul V. Lade
Aalborg University, Denmark

Chi-Tseng Liu
LAW/Crandall Incorporated, Los Angeles, Calif., USA

ABSTRACT: A systematic series of creep tests on sand under triaxial stress states has been performed. The experiments showed that the nature of inelastic creep strains is similar to that of plastic strains. The time-dependent inelastic work may be normalized on values of reference plastic work and is independent of confining pressure and stress level. It varies linearly with normalized time when plotted on log-log scales. To capture the time-dependent inelastic strains, the existing single hardening model is extended by multiplying the plastic work in the hardening function by a power function of time, which requires determination of one parameter. Thus, the predicted time-dependent behavior is based on the modeled time-independent behavior. The time-dependent behavior is predicted and compared with results of creep experiments.

1 INTRODUCTION

In geotechnical engineering projects, soil creep and relaxation often cause long-term deformation and pressures on buildings, bridge abutments, earth retaining structures, and earth slopes. Excessive deformation with time may cause structures to fail. Another mode of soil collapse is instability and liquefaction: Granular materials may become unstable when exposed to certain stress paths inside the failure surface. In nature, especially fine silty sands are susceptible to liquefaction under static and cyclic loading conditions, but these fine silty sands also exhibit considerable creep behavior, and they are therefore affected by time effects. Drained creep at a given stress state causes the plastic yield surface to move out and away from the current stress state, and the sand consequently becomes stiffer in the region immediately outside the stress point. Increasing amounts of drained creep therefore require increasingly high stress differences or decreasingly low effective confining pressures to render the sand unstable. Thus, time effects play significant roles in the behavior of soils.

Prediction of creep, relaxation and loading rate effects in granular materials requires a constitutive model with a capability to account for time-dependent behavior. In this presentation the time-dependent behavior is incorporated in an existing single hardening plasticity model, whose capabilities have been demonstrated in the literature. The advantage of adding this capability to the existing model is that prediction of creep under three-dimensional conditions is immediately incorporated in the model. Because the time effects modeled here correspond to rates lower than those employed in conventional triaxial testing, emphasis has been placed on time effects at working stress conditions away from failure. Although failure may be affected by rate effects, especially at higher rates, this effect is relatively small at the rates and times considered here (see e.g. Yamamuro and Lade 1993), and only small modifications, within the scatter in the data, would be required from effects on failure. This effect has consequently not been taken into account in the study presented below.

2 PREVIOUS OBSERVATIONS

Considerable insight into the time-dependent behavior of clays has been gained from a number of studies in which drained tests have been performed, and excellent reviews have been presented in the literature (see, e.g., Cristescu 1989; Feda 1992; Mitchell 1993; Leroueil and Marques 1996; Sheahan and Kaliakin 1996).

Sands generally creep less than clays at the same stress states (Murayama et al. 1984; Lade et al. 1997), but the amount of creep increases with confining pressure, particularly after crushing becomes important at high stresses (Yamamuro and Lade 1993). More crushing occurs in loose than in dense sand (Bopp and Lade 1997). However, fine silty sands, whose structure is metastable and may be susceptible to instability and liquefaction

(Yamamuro and Lade 1997, 1998; Lade and Yamamuro 1997), also show considerable time-effects. In this particle structure the silt particles may occupy locations at or near the contact points of the larger sand particles. This particle structure tends to form in the fine range of particle sizes where the attractive and repulsive forces begin to compete with the gravitational forces (Lade et al. 1998). Thus, the finer particles push the larger particles slightly apart, and upon application of normal and shear stresses, the fine particles will tend to slide into the void spaces, allowing the larger particles to move together. Such particle structures formed in loose silty sands are precarious and unstable, and slippage occurs easily in such structures. This generates an inherently contractive volumetric tendency, which under undrained conditions may lead to instability and static liquefaction (Bjerrum 1973; Lade et al. 1997). Thus, time-dependent behavior in granular materials may be caused by crushing of sand grains or by collapse of metastable particle structures.

A study of drained creep and relaxation in Antelope Valley sand was performed by Lade and Liu (1998) to (1) investigate the nature of time effects, (2) study whether time-dependent behavior may be modeled by adding time effects within the framework provided by work-hardening plasticity theory, and (3) provide data for development of a constitutive model for soil stress-strain-time behavior. The structure of the dry deposited fine sand exhibited some collapse upon introduction of water in the pores. The fine sand had sufficient permeability to immediately dissipate pore pressures, thus allowing observation of creep deformations directly after changes in stress. The experiments showed that the nature of creep strains is similar to that of plastic strains in the sense that they may be predicted from the framework provided by hardening plasticity theory. In particular, the potential surface determined from and used for prediction of plastic strains may also be used for prediction of time-dependent creep strains. It also appears that the yield surface and the plastic potential surface move out together during creep, and the point at which to evaluate inelastic strain increment directions is at the current location of the yield surface and the accompanying plastic potential surface. Prediction of the movement and new location of the yield surface was attempted on the basis of the change in inelastic work occurring during creep (Lade 1994).

3 SINGLE HARDENING MODEL

The single hardening constitutive model, to which the time-dependent behavior will be added, was developed on the basis of a thorough review and evaluation of data from experiments on frictional materials such as sand, clay, concrete and rock (Kim and Lade 1988; Lade and Kim 1988 a, b). In this model the total strain increments observed in a material when loaded are divided into elastic and plastic components so that

$$\dot{\varepsilon}_{ij}^{t} = \dot{\varepsilon}_{ij}^{e} + \dot{\varepsilon}_{ij}^{p} \tag{1}$$

Both the elastic and the plastic strain increments are expressed in terms of effective stresses, and all functions for the elastic behavior, failure criterion, flow rule, yield criterion, and work-hardening/softening law are expressed in terms of stress invariants I_1, I_2, and I_3 and the stress deviator invariant J_2. The operative equations in this model were briefly reviewed by Lade (1990) and by Lade and Kim (1995). Only those relating to the yield criterion and the work-hardening relation will be used here, and they are reviewed below.

3.1 Yield Criterion and Work-Hardening Law

The model employs a single isotropic yield surface that is expressed as a contour of constant plastic work as measured from the stress origin:

$$f_p = f_p'(\sigma) - f_p''(W_p) \tag{2}$$

in which the expression for the yield surface in terms of stresses is:

$$f_p'(\sigma) = \left[\psi_1 \frac{I_1^3}{I_3} - \frac{I_1^2}{I_2}\right]\left[\frac{I_1}{p_a}\right]^h \cdot e^q \tag{3}$$

where the parameter ψ_1 (= constant for a given material at given density) acts as a weighting factor between the triangular shape (from the I_3 term) and the circular shape (from the I_2 term), p_a is atmospheric pressure in the same units as the first stress invariant I_1, h is a constant for a given soil, and q varies from zero at the hydrostatic axis to unity at the failure surface.

In Eq. 2 the expression for the yield criterion in terms of plastic work is:

$$f_p''(W_p) = \left[\frac{1}{D}\right]^{1/\rho}\left[\frac{W_p}{p_a}\right]^{1/\rho} \tag{4}$$

where ρ and D are constants for a given material, and W_p is the plastic work calculated from

$$W^p = \Sigma\, dW_p = \Sigma\, \sigma_{ij} \cdot d\varepsilon_{ij}^p \tag{5}$$

Thus, f_p'' varies with plastic work only. The values of ρ and D are given by:

$$\rho = p/h \tag{6}$$

and

$$D = \frac{C}{(27\psi_1+3)^{\rho}} \tag{7}$$

in which

$$W_p = C \cdot p_a \cdot \left(\frac{I_1}{p_a}\right)^p \tag{8}$$

The values of the material parameters C and p are determined from isotropic compression tests, as discussed below.

The resulting yield surface is shaped as an asymmetric teardrop, with smoothly rounded triangular cross-sections in octahedral planes. As the plastic work increases, the isotropic yield surface inflates until the current stress point reaches the failure surface. The relation between the yield criterion and the plastic work is described by a monotonically increasing function whose slope decreases with increasing plastic work, as shown in Figure 1. Beyond failure the yield surface contracts isotropically with increasing amount of plastic work.

4 MODEL FOR CREEP EFFECTS IN SAND

The development consists of modifying the plastic hardening relation to include effects of time-dependent inelastic (or visco-plastic) work. The stress-strain-time relation therefore takes the following form:

$$d\varepsilon_{ij}^{vp} = d\lambda(W_p, t) \cdot \frac{\partial g}{\partial \sigma_{ij}} \tag{9}$$

in which $d\lambda(W_p,t)$ is determined from the hardening relation that includes plastic and time-dependent inelastic work. Once the formulation of the multiplier $d\lambda(W_p,t)$ has been determined, the stress-strain-time relation is given by the equation above. All other portions of the constitutive model have been developed and shown to work well.

The scalar factor of proportionality $d\lambda$ in Eq. 9 may be expressed as:

$$d\lambda = \frac{dW_p}{\mu g_p} \tag{10}$$

in which μ is a material parameter, g_p is the plastic potential function, and the increment in plastic work can be determined by differentiation of the modified hardening equation.

Examples of experimental results from this study were presented by Lade and Liu (1998). To compare time-dependent behavior at different stress levels, the time is normalized as follows:

$$T = \frac{t}{t_0} \tag{11}$$

in which t is the elapsed time since the change in stress, and t_0 is a reference time. Consistent patterns of soil behavior were observed in all experiments after 2 minutes of creep, but most often before. Nevertheless, because of this consistency, the reference time was taken as t_0 = 2 minutes in this study. Thus, T = 1 at 2 minutes of creep.

To express the variation of inelastic work in creep experiments, the inelastic work produced during creep (also referred to as W_p) is calculated and normalized on the inelastic work at T =1, as follows:

$$\bar{W}_p(T) = \frac{W_p(T)}{W_p(T{=}1)} \tag{12}$$

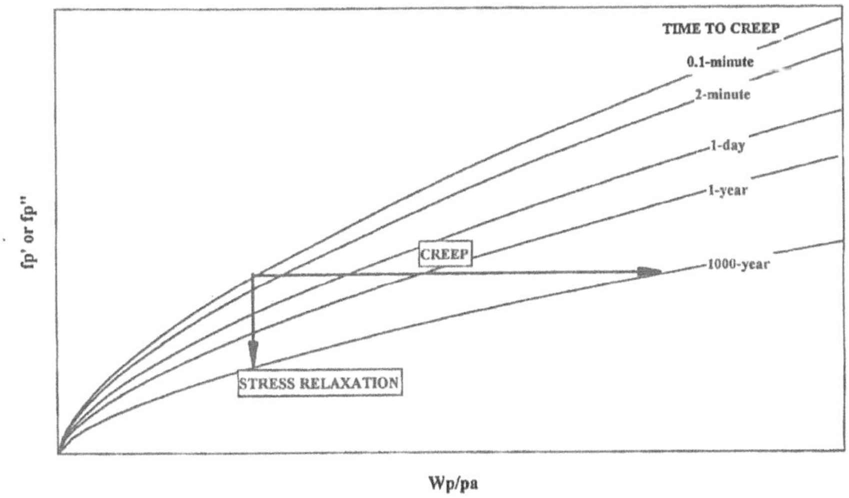

Figure 1. Schematic illustration of creep effects on the change of hardening curves with time.

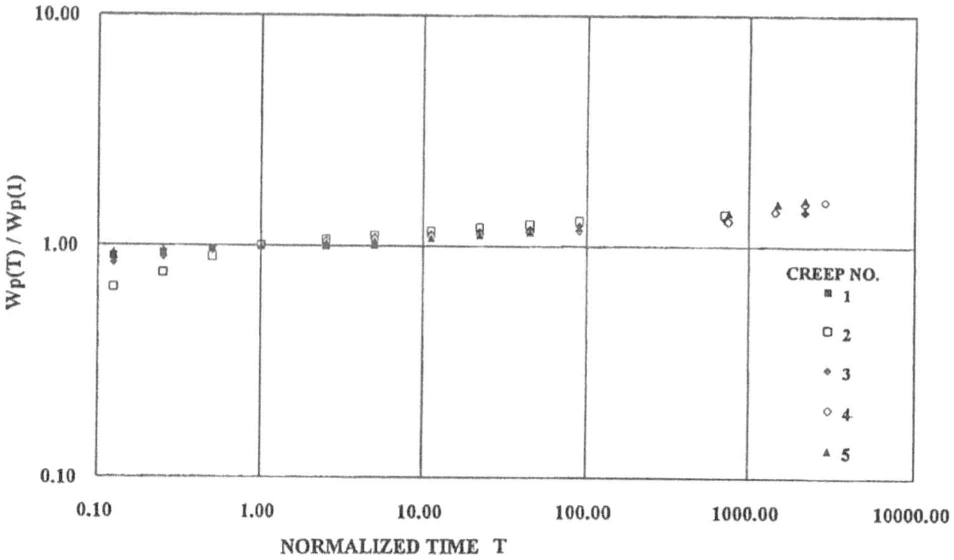

Figure 2. Normalized inelastic work versus normalized time from isotropic compression creep tests.

Figure 2 shows the normalized, accumulated inelastic work plotted versus normalized time on a log-log diagram at five different values of isotropic stress. The results of the five creep tests describe the same straight line within the scatter of the data. The creep at constant isotropic state of stress causes compression and the inelastic work therefore increases with time. Higher stresses produce larger creep strains and more inelastic work. The normalization indicated in Figure 2 captures these effects, and the straight line is expressed as:

$$\frac{W_p(T)}{W_p(1)} = T^{\kappa} \tag{13}$$

Since the state of stress and the yield function is constant during creep, a family of hardening curves for different times can be derived from Eq. 13. The consequence of the variation in plastic work with time expressed in Eq. 13 for creep and relaxation is also indicated in Figure 1.

To determine the dimensionless hardening parameters C and p in Eq. 7 the plastic work during isotropic compression, W_p/p_a, is plotted against the first stress invariant, I_1/p_a, in a log-log diagram and fitting a straight line, as shown in Figure 3. This may be done at different times T, as also shown in Figure 3. The magnitude of C is the value of W_p/p_a at I_1/p_a = 1, and the value of p is the slope of the fitted line. Apart from the initial deviations, the straight lines are essentially parallel and vary in a systematic manner with time for creep. Thus, the slope p is and the parameter C is determined at the reference time of 2 minutes (T = 1). The variation in intercept with I_1/p_a = 1 is then expressed by the relation in Eq. 13, and therefore according to Eq. 7:

$$C(T) = C(1) \cdot T^{\kappa} \tag{14}$$

Creep experiments were performed on Antelope Valley sand in isotropic compression and in triaxial compression tests at various stress levels (proximity to failure) following loading along conventional, constant confining pressure and along proportional loading stress paths. In order to test whether the expression in Eq. 13 is generally correct for creep at all stress states and for all stress paths leading to these stress states, the inelastic work was calculated for various creep tests, normalized on the inelastic work at t_0 = 2 minutes, and plotted versus T. Experiments were performed with creep initiated at up to five different stress states and lasting up to 10 days for each specimen. The different experiments produced the same straight line within the scatter of the results, as shown in Figure 4. The best fitting value of the exponent κ in Eq. 13 is found to be 0.0386. This is the only new parameter added to the model in order to predict time effects under general three-dimensional conditions.

5 PREDICTION OF CREEP STRAINS

All parameters for the single hardening model are independent of time effects except the value of C, whose variation with time is expressed through Eq. 14. Small time effects may also be present on failure,

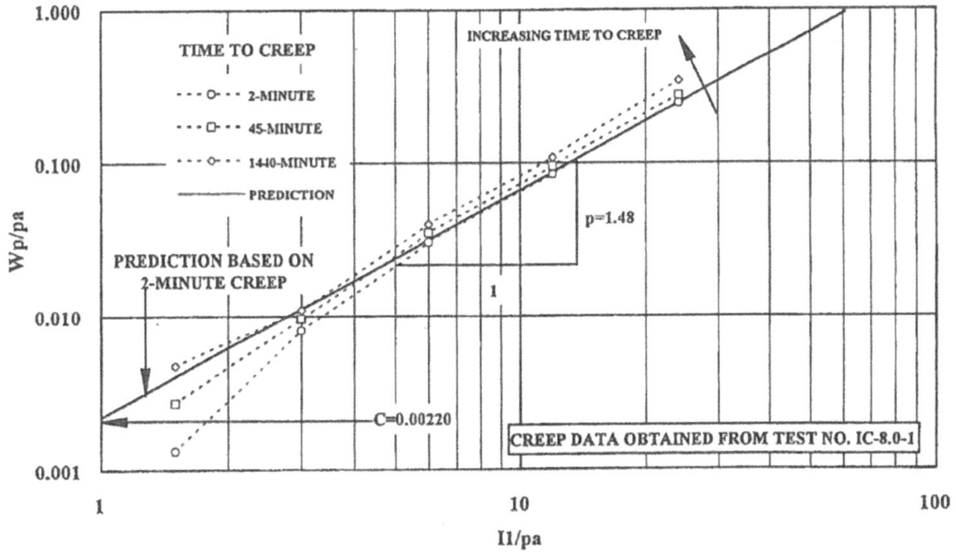

Figure 3. Variation of hardening parameters from isotropic creep tests.

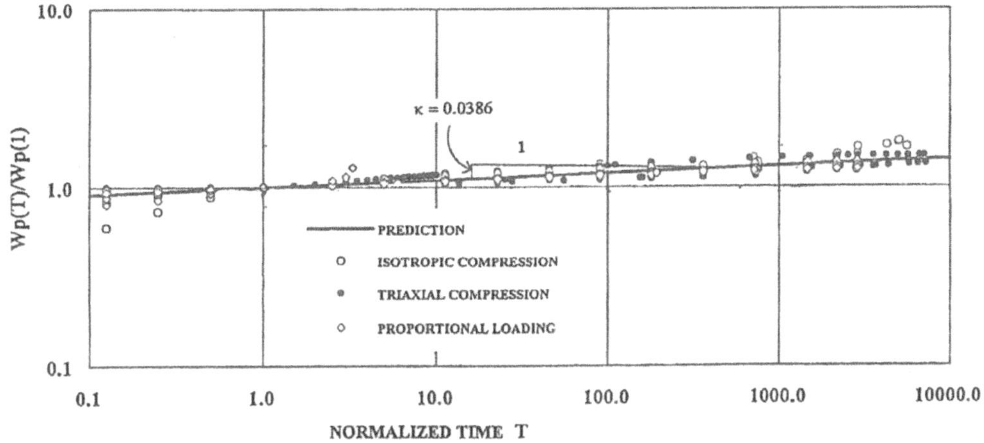

Figure 4. Normalized inelastic work versus normalized time from isotropic, conventional triaxial compression, and proportional loading triaxial compression tests. Determination of parameter κ.

but these are ignored in the present study, because they are relatively insignificant for creep.

Creep strains may be predicted directly from the expression in Eq. 9 with the time variation in plastic work expressed in Eq. 14. Examples of predictions of stress-strain and volume change behavior for a triaxial compression test on Antelope Valley sand are shown in Figure 5, with the solid curves corresponding to the reference behavior, taken to correspond to 2 minutes of creep at each point. The predictions of creep at four different stress ratios are shown on the upper diagram, while the volume change behavior in the lower diagram follows the primary loading curve. This is because the potential for plastic strains and the potential for inelastic, time-dependent strains is the same, thus producing the same ratio between linear and volumetric strains, whether plastic or inelastic.

Figures 6 and 7 show the predicted axial and volumetric strains with time. The strains are shown accumulated for the various stages of elastic, plastic, and creep behavior as predicted by the model. Thus, creep strains are not initialized at each new stage of creep. Deviations between predicted and observed behavior are therefore accumulated as the test progresses. The majority of these deviations are caused by experimental scatter and by failure of the single hardening model to predict the time-independent, plastic strains with high accuracy. If the predicted and the observed creep behavior

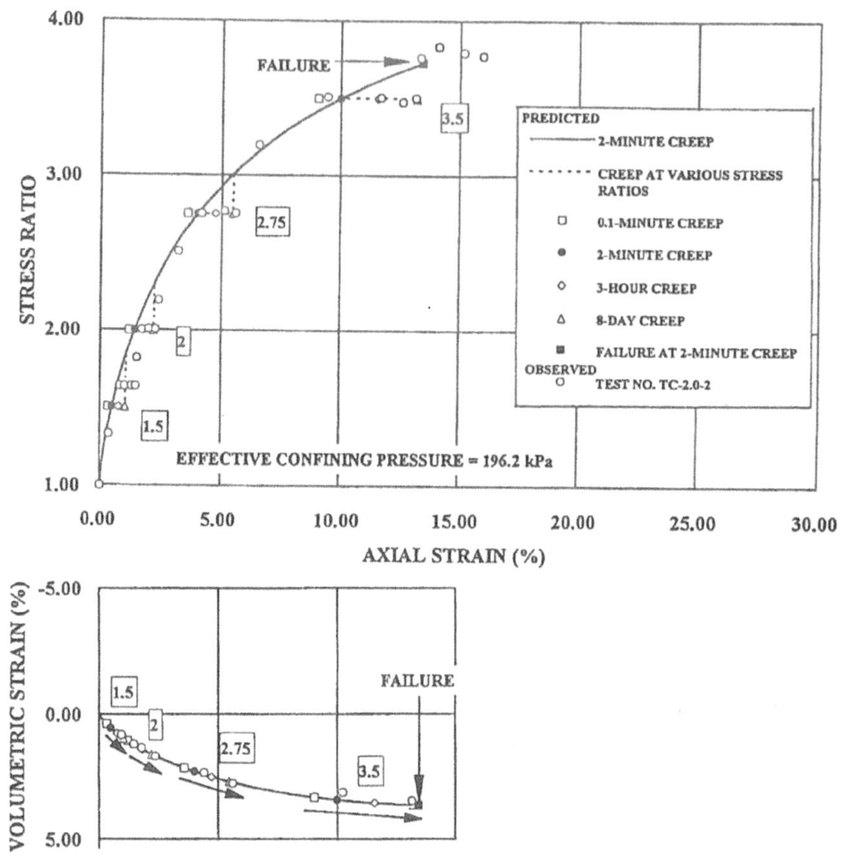

Figure 5. Comparison of observed and predicted stress-strain and volume change relations in triaxial compression test with creep stages at four different stress ratios.

were initialized at the same points at the beginning of each new creep stage, the comparisons would look quite attractive.

6 DISCUSSION

Several advantages are present when incorporating a single yield surface in a constitutive model: Using the plastic work contours as yield surfaces takes into account effects of shear as well as volumetric strains. Furthermore, the direct association between yielding and plastic work results in a unique hardening curve, and it produces a model that is mathematical consistent. Comparison with experimental evidence shows that this formulation of the single yield surface fits very well with observed behavior of frictional materials, i.e. it captures the conditions for yielding correctly for various stress paths. In addition, employment of a single yield surface makes the model more efficient for applications involving large computations.

Employment of plastic work as the hardening parameter (rather than e.g. effective strain or volumetric strain) makes the model particularly suitable for addition of time effects such as loading rate effects, creep, and relaxation. All types of strain increments (normal, shear, volumetric) and all stresses are taken into account in calculating the plastic work, and such strain increments and stresses may also be produced in the process of predictions. Thus, if current stresses and stress increments are given, then strain increments may be predicted. Or if current stresses and strain increments are given, then stress increments may be predicted from the model.
In connection with time effects, this capability comes in conveniently. Pure creep, which occurs under constant stresses, produces inelastic strains and therefore inelastic work, which is very similar to plastic work. Thus, the plastic work increases with time, while the stresses stay constant, as indicated in Figure 1. Conversely, once the model has been modified to include time effects, then it is possible to keep the stresses constant, let the time pass, thereby

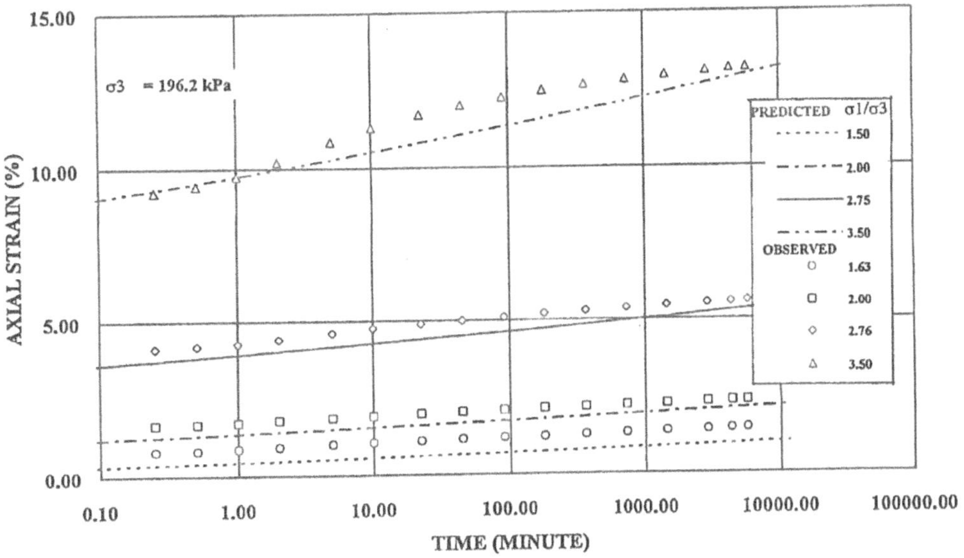

Figure 6. Comparison of observed and predicted axial strain versus time relation in triaxial compression test with creep stages at four different stress ratios.

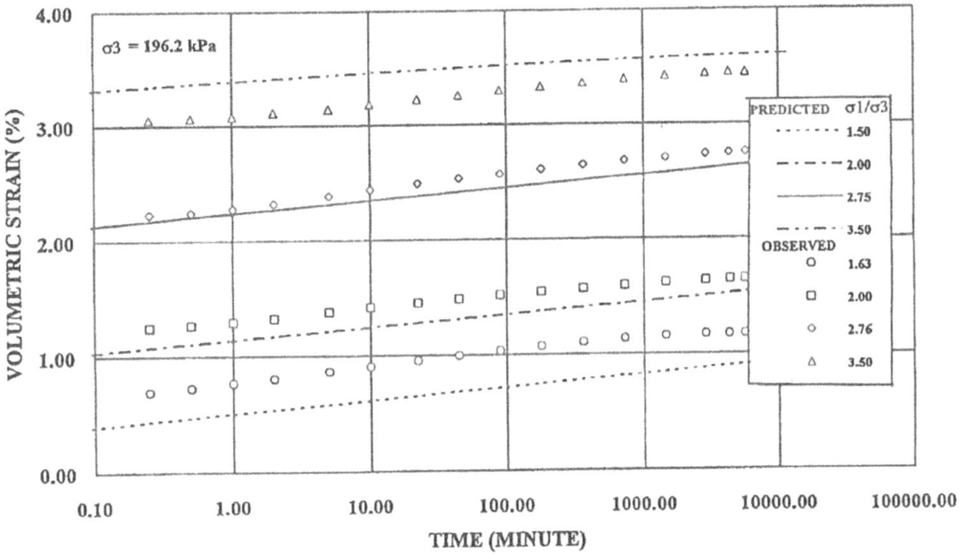

Figure 7. Comparison of observed and predicted volumetric strain versus time relation in triaxial compression test with creep stages at four different stress ratios.

producing inelastic (i.e. plastic) work, and use the model to predict the corresponding inelastic (i.e. creep) strains.

If, on the other hand, strains are prevented from taking place, then stress relaxation will occur. Without strains, inelastic or plastic work will not occur, but since the time passes, then the current stress point moves on to a lower work-hardening relationship, as indicated in Figure 1. Conversely, this feature of the model allows calculation of the relaxed stresses when the strains are fixed. Any combination of creep and relaxation may also be predicted by changing both time and stress state or work input.

The same formulation also allows the calculation of loading rate effects. As both the load and the time changes, the current point moves to another work-hardening curve, and effects of lower as well as higher strain rates may be handled by this formulation.

Thus, using the plastic (inelastic) work as the hardening parameter in the model allows incorporation and calculation of all types of known time effects in the model. The model then involves a hardening relation that is time dependent, as indicated schematically in Figure 1.

7 CONCLUSION

Modeling time-dependent behavior such as creep in sand has been incorporated into a time-independent constitutive model for frictional materials. To capture the creep occurring at various points in stress space, the existing single hardening model has been extended by multiplying the plastic work in the hardening function by a power function of time, which requires determination of one parameter. Thus, the predicted time-dependent behavior is based on the modeled time-independent behavior.

The modified model was shown to predict the observed creep behavior of Antelope Valley sand at various stress ratios with good accuracy.

REFERENCES

Bjerrum, L. 1973. Problems of soil mechanics and construction on soft clays and structurally unstable soils. *Proc. 8th Int. Conf. Soil Mech. Found. Engrg.*, 3: 111-159.

Bopp, P.A. & P.V. Lade 1997. Effects of initial density on soil instability at high pressures. *J. Geotech. and Geoenv. Engrg.*, ASCE, 123(7): 671-677.

Cristescu, N. 1989. *Rock Rheology*, Kluwer Academic Publishers, Boston.

Feda, J. 1992. *Creep of Soils and Related Phenomena*. Development in Geotechnical Engineering 68, Elsevier, Amsterdam.

Kim, M.K. & P.V. Lade 1988. Single hardening constitutive model for frictional materials, I. Plastic potential function. *Computers and Geotechnics*, 5(4): 307-324.

Lade, P.V. 1990. Single Hardening Model with Application to NC Clay. *J. Geotech. Engrg.*, ASCE, 116(3): 394-414.

Lade, P.V. 1994. Creep effects on static and cyclic instability of granular soils. *J. Geotech. Engrg.*, ASCE, 120(2): 404-419.

Lade, P.V. & M.K. Kim 1988 a. Single hardening constitutive model for frictional materials, II. Yield criterion and plastic work contours. *Computers and Geotechnics*, 6(1): 13-29.

Lade, P.V. & M.K. Kim 1988 b. Single hardening constitutive model for frictional materials, III. Comparisons with experimental data. *Computers and Geotechnics*, 6(1): 31-47.

Lade, P.V. & M.K. Kim 1995. Single Hardening Constitutive Model for Soil, Rock and Concrete. *Int. J. Solids Struct.*, Elsevier Science Ltd., 32(14): 1963-1978.

Lade, P.V., C.D. Liggio, Jr. & J.A. Yamamuro 1998. Effects of Non-Plastic Fines on Minimum and Maximum Void Ratios of Sand. *Geotech. Test. J.*, ASTM, 21(4), 336-347.

Lade, P.V. & C.-T. Liu, 1998. Experimental Study of Drained Creep Behavior of Sand. *J. Engrg. Mech.*, ASCE, 124(8): 912-920.

Lade, P.V. & J.A. Yamamuro 1997. Effects of nonplastic fines on static liquefaction of sands. *Can. Geotech. J.*, 34(6): 918-928.

Lade, P.V., J.A. Yamamuro & P.A. Bopp 1997. Influence of Time Effects on Instability of Granular Materials. *Computers and Geotechnics*, 20(3/4): 179-193.

Leroueil, S. & M.E.S. Marques 1996. State of the Art: Importance of strain rate and temperature effects in geotechnical engineering. in *Measuring and Modeling Time Dependent Soil Behavior*, Geotech. Spec. Publ. 61, ASCE, ed. by T.C. Sheahan and V.N. Kaliakin, 1-60.

Mitchell, J.K. 1993. *Fundamentals of Soil Behavior*. 2nd edition, John Wiley and Sons, New York.

Murayama, S., K. Michihiro & T. Sakagami 1984. Creep characteristics of sands. *Soils and Foundations*, 24(2): 1-15.

Sheahan, T.C. & V.N. Kaliakin 1996. *Measuring and Modeling Time Dependent Soil Behavior*, ASCE Geotech. Spec. Publ. 61, New York.

Yamamuro, J.A. & P.V. Lade 1993. Effects of strain rate on instability of granular materials. *Geotech. Test. J.*, ASTM, 16(3): 304-313.

Yamamuro, J.A. & P.V. Lade 1997. Static liquefaction of very loose sands, *Can. Geotech. J.* 34(6): 905-917.

Yamamuro, J.A. & P.V. Lade 1998. Steady-State Concepts and Static Liquefaction of Silty Sands, *J. of Geotech. and Geoenv. Engrg.*, ASCE, 124(9): 868-877.

Postscript

Additional information regarding time effects is given in the following articles:

Lade, P.V., (1994) "Creep Effects on Static and Cyclic Instability of Granular Soils", *Journal of Geotechnical Engineering*, ASCE, 120(2): 404-419.

Augustesen, A., Liingaard, M, and Lade, P.V. (2004) "Evaluation of Time Dependent Behavior of Soils," *International Journal of Geomechanics*, ASCE, 4(3): 137-156.

Liingaard, M., Augustesen, A., and Lade, P.V. (2004) "Characterization of Models for Time Dependent Behavior of Soils," *International Journal of Geomechanics*, ASCE, 4(3):157-177.

Karimpour, H. and Lade, P.V. (2010) "Time Effects Relate to Crushing in Sand." *Journal of Geotechnical and Geoenvironmental Engineering*, ASCE, 136(9): 1209-1219.

Karimpour, H. and Lade, P.V. (2013) "Creep Behavior in Virginia Beach Sand," *Canadian Geotechnical Journal*, 50(11): 1159-1178. DOI: 10.1139/cgj-2012-0467.

Lade, P.V. and Karimpour, H. (2015) "Stress Relaxation in Virginia Beach Sand," Canadian Geotechnical Journal, 52(7): 813-835. DOI: 10.1139/cgj-2013-0463.

Lade, P.V. and Karimpour, H. (2016) "Stress drop effects in time dependent behavior of quartz sand," *International Journal and Solids and Structures*, 87(6): 167–182.

Chapter XIV
Effects of Cross-Anisotropy

Preface

All modeling reviewed above is based on the assumption that the soil is isotropic. However, most often the triaxial testing for determination of material parameters is performed on specimens obtained from boreholes drilled in the vertical direction. Deposition of soil particles in water or in air or compacted soils tend to produce soil layers that are stiffer in the vertical direction than in the horizontal direction. Such soil layers are cross-anisotropic with a vertical axis of symmetry. To include cross-anisotropy requires serious modifications in the modeling framework as well as testing of both vertical and horizontal specimens to obtain appropriate material parameters. Presented here is a failure criterion that account for cross-anisotropy, the most important ingredient of such a modified model.

Available online at www.sciencedirect.com
ScienceDirect
International Journal of Solids and Structures 44 (2007) 5146–5162
INTERNATIONAL JOURNAL OF SOLIDS and STRUCTURES
www.elsevier.com/locate/ijsolstr

Modeling failure in cross-anisotropic frictional materials

Poul V. Lade *

Department of Civil Engineering, The Catholic University of America, Washington, DC 20064, USA

Received 21 August 2006; received in revised form 3 December 2006
Available online 28 December 2006

Abstract

Three-dimensional test results for frictional materials such as soils, concrete, and rock show that the shapes of their failure surfaces are influenced by the intermediate principal stress, shear banding, and cross-anisotropy (transverse isotropy). A general 3D failure criterion for cross-anisotropic soils for both non-rotating and rotating stresses is presented and compared with experimental results for sand, clay, and sea ice. The formulation relates the loading direction to the principal directions of the cross-anisotropic microstructure of the material. The criterion is based on a function of stress, previously used as the 3D failure criterion for isotropic frictional materials, which is set equal to a scalar that varies over a sphere. The formulation is specialized for true triaxial tests and torsion shear tests and determination of material parameters is demonstrated. The failure criterion is compared with experimental results from the literature to show that it is able to capture the conditions obtained in three-dimensional experiments without and with stress rotations. The limitations of the criterion are demonstrated by its failure to capture the behavior of some cross-anisotropic materials with particular micro-structures, and the reasons for the shortcomings are discussed.

Keywords: Cross-anisotropy; Failure criterion; Frictional materials; Plasticity; Shear strength; Torsions shear tests; Triaxial tests; True triaxial tests

1. Introduction

Results of three-dimensional tests show that the shape of the failure surface for frictional materials is influenced by the intermediate principal stress, shear banding, and cross-anisotropy (transverse isotropy). It has long been known that the failure condition is influenced by the intermediate principal stress. Experiments have been performed and several expressions for three-dimensional failure criteria have been proposed to include the effect of the intermediate principal stress for isotropic frictional materials such as soils, rock, and concrete. Experiments on sand have shown that shear banding plays an important role in the shape of the 3D failure surface (Wang and Lade, 2001), and analysis of shear banding under 3D conditions indicated that the isotropic failure criterion for homogeneous deformations played an important role in the prediction of shear banding (Lade, 2003). This isotropic criterion describes failure as occurring by a smooth peak in the stress–strain

* Tel.: +1 202 319 5164; fax: +1 202 319 6677.
E-mail address: Lade@cua.edu

0020-7683/$ - see front matter © 2007 Elsevier Ltd. All rights reserved.
doi:10.1016/j.ijsolstr.2006.12.027

relation corresponding to homogeneous deformations. It was found that shear banding occurs in the hardening regime of the stress–strain behavior in the mid-range of b-values [$b = (\sigma_2 - \sigma_3)/(\sigma_1 - \sigma_3)$], in which σ_1, σ_2, and σ_3 refer to the major, intermediate and minor principal stresses] from about 0.18 to approximately 0.85. In this range both the experiments and the analysis showed that the friction angles were lower than those implied by the isotropic failure criterion. Thus, the onset of shear banding occurs before a smooth peak is reached in the stress–strain relation and the peak stress is the result of shear banding which occurs in the mid-range of b-values. For b-values outside this range, the smooth peak is part of the homogeneous constitutive response, and shear banding is triggered in the softening regime, but then plays no role in establishing peak failure. This could only have been predicted correctly by employing an isotropic failure criterion for failure under homogeneous strain conditions (not influenced by shear banding) in the framework for the constitutive model used to predict the shear bands. Thus, the 3D failure surface is not continuous and cannot be described by a single expression, but may be obtained by combining the continuous failure surface with results of predictions of shear banding.

The experiments for which these conclusions were reached were performed in the first sector of the octahedral plane (in which $0° \leqslant \theta \leqslant 60°$) on pluviated sand specimens with horizontal bedding planes. Thus, the major principal stress was always perpendicular to the bedding planes in these tests. Experiments performed in all three sectors of the octahedral plane (in which $0° \leqslant \theta \leqslant 180°$) indicated that pluviated sand with horizontal bedding planes (in which the principal stresses are aligned with and perpendicular to the bedding planes) exhibited cross-anisotropy (transverse isotropy) with lower strength in the horizontal directions than in the vertical direction (Abelev and Lade, 2004). The failure surface is therefore symmetric around the vertical principal stress axis in the octahedral plane. These 3D experiments also showed that the friction angle in triaxial extension at $\theta = 180°$ is substantially lower than the friction angle in triaxial extension at $\theta = 60°$. Further, the experiments performed in all three sectors of the octahedral plane showed that shear banding occurred in the hardening regime for b-values in the mid-ranges of each sector, therefore producing lower friction angles in these ranges than indicated by a continuous failure surface.

It is necessary to formulate an appropriate failure criterion to incorporate in a constitutive model for prediction of shear banding in cross-anisotropic soils. A criterion was proposed by Abelev and Lade (2004) for cross-anisotropic sand. This criterion was formulated by rotating the isotropic failure surface around the stress origin in the principal stress space. This produced a failure surface that captured the homogeneous behavior of cross-anisotropic soil as obtained by true triaxial tests and shown in an octahedral plane in Fig. 1. However, this criterion is not able to deal with effects of stress rotations in a cross-anisotropic soil, i.e., rotation of the principal stress directions relative to the horizontal bedding planes in pluviated sand.

Presented here is a formulation of a completely general 3D failure criterion for cross-anisotropic frictional materials for both non-rotating and rotating principal stresses. It is compared with experimental results from the literature to show that it is able to capture the conditions obtained in true triaxial tests without stress rotations as well as the conditions in torsion shear tests performed to study effects of stress rotation. The limitations of this criterion for materials with certain particular microstructures are also discussed.

In the presentation below, directions and relative magnitudes of principal stresses as well as directions of material axes are indicated by indices as shown in several diagrams. Indices x, y, and z are used for directions of principal stresses when their relative magnitudes are interchangeable, while indices 1, 2, and 3 are used for principal stresses to indicate their relative magnitude (σ_1 = major, σ_2 = intermediate, and σ_3 = minor principal stresses). Indices 1, 2, and 3 are also used to indicate material axes with the 2-direction being vertical and perpendicular to the horizontal (1–3) bedding plane. When double indices are used, they indicate normal and shear stresses, but principal stresses are not implied by e.g., σ_{22} (= normal stress on 2-plane in 2-direction).

2. Cross-anisotropic failure criterion

2.1. Loading direction

To formulate a failure criterion for cross-anisotropic soils it is necessary to specify the direction of loading relative to the microstructure directions of the material. A coordinate system is therefore aligned with the material axes, as shown on Fig. 2, and the loading direction is defined by a unit vector (ℓ_1, ℓ_2, ℓ_3) as proposed

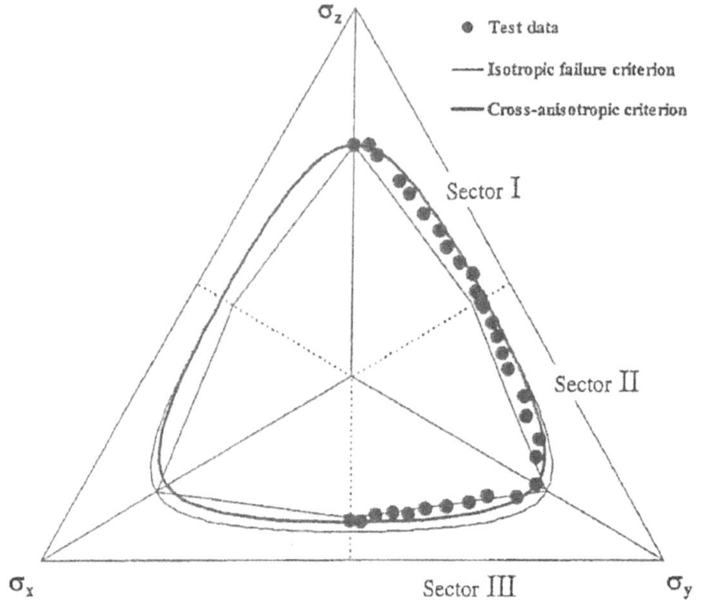

Fig. 1. Experimental strength results of true triaxial tests on dense Santa Monica Beach sand with cross-anisotropic fabric shown in octahedral plane (after Abelev and Lade, 2004). For the experimental results shown, the principal stress directions, σ_x, σ_y, and σ_z, are aligned with the material axes.

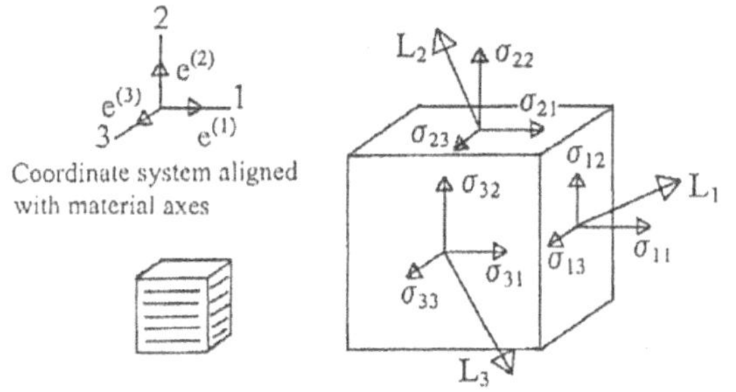

Fig. 2. Depiction of components of loading direction relative to axes of material microstructure.

by Pietruszczak and Mroz (2000, 2001). The magnitudes of resultant of stresses acting on the planes characterized by normals 1, 2, and 3 are calculated as:

$$L_1 = \sqrt{\sigma_{11}^2 + \sigma_{12}^2 + \sigma_{13}^2} \tag{1a}$$

$$L_2 = \sqrt{\sigma_{21}^2 + \sigma_{22}^2 + \sigma_{23}^2} \tag{1b}$$

$$L_3 = \sqrt{\sigma_{31}^2 + \sigma_{32}^2 + \sigma_{33}^2} \tag{1c}$$

The generalized loading vector may then be expressed as

$$L_i = L_1 \cdot e_i^{(1)} + L_2 \cdot e_i^{(2)} + L_3 \cdot e_i^{(3)} = (L_1, L_2, L_3) \tag{2}$$

and the loading direction is then specified by a unit vector relative to the material axes

$$(\ell_1, \ell_2, \ell_3) = \frac{(L_1, L_2, L_3)}{\sqrt{L_1^2 + L_2^2 + L_3^2}} \tag{3}$$

2.2. Failure criterion for anisotropic materials

Based on stress invariants and joint invariants of stress and microstructure, Pietruszczak and Mroz (2000, 2001) formulated a failure criterion for anisotropic materials

$$f = \eta = \eta_0 \cdot (1 + \Omega_{ij}\ell_i\ell_j) \tag{4}$$

Here f is a function of stresses usually employed to describe a failure criterion, and η describes the three-dimensional variation of a scalar over a sphere. The average value of η is equal to η_0 (= the radius of the sphere) and Ω_{ij} describes the deviation in three dimensions from the sphere. For an orthotropic material Ω_{ij} has two distinct eigenvalues, and for a cross-anisotropic material one scalar value is sufficient to describe the rotationally symmetric shape that characterizes such a material. The axis of rotation is oriented in the direction of the normal to the bedding planes for this material, i.e., the 2-direction in Fig. 2. For an isotropic material Ω_{ij} vanishes, and therefore $\eta = \eta_0 = \text{constant}$.

The expression in Eq. (4) may also be written as

$$\eta = \eta_0(1 + \Omega_1\ell_1^2 + \Omega_2\ell_2^2 + \Omega_3\ell_3^2) \tag{5}$$

in which (ℓ_1, ℓ_2, ℓ_3) defines the loading direction relative to the material axes as given in Eq. (3).

2.3. Failure criterion for cross-anisotropic materials

For a cross-anisotropic material the strengths in the directions of the bedding planes are the same and therefore $\Omega_1 = \Omega_3$, and since $\Omega_1 + \Omega_2 + \Omega_3 = 0$ and $\ell_1^2 + \ell_2^2 + \ell_3^2 = 1$, then the expression for a cross-anisotropic material becomes

$$f = \eta_0 \cdot \left[1 + \Omega_1(1 - 3\ell_2^2)\right] \tag{6}$$

where η_0 and Ω_1 are constant material parameters. The average value of η is equal to η_0 = the radius of a sphere, as shown projected on the octahedral plane in Fig. 3, and the factor $\eta = [1 + \Omega_1(1 - 3\ell_2^2)]$, which is controlled by the scalar material parameter Ω_1 and the loading direction ℓ_2, describes the deviation in two dimensions from the sphere (circle in octahedral plane). Thus, the right hand of Eq. (6) describes a shape that is symmetric with regard to the vertical 2-direction, as indicated in Fig. 3. The parameters used to construct the diagram in Fig. 3 are those determined for Santa Monica Beach sand tested in true triaxial tests, as discussed below.

2.4. Failure criterion for cross-anisotropic soils

Combining the expression in Eq. (6) with the expression for the isotropic three-dimensional failure criterion for soils proposed by Lade (1977) produces the following failure criterion for cross-anisotropic soils

$$f = \left(\frac{I_1^3}{I_3} - 27\right)\left(\frac{I_1}{p_a}\right)^m = \eta_0\left[1 + \Omega_1(1 - 3\ell_2^2)\right] \tag{7}$$

in which I_1 and I_3 are the first and the third invariants of the stress tensor, and p_a is atmospheric pressure in the same units as I_1.

For cross-anisotropic materials tested in common laboratory experiments in which up to 3 different, orthogonal normal stresses and one shear stress can be applied (in the 1–2 plane in Fig. 2), the expression for ℓ_2 becomes

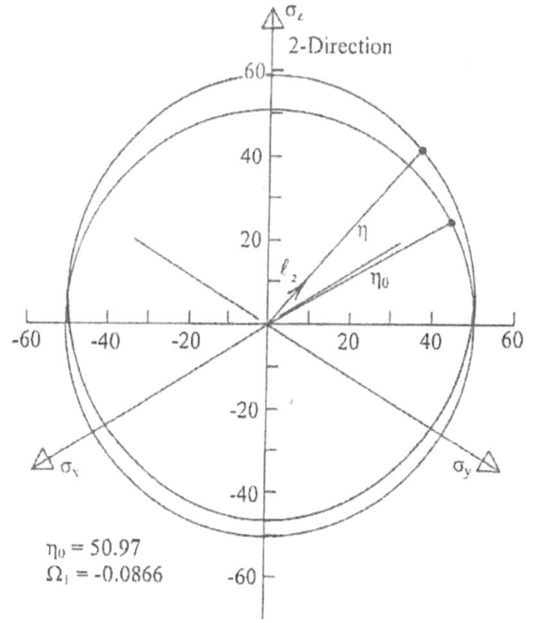

Fig. 3. Variation of factor $\eta = \eta_0[1 + \Omega_1(1 - 3\ell_2^2)]$ indicated by the symmetric shape (around the 2-direction) on the octahedral plane for dense Santa Monica Beach sand tested in true triaxial tests.

$$\ell_2 = \sqrt{\frac{\sigma_y^2 \sin^2\beta + \sigma_z^2 \cos^2\beta}{\sigma_x^2 + \sigma_y^2 + \sigma_z^2}} \tag{8}$$

in which $(\sigma_x, \sigma_y, \sigma_z)$ are principal stresses as indicated in Fig. 4. Practical experimental equipment as employed for torsion shear tests performed on hollow cylindrical specimens allows shearing in the 1–2 plane, but precludes shearing in the 1–3 plane shown in Fig. 2.

3. Formulation for experimental configurations

3.1. True triaxial tests

For true triaxial tests the principal stresses are inclined at $\beta = 0$ and the specialized expressions for the failure criterion for cross-anisotropic materials for each of the three sectors of the octahedral plane are obtained by permutation of the indices. The results are presented in terms of the major principal stress ratio

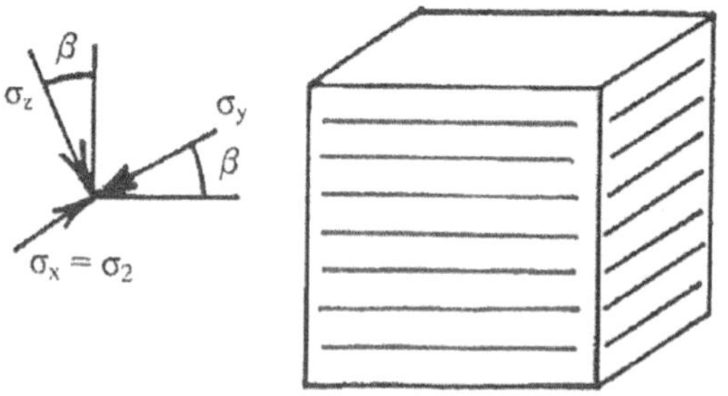

Fig. 4. Principal stress conditions possible in true triaxial and torsion shear tests.

$$R = \frac{\sigma_1}{\sigma_3} \tag{9}$$

and the value of b given by

$$b = \frac{\sigma_2 - \sigma_3}{\sigma_1 - \sigma_3} \tag{10}$$

The b-value varies from $b = 0$ at triaxial compression to $b = 1$ at triaxial extension in each of the 3 sectors shown on the octahedral plane in Fig. 5. In this figure the major, intermediate and minor principal stresses are indicated by σ_1, σ_2, and σ_3, respectively, and shown in orientation relative to the bedding planes of the specimen for each of the 3 sectors.

The expressions for ℓ_2^2 in these sectors are as follows:

Sector I:

$$\ell_2^2 = \frac{R^2}{R^2(1+b^2) + 2R(b-b^2) + (2-2b+b^2)} \tag{11a}$$

$$\text{For triaxial compression}: b = 0: \quad \ell_2^2 = \frac{R^2}{R^2+2} \tag{11b}$$

$$\text{For triaxial extension}: b = 1: \quad \ell_2^2 = \frac{R^2}{2R^2+1} \tag{11c}$$

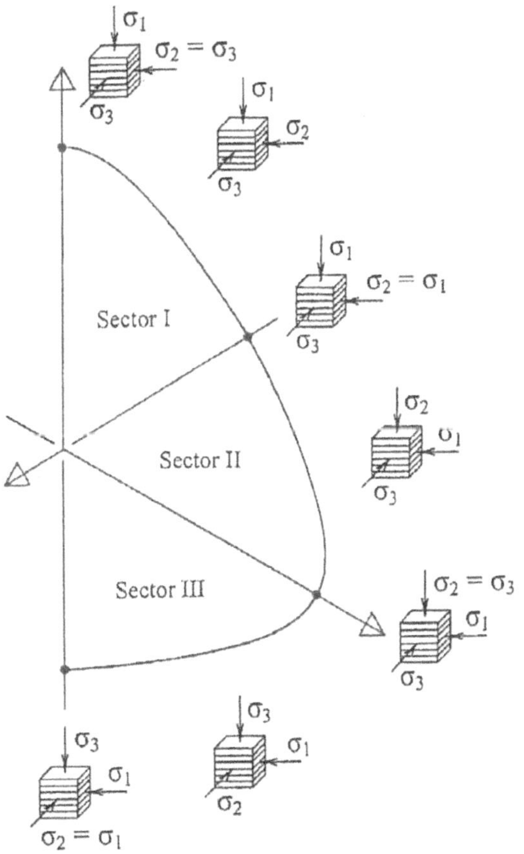

Fig. 5. Application of major, intermediate and minor principal stresses, σ_1, σ_2, and σ_3, to specimen with bedding planes in true triaxial tests to achieve testing in all three sectors of octahedral plane.

Sector II:

$$\ell_2^2 = \frac{b^2R + 2R(b - b^2) + (1 - 2b + b^2)}{R^2(1 + b^2) + 2R(b - b^2) + (2 - 2b + b^2)} \quad (12a)$$

$$\text{For triaxial compression} : b = 0 : \quad \ell_2^2 = \frac{1}{R^2 + 2} \quad (12b)$$

$$\text{For triaxial extension} : b = 1 : \quad \ell_2^2 = \frac{R^2}{2R^2 + 1} \quad (12c)$$

Sector III:

$$\ell_2^2 = \frac{1}{R^2(1 + b^2) + 2R(b - b^2) + (2 - 2b + b^2)} \quad (13a)$$

$$\text{For triaxial compression} : b = 0 : \quad \ell_2^2 = \frac{1}{R^2 + 2} \quad (13b)$$

$$\text{For triaxial extension} : b = 1 : \quad \ell_2^2 = \frac{1}{2R^2 + 1} \quad (13c)$$

Note that Eqs. (11c) = (12c) and Eqs. (12b) = (13b). Thus, the failure surface is continuous across the projected coordinate axes in the octahedral plane shown in Fig. 5.

3.2. Torsion shear tests

For torsion shear tests with the same inside and outside cell pressure, this cell pressure is always the intermediate principal stress, $\sigma_x = \sigma_2$, and the inclination β of the major principal stress relates to the b-value according to $\sin^2\beta = b$. The expression for ℓ_2^2 from Eq. (8) therefore becomes

$$\ell_2^2 = \frac{R^2(1 - b) + b}{R^2(1 + b^2) + 2R(b - b^2) + (2 - 2b + b^2)} \quad (14a)$$

$$\text{For triaxial compression} : b = 0 : \quad \ell_2^2 = \frac{R^2}{R^2 + 2} \quad (14b)$$

$$\text{For triaxial extension} : b = 1 : \quad \ell_2^2 = \frac{1}{2R^2 + 1} \quad (14c)$$

Note that for $b = 0$ Eqs. (14b) = (11b), i.e., conventional triaxial compression in sector I, and for $b = 1$ Eqs. (14c) = (13c), i.e., conventional triaxial extension in sector III. These are the only two comparable stress states for torsion shear tests and true triaxial tests.

4. Parameter determination

The cross-anisotropic characteristics that are present in a given material may change with the stress state as they do in soil deposits. Thus, Lade and Abelev (2005) showed that isotropic compression of initially cross-anisotropic deposits of Santa Monica Beach sand at different densities all lost these characteristics and became isotropic at high pressures. A gradual decrease in the cross-anisotropic behavior occurred with increasing isotopic pressure. If the stresses had been increased under K_0-conditions (in which $K_0 = \sigma_{\text{horizontal}}/\sigma_{\text{vertical}}$ for the condition of zero strain in any horizontal direction, i.e., the usual state of stress in horizontal ground), then another cross-anisotropic state would have been reached at high pressures. Therefore, it is not possible to determine parameters once and for all that characterize the cross-anisotropic behavior of the soil at all states of stress. However, it is possible to determine parameters for the behavior in the vicinity of the stress states at which the given soil deposit has been tested, but these parameters may not apply if the stress state is changed substantially away from those employed in the experiments.

To obtain reliable parameters for the failure criterion, it is important to choose sets of test data that clearly contain the cross-anisotropic characteristics between them. For example, results of triaxial compression tests on specimens with bedding planes in horizontal and vertical directions will clearly embody such characteristics, while results from triaxial compression and extension tests from the first sector will mainly exhibit isotropic characteristics, even if other results indicate that the material is clearly cross-anisotropic. However, results from conventional triaxial compression (in sector I) and conventional triaxial extension tests (in sector III) will clearly represent the cross-anisotropic behavior. Any two sets of data produced with widely different stress states at failure relative to the orientation of the material microstructure will suffice for determination of appropriate parameters.

4.1. Parameters from triaxial compression tests on soils

To determine all three parameters, η_0, m, and Ω_1 in the expression for the failure criterion for soils in Eq. (7), it is preferable to have results from 3 triaxial compression tests on vertical specimens (conventional tests) and 3 triaxial compression tests on horizontal specimens. The results for the dense Santa Monica Beach sand shown in Fig. 1 will be used to demonstrate this approach. In connection with the study of cross-anisotropy of dense Santa Monica Beach sand presented by Abelev and Lade (2004), drained triaxial compression tests were performed on cubical specimens with different confining pressures in the first sector and one test was conducted in the third (and second) sector. The results from the first sector are used to determine the values of η_1 and m, as shown in Fig. 6. These parameters are obtained by plotting $(I_1^3/I_3 - 27)$ versus p_a/I_1 on log–log scales and determining the best fitting line, which is then characterized by the intercept, η_1, with $p_a/I_1 = 1$, and by the geometric slope, m. The value of m characterizes the curvature of the failure surface in meridian planes. It is assumed that the curvature for results from horizontal specimens shown in the third sector is the same as the curvature for vertical specimens shown in the first sector, $m = 0.20$. The point corresponding to the result of the triaxial compression test from the third sector is then employed to locate the parallel line, as seen in Fig. 6. The diagram shows that for vertical specimens, $\eta_{1v} = 78$, while for horizontal specimens, $\eta_{1h} = 62$.

To obtain the values of η_0 and Ω_1 corresponding to the range of stresses in which the experiments were performed, as explained above, the values of η_1 are adjusted to correspond to $I_1/p_a = 4.0$. Thus, for vertical specimens $\eta_v = 59.1$, while for horizontal specimens, $\eta_h = 47.0$. Then two linear equations are solved based on the right hand side of Eq. (7). In this expression the values of ℓ_2^2 are obtained from Eq. (11b) for sector I and from Eq. (13b) for sector III. In these expressions it is necessary to know the principal stress ratios at failure, R. The values of η determined from the vertical and the horizontal specimens are employed for determination of the R-values. This is done by solving the cubical equation based on the left hand side of Eq. (7) for triaxial compression conditions

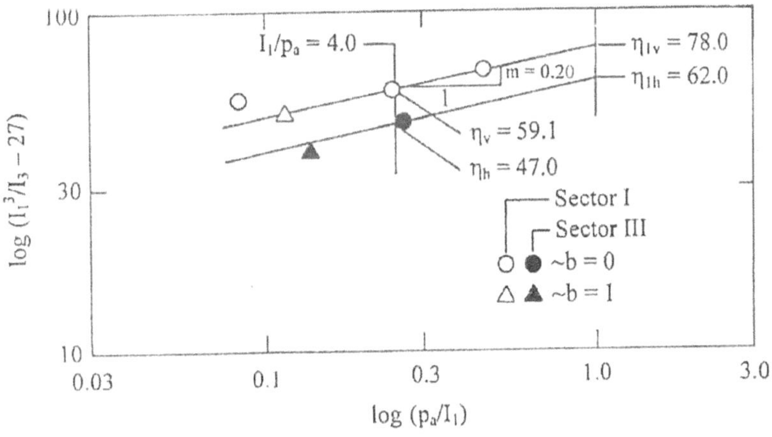

Fig. 6. Determination of parameters η_1 and m from triaxial compression tests on three vertical specimens and one horizontal specimen of dense Santa Monica Beach sand.

$$\frac{I_1^3}{I_3} - 27 = \frac{(\sigma_1 + 2\sigma_3)^3}{\sigma_1 \cdot \sigma_3^2} - 27 = \frac{(R+2)^3}{R} - 27 = \eta \tag{15}$$

The relation between R and η may be determined once and for all for easy determination of $R = f(\eta)$. The values of $R_v = 6.000$ and $R_h = 5.340$ are obtained from Eq. (15) and then substituted into Eqs. (11b) and (13b), which in turn are the bases for the two linear equations from the right hand side of Eq. (7). The two linear equations are then solved for the unknowns, η_0 and Ω_1.

Performing the operations indicated above results in $\eta_0 = 50.97$ and $\Omega_1 = -0.0866$ for the dense Santa Monica Beach sand. Fig. 3 shows the variation of the term on the right hand of Eq. (7) for these parameters.

4.2. Parameters from triaxial compression and extension tests

Parameters may also be determined from conventional triaxial compression tests and a conventional triaxial extension test. To demonstrate this approach the data from a series of undrained true triaxial tests on San Francisco Bay Mud tested in all three sectors of the octahedral plane shown in Fig. 5 will be employed. The specimens were carved from large intact block samples and isotropically consolidated before shearing under three-dimensional conditions (Kirkgard and Lade 1991,1993). Based on effective stresses at failure of the vertical specimens tested in triaxial compression ($b = 0$, sector I) at four different consolidation pressures, Fig. 7 shows the determination of the intercept, $\eta_{1c} = 47.0$, with $I_1/p_a = 1.0$ and the slope $m = 0.64$.

With the assumption that the curvature for results from specimens tested in conventional triaxial extension ($b = 1$, sector III) is the same as the curvature for vertical specimens shown in the first sector, a straight line with slope $m = 0.64$ is drawn through the point corresponding to the available result for the extension test, as shown in Fig. 7. The diagram shows that for horizontal specimens, $\eta_{1e} = 22.5$.

To obtain the values of η_0 and Ω_1 corresponding to the range of stresses in which the experiments were performed, the values of η_1 are adjusted to correspond to $I_1/p_a = 2.0$ which represents the average value at failure for the three-dimensional experiments on San Francisco Bay Mud. Thus, for triaxial compression $\eta_c = 30.1$, while for triaxial extension $\eta_e = 14.5$. Then two linear equations are solved based on the right hand side of Eq. (7). In this expression the values of ℓ_2^2 are obtained from Eq. (11b) for sector I and from Eq. (13c) for sector III. In these expressions it is necessary to know the principal stress ratios at failure, R. The values of η determined from the triaxial compression and the extension specimens are employed for determination of the R-values. This is done by solving the cubical equation based on the left hand side of Eq. (7) for triaxial compression conditions.

The value of $R_c = 4.225$ is obtained from Eq. (15) for $\eta_c = 30.1$, and the value of R_e is obtained by solving the cubical equation based on the left hand side of Eq. (7) for triaxial extension conditions:

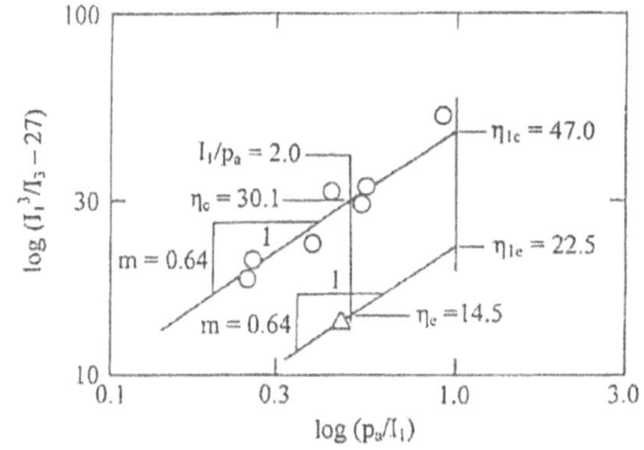

Fig. 7. Determination of parameters η_1 and m from conventional undrained triaxial compression tests and one conventional undrained triaxial extension test on San Francisco Bay Mud.

$$\frac{I_1^3}{I_3} - 27 = \frac{(2\sigma_1 + \sigma_3)^3}{\sigma_1^2 \cdot \sigma_3} - 27 = \frac{(2R+1)^3}{R^2} - 27 = \eta \tag{16}$$

Eq. (16) produces $R_e = 3.460$ for $\eta_e = 14.5$. The two R-values are then substituted into Eqs. (11b) and (13c), which in turn are the bases for the two linear equations from the right hand side of Eq. (7). The two linear equations are then solved for the unknowns, η_0 and Ω_1.

Performing the operations indicated above results in $\eta_0 = 19.82$ and $\Omega_1 = -0.305$ for the San Francisco Bay Mud.

4.3. Parameters from any set of test data containing sufficient information

It is possible to determine the three parameters, m, η_0, and Ω_1 from any type of experiments, as long as the salient information is contained in the two data sets involved in the calculations. It is necessary to follow the procedures demonstrated above for the cases of obtaining the data from (1) conventional triaxial compression tests on vertical specimens and either (2a) triaxial compression tests on horizontal specimens or (2b) conventional triaxial extension tests on vertical specimens. It should also be recalled that the shear strengths from experiments in the mid-ranges of b-values may be affected by shear banding and these strengths are therefore not representative of the homogeneous deformation required for the failure criterion for cross-anisotropic soils. The occurrence of shear banding is a discrete event that cannot be included in a failure criterion for homogeneous deformations and smooth peak failure. The test data employed should therefore be produced near $b = 0$ and/or $b = 1$.

5. Comparison with 3D test results

5.1. True triaxial tests

5.1.1. Sand

Comparisons of test data with failure criterion for the dense Santa Monica Beach sand are indicated in Fig. 8. The material parameters were determined from triaxial compression tests on vertical specimens and one triaxial compression test on a horizontal specimen, as shown in Fig. 6. The comparisons in Fig. 8 are shown as friction angles, φ, versus b-values. The experimental results and the predictions from the failure criterion for cross-anisotropic soils are shown on the same φ–b diagram for all three sectors. The diagram indicates a good fit for the homogeneous results near $b = 0$ and $b = 1$, while the data in the mid-ranges of b-values in all three sectors clearly shows the effects of shear banding in the hardening regime, thus resulting in lower friction angles than obtained from the tests with homogeneous deformations. The criterion indicates a friction angle in conventional triaxial extension ($b = 1$, sector III) that is 1.5° higher than measured, a result that is acceptable in view of the fact that only the triaxial compression tests were used for parameter determination. The proposed failure criterion for the dense cross-anisotropic Santa Monica Beach sand essentially traces the same surface on the octahedral plane as that shown in Fig. 1.

The most noticeable difference between the experimental failure points and the failure criterion, exhibited in Figs. 1 and 8, occurs in the mid-ranges of b where shear banding plays an important role for the shear strength. To analyze the occurrence of shear banding for the dense Santa Monica Beach sand, the homogeneous failure criterion for cross-anisotropic soils presented here is required in the constitutive model to be used for such analyses.

Fig. 3 shows the variation of the term on the right hand of Eq. (7) for the parameters $\eta_0 = 50.97$ and $\Omega_1 = -0.0866$ determined for the dense Santa Monica Beach sand tested in true triaxial tests.

5.1.2. Isotropically consolidated san Francisco Bay Mud

Comparison between the proposed cross-anisotropic failure criterion and the experimental results for the true triaxial tests on San Francisco Bay Mud is shown on the φ–b diagram in Fig. 9. In this case the material parameters were determined from conventional triaxial compression tests ($b = 0$, sector I) and a single conventional triaxial extension test ($b = 1$, sector III), as shown in Fig. 7. The fit between criterion and experimental

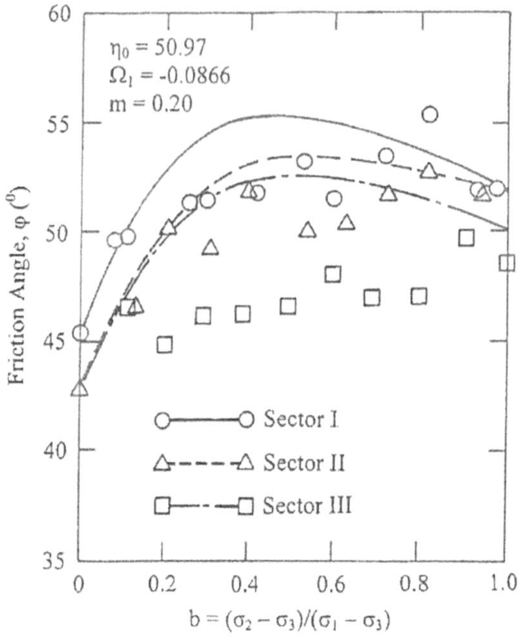

Fig. 8. Comparison of failure criterion with true triaxial test data for cross-anisotropic, dense Santa Monica Beach sand tested in all three sectors of the octahedral plane.

data is reasonably good, with the largest deviations at lower *b*-values in sector II, where differences of 2°–3° occur. This region is furthest away from the points used for parameter determination. Nevertheless, the failure criterion captures the overall cross-anisotropic nature of the San Francisco Bay Mud as tested in three dimensions.

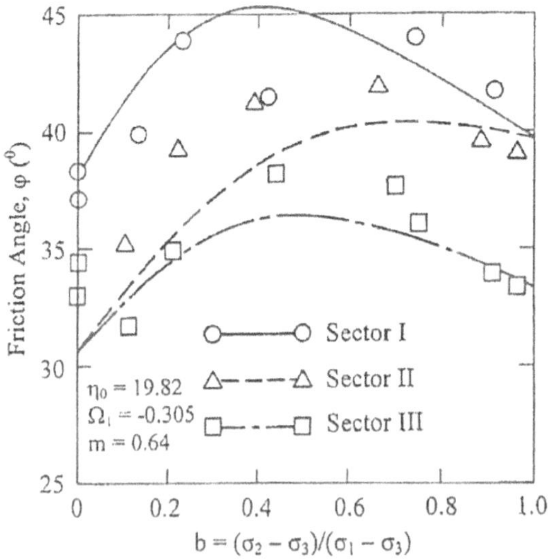

Fig. 9. Comparison of failure criterion with true triaxial test data for isotropically consolidated San Francisco bay mud tested in all three sectors of the octahedral plane.

5.2. Torsion shear tests

Appropriate parameters have been determined by the procedure reviewed above for comparison with the results of 34 drained torsion shear tests on medium dense Santa Monica Beach sand ($e = 0.68$ corresponding to a relative density of 70%) presented by Lade et al. (2006a,b). The hollow cylinder specimens employed in these tests were sheared with the same inside and outside pressure, while vertical deviator stresses and torsion shear stresses were applied at the ends. In these tests the major principal stress is inclined with the axis of the hollow cylinder according to the relation between the deviator and the shear stresses, and the value of b is related to the σ_1-inclination, β, according to $b = \sin^2\beta$. Thus, $b = 0.0$ for $\beta = 0°$, and $b = 1.0$ for $\beta = 90°$. These conditions correspond to those in conventional triaxial compression (sector I) and conventional triaxial extension tests (sector III). To achieve results for b-values between 0.0 and 1.0, experiments were performed with σ_1-inclinations from 0° to 90°. The results of these tests showed considerable scatter, especially at higher b-values, as shown in Fig. 10. It is clear that the strengths correspond to those of a cross-anisotropic material, but they cannot be indicated in the sectors on the octahedral plane. Only the conventional triaxial compression and extension results can be shown on the octahedral plane, and they correspond to the top point in sector I and to the bottom point in sector III.

The parameters for the failure criterion for torsion shear tests were determined from a triaxial compression test in sector I and a triaxial extension test in sector III, i.e., both conventional types of tests performed on the hollow cylinder specimens employed in the torsion shear apparatus. Data were not available for determination of the curvature of the failure surface, and therefore the parameter $m = 0$. The parameters for modeling the cross-anisotropic behavior were determined to be $\eta_0 = 26.55$ and $\Omega_1 = -0.240$. These values are based on friction angles of $\varphi_{comp} = 40.7°$ and $\varphi_{ext} = 38.4°$, i.e., the friction angle in extension is somewhat lower than the friction angle in compression for the torsion shear tests, thus indicating a significant amount of cross-anisotropy in the sand specimens created by dry pluviation.

As may be seen from the φ–b diagram in Fig. 10, the failure criterion fits the data well in the range of b-values from 0.0 to about 0.3 in which homogeneous deformations occurred up to smooth peak failure. Beyond $b = 0.3$, which is approximately the condition in plane strain tests, shear banding in the hardening regime resulted in lower friction angles than indicated by the failure criterion for the cross-anisotropic sand, which models the homogeneous behavior. Thus, with the provision of the occurrence of shear banding in the mid-range of b-values, the proposed failure criterion for cross-anisotropic sand appears to fit the test data well.

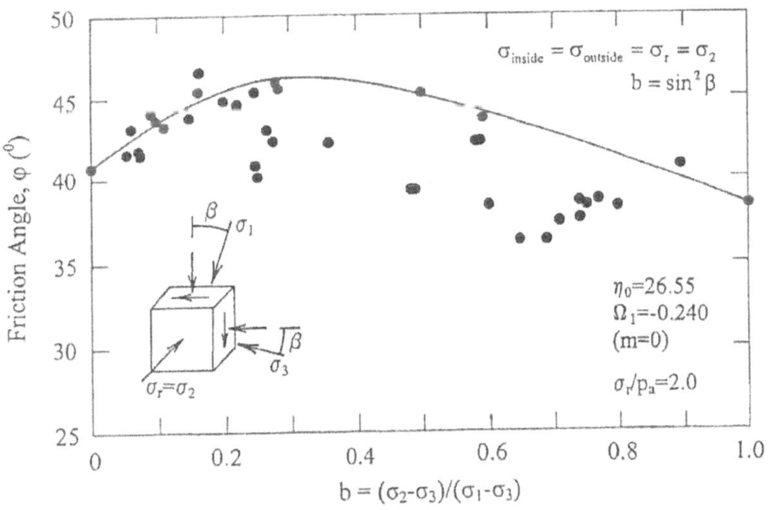

Fig. 10. Comparison of failure criterion with torsion shear test data for medium dense, cross-anisotropic Santa Monica Beach sand. Shear banding in the hardening regime reduces friction angles in mid-ranges of b-values.

6. Limitations

Not all data sets will fit the proposed failure criterion for cross-anisotropic frictional materials, because special features of the material dominate the measured behavior. Two examples are given below.

6.1. Sea ice

Sea ice, formed on the surface of the ocean, consists mainly of closely fitted vertical columnar ice crystals, as shown in Fig. 11. With its columnar arrangement, the sea ice exhibits strong cross-anisotropic behavior with lower strengths in horizontal directions than in vertical directions. The horizontal strength increases with confining pressure, thus revealing its frictional nature, as discussed by Lade (2002). Fig. 12 shows typical biaxial strength envelopes in the horizontal plane (i.e., the plane of the ice floe) and in the vertical plane. The uniaxial strength in the vertical direction is in this case 5–6 times higher than the uniaxial strength in the horizontal direction. The shape of the three-dimensional failure surface is indicated in Fig. 13. This horizontally narrow and vertically oblong shape cannot be modeled by the proposed failure criterion for cross-anisotropic frictional materials. However, the strength is strongly dependent on the strain rate at which it is loaded and the temperature at which it is tested. Fig. 14 shows effects of strain rate and temperature on the shape of the failure surface in the horizontal biaxial plane. While the wide shape observed at low strain rates, shown on Fig. 14(a), may be modeled by the failure criterion, the flexibility required to capture the variable shape indicated in Fig. 14 is simply not present in the criterion and would require a more extensive framework with several additional material parameters. Thus, the strongly cross-anisotropic and variable failure surface of sea ice cannot be modeled by the proposed failure criterion.

6.2. K_0-Consolidated San Francisco Bay Mud

In addition to the experiments presented above for true triaxial compression on isotropically consolidated, intact San Francisco Bay Mud, a series of undrained torsion shear tests were performed on hollow cylinder specimens carved from the large, cylindrical, intact samples taken from the natural clay deposits in the San Francisco Bay (Lade and Kirkgard, 2000). These specimens were consolidated under K_0-conditions, and the results of the torsion shear tests in which stress rotation occurs were substantially different from the results for the isotropically consolidated cubical specimens tested in true triaxial tests without stress rotation. The cubical specimens and the hollow cylinder specimens were carved from the same large samples, but they were consolidated under different stress states. While the K_0-consolidated hollow cylinder specimens appeared to retain their original clay particle fabric, the isotropically consolidated specimens showed signs of fabric alterations and weakening (Lade and Kirkgard, 2000). Consequently, the results were quite different for the two

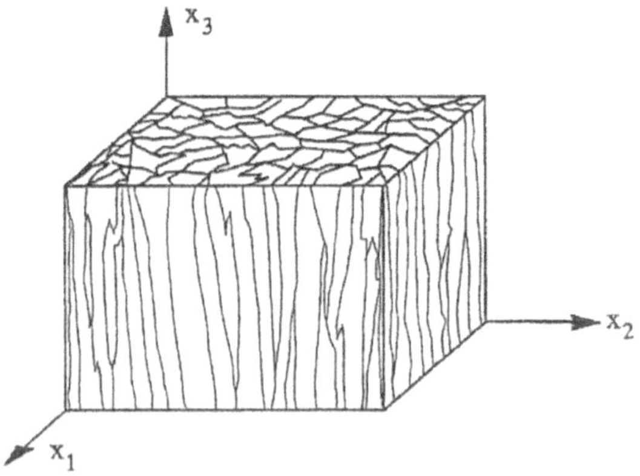

Fig. 11. Macroscopic structure of sea ice with vertical columns of closely fitted ice crystals.

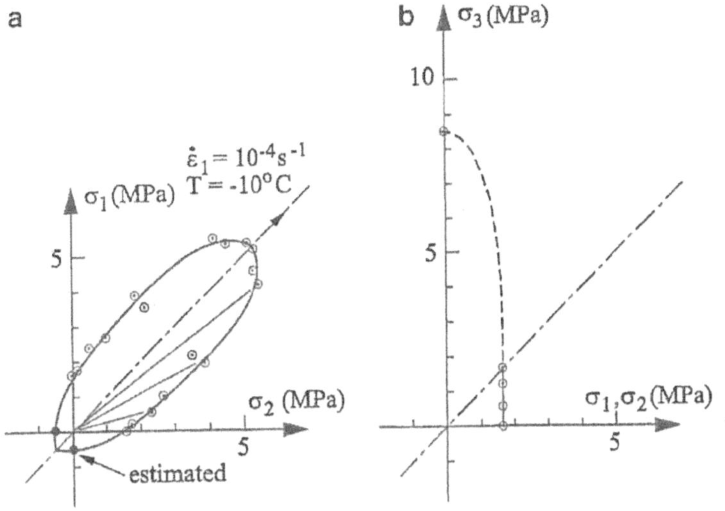

Fig. 12. Comparable biaxial strength envelopes in (a) horizontal X_1–X_2 planes and (b) vertical X_1–X_3 planes (after Lade, 2002; data from Schulson and Nickolayev, 1995).

modes of consolidation. Both exhibited significant cross-anisotropic behavior, but these were opposite in nature, as indicated by the effective stress paths shown on the triaxial plane in Fig. 15 for the triaxial compression and extension tests performed on the San Francisco Bay Mud.

As for the torsion shear tests on Santa Monica Beach sand, the parameters for the failure criterion in Eq. (7) were determined from the friction angles obtained from the conventional triaxial compression and conventional triaxial extension tests performed on hollow cylinder specimens whose effective stress paths are shown in Fig. 15. Insufficient data were available to determine the curvature of the failure surface for the K_0-consolidated Bay Mud, and $m = 0$ was therefore employed. However, the effective mean normal stresses at failure were closely the same and $I_1/p_a = 3 \cdot \sigma_m/p_a = 3.0$ for the triaxial compression and extension tests, as seen in Fig. 15. Thus, for friction angles of $\varphi_{comp} = 34°$ and $\varphi_{ext} = 50°$, the parameters for modeling the cross-anisotropic behavior were determined to be $\eta_0 = 36.62$ and $\Omega_1 = 0.269$.

Note that the value of Ω_1 is positive, indicating that the strength is higher in the horizontal than in the vertical direction. This unusual behavior is supported by the high friction angle in conventional extension in sector III of 50°. This is much higher than the expected value of 38.6° for an isotropic material with a friction angle in compression of 34°. Further support for this high strength in extension is obtained from the other torsion shear tests, which together produce a regular pattern that fits with the high strength in extension, as shown in Fig. 16. Lade and Kirkgard (2000) also showed that a somewhat similar variation in strength as indicated for the Bay Mud was obtained for K_0-consolidated kaolin tested in torsion shear. Thus, it is difficult to dispute the experimental results for the torsion shear tests on San Francisco Bay Mud.

The failure criterion for cross-anisotropic behavior is also plotted on Fig. 16. It is clearly seen that the criterion does not model very well the observed cross-anisotropic behavior of the K_0-consolidated San Francisco Bay Mud. Although fitted to the friction angles at the two ends of the diagram, the criterion simply fails to capture the initial flat strength curve for low b-values, as well as the gradual increase in friction angle at higher b-values. The behavior observed in the torsion shear tests at low b-values may relate to sliding on the plate-like clay particles present in the Bay Mud with very little influence of the intermediate principal stress, while the plate-like structure may buckle up at high b-values and produce higher strengths. Both conjectured phenomena cannot be captured by the proposed failure criterion.

Thus, although the two materials reviewed above clearly exhibit cross-anisotropic behavior, the particular structures of the micro-fabrics inside the sea ice and inside the K_0-consolidated San Francisco Bay Mud cause the materials to behave in ways that are not in accordance with the failure criterion for cross-anisotropic materials proposed here.

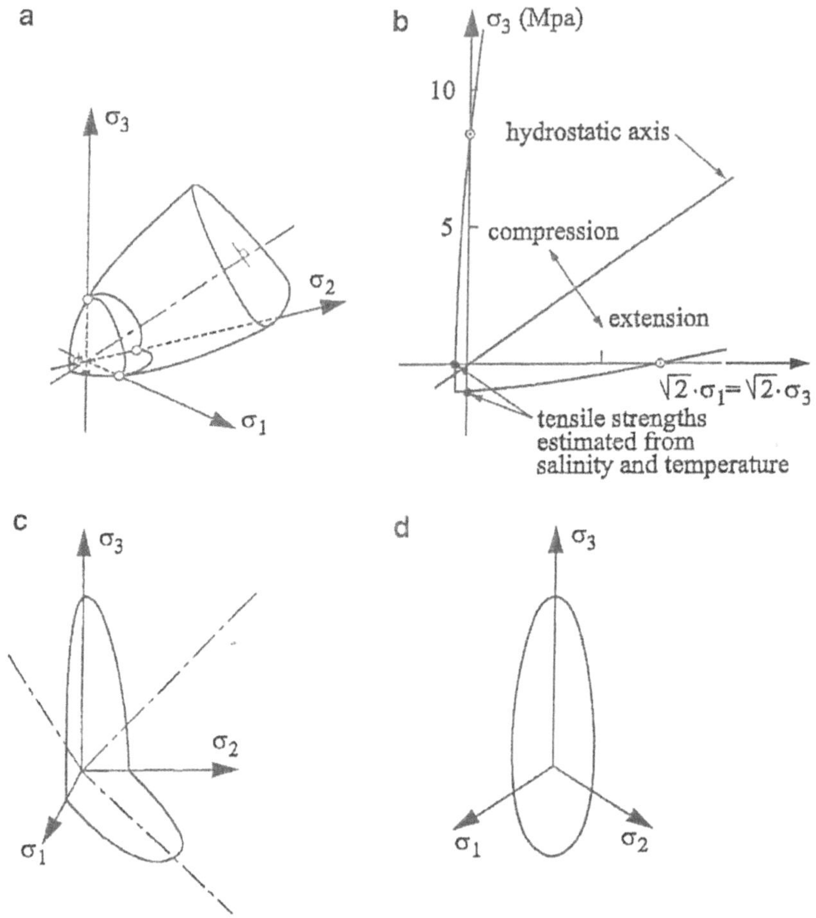

Fig. 13. (a) Overview of principal stress space with failure surface for an isotropic cohesive-frictional material. Experimental results for sea ice shown in (b) triaxial plane, (c) principal stress space with traces in three biaxial planes, and (d) octahedral plane (after Lade, 2002; data from Schulson and Nickolayev, 1995).

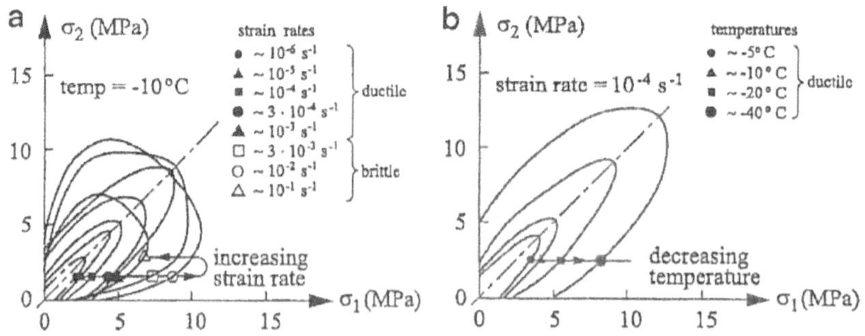

Fig. 14. Biaxial strength envelopes in horizontal X_1–X_2 planes showing (a) effects of strain rate and (b) effects of temperature (after Lade, 2002; data from Schulson and Nickolayev, 1995).

7. Conclusions

A general 3D failure criterion for cross-anisotropic soils for both non-rotating and rotating stresses is presented. The formulation is compared with experimental results from the literature to show that it is able to

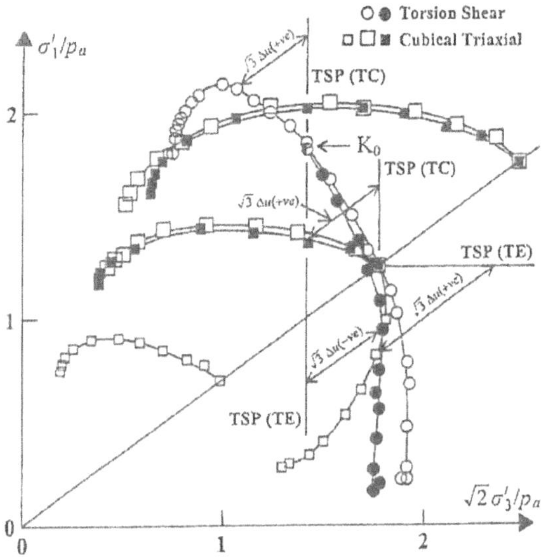

Fig. 15. Effective stress paths in triaxial plane for undrained triaxial compression and extension tests with initial isotropic and K_0-consolidation performed on San Francisco Bay Mud (after Lade and Kirkgard, 2000).

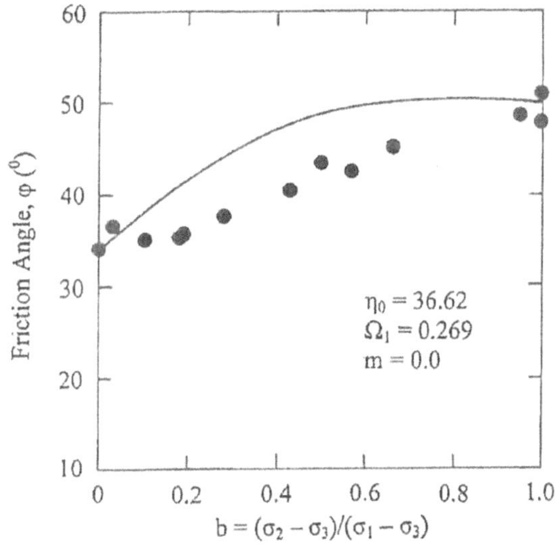

Fig. 16. Comparison of effective friction angles from undrained torsion shear tests on K_0-consolidated San Francisco Bay Mud.

capture the conditions obtained in true triaxial tests without stress rotations as well as the conditions in torsion shear tests performed to study the effects of stress rotation.

The formulation of the failure criterion is based on an isotropic criterion for which the material constant is taken to vary with the direction of loading in three dimensions. The loading direction is defined and included in the formulation of the failure criterion for cross-anisotropic soils.

Determination of the three material parameters, m, η_0, and Ω_1 for the failure criterion is demonstrated for the two most likely conditions, namely based on data from (1) conventional triaxial compression tests on vertical specimens and either (2a) triaxial compression tests on horizontal specimens or (2b) conventional triaxial extension tests on vertical specimens. Other experiments may be employed, but the shear strengths from experiments in the mid-ranges of b-values may be affected by shear banding and these strengths are therefore not representative of the homogeneous deformation required for the failure criterion for cross-anisotropic soils. The test data employed should therefore be produced near $b = 0$ and/or $b = 1$.

The limitations of the criterion are demonstrated by its failure to capture the behavior of some cross-anisotropic materials with particular micro-structures, and the reasons for the shortcomings are discussed.

References

Abelev, A.V., Lade, P.V., 2004. Characterization of failure in cross-anisotropic soils. J. Eng. Mech. ASCE 130 (5), 599–606.

Kirkgard, M.M., Lade, P.V., 1991. Anisotropy of normally consolidated San Francisco bay mud. Geotech. Test. J., ASTM 14 (3), 231–246.

Kirkgard, M.M., Lade, P.V., 1993. Anisotropic three-dimensional behavior of a normally consolidated Clay. Can Geotech. J. 30 (4), 848–858.

Lade, P.V., 1977. Elasto-plastic stress–strain theory for cohesionless soil with curved yield surfaces. Int. J. Solid Struct., Pergamon Press, Inc., New York, NY, 13, pp. 1019–1035.

Lade, P.V., 2002. Characterization of the mechanical behavior of sea ice as a frictional material. J. Geophys. Res. 107 (B12), 2377. doi:10.1029/2001JB000497.

Lade, P.V., 2003. Analysis and prediction of shear banding under 3D conditions in granular materials. Soils Found. 43 (4), 161–172.

Lade, P.V., Abelev, A.V., 2005. Characterization of cross-anisotropic soil deposits from isotropic compression tests. Soils Found. 45 (5), 89–102.

Lade, P.V., Kirkgard, M.M., 2000. Effects of stress rotation and changes of b-values on cross-anisotropic behavior of natural K_0-consolidated soft clay. Soils Found. 40 (6), 93–105.

Lade, P.V., Nam, J., Hong, W.P., 2006a. Shear banding and cross-anisotropic behavior observed in laboratory sand tests with stress rotation. Can. Geotech. J. (submitted for publication).

Lade, P.V., Nam, J., Hong, W.P., 2006b. Interpretation of strains in torsions shear tests. Soils Found. (submitted for publication).

Pietruszczak, S., Mroz, Z., 2000. Formulation of anisotropic failure criteria incorporating a microstructure tensor. Comput. Geotech. 26, 105–112.

Pietruszczak, S., Mroz, Z., 2001. On failure criteria for anisotropic cohesive-frictional materials. Int. J. Num. Analyt. Meth. Geomech. 25, 509–524.

Schulson, E.M., Nickolayev, O.Y., 1995. Failure of columnar saline ice under biaxial compression: failure envelopes and the brittle-ductile transition, J. Geophys. Res. 100 (B11), 22,383–22,400.

Wang, Q., Lade, P.V., 2001. Shear banding in true triaxial tests and its effect on failure in sand. J. Eng. Mech. ASCE 127 (8), 754–761.

Geotech Geol Eng
https://doi.org/10.1007/s10706-023-02553-y

ORIGINAL PAPER

Check for updates

Characteristic Behavior of Cross-Anisotropic Deposits of Granular Soils

Siddharath Singh · Burak F. Tanyu · Poul V. Lade

Received: 29 July 2022 / Accepted: 21 June 2023

Abstract Triaxial compression tests were performed on sand to simulate the cross-anisotropic behavior of soil layers deposited by vertical gravitation. Specimens were created with horizontal and vertical bedding planes, which were obtained by air pluviation in the vertical direction into an upright mold and two halves of the horizontally lying mold respectively. After pluviation, horizontal specimens were water saturated and frozen, which allowed the assembly of the two halves to create a cylindrical specimen for testing. All specimens were isotropically confined at 20, 60, and 180 kPa. The characteristic behavior was controlled by the contact normals between the grains, which favor the vertical direction. Influence of the cross-anisotropy was analyzed by stress–strain, volume change, shear strength, and elastic properties. The behavior of the vertical specimens showed stiffer initial slopes, smaller strain-to-failure, higher shear strengths, less initial contraction and higher rates of dilation than the horizontal specimens. Cross-anisotropy was most pronounced for dense sand and reduces with increasing confining pressure and void ratio and vanishes at very high confining pressures and void ratios. The elastic properties evaluated from unloading–reloading cycles showed slightly higher values of Young's moduli for vertical specimens than for horizontal specimens and they increased with the sand density.

Keywords Cross-anisotropy · Failure · Sand · Shear strength · Stress–strain behavior · Triaxial test

List of Symbols

σ_1	Major principal stress
σ_2	Intermediate principal stress (in triaxial compression test $\sigma_2 = \sigma_3$)
σ_3	Minor principal stress
σ_1/σ_3	Obliquity
ε_v	Volumetric strain ($\Delta V/Vo$, change in volume/original volume of the specimen)
ε_1	Axial strain (($\Delta H/Ho$, change in height/original height of the specimen)
ϕ	Friction angle
ψ	Dilation angle
E	Young's modulus
ν	Poisson's ratio
M	Non-dimensional parameter for elastic component

S. Singh · B. F. Tanyu (✉)
Department of Civil and Environmental Engineering, Sustainable Geo Infrastructure Research Group, George Mason University, 4461 Rockfish Creek Ln, Krasnow Institute, Suite 115, Fairfax, VA, USA
e-mail: btanyu@gmu.edu

S. Singh
e-mail: ssingh26@gmu.edu

P. V. Lade
Sustainable Geo Infrastructure Research Group, George Mason University, 4461 Rockfish Creek Ln, Krasnow Institute, Suite 103, Fairfax, VA, USA
e-mail: ladepoul@gmail.com

Published online: 04 August 2023

Springer

λ	Non-dimensional parameter for elastic exponent
Pa	Atmospheric pressure
I_1	First stress invariant
J_2'	Second stress deviator invariant
R	Non-dimensional constant
e_{max}	Maximum void ratio
e_{min}	Minimum void ratio
D_{10}	Particle size smaller than 10%
D_{30}	Particle size smaller than 30%
D_{60}	Particle size smaller than 60%
Cc	Coefficient of curvature
Cu	Coefficient of uniformity
Gs	Specific gravity

1 Introduction

Many natural and man-made soil deposits exhibit some degree of cross-anisotropy, which is indicated by the material response when loaded in different directions. Granular materials exhibit two types of cross-anisotropy: one is inherent cross-anisotropy in which the particle assembly is in its virgin state before any loading occurs and the other is a result of loading and plastic deformation in one particular direction.

Different studies have been conducted of the characteristics of cross-anisotropy, but few have tested granular material behavior under different loading directions. Rodriguez and Lade (2013) performed torsion shear tests on sand with different stress conditions, as expressed by $b=(\sigma_2-\sigma_3)/(\sigma_1-\sigma_3)$ and different major principal stress directions. In their study, the researchers described that it may also be possible to characterize the three-dimensional failure conditions based on triaxial compression tests if the specimens were created to reflect horizontal bedding planes in vertical and horizontal directions (Fig. 1). Although the researchers conducted triaxial compression tests on vertical and horizontal cubical specimens that were frozen and then installed in the triaxial apparatus for testing, they did not investigate the effects of void ratio and confining pressure on the degree of cross-anisotropy. While Oda et al. (1978) showed that the effects of cross-anisotropic microstructure on strength and deformation characteristics are most often stronger in plane strain compression tests than in triaxial compression tests, the latter were performed in the study described in this manuscript. This is because triaxial compression tests are (1) the

Fig. 1 Specimen orientation **a** vertical and **b** horizontal in cross-anisotropic soil with horizontal bedding planes

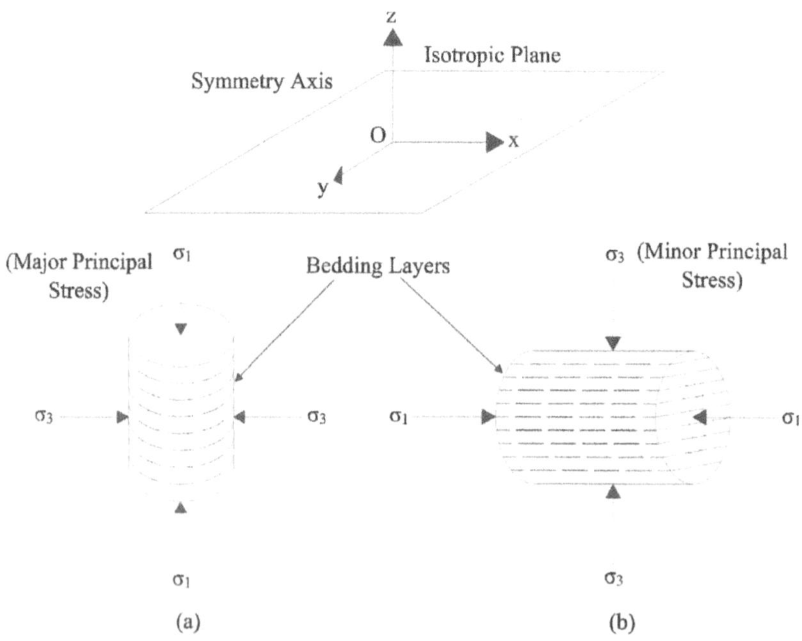

basis for parameter determination in most constitutive models (Lade 2005), (2) easier to perform than plane strain tests and can be performed by most geotechnical laboratories, and (3) shear banding occurs in the softening regime in triaxial compression tests for which $b=0$ as observed experimentally and supported by analysis (Lade 2003), so smooth peak failure is included in the stress–strain behavior, and (4) the loading condition is axisymmetric, as required for definition of Poisson's ratio for the elastic behavior.

The focus of the study presented in this manuscript was to explain the details of the methodology that was used to create cylindrical triaxial specimens with bedding planes in vertical and horizontal directions and investigate the effects of void ratio and confining pressure on the degree of cross-anisotropy.

2 Previous Studies of Cross-Anisotropy

Although this may be true for all sizes of the granular materials, particularly in sand sized particles, the overall granular behavior is affected by fabric anisotropy, which creates a complex granular structure. Such concept is not new as it has been documented by Parkin et al. (1968), Arthur and Menzies (1972), Arthur and Phillips (1975); Oda (1972a, b, c), and El-Sohby and Andrews (1973), when the researchers have investigated the fabric characteristics of natural sands as well as sand sized particles that were randomly assembled in the laboratory to create two and three-dimensional structures. The major findings from these previous research studies indicated that when particles are deposited under gravity, they tend to orient themselves with their long axis in the horizontal direction and with contact normals that favor the vertical direction. Such observations were noted even for perfectly round particles. This is because each particle has a unique preferred orientation as it deposits, and this phenomenon occurs when the sand particles are deposited in air as well as in water (Oda et al. 1978). Ochiai and Lade (1983) also showed that the shapes of the sand particles affect the degree of anisotropy that may occur in a given sand deposit.

Figure 2 has been created to demonstrate the concept of why and how cross-anisotropy occurs in sand deposits (either naturally occurring or man-made fill). Figure 2a shows a schematic assembly of grains deposited by pluviation through air or water

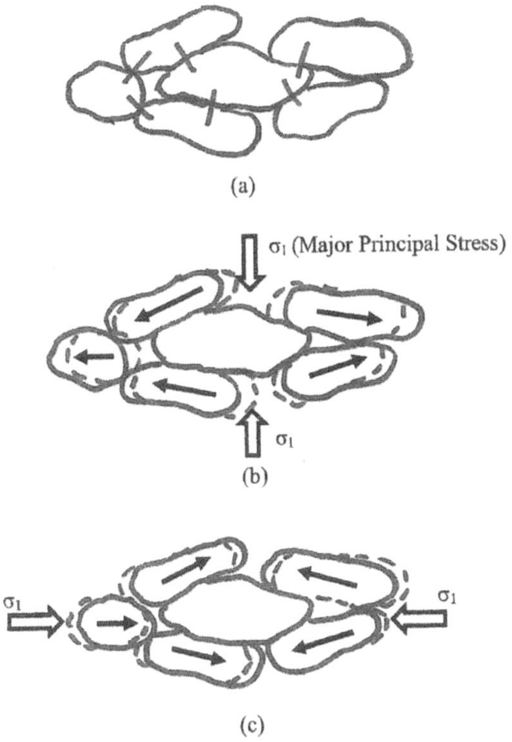

Fig. 2 Pluviated sand deposit **a** with contact normals favoring the vertical direction, **b** loaded in the vertical direction, and **c** loaded in the horizontal direction

(i.e. gravity causes the grains to move in the vertical direction), where the grains tend to settle with their long axes in the horizontal direction and with contact angles favoring the vertical direction (as studied by Ochiai and Lade 1983). When this assembly is loaded, and crushing does not occur, the grains have to slide to the side to pass each other. The friction between grains results in stiffer response in the vertical than in the horizontal direction. Because the contact angles tend to be directed in the vertical direction, vertical loading results in smaller deformation and stiffer response than horizontal loading, as indicated in Fig. 2b, c. Thus, it is the contact angle directions that control the behavior of the granular materials.

Oda et al. (1978) has conducted one of the early studies to demonstrate that cross-anisotropy in sand deposits has implications for the engineering behavior of the granular materials. As part of their research, they designed a laboratory experiment modeling strip foundations on sand with two different relative densities. The model simulating each relative density scenario was loaded in two different directions in relationship to the direction of the pluviation of the sand: one parallel to the direction of the sand deposition and the other perpendicular (normal) to that direction. The results showed that for the dense sand condition, there was 50% difference in the bearing capacity values between the two different loading directions. This observation was noted by Oda et al. (1978) as the importance of cross-anisotropy on the behavior of granular materials.

Previous researchers have used sophisticated equipment such as true triaxial, directional shear, and torsional shear set-ups to investigate and quantify the effects of cross anisotropy on the engineering behavior of sand. Some of these studies even included hollow cylinder specimens where different pressures could be applied inside and outside the specimen. Accumulated research with these sophisticated equipment on cross anisotropy took place over a period of 40 years (Oda 1972a, b; Arthur and Philips 1975; Oda et al. 1978; Yamada and Ishihara 1979; Haruyama 1981; Oda 1981; Hight et al. 1983; Ochiai and Lade 1983; Miura et al. 1986; Tatsuoka et al. 1986; Lade and Wasif 1988; Lam and Tatsuoka 1988; Pradel et al. 1990; Abelev and Lade 2003, 2004; Lade and Abelev 2003, 2005; Naughton and O'Kelly 2007; O'Kelly and Naughton 2009; Chairo et al. 2013; Lade et al. 2014a, b). The results published from these studies overwhelmingly showed that when sand particles are loaded under monotonic conditions, when the direction of the loading and deposition of the sand particles coincide, and when no rotation of principal stresses occurs, the deformation and peak shear resistance of sand is controlled by the initial anisotropic fabric of the sand. The previous research also points out that such condition are especially evident for sands with elongated particles.

Even though there has been a number of studies to prove that the cross anisotropy affects the engineering behavior of granular material, it was Rodriguez and Lade (2013) who developed a three-dimensional constitutive failure criterion for granular soils that was based on the cross-anisotropic behavior. Rodriguez and Lade (2013), pointing out that not all laboratories/investigators/designers have access to sophisticated testing equipment, have derived the parameter values for their failure criterion by using the conventional triaxial compression tests performed on vertical and horizontal specimens. The tests conducted by Rodriquez and Lade (2013) were with fine Nevada sand with relative density of 91% (and a corresponding void ratio of 0.53). Tests for their investigation were conducted at confining pressures of 25, 50, 70 and 130 kPa.

Considering that the fact of cross anisotropy effecting the engineering behavior of granular soils is a proven concept and that there is an existing constitutive failure criterion already developed for cross anisotropic granular soils, it is surprising to see the lack of studies conducted with triaxial compression tests as a means to quantify conditions that are beyond what has been tested by Rodquez and Lade (2013). Therefore, it is desirable to investigate the cross-anisotropy characteristics of a granular material utilizing the conventional triaxial setup as the previous research shows that the degree of cross anisotropy varies with confining pressure under different density. The present study has been geared towards filling up such gap in the literature associated with limited data and to outline details of how the modern day conventional triaxial equipment can be used to conduct consolidated drained tests to evaluate the effects of cross-anisotropy on the shear strengths of granular soils. This is the first study where the work has been performed on one specific soil to evaluate these effects.

3 Material Characterization

All tests conducted in this study were performed on "Play" sand obtained commercially from a construction materials store. The geological make-up of the particles was observed to be of silica (quartz) with particle shapes ranging from rounded to sub-angular as determined from Power's Scale of Roundness system (Powers 1953). The sand was determined to be non-plastic.

Previous literature showed that particles coarser than fine sand will show membrane penetration effects in the volume change measurements when tested in the triaxial cell (Lade, 2016). Similarly, bedding errors may occur in the vertical direction if lubricated ends are employed to reduce friction. These errors may also be minimized if fine sand is used for the experiments. If ignored, this would create an experimental error since the measured volume change is not representative of the soil skeleton compression because it includes the volume of water forced out by the penetration of the membrane into the outer layer of sand grains. Therefore, all specimens used in this study were sieved to be within the range of fine sand (i.e. all sand grains coarser than #40 U.S. sieve or finer than #200 U.S. sieve were removed) as shown in Fig. 3. The properties of the fine sand used in this study are summarized in Table 1.

The maximum (e_{max}) and the minimum (e_{min}) void ratios of the fine sand were determined using the procedures described by Lade et al. (1998) by testing each condition with five replicate tests. The results showed consistently that e_{max} and e_{min} values of this sand were 0.90 and 0.57, respectively. The void ratios of the fine sand specimens created for this study ranged from 0.76 to 0.82 for loose, 0.62 to 0.65 for medium dense, and 0.58 to 0.61 for dense conditions, as determined at the same initial void ratio evaluated at the same partial vacuum at which the specimen dimensions were measured. The corresponding relative densities for these void ratios ranged from 24.3% to 41.8% for loose, 73.3% to 82.9% for medium-dense, and 87.7% to 97.3 for dense sand.

Table 1 Properties of the fine sand used in this study

D_{10} (mm)	0.12
D_{30} (mm)	0.20
D_{60} (mm)	0.29
Cu	2.42
Cc	1.15
USCS classification	SP
Gs	2.64
e_{max}	0.90
e_{min}	0.57

4 Specimen Preparation

All tests were conducted in a triaxial chamber with specimen dimensions of 7.1 cm in diameter and with height in the range of 15.5–16.0 cm (Mahmoudi et al. 2021). Specimens prepared at each relative density were tested under three different confining pressures of 20, 60, and 180 kPa. These pressure ranges were selected because previous studies have shown that all

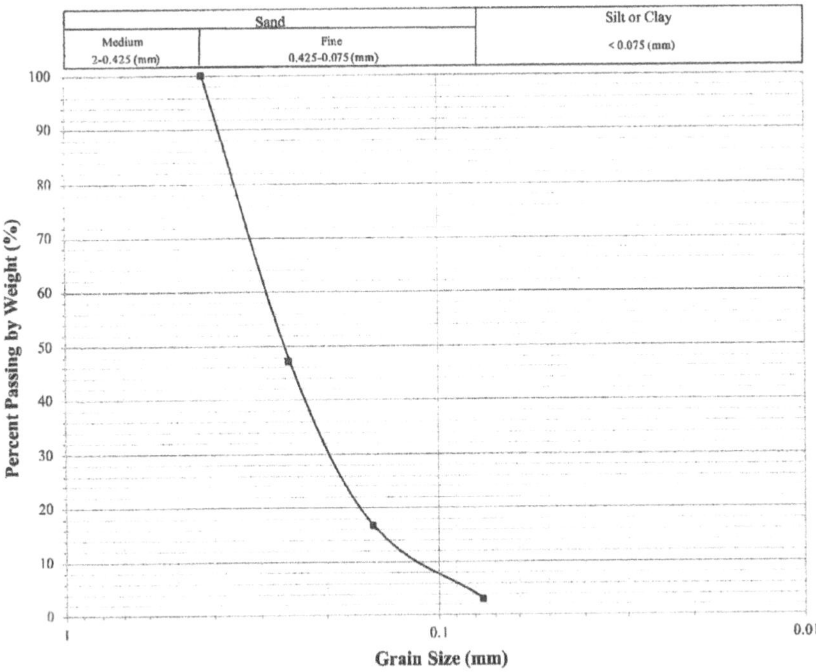

Fig. 3 Grain size distribution of the fine sand used in this study

properties, including the elastic moduli, change with confining pressure and density (Janbu 1963; Abelev and Lade, 2003). Three of the tests (one at each confining pressure) were conducted by just loading the specimens to failure and beyond, and the other three (one at each confining pressure) were performed by loading, unloading and then reloading the specimens. This protocol was followed because without the loading and unloading the true Poisson ratio and true elastic moduli values could not be determined (Lade and Nelson 1987). In addition, results from the tests were compared to evaluate the effects of unloading–reloading rather than just loading of the soil specimens.

All soil specimens were tested vertically under compression loading. However, to evaluate the effects of cross-anisotropy, half of the specimens were prepared vertically and the other half were prepared horizontally. The following sections describe the preparations of these specimens.

4.1 Vertical Specimens

All specimens were prepared by using a split mold and pluviating the sand into the mold that had a rubber membrane already installed. The targeted void ratios were obtained by raining the sand from different heights (Miura and Toki 1982; Vaid and Negussey 1984, 1988; Rad and Tumay 1987). To achieve uniform densities within the sand specimen, the rate at which the sand was rained was kept constant. Drop heights of 0, 15.5, and 50 cm were used to create three different densities. The sand was rained through two 7 cm diameter No. 4 U.S. sieves placed on top of a circular cardboard tube. In the case of 0 cm drop, the sand was gently placed over the base of the triaxial chamber using a conical funnel connected to a plastic tube such that the rate at which the sand was released could be controlled at zero drop height. Void ratios were then calculated based on the measured volumes, masses, and specific gravity (Gs) of the sand.

All sand specimens were initially held stable using a vacuum. The vacuum process started after the sand was deposited in the split mold and the top cap was placed. Before removal of the split mold, the valves of the top cap were shut off and the vacuum line was connected through the drainage line of the bottom base. Only a small amount of vacuum is needed to hold the sand specimen in place. In this study, the maximum applied vacuum was 15 kPa. After assembling the triaxial cell it was filled with water while the specimen was still under vacuum. The vacuum was then slowly reduced while the pressure in the cell water was increased such that the specimen continuously felt the same effective confining pressure.

To saturate the specimen with water, gaseous carbon dioxide (CO_2) was first introduced from the bottom drainage line. The gaseous CO_2 replaced the air inside the sand specimen (Lade and Duncan 1973), and because air is lighter than CO_2, the air was pushed out through the top drainage line and replaced by the CO_2. This replacement process was conducted for about 15–20 min. The top drainage line was connected to a bubble chamber, which was used to confirm that the air and the gaseous CO_2 entering the sand specimen from the bottom was pushed out of the specimen from the top. The bubble chamber is a glass bottle partially filled with water through which the air is passed to create bubbles thus indicating the movement of gas through the specimen. During this process, it was important to monitor the applied CO_2 pressure entering the sand specimen to check that the effective confining pressure within the specimen does not reduce from the desired conditions.

Following the gaseous CO_2, de-aired water is introduced through the bottom drainage line to percolate slowly into the sand specimen while leaving the top drainage valves open and allowing the CO_2 to be pushed out. Because the sand specimen is initially saturated with gaseous CO_2, the de-aired water will replace the CO_2. Any small amounts of gaseous CO_2 still present in the specimen will dissolve in the intruding water because approximately one volume of gaseous CO_2 will dissolve into one volume of water at room temperature and atmospheric temperature. In addition, a small back pressure (20 kPa) was applied to the specimen water to help the dissolution while increasing the cell pressure to 35 kPa allowing the gas bubbles to dissolve under the back pressure. Skempton's B-value was then checked to measure the degree of saturation of the sand specimen by closing the drainage line and increasing the confining pressure of the sand specimen by 70 kPa and recording the increase in pore pressure. The target B-values for these conditions were in the order of 0.95 and above.

4.2 Horizontal Specimens

The horizontal sand specimens were pluviated to void ratios similar to those of the vertical specimens. To construct a horizontal specimen, the triaxial cylindrical split mold was separated into two halves. A frame was used to hold the molds stationary and the semicircular Styrofoam caps were attached to the molds to form the ends of the specimen (Fig. 4a). The sand was pluviated into the half-cylinder cavities through a circular tube with a 7 cm No. 4 U.S. sieve attached to the top. The circular tube was then horizontally moved via a card board that was placed on a smooth track to allow the sand to be pluviated into the cavities of the half-cylinder as shown in Fig. 4b. All of the horizontal specimens were formed by depositing the sand vertically into the split molds at a consistent rate as was done for the vertical specimens. The surface of the pluviated sand was levelled using a spatula to form a flat surface.

The half-cylinder specimens were saturated by introducing de-aired water through the line attached to the sides of the cylindrical split molds. Gaseous CO_2 was not used for this initial saturation procedure. To prevent formation of ice over the top of the sand surface, the wetting procedure was continued only until the specimens showed dry and wet spots on top of the sand surface (as shown in Fig. 5a). The wet sand specimens were then placed in a freezer for about 3 h, which allowed the water to freeze throughout as seen

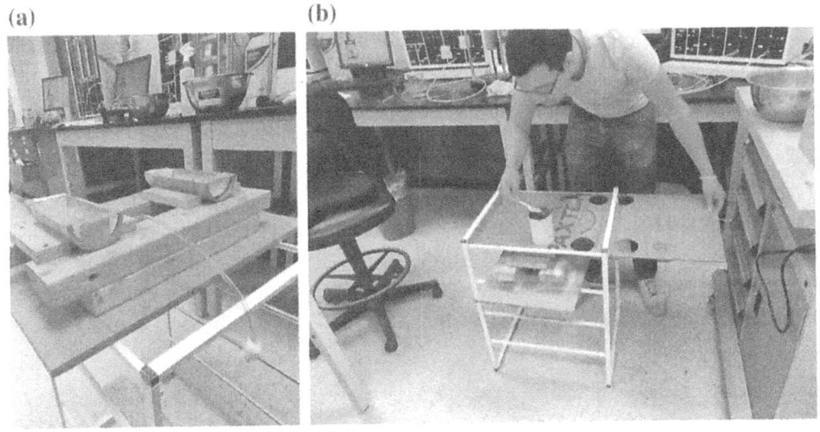

Fig. 4 Preparation of horizontal sand specimens: **a** placement of two split molds, and **b** setup used to pluviate sand into split molds

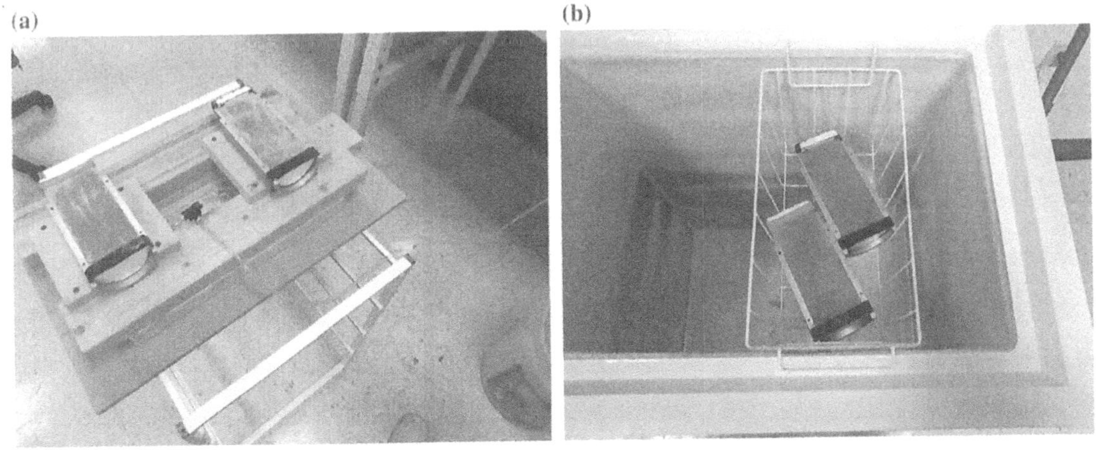

Fig. 5 Horizontal specimens **a** after water permeation up through the sand, and **b** after the specimens are frozen in the freezer

in Fig. 5b. The specimens were targeted to freeze to an extent that when they were removed from the freezer they would keep their integrity but by means of the heat generated by holding the exterior of the cylindrical molds, they could easily be extruded from the two half-cylinder molds. Observations indicated that thin water films did not form on the specimen surfaces, and there were no indications of loose sand on the two halves of the specimen. Once extruded, the two halves were pressed together to form a cylindrical specimen (Fig. 6a). Once completely thawed, the two halves of the cylinder fit together to form a full cylinder with no visible signs of the formation procedure (Fig. 6b). Even though a weak plane may be formed between the specimen halves due to lack of particle interlocking, the volume of sand in this plane is very small compared to the remaining portions of the specimen. Further, in the triaxial setup the two halves are supported between the stiff cap and base such that sliding cannot occur along this plane. Any sliding that may later occur along a shear plane (which in triaxial compression always develops after smooth peak failure), occur at an angle with the vertical weak plane. Thus, the weak plane is considered to have insignificant influence on the overall behavior and strength of the specimen.

After the extrusion from the cylindrical molds, the frozen horizontal sand specimen was quickly positioned into a rubber membrane and placed on the base of the triaxial chamber. The top cap was attached and the membrane was sealed to the cap and base. Once the triaxial cell was assembled and filled with water, a cell pressure of 20 kPa was applied. During the process of thawing the frozen sand specimen, warm water was introduced from the bottom drainage line and percolated through to the top drainage line to help thawing the frozen sand specimen. After the sand specimen had thawed, the water in the specimen was removed by a small vacuum of 15 kPa that was applied through the bottom drainage line while the top drainage line was kept open to allow the air to intrude. Once the removal of water was visibly confirmed, the bottom drainage line was connected to the gaseous CO_2, which was allowed to seep up through the specimen to replace the air and exit through the top drainage line into the bubble chamber. The CO_2 percolated through the specimen for 15 to 20 min as for the vertical specimen. Lade (2016) stated that the CO_2 method works best when the specimen is initially dry, but in this study (as confirmed from B-values) it was observed that the CO_2 method was also effective when the sand particles themselves are slightly wet. Once the horizontal specimen was saturated, the remaining portion of the test proceeded as explained for the vertical specimen.

Fig. 6 Thawing of horizontal specimen after the two halves are joined to create a cylindrical specimen: **a** before and **b** after completion of thawing

5 Triaxial Compression Tests

5.1 Stage 1: Isotropic Compression

The specimens were isotropically compressed to 20, 60 and 180 kPa, respectively, before the triaxial compression tests were performed at these three constant confining pressures. The results of isotropic compression of vertical and horizontal specimens of medium dense sand up to 60 and 180 kPa are shown in Fig. 7. While the two types of specimens are expected to show the same isotropic compression curves, the horizontal specimens indicate slightly larger volumetric compression than the vertical specimens. Linear strains of the specimens were not monitored during isotropic compression. However, during isotropic compression the initial slopes of the volumetric strain curves appear to indicate a small systematic error, but the differences between the remaining portions of the curves are within the experimental scatter of such tests. Similar results were obtained for loose and dense sand specimens.

5.2 Stage 2: Shearing

The shearing stage of the tests on both vertical and horizontal specimens were conducted with an axial strain rate of 10% per hour up to a maximum strain of 20%. To produce results from which the elastic moduli could be determined, half of the triaxial compression tests were sheared to just beyond smooth peak failure after which an unloading–reloading cycle was included. The sand specimens were consequently unloaded at an axial strain of approximately 6%. At this point the fabric of the specimens has changed minimally, and the elastic moduli are therefore expected to be representative of the void ratios of the respective specimens.

6 Results: Stress–Strain Behavior

The stress–strain and volume change curves from the triaxial compression tests are presented in Fig. 8. The stress–strain and volume change behavior shown in Fig. 8a indicates that the vertical and the horizontal medium dense sand specimens loaded to failure and beyond for all three confining pressures. Figure 8b shows similar results for the specimens that were unloaded and reloaded near smooth peak failure.

The shapes of the stress–strain curves for the vertical and horizontal specimens are very different as peak failure is approached. This is due to the grain configuration in the vertical and horizontal specimens. It takes larger strains to mobilize the full shear strength in the horizontal than in the vertical

Fig. 7 Volumetric strains of vertical and horizontal medium dense sand specimens for isotropic compression under minor principal stresses of 60 and 180 kPa

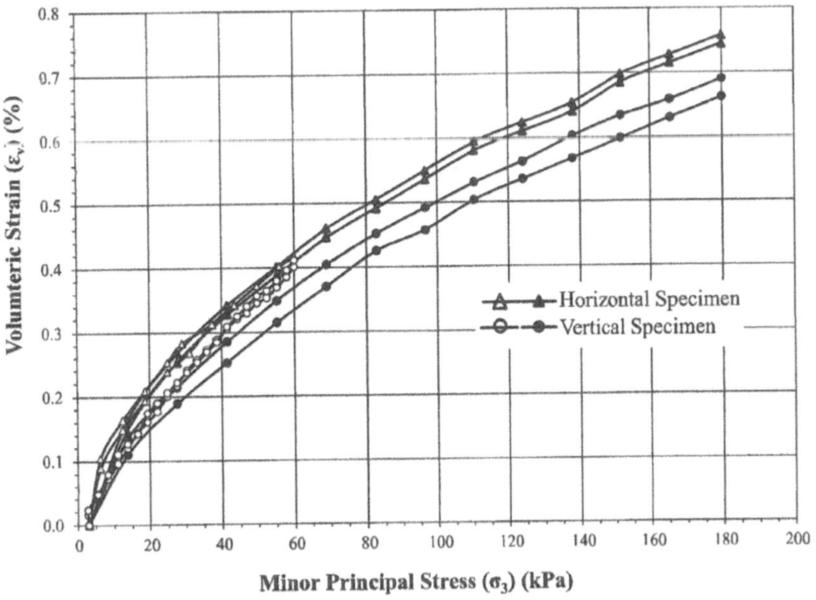

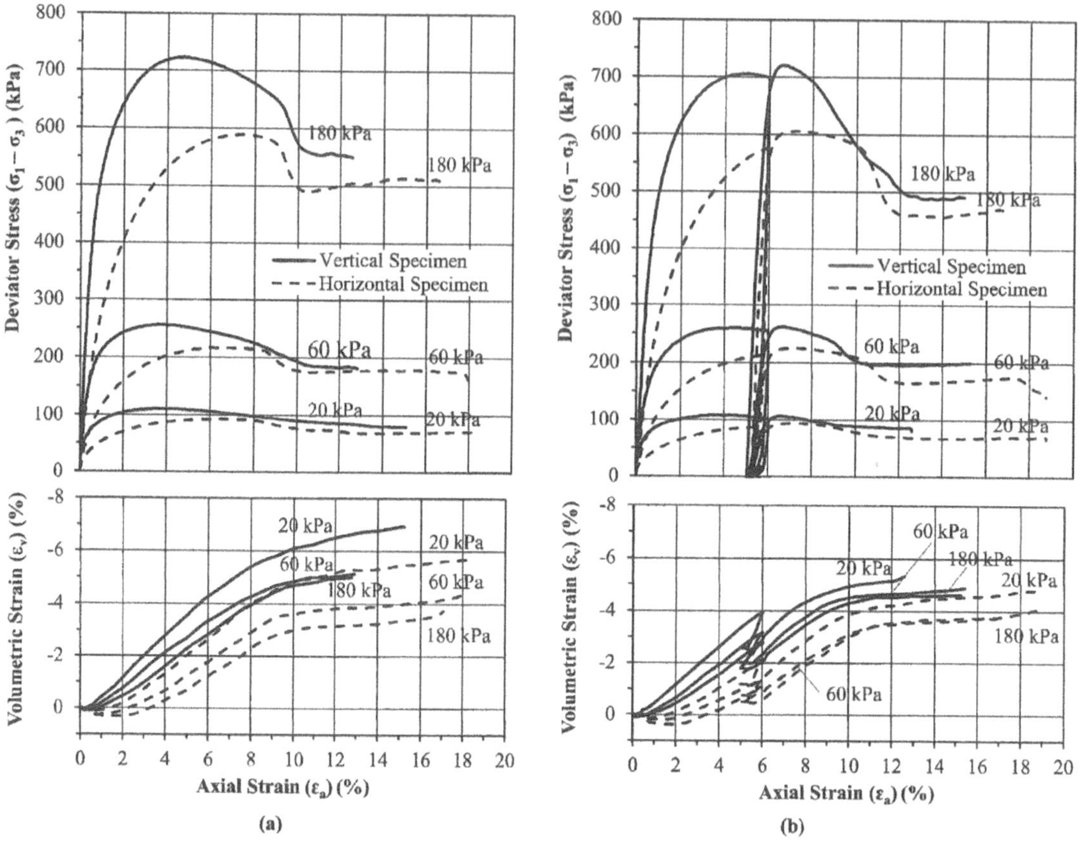

Fig. 8 Changes in axial strain as a function of deviator stress and volumetric strain in medium dense vertical and horizontal sand specimens in triaxial compression tests with **a** loading only and **b** including unloading–reloading cycles

specimens. This is because the grains are pushed together at lower slopes and the resistance to compression and movement of grains is lower for the horizontal specimens than for the vertical specimens.

6.1 Initial Slopes

The initial slopes shown in Fig. 8 for the vertical specimens are consistently stiffer than those for horizontal specimens. This is due to the fabric orientation in the cross-anisotropic deposits, which may be characterized by contact normals that favor the vertical direction. This provides for stiffer response in the vertical direction than in the horizontal direction as explained in connection with Fig. 1. The initial slopes for the horizontal sand specimens do not show a sharp rise similar to those of the vertical sand specimens.

6.2 Strain-to-Failure

The stress–strain behavior of vertical and horizontal specimens is distinctly different due to the fabric orientation. Thus, the vertical specimens show a rapid rise and a higher strength than the horizontal specimens, which exhibit much slower increase in shear resistance. The horizontal specimens also reach peak failure after substantially larger amounts of sliding, and consequently, the horizontal specimens reach smooth peak failure at larger values of axial strain than the vertical specimens.

6.3 Shear Banding

Figure 8 shows that shear banding develops in the softening regime in all the specimens. Most often much larger axial strains than those reached for peak failure are required for shear banding to occur. This has been observed before and it matches the theoretical predictions (Lade 2003). The shear band development is indicated by the break in the stress–strain relations in the softening regime and by the reduction in dilation rate in the volume change curves. While dilation is required for development of shear bands, once they start developing, the void ratio inside the shear band becomes larger and the shear strength becomes lower than in the surrounding sand masses. The following sliding consequently occurs along the shear band which has now exhausted its capacity for dilation, followed by very little or no further dilation in the specimen. This behavior is more pronounced for dense specimens (not shown herein) than for medium dense sand as shown in Fig. 8, but it is not so prominent for the loose sand (not shown herein). This is because shear banding requires the sand to dilate, and loose sand does not dilate or if it occurs, it is much less vigorous than in dense sand.

Figure 8b shows that the vertical sand specimens that included an unloading–reloading cycle developed shear bands in a similar way to the vertical sand specimens that were just loaded, as indicated in Fig. 8a. Note that the vertical specimens in Fig. 8b decreased in shear strength until they developed shear bands around an axial strain of 8–9%, very similarly to those without unloading–reloading cycles. The horizontal sand specimens exhibited shear banding at slightly larger axial strains near 9–10%. The trends in the vertical and horizontal specimens were similar whether or not they included an unloading–reloading cycle.

6.4 Volume Change

Both Fig. 8a, b indicate that the vertical and horizontal specimens of medium dense sand showed increasing initial contraction and decreasing rates of dilation with increasing confining pressure. Further, the horizontal specimens showed more contraction than the vertical specimens and they dilated at a slower rate than the vertical specimens. Similar results were obtained for the vertical and horizontal sand specimens that were unloaded and reloaded, as shown in Fig. 8b.

However, it is readily seen that during reloading all the specimens dilated faster than indicated by the initial dilation rates.

7 Results: Shear Strength

The peak failure points for the two specimen types are different and indicate that the vertical specimens have a higher strength than the horizontal specimens. This is also due to the fabric orientation, which determines their behavior. Inspection of results for dense and loose specimens, which are not shown herein, indicates that this difference increases with increasing density. For loose specimens the difference in strength tends to be very small indicating that the influence of cross-anisotropy is minimal. Cross-anisotropy is therefore most pronounced in dense sand deposits.

Figure 8b shows the behavior of the medium dense sand specimens that include an unloading–reloading cycle for the vertical and horizontal sand specimens. All aspects of the stress–strain behavior up to failure are very similar to those observed for the first series of tests, thus indicating the reproducibility of the tests. During the reloading process both types of sand specimens densify causing the following failure points to increase to slightly higher values.

7.1 Friction Angles

The friction angles (ϕ) were calculated using the maximum obliquity (σ_1/σ_3) (Bolton 1986; Xiao et al. 2014) based on major and minor principal stresses measured during testing of the sand specimens:

$$\phi = \sin^{-1}\frac{\frac{\sigma_1}{\sigma_3} - 1}{\frac{\sigma_1}{\sigma_3} + 1} \qquad (1)$$

where σ_1: major principal stress; σ_3: minor principal stress.

Table 2 gives the friction angles of all sand specimens that were (1) loaded only, and (2) unloaded and reloaded in the vertical and horizontal directions. It is observed that inherent cross anisotropy results in significantly higher friction angle in vertical than in horizontal specimens (Veiskarami et al. 2019).

Figure 9 shows the variation of friction angles with confining pressure for all sand specimens. Friction angles in all cases decrease with increasing confining pressure as previously observed (Lee and Seed 1967). The results also show that vertical specimens

Table 2 Summary of friction and dilation angles for vertical and horizontal sand specimens

	Just loading						Unloading and reloading					
	Vertical			Horizontal			Vertical			Horizontal		
	20 kPa	60 kPa	180 kPa	20 kPa	60 kPa	180 kPa	20 kPa	60 kPa	180 kPa	20 kPa	60 kPa	180 kPa
Friction angle (ϕ)												
Dense	48.6	44.8	43.2	43.4	41.1	39.5	47.9	44.8	42.7	44.1	41.4	39.5
				42.3	41.4	39.7				43.8	40.8	39.8
Medium dense	46.1	42.9	41.8	43.4	39.8	38.3	46.1	42.8	41.7	41.6	39.5	37.9
Loose	39.3	37.5	35.5	37.9	35.7	34.7	39.0	37.2	34.6	36.9	35.7	33.7
	38.3	37.2	35.5				38.9	36.8	34.7			
Dilation angle (ψ)												
Dense	19.3	17.3	15.9	15.8	15.3	13.9	19.3	16.8	15.8	17.3	17.2	15.3
				15.3	15.6	14.2				17.6	16.3	15.7
Medium dense	16.7	14.7	13.8	13.7	12.6	12.3	16.5	14.6	14.0	14.0	13.5	13.3
Loose	4.9	5.0	4.3	5.6	6.4	6.2	6.7	6.7	5.0	7.7	8.1	6.6
	4.1	5.0	4.2				7.6	8.2	5.4			
Difference between friction angles and dilation angles ($\phi - \psi$)												
Dense	29.3	27.5	27.3	27.6	25.9	25.6	28.5	28.0	27.0	26.8	24.3	24.2
				27.0	25.8	25.6				26.3	24.5	24.1
Medium dense	29.5	28.3	28.0	29.7	27.2	26.0	29.5	28.3	27.7	27.6	26.0	24.6
Loose	34.4	32.5	31.2	32.3	29.3	28.5	32.4	30.4	29.6	29.2	27.6	27.1
	34.2	32.3	31.4				31.4	28.7	29.3			

have higher friction angles than void ratio, as also found in horizontal specimens. The ratio of friction angles from vertical and horizontal specimens is relatively independent of the confining pressure and the void ratio similarly as also found by Tatsuoka et al. (1986) for Toyoura sand in drained plane strain tests. Further, the loose specimens show smaller differences than the dense specimens. Thus, cross-anisotropy is most pronounced for dense sand.

7.2 Dilation Angles

The dilation angle (ψ) is determined from the tangent of the volumetric (ε_v) and the axial strain (ε_a) curve measured at peak failure in the triaxial compression tests:

$$\psi = \sin^{-1}\left(\frac{\frac{\Delta\varepsilon_v}{\Delta\varepsilon_a}}{\frac{\Delta\varepsilon_v}{\Delta\varepsilon_a} - 2}\right) \tag{2}$$

where $\Delta\varepsilon_v$: change in volumetric strain; $\Delta\varepsilon_a$: change in axial strain.

Table 2 also shows the dilation angles for the sand specimens tested in this study. The dilation angles have a similar trend as the friction angles where the dilation angles for the vertical specimens being larger than the dilation angles for the horizontal specimens (Xiao et al. 2018). The second series of loose vertical and dense horizontal sand specimens were used to indicate the repeatability of the experimental results, as shown by the very similar friction angles.

Inherent cross anisotropy influences the dilation angles for the sand specimens deposited in two different directions. Thus, the dilation angles are smaller for the horizontal than the vertical specimens. Figure 10 shows the difference between the friction angles and the dilation angles ($\phi - \psi$) for the vertical (solid lines) and the horizontal specimens (dashed lines) that were plotted against the confining pressures. All experiments show that the differences between the friction angles and the dilation angles ($\phi - \psi$) decrease as the confining pressure increases. However, the differences between friction angles and corresponding dilation angles decrease as the density of the sand specimen

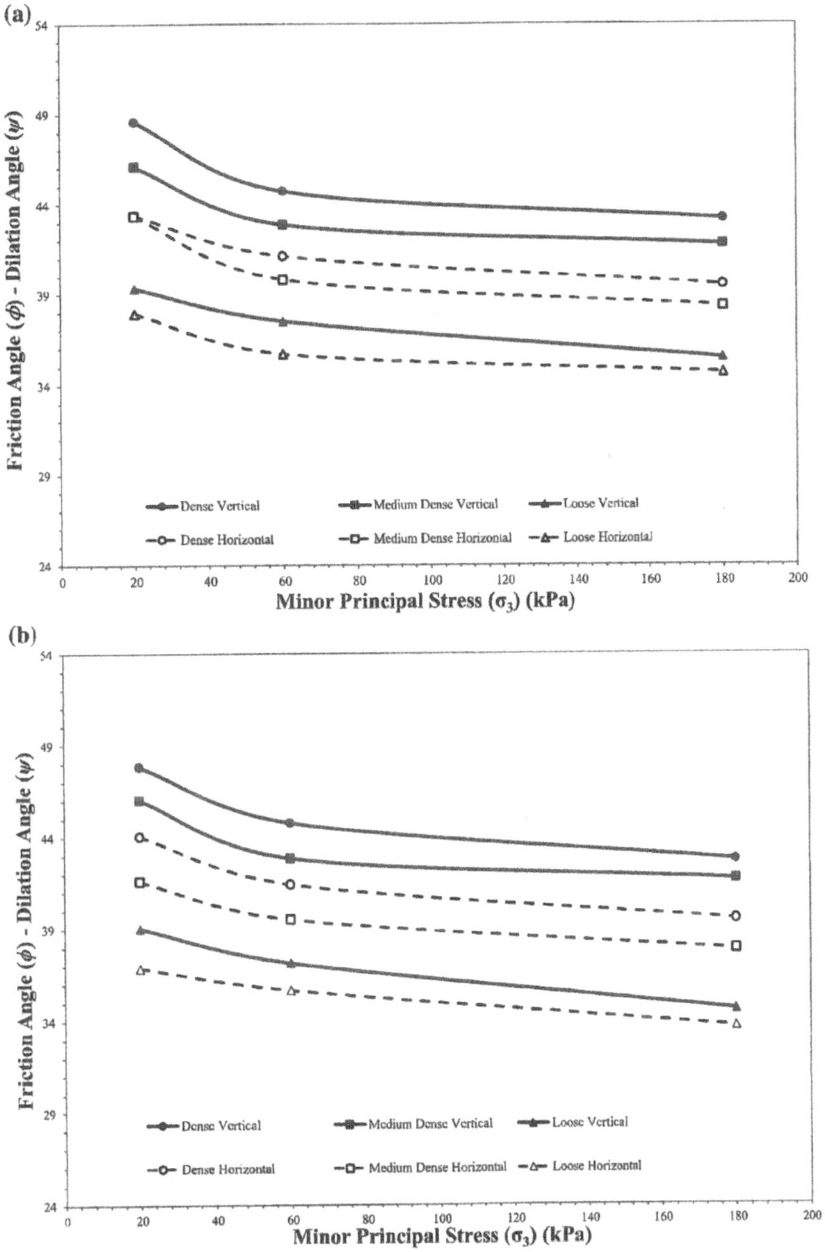

Fig. 9 Variation of friction angles for medium dense vertical and horizontal sand specimens with changes in minor principal stress (σ_3) for tests **a** without and **b** with unloading–reloading cycles

increases, while this difference is highest for the horizontal specimens. The loose sand specimens in both the vertical and horizontal directions have higher values than the medium dense and the dense sand specimens. This observed behavior between the loose and dense sand specimens is similar to what has been observed previously by Lee and Seed (1967) on tests conducted with loose, medium loose, medium dense, and dense Sacramento River sand specimens. The same trend has been observed by Tatsuoka et al. (1986) for drained plane strain tests on Toyoura sand.

Fig. 9 Variation of friction angles for medium dense vertical and horizontal sand specimens with changes in minor principal stress (σ_3) for tests **a** without and **b** with unloading–reloading cycles

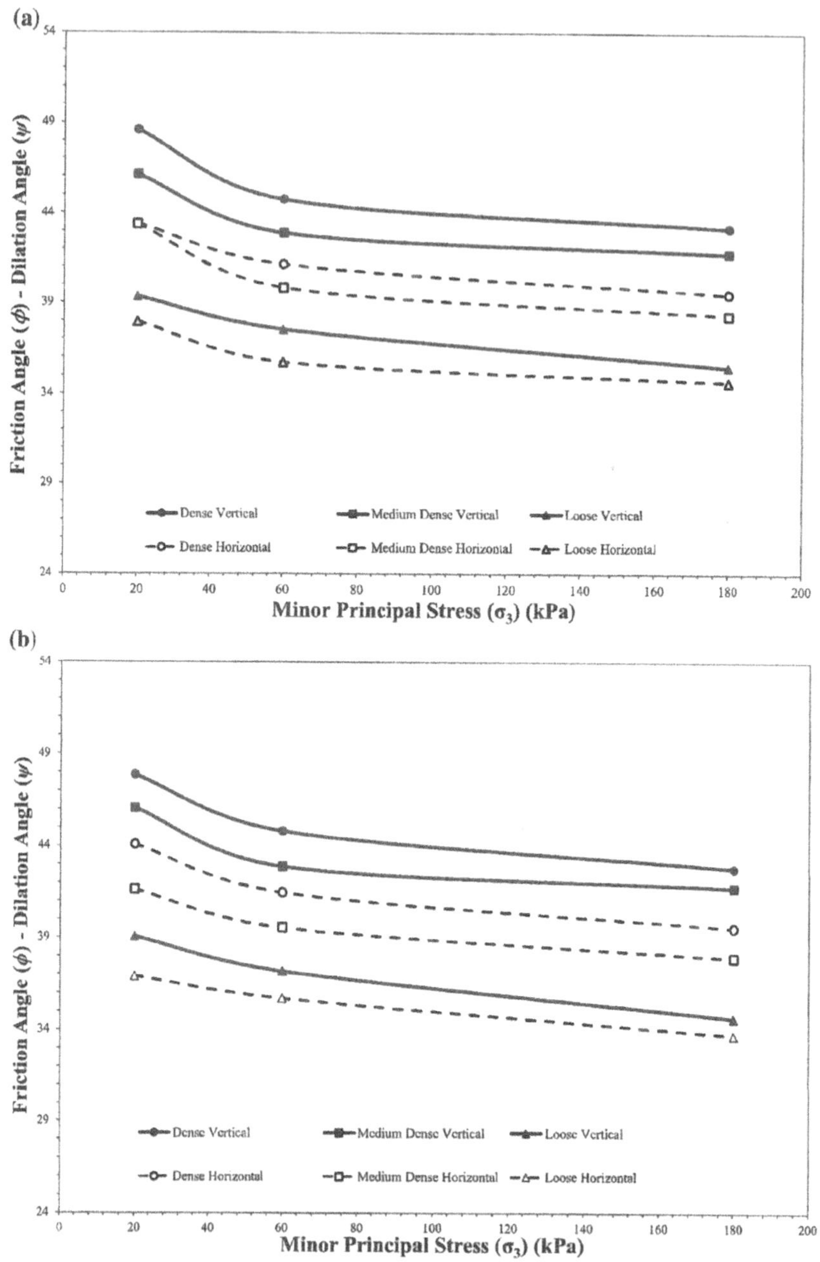

increases, while this difference is highest for the horizontal specimens. The loose sand specimens in both the vertical and horizontal directions have higher values than the medium dense and the dense sand specimens. This observed behavior between the loose and dense sand specimens is similar to what has been observed previously by Lee and Seed (1967) on tests conducted with loose, medium loose, medium dense, and dense Sacramento River sand specimens. The same trend has been observed by Tatsuoka et al. (1986) for drained plane strain tests on Toyoura sand.

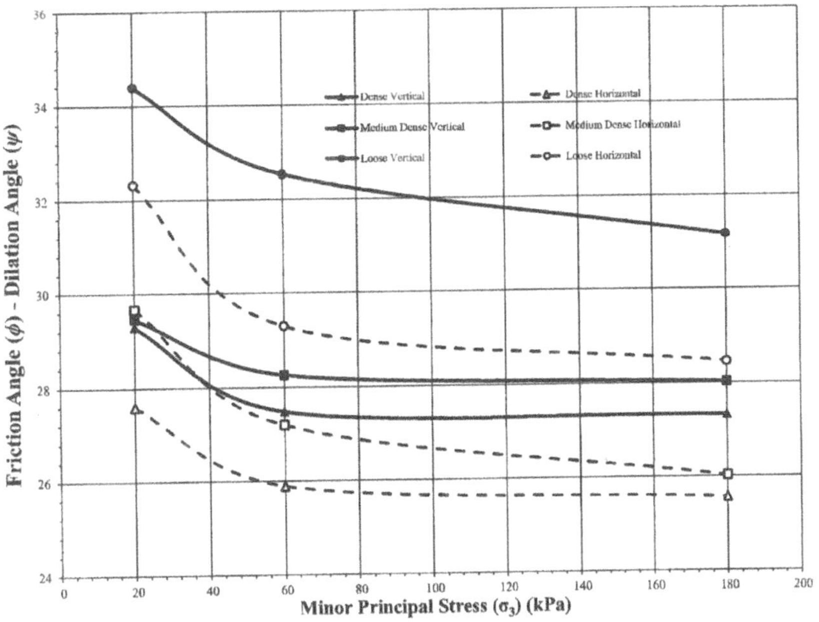

Fig. 10 Differences between friction and dilation angles for medium dense vertical and horizontal sand specimens with confining pressure

8 Results: Elastic Properties

To evaluate the elastic properties of the sand specimens, the slopes of the stress–strain and volume change relations during unloading and reloading were evaluated (Lade and Nelson 1987, Lade 2016). While the volumetric strains near smooth peak failure were in the order of 4–5%, and the particle fabric may have changed a small amount, this change hardly had excessive influence on the values of elastic moduli. The Young's moduli are determined immediately after stress reversals, such that no plastic strains are involved in the evaluation. Thus, Eq. 3 is employed to calculate Young's modulus from axisymmetric loading conditions. Figure 11a shows a schematic diagram in which two values of Young's modulus (E) are determined from the points of unloading and reloading. In both cases, these moduli are associated with the stress points where stress reversal takes place. Poisson's ratio is determined from the slope of the volumetric strain (ε_v) versus the axial strain (ε_a) curve when the

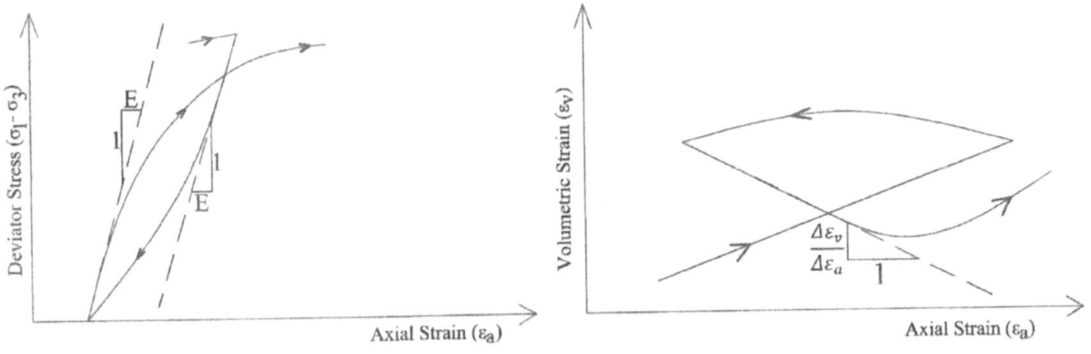

Fig. 11 Schematic diagrams of determination of **a** elastic moduli (E) at stress reversal points and **b** ratio of volumetric to axial strains from volume change relations during reloading of triaxial specimens

specimen is being reloaded as shown schematically in Fig. 11b.

$$E = \frac{\Delta(\sigma_1 - \sigma_3)}{\Delta\varepsilon_a} \quad (3)$$

where $\Delta(\sigma_1 - \sigma_3)$: change in deviator stress; $\Delta\varepsilon_a$: change in axial strain.

Poisson's ratio (ν) is then calculated using Eq. 4.

$$\nu = \frac{1}{2}\left(1 - \frac{\Delta\varepsilon_v}{\Delta\varepsilon_a}\right) \quad (4)$$

where $\Delta\varepsilon_v$: change in volumetric strain; $\Delta\varepsilon_a$: change in axial strain.

To model the elastic strains, a nonlinear variation of Young's modulus with the state of stress was developed by Lade and Nelson (1987). This model was developed on the basis of a constant Poisson's ratio, as it is most often found to be from experimental evidence. The Young's modulus (E) is expressed in terms of stress invariants, as given in Eq. 5:

$$E = M \times p_a \left[\left(\frac{I_1}{p_a}\right)^2 + R \times \frac{J_2'}{p_a^2}\right]^{\lambda} \quad (5)$$

where M and λ: non-dimensional parameters; p_a: atmospheric pressure (in the same units as used to calculate I_1 and J_2').

I_1: first stress invariant and determined as:

$$I_1 = \sigma_1 + \sigma_2 + \sigma_3 \quad (6)$$

σ_2: In triaxial testing, $\sigma_2 = \sigma_3$ representing minor principal stress.

J_2': second stress deviator invariant and determined as:

$$J_2' = \frac{1}{6}\left[(\sigma_1 - \sigma_2)^2 + (\sigma_2 - \sigma_3)^2 + (\sigma_3 - \sigma_1)^2\right] \quad (7)$$

R: non-dimensional constant determined as:

$$R = 6 \times \frac{1 + \nu}{1 - 2\nu} \quad (8)$$

Parameters M, λ and ν are the three material parameters that are required to express the elastic behavior. M and λ are determined by plotting $\left[\left(\frac{I_1}{p_a}\right)^2 + R\left(\frac{J_2'}{p_a^2}\right)\right]$ versus (E/p_a) in a log–log diagram. Figure 12 shows the determination of the values of M and λ for the vertical and horizontal medium dense specimens.

Elastic material parameters determined in this study are presented in Table 3 for the vertical and horizontal sand specimens of loose, medium dense and dense sand. The vertical specimens have slightly higher values of M for all densities than the horizontal specimens. A slight tendency for larger values occurs with increasing density. In view of the scatter in the test results seen in Fig. 12 and the very similar values for ν, M and λ for the vertical and horizontal specimens, it is most practical to assume the elastic properties to be isotropic and that they be represented by one set of ν, M and λ for each density. While the sand tested here shows similar values of Young's moduli and Poisson's ratios for vertical and horizontal specimens, other experiments by e.g. Tatsuoka et al. (1997) and by Hoque and Tatsuoka (1998) indicate that the elastic properties of granular materials can be

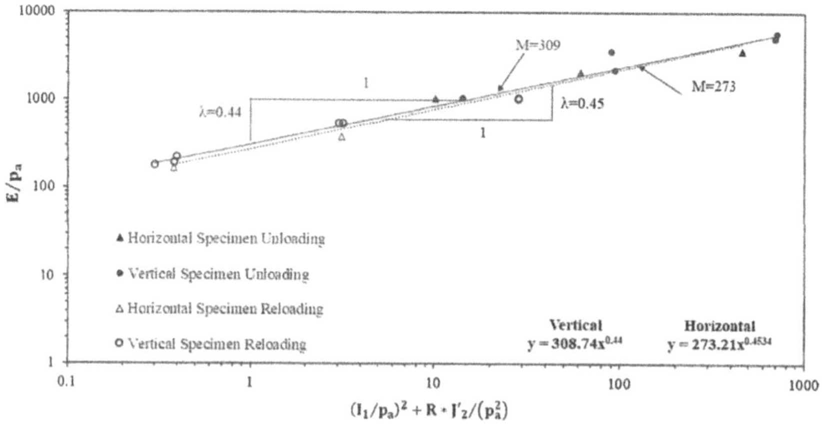

Fig. 12 Determination of parameters M and λ for characterization of elastic modulus variation for medium dense vertical and horizontal sand specimens with changes in stress state (see Eq. 5 for definitions of symbols)

Table 3 Summary of elastic properties of the vertical and horizontal sand specimens

Source of sand	Commercial hardware store					
Description of sand	Playsand					
Void ratio (e)	0.79	0.80	0.65	0.65	0.59	0.60
Relative density	34%	31%	76%	76%	93%	92%
Angularity	Rounded to subangular					
Mineralogy	Consists of mostly quartz					
Minor principal stresses (confining pressures)	20 kPa, 60 kPa, 180 kPa					
Gradation	Poorly graded medium to fine sand					
Density	Vertical loose	Horizontal loose	Vertical medium dense	Horizontal medium dense	Vertical dense	Horizontal dense
ν (poissons ratio)	0.46	0.45	0.38	0.37	0.35	0.33
M (elastic component)	215	255	309	273	408	278
λ (elastic component)	0.42	0.47	0.44	0.45	0.41	0.39

nearly isotropic or highly anisotropic depending on granular soil type and specimen preparation method.

9 Conclusions

A comprehensive experimental investigation was undertaken to study the cross-anisotropy observed in triaxial compression tests on sand specimens deposited into vertical and horizontal specimens. A series of conventional triaxial tests were performed on loose, medium dense and dense air-pluviated specimens of Play sand. The sand specimens were tested under three different confining pressures, 20, 60, and 180 kPa, and two sets of experiments were performed: one set with loading only and one set that included unloading–reloading cycles. The techniques for creating specimens with horizontal layers of sand in vertical and in horizontal specimens were explained in detail.

The volumetric change during isotropic compression was very similar and within the scatter of experiments for the vertical and horizontal sand specimens. The volumetric strains for loose and dense sand specimens resulted in similar trends as observed for the medium dense sand.

The details of the behavior of cross-anisotropic deposits of sands are explained by the contact angle directions between grains. These directions tend to favor the vertical direction and the consequent stress–strain and volume change characteristics can all be explained in view of the consequent cross-anisotropic behavior.

Detailed analyses indicated that:

- the initial slopes of the stress–strain relations for vertical specimens were higher than those for horizontal specimens,
- the strains-to-failure were smaller for the vertical specimens,
- the shear strengths were higher for the vertical specimens,
- shear banding occurred well after smooth peak failure in both vertical and horizontal specimens and were not affected by unloading–reloading cycles,
- the vertical and horizontal specimens of medium dense sand showed increasing initial volumetric contraction and decreasing rates of dilation with increasing confining pressure,
- the horizontal specimens showed more contraction than the vertical specimens and they dilated at a slower rate than the vertical specimens,
- the friction angles were higher for vertical than for horizontal specimens and they all decreased with increasing confining pressure,
- the loose specimens showed smaller differences in friction angles than the dense specimens,
- the angles of dilation showed similar trends as the friction angles,
- the differences between the friction angles and the dilation angles were highest for the horizon-

tal specimens and they were highest for the loose specimens,

- the elastic properties evaluated from the unloading–reloading cycles showed slightly higher values of Young's moduli for vertical specimens than for horizontal specimens and increasing with the sand density, but the elastic behavior may practically be represented by an isotropic elastic model, and
- the essentially constant Poisson's ratio decreased with increasing density.

The results show that conventional triaxial tests on cylindrical specimens can be employed for investigating cross-anisotropy using specimens with bedding planes in horizontal and vertical directions. The overall evaluation of test results is that cross-anisotropy is most pronounced for dense sand and it reduces with increasing confining pressure and increasing void ratio.

Acknowledgements The elastic properties shown in Fig. 12 and listed in Table 3 were determined by Mr. Anil Helvacioglu, former graduate student in the Department of Civil, Environmental and Infrastructure Engineering at George Mason University.

Author Contributions All authors contributed to the study conception and design. Material preparation and data collection were performed by the first author. The analyses and interpretation of the data was conducted by all authors. The first draft of the manuscript was written by the first author and other authors worked on editing and commenting on the previous versions of the manuscript. All authors read and approved the final manuscript.

Funding There is no funding to disclose for this research.

Data Availability All data, models, and code generated or used during the study appear in the submitted article.

Declarations

Conflict of interest The authors have no relevant financial or non-financial interests to disclose.

References

Abelev AV, Lade PV (2003) Effects of cross anisotropy on three-dimensional behavior of sand. I: stress–strain behavior and shear banding. J Eng Mech 129(2):160–166. https://doi.org/10.1061/(ASCE)0733-9399(2003)129:2(160)

Abelev AV, Lade PV (2004) Characterization of failure in cross-anisotropic soils. J Eng Mech 130(5):599–606. https://doi.org/10.1061/(ASCE)0733-9399(2004)130:5(599)

Arthur JRF, Menzies BK (1972) Inherent anisotropy in sand. Géotechnique 22(1):115–128. https://doi.org/10.1680/geot.1972.22.1.115

Arthur JRF, Phillips AB (1975) Homogeneous and layered sand in triaxial compression. Geotechnique 25(4):799–815. https://doi.org/10.1680/geot.1975.25.4.799

Bolton MD (1986) The strength and dilatancy of sands. Geotechnique 36(1):65–78

Chairo G, Kiyota T, Koseki J (2013) Strain localization characteristics of loose saturated Toyoura sand in undrained cyclic torsional shear tests with initial static shear. Soils Found 53(1):23–34. https://doi.org/10.1016/j.sandf.2012.07.016

El-Sohby MA, Andrews KZ (1973) Experimental examination of sand anisotropy. In: Proceedings of the 8th international conference on soil mechanics and foundation engineering, vol 1, Moscow, Soviet Union, pp 103–109

Haruyama M (1981) Anisotropic deformation-strength characteristics of an assembly of spherical particles under three dimensional stresses. Soils Found 21(4):41–55. https://doi.org/10.3208/sandf1972.21.4_41

Hight DW, Gens A, Symes MJ (1983) The development of a new hollow cylinder apparatus for investigating the effects of principal stress rotation in soils. Geotechnique 33(4):355–383. https://doi.org/10.1680/geot.1983.33.4.355

Hoque E, Tatsuoka F (1998) Anisotropy in elastic deformation of granular materials. Soils Found 38(1):163–179. https://doi.org/10.3208/sandf.38.163

Janbu N (1963) Soil compressibility as determined by odometer and triaxial tests. In: Proceedings of the European conference soil mechanics and foundation engineering, vol 1, Wiesbaden, pp 19–25

Lade PV (2003) Analysis and prediction of shear banding under 3D conditions in granular materials. Soils Found 43(4):161–172. https://doi.org/10.3208/sandf.43.4_161Lade,P.V.(2016)TriaxialTestingofSoils,Wiley,Chichester,U.K.10.1002/9781119106616

Lade PV, Abelev AV (2003) Effects of cross-anisotropy on three-dimensional behavior of sand. Part II: volume change behavior and failure. J Eng Mech 129:167–174. https://doi.org/10.1061/(ASCE)0733-9399(2003)129:2(167)

Lade PV, Abelev AV (2005) Characterization of cross-anisotropic soil deposits from isotropic compression tests. Soils Found 45(5):89–102. https://doi.org/10.3208/sandf.45.5_89

Lade PV, Duncan JM (1973) Cubical triaxial tests on cohesionless soil. J Soil Mech Found Div 99(SM10):793–812

Lade PV, Nelson RB (1987) Modelling the elastic behaviour of granular materials. Int J Numer Anal Meth Geomech 11:521–542. https://doi.org/10.1002/nag.1610110507

Lade PV, Wasif U (1988) Effects of height-to-diameter ratio in triaxial specimens on the behavior of cross-anisotropic sand. In: Advanced triaxial testing of soil and rock (ASTM STP 977). ASTM, Philadelphia, pp 706–714. https://doi.org/10.1520/STP29109S

Lade P, Liggio C, Yamamuro J (1998) Effects of non-plastic fines on minimum and maximum void ratios of sand. Geotech Test J 21(4):336–347. https://doi.org/10.1520/GTJ11373J

Lade PV, Rodriguez NM, Van Dyck EJ (2014a) Effects of principal stress directions on 3D failure conditions in cross-anisotropic sand. J Geotech Geoenviron Eng ASCE. https://doi.org/10.1061/(ASCE)GT.1943-5606.0001005

Lade PV, Van Dyck E, Rodriguez NM (2014b) Shear banding in torsion shear tests on cross-anisotropic deposits of fine Nevada sand. Soils Found 54(6):1081–1093. https://doi.org/10.1016/j.sandf.2014.11.004

Lam WK, Tatsuoka F (1988) Triaxial compressive and extension strength of sand affected by strength anisotropy and sample slenderness. In: Advanced triaxial testing of soil and rock (ASTM STP 977). ASTM, Philadelphia, pp 655–666. https://doi.org/10.1520/STP29105S

Lee KL, Seed HB (1967) Drained strength characteristics of sands. J Soil Mech Found Div 93(SM6):117–141

Mahmoudi YC, Taiba A, Hazout L, Belkhatir M (2021) Friction and maximum dilatancy angles of granular soils incorporating low plastic fines and depositional techniques effects. Eur J Environ Civ Eng. https://doi.org/10.1080/19648189.2021.1999334

Miura S, Toki S (1982) A sample preparation method and its effect on static and cyclic deformation-strength properties of sand. Soils Found 22(1):61–77. https://doi.org/10.3208/sandf1972.22.61

Miura K, Miura S, Toki S (1986) Deformation behaviour of anisotropic dense sand under principal stress rotation. Soils Found 26(1):36–52. https://doi.org/10.3208/sandf1972.26.36

Naughton PJ, O'Kelly BC (2007) Stress non-uniformity in a hollow cylinder torsional sand specimen. Geomech Geoeng 2(2):117–122. https://doi.org/10.1080/17486020701377124

Ochiai H, Lade PV (1983) Three-dimensional behavior of sand with anisotropic fabric. J Geotech Eng 109(10):1313–1328. https://doi.org/10.1061/(ASCE)0733-9410(1983)109:10(1313)

Oda M (1972a) Initial fabrics and their relations to mechanical properties of granular materials. Soils Found 12(1):17–36. https://doi.org/10.3208/sandf1960.12.17

Oda M (1972b) The mechanism of fabric changes during compressional deformation of sand. Soils Found 12(2):1–18. https://doi.org/10.3208/sandf1972.12.1

Oda M (1972c) Deformation mechanism of sand in triaxial compression tests. Soils Found 12(4):45–63. https://doi.org/10.3208/sandf1972.12.4_45

Oda M (1981) Anisotropic strength of cohesionless sands. J Geotech Eng Div 107(9):1219–1231. https://doi.org/10.1061/AJGEB6.0001186

Oda M, Koishikawa I, Higuchi T (1978) Experimental study of anisotropic shear strength of sand by plane strain test. Soils Found 18(1):25–38. https://doi.org/10.3208/sandf1972.18.25

O'Kelly B, Naughton P (2009) Study of yielding of sand under generalized stress conditions using a versatile hollow cylinder torsional apparatus. Mech Mater 41(3):187–198. https://doi.org/10.1016/j.mechmat.2008.11.002

Parkin AK, Gerrard CM, Willoughby DR (1968) Discussion on deformation of sand in shear. J Soil Mech Found 94(SM1):336–340. https://doi.org/10.1061/(ASCE)0733-9410(1991)117:11(1810)

Powers MC (1953) A new roundness scale for sedimentary particles. SEPM J Sediment Res 23:66. https://doi.org/10.1306/D4269567-2B26-11D7-8648000102C1865D

Pradel D, Ishihara K, Gutierrez M (1990) Yielding and flow of sand under principal stress axes rotation. Soils Found 30(1):87–99. https://doi.org/10.3208/sandf1972.30.87

Rad NS, Tumay MT (1987) Factors affecting sand specimen preparation by raining. Geotech Test J 10(1):31–37. https://doi.org/10.1520/GTJ10136J

Rodriguez NM, Lade PV (2013) Effects of principal stress directions and mean normal stress on failure criterion for cross-anisotropic sand. J Eng Mech 139(11):1592–1601. https://doi.org/10.1061/(ASCE)EM.1943-7889.0000595

Tatsuoka F, Sakamoto M, Kawamura T, Fukushima S (1986) Strength and deformation characteristics of sand in plane strain compression at extremely low pressures. Soils Found 26(1):65–84. https://doi.org/10.3208/sandf1972.26.65

Tatsuoka F, Uchimura T, Tateyama M (1997) Preloaded and prestressed reinforced soil. Soils Found 37(3):79–94. https://doi.org/10.3208/sandf.37.3_79

Vaid YP, Negussey D (1984) Relative density of air and water pluviated sand. Soils Found 24(2):101–105. https://doi.org/10.3208/sandf1972.24.2_101

Vaid YP, Negussey D (1988) Preparation of reconstituted sand specimens. In: Donaghe RT, Chaney RC, Silver ML (eds) Advanced triaxial testing of soil and rock (ASTM STP 977). ASTM, Philadelphia, pp 405–417. https://doi.org/10.1520/STP29090S

Veiskarami M, Jamshidi Chenari R, Jameei AA (2019) A study on the static and seismic earth pressure problems in anisotropic granular media. Geotech Geol Eng 37:1987–2005. https://doi.org/10.1007/s10706-018-0739-9

Xiao Y, Liu H, Chen Y, Chu J (2014) Strength and dilatancy of silty sand. J Geotech Geoenviron Eng 140(7):06014007

Xiao Y, Meng M, Chen Q, Nan B (2018) Friction and dilatancy angles of granular soils incorporating effects of shearing modes. Int J Geomech. https://doi.org/10.1061/(ASCE)GM.1943-5622.0001289

Yamada Y, Ishihara K (1979) Anisotropic deformation characteristics of sand under three dimensional stress conditions. Soils Found 19(2):79–91. https://doi.org/10.3208/sandf1972.19.2_79

Postscript

Using the strength data for Play sand in Table 2 of the above paper, the values of the intercepts with p_a/I_1 η_v and η_h and the slope m (explained in the first article on cross-anisotropy) for the 3 different densities were determined to evaluate the cross-anisotropy of the sand specimens prepared as described in the paper. The diagrams in Fig. 14.1 clearly shows the systematic behavior across the 3 sand densities. The values of m are practically the same for vertical and horizontal specimens for each of the 3 densities, while the values of η_v and η_h clearly indicate the cross-anisotropic characteristics of these specimens.

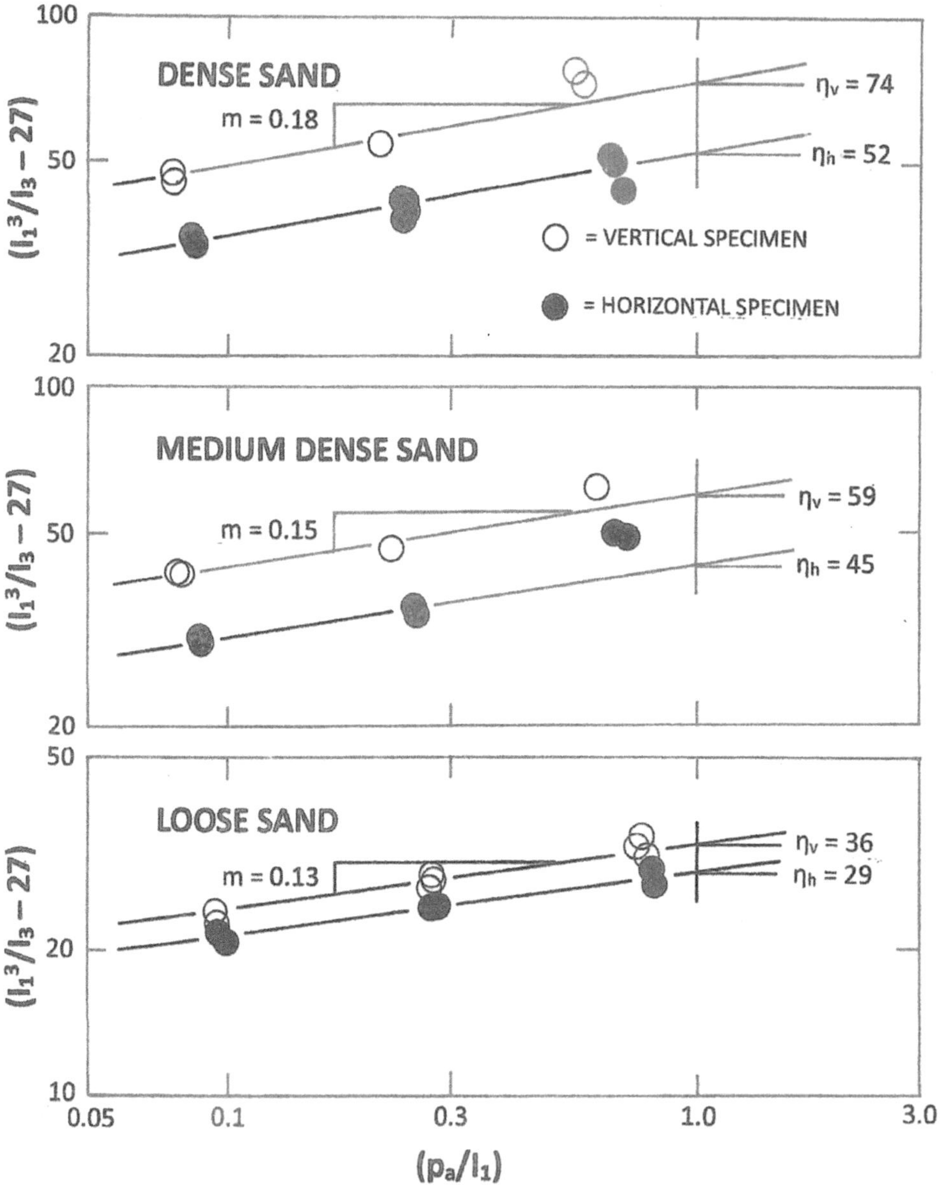

Fig. 14.1. Determination of η_v and η_h and m for triaxial compression tests on Play sand at 3 different densities.

Additional information regarding cross-anisotropy is given in the following articles:

Ochiai, H. and Lade, P.V., (1983) "Three-Dimensional Behavior of Sand with Anisotropic Fabric," *Journal of Geotechnical Engineering*, ASCE, 109(GT10): 1313-1328.

Kirkgard, M.M., and Lade, P.V., (1991) "Anisotropy of Normally Consolidated San Francisco Bay Mud", *Geotechnical Testing Journal*, GTJODJ ASTM, 14(3): 231-246.

Kirkgard, M.M., and Lade, P.V. (1993) "Anisotropic Three-Dimensional Behavior of a Normally Consolidated Clay" *Canadian Geotechnical Journal*, 30(4): 848-858.

Abelev, A.V., and Lade, P.V. (2003) "Effects of cross-anisotropy on three-dimensional behavior of sand. Part I: Stress-strain behavior and shear banding," *Journal of Engineering Mechanics*, ASCE, 129(2): 160-166.

Lade, P.V., and Abelev, A.V. (2003) "Effects of cross-anisotropy on three-dimensional behavior of sand. Part II: Volume change behavior and failure," *Journal of Engineering Mechanics*, ASCE, 129(2): 167-174.

Lade, P.V. (2008) "Failure Criterion for Cross-Anisotropic Soils," *Journal of Geotechnical and Geoenvironmental Engineering*, ASCE, 134(1): 117-124.

Rodriguez, N.M., and Lade, P.V. (2013) "True Triaxial Tests on Cross-Anisotropic Deposits of Fine Nevada Sand," *International Journal of Geomechanics*, ASCE, 13(6), pp. 779-793.

Chapter XV
Summary and Conclusions

Executive Summary

The book presents the development of a realistic, three-dimensional constitutive model for frictional materials as required in finite element and/or finite difference computations performed for advanced analysis, predictions and design of large civil engineering structures. The three-dimensional work-hardening plasticity theory for frictional materials is based on advanced experimental evidence as produced by the author and presented in the literature. The book contains detailed explanations of each component of the model including the general framework for work-hardening plasticity theory, the generation of stress invariants, a model for the elastic behavior, the three-dimensional failure criterion, the yield surface as a surface of constant plastic work, the use of non-associated (rather than associated) plastic flow generated from a plastic potential (different from the yield surface), determination of material parameters from simple triaxial compression tests, incrementalization and implementation of the model, and effects of stress rotation, time effects, and cross-anisotropy.

Summary and Conclusions

The Introduction provides an overview of 31 constitutive models and gives the setting for their use in large finite element of finite difference program for analyses of geotechnical structures. This is followed by the general framework for work-hardening plasticity theory in which it is pointed out that the soil stress-strain behavior consists of both elastic (recoverable) strains and plastic (irrecoverable) strains. In the general framework for work-hardening plasticity there are certain theoretical requirements:

(1) Continuity, i.e. a continuous response must occur at the yield surface.
(2) Uniqueness, for which Prager indicated that associated plastic flow was required, but Mroz showed that it is not required on the basis of fundamental mechanics principles.
(3) Consistency, i.e. changes in stress are countered by corresponding expenditures of plastic work.
(4) Irreversibility, which means that $\sigma \cdot d\varepsilon^p > 0$; Prager derived that the plastic potential is required to be convex, but Mroz indicated that this is not necessary.

The origin of plastic stress-strain relations is normality, as derived for solid metals, which show no plastic volume changes; therefore

$$(\, d\varepsilon_{ij}^p = d\lambda \frac{\partial f}{\partial \sigma_{ij}} \,)$$

in which f is the expression for the yield surface and $d\lambda$ is a proportionality constant. For frictional materials f is replaced with an expression for the plastic potential g, and therefore results in non-associated plastic flow, as observed for frictional materials.

The derivation of the three stress invariants is presented in detail, and the shapes of these invariants in the principal stress space are presented.

A model for the elastic modulus and its variation with stress state is developed based on first principles of no expenditure or generation of work in a closed stress loop. This results in an expression for the elastic modulus in terms of the first and second stress invariants.

Based on true triaxial tests on sand and clay, a three-dimensional failure criterion is developed in terms of the first and the third stress invariants. This criterion is shown to capture the strengths of all frictional materials such as sands, clays, rocks, concrete and metals with pores such as cast iron and powder metals. It also captures the strengths of solid metals.

Experiments were performed in triaxial compression which demonstrated that frictional materials exhibit non-associated plastic flow. The consequence is that stability cannot be guaranteed, and this is demonstrated to be correct for sands: Dense sands that dilate during shearing show stable behavior in triaxial compression well beyond peak failure after which shear banding may occur in the softening regime. On the other hand, loose sands that contracts during shear may become unstable inside the failure surface and this may result in liquefaction.

The Single Hardening Model for Frictional Materials is developed on the basis of experimental evidence and has been tested for a number of conditions that can be developed in laboratory experiments. The model consists of the following components: (1) a failure criterion, (2) a single yield surface, defined as a surface along which the plastic work is constant as measured from the stress origin, and (3) a non-associated plastic potential surface. All required material parameters can be determined from triaxial compression tests.

The Single Hardening Model was enhanced to capture the effects of large three-dimensional stress reversals using a mechanism by which the yield surface rotates around the origin of the principal stress space. This rotational mechanism was shown to work well inside the primary yield surface for up to three reversals after which the sand's memory was erased.

Effects of stress rotation was shown to be contained in the model comprised of stress invariants, and it predicted the effects of stress rotation in torsion shear tests with good accuracy.

Because the expressions for the plastic strain increments in the Single Hardening Model are produced in terms of total stresses, it is necessary to incrementalize the relation to obtain strain increments in terms of stress increments, and vice versa. This is required for the model to be implemented in finite element or finite difference programs. Both the incrementalization and the implementation procedures are presented.

Time effects, such as creep and stress relaxation, play major roles in the behavior of frictional materials, and these have been studied for sands. The effects can be included in the model by enhancing the expression for the plastic work by a factor that account for these time effects. Due to the relation between the yield surface and the plastic work, time effects will cause greater plastic work and therefore movement of the yield surface. Other changes, for example to the plastic potential, are not required.

Real soil deposits tend to be cross-anisotropic due to their mode of deposition in horizontal layers in nature. They therefore have a vertical axis of symmetry, and their strengths and stress-strain behavior are different in the vertical and horizontal directions. This has been known for a long time for clays, for which it is easy to obtain horizontal specimens. Only recently have horizontal sand specimens been produced in the laboratory, and these were tested along with vertical specimens to indicate their differences. A failure criterion for cross-anisotropic sand has been developed on the basis of experiments on vertical and horizontal specimens. However, further developments of the stress-strain model for cross-anisotropy have not been pursued at this time.

The Author

Poul V. Lade received his MS degree from the Technical University of Denmark in 1967 and he continued his studies at the University of California at Berkeley where he received his PhD in 1972. Subsequently his academic career began at the University of California at Los Angeles (UCLA)(1972-1993), and he continued at Johns Hopkins University (1993–1999) in Baltimore, Aalborg University in Denmark (1999–2003), Catholic University of America in Washington, DC (2003–2015) and George Mason University in Fairfax, Virginia (2016-2020).

His research interests include application of appropriate experimental methods to determine the three-dimensional stress–strain and strength behavior of soils and the development of constitutive models for frictional materials such as soils, concrete, and rock. He developed laboratory experimental apparatus to investigate monotonic loading and large three-dimensional stress reversals in plane strain, true triaxial and torsion shear equipment. This also included studies of effects of principal stress rotation, stability, instability and liquefaction of granular materials, and time effects. The constitutive models are based on elasticity and work-hardening, isotropic and kinematic plasticity theories.

He has written nearly 300 publications based on research performed with support from the National Science Foundation (NSF) and from the Air Force Office of Scientific Research (AFOSR). He was elected member of the Danish Academy of Technical Sciences (2001), and he was awarded Professor Ostenfeld's Gold Medal from the Technical University of Denmark (2001). He was inaugural editor of *Geomechanics and Engineering* and he has served on the editorial boards of eight international journals on Geotechnical Engineering.

According to the Stanford/Elsevier tables of paper citations, his H-index (without self-citations) is 53 (2022) and his rank within the subfield of US authors in Geological and Geomatics Engineering is #16 (2022) of 63,200 authors. According to Google Scholar his H-index is 75 (2025). These values are among the highest, if not the highest, in the subfield of Geotechnical Engineering.

Made in United States
Cleveland, OH
05 November 2026

47656824R00285